UNION CARBIDE CORP.
So. Charleston, WV

JUN 2   1999

LIBRARY

Development and Manufacture of Pressure-Sensitive Products

Development and Manufacture of Pressure-Sensitive Products

István Benedek
Consultant
Wuppertal, Germany

Marcel Dekker, Inc.　　New York · Basel · Hong Kong

Library of Congress Cataloging-in-Publication Data

Benedek, István
  Development and manufacture of pressure-sensitive products / István Benedek.
    p. cm.
  Includes bibliographical references and index.
  ISBN 0-8247-0206-9 (alk. paper)
  1. Paper coatings. 2. Adhesives. 3. Adhesive labels. 4. Adhesive tape. I. Title.
  TS1118.F5B45 1998
  668'.3—dc21   98-38672
                                                                                  CIP

This book is printed on acid-free paper.

Headquarters
Marcel Dekker, Inc.
270 Madison Avenue, New York, NY 10016
tel: 212-696-9000; fax: 212-685-4540

Eastern Hemisphere Distribution
Marcel Dekker AG
Hutgasse 4, Postfach 812, CH-4001 Basel, Switzerland
tel: 44-61-261-8482; fax: 44-61-261-8896

World Wide Web
http://www.dekker.com

The publisher offers discounts on this book when ordered in bulk quantities. For more information, write to Special Sales/Professional Marketing at the headquarters address above.

Copyright © 1999 by Marcel Dekker, Inc. All Rights Reserved.

Neither this book nor any part may be reproduced or transmitted in any form or by any means, electronic or mechanical, including photocopying, microfilming, and recording, or by any information storage and retrieval system, without permission in writing from the publisher.

Current printing (last digit)
10 9 8 7 6 5 4 3 2 1

PRINTED IN THE UNITED STATES OF AMERICA

Preface

Over the last two decades the manufacture and application fields of pressure-sensitive products have developed from a largely empirical body of accumulated practical knowledge to an increasingly sophisticated science utilizing the most advanced techniques of physics, chemistry, and engineering. Pressure-sensitive labels, tapes, protective films, seals, business forms, etc., are used in medicine, pharmaceutical applications, electronic circuits, and assembly of machine parts as well as in product promotion, coding, and packaging.

Due to the wide utility and consumer acceptance of these products a high level of basic research and product development has evolved over the last few years and is continuing to grow. Pressure sensitivity, the main performance characteristic of these various products, possesses a common scientific basis in macromolecular chemistry and physics. The ways and means, however, to achieve it differ, and include the technology of adhesives and plastics as well. In the final use, no one asks about the construction of the product, whether it is pressure-sensitive, or with or without adhesive. The technical solution should work. This book is an attempt to integrate the different technologies to give the same result: a pressure-sensitive product.

This is a book about pressure-sensitive products. This work is intended as a companion volume to the book I wrote earlier with L. J. Heymans, *Pressure-Sensitive Adhesives Technology* (Dekker, 1997). My main aim in writing this work was to bridge the gap between theory and practice, between the engineering of plastics, the technology of adhesives, and the conversion and integration of the practical aspects including the engineering fundamentals. This work is a guide to the entire field of pressure-sensitive products, with or without adhesives, and discusses the engineering steps (of paper, plastics, adhesives and other materials) required for their manufacture.

This monograph covers a broad spectrum of knowledge and is designed for production and manufacturing managers, production engineers, material scientists, chemists, new product specialists, and other technologists involved in the efficient production or use of pressure-sensitive products and in new process and product developments. I discuss the whole complex of buildup, manufacture, testing and application of pressure-sensitive products. The focus of the description of the technology is on specific examples of application, rather than on theory. The basic principles of this technical domain, however, are always presented, and the book summarizes our present understanding of the construction and functioning of pressure-sensitive products.

It is not the aim of this book to give a detailed discussion of the science of adhesives or plastics, nor does it constitute a practical vade mecum. It is an attempt to integrate technical domains which belong together, as an aid for those involved in the understanding, design, and use of pressure-sensitive products.

István Benedek

Contents

Preface		*iii*
1. Introduction		**1**
References		4
2. Buildup and Classification of Pressure-Sensitive Products		**7**
1 Construction of PSPs		7
1.1 General Buildup of PSPs		7
1.2 Particular Buildup of PSPs		9
1.3 Components of PSPs		28
2 Classes of Pressure-Sensitive Products		35
2.1 Labels		36
2.2 Tapes		44
2.3 Protective Films		55
2.4 Other Products		66
References		66
3. Physical Basis of PSPs		**73**
1 Rheology of PSPs		73
1.1 Rheology of Carrier Material		77
1.2 Rheology of Adhesives		91
1.3 Rheology of Abhesives		106
1.4 Product Rheology		106

	2	Mechanical Properties of PSPs	116
		2.1 Rheological Background of the Mechanical Properties of PSPs	117
		2.2 Main Mechanical Properties	123
		2.3 Regulating the Mechanical Properties of PSPs	136
		2.4 The Influence of Mechanical Properties on Other Performance Characteristics of PSPs	136
	3	Other Physical Characteristics of PSPs	137
		3.1 Electrical Characteristics of PSPs	137
		3.2 Optical Characteristics of PSPs	140
		References	140
4.	Chemical Basis of PSPs		**149**
	1	Macromolecular Basis and Chemical Composition of Carrier Material	150
		1.1 General Considerations	153
		1.2 Nonadhesive Carrier Material	159
		1.3 Carrier Materials with Adhesivity	165
	2	Macromolecular Basis and Chemical Composition of the Adhesive	169
		2.1 Elastomeric Components	169
		2.2 Viscoelastic Components	202
		2.3 Viscous Components	212
		2.4 Other Components	216
	3	Macromolecular Basis and Chemical Composition of the Abhesive	232
		3.1 Coated Abhesive Components	233
		3.2 Built-in Abhesive Components	236
		References	237
5.	Adhesive Properties of PSPs		**255**
	1	Definition and Characterization of Adhesive Properties	256
		1.1 General Adhesive Performance	256
		1.2 Special Adhesive Performance	263
	2	Regulating the Adhesive Properties	305
		2.1 Regulating the Adhesive Properties with the Adhesive	305
		2.2 Regulating the Adhesive Properties with the Carrier	320
		2.3 Regulating the Adhesive Properties with Manufacturing Technology	322
		2.4 Regulating the Adhesive Properties with Product Application Technology	324
	3	Interdependence: Adhesive Properties and Other Performance Characteristics	325
		References	325

6.	Manufacture of Pressure-Sensitive Products		**333**
	1 Manufacture of Coating Components		336
	1.1 Manufacture of the Adhesive Components		337
	1.2 Manufacture of the Abhesive Components		386
	2 Manufacture of the Carrier Material for PSPs		394
	2.1 Carrier on Paper Basis		395
	2.2 Synthetic Films as Carrier		402
	2.3 Other Carrier Materials		445
	2.4 Manufacture of the Release Liner		445
	3 Manufacture of the Finished Product		448
	3.1 Manufacture of the Finished Product by Extrusion		450
	3.2 Manufacture of the Finished Product by Carrier Coating		462
	References		498
7.	Converting Properties of PSPs		**523**
	1 Coating Properties of PSPs		524
	1.1 Adhesive Coating		524
	1.2 Abhesive Coating		525
	1.3 Printing		526
	2 Confectioning Properties		548
	2.1 Cutting		549
	2.2 Die Cutting		552
	2.3 Perforating, Embossing and Folding		555
	2.4 Winding Properties		557
	3 Dispensing and Labeling		559
	References		560
8.	End Uses of Pressure-Sensitive Products		**567**
	1 General Considerations		568
	2 Labels and Their Application		570
	2.1 Application Conditions		570
	2.2 Application Methods for Labels		575
	2.3 Postmodification of Labels		576
	2.4 Main Types of Labels		576
	3 Tapes and Their Application		584
	3.1 Application Conditions		584
	3.2 Application Method		587
	3.3 Main Tapes		589
	4 Protective Films		602
	4.1 Application Conditions		602
	4.2 Main Protective Films		612
	5 Forms		627
	6 Plotter Films		627

7	Other PSPs	629
	References	630

Index *643*

Development and Manufacture of Pressure-Sensitive Products

1
Introduction

Pressure-sensitive tapes were first used about 150 years ago. Pressure-sensitive labels came on the market 90 years later. About ten years after that, pressure-sensitive protective films were manufactured. Pressure-sensitive products (PSPs) such as pressure-sensitive adhesive (PSA) coated webs have been defined through the special nature of this adhesive although the definition of PSA is not completely clear. The German technical term *Haftkleber* (i.e., adhesive which adheres) supposes that it is possible to differentiate between adhesion and the building up of an adhesive bond. In English the term *pressure-sensitive adhesives* (i.e., adhesives that bond when pressure is applied) admits pressure as an indispensable condition for their function. In reality, as is known from loop tack measurements and touch blow labeling, almost no pressure is required for label application, but high pressures are needed for protective films in coil coating. *Autocollants,* the French term, does not define the application conditions. It refers only to the bonding behavior. The common characteristic of these products is ensured by their special viscoelastic behavior, manifested as permanent cold flow, where the chemistry of the adhesive plays only a secondary role [1].

The development of pressure-sensitive products without a coated pressure-sensitive adhesive layer (in the classical sense known from the converting industry) makes the definition of this product group more difficult. Adhesive-free pressure-sensitive products were developed some decades ago. In this case adhesive-free means that the self-adhesive component is not coated on the product surface. It is included in the carrier; i.e., the carrier per se is pressure-sensitive. In some cases an adhesive-free composition is used and pressure sensitivity is provided by physical treatment of the carrier surface and/or by application conditions (temperature, pressure). According to the definition given in Ref. 2, PSAs are adhesives "which in dry form are aggressively and permanently tacky at room temperature . . . and adhere without the need of more than finger or hand pressure, require no activation by water solvent or heat."

Most pressure-sensitive products do not meet these requirements. However, they manifest self-adhesivity and (under well-defined conditions) can be applied like a PSA-coated classical pressure-sensitive product, i.e., like an adhesive acting via viscous flow and debonding like a viscoelastic compound. Such behavior is achieved by a complex buildup and interaction of the product components and in some cases by special application or deapplication conditions. Obviously, a physically treated hot laminating plastomer film applied under pressure, or a warm laminating film based on a partially viscoelastic olefin copolymer and applied under pressure, cannot have the same chemical basis as a PSA-coated product or a plastic carrier material that includes PSA. As mentioned, their application conditions are quite different also. However, all these products work as viscoelastic bonding elements and are used in application domains of classical, pressure-sensitive adhesive–coated products. Therefore they can be considered pressure-sensitive products. Plastic processing specialists possess the know-how of the manufacture of plastic-based PSPs. Specialists in converting/coating are skilled in the design and testing of pressure-sensitive products. They control the market also. Therefore from an economics point of view the two domains belong together.

Pressure-sensitive adhesives have been used for about a century for medical tapes and dressings. Natural adhesives mixed with natural resins, waxes, and fillers were applied as the first pressure-sensitive products in the form of medical plasters [3]. In 1845 Horace H. Day prepared and patented a plaster composed of a mixture of natural rubber and tackifier resin coated on a cloth [4]. According to Refs. 5 and 6, the first tape patented by Paul Beiersdorf was a zinc oxide rubber–based plaster. At the end of the nineteenth century masking tapes and cellophane tapes were the early nonmedical PSPs. For such products natural rubber was preferred as the raw material. In this period of time, pressure-sensitive adhesives were used for plasters, labels, and tapes [7]. Industrial tapes were introduced on the market in the 1920s and 1930s, and self-adhesive labels in 1935–1936 [8,9]. In the 1930s Stanton Avery developed Kum-Kleen labels [10]. In 1955 the firm Sassions of York was licensed by Avery to produce PSA labels in Europe [11].

Tapes are self-adhesive, self-wound weblike materials, used (mainly) as a continuous web. Generally tapes are produced by coating a nonadhesive web with pressure-sensitive adhesive, but some tapes have a self-adhesive carrier material. The pressure-sensitive layer is protected by the back side of the carrier material.

The permanent weblike character of tapes allows the use of higher forces in their application. The lower converting degree of common tapes permits their design mainly for their adhesive properties. In this case the adhesive properties need not be balanced. Therefore, theoretically, tapes may also be formulated as PSA-free products.

Labels are self-adhesive, laminated carrier materials. Generally they possess a continuous weblike character only during their manufacture. Labels can also be produced as separate items. Quite unlike tapes, labels are used as discrete objects with a well-defined geometry. Because of their adhesive and surface characteristics, the self-adhesive layer of labels must be protected with a supplemental solid-state adhesive material (release liner). The first release material was wax, as used by Stanton Avery [12]. Silicone release coatings have been on the market since the mid-1950s [13,14]. Labels preserve their laminate character until their application. Because of their discontinuous character, limited contact surface, high application speed, and low application pressure, labels have to exhibit well-balanced adhesive characteristics. Therefore most of them are manufactured in the classical way, by coating a nonadhesive carrier material (face stock) with a pressure-sensitive adhesive.

Protective films are removable self-adhesive webs based on a carrier material that possesses built-in or built-on self-adhesive properties. The role of a protective film is to protect a product by adhering to it, covering its surface with a mechanically resistant supplemental layer. This is a time-limited function, i.e., the bond should be removable, allowing the protective sheet to be separated from the protected surface. Protective films are packaging materials, not so much in a legislative sense as in a functional one. Unlike classical packaging materials, where functionality concerns protection of the product during transport and storage and aesthetic, marketing-related design characteristics are determinant, protective films are technological components of a product, attached in many cases to the raw product and passing through the entire manufacturing process up to the finished product, i.e., undergoing the working steps of fabrication. Such products are applied by lamination/delamination of large surfaces. Therefore the resultant bonding/debonding forces are much higher than those employed for labels or tapes. On the other hand, protective films have to be removable. Therefore the instantaneous adhesive performance of a protective film plays only a secondary role. It is evident that in this case adhesive-free constructions may be equivalent to PSA-coated products.

The development of different product classes has been conditioned by the development of the raw materials and of the coating and application technology. Pressure-sensitive labels, stickers, and other products have seen considerable development over the years with the appearance of new materials and combinations of materials as well as new processing technologies. By the end of the 1920s acrylics had been synthesized. Acrylics possess adequate die cutting properties

2
Buildup and Classification of Pressure-Sensitive Products

Pressure-sensitive products (PSPs) may have a simple or sophisticated construction, depending on their end use. More or less expensive products can be used in the same application field; therefore, the buildup of PSPs differs according to their product class and special use. For a better understanding of their function, we first consider the buildup of PSPs.

1 CONSTRUCTION OF PSPs

Generally pressure-sensitive products are sheetlike constructions that exhibit self-adhesion. In principle such products include a component that ensures the required mechanical properties and a component that provides adhesivity.

1.1 General Buildup of PSPs

Supposing that in principle a pressure-sensitive laminate is built up from a carrier material, an adhesive, and a release liner, such a product may be defined as shown in Fig. 2.1. The complex multilayer structure of labels containing separate solid-state carrier and release components bonded by an adhesive can be simplified for tapes. Tapes and adhesive-coated protective films possess only one solid-state carrier component, and it is coated with adhesive. The new generation of protective films is built up from an (adhesive) carrier without a pressure-sensitive adhesive (PSA) layer (Fig. 2.1).

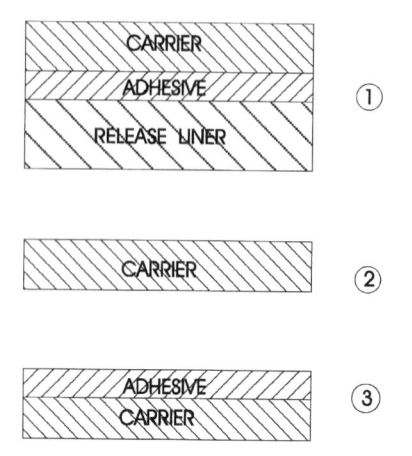

Figure 2.1 Buildup of the main PSPs. (1) Label; (2) extruded protective film; (3) tape.

As can be seen in Fig. 2.1, theoretically such protective films are the simplest PSPs, built up as a one-component pressure-sensitive (self-adhesive) carrier material. Tapes have to be more aggressive; therefore generally they need a PSA layer coated on a nonadhesive carrier material. Labels with balanced adhesive performance and high speed machine application require a separate release liner. Because of the discontinuous (non-weblike) character of labels, their handling and automatic application require a continuous supplemental carrier material, i.e., a release liner. The liner allows labels to be processed as a continuous web and protects their adhesive layer. It can be concluded that PSPs generally have either a carrier and a PSA layer or a pressure-sensitive carrier. Other constructions are known also.

Some PSAs may be used as PSPs per se, without a carrier material [1]. In other cases such as decalcomania (decals) transferable and letters, the pressure-sensitive adhesive layer also plays the role of information carrier. However, because of its discontinuity and for mechanical resistance it also requires a release liner.

The main characteristics of PSPs are their pressure-sensitive bonding and debonding. Their performance has to be ensured by one of the product components or by the assembly as a whole. In classical pressure-sensitive adhesive–coated products, adhesivity was given by the PSA. As discussed above, generally the application of PSPs requires a solid-state carrier material too. The simplest classical PSP can be designed as an adhesive-coated carrier material. Such a product has to adhere on the substrate surface. Theoretically the PSA is the bonding component of the PSP; the carrier should only allow its application.

Buildup and Classification of PSPs 9

Depending on its end use, the nature of the solid-state carrier material and the character of the bond of a PSP can be quite different. Therefore supposing the classical construction of a label as shown in Fig. 2.1, the nature and geometry of both components (solid-state carrier material and pressure-sensitive adhesive) are various. Really because of different manufacture possibilities the construction of PSPs is more sophisticated. The particular buildup is a function of required performance and manufacturing procedure.

For instance, certain tapes have to be primed to ensure good anchorage of the adhesive on the carrier. For some tapes a release layer should be coated on the back side of the carrier. Such a layer is not necessary if the material of the carrier exhibits abhesive properties, like certain nonpolar plastic films. For special tapes a separate release film should be interlaminated [2] (Fig. 2.2).

1.2 Particular Buildup of PSPs

The buildup of labels is described in detail in a companion volume [3]. Historically the production of PSPs started with the manufacture of tapes, i.e., with the production of monowebs coated with PSA. The development of removable PSAs allowed the manufacture of removable labels and tapes and later the manufacture

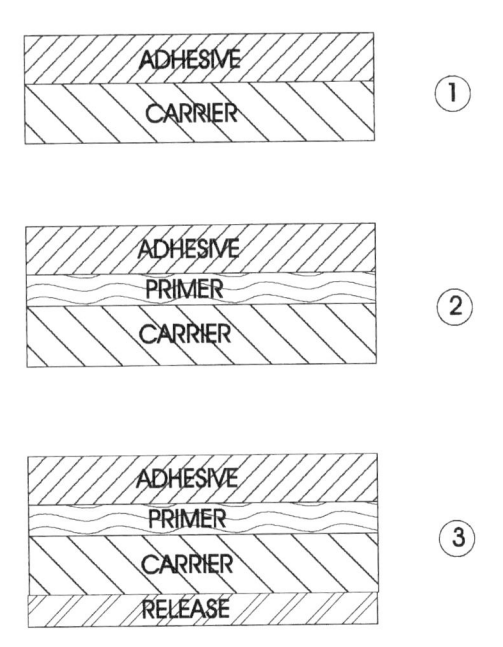

Figure 2.2 Buildup of tapes. (1) Adhesive-coated carrier; (2) adhesive-coated primed carrier; (3) adhesive, primer, and release-coated carrier.

of large surface pressure-sensitive products displaying permanent adhesion during processing of the laminate and removability after. Such products are protective films. Generally the adhesive layer of tapes and protective films is protected by self-laminating (self-winding) only. Such products are manufactured and used as constructions that have only one solid-state carrier material, i.e., as monowebs. Labels, tapes, and protective films are generally designed as products having a continuous solid-state carrier coated with a continuous adhesive.

Special tapes exist that have an adhesive with a carrierlike character. Such behavior can be achieved by crosslinking, filling, foaming, or reinforcing the adhesive layer (Fig. 2.3). The adhesive can contain a metallic network to ensure electrical conductivity [4]. The adhesive (if coated on a carrier) can be discontinuous

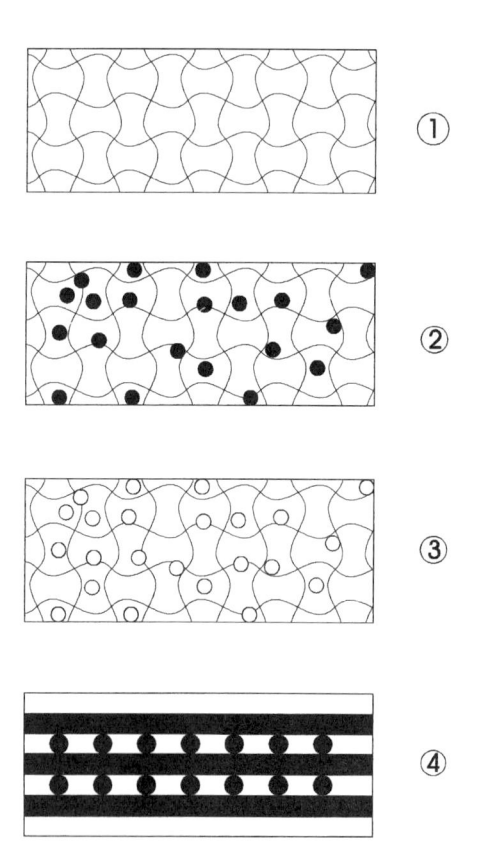

Figure 2.3 Carrierless pressure-sensitive constructions. (1) Crosslinked adhesive; (2) filled and crosslinked adhesive; (3) foamed and crosslinked adhesive; (4) reinforced adhesive.

Buildup and Classification of PSPs 11

also. Producers are looking to produce their own base label stock or special stripe, patch, or spot coated adhesive construction and linerless labels, labels having adhesive-free zones, form and text combinations [5]. Figure 2.4 presents the main pressure-sensitive constructions.

Monoweb Constructions

Classical monoweb constructions (tapes, protective films, etc.) possess a pressure-sensitive adhesive layer coated on a nonadhesive carrier material. As discussed later (see Chapter 6), the adhesive layer can be applied by using various coating techniques such as those used in the converting (coating and/or printing) industry or in plastics manufacture (extrusion). The development of macromolecular chemistry and extrusion technology allowed the manufacture of carrier materials with built-in pressure sensitivity, i.e., PSPs constructed like an uncoated monoweb but behaving, when applied, like a coated web (see also Chapter 4).

Uncoated Monoweb. Pressure-sensitive products that are built up like an uncoated monoweb are composed of a carrier material only. This carrier must have a special chemical nature or undergo special physical treatment to allow self-adhesion under special application conditions (pressure, temperature, and surface treatment). Generally such conformable, autoadhesive monowebs are plastic films (see Chapter 6). According to Djordjevic [6], "films are planar forms of plastics, thick enough to be self supporting but thin enough to be flexed, folded or creased without cracking." The upper dimensional limit for a film is difficult to define but is somewhere between 70 and 150 µm depending on the polymer used as raw material. Uncoated monowebs used as PSPs can have a homoge-

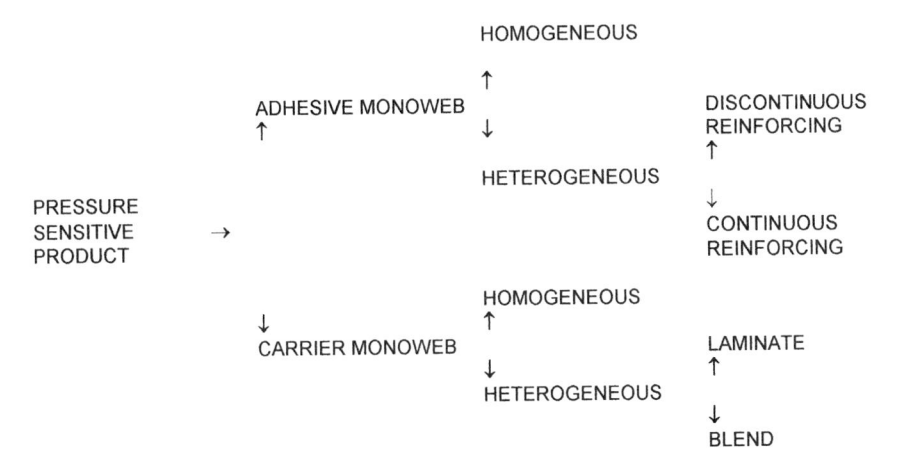

```
                                    HOMOGENEOUS
                                         ↑
                   ADHESIVE MONOWEB      ↓            DISCONTINUOUS
                          ↑                           REINFORCING
                                    HETEROGENEOUS          ↑
  PRESSURE                                                 ↓
  SENSITIVE    →                                      CONTINUOUS
  PRODUCT                                             REINFORCING

                                    HOMOGENEOUS
                          ↓              ↑
                   CARRIER MONOWEB       ↓            LAMINATE
                                    HETEROGENEOUS          ↑
                                                           ↓
                                                      BLEND
```

Figure 2.4 The main pressure-sensitive constructions.

neous or heterogeneous structure. The whole carrier can be autoadhesive (e.g., ethylene-vinyl acetate copolymers or very low density polyolefin-based films), or it may possess an adhesive layer defined by manufacture (coextrusion) and/or by diffusion of a self-adhesive, built-in component (see Fig. 2.5).

The manufacture of an uncoated self-supporting adhesive material is a complex procedure. One possibility is the production of a self-supporting film. It is known that the adhesive bond is the result of chemical attraction as well as physical anchorage. Both require contact surface and interpenetration, i.e., flow. Cold flow depends on chemical basis, product geometry, and application conditions. Unfortunately, mechanically resistant, self-supporting products exhibit only limited flow (see Chapter 3, Section 2). Due to its plasticizing ability, polyvinyl chloride (PVC) is an ideal material to achieve self-adhesive performance and to conserve mechanical strength. Decorative decals (e.g., adhesive films based on PVC with a very high plasticizer level) possess a monoweb construction too. Depending on its formulation and softness, PVC can be used as self-adhesive medical tape (carrier) or as an adhesive coating [7]. "Hardening" of an adhesive can also lead to a carrier-like product. Sealing tapes (without carrier) based on tacki-

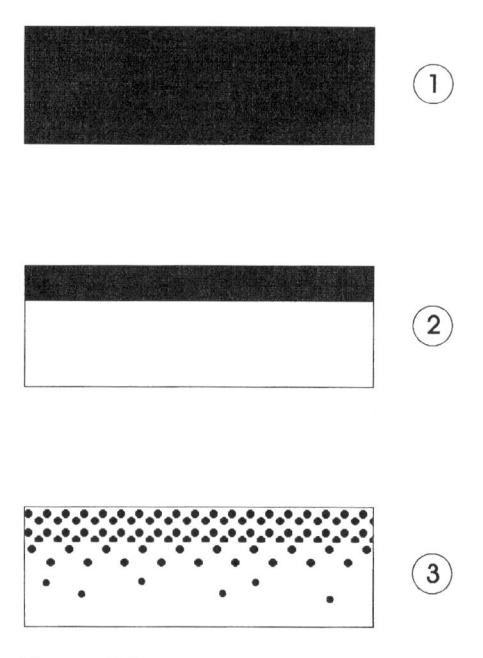

Figure 2.5 Buildup of self-adhesive pressure-sensitive products. (1) Homogeneous adhesive carrier; (2) heterogeneous partially adhesive carrier; (3) heterogeneous adhesive carrier.

fied butyl rubber are applied with an extruder [8]. Sheetlike hot melts are used for thermal lamination of various weblike materials. Ethylene-vinyl acetate and ethylene-propylene copolymers, copolyesters, copolyamides, vinyl chloride copolymers, and thermoplastic polyurethanes (TPUs) have been developed for such applications [9].

Transfer tapes are another class of uncoated and self-supporting monowebs (see also Chapter 8). Such carrierless tapes are prefabricated glues having more than sufficient dimensional stability to permit high speed lamination. In actuality, such products are adhesive layers that have a high mechanical resistance that allows their transfer during application but does not allow their manufacture, storage, and handling (see Chapters 6 and 8). Therefore, before use they must have a laminate structure. An adhesive material in film form (without surface coating) can display many advantages during its application. Additional benefits are cleanliness; controlled, uniform thickness compared to a liquid system; and positionability. Such carrierless glues are also used for non-pressure-sensitive applications. For instance, a modified heat-curable epoxy adhesive based film (25 μm) has been manufactured as a carrierless adhesive sheet for printed circuits [10]. In order to achieve better mechanical performance, the adhesive can be reinforced. As an example, for liquid crystal based thermometers the (repositionable) acrylic adhesive is a 2 mil fiber-reinforced free film [11].

Acrylic foam has been developed as a carrier-free adhesive construction for mounting tapes [12]. Such products are transfer tapes. As mentioned earlier, transfer tapes have a temporary solid-state component that forms the release liner, supporting the adhesive core (Fig. 2.6).

The adhesive core may be a continuous homogeneous or semicontinuous heterogeneous adhesive layer. The heterogeneity of the latter may be due to included solid, liquid, or gaseous particles (holes), i.e., the adhesive layer can be a foam also [13]. Such structural adhesives are used in carrier-free adhesive tapes for bonding dissimilar substrates [14]. Structural tape combines the properties of classical tapes and those of PSA [15,16]. It should be mentioned that virtually carrierless sealing tapes based on butyl rubber have been used in the automotive industry. Such "carrier-free" tapes are reinforced with a metallic wire included in the elastomer.

Transfer tapes without a carrier can have a core of pure adhesive or include a reinforcing material. The reinforcing component can be a continuous web (e.g., a network) or a discontinuous filler-like material. For instance, thick PSA tapes (0.2–1.0 mm) have been prepared by ultraviolet light (UV) initiated photopolymerization of acrylics. In this case the PSA matrix can contain glass microbubbles [17]. The filled layers can be laminated together with the unfilled layers. Such acrylic formulations can be polymerized as a thick layer (up to about 60 mils), or the thick layer may be composed of a plurality of layers, each separately photopolymerized (see Fig. 2.7). The thickness of the layer is a main fac-

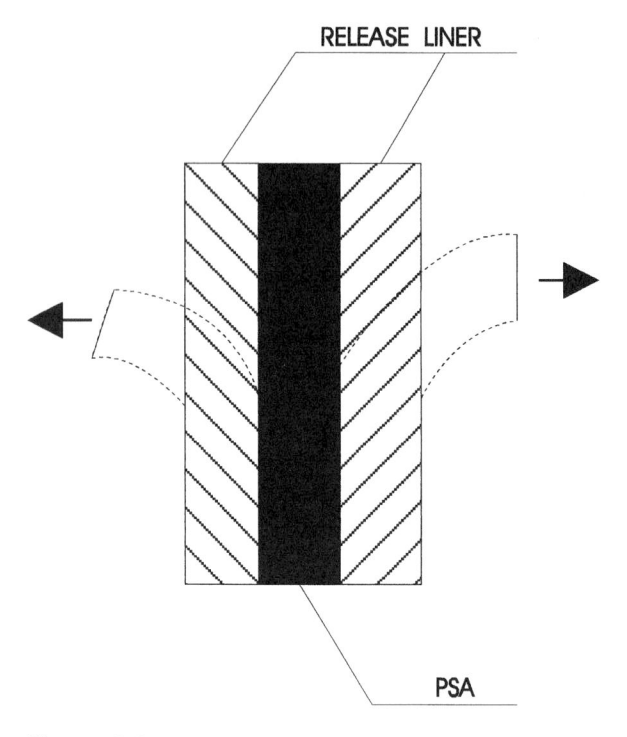

Figure 2.6 Buildup of a transfer tape.

tor; a thicker layer requires a greater degree of exposure. Thick multilayered PSAs or PSPs are made in this way. If the thick layer is sandwiched between two thinner layers, it may be considered a carrier, although it has pressure-sensitive properties also. If the support or carrier layer is 25–45 mils thick, it conforms well to substrates that are themselves not flat. The thick layer may include a filler, such as glass microbubbles as discussed in Ref. 18. The thinner layers are about 1–5 mils thick. Thick, triple-layered adhesive tape can also be manufactured with fumed silica as filler material in the center layer. The viscoelastic properties of the layers are regulated by using different photoinitiators and crosslinking monomer concentrations. A plastomer (polyvinyl acetate) can also be used as filler for the "carrier" layer (see Table 2.1). Such a carrier may be nonadhesive, but for certain applications the carrier itself possesses some pressure sensitivity.

A carrierless, self-sustaining pressure-sensitive film can be manufactured by laminating together an adhesive-based film and a rubber-based film. Such a tape is produced by bonding a 5 mil acrylic adhesive-based film with a 5 mil rubber compound based film (application temperature 180°C) [19]. The adhesive strength of this type of tape was found to have decreased by 10–45% after 1 year.

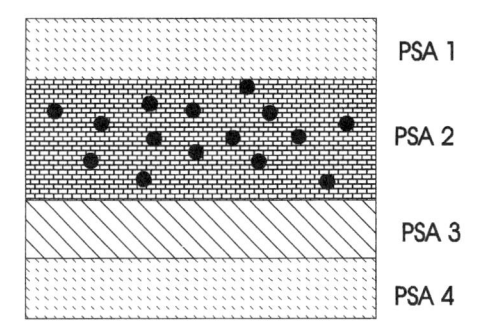

Figure 2.7 Multilayer pressure-sensitive composite manufactured by photopolymerization.

Table 2.1 Fillers for the Pressure-Sensitive Layer

Product	Filler Nature	Function	Ref.
Transfer tape	Glass microbubbles	Mechanical reinforcing	17,18
	Plastomer particles	Mechanical reinforcing	18
	Plastic scrim	Mechanical reinforcing	173
Removable tape	Expanded polymer particles	Stiffening and contact surface reduction	45
	Elastical polymer particles	Contact surface reduction	95
Thermometer label	Continuous fibers	Mechanical reinforcing	11
Medical tape	Inorganic filler	Reinforcing, crosslinking	35
	Crosslinked polysiloxane	Reinforcing	34
	Water	Electrical conductivity	97
	Air	Porosity	37
	Antimicrobial agent	Medical	97
Ironing labels	Plastic powder	Sealability	148

The use of ethylene-vinyl acetate (EVAc) copolymers for carrierless self-adhesive films (tapes) used for bonding roofing insulation and laminating dissimilar materials was proposed by Li [20]. Such formulations contain EVAc-polyolefin blends, rosin, waxes, and antioxidant. They are processed as hot melt and cast as 1–4 mm films. Soft solid pressure-sensitive adhesive has been prepared from high carbon number fatty acid metal soaps containing 20–200% tackifier [21].

In some cases the self-supporting adhesive layer is really a reinforced layer. Such constructions are called *tape prepregs* because the fiber-based reinforcing matrix is impregnated with the adhesive. In such tapes without backing, strength can be enhanced with a tissue-like scrim (cellulose or polyamide) included and coextensive with the adhesive layer. Such tapes behave like classical adhesive-coated carrier-based products. Foam-like carrierless tapes can be considered a development of tape prepregs.

Generally, constructions with porous carrier materials in which the adhesive can penetrate into the carrier, i.e., it can impregnate, can be considered as partially impregnated, partially carrierless, and partially adhesiveless tapes. Such PSPs have layers with both carrier and adhesive characteristics. In reality, transfer tapes with a carrier need the carrier as a technological aid only. During manufacture they are carrier-based; during application they are carrierless. Monoweb labels without a built-in carrier have also been manufactured [22].

In some cases the tape is an adhesive monoweb but behaves before or after use like a nonadhesive product. Such behavior is due to its superficial crosslinking. According to Meinel [23], in this way a carrierless PSP can be obtained (the adhesive layer possesses adequate strength to permit it to be used without a carrier material) that has better conformability. The product exhibits both adhesive and adhesion-free surfaces. The tack-free surface is achieved by superficial crosslink-

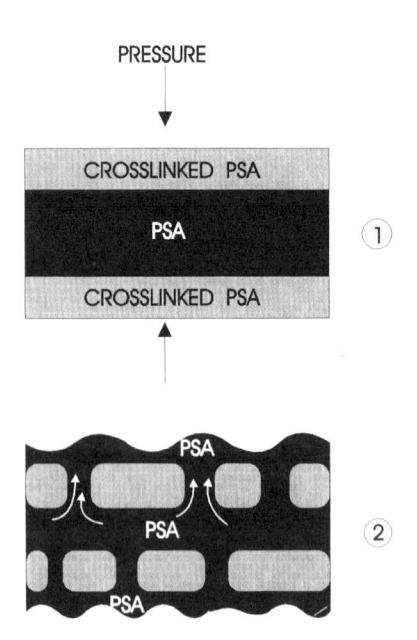

Figure 2.8 Buildup of a virtually adhesive-free tape. (1) Stored tape; (2) applied tape.

Buildup and Classification of PSPs

ing of the adhesive. This crosslinked portion of the adhesive has greater tensile strength and less extensibility. In some cases the polymer can be crosslinked only along the edges of the adhesive layer. The tack-free edge prevents oozing and dirtiness. The superficially crosslinked adhesive layer may be stretched to fracture the crosslinked portions, exposing the tacky core in order to bond it. The surface of the adhesive becomes virtually nontacky under light pressure but becomes tacky when the product is pressed against the adherent surface (Fig. 2.8).

Such monoweb tapes are mostly special or experimental products. It should be emphasized that only the production of superficially soft plastic film carrier materials and the development of thermoplastics having elastic (rubberlike) and viscoelastic properties allowed the manufacture of noncoated pressure-sensitive products.

Coated Monoweb. Coated monowebs are PSPs based on a solid carrier material and an adhesive (Fig. 2.9). For most applications, PSPs must have special adhesive characteristics. As discussed later (see Chapter 4), because of the broad range of raw materials available for PSAs and the sophisticated adhesive coating technologies now in use, the adhesive properties of a PSP can be easily regulated by coating a solid-state carrier material with a low viscosity PSA. Therefore special requirements for aggressive PSPs or removable products can be fulfilled only by coating.

Because of the balanced character of their adhesive performance (see Chapter 5) and their low pressure, high speed application technology (automatic label-

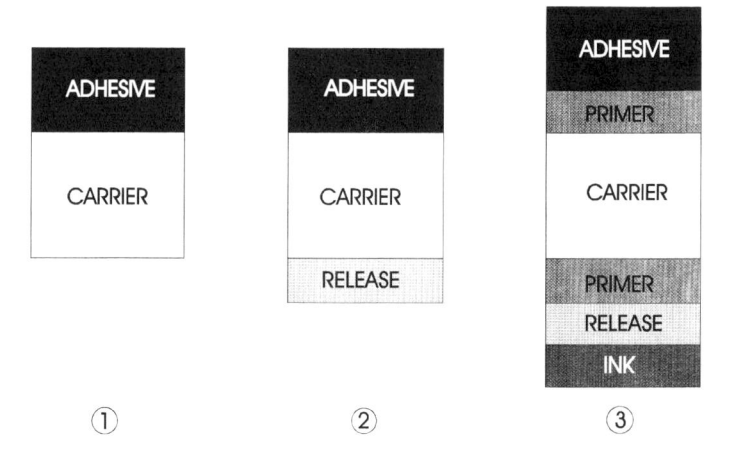

Figure 2.9 Schematic buildup of a coated monoweb. (1) Adhesive-coated carrier; (2) adhesive- and release-coated carrier; (3) primed, printed adhesive- and release-coated carrier.

ing), it is very difficult to manufacture labels without a separate release liner, i.e., labels with a monoweb structure. However, special monoweb labels (roll labels) have also been developed. Linerless labels are supplied as a continuous tape like monoweb material. A special coating on the top surface of the label prevents blocking of the adhesive layer [24]. John Waddington in the United Kingdom launched monoweb production in the 1980s; its application equipment is more expensive and its printing is more complex than those of other types of label manufacture [25].

Tapes were the first PSPs to have a coated monoweb construction. Theoretically, pressure-sensitive tapes have the same construction as wet adhesive tapes. Wet tapes are not adhesive before humidification, i.e., they do not need an abhesive layer. For pressure-sensitive adhesive tapes based on polymeric films, the release layer is usually the surface of the polymeric film remote from the adhesive, thus enabling the tape to be conveniently stored in the form of a spirally wound roll. The development of such products has been enhanced by the use of polyvinyl chloride polymerized in emulsion (EPVC). The surface of films based on EPVC is coated with a thin layer of emulsifier. This layer acts as an abhesive substance and allows low resistance unwinding of the rolls [23]. In practice, for pressure-sensitive tapes with high tack adhesive, a coated release layer is required [26]. This is not necessary if the carrier itself is abhesive like certain nonpolar plastic films or those containing slip or other abhesive additives. It is evident that the abhesivity of the carrier depends on the nature and geometry of the coated adhesive also. As stated by Kuminski and Penn [27], different unwinding

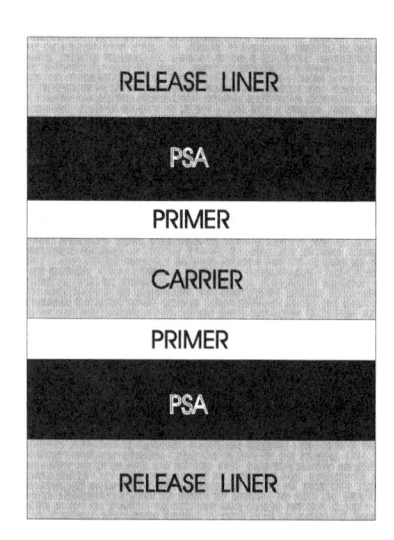

Figure 2.10 Buildup of a double-faced coated tape.

Buildup and Classification of PSPs 19

behavior is to be expected for a tape based on a hard acrylic or a soft hot melt pressure-sensitive adhesive (HMPSA). There is a trend on the market to provide customers with printed packaging tapes. The text or graphics are imprinted on the nonadhesive side of the carrier material before the tape is made. The problem with hot melt PSA is that a release coating has already been applied to the backing material before it is imprinted. Such a coating is necessary because hot melts do not release easily from the nonadhesive side of the carrier material. In some cases a separate release film (paper) should be interlaminated [2]. Double-faced coated tapes are products of this type (Fig. 2.10).

At the beginning of their introduction onto the market, protective films were tape-like products with a special removable adhesive. Masking tapes are really narrow-web masking (protective) films. Such products can also be manufactured as coated monoweb (Fig. 2.11). The buildup of special tapes is more complex. Although most tapes are monowebs, laminated and double-laminated constructions are manufactured also (see later). Transfer PSAs (so-called adhesives from the reel) coated temporarily on a siliconized release liner (double-faced siliconized) can be considered as adhesive-coated monowebs also [28]. In this case their solid-state carrier material is the release liner. Transfer printing materials like Letraset can be considered carrierless or temporary laminates having (after application) a monoweb character. In such cases the printing ink includes

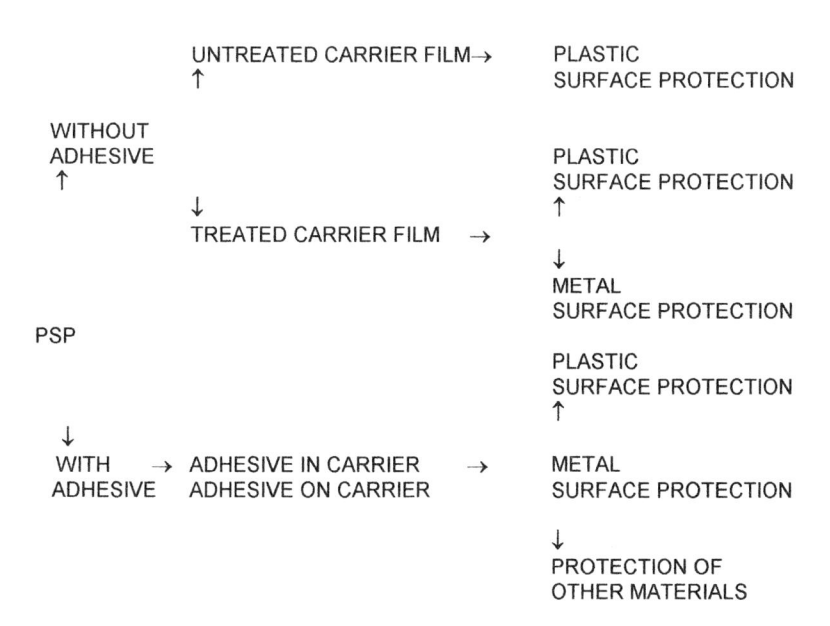

```
                        UNTREATED CARRIER FILM→      PLASTIC
                        ↑                            SURFACE PROTECTION

    WITHOUT
    ADHESIVE                                         PLASTIC
    ↑                                                SURFACE PROTECTION
                        ↓                            ↑
                        TREATED CARRIER FILM  →
                                                     ↓
                                                     METAL
                                                     SURFACE PROTECTION
PSP
                                                     PLASTIC
                                                     SURFACE PROTECTION
                                                     ↑
    ↓
    WITH     →  ADHESIVE IN CARRIER    →     METAL
    ADHESIVE    ADHESIVE ON CARRIER           SURFACE PROTECTION

                                                     ↓
                                                     PROTECTION OF
                                                     OTHER MATERIALS
```

Figure 2.11 Buildup and end use of protective films.

rial, one that has one surface with a series of recesses and another that is smooth [39]. In some cases the cross section of the carrier differs from the usual parallelepiped. Masking tapes have special carrier constructions to allow conformability. According to Lipson [40], a masking tape carrier has a stiffened wedge-shaped, adhesiveless longitudinal section (the thickest side away from the tape centerline) extending from one edge, with a pleated structure to conform to small radii.

There are various possibilities to reduce contact surface between the adhesive and the substrate. Figure 2.13 presents the main ways in which an adhesive surface with reduced contact area can be achieved.

Peel resistance and removability can be controlled by regulating the ratio between adhesion surface and application surface [41] (see also Chapter 5). For a removable product this ratio should be smaller than 1. To achieve such conditions the adhesive layer should be discontinuous or should have a virtually discontinuous surface also. Pattern or strip coating allows the use of nonremovable raw materials for removable applications. Natural rubber latex as an adhesive layer with a contoured surface can be used on paper to give a removable protective web that is adequate for acrylic items [42]. A pressure-sensitive removable adhesive tape useful for paper products comprises two outer discontinuous layers (based on tackified styrene-isoprene-styrene block copolymer) and a discontinuous middle

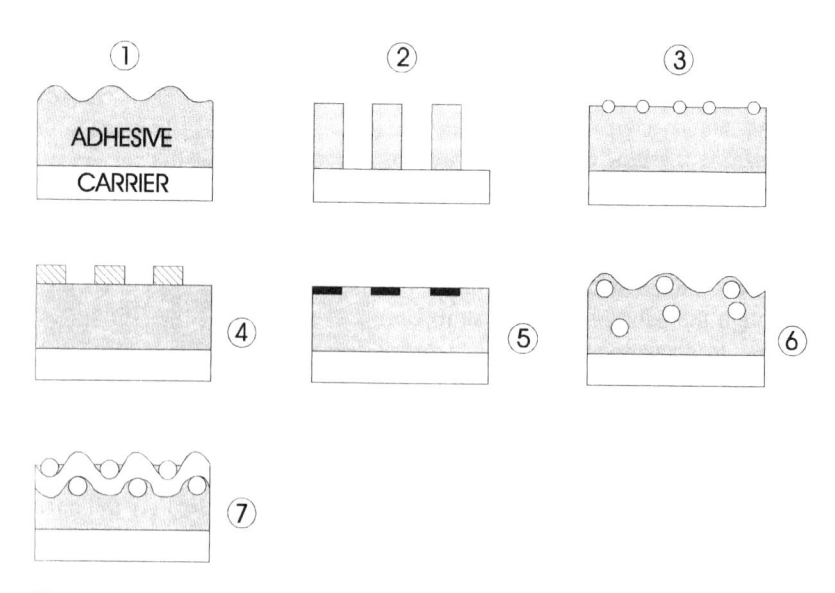

Figure 2.13 Adhesive surface with reduced contact area. (1) Contoured surface; (2) patterned surface; (3) filled surface; (4) release patterned surface; (5) partially crosslinked surface; (6) foamed surface; (7) textile inlay.

Buildup and Classification of PSPs

release layer [43]. The tape can be manufactured with less adhesive than conventional tapes, and its top side can be printed and perforated. Adhesive tapes with a narrower adhesive-coated width than that of the substrate and one edge with an adhesive-free strip are easy to untie and are useful for bundling electronic parts, building materials, vegetables, and other commodities [44].

Removable pressure-sensitive tapes containing resilient polymeric microspheres (20–66%), hollow thermoplastic expanded polymer (acrylonitrile-vinylidene copolymer) spheres having a diameter of 10–125 μm, a density of 0.01–0.04 g/cm^3, and shell thickness of 0.02 μm in an isooctyl acrylate-acrylic acid copolymer, have been prepared [45]. The particles are completely surrounded by the adhesive to a thickness of at least 20 μm. When the adhesive is permanently bonded to the backing and the exposed surface has an irregular contour, a removable and repositionable product is obtained, when the pressure-sensitive adhesive forms a continuous matrix, which is strippably bonded to the backing having a thickness of more than 1 mm. A foamlike transfer tape or foam tape is manufactured. The (40 μm) cellulose acetate–based tape could be removed from paper without being delaminated.

An embossed coating cylinder can coat a PSA pattern on a paper face stock material [46]. A discontinuous adhesive layer can be achieved by spraying also. For instance, polyurethane (PUR) foam alone or with film is used for medical tapes for fractures. For this product, the adhesive is sprayed, achieving a 25–75% adhesive-free back side [47]. The adhesive layer can have a rough surface also, with channels to enclose air flow between the adherent and the sheet. The surface roughness of the sheet has a 50–1000 μm width and a 10–1000 μm height of convexes [48]. An adhesive tape may have an adhesive only along the lengthwise edges (to allow easy perforation) [49]. A removable display poster is manufactured by coating distinct adhesive and nonadhesive strips on the carrier [50]. The adhesive strips are situated on the same plane, the plane being elevated with respect to the product surface. A multilayer adhesive construction can be manufactured also. Here the PSA-coated carrier material functions as a transfer base for a heat-activatable adhesive. After application of the heat-sensitive adhesive the PSA-coated tape can be peeled off [51].

Comparison of Uncoated and Coated Monoweb Constructions. As a result of the continuous development of pressure-sensitive adhesives technology, adhesive-coated monoweb constructions fulfill the most sophisticated end use requirements. Such products can be used for almost all classes of PSPs. Therefore it was mainly economic considerations that led to the development of uncoated self-adhesive PSPs. It is evident that a coated fluid pressure–sensitive layer having a minimum thickness of 1.0–1.5 μm allows better conformability for better contact with a solid surface than a self-supporting plastomer (or plastomer/viscoelastomer) with a minimum thickness of 7–25 μm (see also Chapter 3, Section 1). The more conformable carrierless (transfer) tapes are also manufactured by coating technology. Coating allows the combination of adhesive and nonadhesive areas on the same surface; this is not possible for extruded PSPs.

In some cases a coated PSP can have an enclosed adhesive layer. For certain products an adhesive modification of the bulky carrier material is required to improve the anchorage of the adhesive coating without a supplemental primer coating. In such cases the improved self-adhesivity of the top layer requires the use of an adhesive-repellent (release) layer also [52,53]. Biaxially oriented multilayer polypropylene films for adhesive coating have been manufactured by mixing the polypropylene (PP) with particular resins to improve their adhesion to the adhesive coating. To prepare an adhesive tape that can easily be drawn from a roll without requiring an additional coating on the reverse side, at least two different layers having different compositions are coextruded, with the thin back layer containing an antiadhesive component [54]. In some cases a primer is necessary for anchorage of the adhesive (see Chapter 4). Theoretically, primer thicknesses approaching a single molecular layer are required, but in reality a layer 0.5–1.0 μm thick is formed [55]. According to Ref. 56, tapes are generally 80 μm thick constructions, with 1–1.5 μm accounted for by the primer. Polymer-analogous reactions can be used also to modify the carrier surface before coating. In certain applications the PSA is coated first on a flexible polyurethane, acrylic (AC), or other foam and then is laminated on the carrier surface. The opposite surface of the foam may also be provided with a PSA layer. The polyolefin surface of the carrier can be polarized by graft polymerization using an electron beam (EB) with a higher than 0.05 Mrad dosage level. A special composite PSA construction is made with the adhesive on the lateral portions of the carrier only and with a longitudinal split in the carrier [57]. Such sophisticated constructions can be manufactured using the coating and laminating technology only.

It should be emphasized that generally noncoated monowebs allow lower bond strength than coated products. Because of the bulk monolayer nature of the adhesive and its limited flow in such constructions, the nature of the adhesive break can be easily controlled.

Comparison of Coated Monoweb Constructions. Coated monoweb constructions can have many different application conditions and end uses. Within the same class of labels, tapes, or protective films the end use requirements may also differ. Therefore, coated monoweb constructions are various also. The main representatives are described in Chapters 6 and 8.

Multiweb Constructions—Laminates

Among the PSPs, multiweb constructions were first introduced as labels and special tapes. Labels generally need a separate release liner. Their face stock and release liner are laminated together with PSA to form a multiweb construction. Tapes include both double-faced products and transfer tapes, which need a separate solid-state release or carrier layer [58]. As seen from Table 2.3, a multiweb construction can include multiple solid-state components (face stock and release liner) or multiple adhesive components. For instance, double-faced mounting tapes may have different adhesives on each side of the carrier [59] (see Chapters 6 and 8). Masking tapes may also have a special construction to allow conformability. According to Djordjevic [6], a masking tape has a removable release liner. Abrasion-resistant automotive decoration films also have a separate release liner [60]. Such products are made on a polyurethane or poly(ethylene-vinyl acetate)-polyvinyl chloride (EVAc/PVC) basis. The carrier is coated with a PUR primer. Self-adhesive wall covering consists of a layer of fabric having a visible surface, a barrier paper that has one surface fixed to another fabric layer, a pressure-sensitive adhesive coated on the barrier paper, and a release paper [61]. The release liner itself may also have a multilayer construction. Different degrees of release are achieved for a double-sided liner and a release paper coated on one side with polyethylene and using the same release component.

Multiweb products are not new. They were developed in the packaging materials industry. The introduction of new face stock materials, especially filmlike face stock materials, required the improvement of certain carrier characteristics such as dimensional stability, cuttability, and temperature resistance. This has been possible partially due to the development of laminated face stock materials. Such multiweb constructions have been produced by bonding the solid-state components via adhesive coating or extrusion. Theoretically such adhesive-coated laminates differ from PSPs with respect to the nature of the adhesive and the character of the adhesive bond. They are manufactured using a permanent ad-

Table 2.3 Multiweb PSPs

Product	Multiweb status During manufacture	Multiweb status During application	Components Solid state	Components Coated
Label	Yes	No	Carrier, release liner, overlaminating film	Adhesive, primer, release, ink, lacquer
Double-sided coated tape	Yes	No	Carrier, release liner	Multiple adhesive, primer
Transfer tape	Yes	No	Multiple release liner	Adhesive
Form	Yes	Yes	Multiple carrier, multiple liner	Multiple adhesive, primer, multiple release

hesive. Generally both temporary and permanent laminates are used in the buildup of pressure-sensitive products. Permanent laminates serve as carrier materials; temporary laminates are PSPs (Fig. 2.14).

Multiweb constructions can include solid-state components and adhesives that differ in their chemical nature and buildup (e.g., paper, plastic films, fabrics, nonwovens, foam). For instance, double-faced coated tapes can have a PSA on one side and a reaction mixture of uncrosslinked PUR on the other side. A PVC tape applied on a PUR foam reacts with the adherent and adheres to it chemically [62]. Business forms include permanent and temporary laminates (see Chapter 8). A thermal tag is produced as a reinforced laminated construction for subsequent conversion by label producers. Taking an airline luggage as an example, a central plastic layer gives it strength, the self-adhesive layer allows wrapping around luggage handles, and the upper layer is a top-coated thermal paper onto which the data are printed as a barcode [63].

The PSP laminate manufactured to protect the adhesive-coated surface of a tape or label is a temporary construction. In such a product the release liner may also protect the adhesive or the solvent incorporated in the adhesive layer [64]. Another role of the separate release liner is that of a continuum passing through the coating, converting, and labeling machines carrying the discontinuous label. Decals or labels applied by hand do not need this function. In some cases, such as the manufacture of transfer tapes, the use of a solid-state component is a technological modality only to build up and apply the product. A patent [64] describes the photopolymerization (UV) of acrylic monomers directly on the carrier to manufacture a tape. This patent uses a temporary

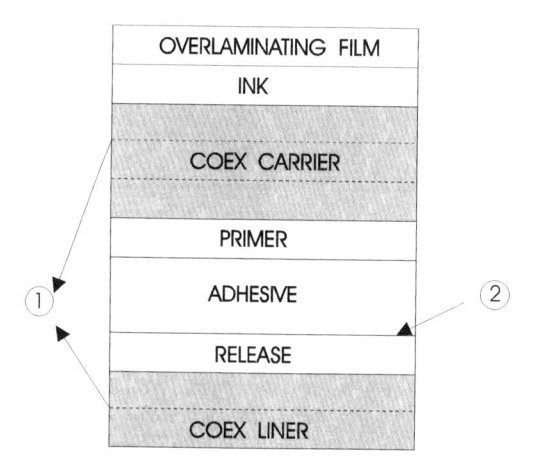

Figure 2.14 Laminates used for PSPs. 1, Permanent laminates; 2, temporary laminate.

carrier, an endless belt, which does not become incorporated in the final product. The tapes have PSA on both sides of the carrier. In this case the laminate is an auxiliary construction built up to allow the construction of the final laminate from the label or tape and the substrate. Labels and certain tapes are built up during manufacture and application as temporary laminates and are used as permanent laminates.

Theoretically, labels are built up by coating a solid-state face stock material with a PSA and covering the adhesive surface with an abhesive one. In practice, transfer coating is common, i.e., the release liner is coated with the adhesive and laminated together with the face stock material. The result is a temporary sandwich. During conversion, the continuous weblike face stock material is die cut, and during labeling the discontinuous face stock material is transferred from the release liner to the substrate and the temporary laminate is destroyed. In the case of double-sided coated tapes, first one of the release liners is taken away, then the other.

Forms are labels that have a multiweb, multilaminate structure, where continuous and discontinuous carrier materials and PSAs with different adhesive properties are laminated together to allow time- and stress-dependent controlled delamination (Fig. 2.15). Their delamination is carried out in several steps during manufacture and application.

Some PSPs are constructed with both types of laminates, temporary and permanent. Security labels may have a complex structure. Such labels are used for hangtags that are printed with advertising text, product information, and a barcode and include an electronic security element [65]. Some forms also include both temporary and permanent laminates.

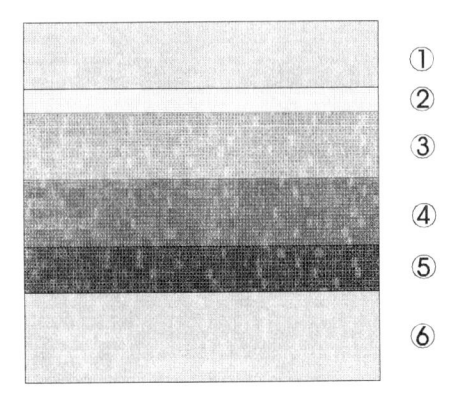

Figure 2.15 Buildup of a form. *1*, Carrier; *2*, adhesive; *3*, carrier; *4*, adhesive; *5*, release; *6*, carrier.

1.3 Components of PSPs

As discussed earlier, during their application, pressure-sensitive products have to be laminated; therefore they need a carrier component. In order to bond they have to display adhesivity; therefore they contain an adhesive component also. Theoretically, such products possess at least a carrier and an adhesive component.

The technical requirements concerning the components of a pressure-sensitive laminate have changed in parallel with the development of PSPs. Table 2.4 lists the requirements for the components of a pressure-sensitive laminate.

It is evident that in the first period of development of pressure-sensitive labels and tapes the most important quality of the adhesive was its adhesion force, i.e., peel resistance. The carrier used for labels had to allow their conversion (printing and cutting). Therefore its mechanical properties, dimensional stability, and surface quality played an important role. For tapes the main requirement was mechanical resistance. Productivity increases forced the improvement of conversion performance. Some performance characteristics are needed during processing of the temporary laminate; others are required during its application. Therefore the components of the PSPs play a special role in their design and manufacture.

Carrier Material

Most pressure-sensitive products are weblike constructions laminated together during their manufacture or application. Exceptions include some special cases where the adhesive layer itself carries information or plays an aesthetic or protective role and the carrier is the laminate component with a packaging and protection function, i.e., the component with special aesthetic and mechanical characteristics. Generally the carrier acts as a face stock material or/and as release liner. In some special cases it also plays the role of adhesive. Table 2.5 summarizes the role of the carrier material in PSPs.

The Role of the Carrier Material. The carrier should have mechanical characteristics that satisfy the end use requirements of the product. In practice, such requirements may vary considerably. Dimensional stability of the carrier material is required for labels and to some extent for certain tapes (e.g., packaging and mounting tapes). For other tapes (e.g., masking or wire wound tapes), hygienic products, deep drawable protective films) carrier deformability is needed [66]. Tamper-evident labels should display excellent conformability and destructibility (low internal strength) [67]. Except for labels and some common packaging tapes, application requires regulating the dimensions and dimensional stability of the carrier material (i.e., control of its plasticity, elasticity, elongation, shrinkage, etc.). If dimensional stability is required, ideally it should not depend on the product geometry. Low gauge carrier materials have to display the same characteristics as thick or reinforced products. Isotropy is another desired property (see Chapter 3, Section 2).

Table 2.4 Requirements for PSP Components

Laminate components	Initial Label	Initial Tape	Actual Label	Actual Tape	Protective film
Adhesive	Peel level	Peel level	Peel level, substrate peel control, time-dependent peel control, removability	Peel level, substrate peel control, time-dependent peel control, removability	Peel level, substrate, removability
Carrier	Printability, mechanical resistance, dimensional stability, die cuttability	Mechanical resistance	High quality printability, mechanical resistance, dimensional stability	Printability, mechanical resistance, deformability	Printability, mechanical resistance, deformability
Release	Removability	Removability	Controlled release	Release	Release

Table 2.5 The Role of the PSP Carrier Material

	Product		
Function	Label	Tape	Protective film
Mechanical strength	x	X	x
Protection	—	x	X
Information carrier	X	x	x
Dosage	x	x	x
Adhesion	—	x	X

X = primary function; x = secondary function.

Cuttability of the carrier material is required for labels during their manufacture. For tapes it is needed during their lamination also. For protective films it is necessary after their application in laminated status (e.g., laser cuttability). For all these products cuttability is also required in the conversion phase of manufacture.

Generally the carrier plays the role of a face stock material. It is the most important solid-state part of the product, functioning as a packaging, protecting, and information carrying component. This is the general role of the carrier material in labels, tapes, and protective films, and there are only a few weblike PSPs without a solid-state face stock material (e.g., transferable letters, decalcomania, transfer tapes) [68].

According to the end use of the PSPs, quite different performance characteristics are required for the carrier used as face stock material (i.e., adhesive-coated permanent top layer) or release (see also Chapter 8). Therefore the nature of the carrier materials differs also. According to Ref. 69, a tape carrier can be any reasonably thin, flat, and flexible material. Thus it may be woven, nonwoven, metallic, electrically resistant, natural or synthetic, tear-resistant or fragile, water-resistant or water-soluble (see also Chapter 6). For instance, carrier materials used for tamper-evident labels should display low mechanical resistance. Fragile pressure-sensitive materials are either very thin or have low internal strength [70]. They are coated with a very aggressive PSA. Slits or perforations improve the tamper-evident performance. For such products the mechanical properties of the liner complement the low mechanical properties of the face stock material during manufacture and application.

The carrier geometry also influences end use performance. For instance, masking tapes can have special carrier constructions to allow conformability [6]. A masking tape carrier has a stiffened, wedge-shaped, adhesiveless longitudinal section (the thickest away from the tape centerline) extending from one edge, with a pleated structure to allow it to conform to small radii. An adhesive carrier material can be perforated along the centerline of the long direction; it is suitable for tying together printer paper [48]. The carrier for tapes can be reinforced; a special buildup (folding) of the reinforcing fibers ensures extensibility [71]. A study in the early 1990s found that reinforced tapes made up 4% of European packaging tape production [72]. Fiber-reinforced tapes include biaxially oriented polypropylene (BOPP), release, adhesive, filaments, and a hot melt adhesive. For a pressure-sensitive label having a wrinkle-resistant, lustrous, opaque facing layer for application to collapsible wall containers (squeeze bottles), the carrier has a thermoplastic core layer, with upper and lower surfaces and voids, a void-free thermoplastic skin layer fixed to the upper surface and optionally to the lower surface of the core layer, and discrete areas of pressure-sensitive adhesive. As the core layer, a blend of isotactic polypropylene/polybutylene terephthalate is used, with polypropylene as the skin layer and circular dots of HMPSA as the adhesive [73]. Woven plastics have been proposed as carrier material for double-layered PSA tape [74]. Paper is suggested as an extensible carrier for medical tapes [75]. Polymer-impregnated paper can be used also as the carrier for tapes [76]. For automotive masking tapes, pleated paper is applied [77]. Woven fiberglass scrim has also been recommended as carrier material for tapes.

The face stock material can repel, absorb, or contain chemicals such as solid-state water-absorbent particles [76]. For such application it serves a dosage and storage function. The carrier materials can also act as protective surfaces. Stain-resistant non-paper face materials help with the staining problem. For temperature-resistant tapes, polyethylene (PE), polyester (PET), and Teflon are suggested [78]. Modified polyethyl methacrylate can be used as a protective film for various substrates such as PVC, acrylonitrile-butadiene-styrene (ABS) and polystyrene, wood, paper, or metal [79]. It is weathering-resistant and UV-absorbent.

It is evident that independently from the special end use requirements, the main surface-related performance required of a face stock carrier material is anchorage of the adhesive. The role of the carrier as the basis for adhesive anchorage is illustrated by the data of Huang Haddock [80]. The same medical adhesive coated on cloth backing gives an average adhesive transfer index of 1.5 in comparison with a vinyl carrier displaying a value of 3.5.

As discussed earlier, labels need a separate solid-state abhesive component (release liner). In this case the abhesive layer is coated on a carrier material. The need for a continuous, solid-state, separate abhesive laminate component for labels arises from their aggressive adhesivity and discontinuous character. Such a liner can also act as an information carrier. A large variety of products, e.g., common paper, satinated craft papers, clay-coated papers, polymer-coated papers, and plastic films, are used as raw materials for liners [81] (see Chapter 6). For tapes and protective films the back side of the self-adhesive (coated or uncoated) materials can act as a liner. This function may be ensured by a supplemental release coating or by the choice of an adequate carrier material with a low level of adhesivity (e.g., nonpolar polyolefins or EPVC) [82]. The abhesive layer can be coated on the back side of the carrier material or included in it, generally as a separate layer [53]. For instance, for a biaxially oriented multilayer polypropylene film used as carrier and/or liner for tapes, the second layer of the film, which faces the adhesive, has a thickness of less than one-third of the total thickness of the adhesive tape and contains an antiadhesive substance [83]. Another construction includes the release substance randomly distributed in the carrier material. Abhesive substances can also be built into the carrier material. For instance, a self-adhesive tape for thermal insulation used to secure a tight connection between heat exchange members (metal foils) and polyurethane foam–based thermal insulation materials is prepared by applying an adhesive that is able to adhere to refrigerated surfaces on an olefin copolymer carrier material (containing fatty acid amides or silicone oils as release agents or lubricants) that is bonded to the polyurethane [84].

The Carrier as Adhesive. The first application domain of adhesive-free laminates was the lamination of packaging films. This development continued in other areas and led to PSPs without a coated adhesive layer. Hot laminated corona-treated common plastic films and special plastic films have been introduced on the market. These films do not contain special viscous components (tackifiers) or viscoelastic components (thermoplastic elastomers or low molecular weight elastomers) and are applied at a temperature above or near the softening range of the products (see also Chapters 6 and 8). Later plastic films having viscoelastic properties (e.g., very low density polyolefins, copolymers of ethylene with polar vinyl or acrylic monomers) or containing viscoelastic additives (e.g., polybutylene copolymers, TPEs) were developed for use at lower or room temperature [85]. It is evident that classical plastic films without an adhesive layer cannot possess the tack of PSA. Plastics can flow, but such flow is a time (stress)- or temperature-related performance (see high pressure hot laminating in Chapter 8, Section 4). It should be mentioned that high application pressure, temperature, and/or special surface treatment alone are not sufficient to allow hot laminating of plastic films. A special chemical or macromolecular composition that ensures improved flow is also required. That means that really adhesive-free hot laminating protective films are products with a special carrier film composition.

The need for an adhesive carrier film having improved adhesive anchorage became evident during the development of polyolefin carrier materials. In this case an adhesive level is required that ensures the anchorage of the adhesive on the carrier. Monolayer and multilayer polypropylene carrier materials with embedded special resins have been developed for this purpose [86,87].

Other Functions of the Carrier. The carrier can function as a storage or dosage system. Its surface characteristics can be modified by built-in additives. In some cases the carrier contains special ingredients for certain purposes. For instance, cover tapes for bathroom use are made with a bacteriostatic agent (which inhibits bacterial growth) incorporated in the vinyl formulation [88]. Tackifier resin can be included in the carrier also [76]. A butadiene-styrene elastomer is mixed with a colophonium-based tackifier and coated (impregnated) on a paper carrier. This carrier is coated with a PSA on natural rubber. Special paper is required for direct thermal coating. Common thermally printable papers include a colorless leuco dye and an acidic color developer [89]. These are coated and held onto the surface of the paper with a water-soluble binder. During printing the two components melt together and react chemically to form the color. Inorganic fillers such as calcium carbonate, chalk, or clay are used to limit this image to the heated area. The problem of image stability (protection again chemicals) is solved by applying a topcoat as a transparent film-forming layer. It is possible to also apply a barrier coat to the underside of the paper to prevent the migration of adhesive or plasticizer from the opposite side.

Coating Components

The classical way to produce pressure-sensitive products uses coating as the main manufacturing step. A carrier material is coated with a pressure-sensitive adhesive. The manufacturing process may include multiple coating, where different surface-coated layers are built up that have various (e.g., adhesive, abhesive, machining, aesthetic) functions.

Adhesive Components. Generally pressure-sensitive adhesivity is provided by a pressure-sensitive adhesive component built into the product. In some special cases (e.g., protective or decorative films) pressure sensitivity can be a characteristic of the carrier material and/or may be the result of special physical treatment or application conditions (pressure and temperature).

In the first stage of their development, PSAs were used to impart pressure sensitivity. Therefore the adhesive was the most important component of PSPs. Classical self-adhesive products are pressure-sensitive. Really good PSAs do not need a measurable application pressure. In blow touch labeling operations, labels fly through the air and land without pressure on the substrate. This is possible because of their high instantaneous adhesivity called tack (see Chapter 5). Tack is the result of molecular mobility and is based on a complex chemical formulation (see Chapter 6). The PSA can be coated on or built into the bulk carrier material. Coated adhesives are such components that are deposited on the surface of a carrier material. As discussed later (Chapter 6), there are various coating methods, depending on the adhesive, the carrier, and end use requirements. Built-in adhesives are chemical compounds that can be mixed in with the components of the carrier material and processed as a homogeneous or heterogeneous (i.e., laminate) composite. Such adhesives are processed together with the main thermoplastic component of the carrier material, and therefore in this case no supplemental adhesive coating is necessary.

A number of raw materials display (alone or in a mixture with viscous components) pressure-sensitive properties. In the classical recipe, PSAs were (at least) two-component formulations composed of elastic and a viscous component. Later one-component viscoelastic raw materials were developed (see Chapter 4). The bonding function of the adhesive is supplemented with other functions. In some cases the adhesive has to also play the role of a porous carrier material. For instance, one such adhesive is manufactured from a thermoplastically deformable composition that is calendered using a release-coated embossed cylinder that perforates the adhesive layer [31]. Fluid-permeable adhesive is required for transdermal applications [34]; such an adhesive must not irritate the skin [90]. The irritation caused by the removal of tape from the skin was overcome by including certain amine salts in the adhesive [91,92]. The role of the adhesive as a storage place for different end use components may be various. Postapplication crosslinking is proposed to improve shear. This is achieved by storing latent crosslinking agents in the adhesive [93]. An antimicrobial agent such as iodine can be incorporated in the skin adhesive as a complex also [94]. As a filler to reduce the contact surface and decrease buildup, elastic polymer particles having a diameter of 0.5–300 μm can be included in the adhesive [95]. To improve their detackifying effect, the particles may include an ionic low tack monomer [96]. The adhesive used for bioelectrodes applied to the skin contains water, which affords electrical conductivity. Electrical performance can be improved by adding electrolytes to the water [97].

The adhesive can have a multilayer structure also. This type of structure may serve as a modality to control creep compliance, according to Gobran [98], with a plurality of superimposed adhesive layers having different gradients of shear creep compliance. Multilayer heterogeneous adhesive is obtained by UV crosslinking of 100% solids acrylics [99]. The radiation is partially absorbed, partially transmitted, and partially reflected. The maximum radiation level is experienced at the top of the adhesive layer; therefore the degree of crosslinking is maximum on this side. Therefore direct and transfer coating give different adhesive characteristics. A uniform, isotropic adhesive layer can be achieved only by using a UV-transparent face stock (backing) and double-sided irradiation. As discussed earlier, a thick, triple-layered adhesive tape (with filler material in the center layer) can be manufactured by photopolymerization also. This is really a carrierless tape that has the same chemical composition for the "carrier" (i.e., self-supporting and adhesive) middle layer and the outer adhesive layers. The second adhesive layer has the same composition as the first one [21]. A tape made according to a patent [22] by superficial crosslinking of the adhesive also has a heterogeneous multilayer adhesive.

Abhesive Components. Because of the high tack of many PSAs or PSPs, an abhesive layer should be used as a protective or intermediate component. In the first

Buildup and Classification of PSPs 35

period of PSA development, paper was used as carrier material. Because of its texture and polarity, most PSAs give a permanent bond with paper. Therefore it was necessary to coat the surface of the paper with an abhesive coating to build up a protective material (release liner). For labels the abhesive properties of the liner have to be exactly controlled. For tapes and protective films, such properties play only a secondary role. In the actual stage of development the precise regulation of the release force is made possible by use of a separate, abhesive-coated solid-state laminate component. This is an expensive and environmentally inadequate technical solution. In addition, it does not satisfy some special requirements (resistance to abrasion, lubrication, etc.) for tapes and protective materials.

Printing Components. To improve the adhesive performance of the PSA layer or the surface characteristics of the carrier material, other nonadhesive or abhesive coatings can also be applied. These include primers, printing inks, antistatic agents, and other materials (see Chapter 4).

2 CLASSES OF PRESSURE-SENSITIVE PRODUCTS

The major product areas in the flexible coatings market are plastic coatings, pressure-sensitive products, coated laminated film composites, and magnetic media. A general classification of the laminates would be possible on the basis of their carrier materials: paper, film, and laminates with other carrier materials. According to Merrettig [100] the most important pressure-sensitive products based on films are:

packaging tapes (based mainly on "hard" films);
office and decorative tapes (based mainly on "hard" films);
insulation tapes (based on "soft" films);
corrosion-protective tapes (based on "hard" and "soft" films); and
decorative films (on printed "hard" film, laminated on paper).

The most important paper- and textile-based pressure-sensitive products are [100]:

labels;
crepe tapes; and
textile tapes (for technical and medical use).

Depending on their bonding characteristics, manufactured laminates may be permanent, removable, or wet removable. Pressure-sensitive products can be divided into classes according to their end use. Some of these classes contain well-known products having a broad range of representatives (labels, tapes, etc.); others are special or one-of-a-kind products. Because of the continuous expansion of their application field, the number of special products is increasing.

Labels and tapes are the main pressure-sensitive products (Table 2.6). Both have been developed by replacing classical wet adhesives with pressure-sensitive products. Pressure-sensitive adhesives have been used since the late 1800s for medical tapes and dressings. Industrial tapes were introduced onto the market in the 1920s and 1930s, self-adhesive labels in 1935 [101]. Another product class including protective, cover, and separation films has been developed to replace carrier-free protective coatings. A comparison of the requirements for these product classes shows that labels and tapes have similarities with respect to the nature of the adhesive, and protective films and tapes exhibit common features concerning the nature of the carrier. In 1983, about 11% (w/w) of European adhesives were used for the primary pressure-sensitive applications (tapes, labels, and postage) and the annual growth (3.2%) of this segment has been higher than the mean value for the adhesives industry (2.5%) [102]. According to one publication [103], in the mid-1980s the most obvious applications for traditional pressure-sensitive polymers were labels, tapes, and decals. A decade later 55% of PSAs were used for tapes, 34% for labels, and 11% for specialties [104].

2.1 Labels

In 1987 labels were the most important pressure-sensitive products in Europe; their proportion in the global pressure-sensitive market was estimated at 72% and it was forecast that by 1990 it would be 76% [105]. As coated carrier material the label/tape ratio attained 7/1. Label production continues to increase, and according to Ref. 106, 7 mrd m^2 are produced each year. As foreseen in Ref. 107, self-adhesive labels will cover about 60% of the label market and the glue-applied market share will continue to decrease.

Table 2.6 The Main Pressure-Sensitive Products

Product	Monoweb	Multiweb	Adhesiveless	With adhesive
		Buildup		
Label	—	x	—	x
Tape	x	x	x	x
Protective film	x	x	x	x
Decalcomania	—	x	—	x
Form	—	x	—	x
Decor film	—	x	x	x
Envelope	x	x	—	x
Separation film	x	—	x	x
Wall covering	x	x	—	x

The definition of this successful product is difficult. Is it a packaging material? Packaging materials have to fulfill the following main functions [108]: They have to (1) contain, (2) preserve, and (3) present.

Containing is a mechanical function. Labels (except for some special dermal dosage products) do not fulfill this requirement. Although there are some anti-theft or closing labels, the function of preserving is not a general requirement for labels. Their principal function is presentation. As stated by Fust [109], the main function of labels is to carry an image such as information for a special product. According to Pommice et al. [110], packaging is used not only as a protection for the products but also as a support for advertisements and as a sales aid. This is also true of labels. The computer label business is one of the fastest growing segments of the label market [111]. Variable data bar-coded labels (routing labels for mailing, inventory control labels, document labels, individual part and product labels, supermarket shelf labels, and health sector labels) are the most important sector of nonimpact printing. Stanton Avery developed Kum-Kleen labels in the 1930s [112]; these were the first pressure-sensitive labels. Other types of labels have been used for almost 100 years. The growth in the use of PSA labels has been a result of substitution [113].

Labels are discontinuous items that serve as carriers of information and can be applied on different substrates. Their geometry, material, buildup, processing, and application technology can vary widely. Materials weighing up to 350 g/m^2 are used [114]. Fan-fold material can be processed. Single labels, strip labels, and label strips can be printed. Labels between 30.2 and 164 mm wide cover almost all areas of application. In the 1970s the major types of labels were wet adhesive labels, pressure-sensitive labels, gummed labels, heat adhesion labels, and shrink labels.

In Europe the traditional labeling techniques are [115] roll feed labeling (wraparound labeling), cut-and-stack labeling, and self-adhesive labeling.

The types of labels now produced include wet glue applied, pressure-sensitive, gummed, heat seal, shrink sleeve, and in-mold. Pressure-sensitive labels have a higher manufacturing cost than wet adhesive labels, but their application is easier. Wet adhesive labels are suggested for application fields with very high speed but low quality requirements (metal cans, glass and plastic bottles, etc.) [116]. Shrink labels use a shrinkable 40–70 μm film. In-mold labeling inserts the label in the mold prior to molding [117]. The insertion of chip cards is also recommended [118]. Heat adhesion labels and papers have been used in the packaging industry [119]. Some heat adhesion labels adhere immediately, and others have a delayed adhesive effect. The advantages of the heat-activatable or heat applied labels are that they can be applied at high speed, they require no preliminary operations before application, they are rapidly activated, their use requires no solvent (no pollution) and no humidity, there is no chemical yellowing, they give good permanent adhesion, and they exhibit no edge splitting. On much the same principle hot stamps, i.e., labels without a solid-state carrier, have been developed. Dormann [120] classified pressure-sensitive labels as cold sealing agents. Generally sealing agents can be divided into hot seal and cold seal systems. Hot seal systems include hot seal dispersions, hot seal varnishes, hot seal film, and extrusion coatings. Cold seal systems include PSAs and cold seal dispersions.

Certain label properties such as removability can be provided only by pressure-sensitive labels. As discussed earlier, labels are not packaging materials per se, but they are used together with packaging materials, to which they are applied to present the product, to advertise, to serve as decoration, and to provide information about the product [121].

Buildup and Requirements for Labels

Labels are temporary laminates composed of a face stock material with an adhesive pressure-sensitive surface that is protected by a separate abhesive component, the release liner. According to Havercroft [122], such products can be considered a special case of soft laminates that allow reciprocal mobility (shrinkage and elongation) of the solid-state components. The nature, geometry, and buildup of the carrier material; the nature and geometry of the adhesive; the nature and geometry of the release liner; and the buildup of the finished product may vary greatly. Labels can be round, square, rectangular, elliptical, or irregular in shape. Their purpose may be to impart essential information, i.e., company address, product description, or weight statement. In addition, depending on the product, they may carry instructions and warning statements. Base labels are almost without exception printed in one color and coated with permanent or removable adhesive. Decorative labels are printed in many colors on various materials. The larger ones are coated with a so-called semipermanent adhesive with low initial tack.

Labels may have a very complex construction. A label construction contains seven layers: release liner, release layer, adhesive, primer, face stock, ink, and top coat [123]. Multilayer self-adhesive labels comprise a carrier label containing silicones, an adhesive layer, a printed message, a carrier layer (e.g., polyester), a release layer (e.g., silicones), a second adhesive layer, a second carrier layer (e.g., polyethylene), and a printed message. Labels may have more than two adhesive–carrier layer–message layer units [124]. Decalomanias can be built up as labels also. Decalomanias are manufactured by using screen printing. First a clear carrier lacquer is printed; this layer provides the mechanical resistance of the product. The image is coated (mirror printing) on this layer with screen or offset printing. The next layer is transparent and is followed by a screen-printed PSA [125]. In another procedure, the PSA is coated on release paper and the PSA-coated liner is laminated together with a printed film [126].

Face Stock Material. Because of the different substrates used and the various requirements concerning the aesthetic, mechanical, and chemical characteristics of the label material, various raw materials have been developed as face stock for labels. Fabric, polyvinyl chloride, cellulose acetate, and polyester films have been suggested for labels [127] (see also Chapter 6). Material combinations are also used. For instance, a polyethylene/paper laminate with the polyethylene layer enclosed between paper layers is recommended for a special application where the product should be partially transparent after delamination [128].

Concerning the face stock, changes in label usage have to be taken into account. The development of reel labels and of nonpaper labels is faster than the average growth [129]. Label form and function can change also. For instance, some special pharmaceutical labels may have take-off parts to allow multiple information transfer; parts of the label remain on the drug or on the patient [130], i.e., the label works like a business form. Advances in document security and identification affect the label market (identification systems, document security, anticounterfeiting, corporate fingerprinting, copyable labels, etc.) and label construction. According to Thorne [113], labels are used mainly to provide information on packs (83%) and on cases (71%), for design purposes (32%), for off-pack promotions (20%), to carry bar codes (20%), and for security (15%). Unlike consumer goods, the promotional function does not have to convince the purchaser of the need to buy.

It is also possible to classify labels according to type of carrier and application domain. For instance, according to Waeyenbergh [115], self-adhesive labels printed on oriented polypropylene (OPP) are used for:

metal cans (preserves, paints, and drinks);
glass bottles (champagne, wine, and cosmetics);
cardboard containers (foodstuffs);
flexible packaging (food and nonfood items);
technical uses (labels, stickers, and rating plates); and
special domains (pocket labels).

Adhesives for Labels. Because of the broad range of bonding forces required in label applications; the differences in types of bonding, debonding speed, and bond break nature; and the various possible types of coatings, a number of different adhesives are used for labels. The main PSAs applied in this field are acrylics and rubber-resin formulations. First rubber-resin PSAs based on natural raw materials (natural rubber and rosin derivatives) were used to produce PSPs. As adhesives for labels they possess the advantage of having well-balanced adhesive properties, easy regulating of adhesive performance, and low cost. Unfortunately they have a limited resistance to aging. Later acrylics and carboxylated styrene-butadiene dispersions and hot melt pressure-sensitive adhesives (HMPSAs) based on styrene-butadiene block copolymer (SBC) were introduced as competitors to acrylics (AC). For special applications a broad range of commercial and experimental low volume products are available (see Chapters 6 and 8). Acrylics are applied as solvent-based, water-based, or hot melt formulations for labels having various buildups and end uses. Labeling is the most important field for water-based PSAs (WBPSAs) [131]. Removable price labels use almost exclusively a paper carrier coated with rubber-resin adhesive [132]. A decade ago in Europe about 30–40% of the HMPSAs produced were used for labels [133]. First SBCs, later acrylic-based HMPSAs were introduced. The manufacture of UV-crosslinkable (100% solids) acrylates is more expensive than that of acrylic dispersions. Such products are an alternative to solvent-based adhesives where without too much capital investment a common hot melt coating line can be equipped with UV curing lamps [99]. Although EVAc-based adhesives are produced as hot melt, solvent-based, or water-based formulations, because of their unbalanced adhesive performance they have been used more in other than web coating applications [134]. Water-based ethylene-vinyl acetate copolymers have been tested for labels as a less expensive alternative to acrylics.

Release Liner for Labels. The release liner for labels is a separate carrier-based component of the laminate. It has to display abhesive characteristics. These are (generally) given by a coated abhesive layer. Therefore, the construction of the release liner is similar to the face stock buildup. A solid carrier bears a coating layer. Various carrier and coating materials and various coating techniques are used for the manufacture of the release liner.

Principle of Functioning of Labels

Labels are sheetlike pressure-sensitive items that carry information and are made to be applied (laminated) on a solid-state adherend surface. In comparison with other PSPs such as tapes or protective films, which are manufactured and applied as a continuous web, labels constitute the temporary component of a laminated web and are applied as discrete, discontinuous items. The main requirements for labels are flexibility, aesthetic value, low cost, quality, ease of application, speed of application, strength, and moisture resistance [112]. It has been stated [112] that, of the technical criteria, ease and speed of application have been the most difficult to fulfill.

It should be mentioned that the technical functions of labels have also changed along with the development of labeled products. As an example, price labels, in addition to communicating such information as product name, weight, composition, expiration date, and manufacturer's name, must be able to capture the potential buyer's attention. For such effects labels have had to become more and more brilliant and bigger. They have had to replace direct conversion of the packaging material. Also, the direct printing of bottles is limited by the quality of the image [117]; here labeling is recommended.

Special Characteristics of Labels

As mentioned above, like other PSPs, labels are manufactured as continuous web but applied as discontinuous items, i.e., discrete parts of this web. Therefore, their construction must allow them to be separated from the rest of the laminate and applied on the adherend surface. To allow them to be separated, labels have to be die-cut and must have an abhesive layer on the back side. They must possess sufficient mechanical stability and adhesivity to be applied. As information carriers, labels have to fulfill aesthetic, coating, and mechanical demands.

Classes of Labels

The diversity of labels is astonishing. According to Ref. 135, thirty years ago more than 100 paper label laminates had been produced. In 1993 the main application domains of labels were computers, foods, cosmetics, pharmaceuticals, and miscellaneous markets. Pressure-sensitive booklets, coupons, and piggybacks can also be considered labels [136]. There are various principles used to categorize labels according to their buildup, dimensions, labeling methods, end uses, and so on. Principally they can be grouped as general labels and special labels. Within each class there are a wide variety of products. The adequacy of a label is judged for a particular use by the customer. According to Ref. 112, the marketing department has a major say in choosing the type of label to be used.

General Labels. Labels are manufactured as weblike products, but they can be applied in other forms. One main classification of labels according to their application technology divides them into roll and sheet labels. Roll labels are those manufactured in roll form and applied from the roll, where their continuous weblike backing allows them to be handled as a continuous material after their confectioning (die cutting) as separate items. Sheet labels are those manufactured in roll form and cut into a discontinuous sheetlike finite product (as finite laminate) that cannot be applied with a common labeling gun. According to Ref. 137, roll labels constituted 66% of the labels produced in 1988 in the United States. The proportion of roll labels is growing [105]. According to a market survey in 1990 [138] about 95% of the label printers in Europe were reel label printers; only 54% also had printed sheetlike products.

Using another general classification based on the nature of the carrier material, labels may be considered as paper or film labels. Labels are also classified as permanent and temporary, according to the character of the adhesive bond. Repositionable labels are a special class of removable labels that stick to various surfaces but remove cleanly and can be reapplied [139]. The final adhesion builds up over a few hours. The major production of labels (75–80%) uses permanent paper label stock [140].

Special Labels. Special labels are manufactured according to one of the above-discussed general label classes, but they have to satisfy certain special end use criteria (water resistance or water solubility, mechanical resistance or loss of mechanical strength, etc.). The special requirements for different labels and their end use characteristics are discussed in Chapter 8. Special labels can also be classified according to the type of adhesive, as permanent, repositionable/ semipermanent, or removable [141]. The hot melt coated products are classified according to label stock manufacture into computer, thermo, price marking, freezer, film, airline luggage, office and retail labels, and shipping documents [141]. Special labels can be classified according to their processibility, application field, function, and so on. The main class of postprocessible labels includes computer labels and copyable labels.

Generally, writability and printability are common features of labels. Postprintability using a computer, i.e., digital printing, is a performance characteristic of a separate label class called computer labels. Other special labels like table and copy labels may belong to this product group also. For table labels, printability and lay flat are required [142]. Such labels are computer-imprintable. Computer-imprintable labels may be transferable or nontransferable and based on paper, Tyvek, PVC, cardboard, aluminum, or other carrier material [143] (see Chapters 6 and 8). There are different digital postprinting methods; therefore, the construction of computer labels can vary also. Laser-printable labels are a special class of computer-printable labels. Computer labels can be classified as label sheets, endless labels, and folded labels (with pinholes for transport) [144]. They are manufactured as printed or bianco (nonprinted) labels. Their printing can be carried out using dot matrix, ink jet, laser, or thermotransfer printers or with a copying machine. In 1995 in Europe about 40% of labels were computer labels. According to their end use they are applied in addressing, marking, organizing, logistics, and other fields.

The original requirements for computer labels included machining and printing quality, application-related properties, and environmental performance. Starting from these requirements the most important development concerns printing quality; modern labels must accept 600 dpi (dots per inch). Release properties have changed also. Some years ago, slow-running machines allowed very slight release forces of 0.1 N/25 mm. At high speed a better mutual anchorage of the laminate components is required, so the release force has increased to more than 0.2 N/25 mm. Copy labels have to allow light to penetrate through the face stock in order to achieve a copy of the image on the liner.

Medical labels can be defined as sterilization labels also [22]. Like medical tapes and bioelectrodes they have a special conformable, removable, porous, skin-tolerant adhesive with well-defined water solubility and electrical properties and a special porous carrier material (see Chapters 6 and 8) or a carrierless construction.

Tamper-evident labels are special permanent labels with a sophisticated construction that does not allow debonding. Tamper-evident cast films ensure that the printed label will fracture if removal is attempted. For this purpose low strength carrier materials are combined with high strength adhesives and optically working printed elements. Their design shows similarities to that of special closure tapes. The optical display of temperature sensitivity is applied in temperature labels also. Common end use areas are airline tags, caution labels, automotive labels, police forces, and gaming machines [145].

Temperature labels are used as thermometers [146]. They possess a conformable heat- or cold-, water-, and oil-resistant carrier material and a special temperature-indicating component. Such products can be applied between 37.8 and 260°C [147].

There are a number of different application fields where labels having water resistance or water solubility are used (see Chapter 8). Solubility or dispersibility in alkaline or neutral aqueous solutions at hot or room temperature and resistance to humid atmospheres or immersion in water are desired. Applicability on wet or condensation-covered surfaces or frozen surfaces is related to water solubility of the PSAs. Although they are more expensive than classical wet glues in such applications, the industry appreciates the fact that PSA labels can be applied at high speed from roll stock, eliminating the need for cut label inventories and glue applicators.

Some special labels are applied as pressure-sensitive products but use other bonding mechanisms (e.g., iron-on labels) or are applied as alternatives to pressure-sensitive labels (e.g., in-mold labels). For iron-on labels, used mostly for marking textile products, the final bonding is achieved via molten plastomer (polyethylene, polyamide, etc.) embedded in the PSA [148]. In-mold labeling (IML) allows the manufacture of the plastic item (via injection molding) and its labeling in the mold with a special label placed in the mold. Form fill and seal cups can also be labeled using this procedure [149]. Classical in-mold labels are not pressure-sensitive. They are coated with a heat-activatable adhesive (210–230°C for injection molding, 100–110°C for deep drawing). New in-mold labels that work like self-adhesive films have also been developed [150].

In-mold labeling is a special case of injection molding with in-mold parts. In-mold fixing techniques [151] include (1) insert molding, (2) outsert molding, (3) back melting of textiles, (4) removable tool technology, (5) in-mold decoration, and (6) in-mold labeling. In these cases, part of the finished product is fixed in or out of the mold by the molten polymer. In-mold decoration and in-mold labeling are used to fix a nonfunctional part of the item. The principle of the method is simple. The label is placed in the mold, and its back side is fixed with the molten polymer. For this fixing no PSA is needed. The handling of the label is difficult. Special apparatus is used to place the labels in the mold. Stacker systems have been developed to feed the label into an in-mold production machine [152]. In-mold labeling is used for injection-molded items, but it can also be applied with thermoformed plastics [153]. No supplemental postlabeling or postprinting of the finished item is necessary. The label is made from the same polymer as the labeled item, and both are recycled together.

Taking into account the increasing number of requirements concerning environmental considerations, labels could also be classified as repulpable and nonrepulpable (see Table 2.7). There are many applications where recyclability is desirable (address and franking labels, magazine supplements, etc.) [154].

It is evident that the existing range of labels—printed circuit board labels, library labels, warehouse labels, floor labels, retroreflective labels, magnetic labels, asset labels, tamper-evident labels, decorative labels, multifunctional labels, bar code labels, imprinted labels, scented labels, rub-off labels, labels with a no-label look, aluminum labels, harsh environment labels, ceramic labels, titanium labels, UV protective labels, textile labels, laboratory labels, etc.—will broaden as the result of new label applications and new engineering concepts [155].

2.2 Tapes

Pressure-sensitive adhesive tapes have been used for more than half a century for a variety of marking, holding, protecting, sealing, and masking purposes. Industrial tapes were introduced in the 1920s and 1930s, followed by self-adhesive la-

Table 2.7 Classification of Labels

Classification criterion	Product classes			
	Main			Special
Form	Sheet	Reel	—	—
Carrier	Paper	Film	Others	—
Adhesion	Permanent	Removable	Repositionable	Permanent destructible; permanent nondestructible Dry removable; wet removable
Recycling	Repulpable	Nonrepulpable	—	—
Web number	Monoweb	Multiweb	—	—
Printability	Classical	Digital	—	Thermal; laser; ink jet; dot matrix
Image	Label look	No-label look	—	—
Labeling	Manual	Automatic	—	—
Application field	General	Special	—	Medical; security; office; food; logistic; computer

bels in 1935 [101]. Single-coated PSA tapes were used in the automotive industry in the 1930s [156]. Self-adhesive packaging tapes replaced gummed tapes on paper basis [157]. According to Ref. 158, the major tape markets in the United States have been packaging tapes (mostly HMPSA-coated), electrical tapes (solvent-based), industrial tapes (HMPSA- and solvent-based), health care tapes (HMPSA- and water-based PSA), masking tapes (water-based and solvent-based), and consumer tapes (HMPSA-, solvent-, and water-based).

Tapes are continuous weblike PSPs applied in continuous form. Generally their role is to ensure the bonding, fastening, and/or assembling of adherend components due to their mechanical characteristics. PSA tape is defined as a pressure-sensitive adhesive-coated substrate in roll form, wound on a core and at least 0.305 m (12 in.) in length [159]. PSTC developed a guide to pressure-sensitive tapes [159] that covers their significance, standard width, labeling, test, units of measurement, and tolerances. The advantages of tapes as an adhesive system are discussed by Bennett et al. [160]. They are positionable, have a controlled coating weight, allow automatic use by having die-cut parts, distribute stress, and exhibit a low level of cold flow. Tapes constitute the largest group of PSPs [161,162].

Buildup and Requirements for Tapes

Generally tapes are PSPs that have a solid-state carrier with a coated (built-in) pressure-sensitive layer. Some tapes also have a release layer. Tapes were the first PSPs to be produced with a coated monoweb construction. Theoretically, pressure-sensitive tapes have the same construction as wet adhesive tapes. For pressure-sensitive adhesive tapes based on polymeric film carrier, the release surface is usually the surface of the polymeric film remote from the adhesive, which allows the tape to be conveniently stored in the form of a spirally wound roll. In practice, pressure-sensitive tapes that have a high tack adhesive (and/or coating weight), for example, double-sided tapes or transfer tapes, need to have a separate solid-state release component.

The development of carrier materials and adhesive components made it possible to manufacture special tapes as a continuous web, which facilitates application, that are not used for bonding in the classical sense.

Double-sided mounting tapes may have a primer coating on the back side to allow the anchorage of a postfoamed layer [101]. They must have a separate release-coated liner. A tape with good adhesion on in situ foamed PUR is described in Ref. 146. This construction includes a PVC carrier coated with a crosslinkable polyurethane adhesive, which cures in use.

Masking tapes may have special carrier constructions to give them conformability. According to Ref. 40, a masking tape carrier has a stiffened, wedge-shaped, adhesiveless longitudinal section extending from one edge, with a pleated structure to conform to small radii. This region is formed of a material different from that of the main tape. The nonadhesive face of the tape is heat-reflective. Conformability is also required for medical tapes. A pleated carrier (which gives extensibility) is needed for some mounting tapes too. Medical tapes require the adhesive, carrier, and the tape construction as a whole to be porous. Electrical conductivity or nonconductivity may be required for insulating, packaging, and medical tapes (see Chapter 8). Temperature-dependent adhesion and temperature-independent bonding are needed for certain tapes. The temperature at which a PSA tape is laminated is critical [163]. Unwinding resistance is also important for tapes [164]. Shrinkage and elongation of the tape during unwinding and deapplication; peel force, fresh and aged; weathering at low temperature; peel from back side; and shear resistance at various temperatures are also measured (see Chapter 8). Water resistance or water transmission and oil resistance are required. Tear and dart drop resistance are needed. Some standards for tapes specify requirements with respect to unwinding resistance and unwinding behavior, water permeation, and water vapor permeation.

Electrical tapes have to possess high dielectric strength and good heat dissipation properties. They are used for taping generator motors and coils and transformer applications, where the tape serves as overwrap, layer insulation, or connection and lead-in tape [165]. The carrier is woven glass cloth impregnated with a high temperature resistant polyester resin or impregnated woven polyester glass cloth. This last variance is used where conformability is required. For some applications transfer tapes are die cut [166]. In other cases (e.g., special packaging tapes and masking tapes), an easy tear carrier or high strength carrier is required. Carrierless mounting tapes may have monolayer or multilayer, filled or unfilled, foamed or unfoamed construction.

Carrier for Tapes. Various weblike materials can be used as carriers for tapes. Films, woven and nonwoven textiles, foams, and combinations of these materials used as carriers (see also Chapter 6). Woven fiberglass can be used for special tapes. Polyurethane-coated PVC can be used also [167]. PUR foams were the first foamlike carriers [168]. Acetate films are used as carrier materials for self-adhesive tamper-evident products. A 50 μm cellulose acetate film has been modified to form a brittle film with a low tear strength, but it has to be die cut [169]. Extensible, deformable paper is required for medical tapes [170]. Well-defined mechanical, electrical, and thermal properties are necessary for insulation tapes [171]. Elongation, shear, and peel resistance are needed also. In some cases shear resistance should be measured after solvent exposure of the PSP. Thermal resistance and low temperature conformability and flexibility are required. Bacteriostatic performance may be necessary also. Cover tapes for bathrooms are made on a special PVC basis, with a bacteriostat (which inhibits bacterial growth) incorporated in the vinyl formulation. They are embossed to produce a secure non-slip surface. The tapes are coated with a repositionable adhesive that builds up adhesion from 26.7 to 36.2 oz/in. [172]. To facilitate immediate adhesion to rough and uneven surfaces, a resilient foam backing can be applied [173]. The use of polyurethanes with an extremely high level of elongation at break will ensure a sufficiently high impact strength.

Adhesives for Tapes. In their first stage of development, rubber-resin-based formulations were used as solvent-based adhesives for tapes. Later synthetic rubbers were introduced. The sealing tape industry had relied heavily on HMPSAs based on SBCs because they offer a good balance of shear, tack, and peel resistance at reasonable price. Water-based acrylics replaced natural rubber and block copolymers for many general applications. Acrylic emulsions represent an attractive alternative to HMPSAs. Waterborne acrylic PSAs do not need tackifiers and exhibit adequate shear and low temperature behavior, which are important in the manufacture of mounting tapes [174]. For special uses, crosslinkable, water-soluble, electrically conductive PSAs have been developed (see also Chapter 6).

Acrylic adhesives are used for general-purpose and special tapes. They are applied as solvent- or water-based or hot melt formulations for tapes of various buildups and end uses. Solvent-based acrylics are used for special medical, insulation, and mounting applications. They allow easy regulation of conformability, porosity, and removability. Hot melt acrylics are expensive products and are used mostly for special medical and sanitary tapes. They are suggested for mounting tapes also [173]. Radiation-cured 100% acrylics are recommended for transfer tapes and medical applications. The development of such crosslinkable and highly filled acrylics allowed the design of carrierless tapes having a foamlike character [93,175–177]. Glass microbubbles have been incorporated to enhance immediate adhesion to rough and uneven surfaces [176]. Such tapes are prepared by polymerizing in situ with UV radiation. Water-based acrylics are preferred for general packaging tapes; they are also formulated as removable PSAs (e.g., application tapes).

The first rubber-resin PSAs were used for tapes with a paper carrier. Because of the relatively simple regulation of the adhesive properties of formulations based on natural rubber (NR) or blends of natural rubber with synthetic elastomers (via crosslinkers and active fillers), recipes were developed for almost every application field (packaging, mounting, and medical tapes) for permanent or removable adhesives. Masticated, calendered compositions have been coated without solvent. The introduction of meltable styrene block copolymers allowed the coating of rubber-resin formulations as HMPSAs. Solvent-based and hot melt rubber-resin formulations are the most common raw materials for inexpensive tape applications. Water-based, carboxylated rubber dispersions replaced acrylics for some packaging applications as a less expensive raw material. According to Wabro et al. [173], for certain tapes rubber-resin formulations cannot be replaced with other raw materials.

Release Liner for Tapes. Tapes generally do not need a separate solid-state adhesive component. Plastic carriers used for tapes may or may not have a coated or built-in release layer, depending on the type of carrier and adhesive (see Chapter 6). Double-sided coated and transfer tapes require a separate solid-state release liner based on a common (uncoated) or coated plastic or paper carrier. Release liners used for double-sided tapes have special requirements [173]. Such materials have to display adequate unwinding performance, controlled adhesion to the tape, dimensional stability, tear resistance, weather and environmental stability, and confectionability/cuttability.

Principle of Functioning of Tapes

Tapes are manufactured as a continuous web and applied as weblike or discontinuous items (portions of this web) by laminating on a solid-state surface.

Special Characteristics of Tapes

Because of their need to be applied as a continuous web and due to their general use as a bonding element of multiple adherends, tapes generally have to possess a mechanically resistant carrier material with a pressure-sensitive adhesive layer. The mechanical characteristics of the carrier material and the adhesive characteristics of the built-in or coated adhesive differ according to the end use of the product. Unlike labels, tapes are not built up as laminates before their use (except for their self-wound character or some special tapes). Because they are unwound at high speed, noise reduction is very important for tapes. For certain tapes the use of a tape applicator is required.

Classes of Tapes

Adhesive tapes are manufactured for packing, masking, office uses, protection, marking of pipes and cables, fixing of carpets, drug delivery, floor marking, etc. Such products can be printed in one or many colors or blank. Tapes may be divided into groups according their buildup and end use (see Table 2.8). End uses include packaging, mounting, construction, medical, decoration, and others. HMPSAs can be incorporated into film tapes, mounting tapes, textile tapes, and insulation tapes [141]. Like labels, a primary classification of tapes is based on their general or special characteristics.

General Tapes. According to their carrier material, tapes may be divided into paper-based and film-based tapes. Some tapes are manufactured with no carrier. Tapes with carrier are classified as single-sided coated and double-sided coated products. Depending on their adhesive characteristics, tapes may be grouped into permanent and removable products. According to their end use there is a broad range of possible classifications. Packaging, masking, protecting, marking, closure, fixing, mounting, and insulating tapes are some examples [169].

Paper-based tapes are manufactured using paper as carrier material. They are adhesive-coated PSPs that have various paper qualities and adhesive coatings de-

Buildup and Classification of PSPs 49

Table 2.8 Classification of Tapes

Classification criterion	Major product classes	Special product classes
Form	Continuous; discontinuous	—
Carrier	Paper; film; other	Textiles; metals; laminates
Adhesion	Permanent; removable; repositionable	Permanent destructible; permanent nondestructible
		Dry removable; wet removable
Buildup	With carrier; carrierless	Transfer
	With adhesive; adhesiveless	—
Recyclability	Repulpable; nonrepulpable	—
Web Number	Monoweb; multiweb	One side coated
		Double side coated
Printability	Classical; digital	Thermal; laser; ink jet; dot matrix
Application method	Manual; automatic	—
Application field	General; special	Packaging; mounting; assembling; office; closure; masking; medical logistic; splicing; electrical

pending on their end use. Extensible, pleated, and conformable paper is required as a carrier for special tapes [169]. Double-sided coated and transfer tapes can have a paper carrier or release liner also (see Chapters 6 and 8).

Film-based tapes are manufactured using plastic films as carrier material. These products can have a coated adhesive, a built-in adhesive, or a virtually adhesive-free construction. The self-adhesivity and conformability of certain plastic films and the nonpolar abhesive surface of some plastic carrier films allow the design of adhesive-free and release layer–free tape constructions.

Carrierless tapes do not have a solid-state carrier material after their application. These products are the so-called transfer tapes or tapes from the reel. Transfer tapes have a PSA layer inserted between two release liners. They are manufactured as a continuous web supported by a solid-state carrier. This allows them to be processed and applied like a common tape, but it serves as a temporary aid only. The adhesive layer of the tape is detached from the carrier during application. Such tapes without carrier can be used on multidirectionally deformed substrates of various shapes, at temperatures up to 150°C because the temperature resistance is limited by the adhesive only [173].

Single-sided coated tapes represent the classical construction of tapes, having a solid-state carrier material coated on one side with a pressure-sensitive adhesive.

Products displaying pressure sensitivity on both sides of the carrier material are double-sided coated. Such tapes may have another coated layer on the back side also, e.g., a primer. A different release degree is achieved for a double-sided coated release liner using the same release component but one-sided polyethylene-coated release paper [58]. The release paper assists in the application of the tape without damaging the coating that is facing the release paper. These tapes are normally used as the bonding agent when combining two materials. The tape is pulled off its roll, and the exposed pressure-sensitive adhesive is placed against the first material. Then the release paper is pulled from the tape and the exposed pressure-sensitive adhesive bonds with the second material. The converting industry uses a large amount of two-sided coated PSA tapes such as flying splicing tapes.

The principle of adhesive detachment from the inserted solid-state release liner is used for transfer tapes. Both surfaces of the carrier may have low adhesion coatings, one of which is more effective than the other. When the tape is used as a transfer tape, when unwound the adhesive layer remains wholly adhered to the more strongly adhesive surface, from which it can be subsequently removed.

Permanent tapes give permanent adhesive bonding. The main representatives of this class are the packaging tapes. High strength carrier materials and aggressive permanent PSAs are recommended for these products.

Certain tapes ensure a removable adhesive bonding, which is required for closure, medical, and masking tapes, among others (see Chapter 8). Like labels, tapes can also be repositionable. For instance, cover tapes for bathrooms are repositionable [88]. Stone impact resistant automotive decor tapes have to be repositionable [60]. Certain closure tapes, e.g., diaper closure tapes, have to be removable [178]. Storable crosslinkable adhesives allowing a built-in controlled removability have been developed. Postapplication crosslinking is proposed to improve shear [175]. Such postcrosslinking is achieved by using free radical [93] or photoinitiated reactions [176]. Postcrosslinking leads to adhesionless surfaces, i.e., easier delamination. According to Bedoni and Caprioglio [72], self-adhesive tapes produced in Europe can be divided into the following classes: packaging (64%); masking (12%); protective (5%); double-sided (5%); dielectric (3%); and stationery and others (7%). Unfortunately, the evaluation criteria for this tape classification are not known, and it is difficult to delimit exactly the difference between masking and protective tapes (and films!), stationery tapes and labels, dielectric (insulating?) and protective, etc. products. Packaging tapes are still the main grade produced, and the productivity developments in the last decade relate mostly to these products. The production speed of such tapes increased from about 250 m/min in 1985 to about 600 m/min in 1995, and the coating machine width increased from about 1400 mm in 1985 to about 2000 mm in 1995. Improvement in the average coating speed is less impressive, from about 100 m/min to about 300 m/min. The average coating machine width increased from 1300 mm to 1500 mm. A new coating machine for packaging tapes having a width of

Buildup and Classification of PSPs 51

1200–2400 mm is running with HMPSA on a 25 μm BOPP film as carrier material, at a speed of 400–600 m/min.

According to Becker [164], technical tapes can be divided according to their end use into closing, bonding, reinforcing, marking, and protection tapes. Closing, bonding, and reinforcing are the main functions of packaging tapes. In this domain, water-resistant tapes form a special category. It is obvious that tapes could also be classified according to their carrier material. For the same carrier class, writeable and printable tapes are considered special products [164]. Table 2.9 summarizes the main tape categories according to their construction.

Table 2.9 The Main Tape Categories According to their Construction

Type of web	Number of carrier layers One	Number of carrier layers Many	Number of adhesive layers One	Number of adhesive layers Many	Components	Tape grade: Application field
Monoweb	Yes	—	Yes	—	Adhesive, primer 1, carrier, primer 2, release, ink	One side coated tape: Packaging, office use, closure, insulating
	Yes	—	—	—	Carrier, ink	Adhesiveless tape: Masking, insulating
Multiweb	—	Yes	Yes	—	Adhesive, primer 1, carrier 1, release, primer 2, carrier 2, ink	One side coated tape: Insulating, medical, assembling, closure
	—	Yes	Yes	—	Adhesive, release 1, primer 1, carrier, primer 2, release 2	Carrierless transfer tape: Mounting, assembling, insulating
	—	Yes	—	Yes	Adhesive 1, adhesive 2, carrier 1, carrier 2, release 1, release 2	Double side coated tape: Splicing, mounting, medical

As can be seen from Table 2.9, there are tapes based on a solid-state carrier and carrierless tapes. Carrierless tapes may be virtually or actually carrier-free. Tapes that are virtually carrierless possess a reinforcing layer (network) embedded in the adhesive mass; this layer ensures the required mechanical strength of the tape.

According to their active (adhesive) surfaces, tapes can be classified as one-sided or double-sided coated. As seen from Table 2.10, double-sided coated tapes are manufactured with or without carrier. Tapes with carrier can be manufactured with a paper, textile, or plastic carrier. The textile carrier used may be woven or nonwoven. Textile carriers ensure nonextensibility and good anchorage [173]. A plastic carrier may be a film or a foam. A film carrier acts as a barrier against chemicals (plasticizers, surfactants, antioxidants, etc.). A foam carrier ensures conformability on uneven surfaces and equalizing of the thermal dilatation coefficients.

Special Tapes. At the beginning of their use, tapes were applied mostly as fastening and bonding elements. Packaging tapes still play this role. The development of new carrier materials made possible the manufacture of special tapes. These products are used as sealants, mounting elements, etc. Some of them are described briefly below.

Textile-based tapes are applied on various substrates. Insulation, medical applications, and packaging are the main end use domains. As stated above, electrical tapes have to possess high dielectric strength and good heat dissipation properties. The carrier for such tapes is woven glass cloth or woven polyester glass cloth impregnated with a high temperature resistant polyester. In the medical domain the textile-based carrier acts as a network that allows the diffusion of air, pharmaceutical agents, vital liquids, etc. Like high strength wet tapes, pressure-sensitive packaging tapes may also have a reinforcing textile carrier.

The plastic web of tapes may be a foamlike product or a combination of film and foam. Double-sided coated tapes using neoprene, PVC, PUR, or PE as carrier have been developed (see Chapter 6). Carrierless foam tapes are manufactured also. Glass microbubble filled mounting tapes imitate the conformability given by cellular (air-filled) structures. Table 2.11 lists the main application domains of special tapes.

As indicated in Table 2.11, such tapes possess two active surfaces; therefore, they can bond in one plane like common packaging tapes as well as between two surfaces. In their early development stage, double-sided tapes were manufactured with a carrier. Later carrierless tapes were produced. In principle such tapes may bond multilaterally, like an overall adhesive profile. As an example, the tapes used for mounting rubber profiles in cars have a special form (see Chapter 8).

Insulation tapes have to ensure thermal, electrical, sound, vibration, humidity, chemical, or other type of insulation. Depending on their application field they

Buildup and Classification of PSPs 53

Table 2.10 Special Tapes

Product	Buildup		Manufacture	Application
Double-sided adhesive tape	Double side coated		Coating/laminating	Splicing, mounting, insulation, assembling
	Self-adhesive	Blend, coex	Extrusion	Masking, insulation
	Carrierless adhesive	Foamed, filled, reinforced, crosslinked	Coating, extrusion	Mounting, insulation, medical
Foam tape	Foamed carrier	Elastomer foam, plastomer foam	Foaming, extrusion	Mounting, insulation, assembling, medical
	Foamed adhesive	Acrylic, rubber	Foaming, coating, extrusion	Mounting, insulation, assembling, medical
Filled tape	Solid state filler in adhesive	Continuous fiber, discontinuous fiber, powderlike filler	Calendering, extrusion, coating	Packaging, medical, electrical
	Liquid filler in adhesive	Electrolyte, antimicrobial agent	Coating	Medical
	Air in adhesive	Foam tapes	Foaming, extrusion	Medical
	Solid-state filler in carrier	Reinforcing fiber, powderlike filler	Extrusion, calendering	Packaging, medical

Table 2.11 Main Application Domains of Special Tapes

Tape	Main characteristics	Application domain
Foam tape	Conformability Damping Multiple adhesive surface (bonding in space)	Mounting Vibration and thermal insulating Sealing Assembling
Carrierless transfer tape	Conformability Damping Multiple, variable adhesive surface (bonding in space)	Mounting Vibration and thermal insulating Sealing Assembling
Double side coated tape	High shear resistance Instantaneous bonding Multiple adhesive surface (laminating)	Splicing Mounting
Self-adhesive tape	Conformability Deformability Removability	Masking Insulating
Discontinuous tape	Automatic applicability	Mounting Assembling
Removable tape	Removability Repositionability	Mounting
Dosage tape	Storage and dosage Conformability Removability	Medical mounting Anticorrosive protection

are made with various carrier materials that have different thermal, electrical, and mechanical properties.

Medical tapes are used as fixing or dosage elements. According to Ref. 179, the earliest plasterlike adhesive medical products were crude mixtures of masticated natural rubber tackified with rosin derivatives and turpentine and filled with zinc oxide as pigment. Cloth served as carrier material. Later, extensible, deformable paper was used. Now plastic films and plastic-based textile materials are used. As adhesive, a broad range of macromolecular compounds (natural and synthetic rubbers, polyurethanes, acrylics, special vinyl polymers) are tested. The special requirements for different tapes and their end use characteristics are discussed in Chapter 8.

Sealing tapes are used as insulating and mounting elements, having sealantlike functions. The three general classes of sealants are hardening, plastic, and elastic sealants [180]. Elastic sealants behave like rubber bands. They deform if a force is exerted on them but return to their original state if this force is removed. Sealing tapes possess intermediate characteristics; they are elastoplastic and plastoelastic. The edge joint of insulating glass is presently manufactured worldwide predominantly using sealants and sealing tapes. During the lifetime of a glazing unit, the edge joint is subjected to many different types of stresses. After installation, mechanical stresses result from wind pressure and suction, flexion of the insulating glass pane due to temperature and pressure fluctuations, and other causes. Chemical and physical stresses appear also [180].

The insulating glass sealant or tape has to satisfy a number of requirements, the most important being (1) optimal adhesion to glass and metals, (2) aging resistance, (3) high flexibility (even at low temperatures), and (4) low water permeability.

2.3 Protective Films

Protective films form a special class of pressure-sensitive products, manufactured as self-wound webs and applied as the laminating component for both weblike and discrete products. They can be used to protect finished surfaces from damage during manufacture, shipping, and handling and to mask areas of a surface from exposure during spray painting. Such products can be formed using a stencil and undergo the same processing steps as the protected item. Taking into account their protective function, they could be considered packaging materials, but actually they cannot be defined as packaging materials because of their aesthetics. According to Djordjevic [6], packaging is a process with a functional definition, playing both a protective and cosmetic role. It should be mentioned that masking tapes may be considered as a special class (narrow web) of protective films.

Table 2.12 summarizes the main types of products used for surface protection. As shown in this table, protective materials can be carrierless (coatings) or be based on a solid-state carrier (protective webs). Coatings display an adhesive contact with the surface to be protected. Protective webs may (e.g., protective films and separation films) or may not (e.g., packaging films and cover films) exhibit an adhesive contact with the surface to be protected. This "adhesive" contact may be the result of an interaction between an adhesive and a substrate (e.g., protective film) or between two adhesiveless surfaces (e.g., cover film). For certain protective webs in adhesive contact with the surface to be protected, this contact is the result of lamination (e.g., laminating and separation films); in other cases (e.g., cover films) no lamination is carried out during application and the contact between protected surface and protective web is not uniform.

The nature of the bond (temporary or permanent) may also vary. Common lacquers build up a permanent coating with the protected surface. The bond between laminating films used for surface protection is permanent too. Other protective materials (the majority of film-like products) are removable.

Table 2.12 The Main Products Used for Surface Protection

	Characteristics				
	Components			Application method	
Protective materials	Carrier	Adhesive	Release	Lamination	Wrapping
Packaging material	x	–	–	–	x
Protective coating	–	–	–	–	–
Protective web					
Cover films	x	–	–	–	x
Separation films	x	x/–	–	x	–
Overlaminating films	x	x/–	x/–	x	–
Protective films	x	x/–	x/–	x	–
Masking tapes	x	x	x/–	x	–
Protective labels	x	x	x	x	–

x = with; – = without.

Table 2.13 lists the main removable products used for surface protection. As can be seen, both carrierless coatings and carrier-based materials are used for surface protection. Carrierless coatings were the first to be produced. Protective films were developed as replacements for masking paper and peelable varnishes (see also Chapter 8).

Peelable varnishes have to fulfill the following requirements [181]: coatability as a monolayer lacquer, weatherability, water and temperature resistance, chemical resistance, nonflammability, transparency, filmlike rigidity and flexibility, and low but sufficient adhesion. According to Zosel [181], such formulations occupy an intermediate position between primers and varnishes with respect to their adhesion and cohesion properties. Primers have to display excellent anchorage on the surface to be coated but medium cohesion. Varnishes have to possess higher cohesion and mechanical resistance. The mechanical resistance and flexibility of removable protective lacquers has to be as high as for varnishes, but their adhesion should be lower (their adhesion is about 1 N/mm^2).

Coating films based on EVAc copolymers and waxes have been used [182,183]. Polyvinylbutyral has been applied for removable coatings [184]. Low molecular weight acrylic polymers have been developed as temporary automotive coatings. They have to exhibit weathering resistance, surface protection, and slight removability with alkali solutions after a long period of application or storage. Low molecular weight solvent-based acrylics have been marketed [181]. Such coatings have to be coatable as monolayer lacquer and should exhibit weather resistance, chemical resistance, nonflammability, mechanical strength and flexibility, and adequate adhesion. For temporary protection of automobiles

Buildup and Classification of PSPs 57

Table 2.13 The Main Removable Products for Surface Protection

Product	Chemical basis	Thickness (μm)	Application method	Deapplication method
Peelable varnish	Wax, low molecular EVAc, acrylics, polyvinylbutyral	5–10	Spray coating, dipping	Mechanical peel-off; washing with water (alkali) or solvents
Peelable low strength films	PVC plastisols, crosslinked polybutadiene, polybutylene/ resin, PP/resin, VC/AC copolymer	10–70	Spray coating, dipping	Mechanical peel-off; washing with water (alkali)
Protective papers	Paper	30–100	Lamination	Delamination
Protective films	PE,PP,PET,PVC	12–120	Lamination	Delamination
Separation films	PE,PP,PVC	25–50	Lamination	Delamination

during transport and manufacture, copolymers of C_{3-4} alkyl methacrylate and methacrylic acid have been proposed [185]. Such coatings sprayed on the surface to be protected give a 6–10 μm film after drying in air (10–15 min), which is removable with a 10% NaOH solution. Crosslinked elastomers have also been used as protective coatings for metals. A patent describes the use of a polyisocyanate cured butadiene polymer [186]. Later, protective papers replaced carrierless protective coatings. Water-removable protective papers have also been manufactured [187].

Buildup and Requirements for Protective Films

Protective films are weblike, self-adhesive, removable materials that are used in intimate contact with the surface to be protected and require a self-supporting carrier in order to display the mechanical resistance necessary for the protective function and for removability. The weblike attribute refers to the solid-state character of the product, which is manufactured as a solid-state continuum (reel). Such products were introduced in the 1960s for the protection of coated coils. As stated in Ref. 188, both cold and hot systems are known. Cold systems use an adhesive-coated film, whereas hot systems apply a meltable film that bonds with the coating layer of the coil (at high temperatures) to produce the desired protection. On the other hand, this bond should allow removability after the coil is processed. It should be mentioned that laminating of coils has been used as a coating method for many years. Such (adhesive) lamination of polyvinyl chloride and polyvinylidene fluoride films has been carried out at 180–200°C.

It is evident that the simplest way (at least theoretically) to build up a protective film is to manufacture an uncoated plastic film that possesses the mechanical strength of a carrier film and has built-in adhesivity (see also Chapter 6).

The monoweb, as a challenge for label producers and a reality for tape manufacturers, is undergoing further development for protective films. To understand the direction of this development, let's have a look at the requirements for protective films (Table 2.14). As shown in Table 2.14, conformability, mechanical strength, balanced adhesion to the protected surface, and removability are the main application criteria for protective films. A number of protective materials fulfill these requirements at least partially. As discussed earlier, there are many types of materials for surface protection. Some, like packaging materials, are self-supporting, mechanically resistant carrier materials that do not need to be in intimate contact with the product to be supported; thus their removability does not affect the product. Other products, like varnishes or lacquers, develop an intimate contact with the surface to be protected. Generally, these are permanent coatings that do not need a self-supporting carrier. In some cases even permanent coatings may display sheetlike character (laminates). Some packaging materials (e.g., cover films) autoadhere to the protected product, although they need no laminating to perform their function.

As shown in Table 2.14, protective materials differ according to their self-supporting or coating-like character (with carrier or without), their contact with the product to be protected (adhesion or overlapping only), and their removability from the product (with or without bond breaking). Protective films include a broad range of products that exhibit pressure-sensitive behavior although they are built up quite differently.

Table 2.14 Requirements for Protective Films

Application requirements	End use requirements	Performance given by Carrier	Performance given by Adhesive
Laminating ability	—	x	x
Cuttability	x	x	—
Mechanical resistance	x	x	—
Machining performances	x	x	—
	Deformability	x	—
	Adhesive bond stability	x	x
	Aging stability and weatherability	x	x
	Delaminating ability	x	x

Buildup and Classification of PSPs

From the history of protective films it is well known that they appeared on the market as improved variants of PSA-coated protective papers. Paper masking tapes were introduced in the 1920s in the automobile industry [173]. Later paper was replaced with the more conformable plastic film. During the 1950s, Japanese polyvinyl chloride–based products were the sole materials used for protective films. Polyvinyl chloride has since been replaced with polyolefins. Taking account of this development from the technical point of view it was easier to manufacture the protective film in a two-step process in which common plastic films (mainly polyolefins) were offline-coated with a pressure-sensitive adhesive. Therefore classical protective films are pressure-sensitive self-adhesive films. They are manufactured by a film conversion process.

Adhesive-free laminating films are not new; they have been used for many years [189]. They have been applied on different substrates. A thermoplastic heat-activated adhesive film has been designed for laminating calendered soft PVC foils to PE foam, on aluminum, fiberboard, and woven fabric. Bonding has been achieved with conventional flame bonding machines at speeds of 10–20 m/min. Coating paper with film-forming (tack-free) ethylene-vinyl acetate copolymers has been carried out to manufacture packaging materials with barrier and sealing properties [190].

In the PSA label and tape industry, the face stock materials used by converters were common materials developed for the packaging industry (PVC, cellulose derivatives, and, later, polyolefins). The main requirements for these packaging materials include mechanical strength, flexibility, printability, and the lack of self-adhesivity (blocking). Only the later development of plastic films for a quite new packaging technology (shrink and cling films), in parallel with the development of tackifiers for hot melt formulations, allowed the manufacture of plastic films with a self-adhesive surface. It should be pointed out that, unlike that of classical PSAs, which are fluid (cold flow) at normal temperature, the adhesivity, i.e., the laminating ability, of the new self-adhesive films (SAFs) can be used only under well-defined temperature and pressure conditions; that is, these materials are adhesive only under exactly defined laminating conditions. Moreover, their removability also depends on the laminating conditions. Another disadvantage of uncoated protective films is their substrate-dependent adhesivity, i.e., they do not have (actually) a general usability for quite different application fields. It is evident that the manufacture of such films requires the skill of specialists from the plastic film production (extrusion) field.

As can be seen from Table 2.14, within the range of protective materials, protective films form a separate class, characterized as a *laminatable monoweb* construction. From the range of monoweb products, cover films (like protecting films) are functional, technological packaging materials, but they are used without laminating [191]. The application of tape to the product to be protected also differs from the use of protective film. It is a low pressure, discontinuous process.

Other products like carrierless protective coatings are not pressure-sensitive. They are tack-free polymer layers, like varnishes, having good adhesion on the coated surface; their removability is mainly chemically controlled.

Carrier for Protective Films. The carrier in adhesive-coated protective films is generally a common polymer used in the manufacture of packaging films. Polyvinyl chloride, polyolefins, and polyester are the most commonly applied materials (see Chapter 6). For self-adhesive protective films, tack-free elastomers (polyolefins) or viscoelastic compositions (based on ethylene-vinyl acetate, ethylene-butyl acrylate copolymers or tackified plastomers) are suggested. Adhesive-coated protective films possess thicknesses of 40–120 μm. Generally it is not possible to coat all thicknesses with every adhesive. Adhesives giving higher peel strength need higher gauge films. It is recommended that films possess enough adhesion to avoid film detachment but not high enough to produce a deposit.

Adhesives for Protective Films. Depending on their construction, adhesive-coated or self-adhesive, PSP protective films may or may not contain an adhesive. Self-adhesive protective films may also include tackifying components (e.g., polyisobutylene copolymers). For coated protective films, crosslinked adhesives are generally used to achieve removability. Various raw materials, elastomers (natural and synthetic rubber) and viscoelastomers (acrylics and polyurethanes), have been suggested for PSA formulations for protective films (see Chapter 6).

Although harder than rubber-resin-based compositions, acrylic adhesives are used as solutions or dispersions for protective films. Generally such formulations are not tackified. Solvent-based acrylics are crosslinked recipes. Acrylic dispersions can be applied in certain cases as non-crosslinked formulations also. Formulations based on rubber-resin adhesives (solutions) are the most important adhesives for protective films. Because of their softness and easy control of removability (via mastication, crosslinking, and tackifying), for certain uses they cannot be replaced with other adhesives (see also Chapter 8).

Release Liner for Protective Films. Generally protective films do not need a separate release liner. In most cases they need no release coating. For products with a higher coating weight or tacky adhesive, special non-silicone-based release coatings are applied (see Chapter 6).

Principle of Functioning of Protective Films

As mentioned before, protective films adhere to the surface to be protected. This is necessary in order to ensure a monoblock character of the product, in order to leave its functional (technological) performance intact. This is a main requirement of protective films, because, unlike classical packaging materials, they are used at the start and not the finish of a technological procedure, i.e., they are ap-

Buildup and Classification of PSPs 61

plied before the product to be protected undergoes its technological cycle. The contact with and adhesion to the surface to be protected is achieved by laminating the web with the product. Thus, the definition of protective films may be that they are self-supporting, removable adhesive webs that build up a laminate with the product to be protected. As discussed above, the main characteristics of a protective film are its self-supporting character and its surface adhesivity.

The self-supporting character is given by the use of a solid-state carrier material (film, paper, etc.). This material has to display the mechanical resistance required during application of the web (bonding and debonding) and during the life of the laminate (protected material/protective web).

Application and deapplication of the protective web refers to its lamination on the surface to be protected and its separation from that surface. As schematically shown in Fig. 2.16, during lamination and delamination the film is stressed by a tension; thus its tensile strength is a very important property.

Quite differently from the laminating of other pressure-sensitive products (labels, tapes, etc.), where low laminating pressure, high laminating speed, and

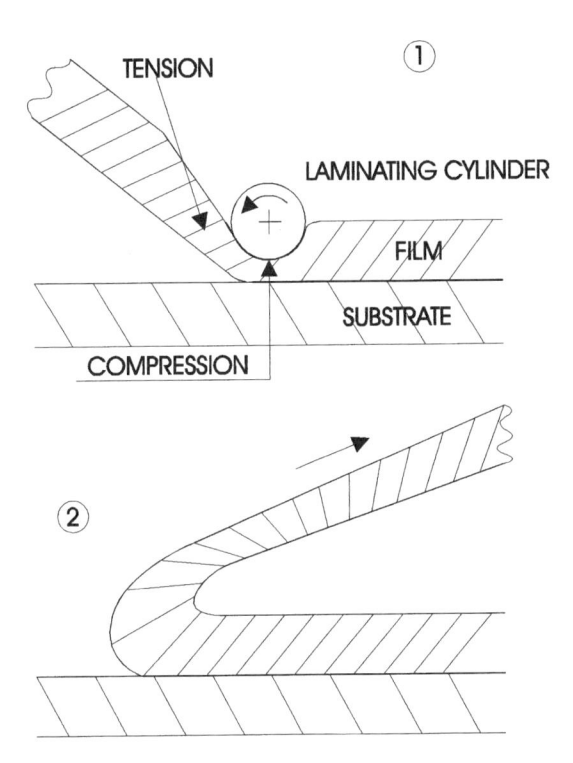

Figure 2.16 (1) Lamination and (2) delamination of a protective film.

(generally) normal temperature are used, the lamination of protective films may require a high laminating pressure and temperature (Table 2.15). Their debonding (delaminating) speed is higher than that of other products. Therefore the mechanical resistance requirements for the carrier material during application/deapplication of protective films are relatively high in comparison with those for labels and tapes. In this operation, the mechanical resistance has to ensure the dimensional stability of the film necessary to allow standard debonding (peel) values (see Removability in Chapter 5, Section 1).

Once applied, the film protects the product against mechanical damage during different working/logistic steps. Depending on the nature of the product, the nature of the applied stresses will differ too. The most common working operations are cutting, drawing, and punching. Therefore the film itself has to resist shear, tensile, and compression stresses (see Chapter 3).

On the other hand, like webless, unsupport protective coatings, the protective web has to conform to the surface profile of the product. Unlike these products, which are liquid during application and exhibit "unlimited" flow, and as a consequence of perfect surface conformance, the solid-state carrier film (protective film) has to allow flow in the solid state. As discussed later in more detail (see Chapter 6), this requirement limits the choice (formulating freedom) of the self-supporting material. In conclusion, the choice of a solid-state, self-supporting carrier as base material for protective films is influenced by the application conditions (laminating/delaminating), by the end use (product working conditions), and by economic considerations.

The application conditions depend on the self-adhesive character of the film. Its influence is very complex. Laminating and delaminating require high mechanical resistance (solid-state-like behavior). Bond formation is a function of the deformability, or plasticity (liquid-like behavior), of the product.

The Self-Adhesive Character. Protective films are laminated on the product to be protected. Unlike the manufacture of classical composite structures, this lamination does not occur simultaneously with the manufacture of the adhesive film

Table 2.15 Laminating Conditions for Protective Films

Protective film	Protected product	Temperature (°C)	Pressure
Hot laminating film	Coil	230	High
Adhesive-coated film	Coil	Room temperature	High
Self-adhesive EVAc film	Cast plastic plate	Room temperature	Medium
Self-adhesive EVAc film	Extruded plastic plate	50–70	Medium

(i.e., converting), but after it, like labeling or the application of pressure-sensitive tape. Protective films are manufactured as preformed materials and undergo laminating after an indefinite storage time. Therefore they have to possess a permanent adhesivity, given by the cold flow of macromolecular compounds above their glass transition temperature T_g (see Chapter 3).

Protective films (like tapes) are monowebs, i.e., pressure-sensitive laminates, where the PSA layer is protected by the back side of the web (carrier material). In some cases, because of the relatively high tack of the PSA layer, protective films need a release layer on their back side. Generally the release layer of a protective film is built up on a nonsilicone basis; therefore it displays a relatively low abhesivity. This is acceptable because of the low tack and peel of the adhesive layer. On the other hand it is necessary because of the increased mechanical requirements for this release coating too. Economic considerations play a role in the choice of the release layer also (see Chapter 6).

In principle, the self-adhesive character of a film may be achieved for the whole material (bulk) or for a built-in or built-on layer of it; i.e., the carrier itself can possess pressure-sensitive properties, or it can be coated with a layer of pressure-sensitive material. Independently from the construction of the self-adhesive layer, the film should display removability.

Removability of Protective Films. The general principles of removability are discussed in Chapter 5. In the case of protective films the protected items are (in comparison with the other components of the laminate) generally mechanically resistant, bulky materials (e.g., profiles, coils, and plates); thus it can be assumed that their internal cohesion (R_s) is much higher than the resistance of the adhesive joint (R_{aj}):

$$R_s \gg R_{aj} \tag{2.1}$$

In working with protective film-coated surfaces, their force of adhesion should ensure a nondestructible, permanent bonding (as in the case of permanent adhesives). On the other hand, delaminating of the protective film after its use requires a very low peel force level, theoretically as low as possible. That means that the real adhesion force is a compromise between the protection-related high peel and the low peel level required for easy delamination.

Generally, relatively thin plastic films are used as carrier materials for protective films. Their mechanical resistance is the result of their formulation and geometrics. It is given by the empirical experience of film and foil manufacturers with common packaging materials. Therefore, in a first instance it may be supposed that the manufacture of protective films requires knowledge of the manufacture of removable film coatings and the mechanics of adhesive joints for plastics.

Special Characteristics of Protective Films

Because of their laminating, end use, and delaminating conditions, protective films have to display some special characteristics. These are given by an adequate physical and chemical basis (see Chapters 3 and 4).

The performance characteristics of protective films should allow their practical use, which includes their application on the product to be protected, performance of their protective function for however long the protected item needs it, and their debonding or separation behavior. Both the application and deapplication (separation) of the protective film depend on its adhesive properties. Therefore protective films have to display special adhesive characteristics manifested as an unbalanced low tack and high cohesion and adhesion. They have to be conformable in order to allow lamination on difficult adherend contours. Protective films have to display plasticity in order to deform in parallel with the protected processed adherend. They have to exhibit permanent adhesion during end use and low force adhesion during deapplication.

Classes of Protective Films

On the basis of their buildup, protective films can be classified as adhesiveless or adhesive-coated films. The adhesiveless products are also called self-adhesive films. Self-adhesive films (SAFs) may work physically or chemically, depending on their principle of functioning, i.e., their buildup. The SAFs include different chemical compounds as pressure-sensitive components. According to the nature of the adhesive polymer, the principal types of self-adhesive films are SAFs based on:

polyolefins;
ethylene-vinyl acetate copolymers;
ethylene-butyl acrylate copolymers;
polyisobutylene copolymers; and
other chemical compounds.

Protective films with adhesive can be classified according to adhesivity, type of carrier film, adhesive properties, bonding/debonding nature, and application domain (Table 2.16).

The self-adhesive protective films can be classified according to their main application fields (and conditions) into SAFs for hot laminating and SAFs for warm or room temperature laminating. The self-adhesive films for hot laminating can be used for metal, plastics, and paper surfaces.

Hot laminating self-adhesive films can be divided into films for uncoated and those for coated adherend surfaces. The films for coated surfaces can be applied on coated lacquered or coated laminated adherends.

Adhesive-coated protective films may be used for metal or nonmetal (glass, plastic, stone, etc.) surfaces. They can be classified according to their function (storage or processing protection), their place of application (face or back side

Table 2.16 Classification of Protective Films

Classification criterion	Product classes			
	Main			Special
Adhesive				
Presence	Adhesiveless	Built-in adhesive	Coated adhesive	—
Nature	Acrylic Crosslinked/uncrosslinked curing agent Solvent-based/water-based	Rubber/resin Primered/unprimered		—
Aggressivity	Low	Medium	High	Easy peel; cleavage peel
Coating weight	Low	Medium	High	—
Recyclability	Yes	No	—	—
Carrier				
Nature	Polyolefin LDPE, LDPE-HDPE, LDPE-PP-LLDPE	PVC	PET	Other materials
Buildup	Homogeneous Clear/pigmented	Heterogeneous Blend/coex	—	Colored
Thickness	Low (25–40 μm)	Medium (40–80 μm)	High (80–120 μm)	—
Release				
Presence	With release liner	Without release liner	—	—
	With release layer	Without release layer	—	—
Nature	Silicone	Carbamate	Fluoropolymers	Other materials; water-based/solvent-based
Laminating				
Temperature	Room temperature	High temperature	—	—
Pressure	Low	High	—	—
Method	Manual	Automatic	—	—
Processibility	Nonprocessible	Processible	—	Deep drawable; laser-cuttable
Weatherability	UV-resistant	Non-UV-resistant	—	—
Protected surface				
Nature	Metal	Plastic	Textile	Copper; anodized aluminum
Roughness	Rough	Polished	Glossy	Soft PVC; enameled metal
Cleanliness	Clean	Oiled	—	—
Delaminating	Manual	Automatic	—	—

protection), and the permanent (mirror tape, label overlaminating film, etc.) or temporary (masking films) character of the bond. Nonmetal surfaces may be film-like surfaces, plates, profiles, and textiles (e.g., carpet).

2.4 Other Products

According to Ref. 173, only about 5% of pressure-sensitive adhesives is used for the manufacture of labels and tapes; the main portion is applied in paper processing (27.5%) and building (28.7%). However, PSPs other than labels, tapes, and protective films having a lower production volume have also achieved economic importance. Business forms, decals, decorative films, and plotter films are the main representatives of these special products.

Business forms are complex, multilayered laminated sheetlike products designed to carry information to be distributed to different end users at different times in different forms. Form–label combinations, cut sheet and laser imprintable forms, and security forms have been developed. A business form with a removable label, with tape pieces adhesively mounted on the paper substrate, is described in Ref. 192. As usual, the business form has die-cut label areas that can be removed from it. This laminate is made of paper, printed continuously in a press. Pieces of adhesive tape are applied at spaced points on the printed paper. The formed binary areas having a greater thickness than the carrier include the tape in relief with respect to the substrate. One of the layers is die-cut.

Plotter films are pressure-sensitive films designed to be used in graphics as signs, letters, etc. [164]. They are cut from special PSA-coated plastic films, using computer programs that allow the design and mounting of letters or text fragments having dimensions of 10–1000 mm [193]. Stickies are carrier-free pressure-sensitive products designed to transfer (usually) graphical information. They use the solid-state carrier material as release liner [194]. According to Refs. 195 and 196, decor films are soft PVC films coated with PSA and printed on the back side with a special design. A detailed description of special PSPs is given in Chapter 8.

REFERENCES

1. *Adhes. Age 9:*8 (1986).
2. K. F. Schroeder, *Adhäsion 5:*161 (1971).
3. I. Benedek and L. J. Heymans, *Pressure-Sensitive Adhesives Technology,* Marcel Dekker, New York, 1997, Chapter 9.
4. T. J. Kilduff and A. M. Biggar, U.S. Patent 3355545; in *Coating, 7:*210 (1969).
5. *Etiketten-Labels 3:*10 (1995).
6. D. Djordjevic, Tailoring films by the coextrusion casting and coating process, Specialty Plastics Conference '87, Polyethylene and Copolymer Resin and Packaging Markets, Dec. 1, 1987, Maack Business Service, Zürich, Switzerland.

7. M. von Bittera, D. Schäpel, U. von Gizycki, and R. Rupp (Bayer AG, Leverkusen, Germany), EP 0147588 B1/10.07.1985.
8. E. L. Scheinbart and J. E. Callan, *Adhes. Age 3:*17 (1973).
9. *Adhäsion 12:*188 (1988).
10. *Coating 12:*390 (1969).
11. *Adhes. Age 9:*30 (1986).
12. *Adhäsion 6:*14 (1985).
13. F. Altenfeld and D. Breker, Semi structural bonding with high performance pressure sensitive tapes, 3M Deutschland, Neuss, 2nd ed., February, 1993, p. 278.
14. *Eur. Adhes. Sealants 6:*37 (1995).
15. G. Bennett, P. L. Geiss, J. Klingen, and T. Neeb, *Adhäsion 7/8:*19 (1996).
16. *Etiketten-Labels 3:*32 (1995).
17. U.S. Patent 4223067, in D. K. Fisher and B. J. Briddell (Adco Product Inc., Michigan Center, MI), EP 0426198 A2/08.05.1991.
18. R. Mast, J. D. Muchin, and J. A. Neri (Lee Pharmaceuticals, Ascutek Adhesive Specialties), EP 262271/06.04.1988, in *CAS 19:*3 (1988).
19. J. Suchy, J. Hezina, and J. Matejka, Czech Patent 247802/15.12.1987; in *CAS 19:*4 (1988).
20. X. Li, Fuaming Zuanly Sheqing Kongkai Shuomingshu, China Patent 86 102255/21.10.1987, in *CAS 19:*4 (1988).
21. D. K. Fisher and B. J. Briddell (Adco Product Inc., Michigan Center, MI), EP 0426198 A2/08.05.1991.
22. F. C. Larimore and R. A. Sinclair (Minnesota Mining and Manuf. Co., St. Paul, MN), EP 0197662A1/15.10.1986.
23. G. Meinel, *Papier Kunstst. Verarb. 19:*26 (1985).
24. United Barcode Industries, Denmark, *Drucker Zubehör,* Technical Booklet, 1996.
25. C. M. Brooke, *Finat News 3:*34 (1987).
26. EP 0251672.
27. F. M. Kuminski and T. D. Penn, *Plastics Process.* 42(2):25 (1972).
28. *Adhäsion 11:*482 (1967).
29. Letraset Ltd., London, U.S. Patent 1545568; in *Coating 6:*71 (1970).
30. R. Higginson (DRG, U.K.), PCT/WO 88 03477/19.05.1988; in *CAS Colloids (Macromol. Aspects) 22:*6 (1988).
31. Mystik Tape Inc., IL, U.S. Patent 3161533; in *Adhäsion 6:*277 (1966).
32. Johnson & Johnson, NJ, U.S. Patent 3161554; in *Adhäsion 6:*277 (1966).
33. Lohmann KG, U.S. Patent 3086531; in *Coating 6:*185 (1969).
34. *Adhes. Age 4:*6 (1983).
35. J. R. Pennace, C. Ciuchta, D. Constantin, and T. Loftus, (Flexcon Co. Inc., Spencer, MA) WO8703477 A/18.06.1987.
36. *Coating 6:*185 (1969).
37. Johnson & Johnson, Br. Patent 799424; in *Coating 6:*185 (1969).
38. *Adhes. Age 4:*6 (1983).
39. D. C. Stillwater, D. C. Koskenmaki, and M. H. Mazurek (3M Co., St. Paul, MN) U.S. Patent 5344681/06.09.1995; in *Adhes. Age 5:*14 (1996).
40. R. B. Lipson (Kwik Paint Products), U.S. Patent 5468533; in *Adhes. Age 5:*12 (1996).

41. U.S. Patent 3364063; in J. C. Pasquali, EP 0122847/24.10.1984.
42. Rohm & Haas Co., Philadelphia, PA, U.S. Patent 3152921; in *Adhäsion* 3:80 (1966).
43. K. Palli, M. Tirkkonnen, and T. Valonen (Yhtyneet Paperitehtaat Oy), Finn. Patent 74723/30.11.1987; in *CAS Adhes.* 14:4 (1988): *Chem. Abstr.* 108:222786.
44. M. Satsuma (Nitto Electric Ind. Co.), Jpn. Patent 63 48381/01.03.1988; in *CAS Adhes.* 14:4 (1988).
45. W. K. Darwell, P. R. Konsti, J. Klingen, and K. W. Kreckel (Minnesota Mining and Manuf., St. Paul, MN), EP 257984/21.08.1986.
46. K. Heinz, Honsel, Bielefeld, Germany, Ger. Patent 2110491; in *Coating* 12:363 (1973).
47. A. D. Little, Inc., U.S. Patent 3039893; in *Coating* 6:185 (1969).
48. Rikkidain K. K., Jpn. Patent 07278508/24.10.1995.
49. H. Nakahata, Jpn. Patent 07316510/05.12.1995; in *Adhes. Age* 5:11 (1996).
50. M. Hasegawa, U.S. Patent 4460634/17.06.1984.
51. Br. Patent 1102244; in *Adhäsion* 10:303 (1972).
52. German Patent Appl. 3144911; in US Patent 4673611/ July 16, 1987.
53. German Patent Appl. 3216603; in US Patent 4673611/ July 16, 1987.
54. G Crass, A. Bursch and P. Hammerschmidt (Hoechst A.G., Frankfurt am Main, Germany), EP4673611/ July 16,1987.
55. Thomas L. Bank and L. Thomas Simpson (Minnesota Mining and Manufacturing Co., St. Paul, MN), EP0120708/ November 19, 1987.
56. J. M. McClintock (Morgan Adhesives Co., Stow, OH), U.S. Patent 4.345678/ 22.04.1986; in *Adhes. Age* 5:26 (1987).
57. *Coating* 11:89 (1984).
58. Arhoco Inc., U.S. Patent 3509991; in *Coating* 5:130 (1971).
59. *Adhäsion* 11:37 (1994).
60. N. W. Malek (Beiersdorf AG, Hamburg, Germany), EP 0095093/30.11.1983.
61. Minnesota Mining and Manuf. Co., St. Paul, MN, Ger. Patent 1486514; in *Coating* 12:363 (1973).
62. E. Borregard, U.S. Patent 3454458; in *Coating* 12:353 (1970).
63. *Converting Today* 7/8:6 (1990).
64. Belg. Patent 675420, in D. K. Fisher and B. J. Briddell (Adco Product Inc., Michigan Center, MI), EP 0426198 A2/08.05.1991.
65. *Packlabel News,* Packlabel Europe 97, Ausgabe 2, Labels and Labelling Packlabelers, The White Lane, England.
66. *Coating* 3:65 (1974).
67. *Adhes. Age* 3:8 (1987).
68. *Adhes. Age* 8:8 (1983).
69. J. H. S. Chang (Merck & Co., Inc., Rahway, NJ), EP 0 179 628/24.10.1984.
70. D. Lacave, *Labels Label.* 3/4:54 (1994).
71. I. P. Rothernberg (Stik-Trim Industries Inc., New York), U.S. Patent 4650704/ 17.03.1987.
72. D. Bedoni and G. Caprioglio, Modern equipment for label and tape converting, 19th Münchener Klebstoff und Veredelungsseminar, 1994, p. 37.
73. G. L. Duncan (Mobil Oil Co.), U.S. Patent 4720416/19.01.1988.

74. Johnson & Johnson, U.S. Patent 3483018; in *Coating* 1:28 (1971).
75. Universum Verpackungs GmbH, Rodenkirchen, Germany, Ger. Patent 12977990; in *Coating* 1:9 (1990).
76. Kimberly Clark Co., U.S. Patent 799429; *Coating* 1:9 (1970).
77. P. Thorne, *Finat News* 3:47 (1988).
78. Mystik Adhesives Inc. IL, U.S. Patent 287842, in Adhesion 6: 277 (1866).
79. Minnesota Mining and Manuf. Co., Saint Paul, MN, DBP 1277483; in *Coating* 7:210 (1969).
80. T. H. Haddock (Johnson & Johnson Products Inc., New Brunswick, NJ), EP 0130080 B1/02.01.1985.
81. *Papier Kunstst. Verarb.* 9:57 (1988).
82. Patent Appl. 3.216.603; in US Patent 4.673.611/July 16, 1987.
83. G. Crass and A. Bursch (Hoechst AG, Frankfurt am Main, Germany), U.S. Patent 4673611/16.07.1987.
84. G. Camerini (Coverplast Italiana SpA.), EP 248771/20.05.1986.
85. I. Benedek, *Eur. Adhes. Sealants* 2:25 (1996).
86. Patent Appl: 3144911; in US Patent 4673611/ July 16, 1987.
87. EP 4673622.
88. *Adhes. Age* 10:125 (1986).
89. E. Park, *Paper Technol.* 8:15 (1989).
90. U.S. Patent 3321451; in S. E. Krampe and C. L. Moore (Minnesota Mining and Manuf. Co., St. Paul, MN), EP 0202831 A2/26.11.1986.
91. U.S. Patent 4260659; in S. E. Krampe and C. L. Moore (Minnesota Mining and Manuf. Co., St. Paul, MN), EP 0202831 A2/26.11.1986.
92. U.S. Patent 4374883; in S. E. Krampe and C. L. Moore (Minnesota Mining and Manuf. Co., St. Paul, MN), EP 0202831 A2/26.11.1986.
93. R. R. Charbonneau and G. L. Groff (Minnesota Mining and Manuf. Co., St. Paul, MN), EP 0106559 B1/25.04.1986.
94. H. Miyasaka, Y. Kitazaki, T. Matsuda, and J. Kobayashi (Nichiban Co., Ltd., Tokyo, Japan), Patent Appl. DE 3544868 A1/18.12.1985.
95. Jpn. Patent 2736/1975, in H. Miyasaka, Y. Kitazaki, T. Matsuda, and J. Kobayashi, (Nichiban Co. Ltd., Tokyo, Japan), Patent Appl. DE 3544868 A1/18.12.1985.
96. U.S. Patent 3691140/12.09.1972, in H. Miyasaka, Y. Kitazaki, T. Matsuda, and J. Kobayashi (Nichiban Co. Ltd., Tokyo, Japan), Patent Appl. DE 3544868 A1/18.12.1985.
97. Engel, U.S. Patent 514950/18.07.1983, in F. C. Larimore and R. A. Sinclair (Minnesota Mining and Manuf. Co., St. Paul, MN), EP 0197662A1/15.10.1986.
98. Gobran, U.S. Patent 4260659, in J. N. Kellen and C. W. Taylor (Minnesota Mining and Manuf. Co., St. Paul, MN), EP 0246 A2/25.11.1987.
99. G. Auchter, J. Barwich, G. Rehmer, and H. Jäger, *Adhes. Age* 7:20 (1994).
100. J. Merretig, *Coating* 3:50 (1974).
101. P. Foreman and P. Mudge, EVA-Based waterborne pressure sensitive adhesives, in *Tech 12,* Technical Seminar Proceedings, Itasca, IL, May 3–5, 1989, p. 203.
102. W. M. Stratton, *Adhes. Age* 6:21 (1985).
103. *Finat News* 3:31 (1994).
104. *Adhes. Age* 7:34 (1983).

105. *Druck Print 10:*32 (1987).
106. *Etiketten-Label 5:*25 (1995).
107. M. Fairley, *Labels Label. Int. 5/6:*76 (1997).
108. G. F. Bulian, Extrusion coating and adhesive laminating: Two techniques for the converter, Polyethylene 93, Maack Business Service, Conference, Zürich, June 6, 1993.
109. K. Fust, *Coating 2:*65 (1988).
110. J. C. Pommice, J. Poustis, and F. Lalanne, *Paper Technol. 8:*22 (1989).
111. A. Prittie, *Finat News 3:*35 (1988).
112. *Etiketten-Labels 3:*8 (1995).
113. P. Thorne, *Finat News 3:*47 (1988).
114. TDI Label Printer, Avery Dennison Deutschland GmbH, Eching, 1997.
115. L. Waeyenbergh, 19th Münchener Klebstoff u. Veredelungsseminar, 1994, München, p. 138.
116. *Etiketten-Labels 3:*9 (1995).
117. H. J. Teichmann, *Papier Kunstst. Verarb. 11:*10 (1994).
118. W. Keller, *Kunstst. Synth. 9:*32 (1995).
119. M. Niemeijer, *Adhäsion 9:*351 (1965).
120. K. Dormann, *Coating 6:*150 (1984).
121. K. W. Holstein, *Neue Verpack. 4:*59 (1991).
122. W. E. Havercroft, *Paper Film Foil Converter 10:*52 (1973).
123. *Paper Film Foil Converter 5:*45 (1969).
124. U. P. Seidl (Schreiner Etiketten und Selbstklebetechnik GmbH), Patent Appl., DE 3625904/04.02.1988; in *CAS Siloxanes Silicones 16:*3 (1988).
125. H. Hadert, *Coating 1:*11 (1969).
126. Johnson & Johnson, Can. Patent 583367; in H. Hadert, *Coating 1:*11 (1969).
127. *Coating 3:*68 (1974).
128. *Verpack.-Rundsch. 9:*994 (1983).
129. *Etiketten-Labels 3:*23 (1995).
130. *Neue Verpack. 1:*156 (1991).
131. G. Bonneau and M. Baumassy, New tackifying dispersions for water based PSA for labels, 19th Münchener Klebstoff u. Veredelungsseminar, 1994, p. 82.
132. H. Mueller and J. Tüerk (BASF AG, Ludwigshafen, Germany), EP 0118726/ 19.09.1984.
133. R. Schieber, *Adhäsion 5:*21 (1982).
134. I. Benedek, *Adhäsion 12:*17 (1987).
135. *Paper Film Foil Converter 5:*45 (1969).
136. *Papier Kunstst. Verarb. 6:*18 (1995).
137. *Adhäsion 11:*9 (1988).
138. 2nd Int. Cham-Tencro Meeting, March 29-31, 1990, Lucern, Switzerland.
139. *Adhes. Age 10:*125 (1986).
140. L. Heymans, Developments in PSA for labels and tapes, Second European Tape & Label Conference, April 28, 1993, Brussels, Belgium, p. 115.
141. Novamelt Research GmbH, Wehr, Germany, *Preselection Guide for Hotmelt Pressure-Sensitive Adhesives.* 1995.
142. *Papier Kunstst. Verarb. 9:*57 (1988).

143. Etilux, *Quelques produits dont vous ne pouvez vous passer,* Booklet, S. A. Etilux N. V., Brussels, Belgium, 1997.
144. *Etiketten-Labels 1:*17 (1995).
145. M. Bateson, *Finat News 3:*29 (1989).
146. Minnesota Mining and Manuf. Co., St. Paul, MN, DBP 1486514, in *Casting 12*:363(1973).
147. *Prodoc,* February 1992.
148. *Etiketten-Labels 5:*136 (1995).
149. *apr 22:*787 (1986).
150. *Etiketten-Labels 1:*10 (1996).
151. W. Keller, *Kunstst. Synthetics 9:*32 (1995).
152. *Finat News 4:*10 (1996).
153. C. Stöver, *Kunststoffe 84* (10):1426 (1994).
154. *Paper Eur. 9:*11 (1993).
155. *Label Buyer International,* Spring 97, pp. 20, 26.
156. F. Altenfeld and D. Breker, Semi structural bonding with high performance pressure sensitive tapes, 3M Deutschland, *Neuss,* Feb. 2, 1993, p. 278.
157. *Klebeband Forum,* Hoechst Films, Hoechst AG, October 1989, No. 27.
158. Second European Tape & Label Conference, Apr. 28, 1993, Brussels, in *Coating 8:*277 (1993).
159. *Adhes. Age 7:*42 (1994).
160. G. Bennett, P. L. Geiss, J. Klingen, and T. Neeb, *Adhäsion 7/8:*19 (1996).
161. *Coating 11:*46 (1990).
162. H. Kniese, *Coating 11:*394 (1990).
163. L. A. Sobieski and T. A. Tangney, *Adhes. Age 12:*26 (1988).
164. H. Becker, *Adhäsion 3:*79 (1971).
165. *Adhes. Age 12:*43 (1984).
166. *Adhäsion 1/2:*29 (1987).
167. *Kunstst. J. 9:*92 (1986).
168. *Converting Today 11:*9 (1991).
169. Johnson & Johnson, U.S. Patent 3403018; in *Coating 1:*24 (1969).
170. *Adhäsion 6:*14 (1985).
171. *Adhes. Age 10:*125 (1986).
172. *Adhes. Age 9:*8 (1986).
173. K. Wabro, R. Milker, and G. Krüger, *Haftklebstoffe und Haftklebebänder,* Astorplast GmbH, Alfdorf, Germany, 1994, p. 48.
174. F. M. Kuminski and T. D. Penn, *Resin Rev. 17*(2):25 (1990).
175. U.S. Patent 4181752; in R. R. Charbonneau and G. L. Groff (Minnesota Mining and Manuf. Co., St. Paul, MN), EP 0106559 B1/25.04.1984.
176. U.S. Patent 2925174; in R. R. Charbonneau and G. L. Groff (Minnesota Mining and Manuf. Co., St. Paul, MN), EP 0106559B1/25.04.1984.
177. U.S. Patent 4286047; in R. R. Charbonneau and G. L. Groff (Minnesota Mining and Manuf. Co., St. Paul, MN), EP 0106559B1/25.04.1984.
178. C. Parodi, S. Giordano, A. Riva, and L. Vitalini, Styrene-butadiene block copolymers in hot melt adhesives for sanitary application, 19th Münchener Klebststoff und Veredelungsseminar, 1994, p. 119.

179. S. E. Krampe and C. L. Moore (Minnesota Mining and Manuf. Co., St. Paul, MN), EP 0202831 A2/26.11.1986.
180. A. Wolf, *Kaut. Gummi Kunstst. 41*(2):173 (1988).
181. U. Zorll, *Adhäsion 9:*237 (1975).
182. *Coating 6:*188 (1969).
183. *Adhäsion 3:*83 (1974).
184. *Coating 8:*200 (1984).
185. O. Cucu, C.Ilie, N. Moga, and M. Popescu, Rom. Patent 92466/30.09.1987; in *CAS Coating Inks & Related Products 11:*6 (1988).
186. Hitco S. A., Fr. Patent 2045508; in *Coating 11:*336 (1972).
187. P. Gleichenhagen and I. Wesselkamp (Beiersdorf AG, Hamburg), EP 0058382B1/25.08.1982.
188. H. Bucholz, *Oberflächentechnik 10:*484 (1973).
189. Thermoplastic adhesive film, *Adhes. Age 12:*8 (1986).
190. C. Bohlmann, *Papier Kunstst. Verarb. 9:*12 (1982).
191. I. Benedek, E. Frank, and G. Nicolaus (Poli-Film Verwaltungs GmbH, Wipperfürth, Germany), DE 4433626 A1/21.09.1994.
192. R. C. Lomeli and G. E. Stewart (Trade Printers), U.S. Patent 4379573 (Apr. 12, 1983).
193. *Siebdruck 8:*14 (1986).
194. Minnesota Mining and Manuf. Co., St. Paul, MN, DBP 1277483; in *Coating 7:*210 (1969).
195. BASF, T1-2.2-21, Teil 3, B.3, Selbstklebende Dekorationsfolien, November 1979.
196. G. Fuchs, *Adhäsion 3:*24 (1982).

3
Physical Basis of PSPs

Pressure-sensitive products are constructed in various ways depending on their application. Their components are manufactured of various raw materials. However, the astonishingly broad range of materials and structures leads to a common characteristic: pressure sensitivity. Why? In order to answer this question, let us make a short examination of the theoretical (physical and chemical) basis of PSPs.

Pressure-sensitive products are viscoelastic materials. As discussed in Chapter 2, such materials are built up as various constructions. The viscoelasticity and flow properties of PSPs play a key role in their application. Their performance results from their intrinsic characteristics and from the interaction of their components. Their mechanical performance characteristics allow their conversion and end use. Such performance characteristics depend on their rheology. Other physical properties influence the performance characteristics of PSPs also. For certain PSPs the polarity and electrical properties of the carrier material affect coatability and processibility. Therefore in this chapter we deal with the rheology, mechanical characteristics, and electrical properties of PSPs.

1 RHEOLOGY OF PSPs

The rheology of pressure-sensitive adhesives is described in detail in Ref. 1. As discussed earlier (see Chapter 2), a non-PSA-related technology has also been developed to manufacture pressure-sensitive products. The first pressure-sensitive products were manufactured from masticated rubber and filler. Later the elastomer was transformed by formulation in a viscoelastic material used as coated PSA. The newest PSPs are adhesiveless films made of plastomer (Fig. 3.1).

The first carrier materials were nonconformable, nondeformable, and nonadhesive. For such products pressure-sensitivity and self-adhesivity were given by a special coated layer. Plastics developed more recently as carrier materials are conformable, deformable, and may exhibit adhesivity. The main components of the classical PSP construction are plastomers (carrier) and modified elastomers (PSA). Among the rubbery products, thermoplastic elastomers (TPEs) can be processed like plastomers. The construction of the new PSPs includes a single

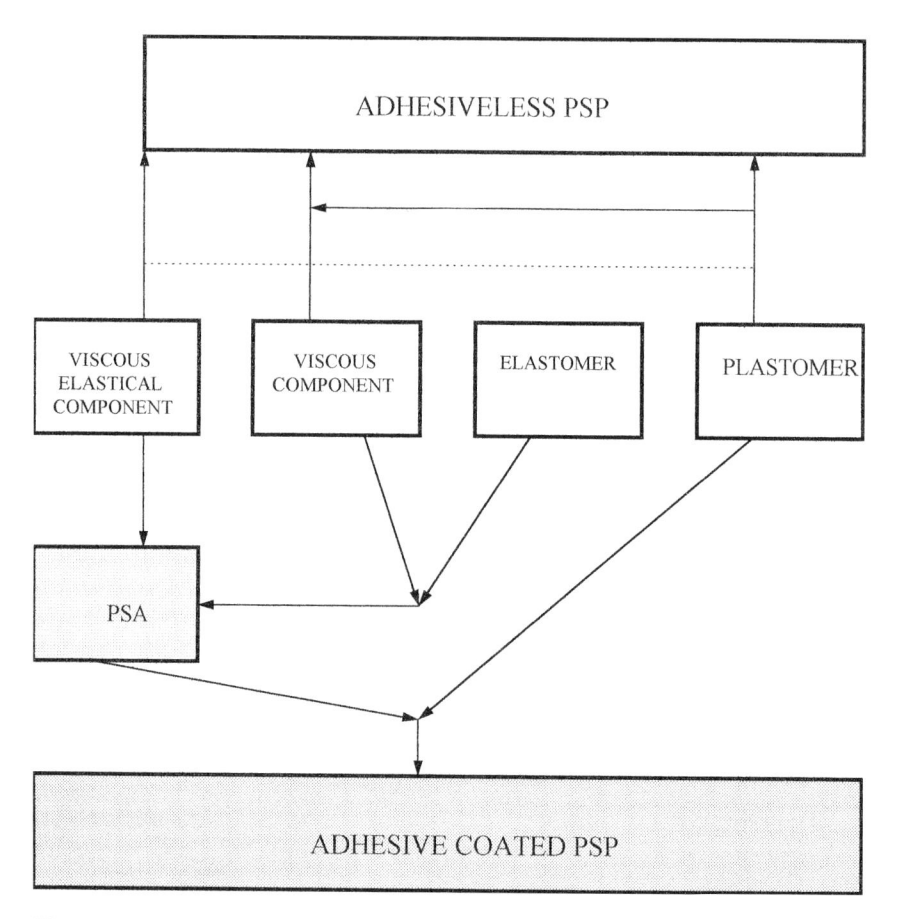

Figure 3.1 Raw material basis and manufacture of pressure-sensitive products.

Physical Basis of PSPs

viscoelastic material processed as a plastomer. It can be concluded that elastomers and plastomers are the base components of PSPs. Both must be modified to achieve adequate viscoelastic properties.

Classical (nonadhesive) plastic carrier materials exhibit viscoelastic properties that are reflected in their mechanical performance during product manufacture and application. Their conformability and deformability do not allow pressure-sensitive adhesion but influence bonding and debonding (stress transfer). Pressure-sensitive adhesives display the viscoelasticity necessary to bond and debond. "One-component" PSPs must also exhibit viscoelastic characteristics that allow them to carry out their mechanical (carrier) and adhesive functions. Pressure-sensitive adhesives exhibit cold flow. It is well known that plastics also display cold flow, but to a different degree. In their common use plastics are virtually cold flow free. Pressure-sensitive adhesives must exhibit a real cold flow. The development of macromolecular chemistry allowed the synthesis of new materials (ethylene-vinyl acetate copolymers, ethylene-butyl acrylate copolymers, etc.) having a PSA-like rheology and the processibility of plastics. Embedded in the carrier film, such products can impart pressure-sensitive properties.

Pressure-sensitive adhesives are based on elastic substances. Rubbery elasticity allows high elastic deformation. Filmlike carrier materials are based on plastics. Their elastic deformation is limited; high force levels or the use of thin gauge materials are required to produce elastic or reversible deformations in them. Pressure-sensitive adhesives undergo continuous self-deformation. Although there are considerable differences between the behavior of PSAs and that of plastomers under stress, it is possible to manufacture plastic carrier materials that will deform like PSAs at a low force level by bonding and debonding. How? What is the theoretical background for the design of such products? Rheological considerations help to answer this question.

For classical adhesives, their chemical nature plays an important role. The nature of the substrate is less important. For PSAs, their physical status is more important, and the surfaces of the carrier material and the substrate determine the type of bond. Because of the mutual influence of the solid-state components and the adhesive, both influence the peel value and the bonding and debonding characteristics. For many classical applications the dimensions and physical/mechanical characteristics of the solid-state components of the joint are less important for the bond. For thin plastic film based pressure-sensitive products, the deformation of the carrier material (its rheology) and consequently its mechanical characteristics play a considerable role. Both the carrier material and the adhesive suffer deformation under stress. Nonreinforced plastics possess mechanical resistance and deformability of the same order of magnitude as the crosslinked adhesive layer [2]. Because of the mutual deformability of the adhesive and the carrier material and their same order of magnitude it is difficult to quantify their participation in the stress transfer (Fig. 3.2).

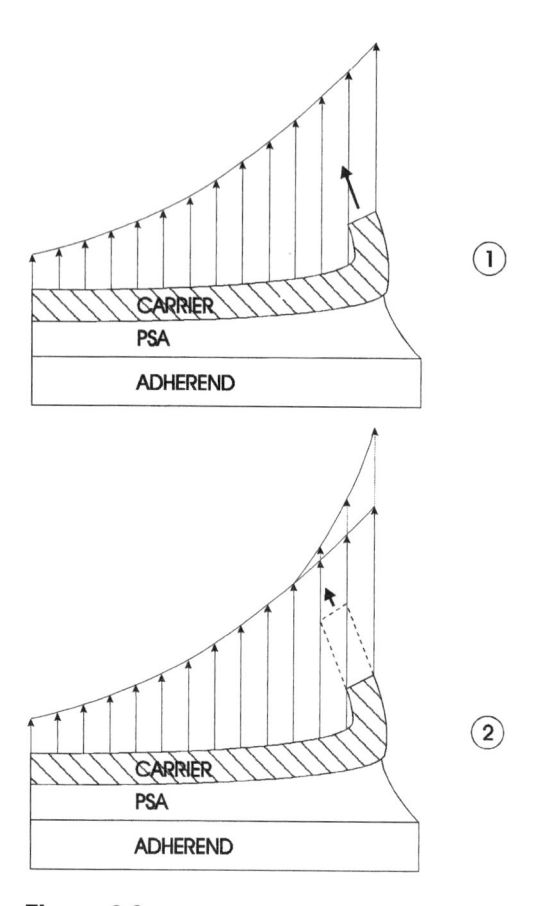

Figure 3.2 Mutual deformation of the adhesive and carrier during debonding of the PSP. (1) Paper carrier; (2) plastic carrier.

Based on industrial requirements the following questions can be formulated:

1. Is it possible to achieve pressure-sensitive behavior for carrier-like materials also?
2. If it is possible to achieve "plastomer-like" pressure-sensitive products or "carrier-like" pressure-sensitive adhesives is their mechanism of functioning the same?
3. For "full plastic" pressure-sensitive products, what are the role (from the rheological point of view) of the carrier material and that of the adhesive?

The manufacturer of pressure-sensitive products can be (and for certain products must be) the manufacturer of the plastic carrier material also. Therefore he is involved in the regulation of polymer rheology during processing at high temperatures and that of the filmlike material at room temperature. The rheology of the molten plastic material is decisive in carrier manufacture. The rheology of the molten or diluted adhesive is important for its coating. The rheology of the softened plastic film is determinative for its application. Therefore these aspects should be discussed separately.

1.1 Rheology of Carrier Material

Some carrier materials and all pressure-sensitive adhesives are viscoelastic materials. They differ only in the ratio of elastic to plastic components and their degree of linearity. A purely elastic (Hookean) material is characterized by a linear relationship between stress and strain, whereas a purely viscous fluid exhibits a stress directly proportional to the strain rate but independent of the strain itself. According to Gerace [3], viscoelasticity means that the bond strength is not constant for varying strain rates and temperatures. Lines of equivalent adhesion can be drawn over all combinations of rate and temperature. Any incompletely elastic body subjected to cyclic strains will respond with an out-of-phase induced stress. If the body is linearly viscoelastic, a simple shift in the phasing of the induced stress with respect to the cyclic strain will occur. Nonlinear viscoelastic behavior is more complex [4]. Pressure sensitivity supposes nonlinear viscoelastic behavior. Pressure sensitivity requires fluidity for low stress rate instantaneous adhesion and solid-state behavior for high speed debonding stress.

In experiments in which bodies are subjected to periodic oscillatory stresses due to sinusoidal strains (γ), the stress and strain rates are never in phase or $\pi/2$ out of phase at the same time but somewhere in between. Such materials when strained store part of the energy and dissipate some of it as heat. This behavior is described as viscoelastic. A viscoelastic body subjected to sinusoidal strain,

$$\gamma = \gamma_0 \sin \omega t \tag{3.1}$$

where γ_0 is the maximum strain amplitude and $\omega/2\pi$ is the frequency, will exhibit a sinusoidal stress (δ) of identical frequency but out of phase with the strain (linear viscoelasticity) [4]. In this case the stress can be expressed as

$$\sigma = \gamma_0[G'(\omega) \sin \omega t + G''(\omega) \cos \omega t] \tag{3.2}$$

where G' is the storage modulus, a measure of the energy stored and recovered in cyclic deformation, and G'' is the loss modulus. The energy loss per cycle hysteresis is given by

$$\pi (\gamma_0)^2 G'' \tag{3.3}$$

The rheological behavior of solid-state materials is characterized by the modulus; the behavior of liquids is described by the viscosity. A classical carrier material is a solid-state component. A classical PSA is liquid. Theoretically, a classical plastic material is a plastomer that does not respond elastically when stressed. Theoretically, rubber (which is the main classical PSA component) is a "pure" elastomer without viscous flow. In reality, ordered structures in both classes of materials allow elastic behavior and excessive stresses cause permanent deformation. Theoretically, the flow properties of a solid-state carrier material and those of a permanently liquid PSA should be discussed separately. In reality, the viscous flow of plastics is temperature-dependent like the viscoelastic properties of PSA. The rheology of both materials is a function of time and temperature. The dynamic modulus and the viscosity depend on the temperature and stress rate. The influence of temperature on viscosity is characterized by the shift factor a_T:

$$a_T = \eta_0(T)/\eta_0(T_0) \tag{3.4}$$

where T_0 is the reference temperature and η_0 is the zero viscosity, i.e., the viscosity that does not depend on the deformation rate (in the linear region, at small shear rates). For amorphous polymers such as PSAs, the dependence of the shift factor on the temperature is characterized by the Williams–Landel–Ferry (WLF) equation including the material-related parameters c_1 and c_2:

$$\ln a_T = -\frac{c_1(T - T_0)}{c_2 + (T - T_0)} \tag{3.5}$$

For partially crystalline polymers (e.g., common carrier materials) the Arrhenius equation is valid:

$$\ln a_T = \frac{E_0}{R(1/T - 1/T_0)} \tag{3.6}$$

where E_0 is the activation energy of the flow related to the material [5]. Pressure-sensitive adhesives are generally amorphous polymers; common plastics are partially crystalline. It is evident that the hot laminating conditions and the bonding behavior of polyisobutylene-based SAFs differ from the laminating conditions of a partially crystalline PVAc. For instance, for a polyvinyl acetate film between −60 and +68°C at a frequency of 1000 Hz there is a change in the dynamic modulus of elasticity of 10^2 dyn/cm^2 [6]. At lower frequencies the difference is more pronounced.

Solid-State Rheology

As a function of the end use of PSPs, a variety of carrier materials have been introduced (see Chapters 2 and 6). Almost all carrier materials are based on macromolecular compounds. Paper and plastics are the main representatives. Paper has a relatively small influence on the rheology of the PSP. It is a polymeric material with a high glass transition temperature (T_g) and crosslinked structure, i.e., low deformability. According to Dunckley [7], for common grades of paper, paper strain appears at 1500 g/in. force during debonding of the pressure-sensitive laminate. The actual forces in the adhesive are higher because they are partially absorbed by the deformation of the paper (Fig. 3.2). In practice it may be supposed that the paper's deformation during the manufacture or application of the PSP is relatively low in comparison with the deformation of other component products. As known the influence of temperature on the rheology of a macromolecular compound can be quantified by the value of the glass transition temperature. Above this temperature the macromolecules have an unordered, free-flowing status, and therefore they can form bonds. Above T_g, the polymer is expanded to the extent that molecular motion is possible. The fluidity of a polymer depends on the position of its T_g relative to the application temperature. The higher the T_g, the higher the fluidity. Carrier materials for PSPs are mainly thermoplasts. Thermoplasts possess a relatively high T_g (Table 3.1). Their end use temperature is situated below their glass transition temperature. They are processed at temperatures above their T_g.

Elastomers (which are the main components of classical pressure-sensitive adhesives) display a much lower T_g than plastomers. They do not melt above this temperature; they are transformed into rubbery elastic materials. They keep this rubbery elastic status due to their crosslinked structure up to their thermal destruction [25]. Thermoplastic elastomers (TPEs) have a relatively low glass transition temperature like elastomers; above this temperature they are also transformed into rubbery elastic materials. This status is, however, less stable than for rubbery materials. As temperature and stresses increase, their stability decreases and a second T_g is displayed. At this temperature the thermoplastic elastomers melt, and above it they are plastic and can be processed like elastomers (Fig. 3.3).

Plastomers are the common carrier materials for PSPs. Elastomers can also be used as carriers, and (reinforced) viscoelastomers play the role of carrier for special PSPs (e.g., transfer tapes). Thermoplastic elastomers have been developed for uses other than PSAs. Their self-adhesivity is low. To be used as PSAs such materials require tackifying like natural rubber. However, their example shows that it is possible to manufacture materials having plastomer-like processibility (as films) that exhibit the properties of elastomers or viscoelastomers. Such performance characteristics have been achieved by development of a broad range of TPEs and pseudo-TPEs (see Chapter 4).

On the other hand, macromolecular compounds that are viscoelastic manifest viscous flow and elasticity. The importance of the viscous or elastic part of their behavior can be characterized by the values of their viscosity or elastic modulus. These parameters are not material characteristics only; they depend on temperature and time, i.e., on the strain rate applied. Paper has a relatively high modulus, i.e., high dimensional stability (at a given atmospheric humidity). Therefore its influence on the rheology of PSPs is manifested more through its interaction with

Table 3.1 Glass Transition Temperature of Main PSP Components

Carrier component	Adhesive component	T_g (°C)	Ref.
—	Polybutadiene	−85	8
Polyethylene	—	−80 to −90	—
—	Natural rubber	−56 to −64	—
Ethylene-propylene copolymer	—	−51 to −59	9
—	Cohesive acrylate	−55	10
—		−53	11
—	Carboxylated styrene-butadiene rubber	−55	12
—	Neoprene	−44	13
—	Acrylate	−43	14
—	Tacky acrylate	−42	15
—	Vinyl acetate-acrylate	−40	16
—	Isoprene-styrene block copolymer, EB cured	−28	17
—	Acrylate	−20	18
—	Vinyl acetate-ethylene, self-crosslinking	−15	19
—	Acrylate, thermally crosslinkable	−7	20
—	Vinyl acetate-dibutyl maleate	−5 to +10	—
—	Very hard acrylate	8	21
Polypropylene	—	−11	9
Polypropylene, cast	—	−3 to +23	22
Vinyl acetate-ethylene copolymer	—	−10 to +15	—
Polyvinyl chloride, plasticized (30–45%)	—	−40 to 0	—
Vinyl acetate, homopolymer, plasticized (40% plasticizer)	—	5	23
Polyvinyl chloride, plasticized (10–30%)	—	0 to +40	—
Vinyl acetate, homopolymer	—	40	23
Polyvinyl chloride	—	75–105	24
Polyimide	—	310–365	24

other laminate components; this interaction is regulated primarily by its surface characteristics.

Plastic films exhibit a different behavior. They display cold flow as a result of their relatively low T_g and modulus. The modulus is a function of chemical composition and macromolecular characteristics (see also Chapter 4). The nature of

Physical Basis of PSPs 81

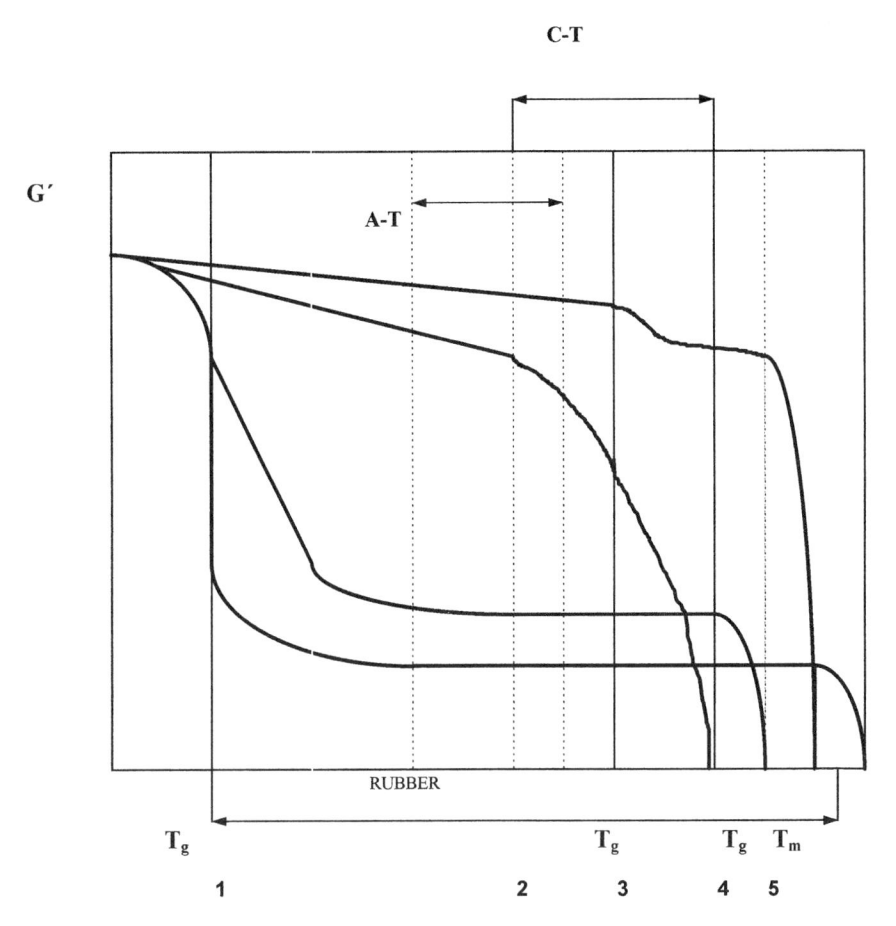

Figure 3.3 Phase transformations of the PSP raw materials. Application domains of rubber, thermoplastic elastomer, amorphous plastomer (*A-T*), and partially crystalline plastomer (*C-T*). *1*, T_g of rubber; *2*, room temperature; *3*, T_g of amorphous plastomer; *4*, T_g of segmented copolymer; *5*, melting point of crystalline domains.

the comonomers, their sequence distribution, and the molecular weight of the polymer influence the modulus. The nature of the comonomers affects the functionality and reactivity of the macromolecules. Functionality refers to the chemistry of the polymer side chain, whereas reactivity refers to crosslinkability. Inter- or intrachain interactions due to functionality or crosslinking due to chain reactivity may reinforce the polymer. The reinforcing effect of the sequence distribution is well known for block copolymers and is taken advantage of in the production of hot melt pressure-sensitive adhesives. The effect of molecular weight (MW) on the cold flow of polymers having the same chemical composition is illustrated by migration, penetration, and blocking. Linear low density polyethylene (LLDPE) contains low molecular weight species and requires more antiblock additive to avoid blocking [26].

The glass transition temperature also has a special influence on aging. A glassy polymer may undergo aging. The temperature range in which this phenomenon occurs has been located by Struik [27] between T_g and the first secondary transition T_β. The data of Struik were obtained from creep measurements. The kinetics of aging has been characterized by the shift rate (M), related to creep (cold flow) at different times according to the following correlation:

$$M = \frac{-\partial \log a}{\partial \log t} \qquad (3.7)$$

where t denotes the aging time and a the horizontal shift factor required to superpose two creep curves related to different aging times. According to Bauwens [28], for the polymers studied the shift rate was found to reach a constant value, near unity, over a more or less wide temperature range. That means that an increase of creep time by a factor of 10 has the same effect on creep compliance as an increase in aging time by a factor of 10. As mentioned earlier, pressure sensitivity requires flow for instantaneous bonding. Let's see what cold flow means for a plastomer, an elastomer, and a viscoelastomer.

Cold Flow and Self-Adhesion. Pressure sensitivity is the result of cold flow. Cold flow (F_c) considered as a deformation (ϵ) of the material under a constant load (σ_0) during a given period of time (t) is expressed by the relation [29]

$$F_c = (1/\sigma_0) \, \epsilon \, (t) \qquad (3.8)$$

Cold flow is a phenomenon that allows bonding and the dissipation of energy by debonding (W_{Fc}), which can be described as a function of the deformation rate v_{Fc} and modulus E [29]:

$$W_{Fc} = \frac{\sigma_0^2}{2E} + \sigma_0^2 \int_0^t V_{Fc}(t) \, dt \qquad (3.9)$$

As stated by Köhler [30], the most important property of a PSA is its permanent liquid status, i.e., cold flow. This allows instantaneous adhesion. Different adhesion theories attempt to explain why one material adheres to another. The best known are the mechanical theory and the adsorption theory. According to the mechanical theory, the adhesive flows and fills microcavities on the surface of the substrate. In general, mechanical anchoring is the prime factor in bonding porous substrates (paper, nonwovens, etc.). It is evident that for this theory the role of adhesive flow is important, but the adsorption theory also supposes intimate contact between adhesive and adherend. The adsorption theory states that adhesion results from molecular contact between two materials and the surface forces that develop. Continuous contact is established between an adhesive and the substrate by wetting. Hot melts, solvent- or dispersion-based adhesives, contact adhesives, and PSAs work physically [31]. It is evident that (generally) the magnitude of the cold flow differs for plastics and pressure-sensitive adhesives. Intermediate values are achieved for self-adhesive films with embedded viscous components and for warm or hot laminated PSPs (see also Chapter 6).

For adhesive-coated PSPs the cold flow of the adhesive depends on its coating weight and viscosity. Therefore for tropically resistant tapes a low coating weight is recommended. For these tapes, a filled adhesive with higher viscosity is used [32]. Special fillers or multilayer structures allow the "combination" of adhesive and carrierlike behavior for transfer tapes. Thick PSA tapes (0.2–1.0 mm) have been prepared by UV photopolymerization of acrylics. The filled layer can be laminated together with the unfilled one. The PSA matrix may contain glass microbubbles [33]. The formulation can be polymerized as a thick layer (up to 60 mils), which may be made up of a plurality of layers, each separately photopolymerized. The middle layer, having a higher degree of polymerization, can be considered the carrier. Such a structure with a plurality of superimposed adhesive layers having different gradients of shear creep compliance may serve as a modality to control the creep compliance, according to Gobran [34].

On the other hand, as is known from the plasticizing of PVC, special plastomers can be formulated as adhesive, and copolymerization may lead to materials having the performance characteristics of plastomers (carrier), elastomers (adhesive component), or viscoelastic compounds (PSA). A vinyl acetate content of 32% in ethylene-vinyl acetate (EVAc) copolymers gives a partially crystalline polymer; at 40% vinyl acetate content, a completely amorphous polymer is synthesized [35]. Polymers with a level of 15% vinyl acetate exhibit polyethylene-like properties. Polymers having 15–30% vinyl acetate exhibit PVC-like characteristics; polymers with more than 30% VAc are elastomers. Copolymerization of ethylene with acrylic acid softens the polymer also. Kirchner [36] showed that the high modulus (of about 1600 kg/cm^2, ASTM 638-58) of high density polyethylene decreases to 430 kg/cm^2 for an ethylene-acrylic acid (EAA) copolymer having 19% polar comonomer. Figure 3.4 illustrates schematically that "softening" of the plastomer carrier material and "hardening" of the adhesive lead to the same result: pressure sensitivity.

The self-deformation (cold flow) of many polymeric materials allows intimate contact to be established between the polymer and the substrate. Adhesion is possible if the flow of the materials allows enough surface contact and the chain segments can interpenetrate [37]. This supposes an increased mobility of the chains. Therefore, crystalline or crosslinked materials are not adhesive above their T_g. As found by the study of mutual adhesion, tack and green strength of ethylene-propylene-diene multipolymer (EPDM) rubber and autoadhesion depend on the

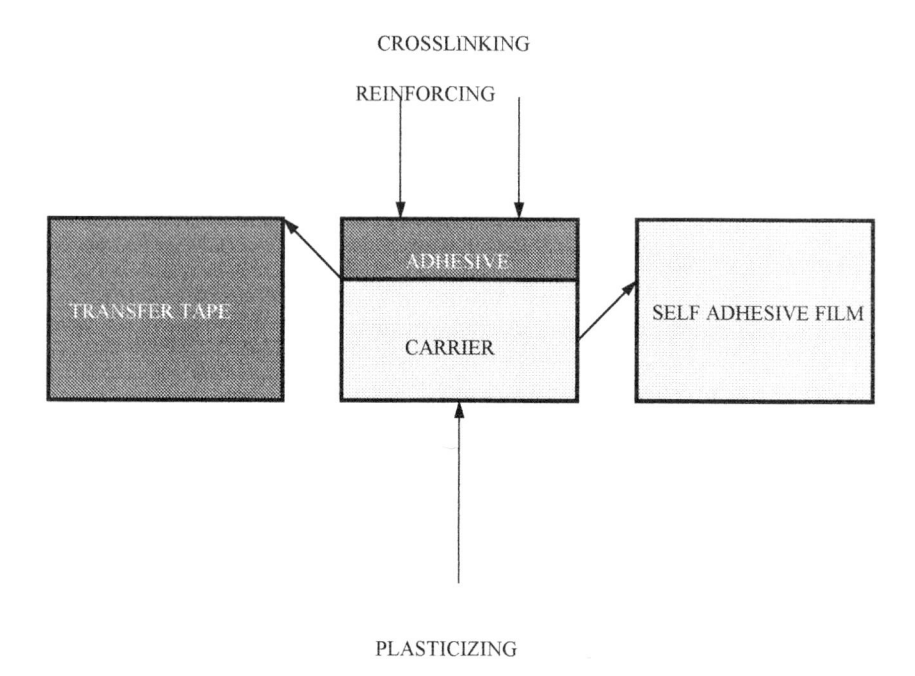

Figure 3.4 Technical possibilities to obtain carrierless or adhesiveless PSP.

(mutual) diffusion of polymer chains and their flexibility and polarity [38]. Studying the contact adhesion of rubber to glass, Carre and Roberts [39] stated that because of energy loss at the border of the contact surfaces, the contact buildup energy is lower than the debonding energy. On wet surfaces, capillary water improves the contact surface, i.e., increases the energy of contact buildup and decreases the debonding energy.

External application conditions may allow better cold flow. Prinz [40] demonstrated that applying a pressure of more than 130 bar to a polyethylene film with aluminum at room temperature (for some seconds) increases the effective contact surface, cold flow of the polymer occurs, and self-adhesion appears. By pressing crosslinked butadiene-styrene copolymer rubber between aluminum plates, a laminate was produced that was used as an adhesive joint model by Carre and Schultz [41]. As stated by Prinz [42], the actual contact surface between two films is very low. (For a polyethylene monocrystal it is about 1–3%.) Therefore the possibility of building up self-adhesion by mutual contact is also low. Corona treatment produces a double layer of electric charges, which causes better contact between the surfaces, i.e., increases the effective contact surface. This increase in the contact surface results in better adhesion. The effect of improving the instantaneous contact surface and adhesion by corona treatment is used for self-adhesive (adhesiveless) films such as hot laminated polyolefins or poly-(ethylene-vinyl acetate) films (see also Chapters 6 and 8).

Cold flow is time- and temperature-dependent. This behavior is illustrated by silicone rubbers. Spontaneous adhesion of slightly crosslinked silicone rubbers on polar substrates builds up linearly over time in the presence of ammonia at sufficiently high humidity and temperature. Reactive groups in the polymer react with the substrate [43]. Cold flow as a surface contact buildup phenomenon and as a component of self-adhesion plays a special role in the manufacture and application of adhesive-free (plastic-based) pressure-sensitive products (protective films, ornamental labels, etc.). Because it allows the bulk deformation of the carrier material and of the pressure-sensitive laminate during application (e.g., tape or protective film lamination) and end use (e.g., processing of protective film/adherend laminates), cold flow plays a very important role for all PSPs that include a plastic carrier material (see also Chapter 5). Cold flow is also important for cold seal adhesives (for both their adhesion and blocking behavior), which are usually mixtures of natural rubber and fillers tackified with vinyl acetate copolymers [44].

There are various theories to explain the adhesion between polymers [45]. New attempts to clarify the role of molecular interdiffusion in self-adhesion of polymer melts draw attention to the effect of labile bonds on the strength of adhesion and of plastic yielding in semicrystalline polymers [46]. Macromolecular diffusion influences the autoadhesion of rubbers. The self-diffusion coefficient of polymers gives information about autoadhesion. The buildup of adhesion is generally proportional to the self-diffusion coefficient (one exception being polybutadiene) [47]. Autoadhesion increases with temperature. It is known from the practice of film tests that measure the coefficient of friction (COF), that a film with low COF at 23°C often displays slip/stick problems at 50°C.

The role of the softness of the carrier material in autoadhesion can be illustrated by cold seal adhesives, where the maximum coating weight (without blocking) recommended for a paper carrier is about 6 g/m^2. The same parameter for a film carrier is only 3 g/m^2 [48]. Here the slow diffusion of the low tack adhesive is facilitated by the pressure used and by the form of the pressing tool. The adhesion of styrene butadiene rubber (SBR) on a glass surface is virtually independent of the degree of crosslinking. However, after the application of pressure the adhesion was found to be related to the degree of crosslinking [49]. The surface characteristics of the adherend determine the softness of the adhesive to be used; for a vinyl adherend, an adhesive with a glass transition temperature of $-10°C$; for fiberboard, $-3°C$; and for steel, $+7°C$ have been suggested [50].

Self-adhesion of plastic films is used industrially for plastic film manufacture via heat laminating. Thick bioriented polypropylene films cannot be produced economically using the common stretching process. They are produced by combining several thin films to form a multilayer film via heat laminating [51].

In industrial practice, cold flow and autoadhesion of plastic films may lead to blocking. According to Schwab [52], blocking is an undesirable adhesive bonding of different surfaces. However, sometimes such adhesion is desired. It is known that the low level of crystallinity of very low density polyethylene (VLDPE) provides an intrinsic cling to extruded films that are used in coextruded structures for industrial stretch wrapping [53]. The examination of the cold flow of common and modified (more viscous) plastomers reveals that although their cold flow is limited (in comparison with the creep of PSA), it can be improved by use of the proper laminating conditions.

Elasticity. Elasticity is related to intramolecular conformational changes. Such changes may occur in elastomers and in the amorphous portion of partially crystalline plastomers. The main problem has to do with how different elastomer structures contribute to such mobility and whether intermolecular interactions are also possible. Common rubber is a crosslinked material and may also be crystalline. It can also be artificially crosslinked. The thermomechanics of the crosslinked networks is considered from the point of view of interchain entropy and energy contributions to the free energy of deformation and the temperature coefficient of the unperturbed chain dimensions. The thermodynamic quantity of primary interest is the energy contribution exhibited by a polymer network. The classical Gaussian theory of rubber elasticity predicts the intrachain entropy and energy contributions at simple deformations of the network and their independence from the deformation (at small and moderate deformations). Analysis of the entropy and energy effects resulting from the simple extension of the stress-softened networks filled with different fillers shows that in many cases the energy and entropy contributions are dependent on the concentration of the fillers, which contradicts the classical theory of rubber elasticity. Early molecular theories of rubber elasticity supposed that the elasticity of a polymer network is exclusively entropic. The classical theory of rubber elasticity takes into account the change in conformational energy by incorporating the front factor (f) into the equation of state for simple elongation and compression [54]:

$$f = \nu \, k \, TL_i^{-1} \, (\langle r^2 \rangle_i / \langle r^2 \rangle_0) \, (\lambda - \lambda^{-2}) \qquad (3.10)$$

where ν is the number of elastically active chains in the network, $\langle r^2 \rangle_i$ is the mean square end-to-end distance of a network in the undistorted state, $\langle r^2 \rangle_0$ is the mean square end-to-end distance of the corresponding free chains, k is a constant, T is the temperature, and L_i is the length of the undistorted sample. From Eq. (3.10) one may derive Eq. (3.11), which relates the intramolecular energy changes to the mean square end-to-end distance of unperturbed chains:

$$\left(\frac{\Delta U}{W} \right)_{V,T} = T \, d \, \ln \, \langle r^2 \rangle \, dT \qquad (3.11)$$

where U is the internal energy and W is the deformational work.

Ordered and Reinforced Systems. Similarities and differences exist between the thermomechanical and thermodynamic behavior of chemically crosslinked polymer networks, filled networks, rubberlike thermoplastics, and crystalline networks. Depending on the composition and dimensions of the reinforcing network, plastomer- or elastomer-like behavior may appear. The proportion of such interpenetrating networks in a material also varies. For instance, polyethylene contains 85–95% crystallinity, EVAc 55–65%, and rubber 0–10%.

The physical reason for the appearance of rubberlike elasticity in SBCs is connected to immiscibility and microphase separation, which lead to a domain structure with domain size on the order of 100 Å. These rigid domains act as crosslinks. Thermomechanical studies of styrene-butadiene-styrene (SBS) and styrene-isoprene-styrene (SIS) block copolymers with a hard block content of less than 40% show that the energy contributions accompanying uniaxial extension are independent of the hard block content and degree of deformation [54]. The energy contributions for diene blocks coincide well with the data on chemically crosslinked diene networks. The energy contributions for polyisoprene and polybutadiene are in good agreement with the results for common chemically crosslinked networks. This seems to indicate that there are no considerable intermolecular changes in the rubbery matrix even at very large deformations. The hard block content also has no influence on the energy contribution. The thermomechanical behavior of segmented polymers with low molecular weight soft blocks and a large content of hard blocks is determined not only by intrachain conformational changes but also by intermolecular changes in both the soft and hard blocks.

The block lengths of segmented polyurethanes and polyesters are much shorter than those of polystyrene-polydiene block copolymers, which tends to limit their extensibility (see also Chapter 4). Their stress-induced softening is accompanied, as a rule, by considerable residual deformation. Such residual deformation is a consequence of plastic deformation of rigid domains and orientation of the domains in the stretching direction. Therefore the energy changes in samples containing less than 50% hard block differ from those with 50% or more. Polymers with a low hard block level behave similarly to typical SBS. Experimental evidence indicates that the free energy of the strained rubberlike block copolymer with a relatively high hard component content cannot originate only within the chains of the network. The interchain effects play a considerable role.

Although it is widely accepted that the free energy of the uniaxial deformation of two-phase crystalline networks is purely intrachain, calorimetric measurements of Godovsky [54] showed that the thermodynamics of the deformation of these networks is controlled by interchain changes in the amorphous region. Crystalline polymers are two-phase systems also, consisting of both crystalline and amorphous regions, and therefore above their T_g they can be considered as networks in which crystallites are the solid filler and act as multifunctional crosslinks. The elastic properties of such systems are related to the conformational changes in the amorphous region. According to this approach the free energy of deformation of two-phase crystalline polymers above their T_g is also purely intrachain. According to the thermomechanics of solids, work (W) is a parabolic function of strain (ϵ) and heat (Q) is a linear function of strain [37]:

$$W = E\,\epsilon^2/2 \qquad (3.12)$$

$$Q = \beta TE\,\epsilon \qquad (3.13)$$

Therefore the heat/work ratio is a hyperbolic function of strain:

$$Q/W = 2\beta T/\epsilon \qquad (3.14)$$

The presence of crystallites in these networks prevents the amorphous chains from deforming, due to conformational rearrangements. The deformation is accompanied by a volume change, with the volume increasing under extension. For undrawn crystalline polymers the thermomechanical properties cannot be related to the conformational changes of the extended tie molecules in the noncrystalline region. Cold drawing of crystalline polymers leads to a change in the sign of the thermal effect accompanying the reversible stretching of a drawn sample; i.e., reversible stretching of drawn polymers is accompanied by evolution of heat. This is a consequence of the stretching of highly oriented tie molecules in amorphous regions like that postulated for crystalline networks. The drawn polymers are able to deform reversibly at 1–30%, and such deformations can be related to the intramolecular conformational changes [54].

As discussed earlier, the rheological behavior of segmented or ordered polymers (e.g., thermoplastic elastomers, partially crosslinked or crystalline polymers) is approximated in many cases by the behavior of filled systems supposing that the hard segments or crosslinking points behave like filler particles embedded in the polymer matrix. In the presence of reinforcing fillers, the elasticity modulus (E) of the elastomers at small and moderate deformation increases in the first approximation according to the Guth–Smallwood [55] equation:

$$E = E_0\,(1 + 2.5\phi + 14.1\phi^2) \qquad (3.15)$$

where E is the modulus of the filled material, E_0 is the modulus of the polymer, and ϕ is the volume fraction of filler. Thus for filled elastomers the mechanical work of deformation, the change of entropy and internal energy, should include the parameter ϕ. The dependence of the energy contribution on the amount of filler and its reinforcing ability demonstrates that in filled elastomers the energy contribution seems to lose its obvious meaning as only a measure of the intrachain effects.

Molten State Rheology

The flow properties of warm and molten polymers depend on their structure and on the forces acting on them. Mobile, fluidlike polymers are the result of the destruction of ordered or segregated structures to achieve chain mobility.

Destruction of Ordered Structures. The flow of the polymer during processing, its extrudability and deformability in the molten state, that allows it to form a tubular or sheetlike film is very important for plastic film manufacturers. The polymers used as raw material for film manufacture are non-Newtonian fluids; their processing viscosity depends on the temperature and pressure and on the deformation rate. The relationships between the processing conditions and flow properties and between the flow properties and the chemical or macromolecular parameters are very important for the film manufacturer [56]. For viscoelastic liquids, the rate of flow along the wall depends on the shear stress on a potential function [57]. The shear rate during the processing of raw materials is quite different during calendering (e.g., 10–10^2 s^{-1}) and extrusion (10^2–10^3 s^{-1}); therefore the residual tension in the materials is also different [58]. In comparison, in the slot die and film forming of hot melts, shear values of 10^2–10^5 s^{-1} are common [59]. According to Ref. 60, a shear rate of 1000 s^{-1} is used in hot melt coating.

Both the plastomers used as carrier material and the elastomers used as carrier and adhesive generally have a low glass transition temperature. The excellent mechanical characteristics of plastomers are due to the presence of ordered structures. The plastomers used as carrier materials possess a multiphase structure given by the presence of crystalline, crosslinked, or filled domains. Such domains disappear above the melting point of the crystallites only. Thermoplastic, segmented elastomers display multiphase portions due to the segregation of their constituents. Such ordered domains can have an amorphous or crystalline structure. Quite different bonding forces (covalent, van der Waals, dipolar, hydrogen, etc.) can hold them together. In comparison, the mechanical characteristics of paper are due to the presence of an ordered structure imposed by hydrogen bonds. In the crystalline region the modulus of paper (E_c) is unaffected by moisture; therefore,

$$E_c = K_c n_c^{1/3} \tag{3.16}$$

where K is the average value of the force constant and n is the number of hydrogen bonds effective in taking up strain under tension at a given moisture content [61]. The mechanical characteristics of such segregated systems depend on the degree of segregation, that is, the degree to which a two-phase construction is built up. Strain may induce ordered structures in elastomers or plastomers. For instance, strain-induced crystallization appears in polyurethane elastomers also. Such crystallization is enhanced by structural crosslinking bimodality [62]. The melting point of the crystalline domains is an important characteristic of such materials. Above their processing temperatures, plastomers exhibit viscous flow and elastomers exhibit viscoelastic flow. Studying the order–disorder transition in styrene block copolymers in comparison with LDPE, Han and Kim [63] evaluated the ratio of loss modulus (G') and storage modulus (G'') of SIS block copolymers over a broad temperature range (140–240°C). The G'/G'' ratio of such copolymers is a function of the temperature (as a result of the order–disorder transitions). Above 165°C the specific volume of SBS polymers increases due to the change in the domain structures [64]. For LDPE there is no such dependence.

The flow properties of a molten polymer are a function of its chemical characteristics. Chemical composition, molecular weight, molecular weight distribution, and branching are the main regulating parameters. These parameters can change the viscosity of the molten polymer and also (because of the non-Newtonian character of its flow) the dependence of the viscosity on the shear rate. Their influence is clearly illustrated by the processing of polyolefins.

Low density polyethylene (LDPE) is highly branched, has a broad molecular weight distribution, and exhibits medium mechanical properties and excellent processability. The extensive branching of LDPE promotes easy extrusion and bubble stability in blown film manufacture (see also Chapter 6). In the early 1970s a linear low density polyethylene (LLDPE) was introduced that had no long branching, had a narrow molecular weight distribution (MWD), and gave better mechanical performance than LDPE [65]. However, its narrow MWD reduced the ease of extrusion and bubble stability. There are fundamental rheological differences between low density polyethylene and LLDPE. Linear low density polyethylene is less viscous at low shear rates and more viscous at higher shear rates. Low density polyethylene hardens as film is drawn off the die surface. The lower viscosity LLDPE does not strain harden and has low bubble stability and little tolerance to high speed air flows [66].

Melt index is a measure of viscosity. High melt index polymers have a lower viscosity, lower molecular weight, and better flow than those of lower melt index. Lower melt index grades are required for melt strength and toughness, i.e., blown film. In spite of its universal use, melt index is a poor measure of processability and molecular structure, being a single-point measurement at low shear rate. Molecular weight distribution has to be considered together with melt index for polymer characterization.

Polymer Blends. The mechanical properties of polymers are basically determined by their mutual solubility. Processability (melt viscosity) depends on the compatibility of the polymer blend also. Interchain interactions responsible for the compatibility in polymer blends tend to reduce the entanglement but increase the friction between dissimilar chains. This friction appears to arise from the local reduction of chain convolution due to segmental alignment, with the latter arising from increased interchain interaction. The entanglement probability and the friction coefficient between dissimilar chains correlate with the strength of specific interactions. The free volume tends to be linearly additive but may deviate according to segmental conformation and packing rather than specific interactions. For instance, the number of branches and branch molecular weight influence both the solid-state morphology of SIS block copolymers and their processing [67]. The reduced entanglement and the free volume additivity tend to reduce the melt viscosity, while the increase in friction tends to increase it. The former two effects are often stronger than the latter, resulting in the reduction of the melt viscosity of compatible polymer blends [68].

1.2 Rheology of Adhesives

The flow properties of the coated adhesive influence the adhesive and end use performance of the PSP. The flow properties of the adhesive during processing influence its coatability.

Rheology of Coated Adhesive

Pressure-sensitive adhesives are macromolecular compounds whose viscoelastic nature is manifested through viscous flow and elasticity and characterized by modulus and glass transition temperature.

Rheological Parameters. The dilatation coefficient increases above the T_g because of the increase of free volume [69]. Adhesive bonding exists only above the glass transition temperature [70]. For a given temperature of application there is an optimum T_g. That means that for PSAs (used generally at room temperature), a T_g value of less than $-15°C$ is required, and values between -40 and $-60°C$ are preferred. As can be seen from Table 3.1, the raw materials for pressure-sensitive adhesives and for certain carrier films possess low T_g values. External and internal plasticizing (copolymerization) allow a decrease in the T_g; crosslinking increases it. From the point of view of chain mobility, pressure sensitivity can be achieved with very different raw materials. Although the T_g is a material constant, its value depends on the purity of the material. As shown by Burfield [70] there is a difference (0.7–0.9 K) between the glass transition temperature of *cis*-polyisoprene and that of natural rubber. As listed by Druschke [71], there are "hardening" (e.g., acrylic acid, acrylonitrile, methyl methacrylate, styrene, vinyl acetate, methyl acrylate) and "softening" (ethyl acrylate, vinyl isobutyl ether, butyl acrylate, isobutylene, ethyl hexyl acrylate, isoprene, butadiene, etc.) monomers. Their use as comonomer influences the T_g, and T_g influences the modulus of elasticity. The T_g showing the temperature domain where chain mobility exists and the modulus indicating the level and nature of material resistance (deformation and change of molecular structures) against external stress are both necessary to characterize the polymer. The hardness and stiffness of polymers increase with T_g [72].

Adhesion is related to the glass transition temperature of the polymer. The adhesive fracture energy exhibits a maximum in the temperature range above the glass transition region. In this temperature range the mechanical behavior is determined by intermolecular interactions, which form entanglements. A network of temporary crosslinks is formed. High tack values require good deformability of the polymer, i.e., a sufficiently low modulus, which means that the material must have an entanglement network with long-chain molecules between two entanglements. For instance, according to Hinterwaldner [73], a radiation-curable base composition for PSA has to possess a T_g lower than $-20°C$ and a controlled crosslinked network. The higher chain flexibility and higher free volume of rubber provides a lower T_g for rubber and a lower modulus [74]. The polymers for PSA must have a T_g 30–70°C below the application temperature [75]. According to Zawilinski [76], room temperature PSAs should possess a T_g between -15 and $+5°C$. The best SBR lattices usable for PSA exhibit a T_g between -60 and $-35°C$. PSAs can be used for tapes, veneers, and wallpapers [77]. The T_g's of such PSAs should be within the range of -35 to $-25°C$. Copolymers of maleic anhydride with acrylics and vinyl acetate have been synthesized for PSAs having glass transition temperatures between -45 and $-65°C$ [78]. The regulation of the T_g can be achieved by mixing base viscoelastic components that have different T_g values. For HMPSAs used for surgical tapes a mixture of acrylic polymers with low T_g (-80 to $-10°C$) and high T_g (10–40°C) is recommended [79]. Elastomers for sealants have a T_g range of -46 to $-50°C$ [80]. The role of the glass transition temperature as a versatility index of raw materials usable for PSAs is illustrated by amorphous propylene copolymers, where long chain comonomers give lower T_g and can be used as one-component HMPSAs also. Amorphous polypropylene (APP) obtained as a by-product has a glass transition temperature of $-14°C$. The polymer produced by direct synthesis exhibits a T_g of $-12°C$ [81]. A propylene-butene APP copolymer displays a T_g of $-16°C$. Even detackifying additives must have a low T_g. The composition of the PSA polymer according to one patent [82] includes a crosslinking agent that is also used for both the matrix polymer and the polymer particles (suspension polymerized) in the matrix, as elastic detackifier. The T_g of the particle should be lower than 10°C.

As discussed in Ref. 1, the T_g of the PSA is the resultant of the components of the recipe. The elastomer or viscoelastic compound and the tackifier resin are the main components of the formulation. Common resins have a T_g situated above room temperature. In certain cases the introduction of a glassy polymer in an elastomer can lead to a product with a broad glass transition [83]. In such cases the rigid polymer component restricts the segmental mobility of the elastomer. Besides the broadening of the tan δ peak, a reduction of the peak value is also readily observable. The T_g allows a forecast for possible tackifier loading

[3]. Therefore certain formulations (e.g., HMPSA) need a liquid plasticizing component also. According to Hughes and Looney [84], fully saturated petroleum resins can be used having a well-defined M_n and a T_g of less than 45°C for an HMPSA formulation based on thermoplastic elastomers without plasticizer oils. It is known that HMPSA formulations incorporate a high level of plasticizer, usually a naphthenic oil or a liquid resin. The use of plasticizers results in a number of disadvantages including long-term degradation of the adhesive bond. Resins synthesized from the C_5 feedstock with a defined diolefin/monoolefin ratio and a molecular weight of 800–960 and M_n of 500–600 (M_w/M_n ratio of at least 1.3), display a T_g of about 20°C and need no oils for formulation with saturated midblock thermoplastic elastomers. Oils can be replaced with other low T_g components also. Low molecular weight polyisoprenes (35,000–80,000) having a low T_g (−65 to −72°C) have been used as viscosity regulators for HMPSA [85]. Used as replacements for common plasticizers, they improve the migration resistance and low temperature adhesion. The elastomer sequence of a rubbery block copolymer must be designed to adequately suppress crystallinity without at the same time raising the T_g. A saturated olefin rubber block has to be obtained with the lowest possible T_g and the best rubber characteristics. The rubbery elasticity is maximized when the rubbery segment is completely amorphous. In commercial styrene-ethylene-butylene-styrene (SEBS) block copolymers, the best rubbery performance is obtained with a rubber block having a T_g of about −50°C and no detectable crystallinity [86].

The flow properties of the adhesive are determinant for its debonding also. It has been shown that for many applications (labels, tapes, protective films, etc.) pressure-sensitive products should be removable (see also Chapter 5). Removability requires a breakable bond at the adhesive/substrate interface. Bond breaking is an energetic phenomenon. For removability, the whole debonding energy should be absorbed by the adhesive itself in such a manner that no failure occurs in the adhesive mass. In this case the energy is used only for the viscous flow and elastic motion of the macromolecules. Therefore a special balance of the plastic/elastic behavior of the PSA is required to allow energy-absorbing flow. The major danger is if the viscous flow is too pronounced, bonds will fail within the adhesive mass. To avoid this danger, the cohesion of the adhesive should be improved in parallel with the reduction of its adhesion to the substrate.

Rheology is a function of the chemical composition and structure of macromolecular compounds. This is illustrated by the investigations of Class and Chu [87], who used low molecular weight polymers of styrene, *t*-butyl styrene, and polyvinyl cyclohexane as tackifiers for natural rubber and styrene-butadiene copolymers. As stated, for optimal rheological behavior the components of the mixture must be compatible. Compatibility depends on their molecular weight and structure. The investigation of the T_g and polystyrene content of thermoplastic elastomers shows that independently of the polystyrene content, (15%, 19%, and

23%) tack and adhesion properties are similar. Cohesion properties and hot melt viscosity differ due to the pronounced dilution. Tack and adhesion properties were found to differ mainly in the midblock phase. The glass transition temperature and the correlation of adhesive properties as a function of plateau modulus allow the prediction of adhesive properties independently of the polymer [88]. The glass transition temperature provides information about the viscoelastic properties, and the value of the modulus quantifies them. The value of the plateau modulus (G_n^0), defined as the storage modulus at the minimum tan δ value of a linear polymer, is an important parameter. Its value depends on the chemical and macromolecular characteristics of the material. For instance, natural rubber latex and natural rubber (even when ground or crushed) exhibit higher plateau moduli ($>10^6$ dyn/cm^2) than carboxylated styrene-butadiene rubber (CSBR) (10^5–10^6 dyn/cm^2) [89]. Degradation of natural rubber does not change the tan δ peak (glass transition temperature). In CSBR, only a high styrene content (46%) gives a high plateau modulus, but such a polymer does not respond well to the addition of resin. Lower styrene content gives a smaller modulus at high temperature.

For a given tensile stress (σ) or shear stress (τ) the interdependence between the deformation force ϵ and the deformation γ is affected by the value of the tension modulus E or shear modulus G:

$$\sigma = \epsilon E \tag{3.17}$$

and

$$\tau = \gamma G \tag{3.18}$$

Plotting the modulus as a function of the temperature reveals a dependence that has the form of the curves in Fig. 3.3, where T_g for an amorphous polymer appears as the transition point from the glassy state to the rubbery elastic state. As can be seen from that figure, for partially crystalline polymers at higher temperatures, above the T_g the crystalline portions of the polymer structure may coexist with the amorphous rubbery parts, reducing the mobility of the chain segments. This is the case for certain pressure-sensitive products made as self-adhesive carriers (e.g., EVAc copolymers or ethylene-butyl acrylate copolymers embedded in a crystallizable plastomer matrix) or based on crystallizable raw materials (e.g., chloroprene or propylene copolymers). It is evident that due to the reduced polymer mobility the tack of such compositions is lower. On the other hand, if the molecular weight is lower than the value required for rubbery performance, the diagram shows no rubbery elastic plateau. This is the case of "postfinished" pressure-sensitive raw materials.

Mechanically resistant plastomers or elastomers need a high modulus value; easy bonding adhesives require a low modulus value. The development of adhesive-free PSPs made softer carrier materials necessary. Making the carrier film thinner decreases its stiffness. There are special tape applications where the conformability of the finished product requires a softness of the same order of magnitude for the carrier and the adhesive. For conformable medical tapes and labels, the films should have a tensile modulus of less than about 4×10^5 psi (in accordance with ASTM D-638 and D-882) [90]. Conformability of PSAs used for medical tapes is measured as creep compliance with a creep compliance rheometer. Acceptable values lie between 1.2×10^{-5} and about 2.3×10^{-5} cm²/dyn [91]. (It has been found that the higher the creep compliance the greater the amount of adhesive residue left on the skin after removal.) Adhesive polymers used for surgical adhesive tapes often exhibited a dynamic modulus too low for outstanding wear performance. A low storage and loss modulus result in a soft adhesive with adhesive transfer. An adhesive suitable for use on human skin should display a storage modulus of 1.0–2.0 N/cm², a dynamic loss modulus of 0.6–0.9 N/cm², and a modulus ratio (tan δ) of 0.4–0.6 as determined at an oscillation frequency sweep of 1 rad/s at 25% strain rate at body temperature (36°C) [92]. On the other hand, crosslinking of the adhesive (necessary in certain applications or for certain coating methods) hardens the adhesive. It can be concluded that there are simultaneous material and product developments that lead to products that barely fall within the definition of carrier or adhesive, i.e., with a well-defined elasticity and/or plasticity.

According to the Newtonian correlation the dependence between shear stress τ and viscosity η as a function of the rate of deformation ($d\gamma/dt$) is given as

$$\tau = \eta \, d\gamma/dt \tag{3.19}$$

Macromolecular compounds undergo relaxation after a force has acted on them. Such compounds are viscoelastic; therefore the force that produces permanent deformation is transformed into viscous energy according to the equation

$$\frac{d\gamma}{dt} = \frac{1}{G}\frac{d\tau}{dt} + \frac{\tau}{\eta} = 0 \tag{3.20}$$

As discussed earlier [see Eqs. (3.1) and (3.2)], in practice there is a shift between the active force and the deformation produced due to the viscoelasticity of the material. Therefore the dynamic modulus is the sum of a synchronous (storage) modulus G' and a delayed (loss) modulus G''. Their ratio (G''/G') is the tangent of the loss angle, i.e., tan δ. The storage modulus is proportional to the average energy storage in a cycle of deformation; the loss modulus is proportional to the average dissipation of energy in a cycle of deformation; tan δ represents the overall behavior of the material with respect to elasticity and plasticity (dissipation). Segregated, ordered structures may exhibit many tan δ peaks. Block copolymers or noncompatible mixtures display two tan δ peaks. The relationship of tan δ to temperature, the tan δ minimum temperature ($T_{\delta min}$), and the tan δ peak temperature ($T_{\delta max}$) describe the rheological behavior of a formulation. The loss tangent versus temperature plot characterizes the cohesive strength. The lower the tan δ value, the greater the cohesive strength. For tackified formulations it is admitted that for a given softening point and resin concentration, the higher the tan δ peak temperature, the better the compatibility [93].

The storage modulus can be used as a measure of crosslinking density. According to Auchter et al. [94], at 50°C a noncrosslinked soft acrylic polymer has a storage modulus of about 10^3 Pa, and this value may be increased to 10^4 Pa by increasing the irradiation level. Such characterization of the flow properties has usually been made by tensile strength measurements [95]. The cohesive (tensile) strength of an uncrosslinked acrylic adhesive is 2500 kPa; that of a crosslinked adhesive can attain 6000 kPa (depending on the crosslinking conditions). The T_g value can also be used to characterize such crosslinked compositions. According to Donker et al. [96], the T_g of electron beam–cured PSA formulations should be situated about 30–70°C below their temperature of use. The cohesive strength is proportional to the concentration of the hard monomer (giving a polymer with T_g higher than $-25°C$).

Admitting that the storage modulus value at low frequency (less than 10^{-2} rad/s) is related to wetting and creep (cold flow) while the value at high frequencies may be related to peel or quick stick properties (i.e., high speed debonding evaluated by tack or peel), both values are necessary for characterizing an adhesive.

For block copolymers the mobility of the midblock can be evaluated by dynamic mechanical analysis (DMA). It is related to the softness or viscous flow, i.e., energy loss modulus [97]. On the other hand, the melting or disappearance of the end block domains (i.e., processibility and temperature resistance) is indicated by the $T_{\delta min}$ value and the crossover temperature of the loss and storage modulus (T_{cross}). Between $T_{\delta min}$ and T_{cross} the end domains soften and disappear. Most tackifiers used for HMPSAs are aliphatic, midblock-compatible resins that do not influence the styrenic domain. Generally a mixture of resins is used for PSAs formulated for tapes. One of the resins is midblock-compatible; the other is compatible with the end blocks. If resins are used that are partially compatible with the styrene domains (e.g., rosin esters, a modified terpene resin, or a modified aliphatic resin), both middle block mobility increase and end block softening are achieved. Such a resin may have an aromatically modified aliphatic hydrocarbon as basis with a higher molecular weight. In a standard HMPSA formulation such tackifiers give similar $T_{\delta max}$ and loss modulus values but a lower T_{cross}, i.e., a lower softening temperature for block domains. According to Donker et al. [96], for optimal PSA packaging tape formulation, $T_{\delta max}$ values of -6 to $0°C$ and loss modulus values at tan δ valleys of 72 and 92 kPa are recommended.

Acrylic block copolymers have soft and hard acrylic phases also. They have a star-shaped, radial structure. The storage modulus has a plateau value of $10^{9.5}$ dyn/cm^2. The soft phase has a T_g of $-45°C$; the hard phase, a T_g of 105°C [97].

As known, because of their more ordered structure, the radial copolymers (SBCs also) generally show a longer linear portion of the rubbery plateau and lower values of tan δ [98]. The two tan δ peaks characteristic of block copolymers are more extreme (−87 and +85°C) for a radial styrene block copolymer than for the linear one (−82 and +75°C). The butadiene used in these types of block copolymers normally has a loss tangent peak at −90°C, the styrene, one at +100°C.

Tack needs low modulus for bonding but high modulus at the strain rates and elongations that obtain during bond breaking. At high frequencies, during debonding the resin stiffens the system. The T_g increases, and the elastic modulus increases also. The lower the modulus at low frequency and the higher it is at high frequency, the better the adhesive. The adhesive strength increases [99]. Energy absorption produces the cold flow of adhesive necessary for bonding. That means that a high loss modulus (peak) at the practical frequency of bonding helps to attain adequate bonding. On the other hand, at the higher debonding frequency the elastic behavior of the adhesive (storage modulus) is required. An index of the energy absorption is the value of the loss modulus. Cold flow (loss modulus) should be maximum during bonding and minimum during debonding (except for removable formulations). Taking into account the different stress frequencies caused by bonding and debonding, dynamic mechanical investigations into the modulus value should be carried out at different frequencies.

In an experiment concerning the PSAs for diaper tapes, the storage modulus ratios of different formulations were determined at high frequency (100 rad/s) and low frequency (0.1 rad/s). A proportionality has been found between the G' ratios at different frequencies and quick stick values (e.g., a $G'_{100/0.1}$ value of 60.9 corresponds to a quick stick value of 24.1, and a $G'_{100/0.1}$ of 5.5 gives a quick stick value of 15.3). Increasing storage modulus ratios lead to higher quick stick values. In a similar manner, decreasing tan δ ratios (at different frequencies) increases the quick stick values. The compliance (J) given by the correlation [100]

$$J = \gamma/\tau \qquad (3.21)$$

where τ is the constant stress for formulations based on radial block copolymers, is lower than 0.2 mi/Pa.

As mentioned with respect to the elevated temperature application of self-adhesive films, the surface temperature of the adherend is another factor to be taken into account. Most general-purpose adhesives are formulated to have tack at room temperature. If the adherend temperature is lower than room temperature, a higher degree of adhesive cold flow is required to provide proper wetout. Sometimes products are labeled at room temperature but subjected to lower temperatures later in their life cycle. Deep freeze labels should display the same viscoelastic properties at −40°C as at +20°C [7]. It is recommended that such labels have a lower storage modulus value than common labels (10^4 Pa). The peel of an untackified acrylic PSA at 0°C is about 210% lower than the peel value at 23°C.

For tackified formulations, peel reduction at 0°C may attain 300% [92]. Due to the increase of the modulus of the adhesive at low temperatures, with a loss of wetting and a jerky debonding are observed with certain formulations.

Tests with different stress frequencies at different temperatures (−50 to +16°C) were carried out by Debier and de Keyzer [100]. Modulus curves plotted at different frequencies and temperatures are superimposed to yield master curves. The logarithm of the dynamic modulus (on the ordinate, in dynes/cm^2) is plotted against that of the reduced frequency, log ω (rad/s). Generally the plot corresponding to test (room) temperatures is used.

As discussed earlier, the majority of pressure-sensitive products are composites with a segregated structure. Their rheology is the result of molecular and macroscopic interactions. Thermoplastic elastomers exhibit a segregated structure on the molecular scale. Their properties depend on their morphology, which is a function of the relative concentrations of the components. The morphology of an SBC polymer in the solid state depends on its polystyrene content. At low polystyrene content (less than 20%), spheres of polystyrene are dispersed in a matrix of elastomer; at higher polystyrene content (20–30%) the spheres begin to interconnect with each other and develop a cylindrical structure. Above 30–35%, a morphology of alternating lamellae is observed. If the polystyrene content is further increased, a phase inversion occurs, and the elastomer becomes the discontinuous phase in a continuum of polystyrene. It is evident that a polymer that has globular polyisoprene "particles" in a continuous polystyrene matrix (i.e., an inverse construction) will display quite different rheological and mechanical properties [101]. The plateau modulus (G_n^0) of a linear polymer is related to the molecular weight between entanglements (M_e), density (ρ), and temperature (T) [102–105]:

$$G_n^0 = (\rho/M_e)RT \tag{3.22}$$

where R is the universal gas constant. The gas constant appears in the correlation due to the hypothesis that the ideal rubber behaves like an ideal gaseous conformation network. In such an ideal "gas" the pressure is given as a function of the number of phantom chains (N), the temperature (T), and the volume (V):

$$p = NkT/V \tag{3.23}$$

where k is the Boltzmann constant. Because the number of phantom chains depends on the density and molecular weight, the equation for the nominal force (f) of the system can be written as a function of the deformation (D) as follows [106]:

$$f = (\rho RT/M_e)D \tag{3.24}$$

Replacing the force with the modulus, Eq. (3.24) becomes Eq. (3.22). Two theories have been proposed to predict the value of the plateau modulus of styrenic block copolymers; both assume that the plateau modulus depends only on the styrene content. According to the filler theory, the polystyrene domains act as a filler dispersed in the continuous polydiene matrix. In this case the plateau modulus is given by the equation [see also correlation (3.15) also]

$$G_n^0 = (\rho/M_e) RT(1 + 2.5\phi + 14.1\phi^2) \quad (3.25)$$

where ϕ is the volume fraction of the filler. In this case the segmented, two-phase block copolymer is considered as a gaseous filled network for which the modulus is given by the combination of correlations derived from the physics of gases and the hydrodynamics of filled liquids (see later). The critical molecular weight M_c, which is normally twice the value of M_e, is extremely dependent on the molecular weight and styrene content, according to Bishop and Davison [107]. At high polystyrene content the morphology no longer resembles that of a spherical filler, and the above equation is no longer valid. According to the structure theory of Lewis and Nielsen [108] for phase-inverted systems, the shear moduli of the filled material (G) and the unfilled matrix (i.e., rubber) (G_1) are related to the fraction of the filler ϕ_2 by the correlation

$$\frac{G}{G_1} = \frac{1 + AB\phi_2}{1 - BU\phi_2} \quad (3.26)$$

where the constant A takes into account the geometry of the filler and the Poisson ratio of the matrix; B is a function of the moduli and A; and U is a function of ϕ_2 and ϕ_m, the maximum packing fraction of filler. The modulus of a thermoplastic rubber between these two extreme cases, where the rubber or the polystyrene acts as matrix, is given by a logarithmic rule of mixtures:

$$\log G = \phi_U \log M_u + \phi_L \log M_L \quad (3.27)$$

where M_u and M_L are the upper (rubber dispersed in polystyrene) and lower (polystyrene dispersed in rubber) modulus, respectively. According to the filler theory, the modulus of a diluted polymer (e.g., thermoplastic rubber in resin and oil) is given by

$$GN^0 = \phi^2{}_2(\rho/M_e)RT(1 + 2.5c + 14.1c^2) \quad (3.28)$$

where ϕ_2 is the fraction of the polymer in the polydiene phase, c is the fraction of the polystyrene block in the entire composition, and ρ is the density of the adhesive. The tackifier resin should be compatible with the elastomer midblock. If the resin is not fully compatible with the elastomer midblock or its concentration is too high, it associates with the polystyrene domain or forms a separate phase, i.e., it does not contribute to the tack. The same can occur with a high oil content. Because of the differences between resin and oil, different exponents are used in the equation for the plateau modulus. For instance, for Cariflex TR-1107-Foral 85 blend the exponent is 1.86, and for Cariflex 1107-Sfellflex 451 blend the exponent is 2.22.

Rheology of Reinforced Systems. The adhesive itself or the adhesive layer or both can have a composite structure. As discussed earlier, segregated, crosslinked polymers (mainly elastomers) are used as base materials for PSAs, carrier materials, and self-adhesive films. Dispersed adhesive systems contain dispersants. Adhesives and carrier materials can include fillers. The rheology of such ordered systems on the macromolecular or product scale is discussed in terms of a filled system.

For carrierless transfer tapes, the use of fillers or fillerlike materials (e.g., voids) is a practical method for strengthening according to Eq. (3.15). The influence of the nature and concentration of the filler on the flow properties of such products has been studied in a detailed manner. Mixtures of glass spheres ranging in size from 10 to 200 µm and suspended in a viscous medium have been investigated [109]. The volume fraction dependence of the viscosity could be fit into an equation developed by using ϕ as a fitting parameter [110]:

$$\eta_r = (1 - \phi/\phi_m)^{-2} \tag{3.29}$$

where η_r is the relative viscosity, ϕ is the volume fraction, and ϕ_m is the volume fraction at maximum packing (i.e., the volume fraction at which the viscosity becomes infinite). This expression is related to the Krieger–Dougherty equation for the relative viscosity [111]:

$$\eta_d = \eta/\eta_s (1 - \phi/\phi_m)_m^{-[\eta]\phi} \tag{3.30}$$

where η_d is the viscosity of the dispersion, η_s is the viscosity of the solvent, and η is the intrinsic viscosity. The displacement force (f_d) in a filled adhesive layer is given by the correlation

$$f_d = 1.7 \times 6\pi\eta\sigma\phi^2 \tag{3.31}$$

where ϕ is the sphere radius, σ is the shear rate, and η is the viscosity.

Rheology—Adhesive Characteristics. As discussed earlier, cold flow and elasticity are the main phenomena influencing the rheological behavior of a PSP. They also influence the adhesive performance characteristics. Toyama and Ito [112] used the data of creep testing (shear resistance) to calculate viscosity according to the relation

$$\eta = \frac{Fh}{A} \left(\frac{1}{dx/dt}\right) \tag{3.32}$$

where F is the load, h is the thickness of the samples, A is the contact area, dx/dt is the slope of the displacement versus time, and η is the steady-state shear viscosity. As seen from the above correlation, sample thickness (coating weight) is a determinant parameter of shear resistance. For a non-Newtonian fluid the above dependence becomes more complex as a function of the time to failure (t_0), the initial overlap (H_0), the reference viscosity at a shear rate of 1 (η_0), and the slope of a plot of log η vs log shear rate [113]:

$$t_0 = \frac{(1-n)H_0^{(2-n)/(1-n)} (w\eta_0)^{1/(1-n)}}{(2-n) (Fg)^{1/(1-n)} h_a} \quad (3.33)$$

where w is the width and h_a is the thickness of the adhesive, F is the load, and g is the gravitational factor. As seen from the above correlation, holding power is a viscosity effect. According to Woo [114], the shift factors used to plot shear measurements at different temperatures on a master curve were almost identical to the shift factors used for modulus curves [correlation (3.5)]. Thus the time to fail in shear can be predicted from viscoelastic functions. As known, the viscosity depends on the loss modulus (viscous flow is the phenomenon that "loses" the energy). It also depends on the shear rate (frequency):

$$\eta' = G''/\omega \quad (3.34)$$

At low frequencies η' and the steady-state viscosity are the same. A static shear test is based on low frequency deformation; therefore creep may be regulated by viscosity.

As stated by Krecensky [113], peel resistance is described by some authors as an elongational flow phenomenon but is viewed by others as shear. Kaelble and Reylek [115,116] found a relationship between a_T (the shift factor from the WLF equation), time (t or $1/\omega$), rate of peel (v_p), and the cleavage stress concentration factor β:

$$a_T/t = \beta v_p a_T \quad (3.35)$$

where

$$\beta = [9G(\omega)/8E_c h_c^3 h_a]^{1/4}$$

In the above correlation, E_c is the Young's modulus of the carrier, $G(\omega)$ is the shear modulus of the PSA at frequency ω, h_c is the half-thickness of the flexible carrier, and h_a is the thickness of the adhesive. Peel data at different rates and temperatures have also been reduced to master curves.

Rheology During Processing

For the manufacturer of PSPs the rheology of the product components before and during their use (production) is important also. It may be different from the rheology of the final product. The manufacture of plastic carrier materials and of certain adhesives (hot melts and heat-curable compositions) entails thermal processing, where elevated temperatures (and pressure) bring about the required flow and forming properties. For other (coated) adhesives, fluidity is achieved via dispersing by the manufacture of polymer solutions or dispersions. A controlled rheology during mixing of the components (formulation) and converting (coating) of the formulated (plastic/adhesive) components is necessary.

Generally the main components of the formulation of solvent-based or water-based adhesives do not influence the processing rheology. This depends more on the choice of the liquid vehicle and dispersing additives. On the other hand, it has been shown that the coating rheology is strongly influenced by the molecular characteristics. The coalescence of aqueous PSAs depends on their molecular weight. For instance, particle coalescence in poly(n-butyl methacrylate) latex films is attributed to interdiffusion of particles. Increasing the molecular weight of the polymer decreases the rate of interdiffusion [117]. The HMPSA formulation can influence both the processing rheology and the rheology of the coated adhesive. The styrene end block of styrene block copolymers provides the cohesion and mechanical properties and influences the processing conditions. Its hardness is reduced by oils that are compatible with the styrene domains. Unfortunately, oils reduce cohesion, and the shear adhesion failure temperature (SAFT) values are mostly influenced by the oil level (see also Chapter 6).

Coating rheology is characterized by wettability. According to Rantz [118], wettability is the extent to which the liquid spreads. Wettability displays a special role for coating systems having a low viscosity (solvent-based, water-based, or radiation-curable compositions) or for coating on difficult substrates (transfer coating and coating on special plastics). The coating rheology of the printing inks

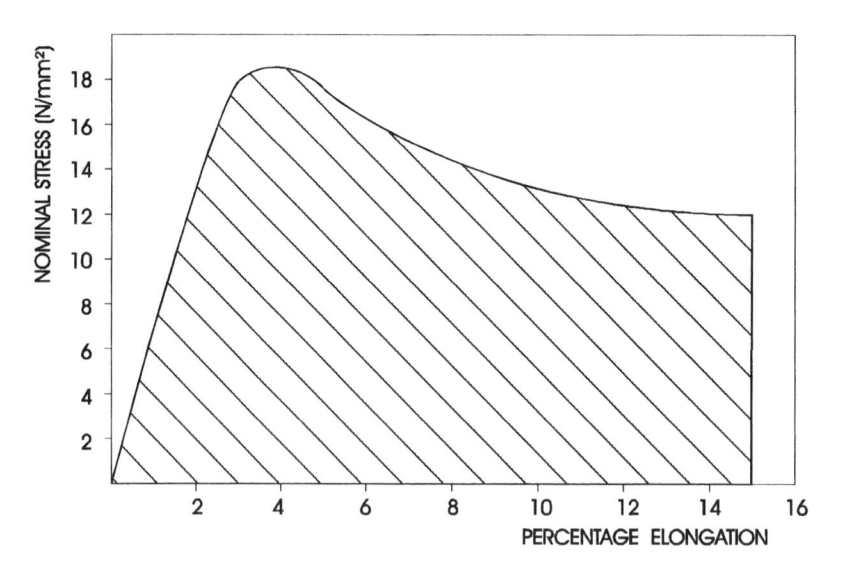

Figure 3.5 Stress–strain plot.

Physical Basis of PSPs 103

is important also. It includes the viscosity, flow limit, thixotropy, and tack of the ink [119]. These parameters influence the transport of the ink on the machine and its transfer on the web to be coated.

Energetic Aspects. As stated by Zosel [120], the separation energy of adhesive bonds can be considered as the sum of the thermodynamic adhesion work (depending on surface characteristics) and a dimensionless term taking into account the viscoelastic properties of the components. At high strain the area between the stress–strain curve and the strain axis is an energy (Fig. 3.5):

$$E = \int_0^{\gamma 0} \sigma \, d\gamma \tag{3.36}$$

A cyclic stress–strain curve for rubber compounds exhibits a hysteresis loop (Fig. 3.6). The area of the loop can be considered as indicative of the energy loss E_L per cycle (hysteresis), the area under the return curve (E_D) as indicative of the energy elastically stored in the system due to the input strain energy E_τ:

$$E_\tau = E_L + E_D \tag{3.37}$$

The change in internal energy during deformation, as determined by measurement of the deformational work and the heat developed, follow the first law of thermodynamics. At small strains, elastomers should react with a positive heat flow above the glass transition point, but at higher strains there is a superimposed

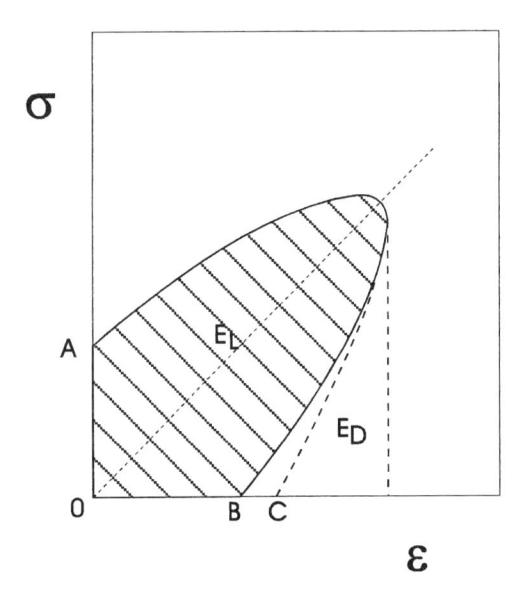

Figure 3.6 Hysteresis by bonding and debonding. See text for explanation.

negative flow that occurs as a result of the decrease in the entropy of stretched polymer chains [121]. This behavior is important for plasticization of rubber and the cuttability of rubber-based coatings. The dissipation energy of polymers is involved in peel energy measurement [122].

Generally for elastomers the dependence of the deformation (ϵ) on the increasing work force (tension σ) and the decreasing work force is not reversible. If viscous flow occurs, permanent deformation appears. Writing the mathematical correlation for the deformation and its dissipation (tensioning and contraction) as the sum of the work of deformation (W_d) and loss of deformation (W_r), their difference (ΔW) gives the residual deformation due to the viscous components:

$$\Delta W = W_d + W_r \tag{3.38}$$

Illustrating this phenomenon as a hysteresis plot where the tension varies between O, A, and B (Fig. 3.6), the corresponding integrals of the positive and negative work of deformation can be written in the form

$$\Delta W = \int_0^{\sigma_A} \sigma_A \, d\epsilon + \int_0^{\sigma_B} \sigma_B \, d\epsilon \tag{3.39}$$

It is known that viscoelastic compounds exhibit viscous flow permanently. For a pressure-sensitive adhesive the viscous deformation due to debonding is very pronounced; therefore,

$$\Delta W \gg 0 \tag{3.40}$$

In a first approximation, admitting that for a removable adhesive the elastic and plastic deformation occur in the adhesive and do not affect the adherend (adhesions break), the starting point of the deformation can be fixed at the adhesive limit, on the surface of the adherend (Fig. 3.7). For a permanent adhesive where debonding destroys at least the surface layer of the adherend also, the starting point of the deformation is situated in the adherend. Both the adherend and the PSA suffer elastic and plastic deformation. For a removable PSA, low forces are generally sufficient to produce debonding. This level of force causes measurable deformation of the adhesive only. It can be supposed that in this case only the PSA suffers plastic deformation. Theoretically for such an adhesive it would be possible to minimize the distance OB (Fig. 3.6) as a function of the stress rate and relaxation time of the adhesive. For a permanent adhesive, where permanent deformation includes the partial destruction of the carrier also (distance BC on Fig. 3.6), this is not possible.

From an energy point of view, the difference between a permanent and a removable adhesive is given by the ratio of areas in which permanent deformation is produced with carrier destruction (OAC) or adhesive flow (OAB) (Fig. 3.6). If this is true, the deformation work that causes permanent deformation is the sum

Physical Basis of PSPs

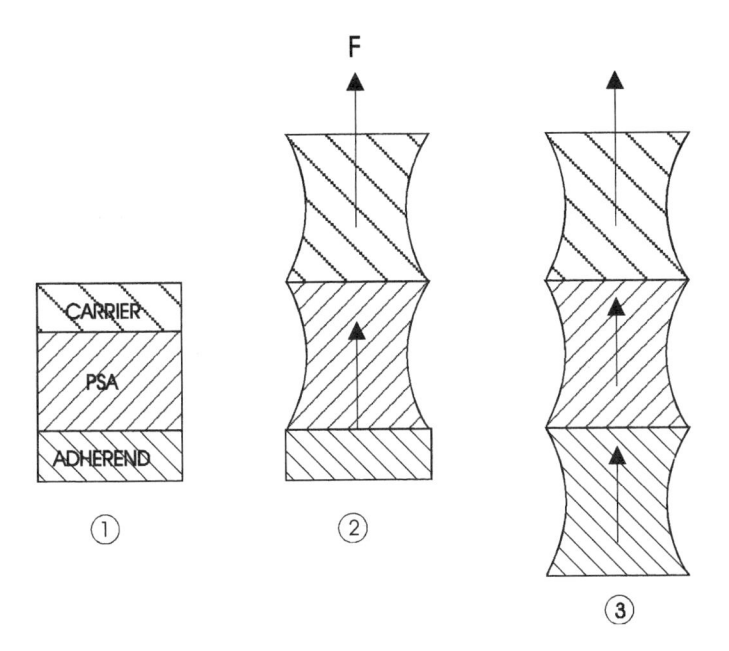

Figure 3.7 Strain in removable and permanent PSPs. (1) Nonstressed PSP; (2) removable PSP; (3) permanent PSP.

of the deformation work that leads to a residual deformation in the adherend (ΔW_{rad}) and the deformation work of the adhesive (ΔW_{psa}):

$$\Delta W = \Delta W_{rad} + \Delta W_{psa} \tag{3.41}$$

$$\Delta W_{rad} + \Delta W_{psa} = \int_0^{\sigma_C} \sigma_A \, d\epsilon + \int_0^{\sigma_B} \sigma_B \, d\epsilon \tag{3.42}$$

Admitting that the debonding energy is absorbed by the permanent deformation, for a removable adhesive the residual deformability must be much higher, i.e.,

$$\Delta W_{psa} \gg \Delta W_{rad} \tag{3.43}$$

During deformation the energy of deformation is transformed partially into heat. For filled PSAs (e.g., transfer tapes) where friction causes supplementary heat, such energy losses are higher. The work of adhesion between two surfaces having surface energies γ_a and γ_b is given by the correlation [123,124]

$$\gamma_a + \gamma_b = \gamma_{ab} + W_{ab} \tag{3.44}$$

This work of adhesion is transformed into heat.

1.3 Rheology of Abhesives

As a function of the product's construction and application requirements, different macro- or micromolecular compounds are used as abhesive materials (see Chapters 4 and 6). The main products are silicones. Unlike adhesives or plastic carrier materials, in which bonding/debonding rheology plays the most important role, for abhesives the coating/processing rheology is more important. For the manufacturer of PSPs the flow properties of the liquid silicone formulation are more important.

1.4 Product Rheology

Because of their various application fields, PSPs have to display different flow properties. Therefore the special aspects of their rheology are discussed separately for each of the main product classes.

Rheology of Labels

Labels are complex products whose various viscoelastic components are built in as thin layers. The introduction of polymeric films as face stock material increased the number of label components with pronounced cold flow. The rheological properties of labels are manifested during their conversion (cutting, printing), application (labeling), and end use (storage, deapplication). Classical (paper-based permanent) labels are products that suffer deformation as a result of the cold flow of the adhesive. Because of the supposed high dimensional stability (i.e., nondeformability) of its solid-state components, the small dimensions of the product, and the low forces acting on it during application, the rheology of the label is that of the coated PSA. Cold flow related phenomena such as bleeding, migration, and cuttability are mostly attributable to the adhesive.

Rheology of Tapes

Tapes are PSPs for which the role of the carrier material as a functional component during application is more important than it is for labels. This is due to the weblike application of tapes where mainly a nonisotropic mechanical character of the web is required. For such applications, except for some (dry or wet) removable tapes, the rheology of the adhesive layer is less important. From the classical point of view, tapes are such products where product deformation is mainly the result of carrier deformability.

Rheology of Protective Films

The industrial production of protective films is based mainly on empirical knowledge. However, future development requires the design of the products to have an engineering basis. This will become possible by studying the physical/chemical basis of the finished product and product components.

As discussed above, a protective film forms a laminate together with the product to be protected. Laminating of the self-supporting film occurs via a pressure-sensitive surface layer. In the classical construction of protective films (and labels or tapes) the PSA is a separate component of the product, coated on the nonadhesive solid-state carrier web (see Chapter 2). Pressure-sensitive adhesives are permanently liquid systems that undergo cold flow. In the case of protective films, at least one of the solid-state components of the final laminated construction (protective film/protected item), namely the weblike, deformable plastic carrier, also exhibits cold flow. Cold flow is the result of the viscoelastic behavior of macromolecular compounds. Some of the polymers used as carrier materials for protective films (e.g., polyethylene) have low T_g values (see Table 3.1). Others (e.g., polyester, polyvinyl chloride) possess higher T_g values, but the interval between their application temperature and their T_g is very narrow. Protective films are applied on plastic surfaces too. As can be seen from Table 3.2, certain plastic substrates to be protected exhibit low T_g also. Such adherends can also exhibit pronounced cold flow.

Therefore cold flow may occur for all components of the laminate, and interpenetration of the solid-state and liquid laminate components is possible. Because the mechanical performance characteristics of the components of a protective film laminate are of the same order of magnitude (Fig. 3.8), such products can be considered PSPs and the deformation of the finished product the result of the deformation of all its components.

In such an interaction the viscoelastic properties of the carrier material influence the rheology of the product. Actually the majority of protective webs are manufactured by using thin plastic films as carrier materials. These are common plastics from the range of materials used as packaging films. In 1974, Toyama and Ito [112] described pressure-sensitive adhesives as being coated "onto rigid (relative to the adhesive) backings without the use of solvents or heat." For

Table 3.2 Glass Transition Temperature of Plastics Used as Substrate

Material	T_g (°C)	Ref.
Polyvinyl acetate	29	125
Polystyrene	100	125
Polyvinyl chloride	−40 to +105	24
Polymethyl methacrylate	105	125
Polycarbonate	150	24
Polyacrylate	190	24
Lacquers and varnishes	−40 to +350	—
Printing inks	−60 to +220	—

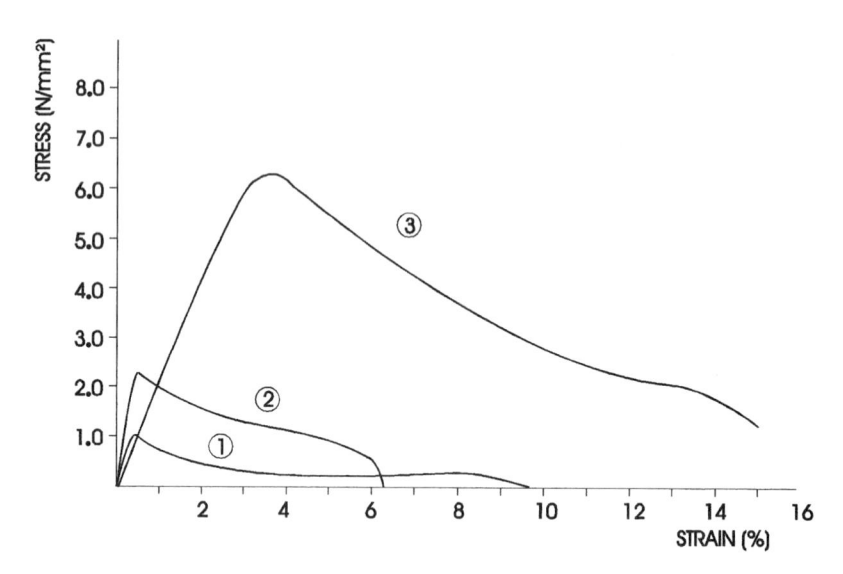

Figure 3.8 Stress–strain plot for PSP components. Curve 1, PSA; curve 2, crosslinked PSA; curve 3, plastomer carrier.

plastic-based protective films this statement is no longer valid. Their cold flow and adhesion are improved by modern application conditions. Table 3.3 lists the application conditions for some protective films.

As illustrated by the data in Table 3.3, in certain applications protective films are laminated on the plastic products to be protected at temperatures higher than room temperature (up to 120–130°C) or are processed with the protected item at elevated temperatures (see also Chapter 8). In some cases such application temperatures are required by the manufacturing technology of the products (e.g., plates of polymethyl methacrylate or polycarbonate). In other applications the high temperature is the processing domain for the protected product (deep drawing, cutting, etc.). In certain uses only the surface layer of the protected item is polymer-based. Uncrosslinked or crosslinked polymers (varnishes and lacquers, inks deposited on metal surface) are protected and undergo thermal treatment. When the working temperature is much higher than the T_g of the protective film components, and (for a short period of time) it can attain the softening/melting range of the base polymer (e.g., hot laminating films), a pronounced viscous flow of the carrier material can occur.

As discussed later (see Chapter 5), protective films possess an unbalanced adhesive character. Because of the relatively low tack of the self-adhesive layer and its low deformability, a high laminating pressure is sometimes necessary to improve the intimate contact between the protective film and the product to be pro-

Physical Basis of PSPs

Table 3.3 End Use Conditions for Protective Films

		End use conditions	
Protective film	Protected product	Laminating	Processing
Hot laminating film	Lacquered coil	120–130°C, pressure	Room temperature
Adhesive-coated film	Lacquered coil	Room temperature, pressure	25–100°C
Adhesive-coated film	Cast PMMA plate	Room temperature, pressure	25–70°C
Adhesive-coated film	Extruded PMMA film	50–70°C, pressure	25–40°C
Self-adhesive EVAc film	Extruded PC plate	Room temperature, pressure	150°C
Self-adhesive EVAc film	Extruded PMMA plate	50–90°C, pressure	25–40°C

tected. The use of high temperature softening (or partial melting) of the carrier film to improve its adherence (surface penetration) allowed the development of so-called hot laminating films without a chemically self-adhesive layer. Here the polarization of the plastic surface and its functionalization by corona treatment (see Chapter 6) provides the adhesive force necessary for the temporary bonding of the protective film to the adherend surface to be protected. The initial requirement to resist elevated temperatures (imposed by the manufacture of the products to be protected) was used later as a high temperature adhesion correction (tack replacement). Classical plastic films without a chemical adhesive layer cannot possess the tack of PSAs. As discussed earlier, such products exhibit flow, but it is time (stress)- or temperature-related (see high pressure hot laminating). It should be mentioned that high pressure, high temperature, and corona treatments are not sufficient to bond hot laminating protective films. A special chemical composition allowing improved flow is also necessary (see also Chapter 4). That means that adhesive-free hot laminating protective films are special products with special carrier film compositions.

If a special carrier composition can be used to simplify the construction of the protective films (and the formulation works under elevated temperature), the next step of development would be the formulation of carrier films displaying both carrier and adhesive properties at normal application temperatures. In this case the rheology of the protective film is more complex because the product should exhibit a more pronounced viscous flow without the loss of mechanical characteristics. It is known that both surface and bulk properties of the carrier material influence the viscoelastic properties of the adhesive layer and vice versa [1]. The adhesive properties related to bonding on the substrate (tack and peel) depend on

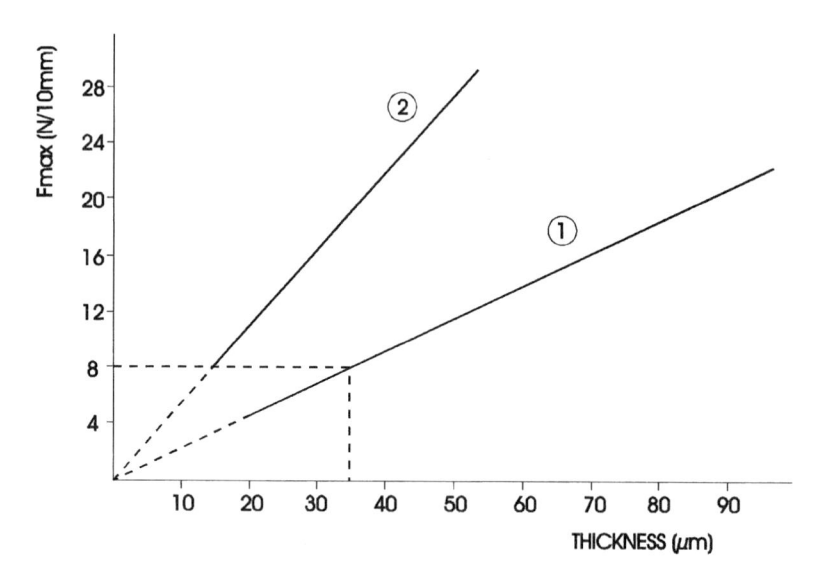

Figure 3.9 Dependence of the carrier film deformability on its thickness. 1, Monolayer LDPE film; 2, coextruded LDPE/LLDPE film.

the anchorage of the adhesive on the carrier. The anchorage is a function of the chemical affinity and physical structure of the surface. On the other hand, migration from the surface into the carrier (or from the bulk plastic material to the surface) will also influence the anchorage and adhesive properties. As known, slip agents from polyolefins, plasticizers and surfactants from polyvinyl chloride, or stabilizers from the plastics will dilute the adhesive, altering its properties. As a secondary effect, the film may harden.

Debonding force is transmitted to the adhesive by the carrier. Generally (for slight removable protective films), the value of the debonding force that causes bond failure at the adhesive/substrate interface is lower than the tensile strength of the plastic film. As known, the mechanical resistance of the film is the result of its own internal characteristics and geometry, which depend on its formulation. Reducing the thickness of the film increases its deformability (Fig. 3.9).

It is well known that for high peel force (permanent film) labels tested in the laboratory, results are degraded by the deformation of the face stock material. Its elongation and relaxation absorb energy and allow the relaxation of the PSA too; therefore peeling off will be delayed in time (the strain and strain rate will be reduced). Protective films display low peel values (see Chapter 5). On the other hand, their mechanical characteristics and thicknesses are also lower (Tables 3.4 and 3.5). Thus, elongation of the film carrier may strongly decrease the measured

Table 3.4 Mechanical Characteristics of Carrier Materials Used for Adhesive-Coated Polyolefin Protective Films

Carrier thickness (μm)	Tensile strength (N/10 mm)				Elongation (%)				Comment
	At maximum load		At break		At maximum load		At break		
	MD	CD	MD	CD	MD	CD	MD	CD	
45	13.6	8.2	13.3	8.1	250	600	259	604	Clear
47	11.4	8.7	11.1	8.6	241	643	244	647	Clear
45	11.9	8.5	11.7	8.3	351	627	353	626	Clear, printed
46	14.7	6.9	14.4	6.7	167	660	171	687	Clear
48	13.4	9.5	13.3	9.3	264	674	265	676	Clear
74	19.3	10.9	19.1	10.4	258	596	261	608	Clear
78	18.6	10.4	18.3	10.4	248	455	248	437	Black/white, printed
80	18.7	9.3	18.3	9.0	239	435	242	440	Black/white, coextruded, printed
111	19.2	15.5	18.9	15.4	352	503	354	504	Black

MD = machine direction; CD = cross direction.

Table 3.5 Mechanical Characteristics of Carrier Materials Used for Common Protective Films[a]

Tensile strength (N/10 mm)		Elongation at break (%)		Modulus (N/mm^2)	
MD	CD	MD	CD	MD	CD
19	11	171	458	19	18
19	11	153	421	20	19
21	12	151	452	21	17
19[b]	15[b]	271[b]	511[b]	13[b]	13[b]

[a]Polyethylene film with a thickness of 50 μm except as noted.
[b]Film thickness 80 μm.

peel value even when the same adhesive, coating weight, substrate, and laminating conditions are used (Fig. 3.10).

As illustrated in Fig. 3.10, there is a critical carrier film tensile strength. Below this limit (of about 6–8 N/10 mm) the adhesion values are degraded (reduced) due to the pronounced deformation of the carrier. The same adhesive and coating weight lead to much lower peel values (0.06–0.10 N/10 mm). For com-

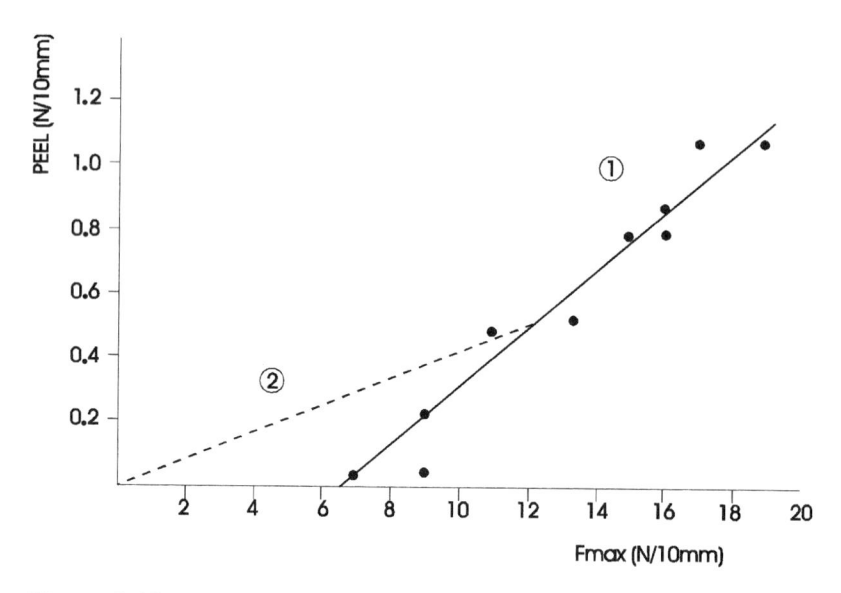

Figure 3.10 Denaturation of the peel values by carrier deformation. Dependence of the peel on stainless steel as a function of the tensile strength (F_{max}) of the plastic carrier material. 1, Measured peel values; 2, theoretical dependence of peel on carrier strength. An LDPE carrier and solvent-based acrylic PSA were used.

mon "nondeformable" carrier materials, values of 0.6–0.8 N/10 mm have been obtained.

In choosing a face stock material for labels, its dimensional stability is a decisive factor. The material's mechanical properties are reported as a stress–strain plot. A criterion for dimensional stability is the value of the force required for a given minimum (5%) deformation (see Table 3.6). For other products, other end uses require other force/deformation values as given in Table 3.6.

For protective films, which exhibit a more pronounced plastic deformation of the carrier film during debonding, it would be more practical to examine the value of the plastic deformation as a criterion of applicability instead of the development of peel force (Table 3.7).

General Considerations

A comparison of the adhesive characteristics of the main classes of pressure-sensitive products (labels, tapes, and protective films) is schematically presented in Fig. 3.11. As Fig. 3.11 (positions 1 and 4) illustrate labels and tapes have balanced adhesive properties. Their tack allows instantaneous adhesion without pressure, their peel resistance ensures the formation of a bond, and their cohesion provides a permanent, mechanically resistant bond. It is evident that for common

Physical Basis of PSPs

Table 3.6 Deformability of Carrier Material for Selected PSPs

Pressure-sensitive product	Tensile stress at a given deformation				Unit	Method	Ref.
	5%	10%	30%	50%			
Harness wrap tape	—	1.5	4.9	7.4	MPa	ASTM-419	126
Easy tear tape	—	3.0–6.5	6.5–13.0	13.0–15.5	MPa	ASTM D-1000	127
Flame-resistant tape	—	1.5–5.4	5.0–12.3	8.5–14.0	MPa	ASTM-412	128
Closure tape	—	>70.0	—	—	N/15 mm	DIN 53455	—
Protective film	8 (5)	—	—	—	N/10 mm	—	—
Label	19 (18)	—	—	—	N/10 mm	—	—
Insulating tape	11 (6)	—	—	—	N/10 mm	—	—
Insulating tape	91	—	—	—	N/10 mm	—	—
Application tape	16 (16)	—	—	—	N/mm^2	—	—
Protective film	13 (13)	—	—	—	N/10 mm	—	—

For tensile stress at 5% deformation values given are MD and (CD).

Table 3.7 Dependence of Peel Resistance on the Carrier Deformation

Carrier elongation (%)				Peel resistance[a] (N/25 mm)	
Nominal		Actual			
MD	CD	MD	CD	Nominal	Actual
220	660	—	—	2.75	—
—	—	680	820	—	0.35
230	600	—	—	3.00	—
—	—	530	800	—	0.38
250	550	—	—	5.00	—
—	—	750	850	—	1.50
300	530	—	—	7.50	—
—	—	730	820	—	2.25

[a] The 180° peel resistance of a crosslinked acrylic adhesive (coated on polyolefin carrier) on stainless steel. Dwell time 1 day.

tapes this balance is shifted toward higher cohesion. Because of their reduced tack, protective films (Fig. 3.11, position 2) have to be applied under pressure. A new class of pressure-sensitive materials is represented by Fig. 3.11, position 4. These are the so-called postfinished products, which need a chemical or physical treatment to complete their synthesis in order to achieve balanced adhesive prop-

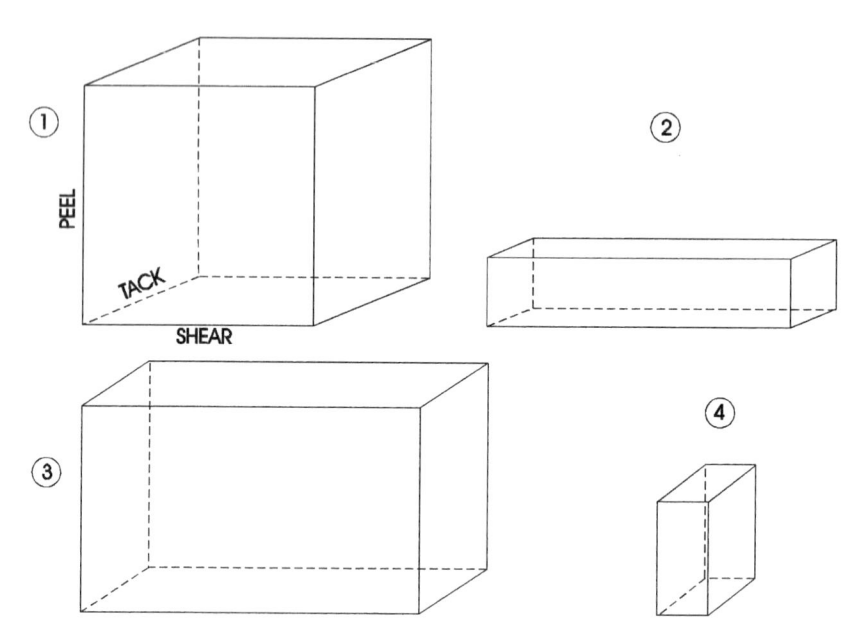

Figure 3.11 Schematic presentation of the adhesive properties of the major PSPs. (1) Label; (2) protective film; (3) tape; (4) postfinished PSA.

erties. This class of products includes mainly low molecular weight 100% solids, coatable as warm melts. As stated in Refs. 94 and 129, even a high molecular weight linear, highly viscous 100% polyacrylate processible as hot melt is not of high enough molecular weight to display balanced pressure-sensitive properties. Pressure sensitivity (sufficient shear resistance) is achieved only by crosslinking. The polymer itself is a highly viscous fluid at room temperature (before crosslinking); it can be processed at 120–140°C. Therefore it should be crosslinked. Special acrylic raw materials for acrylic hot melt PSAs have been developed that allow postcuring. In this way polymacromerization competes with the synthesis of macromolecular compounds having segregated structures (see Chapter 4). The result of both procedures is a new class of "internal" composites having better end use performance characteristics but a more complex rheology.

Pressure sensitivity (Π) is the resultant of two phenomena, viscoelastic bonding (B_{ve}) and viscoelastic debonding (D_{ve}):

$$\Pi = B_{ve} + D_{ve} \qquad (3.45)$$

During bonding and debonding, relaxation occurs [see Eqs. (3.9) and (3.20)]; therefore pressure sensitivity will be the sum of deformation and relaxation by bonding (index b) and debonding (index d):

Physical Basis of PSPs

$$\Pi = \left(\frac{1}{G}\frac{d\tau}{dt} + \frac{\tau}{\eta}\right)_b + \left(\frac{1}{G}\frac{d\tau}{dt} + \frac{\tau}{\eta}\right)_d \qquad (3.46)$$

Writing $1/G$ as J, $d\tau/dt$ as τ^0, and τ/η as f, Eq. (3.44) becomes

$$\Pi = (J\tau^0 + f)_b + (J\tau^0 + f)_d \qquad (3.47)$$

For a pressure-sensitive construction based on carrier and adhesive, the above correlation can be written as the sum of the bonding and debonding components for the carrier (index c) and adhesive (index a):

$$\Pi = [(J\tau^0 + f)_a + (J\tau^0 + f)_c]_b + [(J\tau^0 + f)_a + (J\tau^0 + f)_c]_d \qquad (3.48)$$

Taking into account the various "material characteristics" of the carrier and adhesive for the bonding and debonding and grouping the terms for the carrier and those for the adhesive, Eq. (3.46) becomes

$$\Pi = (J\tau^0_b + f_b + J\tau^0_d + f_d)_a + (J\tau^0_b + f_b + J\tau^0_d + f_d)_c \qquad (3.49)$$

Bonding and debonding occur at different stress frequencies. Because of the dependence of the modulus on the shear rate, correlation (3.47) cannot usually be simplified. The forces acting on the adhesive and are different from those acting on the carrier. Laminating and debonding forces are transmitted to the adhesive by means of the carrier. Force transfer occurs at the same time as carrier deformation and relaxation. According to Eq. (3.47), generally the pressure-sensitive behavior of a PSP is the sum of the bonding and debonding behavior of the components. For a one-component PSP based on self-adhesive carrier (e.g., protective film on EVAc basis), the above correlation can be simplified and written

$$\Pi_{SAF} = (J\tau^0_b + f_b + J\tau^0_d + f_d)_c \qquad (3.50)$$

Taking into account that for this type of product laminating occurs at a higher temperature than delaminating and under pressure and delaminating is carried out with a much higher stress rate at much lower temperatures than laminating, in Eq. (3.48) the terms related to debonding become more important. For such a self-adhesive film, pressure sensitivity should be designed for delamination and relaxation:

$$\Pi_{SAF} = (J\tau^0_d + f_d)_c \qquad (3.51)$$

Most protective films are coated with a hard, crosslinked adhesive and are applied under pressure (and in certain cases with the application of heat). Therefore, Eq. (3.49) is valid for such products also. For a carrierless transfer tape Eq. (3.47) becomes

$$\Pi = (J\tau^0_b + f_b + J\tau^0_d + f_d)_a \qquad (3.52)$$

For such a filled, crosslinked adhesive mass, pressure sensitivity should be designed for bonding:

$$\Pi = (J\tau^0_b + f_b)_a \tag{3.53}$$

Taking into account the differences in the deformability of paper and plastic materials during conversion, application, and deapplication, for practical use it would be more convenient to write the general form of Eq. (3.47) as

$$\Pi = (J\tau^0_b + f_b + J\tau^0_d + f_d)_a + C(J\tau^0_b + f_b + J\tau^0_d + f_d)_c \tag{3.54}$$

where C is a constant taking into account the deformability of the carrier material during debonding and having values between 1 (soft plastic carrier) and 0 (nondeformable carrier).

The relaxation related to bonding and debonding is a complex process. Generally the debonding energy of a composite is described as the reversible energy of adhesion or cohesion with a microscopicol loss factor related to viscoelastic energy losses and a molecular loss factor related to the degree of polymer crosslinking [130]. Relaxation phenomena on a molecular scale are very complex and depend on the buildup of the macromolecules. It has been stated that a relatively simple elastomer such as *cis*-polyisoprene displays three different types of relaxation processes. Near T_g the relaxation is related to crystallization and van der Waals bonds in the polymer network. At higher temperatures, relaxation is related to the interactions between the segments that produce physical contacts of an oversegmental microstructure [131]. After studying relaxation transitions in polybutadiene and poly(butadiene-methylstyrene), Bartenev and Tulinova [132] stated that there existed 17 transitions. As discussed in Ref. 133, in branched polymers of polybutadiene the maximum relaxation time increases with the molecular weight of the branches.

2 MECHANICAL PROPERTIES OF PSPs

The main PSPs are weblike supported or self-supporting materials. Their support has to resist stresses during manufacture. During their application the support also has a mechanical function. The mechanical properties of the carrier material influence bonding and debonding. Therefore for the design and manufacture of PSPs it is necessary to evaluate the mechanical properties of the carrier, the adhesive, and the finished product.

According to Hensen [134], for a polymer used as the raw material for a plastic film carrier, the following characteristics are the most important: mechanical properties (resistance, elongation and shrinkage), chemical resistance, permeability (to oil, gas, and water), sealability, surface quality, stiffness, thermal resistance, adhesion, coefficient of friction, deep drawability, and melting temperature. From these characteristics the resultant mechanical properties (resistance to stress, elongation, and stiffness), the properties related to dimensional stability (shrinkage and deep drawability), the properties related to surface quality (coefficient of friction and sealability), and the properties related to thermal resistance (melting temperature, sealability, and warm deep drawability) have special importance for carrier materials used for PSPs. As a function of their various application fields and manufacturing technology, materials with different mechanical properties and dimensions are processed as PSP components under various stress conditions. Electronic converting machines for labels cut, die cut, punch, and perforate the paper at running speeds of 200 m/min. Carrier materials as different as 15–20 μm thin films and cardboard of 300 g/m^2 are used [135]. Generally the main problem in designing structural bondings with elastic adhesives is the insufficient knowledge about their mechanical behavior. Therefore it is necessary to examine the mechanical properties of the adhesive layer also.

2.1 Rheological Background of the Mechanical Properties of PSPs

The main PSPs are weblike products based on a solid-state carrier or a combination of solid-state carrier materials (laminate) that exhibits pressure-sensitive properties. During the manufacture of PSPs, the solid-state carrier material has to withstand mechanical stresses due to the tensions in the continuous web running at high speed on machine parts having different relative velocities. Stresses are encountered during conversion and application of the material also (see Chapters 6 and 8). Tapes are wound with variable forces; protective films are laminated with high pressure. Labels are applied with high speed as discontinuous items, die cut from a high speed web that has to support (as matrix) the same stresses as the laminated web. The liner plays a weighty role in the functionality and cost of most PSPs. Tear strength, dimensional stability, lay flat, and surface characteristics are the most important features of the release liner [136]. Tear strength is important for label conversion and dispensing. Good caliper control and liner hardness (lack of compressibility) are required. Dimensional stability, the ability to maintain the original dimensions when exposed to high temperature and stresses, is important for print-to-print and print-to-die registration. If the liner stretches under heat and tension, graphics will be distorted. Liner stretch can affect label dispensing also. Lay flat is very important for sheet labels or pin-perforated and folded labels (see Chapter 7). Labels that are butt cut, laser printed, and fan folded must lie flat to give proper feeding and stacking performance. On the other hand, for the application of PSPs on adherends having a complex contour or rough surface, softness and conformability are required. The carrier of tamper-evident labels requires excellent conformability and destructibility [137]. Medical tapes require conformability also [138]. In certain cases for better conformability a carrierless PSP is applied. For such products the adhesive layer possesses adequate strength to permit it to be used without a carrier material (see Chapter 8). Tensile strength, stiffness, tenacity, and Elmendorf tear resistance are important features of sheet labels. Peel resistance lies on both adhesion and dissipation energy. Dissipation energy arises from deformation of the plastic carrier, assuming that the adhesion is high enough to result in deformation during the test [139] [see Eq. (3.47)].

Structural Background of Mechanical Properties

The mechanical properties of PSPs depend on the chemical basis and rheological behavior of the macromolecular compounds. Glass transition temperature, viscoelastic response, and yield behavior of crosslinked systems are explained by extending the statistical mechanical theory of physical aging, taking into account the transition of the WLF dependence to an Arrhenius temperature dependence of the relaxation time in the vicinity of T_g [140] [see Eqs. (3.5) and (3.6)]. For classical PSPs the mechanical properties of the nonadhesive carrier material and of the adhesive are of different orders of magnitude. The carrier is designed to resist stresses without deformation; the adhesive is designed to allow a high degree of deformation without being destroyed. For plastic carrier materials the carrier is considered in a first approximation as nondeformable. Such behavior arises from the nature of plastics. Polymers suffer deformation under stress. The creep modulus E_c is given as a function of the stress (σ, constant in time) and the time-dependent deformation ($\epsilon_{(t)}$) by the correlation

$$E_{c(t)} = \sigma/\epsilon_{(t)} \tag{3.55}$$

At very low elongation (<1‰) a spontaneous reversible elastic deformation exists and Hooke's law is valid. Between 0.5 and 1% elongation, time-dependent viscoelastic deformation takes place. For this domain the laws of linear viscoelasticity are valid [141]. At higher deformation, a time-dependent viscous deformation takes place. Rubberlike materials allow higher elastic deformation.

The mechanical properties of carrier materials are controlled by the choice of raw materials and by film processing conditions. The choice of the film manufacturing procedure (see Chapter 6) and the postextrusion processing affect the film quality. For blown film the mechanical characteristics of the film can be regulated by bubble form [142]. By regulating the blow-up ratio, freezing height, or melt temperature, one can control the film thickness and orientation. The mechanical, thermal, and optical properties of the film are functions of the orientation of the molecules. Orientation occurs during processing and is fixed through freezing. The mechanical properties of plastic films depend on their mono or coex buildup also. As stated by Acierno et al. [143], for semicompatible or almost compatible polymers (e.g., LDPE/LLDPE) the mechanical properties of coextruded films are close to those of films made from blends. In the case of polymers giving rise to

Physical Basis of PSPs 119

incompatible blends, deep minima are observed in many graphs of mechanical performance depending on the composition. In such cases the mechanical properties of coextruded films are expected to be better than those of films made from blends.

The mechanical properties of the adhesive or of a self-adhesive carrier film can be regulated by the choice of raw materials and by their processing. Pressure-sensitive adhesives are based on rubberlike, viscoelastic and viscous products. The mechanical behavior of elastomers is determined by their multiphase network structure. Composite structures allowing controlled mechanical performance can be achieved by synthesis, formulation, or processing of the plastic or elastomer PSP components.

Molecular Order—Mechanical Characteristics. As discussed earlier, the buildup of an ordered structure in the adhesive can change its mechanical performance. Ordered structures are formed via crystallization, association, crosslinking, and bulk fillers (Fig. 3.12). The resulting mechanical properties depend on

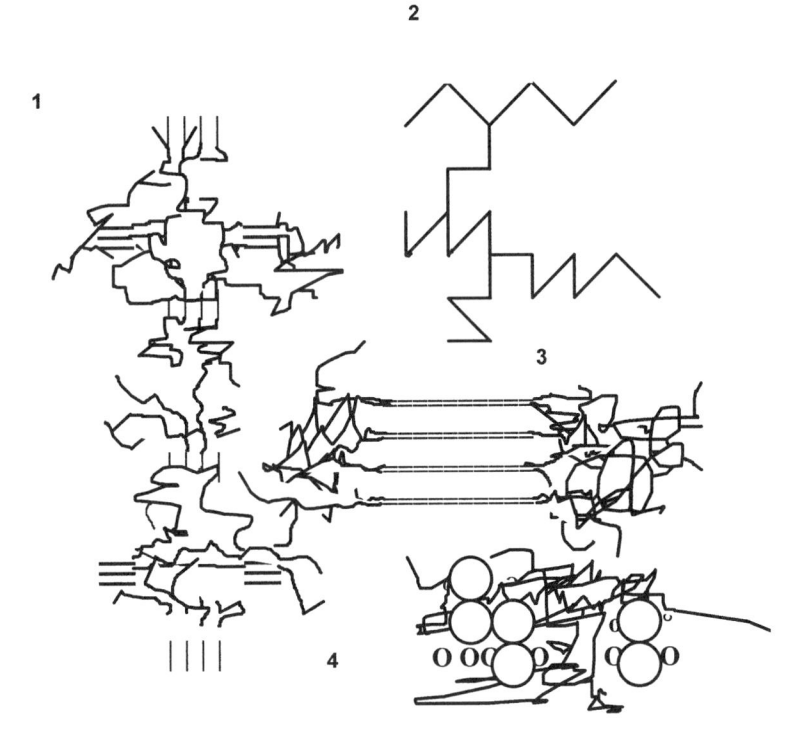

Figure 3.12 Network structures in base PSP polymers. (1) Molecular association; (2) chemical crosslink; (3) crystalline order; (4) filler.

the chain mobility in the "new" construction. This behavior is illustrated by Sun and Mark [144] for polyisobutylene. Viscoelastic polyisobutylene was transformed into an elastomer by crosslinking; the mechanical properties of the material were improved. Tensile strength and elongation increased. Later, by generating in situ reinforcement of this elastomer with inorganic filler particles, other mechanical properties, e.g., stiffness, were improved. The effect of crosslinking and filling on the mechanical performance is complex and depends on the fine structural changes produced. The different methods of building an ordered structure are not interchangeable. Increasing the level of a filler (which does not allow high crosslinking of rubber) decreases the modulus [145]. The hardness of polyurethanes containing acrylonitrile copolymers dispersed (before polymerization) in the oligoether component of a polyurethane is higher than that of a filled (inorganic filler) polyurethane [146]. The existence of a network (composite structure) is the base condition for excellent mechanical properties.

The mechanical resistance ("green strength") of unvulcanized natural rubber is given by the bound cocoons linked together via stress-induced crystal lamellae, providing a stable network structure [147]. For synthetic rubber the type of crosslinking influences the mobility of the chain segments, i.e., the storage modulus. Below the secondary transition temperature for a crosslinked plastic material, the deformation mechanism that governs yield does not depend on the degree of crosslinking. In this region the mechanism of deformation can be regarded as a dislocation glide mode [148]. Above the material's T_g the degree of crosslinking comes into play with respect to yield. Chain length distribution affects the tearing energy of crosslinked networks. Bimodal networks showed a higher tear energy than monomodal constructions [149]. Uncrosslinked and unentangled molecules do not contribute to the mechanical properties of the polymer. It has been shown that a styrene-isoprene (SI) diblock has a tensile strength of 20 psi whereas a fully coupled SIS triblock exhibits 300 psi tensile strength [150]. A shear holding time of 387 days was obtained for a pure triblock polymer in comparison with 86 days for a commercial one.

Mechanical properties are strongly influenced by the reinforcing effect of fillers. Numerous mechanisms have been discussed in relation to their effect. The most significant ones concern the contributions to increased strength, increased stiffness, and increased hysteresis [151].

For filled materials the nature of the filler and its concentration affect the type of structure formed and the resultant mechanical characteristics. Adding EPDM to SBS to improve its ozone resistance, at low filler content (10–20 phr) the particles are dispersed randomly in the polymer matrix; at a higher level (20–40 phr) a quasi-network is built up [152]. In many cases the energetic and entropic changes related to the thermomechanical behavior of filled elastomers depend on the concentration of the fillers [54]. The geometry of the filler particles plays an important role also.

Physical Basis of PSPs

It has been reported that carboxylated diene-based rubbers become strong without reinforcing fillers or covalent crosslinks when the carboxylic acid is neutralized. Their Young's modulus levels off as zinc oxide concentration increases; the Young's modulus of specimens containing carbon black is much higher than that of specimens without carbon black [153]. The effect of the carbon black or zinc oxide loading on the Young's modulus is much greater than the theoretical values calculated using the Guth–Gold equation. This suggests that the effect of surface characteristics is much stronger. The idea of the hydrodynamic or strain amplification effect (which takes into account the concentration of the particles only) is based on the concept that the local strain between the filler particles is amplified because the particles can be considered rigid (compared to the rubbery matrix) [154]. The effect is similar to the enhancement of the viscosity of a fluid due to the incorporation of rigid fillers and has been described by Einstein [155] according to the equation

$$\mu = \mu_0 (1 + 2.5c) \quad (3.56)$$

where μ is the viscosity of the filled liquid, μ_0 that of the unfilled liquid, and c the volume fraction of filler.

According to the paradox of Slayter [156], in a composite structure the mechanical characteristics of the higher strength component are improved. For reinforcing with short fibers, according to the Halpin–Tsai [157] equation, the modulus of the composite (E_C) is a function of the modulus of the matrix (E_M), the volume fraction of the filler (ϕ_F), and the coefficients A and B:

$$E_C/E_M = (1 + AB\phi_F)/(1 - B\phi_F) \quad (3.57)$$

where A is a function of the length L and diameter d of the fibers,

$$A = 2L/d \quad (3.58)$$

and B is a function of the moduli of elasticity of the composite (E_C), fiber (E_F), and matrix (E_M):

$$B = \frac{E_F/E_M - 1}{E_F/E_M + A} \quad (3.59)$$

It should be mentioned that the shape of the filler particles has been taken into account by Guth also with the aspect ratio (f) in the equation for the magnification factor of aspheric fillers [154]:

$$X = 1 + 0.67f\phi + 1.62f^2\phi^2 \quad (3.60)$$

Generally for monodirectional constructions the modulus and the strength of the composite can be calculated from the data of the components. According to the paradox of Griffith [158], a fiberlike material possesses a much higher strength than the same material in another form. Such fillers are used to improve the shear resistance of PSAs. The length of the filler particles is taken into account by the paradox of tensioning or stressed length. It may be supposed that this paradox influences the effect of the thickness of adhesive and of the primer also. According to Klein [159], adhesion decreases if primer thickness increases.

The thermomechanical behavior of segmented block copolymers having a low concentration of soft segments is determined by intermolecular interactions within the soft and hard blocks. First it was supposed that the free energy of monoaxial deformation (elongation) of crystalline two-phase networks influences the internal entropy of the polymer chains. Later it was demonstrated that the deformation produces internal changes in the chains of the amorphous phase [54]. It is supposed that in segmented polyurethane elastomers, during application of a stress the soft polymer matrix is sheared between the small nondeformable "plates" of the hard segments [160], i.e., like an adhesive between the solid-state components of the laminate. The mechanical performance of elastic foams as a function of their blowing degree have been described by considering such systems as two-phase (elastomer phase and cavity phase) systems [161].

Mechanical Properties—Adhesive Performance

The performance of thermoplastic elastomers (TPEs) illustrates that the segregated and ordered structure is a condition sine qua non for adequate mechanical properties. Star-branched polyamides (nylon 6) retained part of their mechanical integrity even after melting, indicating the presence of enhanced entanglements [162]. Bocaccio [163] synthesized a sequenced styrene-isoprene copolymer by depolymerizing natural rubber (MW 50,000) and grafting styrene onto the NR chains, producing a TPE. Because of the only slight segregation in such compounds, their mechanical characteristics are not as good as those of SIS or SBS. On the other hand, the bond strengths of hot melt PSAs formulated with such compounds are comparable to those based on commercial SBCs, illustrating the complex nature of the interdependence of mechanical and adhesive properties. Evaluating the adhesive properties of a pressure-sensitive adhesive tape with a rolling adhesive moment tester, Yoshiaki and Kentaro [164] demonstrated that the mechanical properties of the acrylic adhesive and those of the rubber-based adhesive each had different effects on adhesion. Of block copolymers with different rubbery blocks but similar block molecular weights and about 30% styrene, the SEBS had the highest modulus [153]. The rubbery elasticity of the polymer is maximized when the rubbery polymer segment is completely amorphous and has a low T_g. In commercial SEBS copolymers the best rubbery characteristics are achieved with rubbery blocks having a T_g of $-50°C$. The mechanical performance should be examined as a whole. Macromolecular compounds of quite different chemical compositions or structures can display similarity of some properties but differences in global behavior. As an example, TPUs synthesized with hydroxy-terminated polybutadiene and methylene diisocyanate (MDI) possess the same modulus values as polymers manufactured with poly(oxypropylene glycol)/polybutadiene and MDI. However, their tensile strength and elongation at break are different [165].

2.2 Main Mechanical Properties

Their main mechanical properties allow the transport, processing, application, and deapplication of weblike PSPs. Generally these characteristics are given by the carrier material, but the buildup of the finished product also affects the mechanical characteristics of PSPs. The interface has a significant effect on the structure and properties of the boundary layers, resulting in a change in the packaging density of the macromolecular chains, limitation of their conformational status, a decrease in segmental mobility, and inhibition of relaxation processes near the solid surface [166]. A significant change occurs in the mechanical properties of a polymeric material on a support or in the presence of a solid-state surface of a filler in thin layers when the thickness of the layers becomes comparable to that of the boundary layer. Such effects are "translated" into the adhesive properties of primed PSAs or into the flow properties of filled carrierless transfer tapes. There is a gradient in the segmental mobility and mechanical properties near the phase boundary in moving away from the solid surface. The value of the modulus of the coated material depends on its distance from the solid surface and the nature (high or low energy) of the solid surface. There could also be a shielding influence that decreases the distance through which deformation energy is transmitted from outside to the boundary layer and the more distant layers of the macromolecular coating [167]. The most important mechanical characteristics are tensile strength, tear resistance, and stiffness.

Tensile Strength

Tensile strength, tension yield, and elongation are evaluated by constructing a stress–elongation diagram. If stretched in a stress/strain machine, once a certain degree of elongation is achieved, both elastic and plastic materials tear apart. The force required to rupture the film is expressed as a function of the cross section of the material and is referred to as tensile strength. Direct comparisons cannot be made between the tensile strength values, as they are measured with various elongations at rupture. Comparisons can be made using the moduli, i.e., the forces at specific elongation, reduced to the cross section of the sealant.

For polymers, mechanical properties strongly depend on the chemical composition, macromolecular features, and processing conditions. For polyolefin films, molecular weight and its distribution, crystallinity, the form and order of crystalline parts, chain branching, unsaturation, and polar groups have a weighty influence on mechanical performance. The mechanical properties of plastics are also functions of the structural ordering achieved by postextrusion processing, i.e., orientation [168]. The tensile strength of plastic carrier films depends on their raw materials and manufacturing technology (see Chapter 6). The molecular weight and its distribution affect mainly tensile strength, brittleness, and tear resistance [169]. Crystallinity has a greater influence on the softening point, elastic and flexural moduli, cold flow, and hardness. The texture of the film is given by crystallinity and its buildup conditions. As stated in Ref. 170, the modulus of polyethylene increases with the degree of orientation (achieving values of 10–60 kN/mm^2), but it depends on the molecular weight distribution. The tensile strength of a PE homopolymer increases with its density and generally does not depend on the melt index or molecular weight. For LDPE/LLDPE mixtures processed as blown film, higher LLDPE concentrations increase the degree of orientation of the film and its stiffness. The stiffness of a film containing more than 50% LLDPE is greater than that of the LLDPE [171]. An increase in the molecular weight of a 300 μm PVAc film from 200 × 10^3 to 2000 × 10^3 increases its tensile strength from 7 to 50 kg/cm^2 [172].

Fillers modify the tensile strength and tear resistance of a polymer. As an example, a PUR without filler exhibits an elongation of 360%, but this value decreases to 100% with a filler [173]. Pigment particles alter the characteristics of the film in two ways. If they have a high tensile modulus compared with the polymer, then they will increase the yield stress of the composite, and this will favor brittle fracture. They also flow within the film and act as stress concentrators, reducing the film's overall impact resistance. As is known, the resistance of materials can be evidenced by the buildup of flow regions in amorphous thermoplasts, by the formation of microcrazes in partially crystallized polymers, and by the debonding of adhesive or cohesive joints in fiber-reinforced plastics [174]. The study of the long-range order in the craze microstructure of deformed polystyrene-polybutadiene block copolymers showed that crazes grew parallel to the ordered structures where possible (as in a filled construction) but crossed them when the mismatch between the growth direction and the orientation exceeded a critical value [175].

Tensile strength plays a different role in the evaluation of the quality of carrier material used for different product classes. For labels, adequate tensile strength provides the dimensional stability required mostly during conversion. Tear strength and tensile strength of the face stock for labels depend on the conversion and application methods used [176]. The layout should be in the long grain direction to provide maximum conformability. Tear strength values of 1.6–2.7 in the cross direction (CD) (across the grain) and 2–4 in the grain direction are recommended for paper labels [177]. The characteristics of paper that should be taken into account for its use as release liner [178] are weight, thickness, density, transparency, stiffness (CD/MD), tear resistance, and tensile strength.

For tapes, tensile strength ensures performance of the bonding and assembling function. For protective films, tensile strength should achieve the minimum level

Physical Basis of PSPs

necessary for web processing, lamination, and delamination. For special products like transfer printing elements (e.g., Letraset), a vinyl polymer with high tensile strength ensures the dimensional stability of the letters during high pressure application [179]. Table 3.8 lists some important mechanical characteristics of a number of common PSPs. It can be seen that the tensile strength values required for common packaging tapes are much higher than those for other product classes. The tensile strength values for other special (nonpackaging) tapes are given in Table 3.9.

The mechanical resistance of a PSP depends on the characteristics of the component materials and the geometry of the product. Protective films are based on thin carrier materials (35–120 μm), mostly polyolefins; therefore their downgauging can lead to pronounced elongation of the carrier material during pro-

Table 3.8 Mechanical Characteristics of Selected PSPs

Product	Carrier Material	Thickness (μm)	Mechanical characteristics, MD/CD Tensile strength	Elongation (%)
Automotive protective film	Polyolefin	50	27/13 N/10 mm	550/570
Automotive masking film	Polyethylene	100	19/14 N/10 mm	350/340
Thermoforming protective film	Polyolefin	70	30/27 N/10 mm	680/650
Overlaminating film	Polypropylene	25	8 kg/25 mm	125
Overlaminating film	Polyester	25	10 kg/25 mm	125
Packaging tape	PVC	37	150/70 N/mm^2	—
Packaging tape for carton sealing	PVC	36	7 kg/10 mm	70
Sealing tape for heavy packaging	PVC	70	12 kg/10 mm	70
Packaging tape	Polypropylene, reinforced	50	35 kg/10 mm	—
Palettizing tape	Polypropylene	30	6 kg/10 mm	90
Packaging tape	Polypropylene, oriented	40	250/30 N/mm^2	—
Masking tape	Crepe paper	85–90[a]	3.5 kg/10 mm	—
Label	Polypropylene	50	100/140 N/mm^2	100/40

[a]Weight, g/m^2.

Table 3.9 Mechanical Characteristics of Carrier Materials for Special Tapes[a]

Product	Force at break (N/10 mm)	Elongation (%)	Maximum force (N/10 mm)	Elongation at maximum force (%)
Insulating tape	11	252	14	242
Insulating taping tape	209	26	115	26
Building masking tape	14 (11)	138	19 (19)	14 (11)
Masking tape	11 (8)	374	13 (12)	366 (13)
Application tape	29 (20)	340 (887)	—	—
Application tape	27 (19)	292 (814)	—	—
Application tape	21 (14)	250 (625)	—	—
Medical tape	25 (21)	250 (350)	—	—

[a] Some values are given in machine and cross directions, MD (CD).

cessing and lamination. Such elongation and the relaxation (energy absorption) related to the deformation may decrease the value of the peel force (Fig. 3.10). The use of a coextruded film increases the tensile strength of the film as well as the critical (minimum) thickness (Fig. 3.9).

It must be emphasized that the value of the tensile strength alone does not allow the characterization of the dimensional stability of the carrier material for different product classes. The elongation must also be known. As is known, elastomers display excellent tensile strength, a property associated with rubbery elasticity. Elastic carrier materials do not conform (permanently) to the product surface and may cause delamination. Therefore such products (i.e., linear low density polyethylene alone) are not suggested for PSPs that have large contact surfaces (e.g., protective films). They also cause difficulties in cutting and die cutting (see Chapter 7, Section 2). Other polymers such as high density polyethylene (HDPE) or PET, although they have excellent tensile strength (Table 3.10), are not sufficiently conformable to afford good contact.

A comparable elongation of the carrier material can be achieved with an appropriate formulation. Extensibility may be given by tensioned multilayer carrier materials calendered together [188] or by the use of a filled polypropylene [189]. For many applications the direction of extensibility is very important. For medical tapes, cross-direction elasticity may be imparted by a special nonwoven material [190].

For most PSPs the carrier material is a common film manufactured for other applications also. Theoretically it would be possible to design it specially for PSP applications, knowing the peel value that appears during delamination. Knowing that the mechanical resistance of the carrier material must be higher than that of the adhesive bond, it would be (at least theoretically) possible to design and

Physical Basis of PSPs

Table 3.10 The Main Mechanical Characteristics of Common Plastic Carrier Materials

Material	Tensile strength (N/mm²) MD	CD	Elongation (%) MD	CD	Elastic modulus, MD (N/mm²)	Ref.
HDPE	20–30	—	100–1000	—	600–1400	180
	220–800	—	—	—	—	181
LDPE	8–10	—	300–1000	—	150–500	180
Biaxially oriented	12–13	—	—	—	—	182
MDPE	14–25	—	225–500	—	28–35	169
Biaxially oriented and crosslinked	56–91	—	—	—	—	181
LLDPE	—	—	300–1000	—	400–800	180
PP	21–70	—	200–500	—	630–840	169
			200–600	—	—	181
Biaxially oriented	120–180	300–400	100–200	20–50	—	181
	140–200	280–200	—	—	2000–2500	183
Calendered	>25	25	≥200	≥200	—	184
EVAc	10–20	—	600–900	—	130–700	180
HPVC	50–75	—	10–50	—	2900–3500	180
Oriented	150–170	45–70	—	—	4000	183
	130	50	120	80	—	185
VAc	30–40	—	6–10	—	—	186
Plasticized	10	—	100–1000	—	—	186
SPVC	10–25	—	17–400	—	—	180
PC	60–65	—	80–120	—	2100–2400	180
PET	47	—	25	—	2800–3100	180
Oriented	20	23	110	90	50–130	182, 187
Chemically treated	450	550	125	80	—	187
Filled	450	560	120	90	—	181
Cellulose acetate	37–98	—	—	—	25–45	182

manufacture tailored carrier materials. As known from the production of labels, for common, permanent paper labels a peel value higher than 22–25 N/25 mm causes the paper to tear during label removal. Taking into account a similar dependence between the tensile resistance of the carrier material and the bond strength necessary for a tamper-evident product, films with sufficiently low mechanical resistance can be manufactured for such products. For certain other products, estimation of the required peel value (and of the tensile strength related to it) is more difficult. Some diaper tapes, medical tapes, and protective films have a so-called cleavage or breaking peel value that is higher than the usual so-called continuous peel value (see Chapter 5). For these products the carrier has to be designed for the maximum tensile strength required. Diaper closure tapes are used as refastenable closure systems for disposable diapers, incontinence garments, and similar items [98]. Two- and three-tape systems are known. The two-tape system comprises a release tape and a fastening tape. The fastening tape includes a carrier material such as paper, polyester, or polypropylene. The preferred material is polypropylene (50–150 μm) with a finely embossed pattern on each side. Such systems allow reliable closure and reclosure with a debonding force that depends on the debonding rate. To test a special UV-crosslinked adhesive tape [191], first the force required to start the breaking of the bond (initial breakaway peel) was measured and then the force needed to continue the breaking of the bond (initial continuing peel). The initial continuing peel was found to be about 30% lower than breakaway peel.

The mechanical properties of adhesives and adhesive raw materials are generally measured in order to compare different base materials (Table 3.11). According to a BASF publication [192], such tests "allow a comparison of the order of magnitude" of the mechanical characteristics only. The measured property values depend on the chemical composition, rheology, and geometry of the samples. However, such values are useful for comparison because of the use of standard test methods and their universal character. They allow a comparative test of different components of the pressure-sensitive laminate. For instance, a comparison of the tensile strength of carboxylated butadiene rubber (CSBR) and acrylic PSA shows that CSBR exhibits better tensile strength than acrylics, and

Table 3.11 Mechanical Characteristics of PSA Raw Materials

Adhesive raw material	Tensile strength (N/mm^2)	Elongation (%)	Ref.
Soft acrylate	0.01	>3800	15
Hard acrylate	0.03	2500	14
Hard acrylate	0.15	3000	193
Hard acrylate	0.20	>3800	10
Hard acrylate	0.25	>2000	18
Hard acrylate	0.27	2000	11
Hard acrylate	0.50	1500	194
Hard acrylate	1.00	1340	195
Hard acrylate	1.50	300	196
Hard acrylate	2.70	1100	20
Hard acrylate	7.50	550	197
EVAc, segmented copolymer	11.00	1400	198
SBS, linear block copolymer	15.00	950	98
SBS, radial block copolymer	22.00	620	98

Physical Basis of PSPs 129

there is a good agreement between its tensile strength at 300% elongation and its shear resistance. The addition of styrene block copolymers to natural rubber improves its mechanical properties. Increasing the thermoplastic rubber level in a natural rubber–thermoplastic rubber mixture (0–10 phr) provides a better yield (15–200 kg/cm^2). As can be seen from Table 3.11, the tensile strength of some segmented elastomers attains the level of that of soft plasticized plastomers.

The strength and stiffness of the block copolymer depend on the chemical nature of the segments and their mutual interaction. Block copolymers having the same styrene content and about the same block molecular weight but different types of midblocks display different mechanical properties [199]. The SEBS polymer has the highest strength and modulus. As discussed earlier, the stress–strain behavior of such rubbery polymers can be adequately modeled by the Mooney–Rivlin equation as an elastic network (considering the trapped entanglements as finite crosslink junctions) or using the Guth–Gold equation relating the stress level (modulus) to the hydrodynamic effects of the styrene blocks considered as filler. The mechanical properties of the block copolymers depend on their buildup also. The tensile strength of a radial styrene block copolymer is higher (22 MPa) than that of a linear styrene block copolymer (15 MPa), and the 300% modulus is higher also (5 vs. 2.5). Elongation at break is smaller (620% vs. 950%) [98]. The star-shaped (radial) SBS copolymer Solprene 416x with more styrene has a lower molecular weight but higher tensile strength (200/90 kPa/cm^2) [98]. The copolymer with lower styrene content shows better peel and tack values.

Tensile stress and elongation play an important role for sealants and sealing tapes. Elastomers used for sealants must exhibit good elastic recovery; low temperature ($-30°C$) flexibility; resistance to oil, ozone, and aging; elongation ($>200\%$); and tensile strength (>10 MPa) [80]. Tensile strength, stiffness, density, flexural modulus, flammability, and impact resistance are the most important properties for foams [200]. The mechanical properties of finished products are tested also. The modulus, maximum elongation, and tensile strength are determined for sealants and for carrierless transfer tapes used as sealants.

The plastic anisotropy ratio (R) is used to study the effect of applying a tensile force to plastic films, by considering the change in original width (w_0) and thickness (h_0) resulting from the stress [201]:

$$R = \frac{\ln (w_f/w_0)}{\ln (h_f/h_0)} \tag{3.61}$$

where w_f and h_f are the final width and thickness of the specimen. For most films (especially for monoaxially oriented materials) the R value varies with direction in the material. This variation, ΔR, is called planar anisotropy. Dimensional stability is strongly affected by R. As is known, shrinkage and die cuttability are functions of R (see Chapter 7, Section 2). Considering the planar anisotropy as a sum of the dimensional changes in different directions (angles), the component for normal anisotropy R' is defined as

$$R' = 0.25 \, (R_0 + 2R_{45} + R_{90}) \tag{3.62}$$

According to Ref. 201, if R' is greater than 1 the material could resist thinning, which would lead to better drawability. Drawability is a characteristic known to be required for deep drawable protective films (see Chapter 8).

Tensile strength is related to the shrinkage of plastics also. For heat-shrinkable tapes the nominal value of the shrinkage stress is given. For instance, bilayer heat-shrinkable insulating tapes for anticorrosion protection of petroleum and gas pipelines have been manufactured from photochemically cured low density polyethylene with an EVAc sublayer [202]. The liner dimensions of the two-layer insulating tape decreased 10–50% depending on the degree of curing (application temperature 180°C). The application conditions and physicomechanical properties of the coating were determined for heat-shrinkable tapes with 5% shrinkage, at a curing degree of 30% and shrinkage test tension of 0.07 MPa [202].

Tear Strength Properties

The tear resistance of carrier materials used as face stock plays a pronounced role in their coating or conversion and and use. Web coating and laminating tensions, splitting stresses induced by cutting and die cutting, and forces acting on the matrix after die cutting are very important.

Surface Tear Strength. Paper tear strength is a well-known criterion in evaluating the quality of permanent paper labels. Generally if the adhesion between PSA and substrate is better than the inherent cohesion of the paper during peel-off, the debonding forces destroy the paper face stock material. In some cases only an apparent tear destruction occurs. In these cases the cohesion of the surface layer of the paper (top coat) and its adhesion to the paper fibers are not sufficient. Clay-coated papers tear if the superficial strength of the paper is not sufficient [203]. This phenomenon is also common in offset printing, where it is due mostly to the printing ink and printing conditions [204]. It also occurs for top-coated and uncoated paper. Paper humidity (which reduces the mechanical resistance of paper) can also influence paper surface tear strength, especially for clay-coated papers. Therefore for nonhomogeneous carrier materials (e.g., paper or top-coated or laminated films), both surface tear resistance and bulk tear resistance should be taken into account.

Bulk Tear Strength. Fragile pressure-sensitive materials are either very thin or have low internal strength. They are coated with a very aggressive PSA [205]. Various fragile materials can be used, such as acetate for envelopes, PVC warning labels for electronic equipment, and paper for sterile needle cartridges. For common permanent labels the peel resistance of the adhesive must be higher than the resistance of the carrier. The resultant peel of thermoplastic elastomer based

Physical Basis of PSPs 131

removable formulations should be 2.5–40 oz, with an initial peel of 1.0–16 oz for 1 in. × 6 in. strips (on stainless steel). Values above this can result in a paper tear [206].

For plastic carrier materials, tenacity and tear resistance have to be examined together. These properties depend on the orientation of the molecules. Among the machine parameters, the neck influences the orientation [207]. Elmendorf tear resistance decreases with neck length for bimodal HDPE. At greater neck lengths the machine direction (MD) values are higher [208]. The orientation of the molecules is also a function of blow-up ratio (BUR). The greater the constriction, the higher the real blow-up ratio. The bimodal grades have excellent dart drop impact (DDI) values at high BUR. Tear strength is a function of the density also. As crystallinity decreases, the toughness of the film increases. Elmendorf tear strength increases with the melt flow index (MFI) and decreases with the density. Flexural crack resistance of C_8 LLDPE decreases with density. The isotropy of the tear strength (MD/CD) depends on the polymer and its processing conditions. Because of the tendency of polypropylene to undergo orientation during processing, its tear strength in the machine direction is very low [169].

Tear strength plays an important role for many tapes. Easy tear, breakable tapes are tapes that are easily torn or split during handling. Easy tear breakable hank tapes are used in the same types of finished products as easy tear paper tapes. The tape is hand applied and used for hanking and spot tape when fast and easy removal is required. This tape possesses low elongation and easy break force. Different materials differ in their ability to spread the strain away from the strained zones, thus avoiding the formation of local regions of high strain, which lead to necking and fracture. The strain hardening ability of the material is approximated by the Ludwick–Holloman equation [201],

$$\sigma = K\epsilon^n \tag{3.63}$$

where K is a constant and n is the strain hardening exponent. A high n value is desirable for stretch forming and is required by many tape applications and for deep drawable protective films.

Impact Properties

For special tape applications and for security protective films, the impact properties of the carrier play an important role. Abrasion-resistant automotive decorative films that resist the impact of stones have been developed [209]. These products are made on a PUR or EVAc/PVC basis. They must be bubble-free laminates that are heat-deformable and repositionable. Thick films (80–900 μm) are used as the carrier.

Among the classical polyolefins, LDPE possesses the highest impact strength and lowest modulus [169]. Bimodal PE grades have excellent DDI values at high BUR. Elastic polyolefins and polyolefin copolymers display adequate impact properties. The impact resistance of C_8 LLDPEs is better, increasing with melt flow index and decreasing with density. Ethylene-vinyl acetate copolymers (with 16–30% vinyl acetate) have better impact strength. The use of polyurethanes with an extremely high level of elongation at break ensure a sufficiently high impact strength, which is particularly important in the manufacture of sealing and mounting tapes. For foams, especially "supersoft" foams, a low compression set and rapid recovery from static deformation are required [210].

Stiffness

The stiffness of a carrier (expressed as its modulus) is very important for flatness and machinability. Wrinkle buildup depends on the stiffness of the paper. Wrinkle buildup during printing is due mainly to overdrying of the paper carrier [211]. Its humidity content may decrease to 1–2%, changing its dimensions. Stiffness of the paper is a function of its geometry and grammage, construction, and humidity balance. For lower weight papers (40–80 g/m^2) the buildup of wrinkles is more accentuated. The wrinkles have a shorter wavelength than those formed in papers having a weight of 100 g/m^2. Concerning paper buildup, the choice of paper fiber is strongly influenced by sulfite pulps. Sulfite pulps are used for the manufacture of papers for soft flexible packaging. Machine finish and supercalendering also affect paper stiffness. High stiffness can be obtained by using higher grammage or a more voluminous paper grade. The grade of paper is usually given as substance (mass per unit area of sheet). For labels, paper of at least 70 g/m^2 is used; lower substances make it difficult to provide the necessary strength to enable the paper to be peeled from the substrate. There is no specific functional upper limit of substance for the coating carrier, but paper would not usually have a substance greater than 100 g/m^2, and films not more than 150 g/m^2.

The stiffness of paper and board is a function of the modulus of elasticity and the third power of thickness. Special microspheres used as filler in paper improve the paper modulus if added as a filled middle layer [212]. There is a trend toward the use of high filler loads in fine papers, providing both economic and technical advantages. When the common filler content of a paper increases to 20%, the thickness of the paper decreases by 5% and its stiffness by 15%. With the addition of 0.6% special, expandable filler to the paper layer, its original stiffness is regained.

Tenacity and tear resistance of plastics are examined together and depend mainly on the orientation of the macromolecules. The greater the constriction, the higher the blow-up ratio. For LDPE the maximum possibile stiffness is about 450 N/m^2 [213]. Thermoplastic polyurethane films have moduli of 10–650 MPa [214].

Tape and protective film application depend on the conformability of the film. Stiffness is the main characteristic used for comparison of different carrier materials for tapes [215]. For labels, lay flat in printing is complemented by on-pack

Physical Basis of PSPs

wrinkle-free squeezability. The conformability of carrier materials depends on their stiffness. Stiffness is characterized by the modulus value (Table 3.12). Conformable synthetic films should have a tensile modulus of less than 4,000,000 psi [90] as measured in accordance with ASTM D-638 and D-882, preferably less than about 300,000 psi. Conformable fabric backings should have a tensile modulus of 4,000,000 psi. In comparison, for sealants, modulus values of 0.4–0.6 N/mm^2 are required [216]. Typical examples of conformable carrier materials used for medical labels and tapes include nonwoven fabric, woven fabric, and medium to low tensile modulus plastic films (PE, PVC, PUR, low modulus PET, and ethylcellulose) [90]. Polyethylene terephthalate is considered as nonconformable carrier material.

As shown by Table 3.12, vinyl acetate polymers exhibit a broad range of modulus values. The high values are comparable with the data for common carrier materials; the low ones attain the range of adhesive raw materials. It can be observed also that postprocessing (orientation) has a pronounced influence on the modulus value of films.

As discussed in Ref. 227, stiffness influences cuttability also. According to Ref. 228, the stiffness of the main carrier materials used for tapes decreases as follows:

$$\text{PET} \cong \text{HPVC} > \text{OPP} \gg \text{HDPE} \gg \text{LDPE} \qquad (3.64)$$

Die cuttability of the same materials decreases in a similar sequence:

$$\text{PET} > \text{HPVC} > \text{OPP} \gg \text{HDPE} \gg \text{LDPE} \qquad (3.65)$$

The stiffness of polyethylene increases with its density. Density depends on crystallinity, which is a function of chain length and branching. The crystallinity-related properties decrease with increasing polar monomer content. Stiffness decreases as crystallinity decreases. Bimodality has no effect on stiffness. For polar polymers of ethylene with maleic anhydride, the addition of LDPE to form a terpolymer increases the peel strength because it modifies the stiffness of the film. The greater the density of the PE, the more it increases adhesion.

Stiffness also depends on molecular orientation. A special PE having a high melt flow index, low processing temperature, and high degree of orientation gives high stiffness films. A modulus of 400 N/mm can be achieved for a density of 0.930 g/cm^3 [229]. Most label films are oriented. Orientation gives a 50–100% increase in rigidity [169] (see also Chapter 6). Polypropylene tends to orient during processing. Its stiffness is 3–4 times greater than that of polyethylene [230]. The blow manufacturing process provides high machine direction stiffness and tensile strength for ease of register control, stripping, and dispensing. Balanced orientation makes good die cutting possible during high speed conversion. Films made using a machine direction orientation exhibit outstanding stiffness in the machine direction and flexibility in the cross direction.

Table 3.12 Modulus of PSP Components

	Modulus (N/mm^2)				
Material	Carrier component[a]	Adhesive component[a]	MD	CD	Ref.
PP homopolymer; broad MWD, cast film	x	—	800	—	217
PP homopolymer; narrow MWD, cast film	x	—	700	—	217
Blend of PP homopolymer; broad MWD, low crystallinity	x	—	670	—	217
PP, random copolymer, low comonomer content	x	—	950	—	217
PP, random copolymer, low comonomer content	x	—	800	—	217
PP, random copolymer, low comonomer content	x	—	500	—	217
PP cast film	x	—	670–1400	—	22
PP, bioriented, blown film	x	—	2500	—	183
PP, bioriented, cast film	x	—	2000	—	183
PP, mono-oriented	x	—	200	—	183
Polyethylene (HDPE)	x	—	40–150	—	218
Polyethylene (LDPE)	x	—	40–150	—	218
Polyester	x	—	38–56	—	182
Polyester, oriented	x	—	700	1000	219
Polyester, oriented	x	—	2200	2200	220
Polyvinyl chloride, plasticized	x	x	67	61	—
Polyamide (PA6)	x	—	150–320	—	218
Polyamide (PA66)	x	—	40–150	—	218
Polyurethane	x	—	170	—	171
Polystyrene	x	—	300–360	—	218
Fabric for medical use	x	—	35–40	—	211
EVAc, with 50–70% VAc			20–160	—	222
EVAc, segmented copolymer	—	x	28	—	198
Natural rubber	—	x	0.9	—	223
			1–5	—	74
Natural rubber, cured	—	x	0.9	—	224
Butadiene-styrene copolymer	—	x	7–11	—	225
Butadiene-styrene, radial block copolymer	—	x	5	—	—
Butadiene-styrene, linear block copolymer	—	x	25	—	98
EAA copolymer (hot melt)	—	x	22	—	226
Isoprene-styrene block copolymer	—	x	1.8–7.5	—	—

[a] x = present; — = absent.

Stiffness is a function of crosslinking also. For crosslinked structures the Young's modulus is given by the correlation [231]

$$E = CT\delta_r/M \tag{3.66}$$

where δ_r is the crosslinking density, C a constant, M the molecular weight, and T the temperature. On the other hand, plasticizers added to enable the polymer chains to slide over each other cause a decrease in the modulus (see also Chapter 4). The influence of the degree of crosslinking on the modulus and stiffness can be observed in the practice of protective film application. Because of the differences in the "density" of the crosslinks produced by various crosslinking agents (e.g., isocyanates or aziridine derivatives) and the differences in the mobility of the chain segments between crosslinking points, different degrees of softness of the cured adhesive are achieved. Therefore rubber-based adhesive-coated protective films conform better than cured acrylates, and aziridine-cured formulations are softer than isocyanate-based formulations.

Active fillers serve to stiffen the carrier material or adhesive [83]. Modulus improvement by fillers has been evaluated by a number of authors [232]. The best known equations of Guth and Smallwood were discussed earlier. In a first approximation, tackifier resins were considered as fillers. Such "fillers" possess a much higher modulus than rubber. As stated by Druschke [71] for blends of natural rubber and tackifier resin, the modulus ranges from about 10^{-1} to $2-5 \times 10^{-1}$ N/mm^2 at room temperature. The modulus of a tackifier resin is about 1. Like a filler, the resin level influences the final modulus value.

Flexibility and extensibility of the carrier material influence its energy absorption properties, which help determine the nature of the debonding, i.e., removability (see Chapter 5). Generally the amount of energy (U) that can be stored elastically in a carrier deformed by flexion can be written as a function of the stress (σ) and deformation (ϵ) as [233]

$$U = \int \sigma \, d\epsilon \tag{3.67}$$

For unidirectionally (e.g., filmlike) stressed materials under well-defined conditions, the amount of stored energy may be expressed as a function of the modulus of elasticity (E):

$$U = 0.5 \, E \, \epsilon^2 \tag{3.68}$$

As can be seen from the above relation, the deformability of the material is more important with respect to energy storage than its modulus. Elastic, extensible, low modulus carrier materials can store too much energy during application (e.g., lamination), which causes nonuniform debonding and adhesive deposits. Such materials (e.g., LLDPE or some EVAc copolymers) are not recommended as carriers for protective films (see Chapter 6). It is evident that energy storage may be advantageous for products with high cleavage peel (zip peel) (see Chapter 5).

2.3 Regulating the Mechanical Properties of PSPs

To fulfill the various requirements for different product classes and applications, there is a need to regulate the mechanical properties. This can be accomplished by controlling the formulation of the product components (carrier, liner, and adhesive), the choice of product geometry, and/or the choice of product buildup.

Regulating Mechanical Properties by Formulation

Theoretically the simplest way to regulate the mechanical properties of carrier materials is to control their chemical or macromolecular formulation (see Chapter 5, Section 2). In reality, the range of commercially available raw materials is limited.

Regulating Mechanical Properties by Geometry

The mechanical properties of a PSP are the result of intrinsic chemical and/or structural characteristics and product geometry. Generally product geometry includes the in-plane dimensions of the carrier and its thickness. For labels the product contour (in plane) can influence the die cutting properties. For tapes the cross-sectional contour affects the mechanical characteristics. Special operations such as folding, pleating, punching, and perforating modify mechanical performance also (see Chapter 7).

Regulating Mechanical Properties by Manufacture

As discussed earlier, the mechanical characteristics of a carrier material depend on its chemistry and processing technology. During processing, orientation, tensioning, or relaxation of the material may occur. Such phenomena are partially inherent characteristics of the web buildup and manipulating technology, being affected by the rate differences among the moving parts of the equipment, or they are brought about in a separate technological operation. Tensions are due to thermal effects also. Differential cooling and crystallization may cause tension. The regulation of mechanical characteristics is discussed in Chapter 6.

2.4 The Influence of Mechanical Properties on Other Performance Characteristics of PSPs

Chapter 5 discusses the influence of the mechanical properties of the PSP components on adhesive performance. The mechanical properties also influence the conversion and end use performance characteristics of the PSPs. During conversion, the components of the PSP or the finished product undergo physical, chemical, and mechanical transformations. Conversion is achieved by using high speed, high productivity machines where the machinability of the product or product components plays a special role. The dimensionally stable, weblike behavior of PSPs during conversion is the result of tailored mechanical characteristics. Laminating and delaminating ability is also given by the mechanical properties. The transformability of the web in discontinuous pressure-sensitive products (e.g., labels or special tapes) is ensured by well-defined mechanical features. The influence of the mechanical characteristics on the conversion performance characteristics is described in Chapter 7.

During their end use PSPs have to support increased stresses. For weblike products, unwinding, lamination under pressure and delamination for the labeling of discontinuous items (separating the label from the reinforcing continuous liner) involve stresses. It is evident that the mechanical properties of the carrier material and the adhesive and those of the finished product play an important role in these operations (see Chapter 8).

3 OTHER PHYSICAL CHARACTERISTICS OF PSPs

Among the physical characteristics influencing the manufacture and end use properties of PSPs, polarity and electrical characteristics play an important role.

3.1 Electrical Characteristics of PSPs

In the manufacture of PSPs, coating is one of the main operations. Various components (adhesive, abhesive, primers, printing inks, etc.) are coated on solid-state carrier materials. Their anchorage depends on the polarity of the surface. The polarity of the surface can be improved by physical or chemical treatments (see Chapter 6), which are related to the improvement of electrical characteristics. The electrical characteristics influence the processibility of the laminate components and the application of the finished product. According to Kamusewitz [234], the wetting out of plastic films depends on the temperature, vapor saturation of the atmosphere, and electric charges on the surface. Pitzler [235] studied the transfer coating of water-based and solvent-based polyurethanes and acrylics on siliconized liner to determine the limits of reusability of the temporary carrier. He disclosed a discrepancy between laboratory and industrial results. In practice the reusability of the liner is limited. After a number of uses (five to eight, depending on the nature of the coated liquid), the release force increases. In laboratory tests there is also an increase in the release (peel-off) force (for this increase the polar part of the surface energy is responsible), but at such a low level that it does not affect the reusability (more than 20 times) of the liner. According to Pitzler, under industrial high speed manufacturing conditions the accumulation of electric charges on the silicone surface can lead to sparks that penetrate through the abhesive layer and cause its destruction.

Antistatic Performance—Surface Resistivity

Static electricity can be defined as an excess or deficiency of electrons on a surface. It occurs when two nonconductive bodies rub or slide together or separate from one another. It is a triboelectric effect. Many factors influence the polarity and size of the charge: cleanness, pressure of contact, surface area, and speed of rubbing and separating. As is known from the use of protective films, their delamination from plastic plates produces static electricity. Therefore before processing the plate should be given an antistatic treatment [236].

A second source of static electricity is an electrostatic field formed by charged bodies when they are near each other or near noncharged bodies. This field will induce a charge on a nearby nonconductive object. This voltage can be discharged when the nonconductive body carrying the charge comes in contact with another body at a sufficiently different potential. The discharge may be in the form of an arc or spark. Electric charges on the surface facilitate deposition of suspended particles from the air. To avoid this phenomenon the surface conductivity has to attain a well-defined value. To avoid deposition of dirt (dust) from the air on a polypropylene surface, it should possess a surface resistance of 10^{11} ohms [237]. The prevention of static charge is a requirement for packaging materials for electronic parts [238,239].

The surface resistance R depends on the specific electrical conductivity ϑ as follows [238]:

$$R = (1/\vartheta)(1/h)(a/b) \tag{3.69}$$

where h is the layer thickness, and a and b are the sides of the sample quadrate. Reported as specific surface resistance for a quadratic surface where a and b have the same dimensions,

$$R = 1/\vartheta h \tag{3.70}$$

In practice, ohms per square (quadratic surface resistance) is used. For example, a polyalkoxythiophene has a quadratic resistance of 4×10^4. The common use of antistatic agents includes the domain of 10^6–10^{11} ohms/square. For the evaluation of antistatic agents the maximum surface charge (U) expressed as a function of negative ($U_{(-)}$) and positive ($U_{(+)}$) charges [236]

$$U = \{(1/2)[(U_{(+)})^2 - (U_{(-)})^2[\}^{1/2} \tag{3.71}$$

is used together with the charge decay time τ as a function of the decay time of positive ($\tau_{(+)}$) and negative ($\tau_{(-)}$) charges,

$$U = \{(1/2)[(\tau_{(+)})^2 - (\tau_{(-)})^2]\}^{1/2} \tag{3.72}$$

As a function of the alkylrest and alkoxymethyl group from their constitution, the surface resistance of 1-alkyl-3-alkoxymethylimidazolium chloride–treated LDPE varies between 4×10^6 and 2.5×10^8 ohms, and the charge decay time is between 0 and 0.24 s. An excellent antistatic effect is given by a surface resistivity of less than 10^9 ohms and a shelf life of zero [240,241].

To accommodate the slitting and winding of plastic carrier materials, antistatic films must have a resistance of 10^{11}–10^{13} ohms/square or less. The relationship between surface resistance and charge decay time (in seconds) allows the evaluation of antistatic behavior. A charge decay time of 10–60 s corresponds to a surface resistance of 10^{11}–10^{12} ohms and a medium antistatic behavior.

Antistatic agents consist of ionic or hydrophilic materials that prevent charge accumulation and are either applied to the surface of or compounded with the polymeric raw material (see Chapter 6). To provide the required resistivities, conductive materials can be incorporated also. Reproducible antistatic protection in PE is more difficult to achieve than in PVC because of the semicrystalline nature of PE. A better solution is to incorporate conductive fillers and crosslink the product [242].

Electrical Conductivity

Electrically conductive plastics are used in antistatic products, self-heating plastics, shielding materials (radio-frequency and electromagnetic insulation), and electromagnetic radiation–absorbing materials. Electroconductive adhesives are well known also [243]. Anisotropically electroconductive adhesives are made by dispersing electroconductive particles in an electrically insulating matrix.

The most important parameter characterizing an ionic conducting polymer is the temperature dependence of its ionic conductivity. Plastics can be formulated to fit different levels of conductivity. For the electronic industry, electrostatic discharge protection (ESD) and electromagnetic radio-frequency interference (EMI/RFI) protection are very important.

Modification of thermoplastics by the addition of a conductive material to the resin matrix results in a conductive product that can be used for protection against electrostatic discharge (see Chapters 4 and 6). The surface resistance of the adhesive layer for a cover tape used in the electronics industry is 10^{13} ohms/square or less [244]. Common PE insulating tape possesses an electrical resistivity higher than 10^{14} ohm-cm; a PVC tape has a surface resistivity of more than 10^{12} ohm-cm [169].

Special applications require adhesives and carrier materials with excellent electrical resistivity. Amorphous polypropylene-based HMPSAs have been known for many years [245]. A patent [246] discloses a PSA formulation based on amorphous polypropylene, styrene block copolymers, and tackifier. The formulation discussed by Wakabayashi and Sugii [246] is based on polypropylene, styrene block copolymer, tackifier, and wax and displays special electrical properties as a very low dissipation factor (less than 0.01 at 1 kH) and high volume resistivity of 1×10^{14} ohm-cm. Silicone polymers have excellent electrical properties such as arc resistance, high electrical strength, low loss factor, and resistance to current leakage. These characteristics are very important for electrical insulating tapes [247].

Some applications require electrically conductive films. Such films possess a dielectric constant of 200,000 and electrical resistance of 10^3–10^8 ohm-cm [248]. Conductive fillers such as carbon black, carbon fibers, metal powders, flakes, fibers, and metallized glass spheres and fibers are used to achieve such conductivity. Common plastics contain carbon black to improve their electrical conductivity. Such thermoplastics act as electromagnetic shields up to 30 dB [249]. The structure of carbon black and the polymer morphology influence the conductivity of carbon black–filled plastics. The electrical conductivity of such materials has been explained as a system based on elastomer and carbon black connected in parallel [250]. The electrical characteristics of such systems depend on their degree of mechanical stretching [251]. Different techniques of compounding conductive powders with polymers (mixing the polymer in the plastic or molten state, polymerization or polycondensation of monomer–oligomer mixtures, mixing powdered polymer with conductive powder, suspension mixing, aqueous mixing, etc.) have been developed [252]. Much work has been done in connection with the conductivity of these filled systems. An excellent review of the possible conduction mechanisms (classical, tunnel, etc.) is given by Miyasaka [253].

Very important roles are played by the polarizability of polymer surfaces and by the methods used for polarizability. The majority of self-adhesive films need polarization to achieve instantaneous peel. Polarization methods are discussed in Chapter 6.

3.2 Optical Characteristics of PSPs

Some PSPs are information carriers. For these products, optical properties are basic performance characteristics. These characteristics are discussed in Chapters 6–8.

REFERENCES

1. I. Benedek and L. J. Heymans, *Pressure-Sensitive Adhesives Technology,* Marcel Dekker, New York, 1997, Chapter 2.
2. G. Fauner, *Klebgerechte Gestaltung und Festigkeitsverhalten von Kunststoffklebungen,* Swiss Bonding, 88, May 4, 1988, Rapperswil, Switzerland.

Physical Basis of PSPs

3. M. Gerace, *Adhes. Age* 8:84 (1983).
4. M. Gerspacher, C. P. O. Farrel, and H. H. Yang, *Elastomerics* 11:23 (1990).
5. *Kaut. Gummi Kunstst.* 6:556 (198).
6. H. Gramberg, *Adhäsion* 3:97 (1966).
7. P. Dunckley, *Adhäsion* 11:19 (1989).
8. *Adhes. Age* 4:118 (1974).
9. S. N. Gan, D. R. Burfield, and K. Soga, *Macromolecules* 18:2684 (1985).
10. Condea GmbH, Moers, Germany, *Dilexo AK 691,* Produktinformation.
11. BASF, Ludwigshafen, Germany, *Acronal 80D,* Technische Information.
12. A. Midgley, *Adhes. Age* 9:18 (1986).
13. *Kaut. Gummi Kunstst.* 41(1):34 (1998).
14. Condea GmbH, Moers, Germany, *Dilexo AK 694,* Produktinformation.
15. Condea GmbH, Moers, Germany, *Dilexo AK 680,* Produktinformation.
16. *Adhes. Age* 6:21 (1985).
17. E. E. Ewins, Jr. and J. R. Erickson, *Tappi J* 6:155 (1988).
18. BASF, Ludwigshafen, Germany, *Acronal V303,* Technische Information, August 1985.
19. Hoechst, Frankfurt, Germany, *Mowilith DM 154,* Merkblatt, March 1985.
20. BASF, Ludwigshafen, Germany, *Acronal LA449S,* Technische Information, February 1994.
21. BASF, Ludwigshafen, Germany, *Acronal 330D,* Technische Information, March 1984.
22. Himont Italia, Centro Ricerche "G. Natta," Cast film—New grades for low temperature storage, Technical Information, 1994.
23. *Adhäsion* 10(3):20 (1968).
24. *Coating* 1:12 (1984).
25. H. G. Drössler, *Kaut. Gummi Kunstst.* 1:28 (1987).
26. W. J. Busby, Polyethylene 93, Maack Business Services, Zürich, Switzerland, 1993, p. 3, Session VI, 3–3.
27. L. C. Struik, *Physical Ageing in Amorphous Polymers and Other Materials,* Elsevier, Amsterdam, 1978, pp. 19–22, 22–3, 42–7, 75–8.
28. J. C. Bauwens, *Plast. Rubber Process. Appl.* 7:143 (1987).
29. J. C. Pommier, J. Poustis, and J. J. Azens, *apr* 31:848 (1988).
30. R. Köhler, *Adhäsion* 2:41 (1972).
31. *Coating* 12:432 (1994).
32. B. B. Blackford, Br. Patent 8864365, in *Coating* 6:185 (1969).
33. U.S. Patent 4223067, in D. K. Fisher and B. J. Briddell (Adco Product Inc., Michigan Center, MI), EP 0426198A2/08.05.1991.
34. Gobran, U.S. Patent 4260659, in J. N. Kellen and C. W. Taylor (Minnesota Mining and Manuf. Co., St. Paul, MN), EP 0246352 A2/25.11.1987.
35. *Adhäsion* 3:83 (1974).
36. C. Kirchner, *Adhäsion* 10:398 (1969).
37. *Coating* 3:65 (1974).
38. A. K. Bhowmick, P. P. De, and A. K. Bhattacharyya, *Polym. Eng. Sci.* 27(15):1195 (1987).
39. A. Carre and A. D. Roberts, *J. Chim. Phys., Physico-Chim. Biol.* 84(2):252 (1987).

40. E. Prinz, *Coating* 10:271 (1979).
41. A. Carre and J. Schultz, *J. Adhes.* 18:207 (1985).
42. E. Prinz, *Coating* 10:271 (1978).
43. H. W. Kammer, *Acta Polym.* 34:112 (1983).
44. A. Lamping and W. Tellenbach (Corti A. G.), CH Patent 664971/15.04.1988, in *CAS* 22:4 (1988) 109:130446e.
45. A. N. Gent and P. Vondracek, *J. Appl. Polym. Sci.* 27:4357 (1982).
46. A. N. Gent, Peel strength of model pressure sensitive adhesives, PTSC XVII Technical Seminar, Shaumburg, IL, May 4, 1986.
47. C. M. Roland and G. G. Boehm, *Macromolecules* 18:1310 (1985).
48. L. Placzek, *Coating* 3:94 (1987).
49. M. E. R. Shanahan, P. Schreck, and J. Schultz, *C. R. Acad Sci., Ser. 2* 306(19):1325 (1988).
50. P. P. Hoenisch and F. T. Sanderson, Aqueous acrylic adhesives for industrial laminating, Adhesives 85, Conference Papers, Atlanta, GA, Sept. 10–12, 1985.
51. *Papier Kunstst. Verarb.* 2:18 (1996).
52. U. Schwab, *Coating* 5:171 (1996).
53. A. Barbero and A. Amico, A new performance ULDPE/VLDPE from high pressure technology—Potential applications, Polyethylene 93, Oct. 4, 1993, Maack Business Services, Zürich, Switzerland.
54. Yu. K. Godovski, *Progr. Colloid Polym. Sci.* 75:70 (1987).
55. E. Guth, *J. Appl. Phys.* 16:20 (1945).
56. D. Djordjevic, Tailoring films by the coextrusion casting and coating process, Specialty Plastics Conference '87, Polyethylene and Copolymer Resin and Packaging Markets, Dec. 1, 1987, Maack Business Services, Zürich.
57. H. C. Lau and W. R. Schowater, *J. Rheol.* 30:193 (1986).
58. J. Metall, *Kaut. Gummi Kunstst.* 3:228 (1987).
59. J. Pietschman, *Adhäsion* 5:21 (1982).
60. Exxon Chemical, Adhesives Update, HMPSA for Tape application, Winter 1994/95.
61. L. Salmén, *Tappi J.* 12:190 (1988).
62. C. C. Sun and J. E. Mark, *J. Polym. Sci., Polym. Phys.* 25(10):2073 (1987).
63. C. Han and J. Kim, *J. Polym. Sci., Polym. Phys.* 25(8):1741 (1987).
64. *Kaut. Gummi Kunstst.* 40(10):981 (1987).
65. W. W. Bode, *Tappi J.* 6:133 (1988).
66. S. Wu, *J. Polym. Sci., Part B, Polym. Phys.* 25(12):2511 (1987).
67. D. B. Alward, D. J. Kinning, E. L. Thomas, and L. J. Fetters, *Macromolecules* 19:215 (1986).
68. S. M. Collo and G. Garrabe, *Plast. Mod. Elastomers* 38(8):134 (1986).
69. *Coating* 7:187 (1984).
70. D. R. Burfield, *Polym. Commun.* 29(1):19 (1988).
71. W. Druschke, AFERA, Conference Paper, Edinburgh, October 1986, p. 13.
72. H. G. Bubam and H. Ullrich, *apr* 5:108 (1986).
73. R. Hinterwaldner, *Adhäsion* 10:26 (1985).
74. U. Eisele, *Kaut. Gummi Kunstst.* 40(6):539 (1987).
75. M. C. Chang, C. L. Mao, and R. R. Vargas, Can. Patent 1225792/18.08.1987.

76. A. Zawilinski, *Adhes. Age* 9:29 (1984).
77. P. R. Mudge (National Starch and Chem. Co., Bridgewater, NJ), EP 022541A2/16.06.1987.
78. National Starch and Chem. Co., New York, NY), U.S. Patent 1645063, in *Coating* 7:184 (1974).
79. R. L. Sun and J. F. Kennedy (Johnson & Johnson Products, Inc., USA), U.S. Patent 4762888/09.08.1988; in *CAS, Hot Melt Adhes.* 26:1 (1988).
80. D. Hübsch, *Kaut. Gummi Kunstst.* 3:239 (1988).
81. B. W. Foster, A new generation of polyolefin based hot melt adhesives, Presented at Tappi Hot Melt Symposium, Monterey, CA, June 7–10, 1987.
82. U.S. Patent 3691140/12.09.1972, in H. Miyasaka, Y. Kitazaki, T. Matsuda, and J. Kobayashi (Nichiban Co., Ltd. Tokyo, Japan), Patent Appl. DE 3544868A1/18.12.1985.
83. M. Akay, S. N. Rollins, and E. Riordan, *Polymer* 29(1):37 (1988).
84. V. L. Hughes and R. W. Looney (Exxon Research and Engineering Co., Florham Park, NJ), EP 0131460/16.01.1985.
85. T. R. Mecker, Low molecular weight isoprene based polymers—Modifiers for hot melts, Presented at TAPPI Hot Melt Symposium, 1984; in *Coating* 11:310 (1984).
86. *Kaut. Gummi Kunstst.* 37(4):285 (1984).
87. J. B. Class and S. G. Chu, *Org. Coat. Appl. Polym. Sci. Proc.* 48:126 (1983).
88. M. Dupont and N. Keyzer, Evaluation of adhesive properties of hot melt based on thermoplastic elastomers with different polystyrene contents, Presented at PTSC XVII Technical Seminar, Woodfield Shaumburg, IL, May 4, 1986.
89. K. F. Foley and S. G. Chu, *Adhes. Age* 9:24 (1986).
90. U.S. Patent 3321451, in S. E. Krampe and C. L. Moore (Minnesota Mining and Manuf. Co., St. Paul, MN), EP 0202831A2/26.11.1986.
91. T. H. Haddock (Johnson & Johnson, New Brunswick, NJ), EP 0130080B1/02.01.1985.
92. G. Bonneau and M. Baumassy, New tackifying dispersions for water based PSA for labels, 19. Münchener Klebstoff u. Veredelungsseminar, 1994, p. 82.
93. U.S. Patent 4223067, in R. R. Charbonneau and G. L. Groff (Minnesota Mining and Manuf. Co, St. Paul, MN), EP 0106559B1/25.04.1984.
94. G. Auchter, J. Barwich, G. Rehmer, and H. Jäger, *Adhes. Age* 7:20 (1994).
95. Y. Sasaaki, D. L. Holguin, and R. Van Ham (Avery Int., Pasadena, CA) EP 0252717A2/13.01.1988.
96. C. Donker, R. Luth, and K. van Rijn, Hercules MBG 208 hydrocarbon resin: A new resin for hot melt pressure sensitive tapes, 19th Münchener Klebstoff u. Veredelungsseminar, 1994, p. 64.
97. P. A. Mancinelli, New developments in acrylic hot melt pressure sensitive adhesive technology, in *TECH 12, Advances in Pressure Sensitive Tape Technology*, Technical Seminar Proceedings, Itasca, IL, May 1989, p. 161.
98. C. Parodi, S. Giordano, A. Riva, and L. Vitalini, Styrene-butadiene block copolymers in hot melt adhesives for sanitary application, 19th Münchener Klebstoff und Veredelungsseminar, 1994, p. 119.
99. L. Jacob, New development of tackifiers for SBS copolymers, 19th Munich Adhesive and Converting Seminar, 1994, p. 107.

100. E. Debier and N. de Keyzer, The use of rheometry to evaluate the compatibility of thermoplastic rubber based adhesives, Lecture given at 16th Munich Adhesive and Finishing Seminar, Munich, Oct. 28, 1991.
101. F. Buehler and W. Gronski, *Makromol. Chem. 188*(12):2995 (1987).
102. L. E. Nielsen, *Mechanical Properties of Polymers and Composites,* Vols. 1 and 2, Marcel Dekker, New York, 1974; cited in E. Debier and N. de Keyzer, The use of rheometry to evaluate the compatibility of thermoplastic rubber based adhesives, Lecture given at 16th Munich Adhesive and Finishing Seminar, Munich, Oct. 28, 1991.
103. J. D. Ferry, *Viscoelastic Properties of Polymers,* 3rd ed., Wiley, New York, 1980; cited in E. Debier and N. de Keyzer, The use of rheometry to evaluate the compatibility of thermoplastic rubber based adhesives, Lecture given at 16th Munich Adhesive and Finishing Seminar, Munich, Oct. 28, 1991.
104. J. Kim, C. D. Han, and S. G. Chu, *J. Polym. Sci. B26:*677 (1988).
105. D. W. van Krevelen and P. J. Hoftijzer, *Properties of Polymers, Correlation with Chemical Structure,* Elsevier, Amsterdam, 1972; cited in E. Debier and N. de Keyzer, The use of rheometry to evaluate the compatibility of thermoplastic rubber based adhesives, Lecture given at 16th Munich Adhesive and Finishing Seminar, Munich, Oct. 28, 1991.
106. H. G. Kilian, *Kaut. Gummi Kunstst. 39*(8):689 (1986).
107. E. T. Bishop and S. Davison, *J. Polym. Sci. C 26:*59 (1969).
108. T. B. Lewis and L. E. Nielsen, *J. Appl. Polym. Sci. 14:*1449 (1970).
109. F. Buckmann and F. Bakker, *ECJ 12:*922 (1995).
110. D. Quemada, *Lecture Notes in Physics Stability of Thermodynamic Systems,* Springer, Berlin, 1982, p. 210; cited in F. Buckmann and F. Bakker, *ECJ 12:*922 (1995).
111. I. M. Krieger and T. Dougherty, *J. Trans. Soc. Rheol. 3:*137 (1959).
112. M. Toyama and T. Ito, Pressure sensitive adhesives, in *Polymer Plastics Technology & Engineering,* Vol. 2, Marcel Dekker, New York, 1974, pp. 161–230.
113. M. A. Krecenski, J. F. Johnson, and S. C. Temin, *JMS Rev. Macromol. Chem. Phys. C26*(1):143 (1986).
114. L. Woo, Study on adhesive performance by dynamic mechanical techniques, Paper presented at the National Meeting of the American Chemical Society, Seattle, WA, March 1983; cited in M. A. Krecenski, J. F. Johnson, and S. C. Temin, *JMS Rev. Macromol. Chem. Phys. C26*(1):143 (1986).
115. D. H. Kaelble and R. S. Reylek, *J. Adhes 1:*124 (1969).
116. D. H. Kaelble, *Physical Chemistry of Adhesion,* Wiley-Interscience, New York, 1961; cited in M. A. Krecenski, J. F. Johnson, and S. C. Temin, *JMS Rev. Macromol. Chem. Phys. C26*(1):143 (1986).
117. K. Hahn, G. Ley, and R. Oberthur, *Colloid Polym Sci. 266*(7):631 (1988).
118. L. E. Rantz, *Adhes. Age 5:*15 (1987).
119. R. Dehnen, *Siebdruck 3:*48 (1986).
120. A. Zosel, *Colloid Polym. Sci. 263:*541 (1985).
121. G. Göritz, *Kaut. Gummi Kunstst. 40*(10):966 (1987).
122. M. Bowtell, *Adhes. Age 7:*38 (1987).
123. J. Wiedemeyer, *VDI Beri. 600:*74 (1987).

124. J. Wiedemeyer, *VDI Beri. 600:*3, 171 (1987).
125. *Adhäsion 4:*118 (1984).
126. Packard Electric, Engineering Specification ES M-4037, 1992.
127. Packard Electric, Engineering Specification ES M-2359, 1992.
128. Packard Electric, Engineering Specification ES M-2147, 1992.
129. C. Harder, Acrylic hotmelts—Recent chemical and technological developments for an ecologically beneficial production of adhesive tapes: State and prospects, Presented at European Tape and Label Conference, Brussels, April 28–30, 1993.
130. A. Carre and J. Scultz, *J. Adhes. 17:*135 (1984).
131. *Kaut. Gummi Kunstst. 39*(10):1004 (1986).
132. G. M. Bartenev and V. V. Tulinova, *Vysokomol. Soed. A 29:*1055 (1987).
133. P. M. Toporowski and J. Roovers, *J. Polym. Sci., Polym. Chem. 24:*3009 (1986).
134. F. Hensen, *Papier Kunstst. Verarb. 11:*32 (1988).
135. *Etiketten-Labels 5:*91 (1995).
136. J. R. DeFife, *Labels Label. 3/4:*14 (1994).
137. *Adhes. Age 8:*8 (1983).
138. F. C. Larimore and R. A. Sinclair (Minnesota Mining and Manuf. Co., St. Paul, MN) EP 0197662A1/15.10.1986.
139. J. F. Kuik, *Papier Kunstst. Verarb. 10:*26 (1990).
140. *Cross-Linked Polymers* (ACS Symp. Ser. 367), ACS, Washington, DC, 1988, p. 124; in *CAS Crosslinking Reactions 19:*5 (1988).
141. *Plastverarbeiter 37*(4):48 (1986).
142. BASF Kunststoffe, Lupolen B. 581, d/12.92, pp. 12, 13, 33.
143. D. F. Acierno, F. P. La Mantia, and G. Titomanlio, *Acta Polym. 11:*697 (1986).
144. C. C. Sun and J. E. Mark, *Am. Chem. Soc. Div. Polym. Chem. Prepr. 27:*230 (1986).
145. K. M. Davis, R. Lonnet, and C. R. Stone, *Müanyag, Gummi 22:*233 (1985).
146. V. A. Novak, O. Yu. Krasnova, and R. A. Gommen, *Plast Massy 3:*32 (1987).
147. W. J. McGill et al., A theory of green strength in natural rubber, in *Kaut. Gummi Kunstst. 19:*962 (1987).
148. G. Coulon, and B. Escaig, *Polymer 29*(5):808 (1988).
149. L. C. Yanio and F. N. Kelley, *Rubber Chem. Technol. 1:*78 (1987).
150. K. E. Johnsen, *Adhes. Age 11:*29 (1985).
151. J. B. Donnet, M. J. Wang, E. Papirer, and A. Vidal, *Kaut. Gummi Kunstst. 39*(6):510 (1986).
152. W. Xianglong et al., *Kaut. Gummi Kunstst. 39*(12):1216 (1986).
153. K. Sato, *Rubber Chem. Technol. 56:*942 (1987).
154. E. A. Meinecke and M. I. Taftaf, *Rubber Chem. Technol. 61:*534 (1987).
155. A. Einstein, *Ann. Phys. (Leipzig) 19:*289 (1906); cited in E. A. Meinecke and M. I. Taftaf, *Rubber Chem. Technol. 61:*534 (1987).
156. G. Slayter, *Sci. Am. 206*(1):124 (1962); cited in D. W. van Krevelen, *Kaut. Gummi Kunstst. 37*(4):295 (1984).
157. J. C. Halpin, *J. Compos. Mater. 3:*732 (1969); cited in D. W. van Krevelen, *Kaut. Gummi Kunstst. 37*(4):295 (1984).
158. A. A. Griffith, *Phil. Trans. Roy. Soc. (Lond.) A 221:*163 (1920); cited in D. W. van Krevelen, *Kaut. Gummi Kunstst. 37*(4):295 (1984).
159. H. Klein, *Coating 12:*430 (1986).

160. A. J. Owen, *Colloid Polym. Sci. 263:*991 (1985).
161. I. Skofic and D. Susteric, *Polym. Jugoslav Casop. Plast Gumu.,* in *Kaut. Gummi Kunstst. 41*(8):833 (1988).
162. L. J. Mathias and M. Allison, *ACS Symp. Ser. 367:*66 (1988), in *CAS, Crosslinking Reactions 21:*3 (1988).
163. G. Bocaccio, *Caoutch. Plast 62*(653):83 (1985).
164. U. Yoshiaki and Y. Kentaro, *J. Adhes.* 25(1):45 (1988).
165. T. Yukinari, *Kaut. Gummi Kunstst. 39*(6):545 (1986).
166. R. A. Veselovsky, V. I. Pavlov, and T. P. Muravskaya, Proc. Sixth All-Union Conference on the Mechanics of Polymer and Composite Materials, Riga, *Mekh. Kompozit. Material. 2:*225 (1987).
167. Yu. S. Lipatov, N. P. Pasechnik, and V. F. Babich, *Dokl. Akad. Nauk SSSR 239*(2):371 (1978); cited in R. A. Veselovsky, V. I. Pavlov, and T. P. Muravskaya, Proc. of the Sixth All-Union Conference on the Mechanics of Polymer and Composite Materials, Riga, *Mekh. Kompozit. Material. 2:*225 (1987).
168. *Plastverarbeiter 37*(10):207 (1986).
169. P. Kriston, *Müanyag Fóliák,* Müszaki Könyvkiadó, Budapest, 1976, p. 45.
170. *Kaut. Gummi Kunstst. 40*(10):977 (1987).
171. L. Bayer, Proceedings of Polyurethanes World Congress, Sept. 29–Oct. 2, 1987, p. 373.
172. G. Menges, H. J. Roskothen, and H. Adamczak, *Kunststoffe 62:*12 (1972).
173. C. S. Henkee and E. L. Thomas, *Proc. Annu. Meeting, Electron Microsc. Soc. Am., 45th,* 1987, p. 524; in *CAS Colloids 21:*3 (1988).
174. J. Hansmann, *Adhäsion 9:*251 (1976).
175. A. W. Norman, *Adhes. Age 4:*35 (1974).
176. G. Greiner, *Packung Transport 6:*28 (1983).
177. Letraset Ltd., London, U.S. Patent 1545568, in *Coating 6:*71 (1970).
178. H. Schreiner, *apr 16:*172 (1986).
179. Com-Tech Inc., U.S. Patent 3361609, in *Coating 7:*274 (1969).
180. BASF *Kunststoff Produkte,* BASF, Ludwigshafen, Germany, 1985, p. 53.
181. *Adhäsion 1:*15 (1974).
182. *Adhäsion 5:*208 (1967).
183. P. Hammerschmidt, *apr 7:*223 (1986).
184. Klöckner Pentaplast, Montabaur, Technisches Datenblatt, Pentaform TZ 887/53, Kalandrierte PP Folie.
185. Hoechst Folien, Ausgabe 07/92, Datenblatt, Klebeband Träger und Abdeckfolien.
186. *Coating 3:*65 (1974).
187. Isea Film SPA, Genoa, Italy, *Nu Roll, Film di Poliestere,* Technical booklet.
188. Minnesota Mining and Manuf. Co., Br. Patent 1102296, February 3, 1987.
189. Beiersdorf A. G., Hamburg, Germany, DBP 1667940, in *Coating 12:*363 (1973).
190. D. K. Fisher and B. J. Briddell (Adco Product Inc., Michigan Center, MI), EP 0426198A2/08.05.1991.
191. *Kaut. Gummi Kunstst. 37*(4):285 (1984).
192. BASF, *Acronal 79 D,* Technische Information, BASF, Ludwigshafen, Germany, January 1982.

193. BASF, *Acronal 81 D,* Technische Information, BASF, Ludwigshafen, Germany, October 1983.
194. BASF, *Acronal 85 D,* Technische Information, BASF, Ludwigshafen, Germany, January 1982.
195. BASF, *Acronal 50 D,* Technische Information, BASF, Ludwigshafen, Germany, March 1984.
196. BASF, *Acronal A120,* Technische Information, BASF, Ludwigshafen, Germany, December 1988.
197. BASF, *Acronal 330 D,* Technische Information, BASF, Ludwigshafen, Germany, March 1984.
198. R. Koch and C. L. Gueris, *apr 16:*460 (1986).
199. L. D. Jurrens and O. L. Mars, *Adhes. Age 8:*31 (1974).
200. S. B. Driscoll, L. N. Venkateshwaran, C. J. Rosis, and L. C. Whitney, *Soc. Plast Eng. Annu., Tech. Conf., Tech. Papers 1:*450 (1985), in *Kaut. Gummi Kunstst. 39*(2):161 (1986).
201. Y. W. Lee, N. N. S. Chen, and P. I. F. Niem, *Rubber Process. Appl. 7*(4):222 (1987).
202. V. M. Ryabov, O. I. Chernikov, and M. F. Nosova, *Plast. Massy 7:*58 (1988).
203. *Coating 1:*35 (1988).
204. W. Walenski, *Offset Technik 10:*32 (1988).
205. D. Lacave, *Labels Label. 3/4:*54 (1994).
206. EP 4728572.
207. E. B. Parker, Polyethylene 93, Oct. 4, 1993, Session 3, p. 6. Maack Business Service, Zürich, Switzerland, p. 6.
208. R. Nurse, HDPE applications, PE developments, Polyethylene 93, Oct. 4, 1993, Session 3, Maack Business Service, Zürich, Switzerland, p. 1.
209. N. Wasfi Malek (Beiersdorf AG, Hamburg, Germany), EP 0095093/30.11.1983.
210. S. L. Hager, D. A. Sullivan, B. D. Harper, and L. F. Lawler, *J. Cell. Plast. 22*(11):512 (1986).
211. *Druckwelt 12:*34 (1988).
212. Ö. Söderberg, *Paper Technol. 8:*17 (1989).
213. *Papier Kunstst. Verarb. 11:*28 (1986).
214. *Neue Verpackung 5:*64 (1991).
215. P. Hammerschmidt, *Coating 4:*193 (1986).
216. A. Wolf, *Kaut. Gummi Kunstst. 41*(2):173 (1988).
217. W. Schoene, Speciality Plastics Conference 89, Dec. 4, 1989, Maack Business Service, Zürich, Switzerland, p. 47.
218. M. Heinze, *Kunststoffe 83*(8):630 (1993).
219. R. Kasoff, *Paper, Film Foil Converter 9:*85 (1989).
220. PET Clearyl, Data Sheet, Clearyl Products, Germany.
221. S. E. Krampe and C. L. Moore (Minnesota Mining and Manuf. Co., St. Paul, MN), EP 0202831A2/26.11.1986.
222. T. Matsumoto, K. Nakmae, and J. Chosokabe, *J. Adhes. Soc. Jpn. 11*(5):249 (1975).
223. A. Zosel, *Adhäsion 3:*16 (1986).
224. D. K. Das, R. N. Datta, and D. K. Bash, *Kaut. Gummi Kunstst. 1:*60 (1988).
225. *Kaut. Gummi Kunstst. 40*(1):40 (1987).

226. Primacor Hotmelt Adhesive Resin, 5980, Technical Information, Hot Melt Adhesives, August 1992, Dow Europe.
227. I. Benedek and L. J. Heymans, *Pressure-Sensitive Adhesives Technology*, Marcel Dekker, New York, 1997, Chapter 7.
228. *Adhäsion 9:*19 (1984).
229. *HamLet*, Polyethylene 93, Oct. 4, 1993, Session III, Maack Business Service, Zürich, Switzerland, p. 56.
230. *Polyethylenes, Copolymers and Blends for Extrusion and Coextrusion Coating Markets*, Polyethylene 89, Maack Business Service, Zürich, Switzerland, 1989, p. 107.
231. I. C. Petrea, *Structura Polimerilor*, Ed. Didactica si Pedagogica, Bucuresti, 1971, p. 75.
232. *Adhäsion 7:*242 (1972).
233. N. E. Jung, *Automobiltechn. Z. 90*(3):99 (1988).
234. H. Kamusewitz, Die thermodynamische Interpretation der Adhäsion unter besonderer Beachtung der Folgerungen aus der Theorie von Girifalco und Good, Dissertation, Halle, 1988, in G. Pitzler, *Coating 6:*218 (1996).
235. G. Pitzler, *Coating 6:*218 (1996).
236. Röhm, Technische Information, Produktbeschreibung, Plexiglas GS, XT, Ke. No. 212-1, November 1993.
237. W. Kahle, *Pack Rep. 6:*87 (1991).
238. K. H. Kochem, H. U. ter Meer, and H. Millauer, *Kunststoffe 82*(7):575 (1992).
239. F. Shikata (Teijin Ltd.), Jpn. Patent 63105187/10.05.1988, in *CAS Adhes.* 22:7 (1988); *Chem. Abstr. 109:*130717.
240. J. Pernak, A. Skrzypczak, E. Gorna, and A. Pasternak, *Kunststoffe 77*(5):517 (1987).
241. R. C. Rombouts, Speciality Plastics Conference, Dec. 1–4, 1987, Maack Business Services, Zürich, 1987, p. 401.
242. M. Narkis, *Mod. Plast. Int. 6:*28 (1983).
243. T. Kawagushi, T. Nogami, and K. Nei (Shin-Etsu Polymer Co., Tokyo, Japan), U.S. Patent 4624801 (Nov. 25, 1986).
244. S. Maeda and T. Miyamoto (Sumitomo Bakelite Co. Ltd., Tokyo), U.S. Patent 5456765 (Sept. 13, 1994), in *Adhes. Age 5:*15 (1995).
245. T. Wakabayashi and S. Sugii (Minnesota Mining and Manuf. Co., St. Paul, MN), EP 0285430B1/05.10.1988.
246. U.S. Patent 3686107, in T. Wakabayashi and S. Sugii (Minnesota Mining and Manuf. Co., St. Paul, MN), EP 0285430B1/05.10.1988.
247. L. A. Sobieski and T. J. Tangney, *Adhes. Age 12:*23 (1988).
248. *Adhäsion 5:*211 (1967).
249. D. M. Bigg, *J. Rheol. 28:*501 (1984).
250. K. Baricza, *Müanyag Gummi 24*(11):329 (1987).
251. D. L. Parris, L. C. Burton, and M. G. Siswanto, *Rubber Chem. Technol. 60*(4):705 (1987).
252. T. Slupkowski, *Int. Polym. Sci. Technol. 13*(6):80 (1986).
253. K. Miyasaka, *Int. Polym. Sci. Technol 13*(6):41 (1986).

4
Chemical Basis of PSPs

As discussed in the previous chapter, pressure sensitivity supposes the existence of pronounced flow during bonding and elastic, solid-state-like behavior during debonding. Elasticity requires a reversibly deformable network. The existence of a network does not allow viscous flow. This dilemma can be avoided by synthesizing and formulating viscoelastic compounds that have partially segregated structures. Therefore the chemical basis of PSPs includes raw materials with composite structures. As mentioned earlier, intramolecular or intermolecular segregation is achieved with a crosslinked network, crystalline structure, or reinforcement with fillers. The nature of segregated structures determines their mobility and elasticity. A simplified approach explains the functioning of such networks by comparing them with filled liquids (see Chapter 3). It should be emphasized that this approach assumes that the filler particles are more mechanically resistant than the liquid and that they do not interact with the matrix. This is not always true. A second hypothesis concerns the elastomer. Rubber is considered an ideal network for use as a base elastomer. But rubber-based formulations used for PSPs are tackified and crosslinked; their networks cannot be ideally elastic. Nevertheless, such formulations work well in practice.

"Diluting" the segregated structure with viscous components also allows plastomers to experience pronounced flow. Thus the first requirement for pressure sensitivity is fulfilled. The second requirement is the ability to undergo large elastic deformation. How can that be achieved? To answer this question, the chemical basis of PSPs must be examined. Except for some special materials (e.g., metals, ceramics, glass) used as carrier, the main raw materials of PSPs are organic compounds, mostly polymers. To understand how PSP components work, let's make a short evaluation of their macromolecular basis and chemical composition. Because of the quite different roles of the solid-state carrier material and the liquid adhesive (at least for the classic PSP construction), the raw materials of these components are also different. Therefore, the carrier material and the other components are discussed separately here. The chemical basis of PSAs is described in detail in Ref. 1. These adhesives are often considered the components responsible for pressure sensitivity, which is true for labels. For other products, the PSA–carrier assembly or (in extreme cases) the carrier itself is the component that displays pressure sensitivity. In some cases such behavior can be compared with the "true" pressure sensitivity of the PSA (see also Chapters 5 and 8). For certain products the adhesive properties result from the application conditions only. A detailed examination of the common macromolecular basis of all components of the pressure-sensitive assembly is necessary for its understanding, design, and manufacture.

1 MACROMOLECULAR BASIS AND CHEMICAL COMPOSITION OF CARRIER MATERIAL

The carrier materials used for PSPs have to display special mechanical characteristics (see Chapter 3). Therefore they are manufactured on the basis of high molecular weight polymers. At first paper was used as the carrier; later, organic and inorganic non-paper materials were used. Paper and the main polymeric carrier materials have chemical compositions and macromolecular structures that do not allow enhanced chain mobility at the surface, i.e., surface adhesivity. Paper is a unique carrier based on natural raw material (cellulose derivative) and has its own chemistry and technology. It is not the aim of this book to discuss them. It should be mentioned only that paper contains other inorganic or organic macromolecular compounds, usually as a top coat. The surface layer of top-coated papers is based on about 90% white pigment and 10% binder [2]. The mean coating weight of the top coat is about 15 g/m^2, i.e., the layer has an average thickness of 12 μm. Caseine, starch, polyvinyl alcohol, acrylics, or styrene-butadiene latex is used as binder. The printing and adhesion properties of the top coat depend on its chemical composition.

The development of plastics allowed the manufacture of "full plastic" carrier materials with better mechanical properties. Both paper and common plastic films are nonadhesive carrier materials. Their behavior is due to their partially or-

Chemical Basis of PSPs

dered structure. Although ordered and crosslinked, natural rubber exhibits slight self-adhesion. Its self-adhesivity can be improved by transforming (by tackifying) the crosslinked elastic structure in a mobile construction that exhibits elastic deformation and viscous flow (due to enhanced chain entanglement and molecular motion). As discussed in Chapter 3, the buildup characteristics of the crosslinked network regulate its mobility. The "conformational gas" of ideal rubber with long elastic bridges between the crosslink points allows a high degree of deformation. Crystalline, ordered structures generally do not allow rubberlike high elastic deformations. Their flow occurs in their amorphous domain. Therefore the deformational behavior of plastics is different from that of rubber. Rubber can act as an adhesive; a plastomer cannot (at room temperature). For classic PSP constructions the plastomer is the nonadhesive carrier and rubber is the adhesive component.

However, reducing crystallinity and enhancing chain mobility in the amorphous region can lead to more elasticity and more plasticity for such materials also. The empirics of tackification of natural rubber using liquid resins and the plasticization of cellulose derivatives or PVC demonstrated the possibility of transforming an elastomer or a plastomer into a viscoelastomer (Fig. 4.1). Internal plasticizing competes with external plasticizing. The synthesis of vinyl chloride copolymers, ethylene-vinyl acetate copolymers, propylene copolymers, and branched polyethylenes led to macromolecular compounds with greater chain mobility. Such compounds exhibit viscoelastic flow and are self-adhesive. It should be emphasized that there is a big difference between PSAs and plastomers. The main characteristic of PSAs is their flow; their mechanical strength is generally not sufficient for them to be self-supporting. In some cases their cohesion is not sufficient to ensure acceptable shear resistance. The main characteristic of plastomers is their excellent mechanical resistance; their flow is weaker and allows self-adhesion only with high temperature or pressure lamination.

Various industrial methods can be used to manufacture bulky self-adhesive products (Table 4.1). Synthesis, formulation, and application conditions together ensure their pressure-sensitive adhesivity.

The inverse possibility to stiffen the rubbery network has been studied also. The crosslinking of natural rubber or of viscoelastic compounds such as acrylates made it possible to increase the modulus and to develop carrierless transfer tapes. On the molecular scale, the buildup of segregated structures in "liquid" macromolecular compounds allowed temperature-dependent molecular stiffening, the development of thermoplastic elastomers. Thermoplastic elastomers exhibit rubberlike elasticity at room temperature but behave like thermoplastic resins above their melting point. Such compounds are free of curing agents, accelerators, and reinforcing materials, which reduce the amount of mixing required as well as the amount of materials.

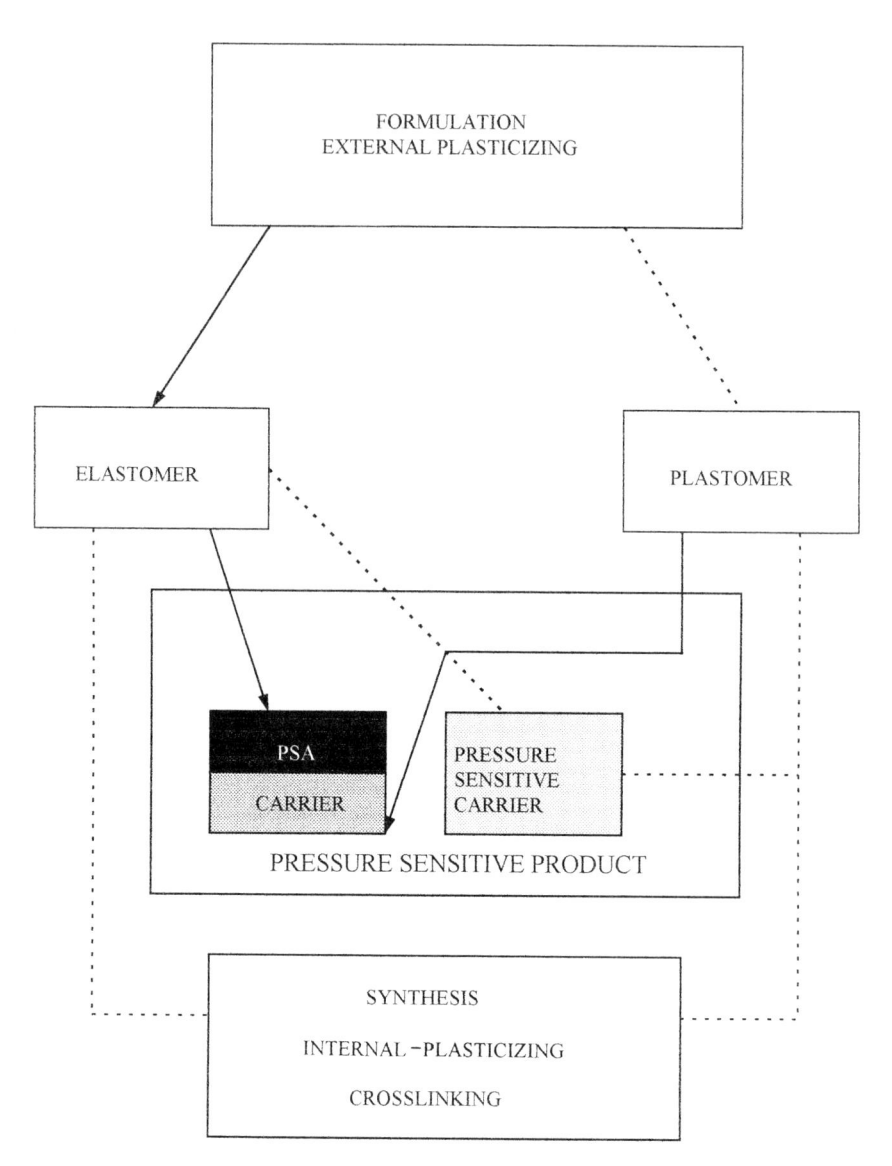

Figure 4.1 Chemical methods used to obtain pressure-sensitive products.

Chemical Basis of PSPs 153

Table 4.1 The Main Manufacture and Application Parameters for Pressure-Sensitive Behavior

Product buildup		Application	
Base component	Additional component/parameter	Product	Working mechanism
Plastomer	Physical treatment; laminating conditions	Hot laminating PE film	Enhanced surface polarity and flow
Plastomer	Viscous additives	Plasticized PVC protective film	Enhanced conformability (lower T_g)
Plastomer	Viscoelastic additives	Polyethylene film tackified with isobutylene copolymer	Surface layer of tacky, viscoelastic compound due to incompatibility
Plastomer	High molecular weight viscoelastic additives; laminating conditions	Polyethylene film tackified with ethylene-butyl acrylate copolymer	Surface layer of tacky, viscoelastic high polymer
Plastomer	Elastic, polar, low T_g polymer as additive; physical treatment	Polyolefin blends with EVAc copolymers	Bulky viscoelastic properties of compatible blends
Plastomer	Physical treatment; laminating conditions	Very low density polyethylene	High conformability due to amorphous structure
Elastomer	Tackification	PSA	Viscoelastic compound
Viscoelastomer	—	PSA	Viscoelastic compound

1.1 General Considerations

The mechanical characteristics and viscoelastic behavior of PSPs depend on their chemical basis (chemical composition and macromolecular characteristics). As is known from work with PSPs, the molecular weight of the adhesive components, coating weight, carrier thickness, debonding angle, and force influence the peel (debonding) resistance [3]. Molecular weight, molecular weight distribution (MWD), even shape and structure (short-chain branching frequency, distribution, and length and long-chain branching and unsaturation), affect the viscoelastic behavior of macromolecular compounds.

Parameters of Molecular Construction

The main parameters that can be manipulated in the polymerization process are the choice of monomers, molecular weight, molecular weight distribution, short-chain branching, long-chain branching, and unsaturation. For instance, common Ziegler–Natta (ZN) catalysts lead to PP with M_w/M_n ratios of 5–6, and special catalysts allow M_w/M_n ratios of 3–4; using the technology of controlled rheology (oxidation in the extruder), M_w/M_n ratios of 2–3 can be obtained [4]. These parameters allow the buildup of multiphase, ordered structures, which are characteristic of rubbery and plastomeric materials also.

Monomers. The most common carrier materials are based on polyolefins. They are generally nonpolar, chemically inert, partially crystalline plastomers. Among the structural characteristics due to the monomers, crystallinity is the most important for mechanical properties and surface adhesivity. To improve their surface wetting properties and elasticity, copolymers of polar monomers (vinyl acetate, acrylics, and maleic derivatives) have been developed. Copolymerization of ethylene with polar vinyl and acrylic derivatives gives rise to polar copolymers whose properties differ greatly from those of LDPE [5]. The term "ethylene copolymers" is restricted to polymers containing more than 50% ethylene. Polymer-analogous reactions have also been used to obtain new copolymers. One such method is grafting. Copolymers of ethylene with vinyl acetate and butyl acrylate and acid-ethylene copolymers have been manufactured to achieve self-adhesion (i.e., low temperature sealability). The comonomer provides polarity in the chain and decreases the crystallinity of the material. Both favor the wettability of the substrate. At the same time its melting point decreases [6]. Butyl acrylate decreases the crystallinity of polyethylene, increases tack, and provides good mechanical properties. Maleic anhydride increases the adhesion on polar surfaces and allows initiation of covalent chemical bonding with some polymers. Terpolymers of ethylene acrylic ester and maleic anhydride have also been synthesized. Ethylene-acrylic acid (EAA), ethylene-maleic anhydride (EMAA), ethylene-butyl acrylate (EBA), and ethylene-methyl acrylate copolymers have been produced [7]. Because of their lower crystallinity these are very low modulus polymers (the value of the flexural modulus of such polymers is situated at 8, respectively 18 MBA). Consequently they also have a greater degree of tack. The presence of functional groups in ethylene copolymers causes them to be more chemically reactive than LDPE. Ethylene-vinyl acetate copolymers have a processing temperature limitation of approximately 230°C. At higher temperatures, acetic acid can split off from the functional group. Ethylene-acrylic acid copolymers can be processed at 330°C [8].

Chemical Basis of PSPs 155

The introduction of a functional monomer into conventional elastomers is a common strategy for altering physical characteristics or designing the macromolecules to fulfill a particular function. The carboxylic acid group is reactive with many agents such as amine and epoxy compounds, which can be used for covalent crosslinks.

Molecular Weight and Molecular Weight Distribution. Higher molecular weight polymers allow the manufacture of stronger films at a penalty in processibility. As known, the tensile strength (S_T) depends on the number average degree of polymerization (DP_n) according to the correlation [9]

$$S_T = S_{T\infty} - C/DP_n \qquad (4.1)$$

where C is a polymer constant. Unfortunately, with high molecular weights, excessive pressures and extrudate melt fracture become problems [10]. Low molecular weight polymers allow better hot seal performance. As stated in Ref. 11, a hot seal PSA formulation may include polyethylene (10–50% w/w) with a (low) molecular weight of 1000–10,000.

Blown film equipment designed optimally for the processing of one kind of polyethylene is generally not suitable for processing other types of polymers. A polymer with a broader MWD will be more processible. On the other hand, plate-out and blocking occurs with low molecular weight species. Linear low density polyethylene sometimes has more low molecular weight species and requires more antiblock additive. Molecular weight and molecular weight distribution influence melt strength and processibility [12]. To avoid blocking with blown HDPE, the best results are obtained by using high blow-up ratios and high frost kine heights; therefore, very broad and preferably bimodal grades are needed [13]. For flat films (monofilaments and tapes), HDPE with a narrow molecular weight distribution is used [14]. Good mechanical properties during and after stretching are the main requirements for these products. For blown film manufacture, a medium MWD is recommended, but for higher melt strength very large MWD grades may be needed too. The blown film is applied for tapes also. Here mechanical strength and elongation are required. On the other hand, broad MWD and low MW (high melt flow index) can result in plate-out and blocking of the film and poorer mechanical properties. Bimodal MWD ensures a combination of excellent mechanical properties and good processibility. Bimodal HDPE can be blended with PP also.

When distinctly different molecular weights are present, the material is described as bimodal. Such materials combine optimum processibility and excellent mechanical properties. The broad molecular weight HDPEs have low melt memory, so the die must be designed to allow time for any melt strain to disappear [13] (see also Chapter 6). The MWD has no effect on the stiffness, which depends only on density [12]. At low film thicknesses and low blow-up ratio, the broad molecular weight distribution increases the dart drop impact strength of the film (see also Chapter 3, Section 2). For special mechanical characteristics, ultra-high molecular weight PE is used [15].

The degree of polymerization and the molecular weight of the plastomers or elastomers used as carrier materials and the additives used to transform them into self-adhesive carriers play an important role in the melting point depression and spherulite growth rate. The melting point depression (ΔT_m) of a crystallizable polymer blended with an amorphous polymer in a compatible mixture according to the Flory–Huggins theory [16,17] can be written as a function of the heat of fusion (ΔH^0), molar volumes of the repeat units of the two polymers (V_1 and V_2), degree of polymerization (m_1, m_2), and volume fractions (Φ_1, Φ_2) of the polymers:

$$-\left[\frac{\Delta H^0\, V_1}{RV_2}(\frac{1}{T_m} - \frac{1}{T_m^0}) + \frac{\ln \Phi_2}{m_2} + (\frac{1}{m_1} - \frac{1}{m_2})\Phi_1\right] = \chi_{12}\Phi_1^2 \quad (4.2)$$

where χ_{12} is the interaction parameter of the components of the mixture. As can be seen from the above correlation, the molecular weight (degree of polymerization) strongly influences the melting point depression. Melting point depression has special importance for hot laminating films, self-adhesive films based on EVAc–PE mixtures or PIB–PE mixtures, and resin mixtures (see also Chapter 6). On the other hand, the volume fraction of the polymers, Φ (which depends on their structure, i.e., branching), and the diffusional processes (which depend on the thickness of the macromolecular layer, b_0) influence the spherulite growth rate (G) of the crystallizable polymer in a mixture according to the correlation [18]

$$\log G - \log \Phi_2 + \frac{U^*}{2.3R(T_c - T_\infty)} - 0.2\, T_m \frac{\ln \Phi_2}{2.3\, \Delta T} = \quad (4.3)$$

$$\log G_0 - \frac{k_g}{2.3 T_c\, \Delta T\, f}$$

where G_0 is a pre-exponential factor, ΔT is the undercooling, $U^*/R(T_c - T_\infty)$ represents the contribution of diffusional processes, f is a correction factor for the heat of fusion (ΔH_0), and k_g is the nucleation factor, which depends on the thickness of the layer:

$$k_g = f(b_0) \quad (4.4)$$

Spherulite growth influences the diffusion of viscous components in self-adhesive films and the decrease of bond performance in semi-pressure-sensitive adhesives (see Chapter 8).

Branching. Long-chain branching provides shear sensitivity, i.e., high processibility [19]. Low density polyethylene is highly branched, gives much entanglement, possesses a broad molecular weight distribution, and displays poor mechanical properties. In the early 1970s, linear low density PE was introduced, with no long branching, little entanglement, narrow MWD, and improved mechanical properties [20]. The entanglement of the short-chain branches (e.g., LLDPE with octene) provides excellent stretchability and flexibility [21].

Segregation

As discussed in Chapter 3, segregated, multiphase structures allow the combination of viscous and elastic, rubberlike and plastomerlike characteristics. In plastomers, multiphase systems are due to crystallinity. The physical status of these polymers is characterized by their orientation, crystallinity, and crystal particle dimensions [8].

Crystallinity. Thermoplasts are generally crystalline materials at their application temperatures. Their excellent mechanical properties are due to crystallinity [22]. Tensile modulus and stiffness decrease as crystallinity decreases [23]. Crystallinity gives the polymer a certain rigidity, and the molecular packaging produces opacity due to the differences in the diffraction indices of the amorphous and crystalline parts. Polyethylene films are either blown or cast. Because of the different degrees of crystallinity of the two types of film, different tackifier loadings are used to achieve the same level of self-adhesivity (see also Chapter 6). For cast films, a level of 1–3% polyisobutylene (PIB) is required, and for blown films a level of 3–6.5% PIB [24].

As mentioned previously, crystallinity also influences the optical properties. LLDPE forms large crystallites that result in surface irregularities and poor optical properties. This can be improved by making the crystallites smaller by using faster cooling. Crystallinity influences density and the modulus. Common HDPE has densities of 0.952–0.960 g/cm^3. Materials of higher density are used when an optimum modulus (rigidity) is required. On the other hand, the impact resistance of C_8 LLDPEs is better. It increases with the MFI and decreases with density. Elmendorf tear strength increases with the MFI and decreases with density. Short chains along the molecules reduce the material's ability to crystallize, and so the more short chains there are, the lower the density.

The formation of nuclei of crystallization is a function of temperature. As the temperature decreases from the melting point, the kinetic effects become smaller and the chances for larger nuclei to form are greater. As the temperature is lowered, the average nucleus size increases and the critical size for crystal growth decreases. For most polymers there is an optimum temperature for crystallization. Polymers may exhibit one or more crystalline transitions at temperatures between T_g and the final melting temperature. The importance of such transformations is due to their possible effects on modulus and thermal, optical, and dimensional properties. Low MW chains possess higher mobility at a given temperature and therefore are more easily crystallized. The crystal grows slowly with the formation of high MW chains with restricted mobility. Plasticizers aid crystallizability; fillers (solid state) have the opposite effect. Low crystallinity may extend the rubbery plateau up to the melting point of the crystallites, giving a much broader application temperature (see Chapter 3, Fig. 3.3). Medium crystallinity (e.g., LDPE) makes the material less flexible; high crystallinity (HDPE, PET) provides stiffness and temperature resistance. Bulky functional groups like rings have a stiffening effect (see SBCs, PET), retaining the glassy plateau up to higher temperatures. Crystallinity due (partially) to incompatibility is displayed by organic antiblocking agents also. Such additives are organic substances incompatible with the polymer carrier. During extrusion these materials tend to create spherulites on the surface of the films, which act as antiblocking agents [25].

Incompatibility between the carrier and an embedded additive related to molecular weight and structure can be used to build up "working" constructions, where the incompatible additive is released from the carrier material. The best known examples are self-adhesive films (SAFs) with embedded tackifying components. Another example is described by Brack [26]. The release agent is incorporated as an incompatible compound in the release binder. This film is cured to a flexible solid layer by irradiation. The abhesive substance migrates to its surface. In a similar manner, combining fluoropolymers with polyamic acid leads to antiadhesive coatings with improved performance, due to the spontaneous separation into two layers, the bottom layer being polyamic acid and the top layer fluoropolymer [27].

Stretching, orienting the macromolecule (see Postextrusion Processing), produces crystallization and improves the mechanical performance of the film. This behavior is exhibited by oriented PET, PP, and PVC [28]. The influence of crystallization on the adhesion of a plastic film is more complex. The proportion of the crystals and their morphology influence the degree of contact and adhesion [29]. Using common industrial methods, it is not possible to manufacture plastic films without inducing order in the macromolecules. This is due to the big differences in the mobility of the macromolecules in molten and "frozen" material. Some raw materials for adhesives are known to exhibit partially plastomerlike structures that become ordered when cooled. These include styrene-diene block copolymers. When they are cooled, rigid polystyrene domains are formed. In this situation, at least theoretically, it would be possible to achieve oriented, ordered molecular structures by high shear, high speed cooling of such materials. This phenomenon was studied by de Jager and Borthwick [30]. To test whether shear-induced orientation of a styrene-diene block copolymer based HMPSA occurs, they measured the shear adhesion failure temperature (SAFT) and rolling ball tack of the films manufactured on samples cut parallel and perpendicular to the machine direction. They did not found shear-induced changes of the adhesive properties. Their results demonstrate the big differences between the rigidity of

Chemical Basis of PSPs 159

the structural order in partially crystalline plastomers and that of thermoplastic elastomers.

Other Structures. Multiphase structures can be built up in plastomers by crosslinking, fillers, or special synthesis. Heterophase polypropylene copolymers have been synthesized that have a continuous polypropylene (PP) matrix and a discontinuous ethylene-propylene rubber (EPR) phase. Such polymers exhibit tensile moduli of 300–1300 N/mm^2 [31]. Traditional thermoplastic rubbers have heterophasic structures and undergo thermally reversible interactions among the polymeric chains. Thermoplastic olefinic rubbers, generally consisting of polypropylene/ethylene propylene elastomer blends, show only weak interactions among the polymer chains; they are not able to develop suitable elastic properties within a wide temperature range. By grafting PP onto EPR or introducing polar groups into the PP and EPR, these interactions have been strengthened. Heterophasic systems have been developed in which PP is the continuous phase and crosslinked ethylene-propylene rubber is homogeneously dispersed in the PP matrix [32].

In rubber resin PSAs, two-phase systems were found by Wetzel [33] and Koch [34] in which the resin was distributed in the rubbery matrix. Later such two-phase systems were synthesized in the same polymer by using block copolymerization. Block copolymers are multicomponent systems that exhibit phase separation on the microscale. Such systems can be manufactured using either common polymerization techniques or special methods. Polymerization of a gaseous monomer in an unswollen, noncrosslinked matrix, crystallization or precipitation of low or high molecular weight materials in situ of polymer systems, and polymerization of a viscous monomer or macromer in the presence of dispersed additives have been described as methods for obtaining segregated polymers [35]. Well-known products include styrene block copolymers. For such polymers the effect of the polystyrene domain is similar to that of vulcanized rubber crosslinks in anchoring the network structure, but in addition it is effective in raising the rubbery modulus of the network by providing a perfectly adhered, hard particle reinforcement. The strength and stiffness of block copolymers depend on a variety of factors, including the chemical nature of the elastomer blocks and the interactions between the different block segments (see later).

1.2 Nonadhesive Carrier Material

As discussed above, in a classic construction and for the major PSPs, the carrier material must ensure the solid weblike character of the product. Therefore it is not adhesive. Nonadhesive carrier materials include natural polymers (paper, cellulose derivatives), synthetic polymers (polyolefins, olefin copolymers, polyesters, polyvinyl chloride, polyurethanes, etc.), metals, and composites with a film-like or textured structure. The best known of them is paper.

Paper-Based Carrier Materials

Paper-based carrier materials have been used since the beginning of the production of PSPs. At first, labels and tapes used paper as the carrier material because it was the best known and most widely available packaging and information-carrying material. The disadvantages of paper are due mostly to the sensitivity of its base polymer, cellulose, to humidity. Paper-based carrier materials differ in paper quality, geometrics, and surface properties. The development of laminating technology led to the use of homogeneous paper/paper and heterogeneous paper/plastic and paper/metal constructions as adhesive carrier materials also. Paper carrier material is used as the face stock and release liner for the main pressure-sensitive product classes. Polymer-impregnated paper can be used for tapes [36]; metallized paper and film/paper laminates are used for labels [37].

Polymeric Carrier Materials

The development of film-forming polymers allowed the use of plastic films as packaging materials. Plastic films were first applied as a homogeneous monolayer, but eventually heterogeneous (plastic/plastic, plastic/paper, plastic/metal, etc., composites) and multilayer constructions were developed.

Homogeneous Plastic Carrier Materials. The use of common packaging films as carrier materials in the early days of PSP manufacture was possible due to the relatively low requirements concerning their mechanical properties and processibility and to the ability of polyvinyl chloride (the best known plastic film carrier material) to be plasticized and thus to allow quite different mechanical properties as a homogeneous, relatively isotropic, polar (coatable) monoweb material.

As mentioned, some years ago polyvinyl chloride was the most common polymeric packaging and carrier material [38]. Labels, tapes, and protective films were manufactured with PVC as the carrier in the form of film or foamlike webs. Both soft (plasticized) and hard forms of PVC were used [39]. Polyvinyl chloride displays excellent mechanical properties. Through plasticizing it has been possible to regulate its mechanical properties, hardness, and conformability so that it can be used as a homogeneous, isotropic, plastic or elastic monolayer material. It also has good thermal resistance. Due to its surface polarity it displays adequate printability. Homo- and copolymers of vinyl chloride (with vinyl acetate, vinyl propionate, vinyl ether, or acrylic ester) have been synthesized [40]. Extruded PVC is the most often used face stock material for screen printing. Therefore PVC has been used for the major PSPs as a universal carrier material. Polyvinyl chloride has been proposed as face stock material for weather-resistant decals [39], for labels in pharmaceutical and electrical use [41], and for tamper-evident labels [42]. Various tapes with a PVC carrier have also been manufactured [43–45]. Polyvinyl chloride foam is used for double-coated tapes [46]. Soft PVC is applied as a carrier material for decorative films [47]. Tapes for freezer use are manufactured with a soft PVC carrier, which needs a primer coating for

Chemical Basis of PSPs

the rubber-based PSA [48]. PVC with silicon carbide as filler is used for electrically conductive tapes [49].

Environmental and economic considerations led to the introduction of polyolefins as carrier materials—first polyethylene and later polypropylene and olefin copolymers [50]. Polyolefins are the most important homogeneous (monolayer) materials used as carrier films. Polyethylene is the most common carrier material for protective films. Recycling is obviously much easier if the materials used for packaging are homogeneous. Therefore in label manufacturing there is a desire to use, to the extent possible, the same base materials for the labels as are used for the products to be labeled. In the early days of plastic carrier development, polyolefins were used as face stock material; later they were used for release liners also. It should be mentioned that the use of nonpolar polyolefin films as adhesive carrier materials sometimes enables PSPs to be constructed without a coated release liner.

It should be mentioned that the introduction of polyolefins as filmlike carrier materials led to diversification of carrier materials having the same chemical composition. This was made possible by the synthesis of polyolefins that had almost the same monomer basis but were polymerized by different methods (see Chapter 6). Later different film manufacturing methods (blowing/casting) and last but not least the orientation of the film (monoaxial or biaxial) allowed further diversification of the product range.

Different grades of polyethylene have been synthesized and applied. Various ethylene polymers are used in films. Low density polyethylene; linear low density polyethylene; high density polyethylene; high molecular weight, high density polyethylene (HMW-HDPE); medium density polyethylene (MDPE); ethylene copolymers; and functionalized ethylene co- and terpolymers are all used in single- and multilayer applications. Choices within polymer types include melt index, MWD, density, and comonomers (C_4, C_6, C_4, MeP-1, C_8, etc.).

Low density PE is a semicrystalline thermoplastic material with a complex chemical structure formed by a repetition of $-CH_2-CH_2-$ groups with some ethylenic side chains. The number and length of the side chains determine the degree of linearity of the structure and therefore the crystallinity of the polymer. The addition of a second bulky or polar monomer, e.g., VAc, destroys the regularity ot the chains and changes the forces existing between them, preventing the free rotation of the segments of the chain around the carbon–carbon bonds and increasing the distance between the chains.

In LLDPE the built-in voluminous side groups do not allow the growth of crystallites. Thus a mainly amorphous structure is formed, having some crystallites also. The amorphous part of the polymer increases the impact resistance of the polymer. The optical properties of LLDPE films and their shrinkage in the cross direction are the main disadvantages. Linear low density polyethylene has a higher melt viscosity than crosslinked LDPE and is difficult to process [51]. It has a melting point situated 10–15 K higher than that of LDPE [52]. Very low density polyethylene (density under 0.918 g/cm^3) was used first together with LLDPE.

New ethylene-octene copolymers with a statistically well defined polymer structure at densities as low as 0.860 g/cm^3 (compared to 0.910 g/cm^3 using conventional technology) and having a fractional melt index (FMI) of more than 100 have been developed using special single-site metallocene catalysts. These are linear polymers that possess a narrow MWD and a large number of long branches. Long-chain branching provides shear sensitivity, i.e., high processibility [53].

New so-called metallocene resins have been synthesized [54] that display a combination of elastomeric and thermoplastic properties. These polymers, with a density of 0.895–0.915 g/cm^3 and a narrower MWD than conventional LLDPE, give films with better tensile strength, excellent puncture resistance, and improved optical quality, but they are difficult to process because of their higher viscosity, lower melting point, and lower melt strength. Due to the presence of long-chain branching, these polymers will process at slightly lower pressure than LLDPE. They are free of low molecular weight oligomers which can produce chill roll plate-out in the manufacture of cast film. They can be processed with conventional equipment for standard LLDPE; however, the die design and downstream equipment could make a big difference. A barrier screw with a grooved feed section, as for LLDPE, should be used. The features of the equipment are yet not finally determined. Metallocene polymers give a tacky film. Very low density polyethylene with a density of less than 0.915 g/cm^3 is not crystalline, is flexible, and exhibits autoadhesion.

Polyethylene is used as film, foam, and fiber-based material or as a laminating component of heterogeneous carrier materials. The following critical characteristics are considered in evaluating polyethylene for various applications: melt index (which is inversely related to MW), density, MWD, and degree of side branching [55]. Bimodal (bimodal distributed molecular weight) HDPE grades allow HDPE levels to be increased in mixtures with LDPE up to 30% and exhibit greater stiffness.

Labels, tapes, and protective films are manufactured with polyethylene carriers. Polyolefin films can also be used for release liners [56]. Single-sided [57,58] and double-sided coated polyethylene tapes have been designed for polyethylene bag manufacturers [59]. Polyethylene foam is suggested for tapes also. Double-coated, crosslinked polyethylene foam tapes have been manufactured [60]. Polyethylene is proposed as carrier for textile labels [61] (see also Chapter 6).

Polypropylene has the lowest density of any commercial plastic material. PP materials can be nonoriented, mono-oriented, or bioriented [62]. Oriented PP film (30–50 μm) is applied for packaging tapes, and oriented LDPE (80–110 μm) for insulation tapes [63]. Bioriented PP is based on cast coextruded film. Polypropylene is used for common tapes [64] and extensible tapes [65]. As a car-

Chemical Basis of PSPs

rier material for tapes, polypropylene can have a silicone release coating [66]. Polypropylene is also used as carrier material for labels for plastic bottles and containers [67] (see also Chapter 8). Copolymerization of PP reduces its tacticity and the mean length of stereoregulated sequences, thus decreasing its melting point [68]. Random and block copolymers have been synthesized. Copolymers with ethylene and butene are more flexible [69]. Special polymers that can be calendered have been manufactured. The raw material has to be ductile, antiblocking, with low depolymerization, low melting point, and adequate MWD. Calendered PP has a higher stiffness modulus, better cuttability, and no slip. Fabrics and nonwovens based on PP are extrusion-coated with EMAA or EVAc [70].

The comonomers for ethylene can be considered as plasticizing (e.g., vinyl acetate, methyl acrylate, isobutyl acrylate, ethyl acrylate, and *n*-butyl acrylate), polar (acrylic acid, methacrylic acid, and maleic anhydride), and reactive (acrylic acid, methacrylic acid, maleic anhydride, and monoethyl maleate) [71]. Such copolymers have the general formula

$$-[-CH_2-CH_2-]_x-[-CH_2-CH-]_y-[-CH-CH-]_z- \quad (4.5)$$
$$|||$$
$$RC=O\ C=O$$
$$||$$
$$OH\ OH$$

where R is vinyl acetate, methyl acrylate, ethyl acrylate, etc.; y has values of 0–40%, z has values of 3–20%, and M is C_1–C_{20}.

Polar copolymers of ethylene were produced in high pressure reactors more than 30 years ago. They can be extruded as conventional films and used as carriers or self-adhesive films. As discussed above, the crystallinity-related properties decrease with increasing polar monomer content. VAc copolymers have low modulus (see Table 3.10). They have been applied as a heat seal layer in extrusion coating and coextrusion. Fire-retardant compositions based on EVAc with special fillers, Al(OH)$_3$ or Mg(OH)$_2$, have been proposed [72]. Ethylene-acrylic copolymers are applied as a tie layer for oriented polypropylene (OPP) and PET film or as heat seal films [73].

Ethylene vinyl acetate copolymers with a VAc contents up to 42% have been grafted with styrene, vinylidene chloride, and styrene-maleic acid. Polyethylene and EVAc have been grafted with VAc. The VAc units have quite different types of influence on the main chain or branches. Short polyvinyl acetate branches are compatible with the amorphous phase of the partially crystalline system; longer PVAc branches build up another amorphous region [74]. EVAc copolymers having more than 70% VAc are polymer plasticizers for PVC [8]. Segmented EVAc has segments with high and low vinyl acetate contents [75].

As raw materials on a natural basis, used first in the packaging industry, cellulose derivatives have been used as carrier materials for PSPs also. Cellulose acetate and hydrate have been used for labels and tapes. Different types of cellulose hydrate (Zellglas) are known. These are generally hydrophilic films plasticized with polyhydroxy derivatives. Their humidity balance is very sensitive to drying conditions. Later top-coated qualities were manufactured to improve water resistance, weldability, and printability [76]. Cellulose acetate has better chemical and water resistance but is sensitive to the plasticizers in printing inks and to electrostatic charging. Clear, destructible acetate film can be used for tamper-evident labels [77]. Acetate films have been used as carrier materials for self-adhesive tamper-evident products. A 50 μm cellulose acetate film modified to be brittle and to have a low tear strength may be used as overlaminate, seal, or label as an alternative to PVC [78]. It is claimed to be useful for applications where labels are designed to cover an existing graphic element or product copy. Insulating tape with a cellulose acetate carrier is used for wire covering by telephone manufacturers. Tapes of cellulose hydrate or PVC have been designed for sealing boxes [44]. Polyvinyl isobutyl ether and polyvinyl alkyl ether can be used for medical tapes on cellophane [79].

As a material that has high thermal resistance, good mechanical properties, and good aesthetic quality, polyester was introduced mainly as a carrier for labels and as a release liner. Polyester is used as a carrier for labels for cosmetics, toiletries, pharmaceuticals, chemical products, and shrink sleeves [80]. Special tapes and protective films also use polyester as a carrier material. Double-faced mounting tapes based on PET are used for electronics products [81]. Polyester is the face stock material for labels for pharmaceutical and electrical applications [41] and for butt splicing of hard-to-stick materials. Metallized polyester has been used for labels and tapes [77]. Polyester fabric is proposed as a label face stock material in Ref. 82.

Polyurethanes are recommended as the carrier material for foam tapes and self-sticking clips [83]. A polyurethane self-sticking, double-backed foam tape should also be fire- and mildew-resistant.

Because of their excellent mechanical properties and thermal resistance, polyamides have been used as a carrier material, mainly for tapes. They are applied as film or as a reinforcing web built into a carrier film. For instance, pipe wrapping tapes and repair tapes are flexible plastic composites reinforced with nylon cord and have to adhere to surfaces of metal, wood, rubber, or ceramic. They are designed to be resistant to punctures and tears [41]. TPEs based on polyamides have also been manufactured. Block amide-ether amide (Peba) copolymers based on combinations of soft, flexible polymer chain segments of polyethers with a high melting point and stiff segments of polyamides possess a T_g of $-50°C$, which allows their use at low temperatures [84]. The polyamide/polyether ratio can be varied between 80/20 and 20/80. Therefore, flexible plastomers can also be obtained [85]. Polybutadiene-aromatic copolyamides give transparent flexible films [86]. The tensile strength and modulus of elasticity of such copolymers increases with the polybutadiene level.

Modified polyethyl methacrylate can be used as a protective film for different substrates such as PVC, ABS, polystyrene, wood, paper, and metals [87]. Polar macromolecular compounds manufactured by polycondensation or polyaddition, e.g., polyesters, polyamides, polyimides, and polyurethanes, are also used as carrier materials [88]. Polyimides have been proposed for transparent pressure-sensitive sheets coated with silicone-based adhesives. The pressure-sensitive laminate resists 8 h at 200°C [89].

Heterogeneous Plastic Films. Composite materials can also be based on plastic films. These are combinations of chemically or physically different polymers that are built together via adhesion. This adhesion is achieved technologically by lamination or coextrusion. In some cases other thermoplasts or duroplasts (e.g., polyurethanes) can be coated on the back side of the adhesive carrier material [90]. Films composed of polyethylene and polystyrene have been manufactured and used as carriers for labels.

1.3 Carrier Materials with Adhesivity

Adhesive carrier materials are self-adhesive films. A self-adhesive film combines the characteristics of a mechanically stable monoweb with the adhesivity of a liquid PSA, i.e., it is a noncoated face stock material that has surface adhesivity. It should be noted that for some carrier materials from this product range, which at normal temperature show a thermoplast-like behavior, surface adhesivity exists only at elevated temperatures or pressures or after a physical pretreatment of their surface. Such carriers are generally films based on nonpolar raw materials. Other carrier materials based on polar polymers exhibit surface adhesivity (with or without surface treatment) at high temperatures. Some carrier materials incorporate viscoelastic or viscous nonpolar or polar components that display surface adhesivity at room temperature.

Self-adhesivity of the carrier material is known from the practice of non-self-adhesive, i.e., adhesive-coated carrier materials too. In some cases, in order to increase adhesive anchorage on the carrier material, its adhesivity has to be improved. This obviates the need for primers; the adhesion of the carrier material is improved by the bulky inclusion of adhesive components. Biaxially oriented multilayer PP films that adhere well to the adhesive coating have been manufactured by mixing the PP with particular resins. Generally, in this case in order to prepare an adhesive tape that can easily be drawn from a roll without requiring an additional coating on the reverse side, at least two different layers of different compositions are coextruded, and the thin back layer contains an antiadhesive component [91]. As a tackifier, nonhydrogenated styrene polymer, α-methylstyrene copolymer, pentadiene polymers, α- or β-pinene polymers, rosin derivatives, terpene resins, and α-methylstyrene-vinyltoluene copolymers are preferred. In this case, self-adhesivity of the plastomer has been achieved chemically, i.e., by tackifying. Physical methods can also be used.

The self-adhesivity of plastic films is useful for other applications too. Biaxially oriented polypropylene (BOPP) films are produced with a thickness of 40–50 μm. Thick BOPP films cannot be manufactured economically with the common stretching processes. For these purposes, several thin films are combined (via heat lamination) to form a multilayer film [92]. As shown in Ref. 93, a polyolefin laminate can be manufactured by laminating together two corona-treated surfaces at a temperature that is lower than the softening temperature of the films.

Nonpolar Carrier Materials

Commercial polyolefins are used for the manufacture of common nonadhesive carrier materials. Special polyolefins can be applied as pressure-sensitive carrier materials. Their adhesivity is due to their chemical composition, to a special surface treatment, and to special application conditions. It should be noted that such materials are not self-adhesive in the classic sense. They behave like a self-adhesive material only at elevated temperatures (near their melting point) and at elevated pressures.

Very low density polyethylene films manufactured by casting (i.e., with a greater amorphous content) and having a low film thickness (conformability) display self-adhesivity at room temperature. Such films are used as cling films (see also Chapter 6). It is well known that VLDPE with a density of 0.885 g/cm^3 is self-adhesive. It is difficult to process for blown film because of its self-adhesivity. The self-adhesivity of polyolefins can be improved by the addition of viscous (nonpolar) and viscoelastic components such as polymers of isobutene. A cold-stretchable self-adhesive film is based on an ethylene–α-olefin copolymer (88–97% w/w) and 3–12% w/w of polyisobutylene, atactic polypropylene, *cis*-polybutadiene, and bromobutyl rubber [94]. The polymer has a density lower than 0.940 g/cm^3. The film exhibits an adhesive force of at least 65 g (ASTM 3354-74). Atactic polypropylene (2–9%) with a molecular weight of 16,000–20,000 has been suggested together with polyvinyl acetate (9–18%) and a tackifier resin for a removable adhesive composition [95]. The adhesion of polybutadiene as an SAF has been improved by modifying it with isopropyl azodicarboxylate [96].

Polar Carrier Materials

Some polar monomers known as adhesion promoters from the manufacture of common viscoelastic raw materials for PSAs can also be used for the synthesis of thermoplasts that display self-adhesion on certain adherend surfaces under well-defined conditions. These monomers are vinyl acetate and acrylics and are used as comonomers with ethylene.

Ethylene Copolymers. As discussed earlier, polar ethylene copolymers have been manufactured by copolymerization of ethylene with polar vinyl and acrylic derivatives. The upper limit for the polar comonomers is about 40% and is limited by the mechanical properties and manufacturing features of the film [5]. Aprotic comonomers (with ester, ether, anhydride, and oxyrane functional groups), protic comonomers (with hydroxy, carboxy, and amide groups), and ionic comonomers (salts) have been polymerized. Polar ethylene copolymers can also be produced by polymer-analogous reactions or grafting. According to Ref. 97, the polyolefin surface of the carrier can be polarized by graft polymerization using electron beam radiation with a dosage level greater than about 0.05 Mrad. Ethylene polymers grafted with a carboxylic reactant or ethylene-vinyl/acrylic copolymers are suggested for adhering propylene polymers to polar substrates [98]. Hydrolysis of EVAc copolymers and neutralization are commonly used to obtain polyvinyl alcohol and ionomers. Acrylic and maleic copolymers are grafted (via processing also) onto the PE backbone. Acid-functionalized polymers may undergo crosslinking. Neutralized ionic polar copolymers differ from acrylic ones by their resistance to solvents, toughness, and heat sealing properties owing to their ionic networks. Both are used as adhesion promoters. Copolymerization of ethylene with acrylic acid leads to softening of the polymer. As shown by Kirchner [99], the high modulus of HDPE (about 1600 kg/cm^2, ASTM 638-58 T) decreases to 430 kg/cm^2 for an EAA copolymer having 19% polar comonomer, and the softening point decreases to 54°C.

Ethylene-vinyl acetate copolymers contain two domains [100]; there is a crystalline region of polyethylene and an amorphous region of EVAc. As the level of VAc increases, the size of the amorphous region increases. The melting point and tensile strength decrease. A correlation has been developed that shows the interdependence of the comonomer content (mole %) and melting point of the copolymer [101]. The number of comonomer molecules is more important than their size. The copolymers of ethylene and vinyl acetate range from typically thermoplastic materials (similar to low density PE) to rubberlike products. The copolymerization of ethylene with vinyl acetate increases the density, clarity, permeability, solubility, environmental stress cracking resistance, toughness (especially at low temperatures), compatibility with other polymers and resins, acceptance of fillers, and the coefficient of friction. In contrast, rigidity, softening point, and surface hardness (modulus) are decreased. Such characteristics are related to film conformability and self-adhesion. As is known, self-adhesion is due to chain mobility (see also Chapter 3, Section 1). Taking as the molecular index for polymer flexibility the freedom of macromolecular chains to freely rotate, given as $(\lambda^2/\lambda_{r,i}^2)^{1/2}$, where λ is the chain length expressed as average quadratic length for a chain in theta solvents and $\lambda_{r,i}$ the chain length capable of rotation, the bonding abilities of different plastomers or elastomers can be compared [102]. As stated, rubber has the lowest value (1.5) for the above ratio compared with 1.8 for polypropylene [103] and 2.3 for polyvinyl acetate [104] or 1.7 for polyisobutylene [105,106]. In self-adhesive films the versatility of these materials is the following: natural rubber > polyisobutylene > polyvinyl acetate.

Ethylene-ethyl acrylate copolymers have been proposed for use as removable clear protective films on metals [107]. In comparison with EVAc, it is possible to reduce the formulating level (and the costs) of EEA by 3–35%.

EVAc copolymers have been used for a number of applications. In the early 1970s they were suggested for heat-activated adhesive bonding [108]. Copolymers of EVAc ensure sealing temperatures 40°C lower than the sealing temperatures for PE. Such polymers were tested for sealable packaging films also. Ethylene-vinyl acetate copolymers have also been used as sealants for PP. A polymer with more than 28% vinyl acetate can be sealed by high frequency. Copolymers having less than 5% w/w EVAc are used in thin films to regulate mechanical properties. Copolymers with 6–12% VAc are used to regulate mechanical properties, for tougher films at low temperatures, and for films with higher impact resistance and greater stretch properties. The sealing temperature of an EVAc copolymer film having 12% VAc is 130–150°C; it decreases as VAc content increases. The melting point of EVAc copolymers with 10–14% VAc content lies between 91 and 96°C [109]. Commercial EVAc types contain 10–40% VAc. A 32% VAc content in EVAc gives a partially crystalline polymer, and at 40% vinyl acetate a completely amorphous polymer is formed [110]. It is to be noted that their properties are strongly influenced by the molecular weight of the polymer and the number of side chains. Copolymers with 15–18% VAc are proposed mainly to improve the sealing properties and stress cracking resistance of polyolefins at low temperatures. Copolymers with 18–30% EVAc are suggested for adhesive and wax-based coatings. EVAc copolymers are also used as plasticizers or flexibilizers for PVC [111]. Copolymers of ethylene with vinyl acetate and hydroxy functional monomers are applied as the sealing layer in coextruded films [112]. Polymers with a level of 15% VAc possess PE-like properties. Polymers having 15–30% VAc display PVC-like properties; polymers with more than 30% VAc are elastomers [113]. Their elongation increases with VAc content, especially up to 15% VAc. The optimum tensile strength is achieved for a content of 20–30% VAc. The clarity of EVAc films increases with the VAc content; however, films with less than 15% VAc are generally manufactured because of their tendency to block. Blocking is reduced by slip agents, antiblocking agents, and cooling during manufacture. Ethylene-vinyl acetate foams are manufactured also. It can be seen that vinyl acetate copolymers are adequate raw materials for nonadhesive carrier films (where the low content of VAc allows the production of a tougher, softer film), for self-adhesive films having a higher VAc content, and for semi-pressure-sensitive and pressure-sensitive adhesives (HMPSAs) with a high polar monomer content. It is evident that for an adhesive formulation the VAc content should be much higher, and generally plasticizers should be included. According to Litz [107], an HMPSA formulation on an EVAc basis contains about 35.5% elastomer, 30–50% tackifier, 0.2% plasticizer, 0.5% filler, and 0.1–0.5% stabilizer. The VAc content does not influence the melt viscosity. The peel value can be increased by increasing the VAc content. As plasticizers, phthalates, phosphates, chlorinated polyphenols, and liquid rosin derivatives have been used. As tackifiers, rosin derivatives, hydrocarbon resins, and low molecular weight styrene polymers have been suggested. Zinc oxide, calcium carbonate, titanium dioxide, barium sulfate, and organic resins have been recommended as fillers. Films based on linear polyurethanes can be hot laminated [88]. Peelable protective films for metals have also been manufactured on a PVC copolymer basis [114].

2 MACROMOLECULAR BASIS AND CHEMICAL COMPOSITION OF THE ADHESIVE

The adhesive used for PSPs has to exhibit pressure sensitivity. As discussed earlier (see Chapter 3, Section 1), pressure sensitivity is given by a special rheology. Viscoelastic behavior allows pressure sensitivity. Viscoelasticity can be the result of a built-in special chemical or macromolecular basis, but it can also be achieved by formulation. In this case the elastomeric and viscous components are mixed to give the desired balance of viscoelastic properties. It is well known that the chemical composition of the adhesive includes nonadhesive components, which are required to ensure the fluid state characteristics of the adhesive and its storage resistance. The adhesive must be fluid for coating. In some cases the adhesive is transferred (during manufacture and application of PSP) onto the final carrier or substrate as a solid-state component (e.g., transfer coating, calender coating, or transfer tapes). The use of transfer tapes allows the choice of adhesive materials that cannot be applied as fluids [113]. The chemical basis of the pressure-sensitive adhesives is described in detail in Ref. 1. Therefore in this chapter only the product-related aspects of the raw materials are discussed.

2.1 Elastomeric Components

The elastomeric components of the adhesive are rubber or rubbery products. Some of them display a low level of adhesivity also. In order to display viscoelasticity they need a viscous component (tackifier). As for plastomers used for carrier materials, the chemical composition, molecular weight, and macromolecular characteristics of the elastomers determine their applicability for PSPs. Polar functional groups, comonomers, side-chain length, crosslinking, molecular weight, and molecular weight distribution influence the rheology of the adhesive [115]. Unlike plastomers, which possess adequate mechanical properties due to their rigidity given by crystallinity, elastomers have a deformable crosslinked elastic construction. Thermoplastic elastomers are elastic structures also where the connection points of the network are given by molecular, segmental associations.

Molecular Weight

The molecular weight, the functionality of the monomers, and the multiphase structure of the macromolecular compunds used for pressure-sensitive products play a primary role in the control of the product characteristics.

The Role of Molecular Weight. It must be emphasized that the role of molecular weight as a macromolecular characteristic is more complex for adhesives than for plastomers used for PSPs. The molecular weight can be decisive for the use of the (chemically) same component as base elastomer or tackifier. For instance, for insulation tapes high molecular weight polyisobutylene (PIB) (MW 100,000) is used as base elastomer and low molecular weight PIB (MW 1500) as tackifier [116]. As known from the practice of sealants, solvent-based low molecular weight acrylates are used as plastic sealants, whereas water-based acrylates (with higher MW) can be used as elastoplastic sealants. They have the ability to absorb vibrations up to 20% [117]. According to Köhler [118], the first pressure-sensitive products were manufactured with relatively low MW polymers. For such pressure-sensitive adhesives, destruction products of rubber together with soft resins and low molecular weight polyvinyl ether have been used.

Increasing the molecular weight by crosslinking can lead to a self-adhesive polymer playing the role of the carrier material in a composite structure manufactured by radiation crosslinking of acrylates [119]. As stated in Ref. 120, the same chemical composition with different molecular weights allows a polymer to be used as carrier and adhesive also. An acrylic acid-ethyl acrylate copolymer with a molecular weight of 40,000 can be used as carrier material for an acrylic acid-ethyl acrylate adhesive (MW < 12,000).

Polymers are not homogeneous on a molecular scale. Generally the macromolecular compounds used for pressure-sensitive products are linear or crosslinked. Linear molecules are typically threadlike structures. In crosslinked polymers separate molecules really do not exist. There are weak and strong regions in the polymers. The weak regions or imperfections may consist of chain ends, which are not entangled, and regions in which chain segments are oriented perpendicular to the direction of the stress. Strong regions include chain entanglements and regions where chain segments are oriented parallel to the stress. When a load is applied, the entangled chains will orient along the stress direction. (For instance, in rubber, stress-induced crystallization may appear.) The weak regions may form submicroscopic cracks.

The segments of the polymer molecule are in constant vibrational motion. This motion together with the imperfect packing of the molecules causes the existence of a free volume in the macromolecular structure. Filling this free volume with low molecular weight substances such as plasticizer changes the polymer's physical properties (e.g., coefficient of linear expansion) and mechanical properties. Thermal motion and free volume depend on the temperature. At a temperature great enough to produce chain mobility (T_g), free volume (and total volume) increase, and the polymer becomes rubbery. The mechanical properties (tensile strength and elastic modulus) decrease above the T_g. Creep, impact resistance, and permeability increase above the glass transition temperature (see also Chapter 3, Section 1). Such molecular mobility above T_g allows a pressure-sensitive adhesive to bond. Molecular flexibility can be controlled by crosslink density, glass transition temperature, fillers, plasticizers, and stabilizers. Molecular mobility causes debonding also. Disentanglement of linear molecules or melting of a crystalline order produces creep, i.e., a slow deformation and slow bond failure. Chain entangling accounts for 75% of the equilibrium modulus, even at a high degree of crosslinking [121].

As discussed earlier, the buildup of ordered multiphase structures can modify the chain mobility. Such structures include fillers, crosslinks, crystallites, and molecular associations. Fillers modify the T_g. The T_g of PVC increases with the level of filler [122].

Crosslinking reduces creep because the polymer segments are immobilized. For systems with limited mobility, weak bonds between or within the chains may rupture. For strongly immobilized structures, rapid bond failure may be manifested as cracks. Such systems do not display creep. As seen from the correlation giving the real stress (σ) in an extended rubbery sample as a function of the lengths of the sample before and after extension (expressed as the ratio λ of the lengths), the temperature (θ), and the end-to-end distance (r) of an uncrosslinked polymer chain [123],

$$\sigma = KT \frac{\langle r^2 \rangle}{\langle r^2 \rangle_0} (\lambda^2 - \lambda^{-1}) \tag{4.6}$$

the stress depends on chain length, and chain length depends on the temperature (the term $\langle r^2 \rangle / \langle r^2 \rangle_0$ takes into account this dependence). That means that molecular weight strongly influences debonding. As shown by Hamed and Hsieh [124], with a short contact time the bond strength (tack) is primarily due to the diffusion of short macromolecular chains. With longer contact times, higher molecular weight chains influence bonding.

Compatibility is a key parameter for the mixing of formulation components. It depends on chemical nature and molecular weight. The mixing enthalpy of macromolecular compounds depends on their chain length, decreasing with increasing chain length. On the other hand, with an increase in temperature the mutual solubility decreases because the entropy and enthalpy effects are working against each other [125]. The free mixing enthalpy ΔG_{mix} is given as a difference of the enthalpy (ΔH_{mix}) and entropy ($\tau \Delta S_{mix}$), where τ is the mixing temperature.

$$\Delta H_{mix} = \Delta H_{mix} - \tau \Delta S_{mix} \qquad (4.7)$$

Elastomers with a broad molecular weight distribution show better performance than materials with a narrow MWD. The concept of tackifying resins functioning as solid solvents for portions of the elastomer has been used to explain the need for a broad MWD. Resins will dissolve the lower molecular weight fractions of the elastomer and increase tack. The undissolved higher molecular weight fractions give the mechanical strength (see also Chapter 6). Natural and synthetic elastomers may be used as elastic components. Their mutual compatibility and their compatibility with the tackifiers depend on their molecular weight. Changes occur in the molecular weight distribution during aging, but coating does not influence it significantly [126]. Compatibility plays different roles for various pressure-sensitive formulations. For classic rubber resin recipes where the resin is dissolved in the elastomer matrix of natural rubber (or vice versa at higher resin levels), the resin should be compatible with the rubber in order to achieve adequate adhesive properties. For block copolymers with segregated structures, resin mixtures are used, each of the components being compatible with one of the polymer blocks. For tackified plastomers, noncompatible viscous components are suggested to allow their migration to the film surface.

Postregulating the Molecular Weight. Regulation of the molecular weight of elastomers during the manufacture of PSPs is a common operation. In order to prepare high solids content solutions, the elastomer is masticated, i.e., its molecular weight is reduced. As discussed in Chapter 3, the properties of the elastic network depend on the crosslinking density and the molecular weight between crosslinked portions. Mechanochemical destruction may change both. Through mastication the Defo elasticity of rubber is decreased, and the flow properties of the elastomer may change also [127]. It is a way to regulate the peel force also. This procedure is usual for tapes and protective films [128]. For such products the rubbery network is rebuilt via crosslinking to achieve high cohesion or low adhesion. A too long mastication leads to excessive degradation of the polymer (dead calendering). Natural rubber for tapes is crosslinked [129]. According to Mueller and Türk [130], rubber resin adhesives have been used almost exclusively for removable price labels. Natural and synthetic rubber and polyisobutylene have been used as the main raw materials. The rubber has been calendered and masticated.

It is well known that crosslinking is also brought about by the synthesis of carboxylated rubber latices. The molecular weight of a crosslinked network cannot be compared with the molecular weight of a linear system. For crosslinked

Chemical Basis of PSPs 173

systems the molecular weight alone is not sufficient for polymer characterization. It is also necessary to evaluate the gel content.

Postsynthesis crosslinking is a common procedure for solvent- and water-based PSAs on different chemical basis using built-in or external (chemical) crosslinking agents. Hot melts are also crosslinked by radiation curing. The crosslinking modifies the molecular weight of a sufficiently high molecular weight polymer in order to limit chain mobility and to achieve a higher modulus according to correlation (3.22) (see Chapter 3). In other cases crosslinking is used to build up a polymer or to achieve a long enough chain for cohesion. At low enough crosslink densities, equilibrium moduli below the the rubbery plateau modulus can be obtained, but such networks are weak due to the presence of long ineffective chain ends. Regulation of the molecular weight via crosslinking is a necessity for low cohesion solid-state acrylics also [129]. Balanced pressure-sensitive properties are achieved by UV-initiated postcrosslinking (after coating) of the formulated adhesive. The curing of monomers as well as of polymers has been carried out to achieve or improve pressure sensitivity. For instance, in Ref. 131 the curing of hot melts using 60–80 kGy radiation (EB) energy dosage is described. Curing of acrylic polymers with UV light is discussed in Ref. 132. Both procedures may lead to higher molecular weight associated with a nonlinear molecular structure. It can be concluded that postregulation of the molecular weight is really a postregulation of the molecular structure also. Pressure-sensitive products need composite structures, and such structures are formed by polymacromerization and polysegregation.

Polymacromerization and Polysegregation

In principle there are two ways to achieve viscoelastic network structures for PSPs: polymacromerization and polysegregation. As discussed earlier for plastomers used as carrier materials for PSPs, the characteristic, ordered multiphase structure is mainly the crystallinity. Crosslinked and filled products can be manufactured also. On the other hand, it is generally assumed that a large number of amorphous polymers feature a definite degree of structural microheterogeneity; elementary domains formed by the macromolecules may associate or form first fibrils and then superdomains from which a three-dimensional network is built up. Such systems are not in equilibrium and remain capable of further ordering, which takes place when the polymer has facilitated mobility.

For pressure-sensitive adhesives more elasticity and molecular mobility are needed. In this case elastomer structures including crosslinked networks, molecular associations, fillers, and in certain cases crystalline portions also have been developed. For such structures the molecular weight between entanglements plays a primary role. As in (crosslinked) plastomers, supplemental crosslinking (by covalent bonds) of the noncovalently segregated structures is possible. Like vulcanization of natural rubber, curing of synthetic elastomers and thermoplastic elastomers can build up new segregated structures. For instance, such crosslinking can be achieved by using radiation curing.

Buildup of Segregated Structures. It has been shown that polymer performance can be enhanced by the formation of heterogeneous systems in which one polymer exists above its glass transition temperature while the other exists below its T_g at room temperature [133]. Thus a composite results at room temperature having one component that is glassy while the other remains rubbery or rubbery elastic. Varying the relative amounts of the polymers in this blend (which can be considered as an interpenetrating network) will alter the properties. The produced material can range from a reinforced rubber to a high impact plastic. Rubber resin adhesive with the high T_g resin and low T_g elastomer or self-adhesive films based on a mixture of plastomer, elastomer, and tackifier resin may be considered structures of this type. Chemical reactions can lead to interpenetrating networks also. Crosslinking reactions in the adhesive layer increase the T_g and build up segregated macromolecular constructions. Crosslinking is responsible for the control of the size of the phase domains in the network and hence for the properties exhibited by the material.

For crystalline polymers the existing order is the main parameter that determines their performance; the chemical structure is secondary. Crystalline structures are nonadhesive. Unlike classic, non-pressure-sensitive adhesives where crystallinity buildup after the formation of bonds improves the joint strength (e.g., polychloroprenes), crystallinity has a negative effect on the usability of PSAs. Molecular weight can influence the tendency to crystallize also, leading to so-called semi-pressure-sensitive products, which have a time-limited self-adhesivity (see also Chapter 6). After crystallization the self-adhesivity of the elastomer disappears. For products where the diffusion of viscous components affects the bonding properties (i.e., tackified self-adhesive films), the time-dependent self-adhesivity is a function of both the molecular weight and crystallinity. In some cases crystallinity is a designed characteristic of the adhesive raw material. For instance, segregated polyurethanes have a crystalline structure.

Thermoplastic elastomers with diblocks and triblocks (SBS, SIS, and SEBS) have ordered amorphous domains and segregated incompatibility, but multiblock systems may also have crystalline segments. This morphology replaces the crosslinked network of common rubbers, but due to its low deformability thermoplastic elastomers like SBCs have only a medium degree of elastic recovery, in comparison with classic rubbers. They can be considered as intermediate to elastoplastic and viscoelastic compounds.

Segregated structures are built up as block copolymers in which molecular associations form amorphous incompatible domains (e.g., SBCs). In such cases the main problem is to find an adequate "blocky" monomer that allows high tem-

Chemical Basis of PSPs

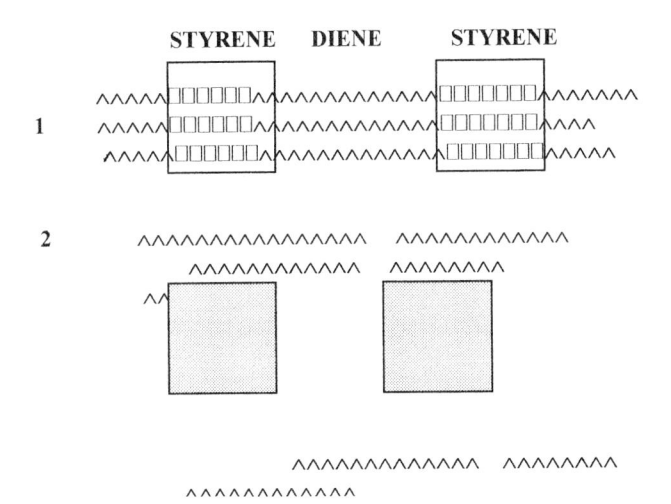

Figure 4.2 Schematic presentation of segmented copolymers. (1) Block copolymer chain; (2) polymer chain considered as blend of homopolymer blocks acting as filler.

perature resistance as well as low temperature processibility. According to Eq. (3.25) (see Chapter 3), in a first approximation such "blocky" units work in an elastomer matrix like a chemically inert filler (Fig. 4.2). Their modification (if possible) with the use of other compatible viscous fillers (tackifiers) allows no further strengthening of the network. Such rubber resin adhesives are considered similar to filled systems having plasticizer or tackifier particles embedded in a rubbery matrix or vice versa (Fig. 4.3). Such systems can also be examined by using Eq. (3.25), i.e., given that the synthesized macromolecular composite is embedded like a filler in another formulated composite. The choice of compatible resins (which are low molecular weight compounds) allows tackifying with a minimum loss of cohesion.

In certain cases segmentation alone does not provide self-strengthening. This is achieved by postcrosslinking. In other cases the thermal limits of strengthening by association have to be improved. For styrene block copolymers with unsaturated midsequences, if strengthening is necessary it can be done by using the reactivity of the rubbery (isoprene, butadiene, etc.) "bridges," i.e., by crosslinking (Fig. 4.4). By such reactions it is possible to create a covalently bonded network within the elastomeric midphase in order to reinforce the physically existing network (which disappears at temperatures above the T_g of polystyrene) [134] (see Fig. 4.4).

As seen in Fig. 4.4 such crosslinking can be carried out without or with a supplemental multifunctional crosslinking monomer. It can be done by classic or

176   *Chapter 4*

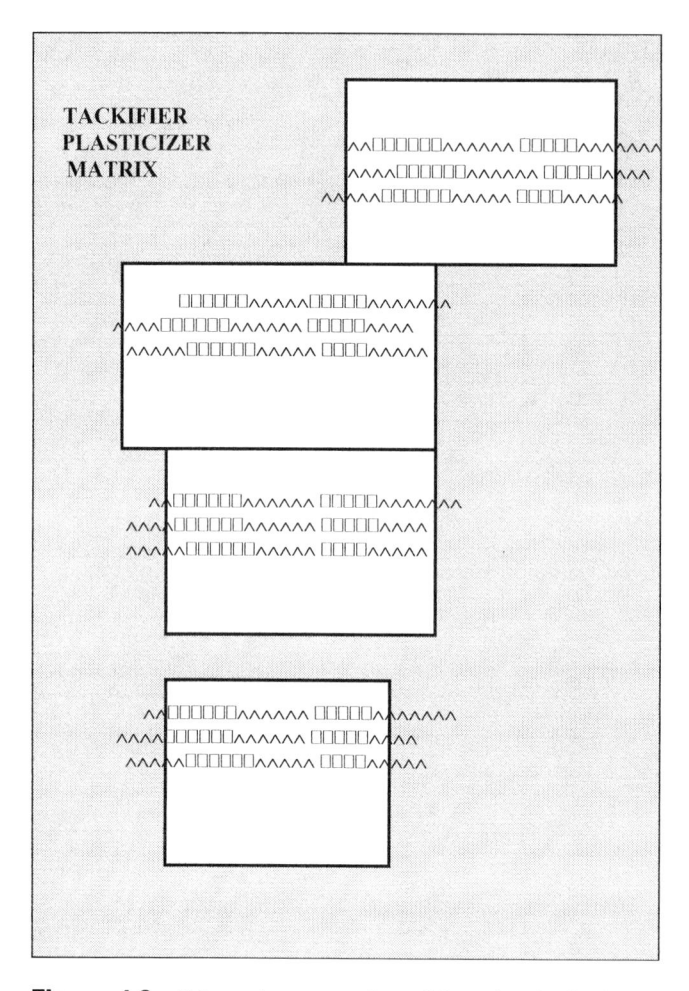

Figure 4.3 Schematic presentation of formulated adhesive based on block copolymers, considered as matrix or filler in a reinforced system.

radiation-induced polymerization. Radiation-induced crosslinking depends on the molecular weight and its distribution. The first radiation-curable styrene block copolymer (Kraton D-1320 X) was processed like a common thermoplastic elastomer, but it was also able to crosslink. This was a star-shaped SIS block copolymer. Because of its high molecular weight, low radiation doses are needed to produce crosslinks. Because of its star structure, partial crosslinking does not modify (essentially) the flexibility of the molecules.

Chemical Basis of PSPs

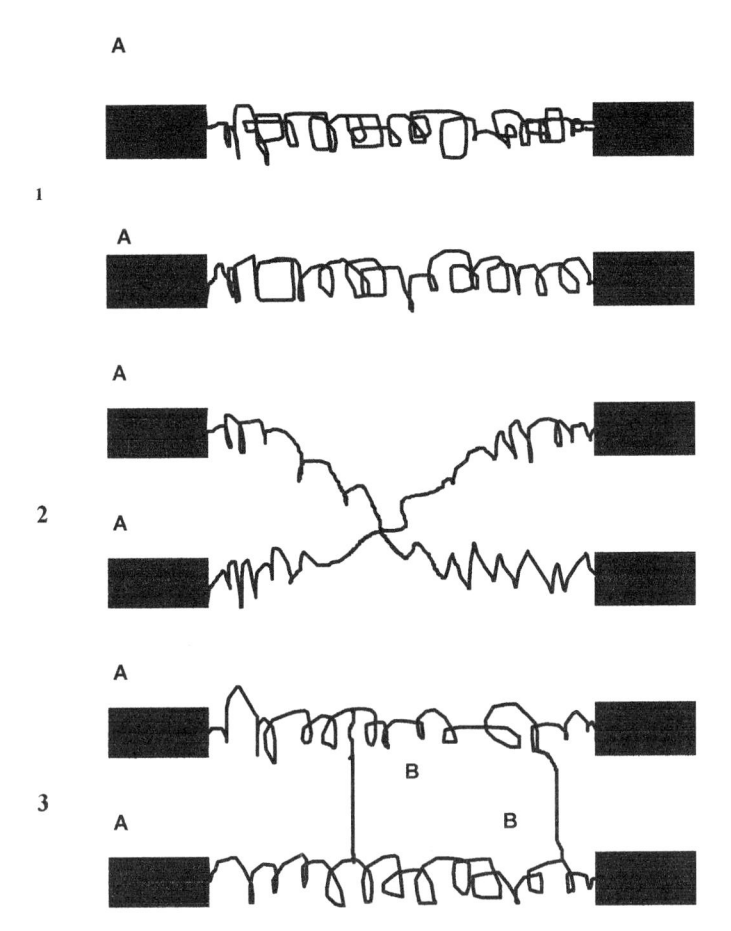

Figure 4.4 Schematic presentation of the uncrosslinked and crosslinked block copolymer. (1) Uncrosslinked chains of a block copolymer (A); (2) crosslinking of a block copolymer by the midblock sequence, without supplemental multifunctional crosslinking agent; (3) crosslinking of a block copolymer using supplemental multifunctional crosslinking agent.

Curing of acrylics is viable too. For instance, one patent [135] describes the curing of copolymers of diesters of unsaturated dicarboxylic acids with acrylics in a composition having a multifunctional crosslinking agent. Such polymers possess a T_g between 30 and 70°C below the temperature of use. In this case curing is used to improve adhesive properties, particularly shear. This composition can be cured chemically or by treatment with any convenient radiant energy. The molecular weight of the formed polymers differs according to their MWD. For a narrow molecular weight distribution, a weight average molecular weight of 100,000, and for a broad molecular weight distribution, a molecular weight of 140,000 is required to enable the desired response to electron beam curing. In general, constituents having a molecular weight of less than 30,000 are nonresponsive to electron beam radiation. An M_w/M_n ratio between 4.2 and 14.3 is preferred. Cohesive strength is proportional to the concentration of the hard monomer (giving a polymer with T_g higher than $-25°C$). When a multifunctional monomer is used, its concentration should preferably be about 1–5% by weight. Its presence enables reduction of the dosage level, at least for EB curing.

Epoxidized star-shaped block copolymers can also be UV cured [136]. Polyisoprene-polybutadiene polymers with a segmented structure have been synthesized as UV-curable adhesive raw materials that are processable as warm melts [137]. Although segmented, the new polymers do not present microdomain phase separation. They are viscous liquids near room temperature. Such behavior is obtained by keeping the polymer molecular weights low and MWD narrow. These polymers are functionalized oligomers that are bi- or multifunctional, and their star-shaped structure is used to increase the functionality, not to allow phase segregation. They possess two different kinds of functionalities: epoxy groups and hydroxyl groups. Epoxy groups may undergo ring-opening polymerization under acidic conditions. When multiple epoxides are located on a polymer, polyether crosslinking results. Epoxy polymerization can be initiated by UV light also when a sulfonium salt is used. The photolysis of the initiator generates a Brönsted acid that protonates the epoxide. The protonated epoxy group may react with another hydroxylated compound to form polyether linkages. Therefore mixtures containing multifunctional oligomers (e.g., epoxidized star polymers, polymers having one or more hydroxy groups, or hydroxy and epoxy groups) allow multiple ways to regulate the reaction and form crosslinks. In such systems the star-shaped polymer is diluted by the monofunctional polymer.

It has been shown that the method of crosslinking influences the flexibility of the polymer network and the flow and mechanical performance of the macromolecular compound [138]. For instance, the modulus of polybutadiene networks made by hydrosilation crosslinking differs from the modulus of radiation-crosslinked polymers. The effect of interaction of the junction points is weaker than for polybutadiene crosslinked by radiation [139]. It is known that in crosslinked systems the rheological and mechanical properties of the system are determined by chain mobility, which is influenced by the molecular weight of the polymer as a whole and the molecular weights of the polymer sequences. The molecular weight between crosslinks, M_c, depends on the chain length between functional groups. In experimental polymers, according to Ref. 137, molecular weights are larger between crosslinkable sites on the polybutadiene blocks than the much smaller molecular weight of the polyisoprene blocks. Both together give the so-called mechanically effective molecular weight M_{cme}. The star-shaped polymer may carry a large number of epoxides per molecule to promote EB curing at low dosage levels. In such polymers there are over 100 epoxy groups located on the polyisoprene endblocks. Such an abundance of functionality leads to overcure. Therefore these polymers exhibit no tack. As stated in Ref. 137, in comparison with polymers that have the longest distance between crosslinks and a molecular weight of about 2000, polymers with PSA properties should have an M_{cme} of at least 3000–4000. A cured blend of mono-ol polymer, epoxidized mono-ol polymer, and hydrogenated tackifying resin exhibits aggressive tack and SAFT values that exceed 175°C without failure. This example illustrates the problem of functionality, i.e., how the balance of linear and crosslinked segments influences the final properties. In this case chemical functionality, and in the case of styrene block copolymers the presence of associative styrene mono- or diblocks (on both ends of the sequence), affects the network buildup. As stated by Holden and Chin [140], a new SEBS with 30% triblock sequences has a tensile strength of 350 psi in comparison with the tensile strength of 142 psi for a 20% triblock copolymer. The chain length used for cross bridges between the main backbones must have a certain minimum length to ensure deformability. Such crosslinking sequences must have a MW greater than 10,000 [141,142]. According to Kerr [143], for UV-curable polyacrylate with styrene side groups, e.g., 2-polystyryl methacrylate, long-chain C_{14} diol diacrylates should be used as crosslinkers. The crosslinking "density" is also important. According to Havranek [144], in slightly crosslinked model networks the T_g for long bridge networks does not depend on the mole ratio of functional groups; in polymers with a high level of short-chain crosslinkings it is a function of this ratio.

All hydrocarbon polymers undergo chain scission during degradation. This reaction competes with crosslinking as the terminal step of the autoxidation process [145]. Since chain scission and crosslinking may occur simultaneously, the oxidative effect on viscosity and adhesivity will depend on the balance of the two competitive reactions. Formulations based on SIS degrade by chain scission; for SBS-based compounds degradation takes the form of crosslinking. Chain scission and crosslinking both occur in the multiblock polymer formulation, and their balance leads to increased bond strength at relatively constant viscosity and service temperature. Electron beam radiation causes the crosslinking of the butadiene–α-methylstyrene block copolymer via the butadiene block and its degradation via the α-methylstyrene blocks [146].

Polymacromerization. Low molecular weight polymers are easily processed. They display the advantages of bulky polymers (no migration, no volatiles, etc.) and low viscosity liquids. Their buildup to a high molecular weight polymer via polymacromerization is a relatively simple process. Classic pressure-sensitive formulations have been based mostly on rubber and resin. Natural rubber has been used as an elastomer. Generally natural rubber has been masticated, i.e., depolymerized to achieve a relatively uniform molecular weight and better solubility. In many cases the molecular weight has been restored by using crosslinkers. The synthesis of ready-to-use, inherently tacky raw materials has allowed control of the molecular weight without the need of depolymerization. However, crosslinking has been used to achieve better cohesion. Developments in macromolecular science have led to the technology of macromer manufacture. Such relatively low molecular weight polymers can be postpolymerized (postcrosslinked) by using various polymerization or curing techniques. Some years ago, the coating of pressure-sensitive raw materials was synonymous with their physical or mechanical transformation. In the future, coating will be associated with the chemistry of macromer transformation (polymacromerization). As discussed later, macromers have been synthesized in almost every class of raw materials. Their starting molecular weights can vary. Their final molecular weight can also vary, depending on how they are "assembled."

As discussed in Ref. 1, formulations with 100% solids have been developed as radiation-curable compositions based on crosslinkable monomers and reactive diluents. Prepolymers with better coatability have been favored. It is possible to manufacture a prepolymer that is polymerized by radiation [147]. A patent discloses the preparation of a low molecular weight spreadable composition to which may be added a small amount of catalyst or polyfunctional crosslinking monomer prior to the completion of polymerization by heat curing. The prepolymer is more viscous and is easily applied to a support. A crosslinkable acrylic formulation is a combination of non-tertiary acrylic acid esters of alkyl alcohols and ethylenically unsaturated monomers that have at least one polar group; it may be substantially in monomer form or may be a low MW prepolymer or a mixture of prepolymer and additional monomers, and it may also contain other substances such as a photoinitiator, filters, and crosslinking agents (such as multifunctional monomers) [135]. The monomers may be partially polymerized (prepolymerized) to a coating viscosity of 1–40 Pa·s before the fillers are added. A prepolymer having a viscosity of (0.3–20 Pa·s) can be synthesized by UV-initiated polymerization. The crosslinking agent is added, and the mixture is coated and polymerized by UV to yield the final product [148,149].

Considering crosslinking as a partial stiffening of the macromolecular sequences, the same effect can be obtained by polymacromerization. It is possible to strengthen the macromolecular structure by partially stiffening it, using macromers to achieve the desired conformability. Polymeric monomers, macromonomers, or macromers are useful as reinforcing monomers. According to Ref. 150, the macromers should have a T_g above 20°C. Representative examples of such macromers are polystyrene, poly-α-methylstyrene, polyvinyltoluene, and polymethyl methacrylate. For instance, a methacrylate-terminated styrene macromer is used for medical adhesives for application to the skin.

Chemical Basis of PSPs *181*

Macromers can be included by the polymerization of vinyl monomers to give better flow, coating, and leveling properties. Ethylene-acrylic acid copolymers have been included in terpolymerization formulations of emulsion-based vinylic monomers for tapes [151]. Unlike the synthesis of segregated polymers, which are ready-to-use raw materials and need no chemical transformation, macromers require chemical reactions to be transformed into pressure-sensitive raw materials or finished products. As presented in Fig. 4.5, thermally initiated or radiation-initiated polymerization of a monomer can be carried out to obtain an oligomer or a high polymer. Either can be postprocessed later. For the oligomer this is a necessity; for the polymer it is a possibility. Generally both are postprocessed as formulated product. The polymerization of the monomer

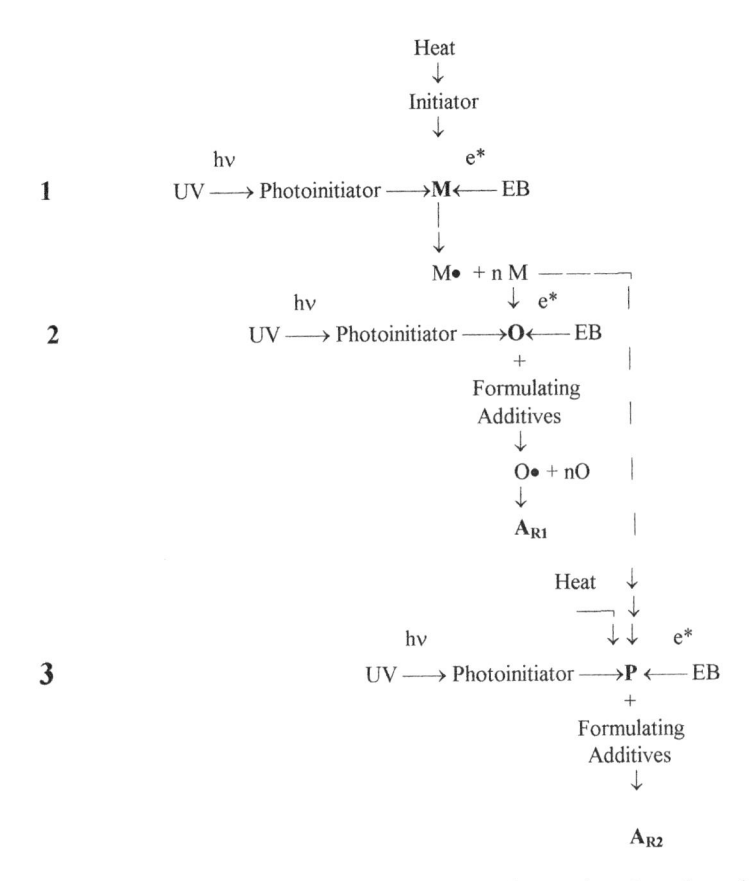

$$
\begin{array}{c}
\text{Heat} \\
\downarrow \\
\text{Initiator} \\
\downarrow
\end{array}
$$

1 UV $\xrightarrow{h\nu}$ Photoinitiator $\longrightarrow \mathbf{M} \xleftarrow{e^*}$ EB

$$M\bullet + nM$$

2 UV $\xrightarrow{h\nu}$ Photoinitiator $\longrightarrow \mathbf{O} \xleftarrow{e^*}$ EB
+ Formulating Additives
↓
$O\bullet + nO$
↓
A_{R1}

3 UV $\xrightarrow{h\nu}$ Photoinitiator $\longrightarrow \mathbf{P} \xleftarrow{e^*}$ EB
+ Formulating Additives
↓
A_{R2}

Figure 4.5 The manufacture of pressure-sensitive products by polymerization, macromerization, and postpolymerization. (1) Polymerization of a monomer to an oligomer (O) or to a polymer (P); (2) polymacromerization of an oligomer; postpolymerization.

(*M*) leads to an oligomer (*O*) or to a polymer (*P*). After its formulation, the oligomer can be polymacromerized to a ready-to-use adhesive product (A_{R1}). The polymerization can lead to a high polymer also, which (after formulating) can be postprocessed by crosslinking to a ready-to-use adhesive product (A_{R2}).

Generally ease of processing is the main argument for preparing oligomers instead of polymers, but the possibility of achieving segregated structures is becoming more and more important. The main problems are related to the nature, molecular weight, and reactivity of the macromers. Their reactivity should be examined in the formulated mixture (see later, UV-Cured Acrylates). The methods used for the synthesis of the macromers are various, and radiation curing is preferred for their postprocessing. Electron beam curing has the advantage that it does not require special initiators or initiating functionalities. Since most of the monomers commonly employed do not produce initiating species with a sufficiently high yield upon UV exposure, for such photopolymerization it is necessary to introduce a photoinitiator to start the polymerization. Among the different possible types of reactions involved in the initiation of photopolymerization, the most important are (1) radical formation by photocleavage and (2) radical generation by hydrogen absorption.

Benzoin derivatives, benzylketals, acetophenone derivatives, hydroxyalkylphenones, acylphosphine oxides, and substituted α-amino ketones belong to the group of initiators that undergo photocleavage. Because of their absorption in the long wavelength domain, acylphosphine oxides allow the use of white pigmented formulations also. Radical generation by hydrogen abstraction is characteristic for benzophenone or tioxanthone. Certain aromatic carbon compounds undergo a Norrish type I fragmentation when exposed to UV light:

$$\text{AR-(CO)-C-X} \xrightarrow{h\nu} \text{AR-(CO)} \bullet + \bullet \text{C-X} \tag{4.8}$$

Efforts have been made to synthesize polymeric compounds that bear side moieties capable of α-cleavage after irradiation with UV light to generate free radicals. Such compounds are polymeric α-hydroxyalkylphenones, structure (4.9), with MW < 2000 [152], e.g., hydroxy-[4-(1-methylvinyl)]-isobutyrophenone, structure (4.10) [153].

$$-[-\underset{\underset{\text{AR(CO)}-\text{C(CH}_3)_2\text{-OH}}{|}}{\overset{\overset{\text{CH}_3}{|}}{\text{C}}}-\text{CH}_2-]_n- \tag{4.9}$$

Chemical Basis of PSPs

$$-[-CH_2-\underset{\underset{O=COCH_2CH_2O-AR(CO)-C(CH_3)_2-OH}{|}}{\overset{\overset{CH_3}{|}}{C}}-]_n- \qquad (4.10)$$

Photoreduction of benzophenone via a charge transfer complex leads to a free radical that reacts with the monomer [154]:

$$(AR)_2 \underset{|\;|}{\overset{|\;|}{C}}=O \; + \; \underset{|\;|}{\overset{|\;|}{:N-CH-}} \; \rightarrow \; (AR)_2 \overset{\bullet}{C}-OH \; + \; \underset{|\;|}{\overset{|\;|}{N-C\bullet}} \qquad (4.11)$$

$$\underset{|\;|}{\overset{|\;|}{N-C\bullet}} + CH_2=CH-COOR \qquad (4.12)$$

According to Scheiber and Braun [155], radiation-cured PSAs can be based on (1) SIS multiarm block copolymers, (2) liquid rubbers with acrylate functionality, and (3) acrylate prepolymers.

The UV curing of "prepolymers" was begun with the curing of PVC plastisols, polymer dispersions in plasticizer. If the plasticizer contains an unsaturated polyfunctional acrylic, it is possible to cure it with UV radiation [156]. A large variety of functionalized prepolymers are now available, making it possible to create networks with tailor-made properties. Acrylic monomers are among the most widely used curable systems because of their high reactivity, moderate cost, and low volatility. Acrylated polyester oligomers have been suggested as raw materials for PSAs. Such products can be processed as HMPSAs [157,158]. For better radiation curing, these products have acrylic side chains. For instance, a UV-curable polystyrene has been developed in which the end groups are methacrylate [159]. In another case a polyacrylate with styrene branching (2-polystyryl methacrylate) has been suggested [143]. For such oligomers a long-chain C_{14} diol diacrylate has been proposed as crosslinking agent:

$$-------C_4H_9-(CH_2-CH_2)_n-CH_2CH_2-O-\underset{\underset{O}{||}}{C}-\overset{\overset{CH_3}{|}}{C}=CH_2 \qquad (4.13)$$

Natural Elastomers

Of the natural elastomers, natural rubber (NR) was the first to be used as a raw material for PSAs. The so-called rubber resin adhesives (mixtures of natural rubber with tackifier resins) were the first commercial PSAs. According to Mueller and Türk [130], removable labels have been formulated on a basis of rubber resin adhesive. The rubber has been tackified with soft resins and plasticizer. Natural rubber based adhesive solutions and styrene block copolymer based HMPSAs are the main formulations for common tapes [160]. To achieve high shear resistance, crosslinked formulations are used. Freezer tapes are coated with an adhesive based on natural rubber, styrene-butadiene rubber, regenerated rubber, rosin ester, and polyether [58]. This adhesive is crosslinked also. Crosslinked rubber resin formulations are the most often used adhesives for protective films. Such formulations are soft, and their adhesivity can be regulated by mastication, tackification, and crosslinking. Generally, using natural rubber as the main formulation component has the advantage that it is easy to regulate the adhesive properties by regulating the molecular weight. Crosslinked natural rubber formulations are softer than certain common noncrosslinked acrylic recipes. In some cases the softness of the masticated elastomer allows "solid-state" mixing and coating (for tapes) of the formulation components. Natural rubber can be formulated together with polar elastomers and reactive resins (phenolic resins). The crosslinked adhesive displays temperature and solvent resistance [161]. The reaction is carried out by mastication of the components. The high molecular weight portion of natural rubber is insoluble in solvents; therefore natural rubber must be milled to a Mooney viscosity of 75 or below for complete solubility (at 10% solids) [162]. The milled sample has a narrower molecular weight distribution. Many natural rubber solvent-based formulations contain a crosslinking agent. The process involves breaking down the natural rubber and then building it back up. It is evident that the amounts of crosslinking agents and tackifiers used depend on the degree of mastication.

Natural rubber latex is applied for low tack removable formulations, especially for tapes and envelopes [21]. A special domain of natural latex based formulations is the manufacture of tapes on paper carrier. Adhesives on a natural rubber latex basis are not compatible with other adhesives [163]. Re-adhering and removable adhesive in a solid stick form (glue stick, applicator crayon) can also be formulated on a natural rubber basis [164]. Such products are removable unless the user applies a very heavy coating. A natural rubber latex, a tackifying agent, and a gel-forming agent are the main components of the formulation. A friction-reducing component and an antioxidant may also be included. Rosin ester and 4,4-butylidene BIS are suggested as tackifier and antioxidant, respectively. Gelling agents are aliphatic carboxylic acid salts. Natural rubber latex used at a level of 10% or less improves the shear resistance of CSBR [165] and also improves the aging characteristics. SBR latex crosslinks and hardens during oxidative aging, whereas natural rubber latex undergoes chain scission and softens; thus a balance is obtained. Pattern or strip coating allows the use of nonremovable raw materials for removable products. Natural rubber latex coated as a contoured surface can be used on paper to give a removable protective web that is adequate for acrylic items [166].

Synthetic Elastomers

Synthetic elastomers are macromolecular products that display rubbery elasticity. This behavior is the result of their special chemical and macromolecular structure. Natural rubber-like stereoregulated polydienes, diene-styrene random and block copolymers, and thermoplastic elastomers on styrene-diene, styrene-olefin, ethylene-propylene-diene, acrylic, etc. basis have been synthesized and tested for use in PSPs. Unlike natural rubber, certain of these products can also be used for noncoated PSPs.

Butene Polymers. Butylene and isobutylene polymers and copolymers are used as primary elastomers or as tackifiers. Polyisobutylene tackified with rosins is suggested as an elastomer for tapes [167,168]. Low molecular weight PIB is used as a tackifier for butyl rubber or acrylic PSA formulations. According to Ref. 169, polyisobutylenes are liquids or solids with molecular weights ranging from 36,000 to 58,000 (liquids) or from 800,000 to 2,600,000 (solids). Nuclear magnetic resonance has disclosed that the isobutylene units are polymerized as head-to-tail sequences [170]. Polyisobutylene (MW 3000) is liquid, whereas high molecular weight PIB (MW 200,000) is a clear solid. Such polymers can be used as solutions or water-based dispersions [168]. There are several tape, label, and protective film applications that use these compounds as the base elastomer or tackifier.

Aluminum film coated with a tackified polyisobutylene-based adhesive is used for street marking. The release layer for this adhesive is PVA (10–40%), silicone (20–50%), mineral oil (20–50%), and sodium acetate (10–12%) [167]. Masking tapes are used during the varnishing of cars [171]. They have to adhere to automotive paint, enamel, steel, aluminum, chrome, rubber, and glass and resist the high temperatures used for the drying and curing of varnishes. For such tapes natural rubber latex tackified with PIB is suggested. Polybutylene has been used in sealing tapes [172] and medical tapes [173]. Polyisobutylene with a molecular weight of 80,000–100,000 is proposed for medical tapes because it does not adhere to the skin [167]. Polyisobutylene is also employed as the base elastomer and tackifier for insulation tapes [116]. Tackified natural and synthetic rubber and polyisobutylene have been used for removable labels [130]. Polybutylene polymers are suggested for hot melts. They display good compatibility with aliphatic resins, atactic polypropylene, and block copolymers. They are not polar, and therefore the adhesives formulated with them have good adhesion on nonpolar surfaces. Polybutylenes are saturated; therefore they exhibit aging resistance. Such materials have a lower rate of crystallization than EVAc [174] (this behavior is very important for self-adhesive films). Low molecular weight butene polymers play a special role as tackifiers for polyolefin carrier materials used as self-adhesive films (see later).

Butyl rubber is a copolymer of isobutylene with isoprene (less than 3% isoprene) and has a molecular weight of 300,000–400,000 [169]. The different grades of elastomers differ in unsaturation, molecular weight, and type of stabilizing agent. Butyl rubber is manufactured as latex also. Halogenated (chlorinated or brominated) grades of butyl rubber have also been synthesized. Butyl rubber (MW 6000–20,000) does not need mastication to dissolve, as many natural and synthetic rubbers do. It can also be crosslinked. Partially crosslinked butyl rubbers (terpolymers with divinylbenzene) are commercially available. The crosslinking improves the flow resistance, and these rubbers are used in sealing tapes. A precrosslinked butyl rubber can also be used instead of partially crosslinking the rubber in the mixer [175]. The degree of partial crosslinking prior to mixing with other components may vary between 35 and 75%. The proportion of partially crosslinked rubber in the composition varies inversely with the percentage of crosslinking. p-Quinone dioxime, p-dinitrosobenzene, phenolic resins, and similar compounds may be added as the crosslinking agent. Butyl rubber is clear and tacky, it is not fragile after aging, and it may be used together with PIB as a tackifier for medical tapes.

Diene Copolymers. The first possibility to achieve natural rubberlike properties is to "copy" the chemical composition and structure of NR, i.e., to synthesize macromolecular compounds having as base monomer isoprene, butadiene, etc. with a (more or less) stereoregulated structure. Such products have been manufactured as solids that can be processed as solvent-based adhesives.

Polar (halogenated) diene monomers have also been polymerized that have special characteristics such as nonflammability. As an example, there are nonflammable tapes with an adhesive based on polychloroprene rubber [176]. Copolymers of chloroprene with other polar monomers are synthesized as water-based dispersions too. Carboxylated neoprene dispersions are used for special temperature-resistant, nonflammable formulations.

Styrene-Butadiene Random Copolymers. Rubberlike properties have been achieved for random styrene-butadiene copolymers (e.g., SBR) also. These products have been used as raw materials for solvent-based PSAs. Styrene-butadiene rubbers contain a residual double bond that enables further polymer chain growth (crosslinking), to form a network with better mechanical properties without apparently affecting the T_g. Unfortunately, they are hard and are difficult to tackify. Styrene-butadiene rubbers possess better aging stability than natural rubber. They harden during aging [56]. Generally polymers with a low level of styrene are suggested for PSAs, mainly for tapes. Freezer tapes are coated with an adhesive based on SBR with 12% styrene [52]. Styrene-butadiene rubber is not recommended for medical tapes [172]. As mentioned above, styrene-butadiene latex is used as a top coat for paper [2].

Carboxylated styrene-butadiene copolymers (CSBRs) have been synthesized and used as aqueous raw materials for PSA. The acid comonomer works as a stabilizing agent for the aqueous dispersion [177]. The styrene content, molecular weight, and gel content are the main quality parameters for such dispersions. Gel and styrene content harden the polymer and make its tackification difficult. With a higher styrene content the latex does not respond well to the addition of resin. Above 30% gel content, CSBR tack decreases rapidly [165]. Shear resistance increases with gel content. Deep freeze properties of formulations are improved if the butadiene level of the polymer is greater than 80%. Natural rubber latex used at a level of 10% or less improves the shear resistance of CSBR and also improves the aging characteristics. Quick stick and high temperature shear are required for tape applications [178]. Such properties may be achieved by using CSBR. A screening formulation for tapes includes CSBR, coumarone-indene resin, hydrocarbon resin, plasticizer, and thickener.

The T_g and modulus of acrylic latices can be adjusted more easily by polymerizing with different monomers at various monomer concentrations. Therefore in comparison with CSBR, acrylates can be used with a low level of tackifier or with none [179]. On the other hand, the multiphase structure given by sequence distribution and crosslinking ensures special properties. Thermoelastic, multiphase styrene-butadiene copolymers have been prepared by radical emulsion copolymerization of styrene with styrene-butadiene copolymers having a molecular weight of 180,000–320,000 and polydispersity of 3–6 [180]. Such graft copolymers exhibit a gel content of 81%, an increase of 30% in tensile strength and 20% in elongation, and a decrease of 80% in the residual elongation in comparison with the ungrafted polymer.

Styrene Block Copolymers. The elastic component of the adhesive is a natural or synthetic elastomer. Natural rubber is a well-defined polymer of isoprene. Isoprene and other diene monomers have been polymerized to form new synthetic rubbers. In a quite different way, block copolymers have been obtained also in which elastic and thermoplastic properties have been combined. Styrene block copolymers have been produced. Styrene-olefin block copolymers with an ABA sequence distribution and styrene endblocks [see (4.14)] have been synthesized since 1965.

$$-[-CH-CH_2-]_x-[-C-CH=CH-C-]_y-[-CH-CH_2-]_x- \quad (4.14)$$

$$\begin{array}{ccc} \text{AR} & & \text{AR} \\ \text{A} & \text{B} & \text{A} \end{array}$$

where $x = 200–500$ and $y = 700–1500$.

Sequenced block copolymers of styrene with different dienes and olefin comonomers have also been produced. First polymers with unsaturated middle segments (butadiene and isoprene) were developed, then products with saturated middle sequences (having about 30% styrene) [181]. According to Gergen [182], styrene-diene block copolymers appeared on the market in 1965, styrene-olefin block copolymers in 1975, and hydrogenated block copolymers in 1983. Styrene block copolymers used for hot melts are based mainly on SIS block copolymers with about 15% styrene [183]. Hydrogenated midblocks were first used in 1972, radial polymers in the mid-1970s, crosslinkable technology in 1979, and interpenetrating networks in 1984 [184]. In bisequenced copolymers, the elastomeric midblock can be polybutadiene (SBS) or polyisoprene (SIS); in trisequenced polymers, polyethylene-butylene (SEBS) and polyethylene-propylene (SEPS) sequences are built in [134]. Commercially available copolymers in this range have a styrene content of 14–25% w/w. These polymers are applied in an amount of 25–50% by weight, preferably 30–40% in the PSA formulation.

According to Rader [185], thermoplastic elastomers can be classified as (1) block copolymers, (2) thermoplastic polyolefins, and (3) thermoplastic vulcanizates. Actually, the styrene block copolymers are the most important TPEs as raw materials for pressure-sensitive products. The main thermoplastic elastomers [186] are styrene block copolymers, olefin copolymers, polyurethanes, copolyesters, and polyetheramides. Segmented EVAc copolymers have also been synthesized [187].

During the development of styrene block copolymers, various products were synthesized. The thermal resistance and cohesion of the copolymers have been improved; their reactivity and functionality have been changed to allow better postprocessing. These modified styrene copolymers have been used as macromers. The degree of crosslinking and chain flexibility have been regulated by controlling the end groups and the molecular weight of the sequences. Functionalizing of the middle and end sequences has been tried also (Fig. 4.6).

Uncrosslinked and unentangled polymer chains do not contribute to the strength of the elastomer. Therefore pure triblock SIS copolymers have also been manufactured. Pure (100% triblock) polymers have a high T_g endblock [188]. Diblock units work as low cohesion diluting agents in the composition and decrease the cohesion. Johnsen [189] discussed the development of pure triblock SBCs with styrene–α-methylstyrene endblocks giving higher T_g and better cohesion. α-Methylstyrene has a higher T_g, 170°C, and therefore provides better thermal properties than styrene. According to Johnsen [189], for SIS a 1/2 ratio of styrene/α-methylstyrene and a molecular weight of 15,000 for the endblock are preferred. Generally the hard segment size influences the melt transition temperature of TPEs [190]. The T_g of polystyrene is a function of its molecular weight up to an MW of about 20,000. The higher MW of SIS allows its use where higher cohesive strength and elevated temperature performance are desired, or

Chemical Basis of PSPs 189

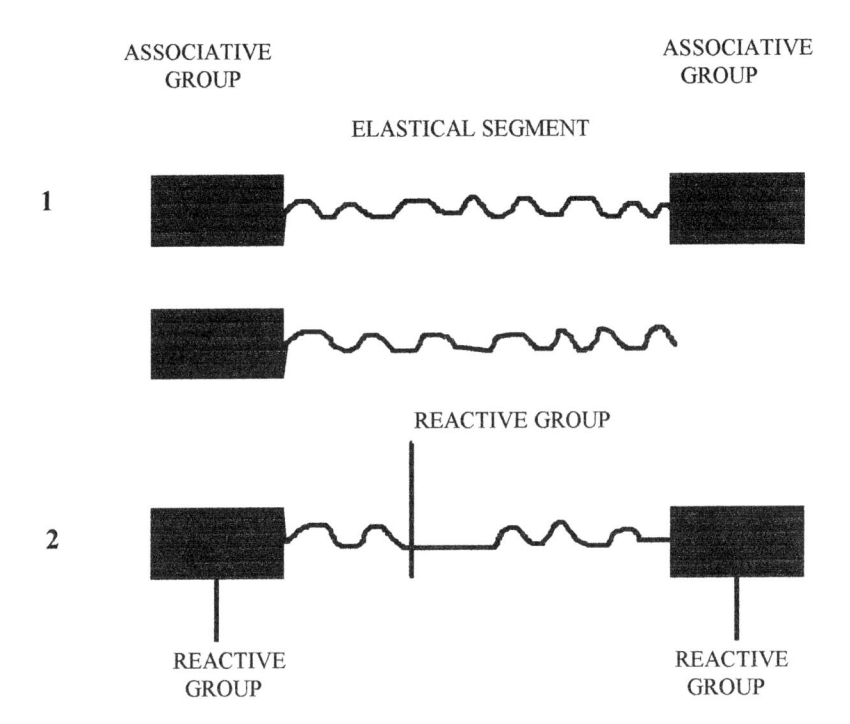

Figure 4.6 Schematic presentation of the macromolecular sites of a styrene-diene block copolymer able to be modified for postprocessing. (1) Unmodified triblock and diblock molecules with associative and elastic segments; (2) functionalized segments.

more tackifier resin should be used (where economy is a prime consideration) [191].

Johnsen [192] examined the influence of the triblock/diblock ratio also. The principles of formulating such polymers have been described by Dehnke and Johnsen [184]. SIS triblocks and SI diblocks have been manufactured separately and mixed together in a desired ratio, which allows the diblock level to be controlled precisely. Common SBCs are manufactured by using the coupling process. Commercial polymers contain 15–20% diblocks. Rubber theory states that uncrosslinked and unentangled polymer chain ends do not contribute to the strength of an elastomer. A pure block exhibits a sharper phase separation than multiblocks and consequently greater heat stability and creep resistance [193]. A range of copolymers with up to 42% diblocks have been synthesized. Such polymers are recommended for label PSAs. Increasing the diblock content improves peel and tack, by a static shear of over 180 h, which corresponds to the label requirements. SAFT decreases with the diblock content. The number of molecules associated decreases at the molecule end. Viscosity decreases, increasing the diblock content also (from about 40,000 to 21,000 mPa.s.) for a standard formulation with an SIS/resin/oil/antioxidant ratio of 100/150/25/1 [194].

The saturated midblock thermoplastic elastomers are commercial ABA-type block copolymers in which the polyisoprene or polybutadiene is hydrogenated. The block copolymers with a saturated midblock segment have an M_n of about 25,000–300,000. As shown in Ref. 195, the elastomer sequence of a rubbery block copolymer must be designed to adequately suppress crystallinity without at the same time raising the T_g. A saturated olefin rubber block has to be obtained with the lowest possible T_g and the best rubbery characteristics. The rubbery elasticity is maximized when the elastomer segment is completely amorphous. In commercial SEBS block copolymers, the best rubbery performance is obtained with rubber blocks having a T_g of about $-50°C$ and no detectable crystallinity. The strength and stiffness of the block copolymer depend on the chemical nature of the segments and their mutual interaction. Block copolymers having the same styrene content, about the same block molecular weight, but different types of midblocks display different mechanical properties. The SEBS polymer has the highest strength and modulus. The stress–strain behavior of such rubbery polymers can be adequately modeled by the Mooney–Rivlin equation as an elastic network (considering the trapped entanglements as finite crosslink junctions) or using the Guth–Gold equation relating the stress level (modulus) to the hydrodynamic effects of the styrene blocks considered as filler (see Chapter 3, Section 1).

In the classic segregated TPE construction, the butadiene (isoprene) units supply the elastic parts of the network and the associated aromatic rings of styrene (or styrene derivatives) provide the connection points of this network. The whole material works as a (spherical particle or short fiber) reinforced composite. It is evident that supplemental bridges between the chains can be formed if the elastic part of this construction is functionalized also. This transformation is made via copolymerization with vinylbutadiene or 3,4-isoprene units.

ABA block copolymers and AB block copolymers of the same type of composition in another geometry, star-shaped block copolymers or radial block copolymers, have also been developed [194]. According to the structure and placement of the side chains, branched radial teleblock and multiarm star block copolymers are known [196]. Various polymer structures are described by the terms "branched," "radial," and "star." "Branched" is a general term indicating a nonlinear structure, which may contain various polymeric subunits appended to various elements of the main polymer chain. "Radial" generally refers to the branched structures obtained by linking individual polymer segments to yield a polymer mixture having four or fewer arms joined centrally. The term "star" describes the structure of a multiarm polymer with copolymer arms that are joined together at a nucleus formed of a linking group. Nonterminating coupling agents are preferred as linking agents for star structures. A desirable melt viscosity and

Chemical Basis of PSPs

shear can be obtained by selecting only star block copolymers with six or more arms [196]. Multiarm styrene-isoprene star diblock copolymers with 18 arms having various styrene contents and styrene block weights have been synthesized also [197]. The effect on the number and molecular weight of the arms influences the solid-state morphology of SIS block copolymers [198]. Polymers with 2–18 arms and molecular weights of 100,000, 23,000, and 33,000 have been synthesized. Structures with more than eight arms are bicontinuous and ordered. Such segregation of the side chains is common for long and flexible chains in linear polymers too (Fig. 4.7).

Radial and teleblock copolymers with isoprene-based branches are used for PSA tapes [199], but butadiene-based armed polymers have been proposed also [200]. The star-shaped (radial) SBS copolymers developed by Phillips (Solprene 416X and 417X) have styrene contents of 30% and 20%, respectively, and molecular weights between 140,000 and 180,000 [201]. The grade 416X with more styrene has a lower molecular weight but higher tensile strength (200 vs. 90 kp/cm^2) and lower elongation (720% vs 920%) than 417X. The copolymer with lower styrene content exhibits better peel and tack values. According to Lau and Silver [196], the formulation with tackified star block copolymer gives a higher

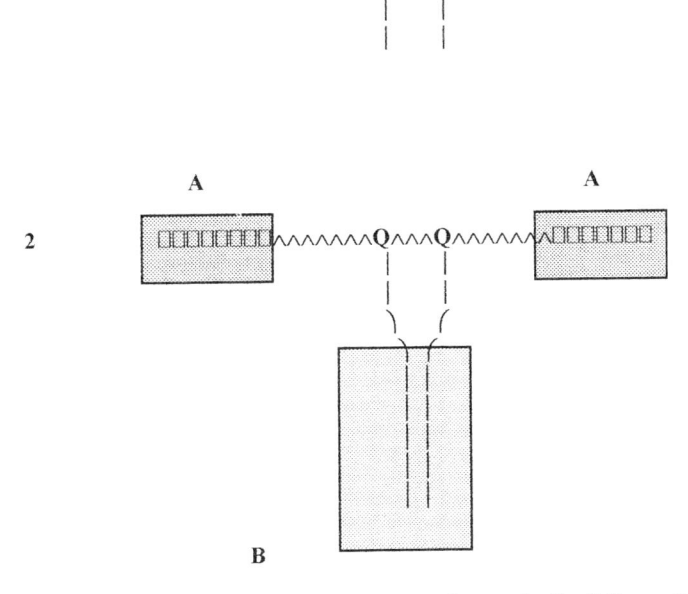

Figure 4.7 Side chain segregation depending on the flexibility and length of branches. (A) Short branches; (B) domain buildup by long branches.

shear resistance (at least 1000 min according to ASTM-D3654, overlap shear to fiberboard at 49°C). The mechanical properties of a radial block copolymer are better. The tensile strength for a radial styrene block copolymer is higher (22 MPa) than for a linear styrene block copolymer (15 MPa), and the 300% modulus is higher also (5 vs. 2.5 MPa). Elongation at break is smaller (620% vs. 950%).

The elastomeric midblocks and the polystyrene end domains are specially designed to be thermodynamically incompatible and to form a two-phase morphology [202]. The polystyrene blocks at the ends of molecules form an associative crosslinked structure. If the polystyrene level is lower than 50%, polystyrene domains are formed because of incompatibility [203]. A physical network is built up that disappears only above the T_g of polystyrene. The compatibility of the polystyrene endblocks and of the elastomeric midblocks depends on their molecular weights and their volume fractions in the mixture. Phase separation in the radial block copolymer is higher owing to the increased organization of the molecule. Therefore physical crosslinking is higher. The better connected pure polystyrene domains ensure higher elasticity of the product [204] (see also Chapter 3). Star-shaped SBCs have an (apparently) higher molecular weight. Therefore they are more reactive to radiation curing. The first radiation-curable styrene block copolymer was processed like a common thermoplastic elastomer, but it was also able to crosslink. This was a star-shaped SIS block copolymer. Because of its high molecular weight, a low radiation dose was needed to produce crosslinking. Because of its star structure, partial crosslinking does not modify (essentially) the flexibility of the molecules.

Styrene-isoprene block copolymers are more expensive than natural rubber. Alternatives are SBS and high styrene content (20–25%) SIS polymers. The adhesive for medical tapes has to be resistant to light and heat, show skin tolerance, and be permeable to water vapor. Such an adhesive may also include styrene copolymers. Common HMPSA systems for tapes are based on SIS, a resin that is partially compatible with the midblock of the SBC, and a plasticizer (oil). Tack properties of SBC are given by the midblock. Its mobility is increased by adding oils and tackifiers to the thermoplastic elastomer. A compatible resin is more effective than oil. The other part of the polymer sequence, the styrene endblock, is responsible for the cohesion and mechanical properties and influences the processing conditions. Its hardness is reduced by oils that are compatible with the styrene domains. Unfortunately, oils reduce cohesion [204]. According to Jacob [205], SIS block copolymers are used exclusively for tapes; as a less expensive alternative, SBSs can be used, but they are less suitable because of their greater rigidity and lower compatibility with tackifier. The branched isoprene midblock is more flexible and more compatible than the butadiene (butene) block. In some cases, the viscosity of SBS-based HMPSA formulations may change during storage; i.e., these products are less aging-resistant. The type of oil used (mineral oil or process oil) also influences the stability of the HMPSA [206].

Carboxylated styrene-butadiene block copolymers can also be used [207]. Styrene-acrylate block copolymers have also been synthesized [208]. They can be crosslinked via methacrylic end groups. Styrene block copolymers are used as compatibilizers (2–8% w/w) for the manufacture of plastic carriers and the recycling of plastics [209]. To improve the radiation curability of SIS block copolymer, products with a low styrene content (less than 15%) have been developed. These products can be crosslinked in a mixture with acrylic monomers and tackifier resins, giving an adhesive with improved temperature resistance and cohesion. Holding power values of more of 30 h have been obtained at 70°C (PSTC-7) [210].

The special segregated structure of SBCs display a crystal-like network of styrene domains. The ordered structure (actually crystalline order) ensures the excellent mechanical properties of some plastomers. Several attempts have been made to copy this structure, i.e., to prepare block copolymers with domainlike structures (see Acrylic Block Copolymers later). The main problems in preparing such polymers are related to the synthesis of blocks and to the ability of the chain segments to build up ordered structures. If it is possible for a polymer chain to remain active ("alive") and to grow for a long time, it should be possible to form long segments of two or more monomer molecules in the same polymer chain. The first example of living polymers was disclosed in 1956 by Szwarc [211] in the copolymerization of styrene with isoprene using sodium naphtalene as anionic initiator. Rather sophisticated block copolymers have since been synthesized using lithium alkyls as initiator. Complex, bifunctionally anionic initiators like 1,4-bis-(4-(1-phenylvinyl)phenyl)butane have also been tested [212].

Isotactic polypropylene has a T_g near 0°C; therefore it becomes brittle when refrigerated. This deficiency was overcome in part by blending isotactic PP with rubber. It has been supposed that incompatibility in blending would be decreased if the rubber molecule were bonded chemically to PP. In some cases such block copolymers have been synthesized with the help of Ziegler–Natta catalysts [213]. Theoretically, if we wish to synthesize tailored block copolymers we have to control certain characteristics of the polymer chain: the molecular weight and molecular weight distribution of each segment; the order of segments of specific monomers; the number of segments in the monomer chain; the composition and order of monomers in each segment; and the linearity, stereochemistry, and microstructure of the olefins and diolefins in each segment. In practice, the usability of such a copolymer depends on the interactions between its segments, which influence its processibility and mechanical properties.

Styrene and alkylstyrenes are unique monomers because they can be polymerized by means of many mechanisms and the bulky aromatic side chain of the vinyl unit may act as a functional group because of its electrons. Associations are also possible. Similar associations based on dipole–dipole interactions have been found in chemically crosslinked polymers. In copolymers of oligomethylenedimethacrylic esters with oligo(ethylene glycol) derivatives, interactions between the soft, elastic "bridges" between the crosslinked portions allow the formation of an ordered structure at 10–20°C above the T_g [214].

Statistical (random) copolymerization of butadiene and styrene has been used as a synthesis procedure for block copolymers also. In this case the segregation in the polymer is achieved by built-in ionic groups. Matsumoto and Furukawa [215] copolymerized styrene and butadiene with maleic acid (2%) and transformed the acid reaction product in salts. The macromolecular ions build up a star-shaped structure that substantially improves the mechanical properties of the polymer. Because of the dissociation of the ionic bonds at higher temperatures only, such polymers display good temperature resistance (up to 150°C). In comparison, a thermally reversible network based on hydrogen bonds (in experimental butadiene block copolymers having urazol groups), although it produces a longer rubberlike plateau of the modulus, is more labile and displays poorer mechanical performance [216]. Star-shaped block copolymers of styrene and ethylene oxide with two arms of polystyrene blocks and two arms of polyoxyethylene have also been prepared [217]. Star polybutadienes with four arms have been included between linear polybutadiene segments [218].

The behavior of SBCs illustrates that the segregated and ordered structure is a necessary condition for excellent mechanical properties. Bocaccio [219] synthesized a sequenced styrene-isoprene copolymer by depolymerizing natural rubber (MW 50,000) and grafting styrene onto the NR chains to produce a thermoplastic elastomer. Because such compounds are only slightly segregated, the mechanical characteristics of the polymers are not as good as those of SIS or SBS. On the other hand, the bond strength of HMPSAs formulated with these TPEs are comparable with those of compounds based on commercial SBCs.

The special characteristics of styrene-based segmented copolymers forced the development of similar structures based on other monomeric segments. Alternating block copolymers have been prepared from aromatic polyethersulfones and polydimethylsiloxane. Phase separation in rubbery polydimethysiloxane and vitreous polyethersulfone domains have been observed [220]. TPEs based on hard fluoroelastomer segments and soft fluororesin segments have also been produced [221]. Star-branched polyamides (nylon 6) retained part of their mechanical integrity even after melting, indicating the presence of enhanced entanglements [222]. In certain cases, properties other than mechanical performance have also been improved. For instance, dotted block copolymers of poly(3,4-diisopropylidene cyclobutene) with polynorbornene exhibit electrical conductivity of 10^{-4} S/cm [223].

Polymers with segregated structures can also be prepared by grafting. Grafting of polytetrahydrofuran chains onto polybutadiene leads to polymers having a two-phase structure of amorphous rubber and crystallites of polyether [224]. Thermoplastic elastomers with hydrogen bonding have been prepared from polybutadiene modified with 4-hydroxyphenyl-1,2,4-triazoline-3,5-dione [225]. The formed urazole groups act via hydrogen bonding, building thermally reversible crosslinks. Polymerization of vinyl chloride in the presence of modified Ziegler–Natta catalysts comprising long-chain organoaluminum compounds (e.g., polystyrene or polyisoprenylaluminum dichloride) leads to block copolymers.

The styrene-diene block copolymers prepared in various ways may have quite different compositions and structures. Because of their hard segments, such polymers are always harder than natural rubber (Table 4.2). Their modulus values are much higher than the common values (1×10^5 and 2×10^4 [235]) for PSAs. They behave like a filled natural rubber. Although the performance of such materials can be easily regulated by changing the molecular weight, styrene content, and sequence buildup, for pressure sensitivity they need to be tackified.

Due to their multiphase structure, the tackification of SBCs supposes special interactions with each of the polymer phases, i.e., unlike natural rubber, SBCs require tackifier mixtures. An additive may be compatible with both the elastomeric phase and the styrenic phase, compatible with only one of these two phases, or compatible with neither phase [140]. In general, resins compatible with only the elastomeric phase or with only the styrenic phase provide the best balance of properties. Resins and oils that are compatible with the elastomeric phase soften the final product, while resins that are compatible with the styrenic phase harden it. Both reduce the viscosity and affect the phase with which they are compatible. It is evident that in such multiphase compounds the concentration of tackifier resin influences the morphology of the polymer also. According to Meyer [236], styrene block copolymers can have five different morphologies, depending on their styrene content. For styrene concentrations of <15%, 15–25%, 35–65%, 65–85%, and >85%, spherical fibrillar and lamellar structures are formed.

For postcrosslinked systems based on block copolymers the behavior of the tackifier resin (absorbancy) toward radiation has to be taken into account also. For such systems the use of an unsaturation index U_f of the formulation has been proposed [134]:

$$U_f = \sum_i^t w_i U_i \qquad (4.15)$$

where i is a particular oligomer in the formulation, w_i is the weight fraction of oligomer i, U_i is the unsaturation index of oligomer i, and t is the total number of ingredients.

Polyacrylate Rubbers. High molecular weight elastomers based on acrylates have been produced by emulsion polymerization. Such polymers do not contain carbon–carbon double bonds as polydienes do. Their backbone includes a satu-

Table 4.2 Mechanical Characteristics of Segregated Elastomers

Type	Characteristics	Tensile strength (N/mm^2)	300% Modulus (N/mm^2)	Shore (A) hardness	Ref.
Natural rubber	Uncrosslinked	1–5	—	—	226
	Uncrosslinked; filled with 50% carbon black	24	10	—	227
Polybutadiene[a]	Uncrosslinked	0.3–0.5	—	—	130
Polybutadiene	Crosslinked	4–5	—	—	130
SBS copolymer	Linear; 31% styrene content	31	3.8	71	228
SBS copolymer	Radial; 30% styrene content	27	4.5	71	229
SBS copolymer	Triblock	30–40	—	—	130
SBS copolymer	Full triblock; 29% styrene content	31	—	65	230
SBS copolymer	Radial; styrene/butadiene ratio 40/60; MW 120,000	22	—	85	203
SBS copolymer	Linear, multiblock; styrene/butadiene ratio 43/57; MW 70,000	15	—	80	203
SBS copolymer	Melt index (g/10 min) <1	35	3	75	231
SBS copolymer	Melt index (g/10 min) 6	35	3	65	231
SBS copolymer	Melt index (g/10 min) 10	25	1	35	231
SBR copolymer	—	15.5–22	7–11	—	232
SEBS copolymer	20% Triblock	282	30[b]	—	140
SEBS copolymer	30% Triblock	490	—	72	140
SEBS copolymer	65% Triblock	34	27	—	140
SEBS copolymer	29% Styrene content	63	—	83	202
SEBS copolymer	28% Styrene content	49	—	—	202
SEBS copolymer	14% Styrene content	34	—	—	202
SIS copolymer	18% Styrene content	27	—	39	233
SIS copolymer	Fully coupled triblock	4500	—	—	192
SI copolymer	Fully diblock	20	—	—	192
MPR[c]	Halogenated polyolefin TPE	12.1	3.7	60	234
MPR[c]	Halogenated polyolefin TPE	13.1	4.5	70	234
MPR[c]	Halogenated polyolefin TPE	13.4	7.2	80	234

[a] Nonsegregated elastomer used as comparison.
[b] 100% modulus.
[c] MPR = melt processed rubber.

rated alkyl and alkoxy acrylate (95–99%) and 1–5%) curing monomers having a chlorine or hydroxy functionality in the side chain. Products with carboxy, epoxy, and isocyanate functionalities have also been developed. The main polymer chain is based on ethylhexyl or butyl acrylate. These polymers can be used for HMP-SAs (reactive hot melts), solvent-based PSAs, pressure-sensitive sealant tapes, and sealant caulks. They are used in tackified (40% resin) or plasticized formulations. Polyisobutylene can also be added. Their quick stick values are low; therefore a common tacky PSA can also be added to the formulation (50 parts PSA to 100 parts acrylic rubber). The peel value of the tackified formulation on aluminum (3–5 mil PSA coated on PET) is about 120–140 oz/in. They can be cured with isocyanates or UV curing agents (e.g., p-chlorobenzophenone). Polyacrylate rubber based formulations can include high loadings of fillers (200 parts hydrated alumina as filler per 100 parts rubber). The mill processibility and rheology of acrylic rubbers depend on the crosslinking agent used for their cure [237]. Acrylic rubbers based on ethyl acrylate-methacrylic acid-allyl glycidyl ether and ethyl acrylate or glycidyl methacrylate can be crosslinked by using onium salt catalysts [238]. Acrylic rubber latices having glass transition temperatures of −20 to +60°C with balanced mechanical characteristics and excellent hysteresis properties contain N-methylol acrylamide as crosslinking monomer and preferably itaconic acid [239]. Peelable protective films (30 μm) comprising methacrylate-(ethylene-methyl methacrylate) copolymers, an organic filler, and slip have been blow molded. Such films are useful for protecting rubber articles and show a peel strength of 165 g/25 mm, in comparison with 500 g/25 mm for ethylene-ethyl acrylate copolymers [240].

Acrylic Block Copolymers. According to Beaulieu et al. [241], the first acrylic hot melts having a common linear, nonsequenced (blockless) structure were introduced in the late 1970s. They were characterized by a modest balance of PSA properties and low cohesion at elevated temperatures [132]. These polymers show no rubbery plateau and exhibit a linear relationship between modulus and temperature. In the early 1980s, thermally reversible acrylic HMPSAs were introduced [242]. Such formulations are harder (Williams plasticity number 2.7) than the early common acrylics and exhibit better cold flow resistance. The new phase-separated acrylic block copolymers display (like SBCs) a thermally reversible crosslinking. As the material cools, the viscosity doubles with a 28°C temperature drop. As the temperature increases, the crosslinking mechanism is deactivated and the modulus drops. Early acrylic HMPSAs exhibited a viscosity of 30,000 cP at 350°F; the new polymers have viscosities of 8000–30,000 cP under the same conditions. The new acrylic HMPSAs exhibit peel values of 62–63 oz/in. (PSTC-1, 25 μm coating thickness on PET), loop tack of 2.5–30 oz, and rolling ball tack of 5–7 in. The shear values (according to PSTC-7, 1.2 in × 1in × 1 kg) are 15–76 h at room temperature. As seen, the rolling ball values are (in comparison with a common value of 1–2.5 cm) too high, i.e., the material is not tacky enough, and the shear values at normal temperature are also low. According to Sanderson [242], such values ensure adequate shear resistance. In reality, they are not high enough to satisfy practical requirements.

The hard blocks of such acrylic block copolymers can contain styrene derivatives also to achieve segregation by incompatibility and/or association. Segmentation and chemical crosslinking are also used. Acrylic hot melt adhesives based on acrylic block copolymers containing zinc carboxylates have been prepared by Enanoza [243]. Such HMPSAs with good melt flow and cohesive strength contain zinc carboxylates and copolymers of methacrylic ester (with C_{1-14} nontertiary pendant group) with polar comonomers and 2-(polystyryl)-ethyl methacrylate macromer. Although theoretically segregation may appear due to the styrenic component, the real elastic network is built up by crosslinking with zinc acetate.

Polyurethane-acrylate hot melt adhesives have been prepared from polyisocyanate-polyol copolymers and unsaturated monomers (alkyl acrylates and vinyl esters) [244]. AB block copolymers containing methacrylic acid and/or methyl methacrylate blocks (methyl methacrylate-styrene, *tert*-butyl methacrylate-styrene, and *tert*-butyl methacrylate-methyl methacrylate block copolymers) have been prepared by selective cleavage of methacrylic esters [245]. Low molecular weight block copolymers have also been prepared from acrylonitrile-methacrylic acid and butadiene [246]. Blocky butyl acrylate-styrene copolymers prepared via step emulsion polymerization showed a broad transition region in differential scanning calorimetry, in comparison with statistical copolymers, but two well-resolved relaxations associated with the glass transitions of styrenic and acrylic sequences [247,248]. The foregoing examples show that there have been many attempts to prepare segregated acrylic block copolymers; however, only a few experimental products promise industrial applicability. According to Ref. 249, in 1988 no commercially available acrylic hot melts had been produced.

Silicones. Elastomers based on silicones can also be used as raw materials for PSPs [250–253]. In comparison with organic polymers, silicones are fluids at high molecular weight (100,000) and crystallize at very low temperatures (−50°C); they exhibit a T_g of −120°C [253]. Silicone adhesives are temperature-resistant and can be coated on apolar substrates [254]. Double-faced mounting tapes used in electronics have silicone adhesive on one side and acrylic adhesive on the other [255]. For temperature-resistant tapes, tackified silicones (water-based dispersions of polydimethylsiloxane) have been suggested [256–259]. High temperature methylsilicone PSAs provide elevated thermal stability [260]. Silicone adhesive with a silicone release layer has been used for freezer tapes [261]. A fluid-permeable adhesive useful for transdermal therapeutic devices to be applied to human skin for periods of up to 24 h is based on an acrylic, urethane, or elastomer PSA mixed with a crosslinked polysiloxane [262]. The therapeutic agent passes through the adhesive into the skin. Pressure-sensitive adhesives made with silicone curable at low temperatures have been developed from rubberlike silicones containing an alkenyl group, a tackifying silicone resin, a curing agent, and a platinum catalyst [263]. Such adhesives for heat-resistant aluminum tapes cure in 5 min at 80°C. The adhesion of silicone pressure-sensitive adhesives can be improved by using high molecular weight dimethylsiloxane or dimethylsiloxane-diphenylsiloxane copolymer gum and resins having R_3Si_{0-5} units ($R = C_{\leq 6}$ hydrocarbyl) and SiO units, with resins having a polydispersity ≤ 2.0 [264]. Emulsified silicone PSAs could be blended with rubber latex and acrylic emulsion PSAs to improve their high temperature shear performance and adhesion to low energy surfaces [265].

Polyurethanes. Polyurethanes are synthesized as plastomers, elastomers, or viscoelastic compounds. Pressure-sensitive products use polyurethanes as rubber, viscoelastic PSA, top coat, and ink or film/foamlike carrier material [266]. Suitable adhesives are based on the chemistry of polyurethanes, i.e., on the chemical reaction between isocyanates and polyesters or polyethers having hydroxylic terminals (see also Chapter 6). The reactivity of isocyanates is used for crosslinking reactions also. The degree of crosslinking determines the hardness, tension yield, and elastic modulus of the polymer. Its thermal stability depends also on its chemical composition. Polyurethane adhesives are built up generally from hard and soft sequences. The soft sequences contain long polyols, polyethers, and polyesterdiols); the hard sequences include polyisocyanates and short-chain polyols or diamines. Thermodynamically these segments are incompatible, i.e., there is a two-phase morphology. The molecular weights of soft sequences are about 600–3000. Primary alcohols and hexamethylene diisocyanate are the preferred base components. Polyetherurethanes are more sensitive to thermal oxidative degradation. The resistance to hydrolysis is more pronounced for polyesterurethanes [267].

Polyurethanes were first raw materials for classic solvent-based adhesives. Later, water-based and hot melt adhesives were formulated. For instance, for PSAs, a glycerine-propylene oxide copolymer (with a molecular weight of 8000) has been suggested for use as the soft segment together with a hard segment of toluene diisocyanate-propylene oxide (with a molecular weight of 850). The reaction of the components occurs in situ, i.e., on paper, after coating [268]. A two-package type of PSA, i.e., an adhesive having two components based on PUR, contains a polyol component (MW 500–100,000) and a polyurethane having a terminal, free isocyanate group obtained by the reaction of an isocyanate component with an oxadiazine-2,4,6-trione ring and a polyol component having an average molecular weight of about 62–500 [269]. Photopolymerizable polyurethane precursors having latent catalysts as salts of organometallic complex cations from Groups IVB, VB, VIB, VIIB, and VIIB have also been prepared [270].

Thermoplastic elastomers based on polyurethanes have also been manufactured. Such thermoplastic polyurethanes (TPUs) are synthesized via polyaddition of polyisocyanates and polyols. [See (4.16).]

$$—[-\underset{\underset{H}{|}}{\overset{\overset{O}{\|}}{C}}-N-X-N-\underset{\underset{H}{|}}{\overset{\overset{H}{|}}{C}}-\overset{\overset{O}{\|}}{O}-(CH_2)_a-O-\underset{\underset{H}{|}}{\overset{\overset{O}{\|}}{C}}-N-X-N-\underset{\underset{H}{|}}{\overset{\overset{H}{|}}{C}}]_x—[-O-R-O]_y—[-\underset{\underset{H}{|}}{\overset{\overset{O}{\|}}{C}}-N-X-N-\underset{\underset{H}{|}}{\overset{\overset{H}{|}}{C}}-\overset{\overset{O}{\|}}{O}-(CH_2)_a-O-\underset{\underset{H}{|}}{\overset{\overset{O}{\|}}{C}}-N-X-N-\underset{\underset{H}{|}}{\overset{\overset{H}{|}}{C}}]_x— \quad (4.16)$$

The hard segments of such "block copolymers" are built up from polyisocyanates and a chain extender, and the soft sequences are polyols. The soft blocks are generally polyether or polyester with low T_g value and weak intermolecular forces, but hydroxy-terminated polybutadienes or polydimethylsiloxanes can also be used. As the polyisocyanate, aromatic and cycloaromatic diisocyanates such as 4,4'-diphenylmethanediisocyanate (MDI) or hexamethylenediisocyanate (HDI) can be used [271]. As the polyol, a polyester (e.g., butanediol-1,4-polyadipate or ethanediol-butanediol-polyadipate) has been proposed [272]. The composition of these TPEs is given by the general formula of multiblock copolymers:

$$—(AB)_n— \quad (4.17)$$

where n should be greater than 10. The soft sequences have a relatively low molecular weight of 600–4000. Such polymers possess a modulus of elasticity of 10–700 MPa. As mentioned earlier, such multiblock TPUs include a spherulitic crystalline phase. The crystalline regions reinforce the polymer. If more voluminous globular parts are synthesized as the hard domain, the mechanical performance of the elastomer can be improved. As reported by Kunii [273], the introduction of polyoxy(tetramethylene glycol) into the polyethylene oxide and polypropylene oxide sequences without modification of the global sequence length of a common propylene oxide/ethylene oxide polymer increases its tensile strength. Polyurethane ureas synthesized with ethylene and propylene oxide, MDI, and ethylenediamine display improved thromboresistance [274]. Thermoplastic polyurethanes based on ring-opening polymerization of δ-valero-lactone and ε-caprolactone are biodegradable [275]. Such polymers possess moduli of elasticity of 0.12–3.83 and elongations of 60–2000%. α-Alkylolefin-urethane block copolymers have also been synthesized [276]. Such copolymers are based on α-alkylolefin polymers bearing NCO reactive groups, polyisocyanates or isocyanate prepolymers, and chain extenders. Generally PUR derivatives are used for medical PSPs.

Propylene Copolymers. Amorphous polypropylene-based HMPSAs have been known for many years [277]. Atactic polypropylene (with a molecular weight of 15,000–60,000) can be used as the base elastomer for PSAs [277]. The selection

Chemical Basis of PSPs

of suitable comonomers and polymerization conditions led to the development of olefin polymers that are pressure sensitive in the neat form [278]. Such polymers have a T_g of $-25°C$. In their formulations no oil is needed (no migration and less skin irritation are given by common tackifiers). They display good thermal stability (less than 10% viscosity break after 100 h at 177°C), low color and odor, lower density, and the ability to be modified with most olefin-compatible ingredients. According to Clubb and Foster [279], amorphous polypropylene obtained as a by-product has a T_g of $-14°C$. Amorphous polypropylene obtained by direct synthesis shows a T_g of $-12°C$. A mixture of 75–85% atactic polypropylene with a molecular weight of 15,000 to 60,000 and 15–25% terpenephenol tackifier resin with a molecular weight of 1200 and a melting point of 115°C has been suggested as an HMPSA [280]. A propylene-butene amorphous polyolefin (APO) copolymer exhibits a T_g of $-16°C$. A patent [281] discloses a PSA formulation based on amorphous polypropylene, styrene block copolymers, and tackifier. The formulation discussed by Wakabayashi and Sugii [277] is based on polypropylene, styrene block copolymer, tackifier, and wax and displays special electrical properties.

Atactic polypropylene has been manufactured as a by-product in the synthesis of isotactic polypropylene. Other products are prepared by the copolymerization of propylene with low molecular weight alkenes such as ethylene or 1-butene. Allen et al. [282] give a raw material formulation that includes ethylene (10–30%), propylene (65–90%), and, optionally, a C_{4-8} olefin ($\leq 15\%$). The tackiest polymers have been synthesized by using hexene and octene as comonomers. The nature of the comonomer influences the crystallization rate of the product, i.e., its shelf life. Copolymers having short-chain comonomers are used for "semi-pressure-sensitive" adhesives. Such products lose their tack after a certain time because of crystallization. Their adhesive properties can be improved by using tackifiers (hydrocarbon, terpene, and rosin resins) and plasticizers (polyisobutylene, polybutene, or mineral oil) and thermoplastic elastomers of styrene (3–7% w/w. SIS or SEBS). HMPSA compositions based on ethylene-propylene copolymers (with 20–50% propylene), aliphatic tackifiers, and naphthenic oil applied on paper (with 20 μm thickness) showed similar tack but better weatherability and better die cutting performance than SBC-based formulations [283]. An HMPSA based on atactic polypropylene includes PP with a molecular weight of 15,000–60,000 and terpene resins with a molecular weight of 1200 [281].

Atactic polypropylene can be crosslinked by using gamma radiation [284]. In a quite different manner from crystallization, where the "condensed" crystalline order, i.e., the crystalline network, causes plastomer-like behavior, i.e., the loss of pressure sensitivity, the crosslinked elastic network exhibits rubberlike properties. Such crosslinked atactic polypropylenes are elastomers. Oxidatively degraded ethylene-propylene copolymers are useful as softeners for rubber resin adhesives for tapes [285].

Polyesters. Thermoplastic elastomers can also be prepared as block copolymers based on esters. A thermoplastic copolyester may have the formula [234]

$$—[-O-(CH_2)_a-O-\underset{\underset{O}{\|}}{C}-AR-\underset{\underset{O}{\|}}{C}-]_x—[-O-(CH_2CH_2CH_2CH_2O)_b$$

$$-O-\underset{\underset{O}{\|}}{C}-AR-\underset{\underset{O}{\|}}{C}-]_x—[-O-(CH_2)_a-O-\underset{\underset{O}{\|}}{C}-AR-\underset{\underset{O}{\|}}{C}-]_x— \quad (4.18)$$

where $x = 1–1.1$, $y = 7–10$, $a = 4$, and $b = 12–16$. Such a copolymer can have hard segments of butylene terephthalate and flexible segments of polyalkylene oxide terephthalate [286]. If polytrimethylene terephthalate blocks are reacted with acrylonitrile-butadiene copolymer blocks, a multiblock copolymer is formed that has crystalline portions due to the hard terephthalate segments. Such polymers resist temperatures up to 125°C. Over 150°C the unsaturation of the polybutadiene segments allows supplemental crosslinking [287]. Pressure-sensitive adhesives based on radiation-curable polyesters include macromers (liquid saturated copolyesters with one terminal acrylic double bond per 3000–6000 molecular weight units) that have glass transition temperatures of -10 to -50°C. These macromers can be coated at elevated temperatures (100°C) and have viscosities of 2000–10,000 mPa.s. [288].

2.2 Viscoelastic Components

The use of elastomers as raw materials for adhesives led to the formulation of viscoelastic materials displaying elastic and viscous properties without the need for supplemental viscous components. A broad range of comonomers can be used to synthesize such polymers. Olefinic, vinylic, acrylic, and maleic comonomers have been used. Water-based PSAs based on polyacrylate latices, ethylene-vinyl acetate copolymer latices, and tackified rubber latices (natural rubber, functionalized or nonfunctionalized styrene-butadiene) have gained a significant share of the total PSA market for labels, tapes, protective films, and miscellaneous products. Benedek and Heymans [1] give a detailed description of the common viscoelastic raw materials used for PSAs. The present discussion concerns the product-related features of these raw materials.

Acrylic Copolymers

The most important viscoelastic polymers used as raw materials for PSAs are copolymers of acrylic and methacrylic esters. Copolymers of acrylic monomers (long side chain derivatives of acrylic and methacrylic acid) have been synthesized as raw materials for PSAs. They are used for almost all PSP classes, mainly for labels and tapes but also for protective films. Generally, common acrylics ex-

Chemical Basis of PSPs

hibit balanced adhesive properties (see also Chapter 5), which allow their use as is or tackified, primarily for labels. Formulations in this application field require relatively small changes in their adhesive properties. Their versatility to be used with low levels of tackifier or additives has been imposed by their coating on migration-sensitive PVC and paper carrier in the first period of their application history. Crosslinking is the main possibility of tailoring them for special applications. As can be seen from the thermomechanical plot of an amorphous polymer (Chapter 3, Fig. 3.3), there is a plateau (of the modulus) corresponding to rubbery/elastic behavior of the material as a function of temperature. The existence of this plateau and its position (T_g) depends on the molecular weight of the polymer. Acrylics possess a very broad range (-85 to $+185°C$) of T_g values [289]. If the molecular weight and the molecular weight between entanglements of the polymer do not attain a critical value (M_c), no rubbery elastic plateau appears. That means that for low molecular weight polymers a "pressure-sensitive" bonding may appear but pressure-sensitive debonding properties do not exist. In this case the tack of the polymer is not sufficient and the cohesion of the material is low. This is the case for low molecular weight acrylics developed for hot melts.

To understand how and why a polymer will behave, one must know the macrostructure as well as the microstructure. Characterization of the macrostructure requires knowledge of the weight average molecular weight (M_w), number average molecular weight (M_n), molecular weight distribution, and branching. Polyacrylates are amorphous polymers. Their properties depend on their molecular weight and chain mobility. For high molecular weight acrylics, the melting point of the polymer is higher than the depolymerization temperature. Therefore their molecular weight is limited. For hot melts, only relatively low molecular weight polymers can be synthesized, and they need a postapplication increase of their molecular weight [132]. In this situation some years ago HMPSAs based on acrylics were considered only as hypothetical products [290], and in a more recent publication Wabro et al. [56] stated that the synthesis of acrylic block copolymers had not yet been carried out.

Acrylics have to be crosslinked to reduce their chain flexibility, i.e., bonding performance. This requirement appears for removability (see also Chapters 5 and 6). Partial crosslinking of water-based acrylics can be carried out during their synthesis. For instance, acrylate-vinyl copolymers with about 3% tetraethylene glycol diacrylate have been polymerized using macromolecular stabilizers (protective colloids) to produce adhesives with better flow (near Newtonian), shear of 500 min, peel strength of 5 pli, and tack of 750 g [291].

For special applications, hydrophilic compositions are needed. Polar functional groups ensure hydrophilicity. Increased polarity and functionality improve bonding performance. To regulate them (i.e., to ensure removability), crosslinking may be necessary. In some cases, porous or water-absorbent macromolecular structures are needed. Acrylic PSAs have been used for many years for medical and surgical applications [292–294]. Acrylic adhesives have been proposed for surgical tapes [295,296]. The acrylic adhesive for medical tapes has to be resistant to light and heat, show skin tolerance, and be permeable to water vapor. Polyacrylate-based water-soluble PSAs are suggested for medical PSPs (operating tapes, labels, and bioelectrodes) [297].

The chemical basis for crosslinking is given by built-in unsaturation or functional groups. The free radical crosslinking of multifunctional monomers or macromers with acrylic groups can lead to network structures (Fig. 4.8). Such functionality is used to enhance the crosslinking ability of various raw materials. Polyfunctional acrylates are suggested for the crosslinking of ethylene-propylene rubber [298]. As stated by Sasaaki [299], the EB curing of SBS using a dosage

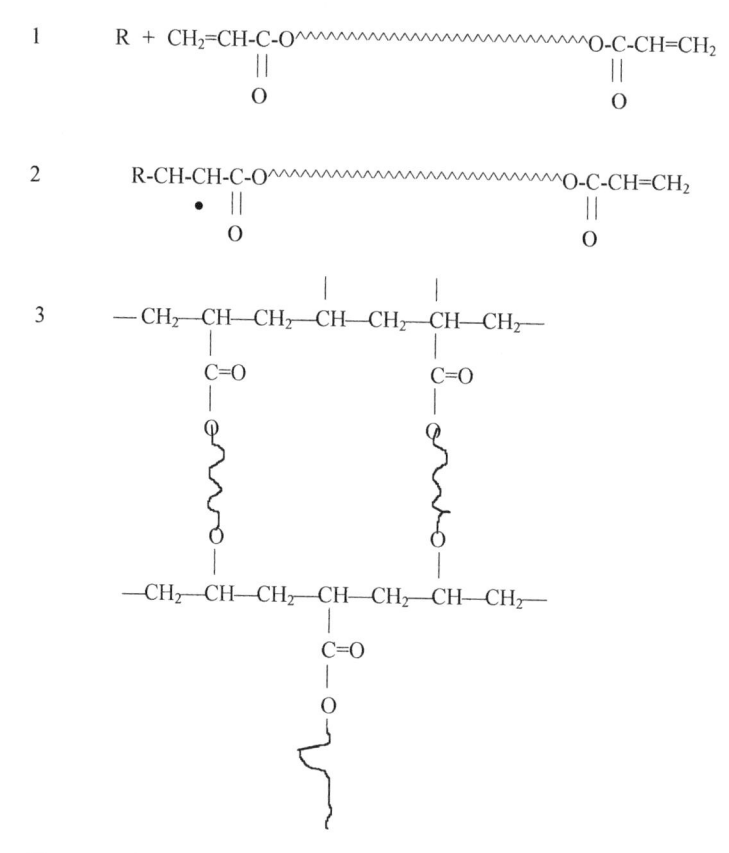

1 R + CH$_2$=CH-C-O~~~~~~~~~~~~~~O-C-CH=CH$_2$
              || ||
              O   O

2 R-CH-CH-C-O~~~~~~~~~~~~~~~~~O-C-CH=CH$_2$
         • || ||
            O   O

3 —CH$_2$—CH—CH$_2$—CH—CH$_2$—CH—CH$_2$—
            | |
            C=O   C=O
            | |
            O   O
            {   {
            }   }
            {   {
            O   O
            | |
   —CH$_2$—CH—CH$_2$—CH—CH$_2$—CH—CH$_2$—
                   |
                   C=O
                   |
                   O
                   {
                   }

Figure 4.8 Buildup of a crosslinked network based on acrylic functionality. (1) Reaction of a free radical with a linear macromer; (2) linear macroradical; (3) molecular network.

of 89 kGy without crosslinker leads to a polymer with 47 h shear at 70°C; with a multifunctional acrylate crosslinker and a dosage of 30 kGy, more than 1440 h of shear is obtained.

It is well known that the main acrylic pressure-sensitive copolymers contain acid groups. The increase in the level of carboxy groups allows a crosslinking mechanism as in CSBR. Special acid-sensitive organic crosslinkers or multivalent metallic salts and oxides may be used for their curing. For instance, a special crosslinkable carboxylated acrylic with a high T_g ($-25°C$) has been developed [300] for high shear, low tack applications and for contact adhesives. Mounting tapes and double faced foam tapes are made using this adhesive.

The (neutralized) carboxylic groups may provide water solubility for solvent-based acrylics. Tack and heat stability together with water solubility or dispersibility are required for splicing tapes in papermaking and printing. For such products water solubility is given by a special composition containing vinyl carboxylic acid (10–80% w/w) neutralized with an alkanolamine [301]. The irritation caused by the removal of a medical tape applied to the skin was overcome by including in the adhesive certain amine salts [131], making it water-washable. As stated by Wistuba [9], carboxylic functionality in such compounds also influences their adhesion. The adhesion coefficient (K_{adh}) depends on the number of carboxy groups according to the correlation

$$K_{adh} = k \, (COOH)^n \tag{4.19}$$

where $n = 0.59$, $k = 0.87$ N/mm, and (COOH) = moles of COOH per 1000 polymer molecules.

Generally the crosslinking formulation also contains other multifunctional monomers. In order to regulate the elasticity of the crosslinked network, difunctional monomers of various chain lengths may be used. Copolymers of n-butyl acrylate with acrylic acid and glycidyl methacrylate, N-vinylpyrrolidone, methacrylamide, acrylonitrile, and methacrylic acid have been tested as PSAs [302]. Such polymers have a molecular weight average of 200,000–1,500,000 and an M_w/M_n ratio of 2.0–4.0 [303], i.e., a broad molecular weight distribution.

Mostly solvent-based or 100% solids acrylics are suggested for such special applications. Their manufacture includes the synthesis of the main polymer and its crosslinking. Thermally induced or radiation-induced polymerization is used to manufacture the main PSA. Thermally induced or radiation-induced curing may be used for its crosslinking also. The adhesive can be synthesized as a thermally noncrosslinkable prepolymer and its recipe can be modified before crosslinking (Fig. 4.5). It has been suggested that a prepolymer be manufactured that is then polymerized by radiation [138]. A low molecular weight spreadable composition is prepared to which may be added a small amount of catalyst or polyfunctional crosslinking monomer prior to completing the polymerization by heat curing.

Postapplication crosslinking is proposed to improve shear or to decrease peel resistance, manufacturing a storable crosslinkable adhesive. Postcrosslinking is achieved by thermally initiated or photoinitiated reactions [304]. A latent crosslinking agent has to be used. As crosslinking agent a lower alkoxylated amino formaldehyde condensate having C_{1-4} alkyl groups (e.g., hexamethoxymethylmelamine) has been proposed. A prepolymer can be synthesized by UV-initiated photopolymerization, then the crosslinking agent is added, and the mixture is coated, and polymerized by UV to the final product.

According to Harder [305], HMPSA on acrylic basis has been prepared via solution polymerization. Relatively low molecular weight, low viscosity polymers have been synthesized and isolated by evaporating the solvent and degassing the polymer as solid-state raw materials for HMPSA. In such low molecular weight polymerizations, a gel effect may appear, allowing a high reaction rate. The gel effect is an autocatalytic phenomenon occurring at high conversion rates and is marked by a reduction in the termination of polymer chain radicals. The termination rate is decreased because the increased viscosity at high conversion makes it more difficult for two polymer chain radicals to come together to produce a termination reaction. Such polymers can be coated at 90–140°C and postcrosslinked. The HMPSA can be crosslinked by using EB and UV curing. Chemical crosslinking (by NCO-containing systems) is possible also.

Both operations, the preparation of the prepolymer and its postpolymerization, can be effected by radiation-induced polymerization also. Photopolymerization of the monomers neat, without any diluent that needs to be removed after the polymerization, provides processing advantages [306]. The liquid monomers are applied to the carrier, which is an endless belt, and the polymerization is completed thermally in an oven or by UV light. A patent [307] describes the polymerization of acrylic monomers directly on the carrier to manufacture a tape. The acrylic ester PSA formulation is a combination of non-tertiary acrylic acid esters of alkyl alcohols and ethylenically unsaturated monomers having at least one polar group, which may be substantially in monomer form or may be a low MW prepolymer or a mixture of prepolymer and additional monomers; it may further contain other substances such as a photoinitiator, fillers, and crosslinking agents (such as multifunctional monomers). The acrylic esters generally should be selected primarily from those that as homopolymers possess some PSA properties. Diacrylates are used as crosslinking agents in amounts of about 0.005 to 0.5% by weight. Typically the crosslinking agent is added after the formulation of the prepolymer. As photoinitiator, 2,2-dimethoxy-2-phenylacetophenone has been proposed also.

If UV-induced photopolymerization is used, there is a choice between external and built-in photoinitiators. Generally benzophenone-type initiators are suggested. The use of benzophenone is known from the light-induced crosslinking of ethylene-propylene copolymers [308]. Benzophenone or deoxybenzoin have been tested for UV photocrosslinking of butadiene-styrene copolymers also [309]. Benzyldimethylketal can be used also [134,310]. For instance, pressure-sensitive adhesive tapes whose adhesion can be decreased by UV irradiation are prepared from an elastic polymer, a UV-crosslinkable acrylate, a polymerization initiator, and a methacrylate photosensitizer containing an NH_2 group [311]. An ethylene glycol-sebacic acid-terephthalic acid copolymer (having a T_g of about 10°C), a tackifier resin, dipentaerythritol hexaacrylate as crosslinking monomer, benzyldimethylketal as photoinitiator, and dimethylaminoacrylate have been polymerized. The resulting adhesive has been coated on a 100 μm PVC carrier primed with a modified acrylate. Such an adhesive displays a peel resistance of 140 g/25 mm before irradiation and 52 g/25 mm after UV irradiation.

The initiating mechanism via benzophenone-containing systems and the main photoinitiating systems have been described in detail by Röltgen [312]. The photoinitiator absorbs light energy, and $n \rightarrow \pi$ or $\pi \rightarrow \pi^v$ electron transfers occur. The excited electron is in the singlet state. It can return to the basic state or go into a more stable triplet state. The photopolymerization involves the participation of an excited (singlet or triplet) form of the monomers or oligomers. Light absorption is needed to achieve this excited state. Chromophoric molecular parts are required to ensure light absorption. The exciting energy can be transmitted from a photosensitizer also. The photoinitiated polymerization occurs via free reactive species (radicals or ions) that initiate thermal polymerization [see expression (4.11)]. Short-lived intermediates (biradicals, carbenes, nitrenes) can also be formed that cause crosslinking. The photolysis of the initiator occurs via intramolecular dissociation (benzoin derivatives, benzyl and acetophenone ketals, peroxides, perhalogenides, benzoyloxime esters, etc.), interchain hydrogen abstraction (aromatic ketones such as benzophenone, thioxanthones, quinones), or photoinitiated transfer of electrons (ketones with amines, aromatic onium derivatives with electron donors, etc). The wavelength of the light should be between 300 and 800 nm. The efficacy of the initiation expressed as liters per mole per centimeter is between 10 and 50,000 (benzophenone has values of 100–200) [313].

The aromatic ketones absorb UV radiation to form a triplet excited state. These molecules can abstract hydrogen radicals from the polymer and create free radical sites on it. Benzophenone may be used together with amines where initiating radicals are generated by photoreduction of benzophenone by the amine, via formation of an intermediate complex between a photoexcited (triplet state) benzophenone molecule and a ground-state polyamine molecule [314]. If many polymer layers are manufactured by photopolymerization, their macromolecular characteristics are regulated by the concentration of the crosslinking agent and that of the photoinitiator (see Chapter 6). The use of external photoinitiators has the disadvantage that they interact with atmospheric oxygen. By quenching, the (excited) photoinitiator reacts with oxygen and will be deactivated; active polymeric radical sites are scavenged by oxygen molecules to give stable peroxides.

As demonstrated by Decker [315], the exposure time for UV curing in air is of 0.2 and in nitrogen, 0.03. Therefore crosslinking by photosensitive monomers with a built-in photoinitiator has also been tested.

Photopolymerization as a generally useful finishing technology has been developed for various applications. Copolymerizable photoinitiators have been studied for the UV curing of epoxy acrylates and hexanediol diacrylates [316]. Photoinitiators for waterborne UV-curable coatings can contain oligomeric hydroxyalkylphenyl ketone as photoinitiator [317]. Monoethylenically unsaturated aromatic ketones can be used for crosslinking PSA compositions with UV light [318]. Such monomers are copolymerized in the adhesive [132]. Preferred monoethylenically unsaturated aromatic ketone monomers have the general formula

$$R - \underset{\underset{O}{\|}}{C} - \underset{\underset{(X)_n}{|}}{Ar} - Y - Z \qquad (4.20)$$

where R is a lower alkyl or phenyl, provided the R may be substituted with one or more halogen atoms, alkoxy groups, or hydroxy groups, X is halogen, alkoxy, or hydroxyl; n is an integer from 0 to 4; Y is a divalent linking group; and Z is alkenyl or ethylenically unsaturated acyl. The aromatic ketone monomer is free of ortho aromatic hydroxyl groups. Particularly preferred monomers are the acryloxybenzophenones, e.g., *para*-acryloxybenzophenone. Unfortunately, the polymerization of such monomers may be accompanied by side reactions (see Chapter 6).

Crosslinking via polymer-bonded benzophenone groups has been studied on model polyimides containing benzophenone as well as alkyl-substituted diphenylmethane groups in the main chain. As stated, the crosslinks are formed upon UV irradiation through hydrogen abstraction by triplet benzophenone from the alkyl groups acting as hydrogen donors and subsequent coupling of the formed radicals. The quantum yield of the process is low. This is attributed to a specific energy dissipation process that operates at the reactive site, namely, reversible hydrogen exchange between benzophenone and the hydrogen donor [318]. There is a correlation between the activity and structure of polymeric photoinitiators containing side chain benzophenone [e.g., poly(4-vinylbenzophenone) or 4-acryloyloxybenzophenone] chromophores. It has been confirmed that the main activation mechanism in the UV polymerization of acrylic monomers by benzophenone-containing polymeric photoinitiators is intramolecular hydrogen abstraction by excited benzophenone moieties [319]. Those photoinitiator systems show high activity, with both benzophenone and amine groups displaying a large conformational mobility.

Such built-in photosensitizers have been tried for styrene-butadiene copolymers also. A styrene-butadiene copolymer has been modified by the addition of benzoyl or phenylacetyl groups. The built-in functional groups served for the postcrosslinking of the polymer [320]. Mannich polymers can be used as photosensitizers for onium salts for the polymerization of vinyl monomers too [321].

Electron beam crosslinkable acrylics contain a multifunctional monomer such as pentaerythritol triacrylate to allow curing with low energy use [322]. Such adhesives display good high temperature shear strength. As stated by Braun and Brügger [323], most EB-curable PSAs are AC functionalized; they are based on (1) postcrosslinking of SIS HMPSA; (2) acrylate functionalized polyester prepolymers; (3) acrylate functionalized liquid rubber; or (4) acrylate prepolymers.

Acrylate-based PSAs containing acid or acid anhydride groups and glycidyl methacrylate coated on a flexible carrier can be crosslinked thermally between 60 and 100°C using a zinc chloride catalyst [306]. Instead of acrylate copolymers with monomers containing epoxide groups and compounds containing an acid anhydride group, mixtures of two separately prepared copolymers can also be used, for example, an acrylate copolymer containing epoxide groups and an acrylate copolymer containing anhydride groups. Crosslinking catalysts (0.05–5% w/w) are added to the polymer. The polymer coated on a carrier is crosslinked (after solvent evaporation) for 2–7 min at 60–100°C. Zinc or magnesium chloride, monosalts of maleic acid, organic phosphoric or sulfonic acid derivatives, or oxalic or maleic acid can be used as the catalyst. Tertiary amines and Friedel–Crafts catalysts accelerate crosslinking. Compared to common uncrosslinked rubber-based PSA (with a holding power of some minutes at 100°C), the holding power of these adhesives attains 20 h at 150°C, and they are more resistant to water and solvents.

Principally the same comonomers can be used as crosslinking agents for water-based formulations. Adhesives for diaper tapes have to display high shear. Therefore crosslinking monomers are added (e.g., N-methylolacrylamide) to increase their shear resistance [324]. Unsaturated dicarboxylic acids may be included as crosslinking agents [325] together with polyfunctional unsaturated allyl monomers [326], and di-, tri-, and tetrafunctional vinyl crosslinking agents [327]. The polyolefinically unsaturated crosslinking monomer enhancing the shear resistance should have a T_g of −10°C or below. Shear resistance is improved by using a special stabilizer system comprising a hydroxypropyl methyl cellulose and an ethoxylated acetylenic glycol.

Crosslinking leads to a blend of polymers having different molecular weights. Pressure-sensitive macromolecular compounds of different molecular weights can be prepared by other methods also. According to numerous patents (see later), PSAs synthesized via suspension polymerization are used together with solution- or emulsion-polymerized adhesives. The PSA surface can have a special discontinuous shape ensured by the PSA itself if it is a suspension. Thus spherical contact sites are formed [326]. The anchorage of the spherical PSA particles needs a continuous matrix. In this case the PSA particles synthesized by suspension polymerization and the adhesive used as matrix are of different molecular weights. Suspension polymerization leads to lower degrees of polymerization than the emulsion procedure [328].

Vinyl Acetate Copolymers

Polyvinyl acetate homopolymer has been used in solution- and water-based dispersions for classic adhesives. Its copolymerization with comonomers that allow a lower T_g leads to raw materials for PSAs. Vinyl acetate-acrylate, vinyl acetate-maleinate, and ethylene-vinyl acetate copolymers are the main representatives of this class of raw materials. Copolymers of vinyl acetate have been used as raw material bases for classic adhesives. Vinyl acetate-acrylate copolymers and vinyl acetate-ethylene copolymers have been manufactured as raw materials for solvent- and water-based pressure-sensitive adhesives. Ethylene-vinyl acetate copolymers have been developed for semi-pressure-sensitive adhesives also and also as primers for extrusion coating [329].

As discussed by Benedek [330] and by Benedek and Heymans [1], because of their unbalanced adhesive properties and limited tackification ability, vinyl acetate copolymers are not competitive with acrylics for use in labels. However, copolymers of vinyl acetate with long-chain acrylic comonomers are raw materials for PSAs. Vinyl pyrrolidone-vinyl acetate copolymers are used for water-soluble PSAs [331]. Adhesives for medical tapes have also been formulated on vinyl acetate copolymer basis [332]. Ethylene-vinyl acetate-vinyl chloride copolymers can also be used for medical tapes [331]. The use of ethylene as an inexpensive low T_g comonomer allows the synthesis of new raw materials based on vinyl acetate, maleinates, and acrylics for PSAs. Ethylene-vinyl acetate (EVAc) copolymers were first marketed in the early 1960s [333]. The first European EVAc-based pressure-sensitive aqueous dispersion was Vinnapas EAF 60, manufactured by Wacker [334]. The so-called combination adhesives are made as blends of PSA and non-pressure-sensitive EVAc dispersions (70/30) [335]. Such formulations used as flooring adhesives contain a PSA on EVAc basis and a non-pressure-sensitive EVAc dispersion, resins, fillers, and additive. They can be applied as one-sided flooring adhesives after a drying time and display excellent thermal stability [336]. So called legging adhesives [337] are also formulated. Here the addition of a resin allows the regulation of the legging (see Chapter 3, Section 1). Removable flooring adhesives are also required. Such adhesives can be removed with or without water (see Chapter 8). The PVA content of EVAc dispersions allows their easy removal with water. Systems that can be removed without water are also available. Here an EVAc-based primer is applied on the substrate to facilitate removability. High shear pressure-sensitive EVAc emulsions resistant to high temperatures are used for labels in the automotive industry and for tiles and packaging tapes [337]. Generally EVAc dispersions are used for flooring adhesives and as tackifiers for contact adhesives for tapes, labels, and adhesive films [338]. Slightly pressure-sensitive ethylene-vinyl acetate copoly-

Chemical Basis of PSPs

mers with a minimum film-forming temperature (MFT) of about 0°C and T_g of −20°C have been suggested as one-sided flooring adhesives also [339]. Ethylene-vinyl acetate copolymer emulsions useful as carpet adhesives have good compatibility with SBR emulsions [340]. EVAc copolymers with improved plasticizer, temperature, and shrinkage resistance, and good cohesion have been proposed for flooring adhesives [341].

Special segmented EVAc copolymers have been developed as raw materials for HMPSAs. They have a higher VAc content (36%) and can be considered elastomers [342]. They exhibit improved aging stability and adhesion on soft PVC [343]. Their adhesive characteristics are not as good as those of SBCs. A peel adhesion of 14.5–24.0 N/25 mm (PSTC-1, 180°C), rolling ball tack of 11–50 cm, Polyken tack of 2.5–5.6 N (300–600 g), and shear adhesion of 1–7 h (23°C) have been achieved on paper with an adhesive coating thickness of 25.4 μm. The comparative examination of the adhesive characteristics of HMPSAs based on different segregated copolymers shows that SBCs exhibit the best performance (Table 4.3) although they are less tacky than PSAs based on natural rubber.

A hot melt PSA on EVAc basis contains about 35.5% elastomer, 30–50% tackifier, 0.2% plasticizer, 0.5% filler, and 0.1–0.5% stabilizing agents [109]. The VAc content does not influence the melt viscosity. The peel value can be increased by increasing the VAc content. As plasticizer, phthalates, phosphates, chlorinated polyphenols, and liquid rosin derivatives have been used. As tackifier, rosin derivatives, hydrocarbon resins, and low molecular weight styrene polymers have been suggested. Zinc oxide, calcium carbonate, titanium dioxide, barium sulfate, and organic resins have been recommended as filler. High mo-

Table 4.3 Adhesive Characteristics of Pressure-Sensitive Formulations Based on Thermoplastic Elastomers[a]

Chemical composition				Adhesive characteristics		
Elastomer		Tackifier		Peel resistance	Shear	
Type	Level (pts)	Level (pts)	Tack (cm)	(N/25 mm)	resistance (h)	Ref.
SEBS	100	115	4	15	100	202
SEBS	100	120	7	12	>83	344
SBS	100	125	2.5	6	>150	205
SIS	100	125	1.5	5	>150	205
EVAc	100	217	21	8	5	345
EVAc	100	150	13	24	7	338
AC	100	—	12	62	15	242
NR	100	125	1.5	7.5	1	346

[a] 180° peel on stainless steel and room temperature static shear were measured.

lecular weight PVAc dispersions are used for hot sealing at elevated temperatures (120°C), low molecular weight ones at temperatures lower than 60°C [342]. A polyvinyl acetate polymer can be used as filler in UV-curable formulations [302]. Water-soluble hot melt adhesives can be formulated with vinyl pyrrolidone-vinyl acetate copolymers [347,348].

Pressure-sensitive cold sealing adhesives have been manufactured from EVAc copolymers with 55–85% ethylene, tackifiers, and special microsphere fillers [349]. Such adhesives can include natural rubber latex (2–32 parts and vinyl acetate copolymers (8 parts) [350]. As filler, micronized pyrogenic SiO_2 has been proposed. Mixtures of EVAc (30–45% with 15–50% VAc content) with 55–70% chlorinated polypropylene (73–85% chlorine) have been suggested as adhesives for removable protective films [351].

Polyvinyl Ethers

Polyvinyl methyl, ethyl, and isobutyl ethers are used as viscoelastic raw materials for PSAs. They are soluble in common solvents such as alcohol, acetone, butyl acetate, ethyl acetate, and toluene, and some of them are also soluble in water [352]. Polyvinyl ethers with acrylics are used for tapes [353]. Polyvinyl ethers are known as raw materials for wet tapes also [354]. An adhesive for medical tapes contains polyvinyl ether with a K value of 50–130 [161]. Polyvinyl ether is recommended for medical tapes alone or with polybutylene and titanium dioxide as filler. Such compounds are coated warm without solvent. A porous adhesive is achieved by cooling [169]. Polyvinyl isobutyl ether is a component of skin-compatible acrylic PSAs also [355]. Polyvinyl ether can be used for masking tapes during the varnishing of cars [161]. The aging resistance of PVE can be improved by using phenol and pinene derivatives [356]. Polyacrylates tackified with water-soluble PVE have been proposed for water-soluble splicing tapes [357]. Because of its lack of sensitivity to atmospheric humidity, polyvinyl methyl ether has been added to tape formulations to avoid curling [346].

2.3 Viscous Components

At first, rubber resin formulations were used for PSAs. To achieve viscoelastic behavior, the elastic component (rubber) had to be mixed (formulated) with a viscous component. Viscoelastic raw materials can also be blended with such viscous components to regulate their viscosity. Either macromolecular or micromolecular compounds can be used as viscous components. The best known are resins and plasticizers; both are used as tackifiers.

Tackifiers

Tackifiers are raw materials used in PSA formulations to improve the tack (see Chapters 5 and 6). However, they also change other adhesive properties. Generally they improve the peel resistance and decrease the shear resistance (cohesion) of the adhesive. Other nonadhesive properties (e.g., coatability, cuttability, aging resistance) are influenced by tackifiers also (see Chapter 8). Tackifiers have to be viscous. Therefore their molecular weight is limited. Generally low molecular weight resins and relatively high molecular weight solvents are used as tackifiers in PSA formulations. Tackification, tackifiers, and plasticizers for pressure-sensitive adhesives are described in detail in Ref. 1. This chapter discusses some of their PSP-related features only. Tackification of PSPs is more than tackification of the adhesive. It can provide an adhesive character for the carrier too. As described in previous chapters, certain PSPs contain built-in pressure-sensitive components. Such (extruded) products are tackified plastics. For their tackification, high polymers as well as common tackifiers are used (see also Chapter 5).

Resins. The range of resins used as tackifiers has been continuously expanded over recent decades. At the beginning of the manufacture of pressure-sensitive products, mostly natural raw materials were used. Natural rubber was blended with natural (rosin) resins. Later, modified natural resins (rosin derivatives) with better aging resistance were used as the viscous component. Some decades ago, hydrocarbon-based synthetic resins were produced. The addition technology of these resins has been developed also. First resin solutions were developed, later water-based resin dispersions. Such products have to display tackifying ability, but other properties are also needed. For water-soluble formulations, the resin also has to be water-soluble [358].

Molecular weight and molecular structure have a special importance in processes where mutual compatibility of the components is required. Compatibility depends on the molecular weight and chemical structure of the tackifier. The compatibility of the resin plays a special role for block copolymers where the polymer sequences have a quite different chemical nature (and structure), and depending on the choice of tackifier, on its compatibility with one of the sequences or both, different adhesive (and processing) performances can be obtained. Resins are defined by generic type, softening point, color, and molecular weight. Tackifier resins are situated in the molecular weight range 300–3000, where T_g strongly depends on molecular weight [359].

Crystallization, oxidative destruction, or polymerization can lead to changes in the tackifying or adhesive properties of the resin. Non-esterified rosin is recommended as tackifier for hot melt adhesives, providing it is stabilized against crystallization and oxidation [360]. Freezer tapes are coated with an adhesive based on natural rubber, styrene-butadiene rubber, regenerated rubber, or rosin ester [129].

According to Donker [183], styrene block copolymers used for hot melts are mainly SIS block copolymers with about 15% styrene. Alternatives are SBS and high styrene (20–25%) SIS polymers. Special solvent blends are suggested to identify resin compatibility for such elastomers. For SIS block copolymers, modified C_5 hydrocarbon resins, hydrogenated rosin esters, and alkylaromatic hydrocarbon resins are suggested [360]. Common HMPSAs for tapes are based on SIS plus a resin that is partially compatible with the midblock of the SBC and a plasticizer (oil). Tack properties of SBC are given by the midblock. Its mobility is increased by adding oils and tackifiers to the thermoplastic elastomer. A compatible resin is more effective than oil. The other part of the polymer sequence (the styrene endblock) provides the cohesion and mechanical properties and influences the processing conditions. Its hardness is reduced by oils that are compatible with the styrene domains [361]. One way to reduce the processing viscosity of the HMPSA is to increase the processing temperature (upper limit 190°C) or to decrease the melting temperature of the polystyrene domains. According to Hollis [360], 80 Pa.s is the preferred maximum viscosity for high speed coating. Using a mid- and endblock-compatible resin for SIS, this viscosity value can be obtained with a lower resin level (a rubber/resin ratio of about 1:1).

It is known that common HMPSA formulations incorporate a high level of plasticizer, usually a naphthenic oil or a liquid resin. Most tackifiers used for HMPSA are aliphatic, midblock-compatible resins that do not influence the styrenic domain. Oils are present to act on the polystyrene domain (being compatible with these groups) and thus soften it. For instance, a partially fumarized or maleinized, disproportionated rosin ester is used as tackifier together with a plasticizer and a thermoplastic rubber for PSA [362]. The use of plasticizers results in a number of disadvantages including long-term degradation of the PSA. For tapes a mixture of two resins is suggested, with one being midblock-compatible and the other compatible with the endblocks. If resins are used that are partially compatible with the styrene domains (e.g., rosin esters, modified terpene resin, or a modified aliphatic resin), two effects—midblock mobility increase and endblock softening—are achieved. Such a resin can have an aromatically modified aliphatic hydrocarbon as its basis with a higher molecular weight. Fully saturated petroleum resins with a well-defined number average molecular weight of 400–800, a softening point of 40–70°C, and a T_g lower than 45°C have been developed for use in HMPSA formulations based on thermoplastic elastomers without plasticizer oils. According to Donker et al. [362], resins synthesized from the C_5 feedstock having a major portion of piperylene and 2-methylbutenes with a defined diolefin/monoolefin ratio and a weight-average MW (M_w) of 800–960 and M_n of 500–600 (M_w/M_n ratio at least 1.3) are clear, exhibit a softening point of about 57°C and a T_g of about 20°C, and need no oils in the formulation with saturated midblock thermoplastic elastomers. The saturated midblock thermoplastic elastomers are commercially ABA block copolymers where the polyisoprene or polybutadiene is hydrogenated. The block copolymers with the saturated midblock segment have an M_n of about 25,000–300,000.

Rosin derivatives have to be disproportionated to decrease the conjugated double bond concentration. Such bonds are sensitive to oxygen attack [363]. For instance, a common acid resin used for water-based tackifier dispersions displays a narrow molecular weight distribution (M_w/M_n = 1.19) and a T_g of 10°C. Ester resins have a broader MWD (1.19–2.48) and somewhat lower T_g (4–6°C); they display better compatibility and aging resistance. At low temperatures, loss of wetting and a jerky debonding are observed for formulations with the acid resin. The molecular weight of the acid resin is about 400; that of the resin esters is situated between 1000 and 1300. Such (small) differences in the molecular weight and the greater tendency of acid resin to crystallize can cause incompatibility at low temperatures. The crystallization of rosin acids can also result in the formation of grit [364]. Such crystals are insoluble in water and alkalis. Crystallization depends on the rosin type. Tall oil rosin has the highest tendency to crystallize. Gum rosin (from living trees) crystallizes due to its higher acid content.

The nature of the resin affects radiation curing of PSA formulations. Curing by UV light is influenced by the unsaturation index of the tackifier resins [132]. The formulation of EB-crosslinked, SIS block copolymer based PSAs has been discussed by Ewins and Erickson [365]. The effect of resins and plasticizer in a standard formulation having 100/100 phr of rubber (Kraton D1320X) resin or plasticizer has been evaluated. Electron beam cured thermoplastic elastomer-based tackified HMPSA formulations have been studied by Nitzl and coworkers [366] also. They stated that when radiated at a high dosage level of 20 kGy, unsaturated resins give adhesives with increased temperature resistance. Resins can be used as curing agents in classic formulations also. Butyl rubber based formulations use brominated phenolic resins as the crosslinking agent.

Plasticizers. Plasticizers were originally used as viscous components for plasticizing PVC. Their use made it possible to regulate the hardness and mechanical properties of PVC. As an (undesired) side effect, its surface adhesivity was also changed. Such chemical compounds have been used for adhesives too, to give them permanent adhesive characteristics (pressure-sensitive adhesivity) or to regulate their adhesion–cohesion balance to achieve removability. Plasticizers can be incorporated in the peelable pressure-sensitive adhesives to soften the adhesive and thus to improve peelability. However, care may be needed, as some plasticizers can have a tackifying effect on adhesive polymers and this may limit the amount that can be used. The use of plasticizers in printing inks makes the dried film more flexible and pliable [367].

The chemical nature of plasticizers differs according to the type of elastomers or plastomers to be plasticized. For polar hydrophilic macromolecular compounds, polar hydrophilic plasticizers are chosen. Macromolecular polyester-based plasticizers have been suggested for soft PVC to be used as carrier for tapes. For a tape carrier the adhesive has been formulated with natural rubber, butadiene-styrene rubber, and terpene or phenolic resin as tackifier [368]. Dioctylphthalate (a common micromolecular plasticizer for PVC) has been tested as plasticizer or tackifier for tape adhesives on natural rubber and chlorinated rubber [161] and has also been used for medical tapes [369]. In SBCs aromatized plasticizers like dioctylphthalate associate with the polystyrene domains and work like aromatized oils, softening the polystyrene domains. Therefore if plasticized PVC is used with such adhesives, it should be coated with a barrier layer.

Plasticizers can be used in water-soluble formulations also [161,370]. According to Ref. 371, water-soluble compositions used for splicing tapes contain polyvinylpyrrolidone with polyols or polyalkyglycol ethers as plasticizer. Ethoxylated alkylphenols, ethoxylated alkylamines, and ethoxylated alkylammonium derivatives have also been suggested as the water-soluble plasticizer.

The molecular weight of the plasticizer influences its water dispersibility. A maximum or minimum molecular weight is desired depending on the use of the plasticizer. Water-soluble plasticizers whose molecular weight does not exceed 2000 (e.g., polyethylene glycols with a molecular weight of about 200–800) are preferred for water-soluble splicing tapes [348]. For resin dispersions with softening points above the boiling point of water, plasticizers with a well-defined molecular weight should be added [364].

Low molecular weight (35,000–80,000) polyisoprenes with a low T_g (−65 to −72°C) have been used as viscosity regulators for EVAc-based HMPSAs [372]. They replace common plasticizers and improve the migration resistance and low temperature adhesion. Dibutylmethylene-bisthioglycolate has been suggested as plasticizer for special rubbers [373]. Special plasticizers such as lanolin, sesame oil, petroleum jelly, and ricinus oil are applied as adhesives for medical tapes [161]. Plasticizer oils for block copolymers have to fulfill the following requirements [374]: no solubility in the polystyrene domains, compatibility with the elastomer segments, low volatility, low density, and good aging performance.

2.4 Other Components

The processing of the PSP components and of the finished product and its end use or recycling require different additives. Such products can have various chemical bases. The right selection of fillers, antiblocking agents, antioxidants, antistatic agents, and processing aids depends on the type of processing as well on the properties that must be achieved. Some of these additives are used as technological aids whereas others are constituents of the finished product.

Final Constituents

Various materials (solubilizers, fillers, antioxidants, etc.) are needed as additives for PSP formulations. Some are adhesive, but others are not. Most are inherent components of polymer formulations; therefore they are used for both the carrier material and the adhesive.

Solubilizers. The achievement of water solubility or dispersibility requires special additives. Generally such substances have hydrophilic functional groups that increase water solubility and water absorption. Some are macromolecular derivatives (e.g., plastomers or resins), and others are micromolecular products. Solubilizers can be elastomers, plastomers, viscoelastic materials, or viscous additives. Depending on their influence on the pressure sensitivity of the product, they can be considered as main adhesive components or as special fillers.

Macromolecular solubilizers have polar atoms (heteroatoms) in the main polymer backbone or as branches. They may possess the usual hydroxy, amine, carboxy, etc. polar groups that provide the hydrophilicity. Vinyl ether copolymers [375] and neutralized acrylic acid-alkoxyalkyl acrylate copolymers [376] have been suggested as macromolecular pressure-sensitive solubilizers. Polyacrylate-based water-soluble PSAs can be used for medical PSPs (operating tapes, labels, and bioelectrodes) [297]. Suitable water-soluble plasticizers include polyoxyethylenes and rosin derivatives that contain carboxyl groups. Compounds like C_4–C_{12} alkyl acrylates give adhesion by having a low T_g; vinyl carboxylic acids and hydroxyacrylates display hydrophilic properties. Their combination possibilities are numerous. About 30 patents are mentioned by Czech [297] that concern only water-soluble acrylics.

As mentioned earlier, some of these components exhibit adhesive properties whereas others do not. Water-soluble waxes are not pressure-sensitive [377]. Other macromolecular compounds are only slightly adhesive. Water-soluble hot melt adhesives can be formulated with vinylpyrrolidone and pyrrolidone-vinyl acetate copolymers [347]. Polyalkylene oxides and polyalkylenimines can be used as solubilizer also. For instance, a hydroxy-substituted organic compound (alcohols, hydroxy-substituted waxes, polyalkylene oxide polymers, etc.) and a water-soluble N-acyl-substituted polyalkylenimine obtained by polymerizing alkyl-substituted 2-oxazolines can be added also. Such water-releasable HMPSAs contain 20–40% polyalkylenimine, 15–40% tackifying agent and 25–40% hydroxy-substituted organic compound [348].

A repulpable splicing tape specially adapted for splicing carbonless paper contains a water-dispersible PSA based on acrylate-acrylic acid copolymer, alkali, and ethoxylated plasticizers. A polar hydrophilic polyamide-epichlorohydrin crosslinker may also be included in its formulation [378]. Polyfunctional aziridine derivatives (1–3%) can be used as crosslinker also. Such compounds work at room temperature [379]. Polyacrylic acid blended with polypropylene glycol has been suggested too [380]. Amine-containing soluble polymers and soluble plasticizers are proposed in Ref. 381 (see also Chapter 6).

Fillers (Pigments and Extenders). Common fillers are used mainly as inexpensive PSP components. They are used for carriers as well as for adhesives. In some cases fillers are included in the recipe to achieve opacity or special mechanical properties. Fillers in adhesive formulations increase viscosity and improve tensile strength but can cause problems due to their abrasiveness [382].

Generally fillers are inorganic, low molecular weight materials. Fillers are described in detail in Refs. 383–385. Fillers in carrier materials influence the bulk and surface characteristics. Regulation of the mechanical properties is the most important function of fillers in carrier materials. Optical and electrical characteristics are affected also. Surface absorptivity, porosity, and polarity can be controlled by fillers. In the range of mechanical characteristics, stiffness, tensile strength/extensibility, and tear resistance are the main properties regulated by fillers (see also Chapter 6 and Chapter 3, Section 2).

The influence of fillers on the mechanical characteristics of polymers has been the subject of extensive study. Filled structures are used on the macromolecular scale also (see Chapter 3, Section 1). Plastomer–plastomer, plastomer–elastomer, and elastomer–viscous component mixtures have been examined as filled systems. The effect of fillers on the Young's modulus of filled formulations can be calculated by using the Guth–Gold, Smallwood, Brinkmann, Zosel, or other equations [386]. The most important effect of excessive filler levels in carrier or adhesive formulations is the loss of the main mechanical or adhesive characteristics. Such effects are used to regulate the tear resistance of carrier materials or to reduce the adhesivity of pressure-sensitive products. For instance, easy tear breakable hank tapes are used in the same finished product as easy tear paper tapes. The tape is applied by hand. Such tapes are easily torn or split during handling. The carrier is highly filled and has a low elongation and easy break force. Highly filled pressure-sensitive layers ensure removability or repositionability.

The main natural white pigments for paper are kaolin, limestone, and talcum. Synthetic products such as calcium carbonate, calcium sulfoaluminate, titanium dioxide, calcium silicate, sodium aluminum silicate, barium sulfate, and zinc oxide have been developed also. The most used filler for paper is kaolin. Calcium carbonate is less expensive and allows fast ink penetration. Calcium carbonate can be used to increase the porosity of the paper, which is very important for heat set rotary offset printing [387]. Precipitated silicates regulate the ink absorption and increase opacity [388].

Special fillers ensure that the plastics in which they are used will be nonflammable. Aluminum hydroxide, silica, antimony oxide, halogenated compounds, and phosphor derivatives are the main additives used to impart nonflammability. Aluminum hydroxide can be used up to 180°C and magnesium hydroxide up to 340°C [389]. Barium metaborate can replace 50–75% of the antimony trioxide used [390].

PVC with silicon carbide as filler has been suggested for electrically conductive tapes [391]. Metallized ceramic bubbles can be used for electrically conductive plastics or adhesives. Such materials are proposed for protection against EMI (electromagnetic interference) or RFI (radio-frequency interference) [392]. Metallic fillers have been developed for medical tapes (embedded on one side of the carrier material) to avoid the appearance of static electricity and sparks when the tapes are slit or cut [393]. Highly conductive silver-clad hollow microspheres exhibit low resistivity at loading levels of slightly more than 0.1 g/in.^{2} [394].

Carbon black fillers are used to impart electrical conductivity (see Chapter 3). A high carbon black content may make processing difficult and requires previous drying. The conductivity achieved with carbon black depends on the additive concentration and on the morphology of the product [395]. Special composite systems consisting of electrically conductive carbon black and rubber are also known as pressure-sensitive rubbers, because their conductivity depends on compression [396]. Antistatic webs that prevent the accumulation of dust are manufactured by using carbon black filled polyurethanes [397]. In some cases the filler should ensure electrical conductivity and nonflammability without affecting the conformability and extensibility of the carrier material. Such formulations are required for automotive electrical tapes. A patent [398] describes the use of special fillers for such formulations. Photochromic and thermochromic master batches can be applied also. Such fillers cause the finished article to change color when exposed to sunlight or a change of temperature, respectively [399].

As in plastomer formulation (see above), in classic adhesive formulations, fillers are used mainly as inexpensive recipe components. As stated in Ref. 400, extenders or fillers have been regarded as the least important ingredients in the adhesive or sealant formulation, and in the past almost any type of calcium carbonate was used as filler. Active fillers served primarily to stiffen the material; inactive fillers were added to reduce the cost of the formulation, but they can also act as "builders," i.e., they may have a decisive influence on stiffness [117]. The importance of fillers as raw materials is quite different for labels, tapes, and protective films. As discussed in Ref. 1, PSA formulations for labels contain only a low level of fillers. Because of the negative effect of chemically inert fillers on pressure sensitivity, this statement is also valid for tapes and protective films. Because of their generally negative influence on adhesive properties, fillers can only be used for well-balanced formulations or for products having a high coating weight. Therefore they are suggested mainly for the formulation of PSAs for tapes. On the other hand, there are several active fillers that positively modify some features of the adhesive or end use behavior such as shear resistance, cuttability, or tear strength. These requirements are important mainly for tapes. Therefore, fillers are used as adhesive components mostly for tapes. Fillers can improve the processing characteristics of an adhesive raw material blend. As stated in Ref. 401, adding CaO to a polychloroprene-based raw material formulation reduces the mixing (kneading) time by 90%.

As mentioned earlier, fillers generally do not possess adhesive properties. They have the disadvantage of being inert. In some cases they can contain moisture, which is not inert, and may interact with other components of the formulation (e.g., moisture cure polyurethane formulations). Most good quality fillers have a moisture content of less 0.2% [399]. In order to achieve this level, special calcium carbonate fillers can be given two kinds of surface treatments, one to make them opaque and the other to impart hydrophobicity. According to Ref. 205, a level of 15% $CaCO_3$ is used for carpet tapes coated with HMPSA on SBC basis. For HMPSA on EVAc basis, zinc oxide, calcium carbonate, titanium dioxide, barium sulfate, and organic resins have been recommended as filler. These adhesives require only a low level of filler. They contain about 35% elastomer, 30–50% tackifier, 0.2% plasticizer, 0.5% filler, and 0.1–0.5% stabilizing agents. As plasticizers, phthalates, phosphates, chlorinated polyphenols, and liquid rosin derivatives have been used. As tackifiers, rosin derivatives, hydrocarbon resins, and low molecular weight styrene polymers have been suggested [107]. In so-called combination adhesives made as blends of PSAs and non-pressure-sensitive EVAc dispersions (70/30), higher levels of filler (up to 40%) and additives are used [335]. Filled ethylene-vinyl acetate copolymer emulsions useful as carpet adhesives can have a solids content as high as 85% [341].

Fillers have been proposed as viscosity-regulating agents also. The cold flow of the adhesive in adhesive-coated PSPs depends on its viscosity and coating weight. Therefore, for tropically resistant tapes a low coating weight and high filler loading are recommended. For these tapes, the use of silica (0.01–0.03 µm) as filler is suggested to increase the viscosity of the adhesive [402]. Polyvinyl ether is used for medical tapes with titanium dioxide as filler for coating as a highly viscous warm mass without solvent. Since tapes used for automobile interiors have to resist high temperatures, it is recommended that they be manufactured of crosslinked acrylics with filler [403].

The adhesive properties of polychloroprene-based PSAs can be improved with chloroparaffins and fillers [404], for which 5% w/w zinc oxide and 12% w/w magnesium oxide, respectively, are used as filler. In this case the filler interacts with the base elastomer [405] and improves the cohesion of the adhesive. In a quite unusual manner, zinc oxide may increase both tack and cohesive strength in formulations with butyl rubber [406]. Zinc oxide has been used as a reactive filler in PSAs for medical tapes [147]. It is also suggested for use as an active filler for medical and insulating tapes because of its good resistivity. In such formulations zinc soaps are formed. For instance, 15–55 parts butyl rubber untackified or tackified with 7.5–40 parts PIB and crosslinked with 45 parts zinc oxide is a suggested recipe. Such formulations give 180° peel values of 0.73–1.81 kg/25.4 mm. The zinc oxide induced crosslinking is also of interest in FDA-sensitive applications. Silica and ZnO increases the aging stability of adhesives on a rubber (natural or synthetic) basis [407,408]. They also affect the coating viscosity. As absorbing pigments they can change the mechanical properties of the product by the photooxidation of ethylene-propylene copolymers [409].

There are special formulations in which the filler improves the application performance of the product and allows a carrierless construction. Such fillers are needed for transfer tapes used as carrierless high cohesion mounting tapes. Foamlike pressure-sensitive tape can be manufactured with glass microbubbles as filler [410]. Dark glass microbubbles are embedded in a pigmented adhesive matrix. The glass microbubbles have an ultraviolet window that allows UV polymerization of the adhesive composition. The pressure-sensitive tape filled with glass microbubbles has a foamlike appearance and character and is useful for purposes that previously required a foam-backed PSA tape. The adhesive matrix may include 0.1–1.15% w/w carbon black also without affecting the UV polymerization. Glass microbubble fillers have the advantage of higher distortion temperatures.

Transparent microbubbles are included in the composition of one foamlike tape [411]. In this case the glass microbubbles are embedded in a polymeric matrix, 5–65% by volume. The thickness of the pressure-sensitive layer should exceed three times the average diameter of the microbubbles in order to enhance the movement of the bubbles in the matrix. Optimum performance is attained if the thickness of the PSA layer exceeds seven times the average diameter of the bubbles. The bubbles may be colored by adding metallic oxides. The average diameter of stained glass microbubbles should be of 5–200 µm. Bubbles with a diameter greater than 200 µm would make UV polymerization difficult. The die embedded in the PSA matrix should have a UV window also. Thick, triple-layered adhesive tape with filler material in the center layer can also be manufactured. A fumed silica may be used as filler [411]. Microspheres made from borosilicate glass have been suggested as a filler with a good stiffening effect [412]. Special expandable microspheres based on a copolymer of vinylidene chloride and acrylonitrile that encapsulate a liquid blowing agent have been used to improve paper stiffness [413]. Similar products can be used for the adhesive. Pressure-sensitive cold sealing adhesives have been manufactured from EVAc copolymers with 55–85% ethylene, tackifiers, and special microsphere fillers [414]. A closure tape with crosslinked acrylate terpolymer is based on a reinforced web and a reinforcing filler. This type of adhesive is useful for bonding the edges of heat-recoverable sheets to one another to secure closure. An amine-formaldehyde condensate and a substituted trihalomethyltriazine are used as additional crosslinking agents [415].

The influence of fillers on the tack or peel and on peel buildup has to be taken into account also. Special fillers are used to minimize the adhesive–adherend contact to achieve removability. These fillers can be rigid, nonadhesive, elastic, or viscoelastic. As an example, microbubbles of glass have been used as fillers for a repositionable tape [416] and for sheet labels [417]. Removable pressure-sensitive tapes containing resilient, hollow polymeric microspheres (20–66.7%) of a thermoplastic, expanded acrylonitrile-vinylidene copolymer having a diameter of 10–125 µm, density of 0.01–0.04 g/µm^3, and shell thickness of 0.02 µm in an isooctyl acrylate-acrylic acid copolymer have been prepared [418]. The particles are completely surrounded by the adhesive to a thickness of at least 20 µm. When the adhesive is permanently bonded to the backing and the exposed surface has an irregular contour, a removable and repositionable product is obtained. When the pressure-sensitive adhesive forms a continuous matrix that is strippably bonded to the backing (see Transfer Tape), having a thickness higher than 1 mm. This product is a foamlike transfer or foam tape. The (40 µm) cellulose acetate based tape could be removed from paper without delaminating it.

As detackifying filler, polymer (PSA) particles having a diameter of 0.5–300 µm can be used also [419]. Such pressure-sensitive filler particles added as detackifiers are synthesized via suspension polymerization. The T_g of the particles should be below 10°C. They allow the use of special inexpensive release coatings. Modified starch or starch and a water-soluble fluorine compound (water-soluble salts of perfluoroalkyl phosphates, perfluoroalkyl sulfonamide phosphates, perfluoroalkoxyalkyl carbamates, perfluoroalkyl monocarboxylic acid derivatives, perfluoroalkylamines, and polymers having as skeleton epoxy, acrylic, methacrylic, fumaric acid, or vinyl alcohol, each of which contains a C_6–C_{12} fluorocarbon group on a repeating unit of the polymer) can be used as the release layer on paper for an adhesive containing a minute spherical elastomeric polymer [420]. The adhesion buildup can be reduced according to Pagendarm [421] by using a powderlike surface coating of the PSA layer with particles having a diameter of less than 10 µm, where the particles are embedded in the PSA to one-third of their diameter. Special fillers are used as detackifier agents mixed with the PSA. Their chemical compositions vary. For instance, a pressure-sensitive adhesive for mounting tapes includes a detackifying resin comprising a caprolac polymer (1–30% w/w) [422].

Common fillers for electrical conductivity are carbon black, graphite, carbon fibers, metallic powders, metal oxides, metallic fibers, and metallized glass bubbles. Metallic powder is used as filler for electrically conductive adhesives. Electrically conductive adhesive tapes have been manufactured by coating an acrylic emulsion containing 3–20% nickel particles onto a rough flexible material (e.g., Ni foil) [423]. Medical tapes with gelled PUR adhesive layer can include metallic powders [424]. Fine powders of tin oxide, zinc oxide, titanium oxide, and silicon-based organic compounds have been used as conductive fillers [425]. It should be mentioned that the notion of coagulum or "grit" is used in connection with fillers also. Such particles are included in carbon black used as filler [22].

Water can also be used as a conductive filler [426]. Compositions based on acrylic acid polymerized in a water-soluble polyhydric alcohol and crosslinked with a multifunctional unsaturated monomer contain water, which affords electrical conductivity. Electrical properties can be improved by adding electrolytes

Chemical Basis of PSPs

to the water. Such products can be used as biomedical electrodes applied to the skin. Starch is used also as a water-absorbing filler. As shown by Eifert and Bernd [427], an electrically conductive (higher than 10^{-5} S) transparent PSA layer can be achieved by using a special electrolyte. Electrically conductive PSAs can be manufactured as gels containing water, NaCl, or NaOH [428]. These gels are useful in detecting voltages in living bodies. Such polymers based on maleic acid derivatives have a specific resistivity of 5 k Ohm (1 Hz).

Flame-retardant adhesives are prepared by using halogen derivatives. A chlorine or bromine flame retardant with 65% or more chlorine and or 80% or more bromine has been suggested [429]. A formulation with 1–30 parts by weight of the inorganic flame retardant and 0.1–20 parts by weight of silica powder has been recommended. As known, decabrominated compounds are not allowed in food products. Common inorganic fillers can be used as flame retardants also. Sealant tapes can comprise 200 parts hydrated alumina as filler per 100 parts rubber, 50 parts carbon black, and 50 parts plasticizer. The efficacy of inorganic fillers as flame retardants decreases as follows [430]: $Al(OH)_3 > Mg(OH)_2 >$ kaolin $>$ limestone $>$ zinc oxide. Their activity is based on the increase in the enthalpy of gases built up [431].

Pigmented adhesives for labels offer better opacity than those without pigments [432]. For such pigments light absorption up to 385 nm is required to avoid damage to the adhesive. This degree of protection is generally given for 1 year in the Middle European climate. Because the release solution is transparent, converters may add an ultraviolet-detectable agent [433]. Special paints, inks, and light-sensitive additives can be incorporated in the adhesive. For instance, squarylium-type sensitizing dyes have been developed for (laser-activated) anti-theft labels sensitive to visible light [434].

Thermally activatable PSAs contain a thermoplastic resin [435]. The formulation of hot seal PSA may include polyethylene (10–50% w/w) with a molecular weight of 1000–10,000 [436].

Fillers influence the crosslinking of both adhesives and carrier materials. The small difference in crosslink density of certain filled and unfilled polyurethanes is attributed to the presence of an active pigment, which may affect the distribution of crosslinking by localizing crosslinks in the network system [138]. Fillers also influence drying time [191]. Coatings containing a higher level of fillers may be dried more quickly.

Antioxidants. Generally each PSP component (elastomer, resin, plastomer) contains an antioxidant to improve its weathering, thermal, and light resistance. The chemical composition of the antioxidant may vary according to the product component, its end use, and formulation technology. Antioxidants react with free radicals formed during the oxidation process. Their role is to delay the oxidative reactions. The most common antioxidants are naphthol and phenol derivatives, oximes, and aromatic amines [40]. For instance, for natural rubber based adhesives, 0.05–2% Neozone may be used [111]. Generally an HMPSA formulation contains 0.2–2.0% w/w of an antioxidant. High molecular weight hindered phenols and multifunctional phenols such as those containing sulfur and phosphorus are suggested. Antioxidants such as 2,2-thiodiethyl-bis [3-(3,5-di-*tert*-butyl-4-hydroxyphenyl)]propionate (known under the commercial name of Irganox 1035) deactivates the singlet oxygen and reacts with alkoxy radicals. Aging of SBC-based PSAs may be related to the incompatibility of the two-phase system and to the unsaturation that exists in styrene midblocks [126]. For such compounds, stabilizers based on salts of di(alkyl, phenyl, or benzyl)thiocarbamic acid have been recommended [437]. The performance of such stabilizers may be further enhanced by adding known synergists such as thiodipropionate esters and phosphites. Antioxidants for hot melts should have low volatility, provide better maintenance, reduce viscosity changes during processing, delay skin formation, and not impart color. For SIS block copolymers in HMPSAs, amines, phenols, phosphites, and thioesters are used as antioxidants. Oxybenzoquinones and oxybenzotriazoles are added as UV stabilizers [438]. For saturated midblock thermoplastic elastomers (commercial ABA-type block copolymers), the formulations include a phenolic antioxidant 1–5% by weight.

The photoinitiated degradation of polymers is a radical reaction caused by light of 290–400 nm [439]. As protective agents, UV absorbers, quenchers, hydroperoxide-destroying additives, and radical scavengers are used. According to Jacob [205], for tapes one of the most important properties is open face aging. (This is not a crosslinking phenomenon, according to the author [205].) Common tackifiers used for SBS show after a week of aging an important loss of tack and later a loss of peel. For screening of resins for SBS, a standard formulation with 200 parts SBS, 125–150 parts resin, 25–50 parts oil, and 1 part antioxidant has been used. For SIS a 100/100/10/2 resin/SIS/oil/antioxidant composition has been suggested [440]. Special antioxidants should be used for EB-cured SBCs, because BHT and ZBTC inhibit the crosslinking. Phosphites work better [365].

As known, storage stability of 6 months is expected for tackifier dispersions [263]. A typical resin dispersion contains 40–60% resin, 2–20% plasticizer, 2–10% surfactant, 0–3% stabilizer/thickener, 0–0.5% antioxidant, and 0–0.01% biocide. BHT (2,6 di-*tert*-butyl-*para*-cresol) or tetrakis(methylene(3,5-di-*tert*-butyl-4-hydroxyhydrocinnamate)methane can be added to the resin dispersions to avoid color degradation and prevent oxidation [441]. As antioxidant for polyvinylpyrrolidone-based formulations, distearylpentaerythritol diphosphite is suggested; for acrylics, 4,4-butylidene-BIS has been recommended [348,441].

Antioxidants are also used to prevent chemical denaturation of the carrier material over time under environmental influences. They prevent its oxidation. However, side effects can appear. Some antioxidants can cause gel sensitivity due to stabilization of free radicals that are released. Thus they can produce large gel

Chemical Basis of PSPs 225

particles in the die [8]. Another problem is discoloration. No antioxidants should be added after formulation because of possible yellowing of LDPE. If LLDPE is used as the outer layer for a coextruded product, more antioxidant has to be added in order to avoid material destruction due to prolonged residence time [13]. The level of antioxidants depends on the film thickness also. For thinner films, more antioxidant should be used.

Antistatic Agents. To avoid accumulation of electric charges on the carrier surface, antistatic agents may be coated on or built in as additives [442] (see also Chapter 6). Antistatic additives are classified as internal or external according to their application [443]. Such compounds possess apolar and polar parts. This one may be a nonionic functional group (OH) such as glycerine monostearate, an anionic group (e.g., an alkylsulfonate) or a cationic one (e.g., a quaternary ammonium salt). Fatty acid choline ester chloride is BGVV allowed. Antistatic agents reduce the usual 10^{16}–10^{17} ohm resistance of plastic films to 10^{10} ohms [444] (see Chapter 3). External antistatic agents are used at a level of 10–100 mg/m^2, depending on the plastic. Internal antistatic agents may have different levels of application (0.05–8%) according to the type of plastic; conductive fillers are added at a level of 5–30% [443,445].

External antistatic agents pose the disadvantage of requiring a supplemental processing (coating) step. On the other hand, they display the following advantages: They work instantaneously; the influence of air humidity is less important; only a low level (0.01 g/m^2) is needed; one-sided application is possible.

Internal antistatic agents must be soluble in the molten polymer. Ethoxylated alkylamine and other agents migrate on the surface of the polymer and together with the humidity of the air build up a thin conductive layer [446]. They also have certain disadvantages: The crystallinity of the polymer, pigments and migratory additives, and the temperature all influence their migration; heat and chemical reactions influence their concentration and stability. They have to be used in higher concentrations than the external agents, and their antistatic properties decrease when they come in contact with humidity and oxygen [447]. A too high level of antistatic agents has an adverse effect on the anchorage of printing inks. Generally the resistivity that can be obtained when migrating additives are used is not lower than 10^9 ohms.

Ethoxylated amines are the best known compounds [25]. Unfortunately, they have a very low compatibility in master batches and a strong amine odor. For good antistatic behavior they need a relative humidity of 40%. Cationically active compounds with imidazole as quaternary group also display antistatic effects [448]. Derivatives of 1-alkyl-3-alkoxymethylimidazolium chlorides have been tested as antistatic agents for LDPE. Antistatic PP films contain *N,N*-bis-(β-hydroxyethyl)stearylamine, or stearylamide, or glycerine stearate [449].

Soluble, electrically conducting polyalkoxythiophenes may be used as antistatic agents also. Due to the sp^2pz hybridization of the orbitals in their aromatic rings, the free-moving electrons ensure (if doped) sufficient conductivity. Polyethoxythiophene is used together with bonding agents (5% w/w active substance). Antistatic agents like polyethoxythiophene can be coated onto the film surface [447]. Polyethoxythiophene is coated as a solution in organic solvents (4–8% solids content) by using a gravure roller (with a coating weight of 1–4 g/m^2). Polypyrrole has been coated onto plastic surfaces displaying a surface resistance of 10^6 ohms [450]. Polyanilines have been suggested as built-in or coated (100–400 nm) layers for improving the electrical conductivity of polymers [451].

Antiblocking Agents. Blocking is an undesirable adhesive bonding of different surfaces [452]. Antiblocking agents are special fillers (see also Chapter 6). Antiblocks prevent blocking, or autoadhesion between layers. They are also used in carrier and adhesive formulations. They are added to an adhesive (or carrier) formulation to prevent the adhesive coating (or carrier) from adhering to its backing when the adhesive-coated or uncoated carrier material is rolled or stacked at ambient or elevated temperatures and relative humidities [453]. Blocking can be caused by heat or moisture, which may activate latent tack properties of the adhesive/carrier composition. Excessive surface treatment can cause blocking also. Excessive tackiness can require the use of additives or special processing techniques to facilitate handling. For instance, in film manufacture, good chill roll release is needed [454].

Antiblocking additives are small (a few micrometers) particles of hard materials such as silica and talc that protrude from the film and prevent the surface from making close contact with another surface, which would cause blocking. Their nature and concentration depend on the nature of the base plastomer. In some cases the additives used as opacifying agents can be used as antiblocking agents also [455]. Matting agents have been suggested as antiblocking additives for PVC. For LDPE, silica is almost always added as the antiblock [456]. Ultrahigh molecular weight PE may be used as an antislip additive for LDPE also [457]. Linear low density polyethylene sometimes includes a number of low molecular weight species and will require more antiblocking additive. The processing aid concentrations needed depend on the levels of antiblocking additive and other additives. Processing additives require a higher percentage of antiblocking agents and may influence the adhesion on the film surface [458].

Slip Agents. Slip agents are additives that reduce friction during the transport and processing of plastic carrier materials. When films are blown, slip tends to bloom more to the outside of the bubble [459]. According to Grünewald [458], slip agents affect adhesion and surface treatment. They also influence printability. Their presence as an undesirable boundary layer can interfere with printing and coating [40]. Recommended levels of slip agents are 1000–2000 ppm.

In LDPE and LLDPE the the most commonly used slip agents are unsaturated C_{18} primary amide (oleamide, MW 275), unsaturated C_{22} primary amide (erucamide), unsaturated C_{36} secondary amide, and unsaturated C_{40} secondary amide, the latter having a molecular weight of about 600. It is known that migration is inversely proportional to molecular weight [25]. The molecular weight of slip agents is a compromise between the need to migrate to the surface within a reasonable period of time and their blocking tendency. Amides with higher melting points have less effect on wettability after corona treatment.

The level of slip additives needed depends on the application, film thickness, and treatment intensity [460]. Thin glossy films generally have poorer slip and block more. Friction increases with the density of the base polymer. For cast film, the surface quality of the chill roll also affects the antislip concentration. Secondary amides with better thermal resistance are particularly suitable for cast film. Slip agent migration is influenced by the laminate components. Compatibility of amides with EVAc, EBA, or EEA copolymers influences their efficacy. For such polymers metallic soaps are added as dispersing aids [25]. As the polarity of co-laminated extruded films increases, slip agent migration increases [458]. Printing inks without a slip agent may also influence slip agent migration.

The temperature resistance of slip agents is important also. Erucamide (ESA) is more efficient and has higher temperature resistance than oleamide (OESA). Its efficiency is less affected by the adhesive. For polymers processed at higher temperatures, ESA is suggested [458]. Slip, antiblocking, and antistatic agents are interactive. Their combination can have effects that are difficult to predict. Other additives such as UV absorbers and UV stabilizers, antioxidants, processing aids, and some pigments can also strongly interfere with the slip agent. Fluorocarbon elastomers are the main processing aids used for polyethylene [461–463]. Such compounds work as slip agents between the molten polymer and the metallic surface of the processing equipment.

Printing Inks. Printing inks can have various chemical compositions. Their composition depends on the type and geometry of the carrier (web or sheet base material); type of printing process (flexography, lithography, gravure, etc.); type of printing press (production speed and drying characteristics); finish required (matte or glossy); end use application (color detail required and other processing steps such as cutting, thermoforming, laminating); run length; and sequence of ink application. Ink properties such as the coefficient of friction, scuff resistance, gloss, and pigment light fastness are important characteristics for printing and film processing [40].

Special inks and coatings are used to combat theft and forgery; examples include thermochromic inks, photochromic inks, optically variable inks, and reactive, metameric, infrared, and luminescent inks [465]. Thermochromic inks change color at certain temperatures. This change is normally irreversible, but reversible inks also exist [462]. Chiral nematics (liquid crystals) are applied by screen printing. These are water-transparent inks applied over a black background. They allow an invisible printed subject to become visible simply by being touched. The most used temperature range is 27–33°C.

Photochromic inks change color with light. Optically variable inks display different colors when viewed from different angles. Reactive inks change color when reacted with special chemicals. Metameric inks change color under different light sources. Metallic inks change color when photocopied. Infrared inks either show or do not show their color under infrared light. Luminescent inks continue to glow in the dark.

As stated by Nitschke [466], there is also a trend to use water-based compositions for printing inks. New printing ink systems without solvents, with reduced solvent levels, and with "new" solvents have been developed. UV-cured printing ink systems, high solids ink systems, and water-based inks have been manufactured [467]. It should be emphasized that, because printing and drying properties of the inks are controlled with solvents, a general (full) replacement of solvents with water is not possible. Water-based printing inks are really water-containing or water-dilutable systems. This is a big difference in comparison to water-based adhesives. Water-based ink systems have lower running speeds [466]. Water-based printing inks bond according to a physical or chemical process [467]. Physically bonding systems are based mainly on styrene-acrylic copolymers. Chemically bonding binder systems are based on acrylated polyester prepolymers and acrylics. Vehicle systems embody the properties required by printing process, drying, and end use [40]. The polymer is the main part of the vehicle. Low molecular weight, narrow MWD polymers are used as vehicles. Solvent release and good block resistance are required. For rotary screen printing, PVC-based inks (plastisols, foamed plastisols) and water-based dispersions are used [468]. Water-based screen printing inks have excellent anchorage on PVC-based pressure-sensitive products [469]. Binders for rotogravure printing are based mostly on acrylics. Special clays are designed for rotogravure with coarse pigments, calcinated clays, talcum, and montmorillonite are used with styrene-butadiene rubber latex [470]. The water-based systems are stabilized with surfactants and protective colloids, but surfactant-free systems have been developed also [471].

Reactive printing inks such as hot melts and UV-curable, water-dilutable printing inks and lacquers have been developed also [472]. They are curable after drying. Generally, radiation-curing systems are based on binders (vehicles) that cure with UV light or electron beam radiation [473]. Electron beam curing is too expensive for printing inks. Fewer than 50 such (offset) printing units have been installed in the 1990s throughout the world [474]. Less expensive combined presses with UV flexo and rotary screen printing and cold UV systems have been introduced.

Ultraviolet-crosslinkable systems can be cured with free radical or ionic mechanisms. Free radical UV systems contain acrylics as the base macromo-

Chemical Basis of PSPs

lecular compounds. Cationic systems include low viscosity epoxies. Both are 100% solids, whereas common solvent-based inks are 30% solids. Free radical systems react by light and during light action only and can lead to almost total crosslinking [475]. Systems crosslinked by free radical mechanisms use a liquid vehicle based on multifunctional acrylics. Allyl and vinyl derivatives are applied also. These are di- and trifunctional monomers or oligomers. Tetrafunctional derivatives give films that are too rigid due to the advanced crosslinking. Polymers modified with acrylics are used also. They exhibit the advantages of the base compound (epoxies, polyesters, or polyurethanes) and the reactivity of acrylates. Epoxy acrylates, amine-modified polyether acrylates, and urethane acrylate oligomers have been developed [473]. Modified natural raw materials (oils, fatty acids) have also been introduced. The viscosity of the reactive prepolymers must be reduced by using reactive diluting agents (di- and trifunctional acrylics). Tripropylene glycol diacrylate and dicyclopentenyl oxyethyl acrylate are the most often recommended difunctional reactive monomers. Trimethylolpropane triacrylate is used as a monomer with more functionality. The adhesion between carrier and ink is improved by hydroxyl groups.

Seng [476] discussed the printing of polypropylene tapes using UV-cured systems. Ionic and free radical polymerization are compared. Cationic initiators are onium salts that produce an acid when photolyzed. Such onium salts include a diazonium, oxonium, sulfonium, or iodonium cation, and the anion is ClO_4^-, BX_4^-, PX_6^-, AsX_6^-, or SbX_6^-, where X is chlorine or fluorine [477]. Good results are obtained using cycloaliphatic epoxies (chain-opening polymerization via oxirane rings). The ionic UV systems have the following advantages: They are not inhibited by oxygen; they exhibit little shrinkage; they adhere well on metals; and they are less toxic. Unfortunately, such systems are less reactive, cure more slowly (curing is not finished after radiation; postcuring is necessary) and have lower penetration. Therefore dual cure systems are proposed. First a peroxidic initiation, then UV curing are carried out. As UV sensitizers, benzophenone derivatives are used. As discussed earlier, such compounds have been suggested for pressure-sensitive adhesives also.

Technological Additives

Materials used to facilitate the mixing of formulation components and the coating and drying of adhesives (or other PSP components) are called technological additives. Some of them (like solubilizers) may be constituents of the finished product also.

Solvents. Solvents are temporary components of adhesives, primers, release coatings, and printing inks. Evaporation rate and solvent retention influence the properties of the coating. Solubility is a fundamental property of chemical compounds that can be calculated or measured. Such parameters are of great help in choosing a solvent.

Some years ago, toluene, xylene, perchloroethylene, and naphtha blends were the most common solvents for adhesives [477]. Special petroleum fractions were recommended for rubber-based adhesives [478]. Rubber resin PSAs on solvent basis are manufactured with hydrocarbon solvents. Pressure-sensitive adhesives based on synthetic rubber use special petroleum fractions as solvents also [478]. The common manufacturing procedure for tapes applies a petroleum fraction with a boiling range of 60–100°C as the solvent [479]. Although generally aliphatic solvents are suggested, aromatic compounds can be used also. Such derivatives influence viscosity. Adding aromatic solvents to SBR-based PSA formulations results in increased viscosity on aging [480]. For printing inks the use of hydrocarbons is limited to rotogravure; for solvent-based silicones, hexane, toluene, or xylene can be used. According to Mueller and Türk [130], rubber resin adhesives are recommended for removable price labels. For such solvent-based adhesives, toluene, acetone, and ethyl acetate have been suggested as solvents. Turpentine has been applied for medical tapes as plasticizer solvent [292]. Toluene replacement is a special problem [481]. High solvency hydrocarbon solvent systems with a high content of naphthenic species provide a good combination of solvency and evaporation characteristics suitable for PSA formulation, using either styrene block copolymer or solvent-based acrylics. Because the solvent system has a low affinity for water, it may be an ideal candidate in current solvent capture and recycling equipment. Isoparaffinic solvents are suggested for natural rubber, butyl rubber, polyisobutylene, EPR, and polyisoprene [482].

Ethyl acetate, special boiling point gasoline, n-hexane or n-heptane, toluene, acetone, and isopropyl acetate are used as solvents for PSAs [483]. Typical solvent blends for acrylics contain (1) ethyl acetate (54%), isopropanol (35%), and hexane (11%), or (2) toluene (60%), hexane (39%), and methanol (1%) or pure ethyl acetate (100%).

For printing inks the rate of development of internal stresses depends on the type of solvent used and its rate of evaporation. The slower the evaporation of the solvent, the lower the internal stress of the dried layer [40]. This influences plastic film shrinkage and curl also (see also Converting Properties). Alcohols are the primary solvents for flexographic inks; they are used in crosslinkable solvent-based acrylic PSAs also and may improve the solubility of sulfosuccinates in water-based PSA dispersions. Esters are used because of their solubility as the main solvents for solvent-based acrylic PSAs, for crosslinking agents, and as components in solvent blends for printing inks.

The addition of special solvents to water-based acrylic dispersions improves their adhesive characteristics [484]. Solvents with a dielectric constant higher than 80 have been added to copolymers of C_{1-14} methacrylates with ethylenic unsaturated carboxylic acids. Double-sided coated PET tapes with such formulations have shown an increase of about 10–20% of peel.

Chemical Basis of PSPs

Adequate choice of the solvent can increase the pot life of two-component crosslinked adhesives. Using *tert*-butanol as solvent instead of ethyl acetate, the pot life of formulations based on copolymers of C_{4-12} of alkyl(meth)acrylates and unsaturated monomers having functional groups reactive to NCO crosslinked with hexamethylene diisocyanate, isophorone diisocyanate, hydrogenated MDI, hydrogenated xylene diisocyanate, or lysine di- or triisocyanate and their derivatives has been increased to 8 h [485].

For solvent-based PSAs, the choice of solvent is determined by the type of polymerization. Blends of acrylic and vinyl esters can be copolymerized to produce medical PSAs. Polymer solutions in isopropyl acetate (with a viscosity of 300–10,000 cP) are produced [486]. Cycloaliphatic, hydrocarbon-based solvents, ketones, and alcoholic mixtures can be used also [292].

In certain cases the choice of solvent is imposed by coating problems. Wetting problems can be corrected by the addition of cosolvents. Water–solvent mixtures have lower surface tension and may swell the dispersion particles, giving a more continuous dried film [487].

In some cases the solvents are needed after coating and finishing of the PSP, if a posttreatment is carried out. According to Krampe and Moore [292], a tack-free surface is achieved by superficial crosslinking of the adhesive by polyvalent cations (Lewis acid, polyvalent organometallic complex, or salt). The crosslinking penetration depends on the type of polymer and the solvent. Crosslinking can be achieved by dipping also. Water or water-miscible solvents should be used.

Primers. The common manufacture of tapes includes in-line primer coating, release, and adhesive coating [479]. As primer, solutions (in toluene) of butadiene-acrylonitrile-styrene copolymers have generally been preferred. For solvent-sensitive SPVC films, primer dispersions based on Hycar latex and Acronal 500 D (1/1) were suggested by Reipp [479]. For polypropylene, Tatsumo et al. [488] suggested as primer a mixture of maleated polypropylene and acrylic resin. The use of such a primer increases the bonding strength from 0.5 to 2.4 kg. Chlorinated polyolefin/polyurethane adhesives have been proposed as primer for tapes also [489]. Epoxy-containing silanes and vinyltriacetoxysilanes have been suggested as primer for silicone coating on paper [490]. As solvent-based primer, a mixture of 50–90% chlorinated polyethylene (with 40–75% chlorine) and 1–50% chlorinated polypropylene (with 25–35% chlorine) has been proposed [491]. Tackified cyclized rubber has been suggested as primer for polypropylene [492]. For polyester films, primers based on a reactive component (e.g., chlorinated and fluorinated acetic acid or benzophenone) and a film-forming polymer (PVC, polyvinylformal, etc.) have been proposed [493]. For polypropylene, a mixture of 20–60% chlorinated rubber (60% chlorine), 40–80% EVAc copolymer (25% VAc), and 1–15% chlorinated polypropylene has been suggested [494]. For a polymeric film for silicone release applications, the primer composition con-

Table 4.4 Chemical Basis of Primers

Primer Base components	Type	Carrier	Application	Ref.
Butadiene-acrylonitrile-styrene copolymer	SB	Paper, film	PSA coating	479
Polypropylene graft copolymer-polyacrylate	SB	PP	PSA coating	488
Chlorinated polyolefin-polyurethane	SB	Film	Tapes	489
Chlorinated polyethylene-chlorinated polypropylene	SB	Film	Tapes	491
Chlorinated polypropylene-EVAc blends	SB	PP	Tapes	494
Tackified cyclized rubber	SB	PP	Tapes	492
Blends of PVC with chemical reagents	SB	PET	Tapes	493
Carboxylated rubber-polyacrylate	WB	SPVC	PSA coating	479
Epoxidized silanes-vinyltriacetoxysilane	SB	Paper	Siliconizing	490
Polyester-glycydoxysilane	SB	Film	Siliconizing	495

tains 25–75% glycidoxysilane and 25–75% polyester [495]. Solvent-based one-component polyurethanes or polyimines can be used as primer ("tie-coat" adhesive) too [495,497]. Maleic anhydride and vinyl acetate grafted butyl rubber have been suggested as primer for cellophane tape [498]. As illustrated by the above examples (see Table 4.4), the chemical basis of primers varies. Because of the role of primers as anchoring intermediate between two chemically different layers, their composition assumes reciprocal affinity and reactivity toward both components.

3 MACROMOLECULAR BASIS AND CHEMICAL COMPOSITION OF THE ABHESIVE

As discussed earlier, most PSPs require an abhesive layer between the pressure-sensitive layer and the (self-wound) back side of the product. In some special cases the low level of tack or peel and the nonpolar character of the carrier material allow construction without a release layer, but in the most cases a release layer is required. This layer may be built into or coated on the product.

3.1 Coated Abhesive Components

Coated abhesive layers are based on special chemical compounds that have a nonpolar or closed surface. Except for some low molecular weight fatty acid derivatives, the chemicals most often used as release coatings are macromolecular compounds (silicones, polyvinylcarbamates, etc.). They can be classified according to their raw materials into non-silicone-based and silicone-based compounds. Because of their excellent release characteristics, silicones are the best known abhesive materials. Because of the need for a separate solid-state liner for labels and certain tapes, the manufacture of siliconized carrier materials is an industry in itself. It is not the aim of this book to discuss the complex problems of this domain. Some of them are described in Ref. 1. The number of products and applications where such very pronounced de-adhesive characteristics may disturb the manufacture and application of the product or where they are unnecessary is increasing. On the other hand, as discussed earlier, some plastomers used as solid-state laminate components perform well as release agents. There are processing aids for plastomers (e.g., slip, antiblocking, matting agents) that display abhesive properties too. There are special contoured or textured carrier materials (e.g., pleated (crepe) paper, embossed PVC) that possess abhesive characteristics. Therefore the importance of non-silicone-based release materials is growing.

Nonsilicone Release Agents

As mentioned, for some products there is no need for the high level of abhesive performance afforded by silicones. Also, for adhesive coatings with very low tack and peel the use of silicones would give laminates having insufficient bond strength. In such cases the abrasion sensitivity and low anchorage of silicones would also influence the peel level. For these applications it is recommended that release layers be produced with nonsilicone materials. Various products have been developed for release layers using CMC, alginates, and shellac [499]. As a release with good printability for tapes having siloxane, polyurethane, acrylic, or modified rubber based PSA coated on polyamide or glass fiber mats, polyesters, nitrile rubber, neoprene, and chlorosulfonated polyethylene have been suggested [500]. For medical tapes, nitrocellulose, polyethylene, rubber, proteins, and silicones have been used as release layer [128]. Copolymers of maleic acid with acrylic and vinyl monomers, styrene, vinylpyrrolidone, and acrylamide have been proposed as release coating for tapes [501]. Aluminum film is used as street marking tape coated with a tackified polyisobutylene-based adhesive. The release layer for this adhesive is polyvinyl alcohol [502].

The release mechanism involves low polarity surface or low surface tension and incompatibility of the release surface and adhesive surface [503]. Generally the abhesive raw materials are high polymers bearing nonpolar or/and voluminous side chains or functional groups. The abhesive effect of low polarity is well demonstrated by silicones and fatty acid derivatives. Steric and low molecular mobility effects are illustrated by fluorine, phosphorus, or carbamide derivatives.

Werner complexes, complexes of basic chromium chloride with a fatty acid rests, e.g., stearine, palmitine, or lauric acid, and metal complexes of fluorinated compounds have been tested as release agents. Chromium complex salts in isopropyl alcohol are suggested; film-building substances can be added also, as can mixtures of silicones with chromium complex salts. Such compositions contain chromium complex salt and silicone resin emulsion (1:1) or chromium complex salt and silicone rubber emulsion plus hardener; or silicone resin, silicone rubber, and crosslinking agent [504].

A silicone release coating that stratifies when backed comprises a hydroxy functional polyester containing fatty acid, crosslinking agent, and silicone emulsion [505]. Polyester coated with stearic acid solutions in chloroform and treated with aqueous copper chloride solution displays release properties (peel strength of about 10% for an acrylic adhesive tape).

Omote et al. [506] describe a release layer for paper carrier based on modified starch or starch and a water-soluble organic fluorine compound, such as a perfluoroalkylphosphate, fluoroalkylsulfonamide phosphate, perfluoroalkoxycarbamate, perfluoroalkyl monocarboxylic acid derivative, perfluoroalkylamine, or a polymer having as a skeleton acrylic acid, methacrylic acid, vinyl alcohol, etc. containing C_6–C_{12} fluorocarbon groups on a repeating unit of the polymer (see Fig. 4.9).

Vinyl esters having long pendant groups containing tetrafluoroethylene and hexafluoropropylene oligomers can be used as release coatings [507]. In a similar manner, combining fluoropolymers (hexafluoropropylene-tetrafluoroethylene copolymers) with polyamic acid leads to antiadhesive coatings with improved performance due to the spontaneous separation into two layers, the bottom layer being polyamic acid and the top layer fluoropolymer [27]. Favorable conditions for stratification have been created by adjusting the concentration of fluoropolymers to 32–40 vol % and increasing their particle size (water-thinned dispersion) to 1 μm. Nonporous, thin $(CF_x)_n$ coatings, where x has values of 0.3–2.0, with controlled wettability and good adhesion to plastics have been manufactured by plasma polymerization deposition of C_2F_5–H, C_2F_6–C_2F_4, CF_3–H, and CF_4–C_2F_4 mixtures or C_2F_4. Such deposits display a surface tension of 18 dyn/cm.

Polyvinylcarbamate is the main product used for non-silicone-based release coatings for films (see also Chapter 6). For special applications, vinyl acetate polymers can be used as release agents. For tapes an abhesive layer based on cellulose derivatives and PVAc has been suggested [44]. In certain cases acrylic

1
~-CH$_2$-CH-CH$_2$-CH-CH$_2$-CH-~
    | | |
  OC-OH   OC-OH   OC-OH

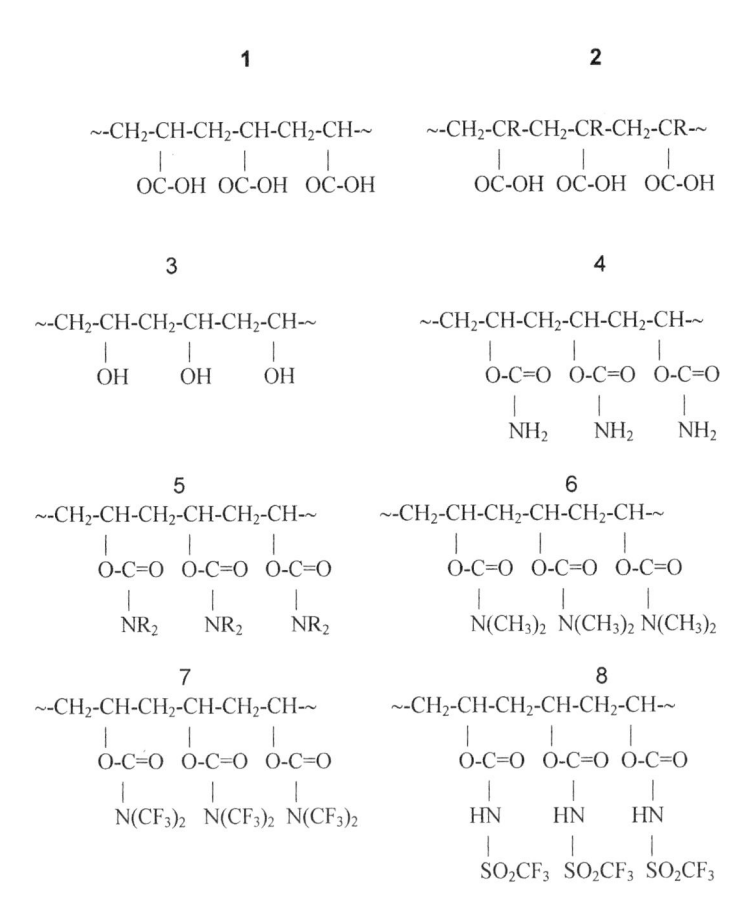

2
~-CH$_2$-CR-CH$_2$-CR-CH$_2$-CR-~
    | | |
  OC-OH   OC-OH   OC-OH

3
~-CH$_2$-CH-CH$_2$-CH-CH$_2$-CH-~
    | | |
   OH   OH   OH

4
~-CH$_2$-CH-CH$_2$-CH-CH$_2$-CH-~
    | | |
  O-C=O   O-C=O   O-C=O
    | | |
   NH$_2$ NH$_2$ NH$_2$

5
~-CH$_2$-CH-CH$_2$-CH-CH$_2$-CH-~
    | | |
  O-C=O   O-C=O   O-C=O
    | | |
   NR$_2$ NR$_2$ NR$_2$

6
~-CH$_2$-CH-CH$_2$-CH-CH$_2$-CH-~
    | | |
  O-C=O   O-C=O   O-C=O
    | | |
 N(CH$_3$)$_2$ N(CH$_3$)$_2$ N(CH$_3$)$_2$

7
~-CH$_2$-CH-CH$_2$-CH-CH$_2$-CH-~
    | | |
  O-C=O   O-C=O   O-C=O
    | | |
 N(CF$_3$)$_2$ N(CF$_3$)$_2$ N(CF$_3$)$_2$

8
~-CH$_2$-CH-CH$_2$-CH-CH$_2$-CH-~
    | | |
  O-C=O   O-C=O   O-C=O
    | | |
   HN   HN   HN
    | | |
 SO$_2$CF$_3$ SO$_2$CF$_3$ SO$_2$CF$_3$

Figure 4.9 Buildup of some nonsilicone release agents. (1) Acrylic base polymer; (2) alkyl acrylic base polymer; (3) polyvinyl alcohol base polymer; (4) polycarbamate based on polyvinyl alcohol; (5) alkyl-substituted polycarbamate; (6) methyl-substituted polycarbamate; (7) perfluoromethyl polycarbamate; (8) perfluoroalkylsulfonamide polycarbamate.

polymers are adequate release agents. Methyl methacrylate, styrene, butyl acrylate, and methacrylic acid have been polymerized to a copolymer with a T_g of 60°C. The polymer has been grafted with the same monomers, giving a polymer with T_g of −80°C. The neutralized product showed release properties [508,509].

Silicone Release Agents

The most used release coatings having general applicability are those based on silicone polymers. Such products were developed many years ago for greaseproof and release papers and antiblocking coatings. Polymethylsiloxanes are used as soft block segments in high polymers or as oligomers in surface-active agents. Silicone release is used mostly for labels, but it has been tested for other products too (see also Chapter 6). Such formulations include silicones per se as well as other abhesive or nonabhesive polymers or silicone-modified polymers. For instance, dimethylpolysiloxane and methyl hydrogen polysiloxane based coatings are used as release liners for stickies [46]. Diorganopolysiloxanes have been suggested as water-based release dispersions [510]. Such release coatings contain 10–70% w/w of polyvinyl acetate and 2–20% of polyvinyl alcohol. The T_g of the polymer is situated between -10 and $+35°C$. Mixtures of silicones with vinyl polymers [511] and ionomers [512] have been proposed as antiblocking surface treatment solutions for elastomers.

Ultraviolet-cured silicones have been proposed for narrow webs; for broad webs, EB curing has been recommended [513]. Because EB curing occurs at low temperatures, these types of silicone release coatings could be applied and cured on heat-sensitive liners such as polyester, polypropylene, and polyethylene films and polyethylene-coated papers without distorting the liners. Radiation (UV)-curable release materials include reaction products of esters of pentaerythritol and methacrylic acid containing hydroxy groups and alkoxysilyl-terminated siloxanes [514]. Tapes with good adhesion and with release properties that decrease after irradiation have been prepared by treating substrates with silicone release agents (containing SiH and vinyl groups) on one side and butyl acrylate-acrylic acid copolymer on the other side of a 60 μm corona-treated PP film. After UV irradiation of the release side (30 s, 1000 W/m^2), the adhesive strength was 150 g/50 mm [515].

3.2 Built-in Abhesive Components

As discussed earlier, some nonpolar carrier materials exhibit abhesive properties. Their performance can be improved if special chemicals are built into the bulk carrier material. These compounds may be macro- or micromolecular. It is well known from the production of packaging materials that their processing and use on packaging machines require a low friction coefficient. Therefore slip agents are built in. Tapes have to have low resistance to unwinding. Protective films laminated on the surface to be protected must be able to endure high speed processing. Therefore in certain cases during film manufacture, macro- or micromolecular compounds having abhesive properties are added to the base polymer (see earlier discussions of slip and antiblocking agents).

Micromolecular abhesive components are chemically similar to some low molecular weight compounds used as abhesive for web coating. For instance, adhesive tapes with silicon PSA are coated with a release layer based on secondary amines, long-chain fatty acid, and vanadium oxytrichloride [516]. Such formulations can be embedded in the carrier also. Macromolecular abhesive components are polymers having a nonpolar closed surface. It is well known that highly crystalline polymers like HDPE are difficult to coat. Macromolecular abhesive components can be embedded in common plastomers or elastomers also, to improve their abhesivity. The best known dehesive macromolecular compounds are silicone derivatives. They can be used as additives also. For instance, polydiorganosiloxanes are suggested as additives in coextruded polypropylene carrier materials with a tackified top layer [91]. A dimethylpolysiloxane having a kinematic viscosity of at least 100 mm^2/s at a temperature of 25°C is suggested (0.3–2.0% w/w) as such an additive. The release layer of the coextrudate has a thickness of 0.5–10 μm. The total thickness of the tape carrier film is about 15–50 μm. The removability of acrylic adhesives for tapes is improved by adding small amounts of organofunctional silanes (e.g., methacryloxypropyltrimethoxysilane) [517].

REFERENCES

1. I. Benedek and L. J. Heymans, *Pressure-Sensitive Adhesives Technology,* Marcel Dekker, New York, 1997, Chapter 5.
2. R. Stockmeyer, *Deut. Drucker 13:*42 (1988).
3. D. W. Aubrey, G. N. Welding, and T. Wong, *J. Appl. Chem. 19:*2193 (1969).
4. E. Seiler and B. Göller, *Kunststoffe 10:*1085 (1990).
5. H. Hub, Composition, characteristics and application of polar ethylene copolymers, Presented at Polyethylene and Copolymer Resin and Packaging Markets, Dec. 1–4, 1990, Zurich, Maack Business Service, Zurich, Switzerland, 1990, p. 101.
6. J. F. Kuik, *Papier Kunstst. Verarb. 10:*26 (1990).
7. R. M. Ward and D. C. Kelley, *Tappi J. 6:*140 (1988).
8. E. B. Parker, PE copolymers and their application, Presented at Polyethylene and Copolymers Resin and Packaging Market, Speciality Plastics Conf. 87, Dec. 1, 1987, Maack Business Service, Zurich, Switzerland, p. 259.
9. E. Wistuba, *Kleben und Klebstoffe,* Sonderdruck, TI/ED 1665d, September 1993, BASF A. G., Ludwigshafen, Germany.
10. W. A. Fraser, Novel processing aid technology for extrusion grade polyolefins, Presented at Speciality Plastics Conference '87, Polyethylene and Copolymer Resin and Packaging Markets, Dec. 1, 1987, Zurich, Switzerland.
11. R. Hauber (Hans Neschen GmbH, Bückeburg, Germany), DE 42 02 070/29.07.1993.
12. Polyethylene 93, Oct. 4, 1993, Maack Business Service, Zurich, Switzerland, Session III, p. 6.
13. W. J. Busby, Processing problems with PE films, Polyethylene 93, 3-3 Maack Business Service, Zurich, Switzerland, Session VI, p. 3.
14. R. Nurse, HDPE applications, PE developments, Polyethylene 93, Oct. 4, 1993, Zurich, Switzerland, Session 3, p. 1.
15. Packard Electric, Engineering Specification ES-1881, 1992.
16. P. J. Flory, *Principles of Polymer Chemistry,* Cornell Univ. Press, Ithaca, NY 1953; in T. Nishi and T. T. Wang, *Macromolecules 8:*908 (1975).
17. T. Nishi and T. T. Wang, *Macromolecules 8:*908 (1975).

18. M. Avella and E. Martuscelli, *Polymer* 29(10):1731 (1988).
19. *Eur. Plast. News* 19:19 (1993).
20. W. W. Bode, *Tappi J.* 6:133 (1988).
21. Kimberly Clark Co., U.S. Patent 799429; in *Coating* 1:9 (1970).
22. H. G. Drössler, *Kaut. Gummi Kunstst.* 1:28 (1987).
23. G. Bodor, *A Polimerek Szerkezete,* Müszaki Könyvkiadó, Budapest, 1982, pp. 85, 147.
24. *Deut. Papierwirtsch.* 3:133 (1987).
25. R. C. Rombouts, Speciality Plastics Conference, Dec. 1, 1987, Maack Business Services, Zurich, Switzerland, p. 401.
26. K. Brack (Design Coat Co.), U.S. Patent 4288479/08.09.1981; in *Adhes. Age* 12:58 (1981).
27. V. D. Babayants, V. V. Kolesnitschenko, and S. G. Sannikov, *Lakokras. Mater. Ikh. Primen.* 3:45 (1988); in *CAS Coatings, Inks Related Products* 17:9 (1988).
28. E. Steffens, *Plaste Kaut.* 3:171 (1969).
29. J. Patschorke, *Adhäsion* 2:49 (1970).
30. D. de Jager and J. B. Borthwick, *Thermoplastic Rubbers for Hot Melt Pressure Sensitive Adhesives—The Processing Factors,* Shell Elastomers, Thermoplastic Rubbers, Tech. Manual, TR 8.11, 1989. p. 5.
31. W. Neissl and H. Ledwinka, *Kunststoffe* 8:577 (1993).
32. S. Danesi and E. Garagnani, *Kaut. Gummi Kunstst.* 3:195 (1984).
33. J. Wetzel, *ASTM Bull. 221,* 1957.
34. C. W. Hock, *J. Polym. Sci.* C3:139 (1963).
35. M. Kriszewski, *Acta Polym.* 1/2:37 (1988).
36. *Etiketten-Labels* 3:10 (1995).
37. Minnesota Mining and Manuf. Co, St. Paul, MN, U.S. Patent 3129618; in *Adhäsion* 6:14 (1985).
38. *Adhäsion* 11:482 (1967).
39. R. Davis, *Fasson Facts Int.* 1:2 (1969).
40. R. M. Podhajny, *Convert. Packag.* 3:21 (1986).
41. *Adhes. Age* 11:10 (1986).
42. *Labels Label.* 3/4:82 (1994).
43. *Adhäsion* 6:14 (1985).
44. B. Hanka, *Adhäsion* 10:342 (1971).
45. *Paper, Film, Foil Converter* 5:31 (1973).
46. Minnesota Mining and Manuf. Co, St. Paul, MN, U.S. Patent 3129618; in *Adhäsion* 2:79 (1966).
47. G. Fuchs, *Adhäsion* 3:24 (1982).
48. H. K. Porter Co. Inc., Pittsburgh, PA, U.S. Patent 3149997; in *Adhäsion* 2:79 (1966).
49. Allmänna Svenska Elektriska AB, DBP 1276771; in *Adhäsion* 2:79 (1966).
50. *Etiketten-Labels* 3:9 (1995).
51. *Kaut. Gummi Kunstst.* 6:564 (1985).
52. H. Münstedt and H. J. Wolter, *Kunststoffe* 10:1076 (1990).
53. *Eur. Plast. News* 19:19 (1993).
54. *Eur. Plast. News* 19:22 (1996).

55. S. Füzesséry, West European PE film market and applications, by melt index, density and comonomers type, Polyethylene '93, Oct. 4, Maack Business Service, Zurich, Switzerland.
56. K. Wabro, R. Milker, and G. Krüger, *Haftklebstoffe und Haftklebebänder,* Astorplast GmbH, Alfdorf, Germany, 1994, p. 48.
57. E. Djagarowa, W. Rainow, and W. Dimitrow, *Plaste Kautschuk 1:*28 (1970).
58. E. Djagarowa, *Plaste Kautschuk 9:*678 (1969).
59. *Adhes. Age 12:*8 (1986).
60. *Adhes. Age 1:*43 (1985).
61. *Etiketten-Labels 5:*136 (1995).
62. D. Djordjevic, Tailoring films by the coextrusion casting and coating process, Speciality Plastics Conference '87, Polyethylene and Copolymer Resin and Packaging Markets, Dec. 1, 1987, Maack Business Service, Zurich, Switzerland, p. 67.
63. B. Kunze, S. Sommer, and G. Düsdorf, *Kunststoffe 84*(10):1337 (1994).
64. *Handling 5/6:*12 (1994).
65. Minnesota Mining and Manuf. Co., St. Paul, MN, Br. Patent 1102296; in *Coating 7:*210 (1969).
66. K. Nakamura and Y. Miki (Nitto Electric Industrial Co. Ltd.), Jpn. Patent 6386786/18.04.1988; *CA Selects Adhes. 21:*5 (1988).
67. *Etiketten-Labels 5:*6 (1995).
68. P. Galli, F. Millani, and T. Simonazzi, *Polym. J. 17:*37 (1985).
69. *Kunststoffe 7:*611 (1992).
70. B. H. Gregory, Extrusion coating advances—Resins, processing, applications, markets, Polyethylene '93, The Global Challenge for Polyethylene in Film, Lamination, Extrusion, Coating Markets, Oct. 4, 1993, Maack Business Services, Zurich, Switzerland.
71. E. Eastmann, Ethylene polymer compositions for hot melt adhesives, *Coating 10:*368 (1988).
72. M. Gebauer and K. Bühler, *Kunststoffe 1:*21 (1992).
73. *Kunststoff-J. 4:*60 (1985).
74. A. Barbero and A. Amico, A new performance ULDPE/VLDPE from high pressure technology—Potential applications, Polyethylene 93, Oct. 4, 1993, Zurich, Switzerland, p. 3.
75. F. R. Baker, A unique ethylene/vinylacetate copolymer for the adhesives industry, *Tappi,* Hot Melt Adhesives and Coating Short Course, May 2, 1982, Hilton Head, SC.
76. K. Taubert, *Adhäsion 10:*379 (1970).
77. *Adhes. Age 8:*8 (1983).
78. *Converting Today 11:*9 (1991).
79. DBP 1079252; in *Coating 6:*185 (1969).
80. *Etiketten-Labels 35:*21 (1995).
81. *Adhäsion 11:*37 (1994).
82. *Etiketten-Labels 5:*136 (1995).
83. *Adhes. Age 12:*38 (1987).
84. Ph. Barot and J. Goletto, *Kaut. Gummi Kunstst. 10:*967 (1986).
85. G. Deelens, *Kaut. Gummi Kunstst. 10:*967 (1986).

86. D. Ogata and M. Kakimoto, *Macromolecules* 18:851 (1985).
87. *KI, 1020:*2 (1991).
88. *Coating 6:*154 (1974).
89. T. Nakajima, K. Oda, K. Azuma, and K. Fujita (Nitto Electric Ind. Co.), Jpn. Patent 6327579/05.02.1988; in *CAS Siloxanes Silicones* 17:5 (1988).
90. Minnesota Mining and Manuf. Co., St. Paul, MN, DBP1486514, in Coating 12:263 (1973).
91. G. Crass, A. Bursch and P. Hammerschmidt (Hoechst AG, Frankfurt am Main, Germany), U.S. Patent 4673611/June 16, 1987.
92. *Papier Kunstst. Verarb.* 2:18 (1996).
93. Milprint Overseas Corporation, Milwaukee, WI, U.S. Patent 1504556; in *Coating 8:*240 (1972).
94. A. Haas (Societe Chimique des Charbonnage-CdF Chimie, France), U.S. Patent 4624991/25.11.1986; in *Adhes. Age* 5:26 (1987).
95. Sun Oil Co., U.S. Patent 3342902; in *Coating 8:*244 (1969).
96. J. C. Chen and G. R. Hamed, *Rubber Chem. Technol.* 60(2):319 (1987).
97. T. J. Bonk and J. T. Simpson (Minnesota Mining and Manuf. Co., St. Paul, MN), EPA 0120708/ November 19, 1987.
98. S. Schmukler, J. Machonis, Jr., and M. Shida (Chemplex Co., Rolling Meadows, IL), U.S. Patent 4472555/18.09.1985.
99. C. Kirchner, *Adhäsion 10:*398 (1969).
100. C. L. Gueris and E. McBride, Ethylene copolymers for hot melt adhesives for adhesion to difficult plastics, 16th Münchener Klebstoff und Veredelungsseminar, 1991, p. 72.
101. J. Brandup and E. H. Immergut (Eds.), *Polymer Handbook,* 3rd ed., Wiley, New York, 1990.
102. H. Wagner and P. Flory, *J. Am. Chem. Soc. 74:*195 (1952); in I. C. Petrea, *Structura Polimerilor,* Ed. Didactica si Pedagogica, Bucuresti, 1971, p. 75.
103. F. Danusso, G. Moraglio, and G. Gianotti, *Rend. Ist. Lombardo, Sci. P.I., Classe Sci., Mat.e Nat. A93:*666 (1959); in I. C. Petrea, *Structura Polimerilor,* Ed. Didactica si Pedagogica, Bucuresti, 1971, p. 75.
104. A. Schultz, *J. Am. Chem. Soc.* 76:3422 (1954); in I. C. Petrea, *Structura Polimerilor,* Ed. Didactica si Pedagogica, Bucuresti, 1971, p. 75.
105. M. Kuwahara, M. Kaneko, and J. Furuichi, *J. Phys. Soc. Jpn. 17:*568 (1962); in I. C. Petrea, *Structura Polimerilor,* Ed. Didactica si Pedagogica, Bucuresti, 1971, p. 75.
106. I. C. Petrea, *Structura Polimerilor,* Ed. Didactica si Pedagogica, Bucuresti, 1971, p. 75.
107. R. J. Litz, *Adhes. Age 8:*38 (1973).
108. *Coating 8:*246 (1972).
109. J. Verseau, *Coating 11:*330 (1972).
110. *Adhäsion 3:*83 (1974).
111. J. Gerecke, F. Zachaeus, and R. Wintzer, *Plaste Kaut. 32:*332 (1985).
112. Neste Chemicals International NV-SA, PE Marketing Department, *Technical Information, Polyethylene,* SP0237, 1991, p. 11.
113. K. F. Schroeder, *Adhäsion 5:*161 (1971).

Chemical Basis of PSPs 241

114. Takdust Products Co., *Rubber Plast. Age (Lond.)* 6:559 (1968); in *Coating* 1:14 (1987).
115. M. A. Krecenski, J. F. Johnson, and S. C. Temin, *J. Macromol. Sci., Rev. Macromol. Chem. Phys.* C26:143 (1986).
116. A. B. Wechsung, *Coating* 9:268 (1972).
117. A. Wolf, *Kaut. Gummi Kunstst.* 41(2):173 (1988).
118. R. Köhler, *Adhäsion* 6:147 (1968).
119. U.S. Patent 4223067; in D. K. Fisher and B. J. Briddell (Adco Product Inc., Michigan Center, MI), EP 0426198 A2/08.05.1991.
120. K. S. Lin (Avery Int. Co.), PCT WO 88 02014/24.03.1988; in *CAS Adhes.* 20:4 (1988).
121. *Kaut. Gummi Kunstst.* 40(2):110 (1987).
122. *Kaut. Gummi Kunstst.* 39(5):450 (1986).
123. W. Gleim, W. Oppermann, and G. Rehage, *Kaut. Gummi Kunstst.* 6:516 (1986).
124. G. R. Hamed and C. H. Hsieh, *J. Polym. Phys.* 21:1415 (1983).
125. *Chemie in unserer Zeit* 21(2):52 (1987).
126. D. J. P. Harrison, J. F. Johnson, and W. R. Yates, *Polym. Eng. Sci.* 22(14):865 (1982).
127. *Kaut. Gummi Kunstst.* 40(8):34 (1987).
128. *Coating* 6:184 (1969).
129. Minnesota Mining and Manuf. Co., U.S. Patent 1594178; *Coating* 8:240 (1972).
130. H. Mueller and J. Türk (BASF AG, Ludwigshafen, Germany), EP 0118726/19.09.1984.
131. U.S. Patent 3321451; in S. E. Krampe and C. L. Moore (Minnesota Mining and Manuf. Co., St. Paul, MN), EP 0202831A2/26.11.1986.
132. G. Auchter, J. Barwich, G. Rehmer, and H. Jäger, *Adhes. Age* 7:20 (1994).
133. U.S. Patent 3725115; in Y. Sasaaki, D. L. Holguin, and R. Van Ham (Avery International Co., Pasadena, CA), EP 0252 717A2/13.01.1988.
134. M. Dupont and N. De Keyzer, Ultra violet light radiation curing of styrenic block copolymer based pressure sensitive adhesives, 19th Münchener Klebstoff und Veredelungsseminar, 1994, p. 120.
135. U.S. Patent 4069123; in Y. Sasaaki, D. L. Holguin, and R. Van Ham (Avery International Co., Pasadena, CA), EP 0252 717A2/13.01.1988.
136. R. Hinterwaldner, *Coating* 12:477 (1995).
137. J. R. Erickson, E. M. Zimmermann, J. G. Southwick, and K. S. Kibler, *Adhes. Age* 11:18 (1995).
138. M. Akay, S. N. Rollins, and E. Riordan, *Polymer* 29(1):37 (1988).
139. M. I. Aranguren and C. W. Macosko, *Macromolecules* 21(8):2484 (1988).
140. G. Holden and S. Chin, *Adhes. Age* 5:22 (1987).
141. J. Weidenmüller, *Deut. Papierwirtsch.* 3:T78 (1987).
142. R. Hinterwaldner, *Coating* 7:252 (1991).
143. S. R. Kerr, Heat seal macromolecular compounds, in F. Th. Birk, *Coating* 9:238 (1985).
144. A. Havranek, *Kaut. Gummi Kunstst.* 39(3):238 (1985).
145. J. L. Haldeman, *Adhes. Age* 9:35 (1984).
146. A. T. Govorkov, D. L. Muriskin, and Yu. N. Safonov, *Plast Massy* 3:28 (1987).

147. Y. Sasaaki, D. L. Holguin, and R. Van Ham (Avery International Co., Pasadena, CA), EP 0252717A2/13.01.1988.
148. Lehmann et al., U.S. Patent 3729338; in D. K. Fisher and B. J. Briddell (Adco Product Inc., Michigan Center, MI), EP 0426198A2/08.05.1991.
149. U.S. Patent 4181752; in R. R. Charbonneau and G. L. Groff (Minnesota Mining and Manuf. Co., St. Paul, MN), EP 0106559 B1/25.04.1984.
150. M. H. Mazurek (Minnesota Mining and Manuf. Co., St. Paul, MN), EP 4693935/ March 15, 1987.
151. G. R. Frazee (S. C. Johnson and Son, Inc.), EP 259842/16.03.1988; in *CAS Adhesives 20:*4 (1988).
152. *Coating 8:*268 (1987).
153. H. Bayer, *Adhäsion 5:*17 (1987).
154. K. Fuhr, *Defazet 31*(6–7):259 (1977).
155. M. Scheiber and H. Braun, UV Vernetzbare Haftklebstoffe, 19th Münchener Klebstoff u. Veredelunsgseminar, 1994, p. 141.
156. C. R. Morgan, *Adhäsion 12:*25 (1983).
157. H. Huber and H. Müller, *Coating 9:*328 (1987).
158. J. Weidenmüller, *Deut. Papierwirtsch. 3:*T78 (1987).
159. S. Kerr, *Polygraph 5:*366 (1986).
160. E. Djagarowa, *Plaste Kautschuk 19:*748 (1969).
161. P. Beiersdorf & Co. AG, Hamburg, U.S. Patent 1569882; in *Coating 11:*336 (1972).
162. A. Zawilinski, *Adhes. Age 9:*29 (1984).
163. Sealock, *Product Data, Adhesive L1233,* Andover, Hampshire, UK 1994.
164. R. J. Shuman and B. Burns (Dennison Manuf. Co., Framingham, MA), PCT, WO 88/01636/28.10.1986.
165. P. Green, *Labels Label. 11/12:*38 (1985).
166. Rohm & Haas Co., Philadelphia, PA, U.S. Patent 3152921; in *Adhäsion 3:*80 (1966).
167. TN II Bumagi, SSSR Patent 300561; in *Coating 1:*14 (1987).
168. *Coating 6:*185 (1969).
169. *Coating 1:*24 (1969).
170. Y. C. Chu and R. Vukov, *Macromolecules 18:*1423 (12985).
171. *Coating 12:*455 (1969).
172. *Eur. Adhes. Sealants 6:*23 (1995).
173. *Eur. Adhes. Sealants 6:*36 (1995).
174. W. H. Korcz, *Adhes. Age 11:*19 (1984).
175. E. G. Huddleston (The Kendall Co., Boston, MA), U.S. Patent 4692352/ 08.09.1987.
176. Johns Manville Corp., U.S. Patent 3356635; in *Coating 7:*210 (1969).
177. A. J. Liebermann, A. J. Kotova, N. Vedenov, V. V. Verchoyancev, and R. J. Mirkina, *Lakokras Mater., Ih. Primen. 3:*12 (1983).
178. R. G. Jahn, *Adhes. Age 12:*35 (1977).
179. *Adhes. Age 9:*24 (1986).
180. H. J. Neupert, M. Arnold, S. Klodt, U. Schoenrogge, B. Rothenhauser, and U. Ros-

sow (VEB Chemische Werke, Buna,) German East Patent 247903/22.07.1987; in *CAS Emulsion Polymerization 18:*2 (1988).
181. D. J. St. Clair and J. T. Harlan, *Adhes. Age 12:*39 (1975).
182. W. P. Gergen, *Kaut. Gummi Kunstst. 37*(4):284 (1984).
183. C. P. L. C. Donker, Characterisation of resins for hotmelt pressure sensitive adhesives, 16th Münchener Klebstoff und Veredelungsseminar, 1991, p. 41.
184. M. K. Dehnke and K. E. Johnsen, Formulating flexibility of pure triblock styrene/diene thermoplastic elastomers, *Coating 5:*176 (1988).
185. C. P. Rader, *Kunststoffe 83*(10):777 (1993).
186. C. N. Smit, *Kunstst. Ruber 41*(2):16 (1988).
187. R. Hinterwaldner, *Adhäsion 6:*11 (1984).
188. *Adhes. Age 11:*32 (1985).
189. K. E. Johnsen, New developments in thermoplastic elastomers, *Tappi* Hot Melt Symposium 85, June 16–19, Hilton Head, SC, 1985.
190. S. V. Canewarolo and A. W. Birley, *Br. Polym. J. 19:*43 (1987).
191. I. Jagisch, Advances in styrene block-copolymer technology, Paper presented at the PSTC Technical Seminar, Tech XIV, May 2, 1991.
192. K. E. Johnsen, *Adhes. Age 11:*29 (1985).
193. E. Diani, A. Riva, A. Iacono, and E. Agostinis, Styrenic block copolymers as base material for hot melt adhesives, 16th Münchener Klebstoff und Veredelungsseminar, 1991, p. 77.
194. J. A. Miller and E. von Jakusch (Minnesota Mining Manuf. Co., St. Paul, MN), EP 0306232B1/07.04.1993.
195. *Kaut. Gummi Kunstst. 37*(4):285 (1984).
196. F. F. Lau and S. F. Silver (Minnesota Mining and Manuf. Co., St. Paul, MN), EP 0130087B1/02.01.1985.
197. L. J. Fetters, R. W. Richards, and E. L. Thomas, *Polymer 28*(13):2252 (1987).
198. D. Alward, D. J. Kinning, E. L. Thomas, and L. J. Fetters, *Macromolecules 19:*215 (1986).
199. U.S. Patent 4163077; in F. F. Lau and S. F. Silver (Minnesota Mining and Manuf. Co., St. Paul, MN), EP 0130087B1/02.01.1985.
200. Fr. Patent 2331607; in F. F. Lau and S. F. Silver (Minnesota Mining and Manuf. Co., St. Paul, MN), EP 0130087B1/02.01.1985.
201. L. D. Jurrens and O. L. Mars, *Adhes. Age 8:*31 (1974).
202. E. H. Otto, *apr 16:*438 (1986).
203. C. Parodi, S. Giordano, A. Riva, and L. Vitalini, Styrene-butadiene block copolymers in hot melt adhesives for sanitary application, 19th Münchener Klebstoff und Veredelungsseminar, 1994, p. 119.
204. C. Donker, R. Luth, and K. van Rijn, Hercules MBG 208 hydrocarbon resin: A new resin for hot melt pressure sensitive (HMPSA) tapes, 19th Münchener Klebstoff und Veredelungsseminar, 1994, p. 64.
205. L. Jacob, New development of tackifiers for SBS copolymers, 19th Munich Adhesive and Finishing Seminar, 1994, p. 107.
206. *Coating 12:*480 (1995).

207. E. J. Lutz, A. Lepert, and M. L. Evans (Exxon Research and Eng. Co. Florham Park, NJ), U.S. Patent 4623698/18.11.1986.
208. P. A. Mancinelli, J. A. Schlademann, S. C. Feinberg, and S. O. Norris, Advancement in acrylic HMPSAs via macromer monomer technology, PTSC XVII Technical Seminar, May 4, 1986, Woodfield, Shaumburg, IL; in *Coating 1:*12 (1986).
209. G. Obieglo and K. Romer, *Kunststoffe 83*(11):926 (1983).
210. J. R. Erickson, Experimental thermoplastic rubber with improved radiation curing performances for HMPSA, *TAPPI* Hot Melt Symposium '85, June 16–19, 1985, Hilton Head, SC; in *Coating 1:*6 (1986).
211. M. Szwarc, *Nature 178:*1168 (1956).
212. F. Bandermann, H. D. Speikamp, and L. Weigel, *Makromol. Chem. 186:*2017 (1985).
213. Z. Mo, L. Wang, H. Zhang, P. Han, and B. Huang, *J. Polym. Sci., Polym. Phys. 25*(9):1829 (1987).
214. N. G. Matveeva, A. G. Kondratieva, E. S. Pankova, O. G. Selskaya, and E. S. Mamedova, *Kinetics and Mechanism of Polyreactions,* Akadémiai Kiadó, Budapest, 1969, Vol. 3, p. 161.
215. Y. S. Matsumoto and J. Furukawa, *Kaut. Gummi Kunstst. 6:*541 (1986).
216. L. L. de Lucca Freitas and R. Stadler, *Macromolecules 20*(10):2478 (1987).
217. H. Xie and J. Xia, *Makromol. Chem. 188*(11):2543 (1987).
218. J. Roovers, *Macromolecules 20*(1):148 (1987).
219. G. Bocaccio, *Caoutch. Plast. 62*(653):83 (1985).
220. B. C. Auman, V. Percec, H. A. Schneider, and H. J. Cantow, *Polymer 28*(8):1407 (1987).
221. S. Kawachi, *GAK 4:*162 (1986).
222. L. J. Mathias and M. Allison, *ACS Symp. Ser. 367:*66 (1988); in *CAS, Crosslinking Reactions 21:*3 (1988).
223. TZ. M. Swager and R. H. Grubbs, *J. Am. Chem. Soc. 109*(3):894 (1987).
224. *Kaut. Gummi Kunstst. 41*(1):102 (1988).
225. L. de Lucca Freitas, J. B. Urgert, and R. Stadler, *Polym. Bull. 17:*431 (1987).
226. U. Eisele, *Kaut. Gummi Kunstst. 40*(6):539 (1987).
227. R. Jordan, *Adhäsion 5:*256 (1980).
228. *Vector 2518D,* Product description, Dexco Polymers, November 1992.
229. *Vector 2411D,* Product description, Dexco Polymers, November 1992.
230. *Vector 8508D,* Product description, Dexco Polymers, November 1992.
231. Shell Elastomers, *Cariflex TR-1000 Polymere für Klebstoffe, Beschichtungen und Dichtungsmassen,* Technische Broschüre, RBX/73/8(G), 1973.
232. *Kaut. Gummi Kunstst. 40*(1):40 (1987).
233. *Vector 2514D,* Product description, Dexco Polymers, November 1992.
234. R. Koch, *Kaut. Gummi Kunstst. 39*(9):804 (1986).
235. D. W. Bamborough and P. M. Dunkley, *Adhes. Age 11:*20 (1990).
236. W. H. Meyer, *Chem. unserer Zeit 21*(2):59 (1987).
237. N. Nakajima, R. A. Miller, and E. R. Harrel, *Int. Polym. Process. 2*(2):88 (1987).

Chemical Basis of PSPs 245

238. K. Tanakagawa, S. Yagishita, K. Hosoya, and N. Inagami, *Am. Chem. Soc. Div. Polym., Chem. Prepr. 26*(2):32 (1985).
239. V. Stanislawczyk (B. F. Goodrich Co.), EP264903/27.04.1988; in *CAS Emulsion Polymerization, 18:* 2 (1988).
240. K. Yamada, K. Miyazaki, Y. Oowatari, Y. Egami, and T. Honma (Sumitomo Chem. Co., Ltd.), EP257803/02.03.1988; in *CAS, Coatings Inks Related Comp. 17:*6 (1988).
241. A. Hecht Beaulieu, D. R. Gehman, and W. J. Sparks, *Tappi J. 9:*102 (1984).
242. F. T. Sanderson, Acrylic hot melt pressure sensitive adhesives, Hot Melts—The Future is Now, Tappi Hot Melt Symposium, June 2–4, Toronto; in *Coating 7:*175 (1980).
243. R. M. Enanoza (Minnesota Mining & Manuf. Co.), EP 259968/16.03.1988.
244. National Starch and Chem Co., Jpn. Patent 6306076/12.01.1988; in *CAS Adhesives 19:*2 (1988).
245. D. E. Bugner (Eastmann Kodak Co., Rochester, NY), *ACS Symp. Ser. 364 (Chem. React. Polym.):*276 (1988); in *CAS Polyacrylates 9:*2 (1988).
246. T. Hamaide, A. Revillon, and A. Guyoz, *Eur. Polym. J. 23*(10):787 (1987).
247. E. V. Gruzinow, V. P. Panov, V. V. Gusev, V. N. Frosin, and V. S. Rytbchinskaya, *Vysokomol. Soed. B29:*373 (1987).
248. P. Cebeillac, D. Chatain, C. Megret, C. Lacabanne, A. S. Bernes, and P. Dupuis, *Makromol. Chem., Macromol. Symp. 20/21:*335 (1988).
249. *Druck Print 4:*18 (1988).
250. G. R. Homan and H. L. Vincent (Dow Corning Corp., Midland, MI), EP0183378/29.10.1984.
251. J. D. Blizzard and T. J. Swihart (Dow Corning Corp., Midland, MI), EP 0183379/29.10.1984.
252. J. D. Blizzard and D. Narula (Dow Corning Corp., Midland, MI), EP0183377/29.10.1984.
253. A. Tomanek, *Kunststoffe 11:*1277 (1990).
254. F. Hufendiek, *Etiketten-Labels 1:*9 (1995).
255. *Adhäsion 11:*37 (1994).
256. *Adhes. Age 9:*8 (1986).
257. Dow Corning Corp., U.S. Patent 2814601, in Coating, (1)24(1987).
258. Minnesota Mining and Manuf. Co. St. Paul, U.S. Patent 2882183, in *Coating*, (1)24(1987).
259. Midland Silicones Ltd., Br. Patent 792041, in *Coating*, (1)24(1987).
260. S. R. Kerr and S. B. Lin, Recent advances in silicone PSA technology, PTSC, XVII Technical Seminar, May 4, Woodfield, Shaumburg, IL, 1986.
261. Mystik Tape Inc., U.S. Patent 3161533; in *Adhäsion 6:*277 (1966).
262. J. R. Pennace, C. Ciuchta, D. Constantin, and T. Loftus, WO8703477A/18.06.1987.
263. Y. Hamada and O. Takuman (Toray Silicone Co., Ltd.), EP 253601/20.01.1988.
264. B. Copley and K. Melancon (Minnesota Mining and Manuf. Co., St. Paul, MN), EP255226/03.02.1988.
265. D. F. Merril, *Int. SAMPE Symp. Exhib., 1988:*33; in *CAS Adhesives 14:*2 (1988).

266. R. Milker, PUR PSAs, Pressure Sensitive Adhesives and Adhesive Coatings, Cowise, Management and Training Service, 24.04.1996, Amsterdam.
267. B. H. Edwards, Polyurethane structural adhesives, Adhesives '85, Conference Papers, Sept. 10, 1985, Atlanta, GA.
268. *Coating 8:*242 (1972).
269. M. K. Yamazaki and S. Kamatani (Takeda Chemical Industries, Ltd., Osaka, Japan), U.S. Patent 4471103/September 11, 1984.
270. R. J. DeVor and C. D. Lynch (Minnesota Mining and Manuf. Co., St. Paul., MN), U.S. Patent 474577/26.04.1988.
271. C. Hepburn, *Progr. Rubber Plast. Technol. 3*(3):33 (1987).
272. B. Krüger, *Kaut. Gummi Kunstst. 10:*967 (1986).
273. N. Kunii, *Kaut. Gummi Kunstst. 39*(6):541 (1986).
274. N. Yamazaki, *Kaut. Gummi Kunstst. 9:*836 (1986).
275. C. G. Pitt, Z. W. Gu, P. Ingram, and R. W. Hendren, *J. Polym. Sci., Polym. Chem. 25*(4):995 (1987).
276. J. Franke, K. P. Meurer, P. Haas, and J. Witte (Bayer A. G. Leverkusen, Germany), DE 3622825/21.01.1988.
277. T. Wakabayashi and S. Sugii (Minnesota Mining and Manuf. Co., St. Paul, MN), EP 0285430B1/05.10.1988.
278. Sun Oil Co., U.S. Patent 3356766; in *Adhäsion 10:*303 (1972).
279. C. N. Clubb and B. W. Foster, *Adhes. Age 11:*18 (1988).
280. B. W. Foster, A new generation of polyolefin based hot melt adhesives, Tappi Hot Melt Symposium, June 7–10, 1987, Monterey, CA.
281. U.S. Patent 3686107; in T. Wakabayashi and S. Sugii (Minnesota Mining and Manuf. Co., St. Paul, MN), EP 0285430B1/05.10.1988.
282. G. C. Allen, J. B. Pellon, and M. P. Hughes (El Paso Products Co.), EP 251771/07.01.1988.
283. Y. Mizutani, T. Noguchi, H. Kuroki, and T. Imahaba (Tosoh Co.), Jpn. Patent 62265379/18.11.1987; in *CAS, Hot Melt Adhes. 11:*1 (1988).
284. L. C. DeBolt and E. Riande, *Makromol. Chem. 187:*2497 (1986).
285. H. Takao, H. Kuribayashi, and E. Usuda, EP 254002/27.01.1988.
286. G. A. Luscheyki, F. M. Medvedeva, L. I. Voytesonok, M. K. Polevaya, and L. D. Pin, *Plast Massy 10:*16 (1985).
287. B. M. Mahato, S. C. Shit, and S. Maiti, *Eur. Polym. J. 21:*925 (1985).
288. H. F. Huber and H. Müller, Radcure '86, Conf. Proc., 10th, 12.01.1986.
289. P. P. Hoenisch and F. T. Sanderson, Aqueous acrylic adhesives for industrial laminating, Adhesives '85, Conference Papers, Sept. 10–12, 1985, Atlanta, GA.
290. Minnesota Mining and Manuf. Co., St. Paul, MN, U.S. Patent 2926105; in *Coating 6:*185 (1969).
291. R. G. Frazee (S. C. Johnson and Son, Inc.), EP 258753/09.03.1988; in *CAS 19:*3 (1988).
292. S. E. Krampe and C. L. Moore (Minnesota Mining and Manuf. Co., St. Paul, MN), EP 0202831A2/26.11/1986.
293. U.S. Patent 288416/RE 24906; in S. E. Krampe and C. L. Moore (Minnesota Mining and Manuf. Co., St. Paul, MN), EP 0202831A2/26.11.1986.

294. U.S. Patent 3121021; in S. E. Krampe and C. L. Moore (Minnesota Mining and Manuf. Co., St. Paul, MN), EP 0202831 A2/26.11.1986.
295. British United Shoe Machinery Co., Br. Patent 761840; in *Coating 6:*185 (1969).
296. *Druckprint 4:*19 (1988).
297. Z. Czech, *Eur. Adhes. Sealants 6:*4 (1995).
298. *Kaut. Gummi Kunstst. 40*(4):313 (1987).
299. Y. Sasaaki, European Tape and Label Conference, Apr. 28, 1993, Brussels, p. 133.
300. D. G. Pierson and J. J. Wilczynski, *Adhes. Age 8:*52 (1990).
301. U.S. Patent 3865770; in F. C. Larimore and R. A. Sinclair (Minnesota Mining and Manuf. Co., St. Paul, MN), EP 0197662A1/15.10.1986.
302. P. K. Dhal, R. Murthy, and G. N. Babu., *Org. Coat. Appl. Polym. Sci., Proc. 48:*131 (1983).
303. Sekisui Chemical Ind. Co. Ltd., Jpn. Patent 072785513/24.10.1995; in *Adhes. Age 5:*12 (1996).
304. U.S. Patent 2925174; in R. R. Charbonneau and G. L. Groff (Minnesota Mining and Manuf. Co., St. Paul, MN), EP 0106559B1/25.04.1984.
305. C. Harder, Acrylic hotmelts—Recent chemical and technological developments for an ecologically beneficial production of adhesive tapes: State and prospects, European Tape and Label Conference, Brussels, Apr. 28–30, 1993.
306. D. K. Fisher and B. J. Briddell (Adco Product Inc., Michigan Center, MI), EP 0426198 A2/08.05.1991.
307. Belg. Patent 675420; in D. K. Fisher and B. J. Briddell (Adco Product Inc., Michigan Center, MI), EP 0426198A2/08.05.1991.
308. G. Geuskens and M. S. Kabamba, *Polym. Degrad. Stabil. 19*(4):315 (1987).
309. R. Schaller, M. G. Mertl, and K. Hummel, *Eur. Polym. J. 23:*259 (1987).
310. J. N. Kellen and C. W. Taylor (Minnesota Mining and Manuf. Co., St. Paul, MN), EP 0246826A2/25.11.1987.
311. H. Kuroda and M. Taniguchi (Bando Chem. Ind. Ltd.), Jpn. Patent 6343987/25.02.1988; in *CAS Adhes. 20:*3 (1988).
312. H. Röltgen, *Coating 5:*124 (1985).
313. H. J. Timpe and H. Baumann, *Adhäsion 9:*9 (1984).
314. P. Gosh and A. R. Bandyopadhyay, *Eur. Polym. J. 11:*1117 (1984).
315. C. Decker, Radcure 85; in F. Th. Birk, *Coating 12:*278 (1985).
316. W. Baueumer, M. Koehler, and J. Ohngemach, Radcure '86, Conf. Proc. 10th, 4/43 (1986); in *CAS Coatings, Inks Related Products 22:*3 (1988).
317. G. Li Bassi and F. Broggi (Fratelli Lambert S.p.A., Albizzate, Italy), *Polym. Paint Colour J. 178*(4210):197 (1988); in *CAS Coatings Inks Related Products 18:*2 (1988).
318. A. A. Lin, R. V. Sastri, G. Tesoro, E. Reiser, and R. Eachus, *Macromolecules 21*(4):1165 (1988).
319. C. Carlini, L. Toniolo, P. A. Rolla, P. Barigeletti, P. Bortolus, and L. Flamigni, *New Polym. Mater. 1*(1):63 (1987).
320. K. Hummel, R. Schaller, and M. G. Martl, *Polym. Bull. 17:*369 (1987).
321. R. J. DeVoe and S. Mitra (Minnesota Mining and Manuf. Co.), EP 260877/23.03.1988.

322. B. K. Bordoloi, Y. Ozari, S. S. Plamtottham, and R. Van Ham (Avery Int. Co.), EP 263866/13.04.1988; in *CAS 19:*4 (1988).
323. H. Braun and Th. Brugger, *Coating 9:*307 (1993).
324. G. W. H. Lehmann and H. A. J. Curts (Beiersdorf AG, Hamburg, Germany), U.S. Patent 4038454/26.07.1977.
325. U.S. Patent 3257478; in W. E. Lenney (Air Products and Chemical Inc., USA), Can. Patent 1225176/04.08.87.
326. (a) U.S. Patent 3697618; in W. E. Lenney (Air Products and Chemical Inc., USA), Cand. Patent 1225176/04.08.87. (b) U.S. Patent 3971766; in W. E. Lenney (Air Products and Chemical Inc., USA), Cand. Patent 1225176/04.08.87.
327. U.S. Patent 3998997; in W. E. Lenney (Air Products and Chemical Inc., USA), Cand. Patent 1225176/04.08.1987.
328. H. Gunesch and I. A. Schneider, *Makromol. Chem. 125:*213 (1969).
329. R. N. Henkel, *Paper Film Foil Convert. 12:*68 (1968).
330. I. Benedek, *Adhäsion 12:*17 (1987).
331. H. Monsey and A. Malersky, U.S. Patent 4331576/25.05.1982; in *Adhes. Age 12:*53 (1982).
332. Protective Treatments Inc., Br. Patent 925007; in *Coating 6:*185 (1969).
333. F. M. Rosenbaum, *Adhes. Age 6:*32 (1972).
334. *Coating 2:*123 (1984).
335. H. Hintz, Kunststoff Dispersionen für das Verkleben von PVC Bodenbelägen und Textiler Auslegeware, *Kunstharz Nachr. (Hoechst) 2:*18 (1983).
336. W. M. Stratton, *Adhes. Age 7:*21 (1985).
337. *Adhäsion 12:*283 (1983).
338. F. R. Baker, A unique ethylene/vinyl acetate copolymer for the adhesives industry, Tappi Hot Melt Adhesives and Coating Short Course, May 2–5, 1982, Hilton Head, SC.
339. A. A. Drescher, *Coating 5:*113 (1974).
340. Hoechst Aktiengesellschaft, F+E/Polymerisate II, Mowilith VDM 1360 ca 55%, Data Sheet, 1988.
341. J. G. Iacoviello (Air Products and Chem. Co.), U.S. Patent 4735986; in *CAS Adhesives 20:*3 (1988).
342. *Coating 2:*44 (1984).
343. W. Hoffmann, *Kunststoffe 78*(2):132 (1988).
344. *Coating, 7:*186 (1984).
345. R. Koch and C. L. Gueris, *apr 16:*460 (1986).
346. *Coating 1:*15 (1984).
347. Colon et al., U.S. Patent 4331576; in W. L. Bunelle, K. C. Knutson, and R. M. Hume (H. B. Fuller Co., St. Paul, MN), EP 0199468A2/29.10.1986.
348. Morrison, U.S. Patent 4052368; in W. L. Bunelle, K. C. Knutson, and R. M. Hume (H. B. Fuller Co., St. Paul, MN), EP 0199468A2/29.10.1986.
349. Y. Aizawa (Cemedine Co. Ltd.), Jpn. Patent 6317945/25.01.1988; in *CAS Adhes. 14:*4 (1988).
350. A. Lamping and W. Tellenbach (Nyffeler Corti A. G.), CH Patent 664971/15.04.1988; in *CAS Colloids (Macromol. Aspects) 21:*6 (1988).

351. Nitto Electrical Co., Jpn. Patent 24229/70; in *Coating* 2:38 (1972).
352. *Coating* 2:45 (1969).
353. Beiersdorf AG, Hamburg, U.S. Patent 15698888; in *Coating* 8:240 (1972).
354. *Allgem. Papier Rundsch.* 18:664 (1970).
355. R. Hauber (Hans Neschen GmbH, Bückeburg, Germany), DE 43 03 616/04.08.1994.
356. Minnesota Mining and Manuf. Co., St. Paul, MN, Can. Patent 588869; in *Coating* 6:185 (1969).
357. U.S. Patent 3441430; in F. D. Blake (Minnesota Mining and Manuf. Co., St. Paul, MN), EP 0141504 A1/15.05.1985.
358. Z. Czech (Lohmann GmbH, Neuwied, Germany), DE 44 31053/01.09.1994.
359. M. F. Tse and K. O. McElrath, *Adhes. Age* 9:32 (1988).
360. S. D. Hollis, Non-crystallizing rosin. A tackifier for hot melt adhesives, Tappi Hot Melt Adhesives and Coatings, Short Course, Hilton Head, SC, June 1982.
361. J. Rustige, *Adhäsion* 10:275 (1977).
362. C. Donker, R. Luth, and K. van Rijn, Hercules MBG 208 hydrocarbon resin: A new resin for hot melt pressure sensitive (HMPSA) tapes, 19th Münchener Klebstoff u. Veredelungsseminar, 1994, p. 64.
363. G. Bonneau and M. Baumassy, New tackifying dispersions for water based PSA for labels, 19th Münchener Klebstoff u. Veredelungsseminar, 1994, p. 82.
364. G. J. Kutsek, *Adhes. Age* 6:24 (1996).
365. E. E. Ewins, Jr. and J. R. Erickson, Formulation to inhance the radiation crosslinking of thermoplastic rubber for HMPSAs, in Tappi Hot Melt Symposium 85, June 16–19, Hilton Head, SC.
366. K. Nitzl, Th. Horna, and U. Hoffman, Strahlenhärtbare hotmelts, 16th Münchener Klebstoff u. Veredelungsseminar, 1991, p. 100.
367. S. Tsuchida, Y. Kodama, and H. Hara (Arakawa Kagaku Kogyo Kabushiki Kaisha, Osaka, Japan), U.S. Patent 4622357/11.11.1986; in *Adhes. Age* 5:15 (1987).
368. V. L. Hughes and R. W. Looney (Exxon Research and Engineering Co., Florham Park, NJ), EP 0131460/16.01.1985.
369. R. M. Podhajny, *Convert. Packag.* 3:21 (1986).
370. Minnesota Mining and Manuf. Co., U.S. Patent 3129816; in *Adhäsion* 6:271 (1965).
371. Protective Treatments Inc., Br. Patent 814001; in *Coating* 6:185 (1969).
372. T. R. Mecker, Low molecular weight isoprene based polymers—Modifiers for hot melts, Tappi Hot Melt Symposium, 1984; in *Coating* 11:310 (1984).
373. *Kautschuk Gummi Kunstst.* 41(7):662 (1988).
374. R. Jordan, *Coating* 2:37 (1986).
375. Br. Patent 941276; in P. Gleichenhagen and I. Wesselkamp (Beiersdorf A. G., Hamburg), EP 0058382B1/25.08.1982.
376. U.S. Patent 3441430; in P. Gleichenhagen and I. Wesselkamp (Beiersdorf A. G., Hamburg), EP 0058382B1/25.08.1982.
377. U.S. Patent 3152940; in P. Gleichenhagen and I. Wesselkamp (Beiersdorf A. G., Hamburg), EP 0058382B1/25.08.1982.

378. F. D. Blake (Minnesota Mining and Manuf. Co., St. Paul, MN), EP 0141504 A1/ 15.05.1985.
379. Zeneca Resins, Waalwijk, Netherlands, *Crosslinker CX-100,* Data Sheet, 1995.
380. U.S. Patent 2838421; in F. D. Blake (Minnesota Mining and Manuf. Co.), EP 0141504 A1/15.05.1985.
381. U.S. Patent 3661874; in F. D. Blake (Minnesota Mining and Manuf. Co.), EP 0141504 A1/15.05.1985.
382. P. Dreier, *Adhes. Age 6:*32 (1996).
383. J. Hansmann, *Adhäsion 10:*360 (1970).
384. B. Mayer, *Coating 4:*109 (1969).
385. *Eur. Plast. News 4:*24 (1992).
386. K. Sato, *Rubber Chem. Technol. 56:*942 (1984).
387. A. T. Franklin, *Printing Trades J. 1044:*46 (1974).
388. O. Huber, *Wochenbl. Papierfabrik. 17:*657 (1973).
389. *Kaut. Gummi Kunstst. 12:*1160 (1986).
390. *Kaut. Gummi Kunstst. 1:*6 (1986).
391. Allmänna Svenska Elektriska AB, DBP 1276771/1987.
392. *Coating 10:*373 (1986).
393. D. R. M. Harwey, U.S. Patent 2822509; in *Coating 6:*184 (1969).
394. *Adhes. Age 5:*6 (1985).
395. A. I. Medalia, *Rubber Chem. Technol. 59:*432 (1986).
396. G. Langsley, *Int. Polym. Sci. Technol. 13*(1):44 (1986).
397. F. Shikata (Teijin Ltd.), Jpn. Patent 63105187/10.05.1988; in *CAS Adhes. 22:*7 (1988).
398. I. Benedek and E. Frank, DE 4421420A1/18.06.1994.
399. *Eur. Adhes. Sealants 9:*22 (1987).
400. *Eur. Plast. News 11:*67 (1993).
401. T. Aryoshi, G. Fujiwara, T. Hayashi, and Y. Kaneshige (Tosoh Co.), Jpn. Patent 6341648/22.02.1988; in *CAS Adhes. 22:*7 (1988).
402. B. B. Blackford, Br. Patent 8864365; in *Coating 6:*185 (1969).
403. U.S. Patent 2925174; in EP 0120708, p. 2.
404. V. G. Raevski et al., SSSR Patent 304264; in *Coating 6:*185 (1969).
405. SSSR Patent 349782, TNII Bumagi; in *Coating 5:*122 (1974).
406. G. Grove, *Adhes. Age 5:*39 (1971).
407. Ankerwerk Rudolfstadt, DBP 1056787; in *Coating 6:*185 (1969).
408. Johnson & Johnson, U.S. Patent 2909278; in *Coating 6:*185 (1969).
409. R. P. Singh, J. Lacoste, R. Arnaud, and J. Lemaire, *Polym. Degrad. Stabil. 20*(1):49 (1988).
410. G. F. Vesley, A. H. Paulson, and E. C. Barber, EP 0202938 A2/26.11.1986.
411. U.S. Patent 4223067; in G. F. Vesley, A. H. Paulson, and E. C. Barber, EP 0202 938 A2/26.11.1986.
412. *Adhäsion 12:*12 (1983).
413. Ö. Söderberg, *Paper Technol. 8:*17 (1989).
414. Y. Aizawa (Cemedine Co. Ltd.,) Jpn. Patent 6317945/25.01.1988; in *CAS Adhes. 14:*4 (1988).

415. T. J. Bonk, T. I. Cheng, P. M. Olso, and D. E. Weiss, PCT, WO 87/00189/ 15.01.1987.
416. Vitta Corporation, U.S. Patent 3598679, in *Coating 6:*388 (1969).
417. *Druck Print 9:*65 (1986).
418. W. K. Darwell, P. R. Konsti, J. Klingen, and K. W. Kreckel (Minnesota Mining and Manuf. Co., St. Paul, MN), EP257984/21.08.1986.
419. Jpn. Patent 2736/1975; in H. Miyasaka, Y. Kitazaki, T. Matsuda, and J. Kobayashi (Nichiban Co., Ltd., Tokyo, Japan), Offenlegungsschrift, DE 3544868 A1/ 18.12.1985.
420. T. Shibano, I. Kimura, H. Nomoto, and C. Maruchi (Sanyo Kokusaku Pulp Co., Ltd., Tokyo, Japan), U.S. Patent 4624893/25.11.1986; in *Adhes. Age 5:*24 (1987).
421. E. Pagendarm, Hamburg, Germany, Offenlegungsschrift, DE 3632816 A1/ 26.9.1986.
422. J. W. Otter and G. R. Watts (Avery International Corp., Pasadena, CA), U.S. Patent 5346766/13.09.1994.
423. R. Shibata and H. Miyagawa (Hitachi Condenser Co. Ltd.), Jpn. Patent 63 86785/ 18.04.1988; in *CAS Adhes. 20:*4 (1988).
424. M. von Bittera, D. Schäpel, U. von Gizycki, and R. Rupp (Bayer AG, Leverkusen, Germany), EPA 0147588 B1/10.07.1985.
425. S. Maeda and T. Miyamoto (Sumitomo Bakelite Co. Ltd., Tokyo), U.S. Patent 5456765/13.09.1994; in *Adhes. Age 5:*15 (1995).
426. Z. Czech and H. D. Sander (Lohmann GmbH, Neuwied, Germany), DE 4219368/ 28.07.1994.
427. H. Eifert and G. Bernd (Fraunhoffer Gesellschaft, München, Germany), DE 4228608/28.08.1992.
428. K. Takashimizu and A. Suzuki (Advance Co. Ltd.), Jpn. Patent 63 92683/ 23.04.1988; in *CAS Adhes. 19:*5 (1988).
429. N. Hosoi and S. Azuma (Sumitomo Electric Industries Ltd., Japan), U.S. Patent 5346539/13.09.1994.
430. V. A. Uskov, V. M. Lalayan, A. K. Abysev, and N. Yu. Morozova, *Kauch. Resina 3:*8 (1986).
431. *Neue Verpackung 6:*72 (1991).
432. *Finat Labelling News 3:*29 (1994).
433. D. Lacave, *Labels Label. 3/4:*55 (1994).
434. Oji Paper Co., Jpn. Patent 07325391/12.12.1995; in *Adhes. Age 5:*12 (1996).
435. Toyo Ink Manuf. Ltd., Jpn. Patent 07292344/07.11.1995.
436. R. Hauber (Hans Neschen GmbH and Co. KG, Bückeburg, Germany), DE 42 02 070/29.07.1993.
437. S. Toshiki, M. Nagano, H. Ito, and T. Miyaji (Japan Synthetic Rubber Co., Ltd.), Jpn. Patent 6351443/1988; in *CAS Adhes. 20:*3 (1988).
438. S. Milton and C. Max, *Adhes. Age 1:*12 (1983).
439. K. Berger, *GAK 6:*318 (1985).
440. M. E. Ahner and M. L. Evans, Light color aromatic containing resins, Tappi Hot Melt Symposium '85, June 16–19, 1985, Hilton Head, SC; in *Coating 1:*6 (1986).

252   Chapter 4

441. R. J. Shuman and B. D. Josephs (Dennison Manuf. Co., Framingham, MA), PCT, WO 88/01636.
442. U. Storb, *apr 9:*244 (1986).
443. W. Kahle, *Pack Rep. 6:*87 (1991).
444. *Neue Verpackung 5:*63 (1991).
445. E. Fuchs, *Coating 3:*78 (1969).
446. BASF Kunststoffe, *Lupolen B 581,* d/12.92, Technical booklet, p. 33.
447. K. H. Kochem, H. U. ter Meer, and H. Millauer, *Kunststoffe* 82(7):575 (1992).
448. J. Pernak, A. Skrzypczak, E. Gorna, and A. Pasternak, *Kunststoffe* 77(5):517 (1987).
449. Jpn. Patent 55151046/15.05.79; in *Hochmolekularbericht* 57(1):94 (1982).
450. K. Sirinyan and F. Jonas (Bayer A. G.), Ger. Offen., DE 3625272/04.02.1988; in *CAS Coatings, Inks Related Products 17:*9 (1988).
451. Zippeling, Kessler, Ahrtensburg, Germany, *Incoblend;* Technical Information.
452. O. Schwab, *Adhäsion 5:*155 (1972).
453. M. W. Uffner and P. Weitz (General Aniline and Film Co., New York, NY), U.S. Patent 3345320/03.10.1967.
454. R. M. Ward and D. C. Kelley, *Tappi J. 6:*140 (1988).
455. *Coating 11:*274 (1985).
456. E. Haberstroh, *Papier Kunstst. Verarb. 8:*14 (1986).
457. C. Gondro, *Kunststoffe 19:*1084 (1990).
458. N. Grünewald, Additives for the improvement of polymers, SP '89, Maack Business Service, Zürich, Switzerland, p. 148.
459. F. T. Kitchel, *Tappi J. 12:*156 (1988).
460. Dow Chemicals, Horgen, Switzerland Films Dowlex. For Lamination. CH-254-052-E-288, Data Sheet, 1995.
461. A. Rudin, A. T. Worm, and J. E. Blacklock, *J. Plast Film Sheeting 1:*189 (1985).
462. C. De Smedt and S. Nam, *Plast Rubber Process. Appl.* 8(1):11 (1987).
463. W. N. K. Revolta, *Kunstst. J. 9:*72 (1986).
464. *Labels Label. 3/4:*44 (1994).
465. G. Joannou, FINAT Seminar, March 1987, p. 126.
466. H. Nitschke, *Papier Kunstst. Verarb. 10:*8 (1987).
467. *Etiketten-Labels 35:*28 (1995).
468. K. Heger, *Coating 5:*180 (1987).
469. R. Hinterwaldner, *Coating 3:*98 (1993).
470. J. Dietz, *Coating 11:*296 (1982).
471. G. Bosch, *Siebdruck 6:*58 (1988).
472. J. F. LaFaye, J. P. Maume, J. M. Schwob, and R. Chiodi, *Tappi J. 12:*63 (1988).
473. *Coating 7:*246 (1993).
474. *Coating 3:*101 (1993).
475. A. Dettling and J. Burri, *Etiketten-Labels 3:*4 (1995).
476. H. P. Seng, *Coating 9:*324 (1993).
477. P. Penczek, *Adhäsion 1/2:*32 (1988).
477. *Adhes. Age 6:*18 (1987).
478. *Coating 12:*362 (1970).

479. H. Reipp, *Adhes. Age 3:*17 (1972); in *Coating 6:*187 (1972).
480. G. L. Burroway and G. W. Feeney, *Adhes. Age 7:*17 (1974).
481. R. A. Tait, A. Salazar, W. C. Chung, R. J. Skiscim, and D. R. Hansen, Toluene replacement in solvent borne pressure sensitive adhesive formulations, PTSC XVII Tech. Seminar, May 4, Woodfield Shaumburg, IL, 1986.
482. Esso Research and Eng. Co., U.S. Patent 3351572; in *Coating 7:*210 (1969).
483. R. Milker and Z. Czech, Lösungsmittelhaltige Akrylhaftklebstoffe, 19th Münchener Klebstoff u. Veredelungsseminar, 1994, p. 136.
484. Y. Moroishi, T. Sugii, and K. Noda (Nitto Electric Ind. Co. Ltd.), Jpn. Patent 6381183/12.04.1988; in *CAS Adhes.* 22:6 (1988); *Chem. Abstr. 109:*130548q.
485. T. Sugiyama, N. Miyaji, Y. Ito, and T. Tange (Nippon Carbide Industries Co., Inc.), Jpn. Patent 62199672/03.09.1987; in *CAS Adhes.* 14:4 (1988).
486. Pittsburgh Plate Glass Co., U.S. Patent 3355412; in *Coating 7:*274 (1969).
487. N. C. Smith, *Tappi J.* 7:106 (106).
488. T. Tatsuno, K. Matsui, T. Masaharu, and W. Mitsui (Kansai Paint Co., Ltd.), Jpn. Patent 62285927/11.12.1987; in *CAS Adhes.* 14:4 (1988).
489. T. Murachi (Toyoda Gosei Co., Ltd.), Jpn. Patent 63 61075/130388; in *CAS Adhes.* 20:4 (1988).
490. H. J. Northrup, U.S. Patent 3691206; in *Coating 5:*123 (1974).
491. Showa Denko K. K. Tokyo, Japan, U.S. Patent 1900521; in *Coating 3:*60 (1974).
492. Mitsubishi Petrochemical Co. Ltd., Tokyo, Japan, DBP 1277217; in *Coating 9:*272 (1969).
493. Keuffel and Esser Co., Morristown, NJ, U.S. Patent 1569086; in *Coating 1:*31 (1978).
494. Kjin Co., Jpn. Patent 28520/70; in *Coating 2:*38 (1972).
495. Hoechst Celanese, Sommerville, NJ, EP 484819/30.10.1991; in *Coating 19:*354 (1992).
496. Morton Adhesives Europe, Adhesives and Coatings, *Pentacoll ET 691A,* Data Sheet, 1994.
497. Dow Chemical Co., *Montrek Polyethylenimine Products,* Data Sheet, 1995.
498. T. Kishi (Sekisui Chem, Co. Ltd.), Jpn. Patent 6369 879/29.03.1988; in *CAS Adhes.* 19:6 (1988).
499. D. H. Teesdale, *Coating 6:*243 (1983).
500. A. K. Agarwal and C. Balian (CMR Ind. Inc.), EP 258974/09.03.1988.
501. Br. Patent 1123014; in *Coating 7:*274 (1969).
502. TNII Bumagi, SSSR Patent 300561; in *Coating 7:*368 (1969).
503. L. Bothorel, *Emballages 278:*372 (1972).
504. L. Dengler, *Coating 6:*143 (1974).
505. K. Kriz and T. H. Plaisance (De Soto Inc.), U.S. Patent 15.12.1987; in *CAS, Coating Inks Related Products 10:*2 (1988).
506. M. Omote, I. Sakai, and T. Matsumoto (Nitto Electric. Ind. Co.), Jpn. Patent 6386788/18.04.1988; in *CAS 19:*4 (1988).
507. H. Shimitzu, F. Nemoto, I. Okamura, and K. Ito (Neos Co. Ltd., Toray Industries Inc.,) Jpn. Patent 62270638/5.11.1987; in *CAS Coatings, Inks Related Products 11:*11 (1988).

508. T. Kikuta and H. Hori (Nippon Shokubai Kagaku Kogyo Ko. Ltd.), Jpn. Patent 6333482/13.02.1986; in *CAS Coatings, Inks Related Products* 17:10 (1988).
509. T. Kikuta and H. Hori (Nippon Shokubai Kagaku Kogyo Ko. Ltd.), Jpn. Patent 6351478/21.08.1986; in *CAS Coatings, Inks Related Products* 17:11 (1988).
510. A. Fau (Rhone Poulenc Chimie, Courbevoie, France), EP 0169098 B1/22.01.1986.
511. T. Horichi and M. Tanaka (Uchiyama Kogyo Kaisha Ltd.), Jpn. Patent 6308429/14.01.1988; in *CAS Siloxanes Silicones* 17:7 (1988).
512. H. Yasui and M. Okamoto, Jpn. Patent 6308430/14.04.1988; in *CAS Siloxanes Silicones* 17:7 (1988).
513. *Coating* 12:344 (1984).
514. N. Hayashi (Toshiba Silicone Co. Ltd.), Jpn. Patent 62292867/19.12.1987; in *CAS Coatings Inks Related Products* 11:11 (1988).
515. K. Nakamura, Y. Miki, and Y. Nanzaki (Nitto Electric Ind. Co. Ltd.), Jpn. Patent 63 86787/18.04.1988; in *CAS Adhes.* 19:5 (1988).
516. Norton Co., U.S. Patent 3508919; in *Coating* 5:130 (1974).
517. J. L. Walker and P. B. Foreman (National Starch and Chem. Co.), EP224795/10.06.1987; in *CAS Colloids (Macromol. Aspects)* 1:5 (1988).

5
Adhesive Properties of PSPs

In principle, PSPs are characterized by their adhesive, converting, and application performance characteristics. Table 5.1 summarizes the main performance characteristics of PSPs and their components. It is evident that for different pressure-sensitive products the converting and end use behavior have a specific character. It is astonishing that the general adhesive properties of various product classes must be examined as specific properties also. For different PSPs the level of the instantaneous and dwell time dependent peel force and the permanent or removable character of the bond and/or its deposit-free destruction are of quite different degrees of importance and are evaluated in different ways. For labels and tapes, instantaneous adhesion, tack, and peel are the most important adhesive properties. For the major part of these products, removability is a special characteristic. For protective films it is essential.

With the development of PSP applications, the requirements for different product classes became more complex. Aesthetic and chemical resistance considerations imposed the use of plastics as solid-state carrier materials. Environmental considerations forced changes in film carrier composition. Polyvinyl chloride has been replaced, mainly by polyolefins. Recycling required improvements in paper quality too. Short fiber papers were introduced. Due to the development of plastics, especially thin polyolefin films, for the packaging industry, a need developed for adhesion and anchorage of PSPs on nonpolar, nongeometric, and nonstatic items. The broad range of new, unknown substrates requires the PSA adhesion to be substrate-independent. For different applications, peel regulation

Table 5.1 Major Performance Characteristics of PSPs

Adhesive properties
  Adhesion: Tack, peel resistance
  Cohesion: Shear resistance
Converting properties
  Coating properties: Printability, laminability
  Confectioning properties: Cuttability; die-cuttability
End use properties
  Aging resistance
  Chemical and water resistance
  Bonding/debonding properties
  Processibility

should ensure instantaneous permanent adhesion (e.g., permanent labels and tapes) as well as a controllable permanent adhesion (e.g., repositionable labels) or nonpermanent adhesion (e.g., removable labels, tapes, and protective films) (Fig. 5.1).

1 DEFINITION AND CHARACTERIZATION OF ADHESIVE PROPERTIES

Adhesion onto the substrate surface under either slight or high pressure is a general characteristic of PSPs. It is achieved with products having quite different buildup, under different application conditions for various end use times, and under different deapplication conditions and methods. Generally adhesion is evaluated in terms of adhesive properties. The adhesive properties are characterized by general adhesive performance characteristics and special product-dependent properties.

1.1 General Adhesive Performance

The characterization of the adhesion of PSAs using common, standard adhesive properties is described in detail by Benedek and Heymans [1]. As is known, adhesive properties are defined by tack, peel, and shear resistance of the joint. Tack is the ability to bond instantaneously. Peel is the strength of the joint, tested by debonding, using a stress direction other than that of the laminate. Shear is the internal cohesion of the adhesive tested by shear stresses parallel to the joint direction. Tack, peel, and shear are interdependent. They are controlled by the adhesion–cohesion balance. This balance depends on the material and application or test conditions. Theoretically it is a function of only the material. In practice,

Adhesive Properties of PSPs

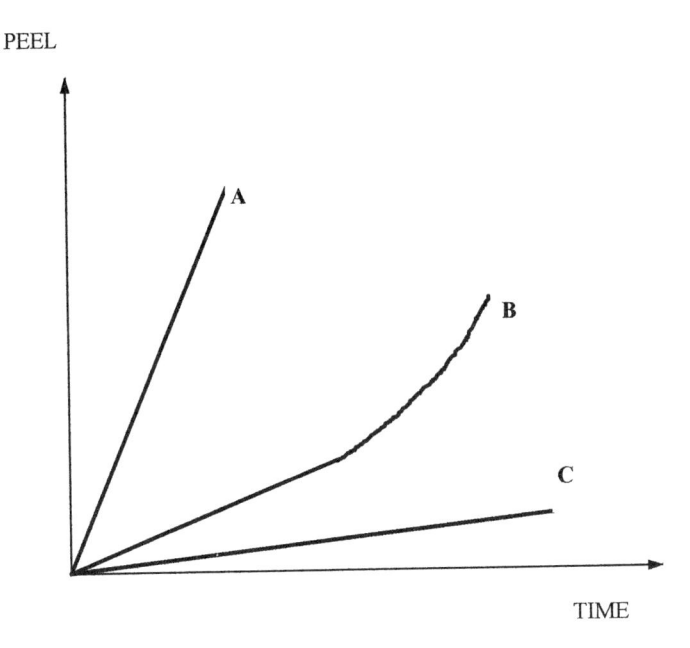

Figure 5.1 Schematic presentation of the adhesion buildup over time for the main types of PSPs. A, Permanent PSP; B, repositionable PSP; C, removable PSP.

because of the non-Newtonian flow of the macromolecular compounds used as chemical bases for PSAs, this balance depends on time and temperature, that is, on actual test and application conditions (see also Chapter 3, Section 1). In the end use of the product the adhesion–cohesion balance is a function of the rate of product application and deapplication. Moreover, because of the influence of the solid-state laminate components (substrate and carrier) on the rheology of the PSA and vice versa, the adhesion–cohesion balance depends on the buildup and manufacturing parameters of the PSP and also on the substrate. According to an old definition [2], "pressure-sensitive adhesives give instantaneous, mechanically removable, nonspecific adhesion, generally without damaging the substrate." According to Chang [3], PSAs are adhesives "which require no activation by water, solvent or heat . . . they have a sufficiently holding and elastic nature." Except for labels, there are numerous pressure-sensitive products that do not give sufficient instantaneous adhesion. Such products do not have balanced adhesive properties and may require activation. Nonspecific adhesion can be given by PSAs applied to one side (e.g., labels), double side (carrier and substrate) coated PSAs, (e.g., envelopes), or double side (carrier and carrier) coated cold sealing adhesives (Fig. 5.2). Pressure-sensitive cold sealing adhesives have been manufac-

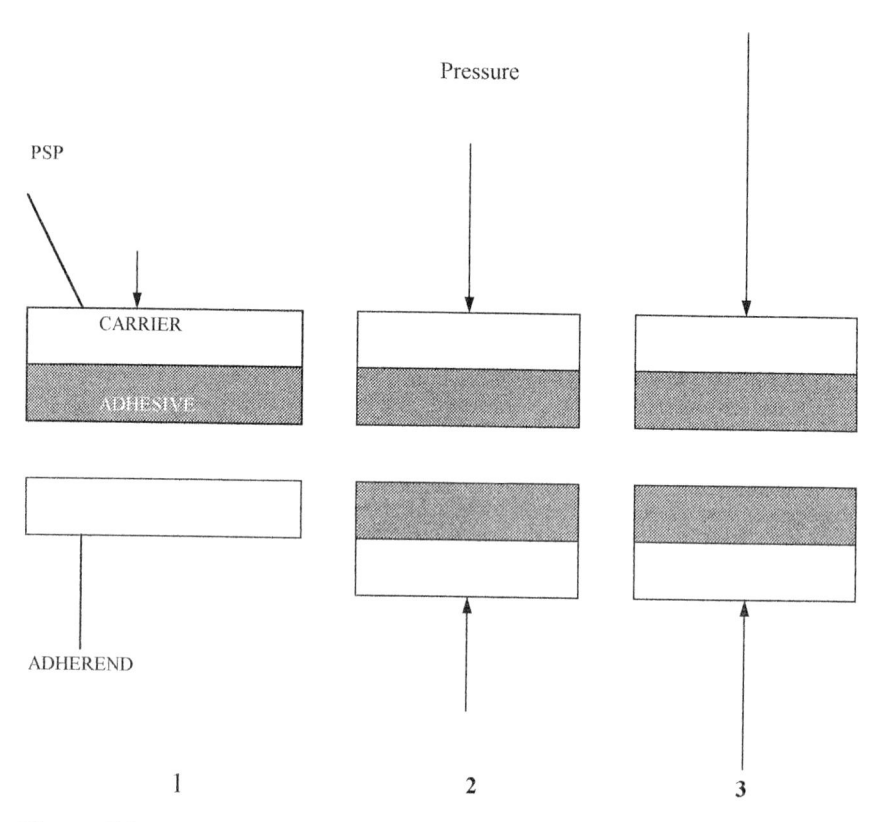

Figure 5.2 Pressure-dependent contact buildup for different products. (1) Bonding of PSA-coated PSP on different adherend; (2) bonding of PSA-coated PSP on itself; (3) bonding of cold sealing adhesive coated carrier on itself.

tured from tackified EVAc copolymers with 55–85% ethylene, tackifiers, and special microsphere fillers [4]. It is evident that autoadhesion as a special case of adhesion presents a simplified model for the study of pressure sensitivity. For such a model, contact building is controlled by diffusion of the same material on both sides of the interface. Physical adhesion can develop from adequate wetting at the interface or by diffusional interpenetration of segments across the interface when this is thermodynamically possible [5]. For noninteracting polymer–polymer pairs, the only driving force for diffusion is entropic in origin and very small. According to Dormann, [6], cold seal is "the use of a soft adhesive sealing medium with or without temperature and pressure." Autoadhesive tack is "the ability of two similar materials to resist separation after their surfaces have been brought into contact for a short time under a light pressure" [7]. Autocontact (self-adhesion) is characteristic of contact adhesives. Common contact adhesives are applied under pressure. Lap shear adhesion of acrylic contact adhesives increases with pressure [8].

A common PSA does not have to bond to itself. It has to be xenophilic. Taking into account the above definition, current (one-side coated) PSA applications must have another bonding mechanism. A more complex adhesion mechanism is given by macromolecular compounds that exhibit diffusion but are chemically different. Studying the fracture energy of polystyrene and PMMA adhesive joints attained by contact molding of polystyrene and PMMA sheets, it has been stated that in such cases the van der Waals interaction was sufficient to create strong adhesive joints [9]. Such cases do not take into account the influence of the surface and bulk characteristics of the solid-state components. According to another definition, pressure-sensitive adhesives are coated "onto rigid (relative to the adhesive, N.A.) backings" [10]. This statement excludes autoadhesive tack and self-adhesive soft carriers.

The definitions given above for pressure-sensitive adhesives illustrate the fact that "classic" pressure-sensitive products (e.g., labels) that have excellent instantaneous tack or peel on other substrates can be considered only as special cases. Certain PSAs have to be applied via autoadhesion and under pressure. On the other hand, pressure-sensitive products are not the same as pressure-sensitive adhesives. Pressure sensitivity of PSPs is the result of a complex structure with or without a PSA. Therefore, its characterization is more difficult. Standard test methods for characterizing adhesive properties have been developed for PSAs. In certain cases such methods do not fulfill the practical requirements for PSPs. Therefore, in this chapter we examine whether it is possible to evaluate the adhesive performance characteristics of PSPs with unbalanced adhesive–cohesive properties and with or without PSA, and if so, how.

As mentioned earlier, tack, peel, and shear resistance are the main parameters of adhesion (Fig. 5.3). Normally, tacky pressure-sensitive adhesives are used in the manufacture of a variety of articles such as adhesive labels, tapes, and other materials that are intended to be easily attached to another substrate by the application of slight pressure. Such PSAs are adhesives that do not require specific activation to render them adhesive.

In practice, depending on the end use requirements, i.e. high application speed (e.g., label) or durable high strength joint (e.g., mounting tape, transfer tape) and removable adhesion on large surface items (e.g., protective films) the balance of the tack, peel, and shear may be different (see Fig. 5.3). Problems may also arise from the evaluation of adhesives having balanced tack–peel–shear adhesive. The definition of "adequate" peel, tack, or shear resistance is relative, and the test methods used are more or less versatile for a given PSP class.

According to Pasquali [11], common PSAs exhibit a peel level of 50–800 N/m (g/cm). As is known, reducing the peel value makes PSP application difficult.

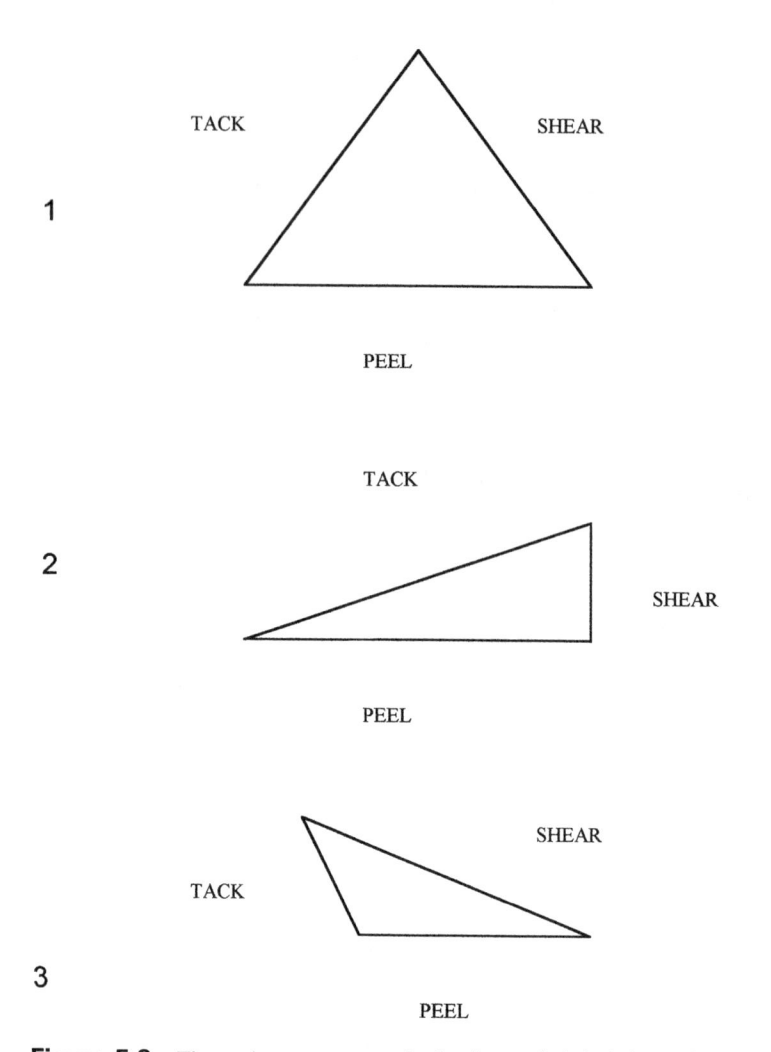

Figure 5.3 The main parameters of adhesion and their balance for different PSPs. (1) Label; (2) tape; (3) protective film.

Adhesives with peel values lower than 150 N/m give almost no adhesion (except for some special adherends such as marble, glass, or metal). On the other hand, it is also difficult to delaminate large adhesive-coated surfaces. Products having a common, medium peel value of 350 N/m would require a 350 N force for delamination of a sample 1 m wide. According to Ref. 12, highly aggressive PSAs are those having peel adhesion values in the range of 60–100 N/100 mm (15–25

N/25 mm). Tackified acrylic block copolymers for hot melts show excellent peel values of 17–30 N/25 mm [13]. The peel of thermoplastic elastomer based removable formulations should be 2.5–40 oz, and the initial peel 1.0–16 oz (using 1 in. × 6 in. strips on stainless steel). Values above these limits can result in paper tear [14]. Tackified formulations based on amorphous polypropylene exhibit peel resistance of 31 N/in. (PSTC-1, steel), loop tack of 25 N/in.^{2}, and shear resistance of 2.5 h (PSTC-7) [15]. Adequate selection of the comonomers and polymerization conditions led to the development of olefin polymers that are pressure-sensitive in the neat form. Such polymers possess a peel value of 2.5 N/in. (PTSC-1), quick stick of 1.2 g/in. (PTSC-5), and shear resistance of 20 h [16]. Preferred skin adhesives exhibit initial peel values of 50–100 g and final peel values (after 48 h) of 150–300 g/25 mm [17]. A cold-stretchable self-adhesive film based on a tackified ethylene–α-olefin copolymer displays an adhesive force of at least 65 g (ASTM 3354-74) [18].

Tack values of 16–24 N/25 mm are acceptable for use on diapers [19]. For CSBR, holding time values of 116 h at 75°F and 120°F (PSTC-7) are, according to Ref. 20, excellent. Mueller and Türk [21] proposed a water-based acrylic formulation having physically or chemically crosslinkable comonomers like acrylo- or methacrylonitrile, acrylic or methacrylic acid, N-methylolacrylamide, and plasticizer (10–30% w/w). Such formulations give a peel resistance of 1–6 N/20 mm on different substrates (at debonding speed of 75 mm/min). An acrylate-based PSA crosslinked between 60 and 100°C using a zinc chloride catalyst shows holding power of 20 h at 150°C compared to common uncrosslinked, rubber-based PSAs with holding power values of some minutes at 100°C [22]. According to Midgley [23], an arbitrary shear failure time of 20 h (1.25 cm overlap, 2.5 cm width, 1 kg at room temperature) was adopted as an adequate value in practice. Shear values of 12–680 min (on steel, according to PSTC) are acceptable only for labels. Packaging tapes need higher shear [24]. A special crosslinkable, carboxylated water-based acrylic displays room temperature shear values of 5000–60,000 min at 8.8 psi (0.5 × 0.5 in., 1000 g). A SAFT value of 240°C has been obtained, which compares with the shear performance of the highest quality commercial solvent-based acrylics [25]. Unfortunately, these values were obtained without tackifier, and the corresponding peel and tack values are very low (180° peel, PSTC-1, on PET, 1.1 lb/in.; quick stick 0.7 lb/in.). Formulations with acrylic PSAs for permanent label application show (minimum) shear values of 8 min (tackified) to 380 min (nontackified); for tackified SBR, values of 60–100 min are obtained [26]. As illustrated by the above test values, although standard methods exist, such values must always be evaluated with respect to the specific formulation.

Different tack methods are adequate for different PSPs [27]. With the probe tack test the load is regulated, as in the lamination of protective films. Polyken tack is less strongly affected by the resin softening point (it is applied under load!) during tackifying. Loop tack simulates the actual application conditions of labels, which are blown onto a surface by using air pressure. Rolling ball tack is more complex. It is friction-related, but the correlation between polymer friction and its viscoelastic properties during bonding and debonding are not clear.

A well-founded study of the dependence of adhesive performance on the experimental conditions was carried out by Bonneau and Baumassy [28], who used various water-based tackifier dispersions (Dermulsene, DRT). As stated previously, the optimum resin concentration for the best peel value depends on the peel angle. For 90° peel on polyethylene, the highest value is obtained at the highest resin loading ($\approx$35%); for 180° peel the best values are obtained with less resin (30 phr). That means that the common peel test methods used for tapes (90° peel) and labels (180° peel) give quite different evaluations of the same formulation.

Tack and peel are characteristics used for evaluation of instantaneous or short-term behavior. They can be estimated or exactly measured and if necessary rapidly improved. Shear measurement can be used to test the internal cohesion of the adhesive. Quick stick and high temperature shear are required for PSAs for tapes [20]. For labels, creep is not so important, because high load-bearing capability is not required [23]. Except for some special applications for PSPs where a high level of shear resistance is necessary, in most cases the shear resistance is taken into account as a cohesion-dependent component of the adhesion–cohesion balance only, which can be evaluated more easily than other cohesion-dependent converting or application properties such as cuttability and die cuttability.

Cuttability and die cuttability depend on the adhesive also. Cohesive, tack-free elastomers or duromers can be more easily cut than tacky, low viscosity materials. Therefore one can suppose that the cohesion of the adhesive and its shear resistance allow characterization of cuttability. Practically, shear tests and cutting or die-cutting operations are carried out at quite different stress rates and temperatures. Supplemental problems arise from the lack of correlation between the shear values measured at room temperature and cuttability. Room temperature tests are too slow, and the dispersion of the test results is too high. Hot shear measurements give better results, but there is no linear correlation between the measured values and cuttability. The situation is more complex for water-based dispersions where moisture content of the laminate (degree of drying) influences the shear resistance also, or for thermally crosslinking compositions, where hot shear results are denaturated by the crosslinking of the product. Generally, high surfactant concentrations in the formulation lower the peel, and in many cases they lower the shear also [23]. Excess moisture in the carrier paper can be troublesome in tests of the release force, and it is preferable for the papers to have a moisture content of about 4% or less by weight [27].

As discussed in Chapter 3, the T_g and the modulus are base characteristics that make it possible to predict adhesive performance. Equations can be derived that express the effect of molecular weight, plasticization, degree of crosslinking, and copolymerization on the second-order transition temperature. Such equations may be reduced in form to equations derivable from free volume theory [29]. Although the possibility of computer-based tape and label design was demonstrated by Kaelble [30] a decade ago, the evaluation of adhesive performance needs a thorough experimental study. DMA measurements are useful for the base elastomer but can be denaturated by the composite structure of the adhesive and adhesive–carrier interactions. Such investigations were proposed in the 1960s by Gramberg [31]. Later the value of the storage modulus at the tan δ peak as a function of the tan δ peak value was successfully used by Bamborough and Dunckley [32] to predict the application window for bookbinding hot melts. Bamborough [33] proved the applicability of the application window values stated by Chu [34] for solvent-based, hot melt, and water-based PSAs. It has been shown that for natural rubber and styrene-butadiene copolymers the addition of a compatible resin allows us to bring the system's values into the application window. For SBCs at room temperature the loss tangent peak temperature is lower than that of natural rubber or SBR and the storage modulus is greater by two decades than that of natural rubber (see also Chapter 3). Therefore the concentration required to bring the system into the application window is much higher but is possible if an oil is added as well. Unfortunately, for WBPSAs, such simple prediction of adhesive performance is is impossible. As stated by Bamborough [33], the loss tangent peak temperature (i.e., the interdependence of storage modulus and loss tangent peak temperature) for water-based adhesive systems is not a reliable predictor of PSA performance.

1.2 Special Adhesive Performance

Table 5.2 lists the main adhesive characteristics of PSPs and (standard or special) test methods for evaluating them. Examination of these methods on the basis of industrial experience shows that the actual definition of and test methods for the adhesive properties (developed originally for labels and tapes produced according to the classic PSA coating technology) are not adequate for the whole PSP domain. The standard adhesive characteristics describe the adhesive performance of PSPs having a balance between tack or peel and cohesion. Labels, tapes, and some protective films belong to this group of products. On the other hand there are some special application fields where bonding and debonding are influenced more by the application/deapplication conditions and standard test methods cannot characterize the real behavior of the PSP. In such cases it would be better to describe the end use tack and the end use peel, which are complex functions of the application conditions.

Table 5.2 Main PSP Adhesive Properties and Test Methods for Their Evaluation

Characteristic	Principle of measurement	Method
Adhesion		
Tack	Rolling adherend	Rolling substrate (rolling ball, rolling cylinder) or rotating substrate (toothed wheel)
	Peel measurement	Flexible carrier (loop tack) or rigid carrier (Polyken tack)
Bond strength	Peel resistance	Standard carrier and substrate quality (adhesion peel, self-adhesion peel); standard peel angle (90° and 180°) and rate
Cohesion	Shear resistance	Static, standard debonding angle, standard debonding force, and standard contact surface; room temperature and high temperature; dynamic

Application Peel and Tack

As discussed above, an ideal PSA, by definition, gives instantaneous tack and peel on a rigid substrate. Certain pressure-sensitive products must adhere to soft substrates and provide enough adhesion after a given time under special application conditions. Under such conditions, "dwell time" means time of forced contact and strongly depends on the coating weight. The importance of peel buildup for the strength of the adhesive joint is well known. Generally such an increase in the adhesion occurs under well-defined static (storage) conditions having time as the sole parameter. For certain products (at least in the first period of this time), supplemental parameters can also enhance contact and peel buildup.

Influence of Coating Weight. The role of coating weight for adhesive-coated PSPs is discussed in detail in Refs. 1 and 35. It has been stated that adhesion generally increases with coating weight and a critical coating weight is necessary to achieve adequate adhesive properties. On the other hand, an excessive coating weight can negatively influence removability, cuttability, and shear resistance. As discussed in Ref. 1, the adhesive properties of a PSA depend on the coating weight. Peel resistance increases with coating weight (Fig. 5.4) up to a certain value. The form of this dependence is a function of the adhesive, substrate, and application conditions.

As presented in Fig. 5.5 in a simplified form, a plot of peel vs. coating weight can be considered as a diagram having a point of inflection. The dependence of peel resistance on coating weight before the point of inflection can be considered as a linear one (Fig. 5.5).

Adhesive Properties of PSPs 265

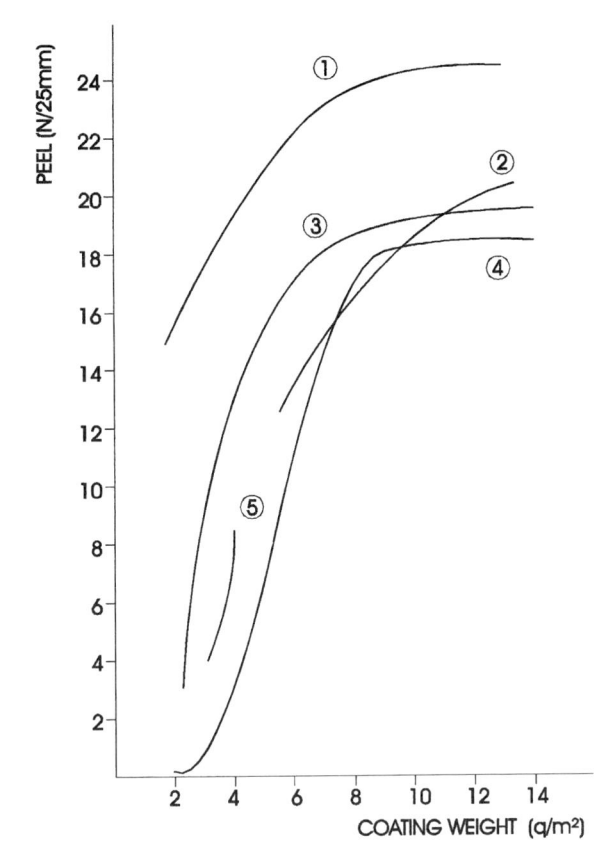

Figure 5.4 Dependence of peel resistance on coating weight. *1,* Water-based, tackified acrylic PSA, 180° peel on glass; *2,* water-based tackified acrylic PSA, 180° peel on glass; *3,* water-based, tackified acrylic PSA, 180° peel on polyethylene; *4,* water-based, tackified acrylic PSA, 180° peel on polyethylene; *5,* tackified CSBR dispersion, 180° peel on glass.

As a first approximation, in the linear domain of coating weight values (*OL*), the peel resistance (*P*) increases with coating weight (C_w). This increase can be characterized by the angle α of the plot and/or the critical coating weight (C_{wcr}), according to the correlations

$$P = f(C_w) \tag{5.1}$$

and

$$P = \alpha C_w \tag{5.2}$$

PEEL

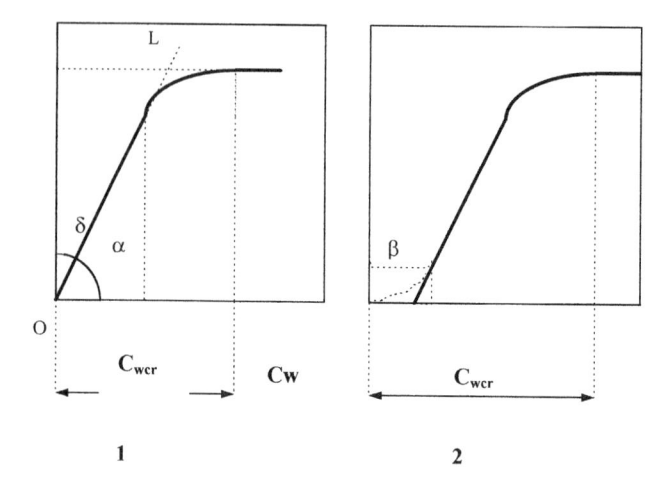

Figure 5.5 Schematic presentation of peel dependence on coating weight. (1) Dependence function composed of coating weight dependent and coating weight independent parts at high coating weight values. (2) Dependence function composed of a coating weight "independent" part at low coating weight values and a coating weight dependent part and a coating weight independent part at high coating weight values.

Generally α depends on the adhesive and substrate and application conditions:

$$\alpha = f(\text{adhesive, substrate surface, application conditions}) \quad (5.3)$$

The application conditions are characterized by time and temperature. Therefore,

$$\alpha = f(\text{adhesive, carrier, substrate, time/temperature}) \quad (5.4)$$

For an ideal adhesive, α should be zero, i.e., the adhesion should not depend on the coating weight. On the other hand, an examination of Figs. 5.4 and 5.5 shows that the peel resistance attains a measurable value only above over a certain coating weight. Contact buildup needs adhesive flow. For given application conditions, surface roughness, carrier conformability, and adhesive cold flow are the main parameters of contact buildup. Therefore the dependence of the peel force on the coating weight can be described by a simplified plot, with the linear part of the dependence shifted from the origin to higher coating weight values (Fig. 5.5) according to

$$P = \alpha C_w - \beta \quad (5.5)$$

where β is a conformation factor. The value of this factor is a function of the parameters influencing the contact buildup between the adhesive and substrate sur-

Adhesive Properties of PSPs

face and the so-called free flow region of the PSA. That means that for too low a coating weight or too rough a surface (see Application of Protective Films) no adhesion builds up. It is evident that rough, nonpolar surfaces and harder adhesives on rigid nonconformable carrier materials lead to higher values of β; in such cases the critical coating weight is higher (see Fig. 5.4 also). It is well known from industrial practice that the coating weight depends on the mechanical characteristics of the carrier (Table 5.3).

As can be seen from Table 5.3, greater carrier thickness requires the use of higher coating weights to achieve better contact buildup. Therefore it can be supposed that

$$\beta = f(\text{adhesive, carrier, substrate, time/temperature}) \qquad (5.6)$$

Figure 5.4 illustrates that a permanent acrylic PSA is coated for labels with a coating weight situated above the critical value. Table 5.4 presents the coating weight domains for the main PSP classes. As can be seen from this table, tapes may have very high coating weights. On the other hand, separation and protective films may have coating weights that are lower than common coating weights

Table 5.3 Dependence of Coating Weight Values[a] (g/m^2) on Carrier Thickness

Code	Carrier thickness (μm)							
	50		70		80		100	
	TH	PR	TH	PR	TH	PR	TH	PR
1	2.80	3.60	2.40	2.70	—	—	—	—
2	2.80	2.86	2.80	2.83	—	—	—	—
3	4.50	3.97	—	—	4.50	4.13	—	—
4	4.50	4.80	—	—	—	—	4.50	3.86
5	4.50	3.95	—	—	—	—	4.50	3.84
6	4.50	4.10	—	—	—	—	4.50	4.02

[a] TH = theoretical values; PR = measured values.

Table 5.4 Coating Weights of Selected PSPs

Pressure-sensitive product	Coating weight range (g/m^2)
Label	5–30
Tape	1.5–500
Protective film	0.5–20
Separation film	0.5–2.5

for lacquers or printing inks. It is evident that the coating weight values strongly depend on the type of adhesive. As stated in Ref. 36, solvent-based acrylics are coated for different product classes with the following coating weight values: 5–10 g/m^2 for protective films; 25–40 g/m^2 for film-based tapes; 50–100 g/m^2 for foam tapes; and 20–30 g/m^2 for labels and tags. For labels, peel is regulated by formulation, not by coating weight. Labels can be considered a special case of PSPs where coating weights are relatively high (above the critical domain, see Fig. 5.6).

Higher coating weights are used for tapes. These weights are needed for high tack and instantaneous peel for adhesives having high cohesion. The high coating weight of tapes can be apparently reduced for applications where conformability and cohesion are required. Such an apparent reduction is made by using special fillers (see later). For an adequate peel value, a high coating weight (above the critical value) is needed. The critical coating weight value depends on adhesive and surface qualities and application conditions. For rough surfaces, high coating weights are generally needed. Although these values are necessary for rapid peel buildup, later they may disturb the application of the adherend–PSP assembly. Therefore a postreduction of the initial coating weight may be necessary. Such an apparent reduction of the coating weight, i.e., of the mobility of the adhesive layer, can be achieved by modifying the rheology of the adhesive. As is well known, one way to achieve this modification is by crosslinking (see also Chapter 3, Section 1). Another possibility is to use fillers.

Generally, crosslinking of the adhesive is carried out before and during its coating, but it can also be achieved after the application of the adhesive coat. Peel

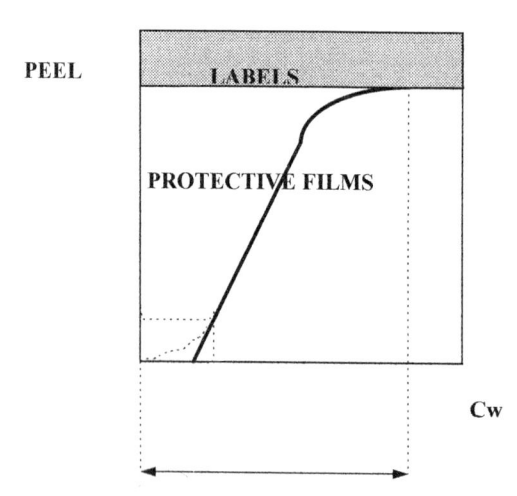

Figure 5.6 Typical coating weight domains for labels and protective films.

can be postregulated by crosslinking after use of the laminate [37]. In this case the storable, crosslinkable adhesive is formulated for postapplication crosslinking, which is proposed to improve shear resistance and/or to allow easier delamination of the applied product. Postcrosslinking is achieved by using free radical initiated [38] or photoinitiated reactions [39]. Postcrosslinking leads to adhesionless surfaces, i.e., to easier delamination. This is a development of the adhesive properties as in the case of the so-called semi-pressure-sensitive adhesives, where crystallization causes the time-dependent loss of adhesivity (see later). According to Charbonneau and Groff [37,40], the procedure allows delicate electronic components to be removed from the tape even though they were nonremovable before crosslinking. A latent crosslinking agent has to be used. The tape has a high coating weight and is conformable to rough, uneven surfaces (coating thickness of 0.5–1.5 mm). To facilitate immediate adhesion to rough and uneven surfaces a resilient foam backing can be used (see also End Use of Tapes).

Such foam carriers coated on either one or both sides can have an overall thickness of 0.1–2.0 mm [41]. Glass microbubbles can be incorporated also to enhance immediate adhesion to rough and uneven surfaces [42]. As is known from formulating practice, microbubbles have been suggested as fillers for the adhesive layer (see also Chapter 4). They can be applied to reduce the contact surface, i.e., for removability (see later), or to improve the cohesion of the adhesive (see Chapter 3, Section 1). Such improvements are necessary for tapes for rough surfaces where adhesive conformability needs a very high coating weight (one that shows pronounced cold flow) or/and a soft foamlike carrier. A foamlike pressure-sensitive tape is manufactured according to Vesley et al. [43] by using glass microbubbles as filler. The microbubbles are embedded in an adhesive matrix. Pressure-sensitive tapes filled with glass microbubbles have a foamlike appearance and character and are useful for purposes previously requiring a foam-backed PSA tape.

Tapes with a thick adhesive layer (0.2–1.0 mm) have been prepared by photopolymerization (UV) of acrylics. The PSA matrix may contain glass microbubbles [42]. Microbubbles are included in the composition of a foamlike tape according to Ref. 41. They are embedded in a polymeric adhesive matrix containing at least 5% (5–65% v/v) filler. The thickness of the pressure-sensitive layer should exceed three times the average diameter of the microbubbles to enhance the migration (flow) of the bubbles in the matrix under applied pressure instead of breaking. Such a formulation enables flow and the buildup of intimate contact with rough and uneven surfaces while retaining the foamlike character. Optimum performance is attained if the thickness of the PSA layer exceeds seven times the average diameter of the bubbles. This construction ensures an apparent reduction of the coating weight and an apparent improvement of the cohesion of the adhesive layer. To increase cohesion of the adhesive layer, polar monoethylenically unsaturated monomers (less than 20%) are included in the acrylic polymerization recipe and photoactive or heat-activatable crosslinking agents are added [44,45]. It is evident that the reduced flow properties influence the peel values negatively. This can be seen from the peel measurement method (T-peel is tested) and the peel values. The cohesive (tensile) strength of the uncrosslinked adhesive is 2550 kPa (elongation 840%); that of the crosslinked formulation may attain 6000 kPa (depending on the crosslinking conditions).

In some cases the conformability of the PSP is achieved by the manufacture of a carrierless construction [46]. The adhesive can be crosslinked with a multi-functional, unsaturated monomer. A prepolymer (precursor) is synthesized that is polymerized to the final product by using UV radiation [47]. According to Larimore and Sinclair [46], a carrierless PSP can be obtained (the adhesive layer has adequate strength to permit it to be used without a carrier material) that is more conformable. The product exhibits an adhesive surface and an adhesion-free surface. The tack-free surface is achieved by superficial crosslinking of the adhesive by polyvalent cations (Lewis acid, polyvalent organometallic complex, or salt). The crosslinked portion of the adhesive has greater tensile strength and less extensibility. The crosslinking penetration depends on the type of polymer and the solvent, i.e., a crosslinking gradient can be achieved as for the UV photopolymerized products (see later). Thick PSA tapes (0.2–1.0 mm) have been prepared by UV photopolymerization of acrylics [42]. The formulation can be polymerized as a thick layer (up to about 60 mils), which may be composed of a plurality of layers, each separately photopolymerized. The length of polymerization zones and the density of lamps in these zones affect the manufacture. When the thick layer is sandwiched between two thinner layers, the thick layer is often referred to as a carrier, but it has pressure-sensitive properties also. The support or carrier layer is 25–45 mils thick, and it conforms well to substrates that are not flat.

As stated in Ref. 48, even a high molecular weight, highly viscous 100% polyacrylate processible as hot melt is not of high enough molecular weight to display balanced pressure-sensitive properties. For such a product the PSA definition given by Chang [3] is not valid. Therefore, the PSP should be crosslinked. A special acrylic raw material for acrylic HMPSAs has been developed that can be cured photochemically. The photoinitiator is built into the polymer. The uncrosslinked product does not have balanced adhesive properties. Because of its low molecular weight it is very tacky but does not have sufficient shear. The pressure-sensitive properties are achieved only by crosslinking. Another aspect of its crosslinking is its anisotropy. The radiation is partially absorbed, partially transmitted, and partially reflected. The maximum radiation level is given at the top of the adhesive layer. The crosslinking degree is the maximum on this side. Therefore direct and transfer coating give different adhesive characteristics. A uniform, isotropic adhesive layer can be achieved by using a UV-transparent face stock (backing) and irradiation on both sides. A transfer coated PSA (with 15% tackifier resin) gives 25 N/25 mm peel resistance (on steel) after irradiation and

27.5 N/25 mm quick stick compared to 18.7 N/25 mm peel on polyethylene, and 44 min shear resistance (1.2 × 1 in.^{2}, 500 g). The adhesive layer irradiated on the face stock material gives 20 N/25 mm peel on steel, 16.8 quick stick, and 6.2 N/25 mm peel on polyethylene and (surprisingly) a quite different shear value also (77 min). Unfortunately, the shear values are not comparable for the given experiments. These examples illustrate that balancing of the adhesive properties is achieved in many cases by product buildup.

Protective films use low coating weights (Table 5.5). Such products are always situated in the critical domain (see Fig. 5.6). Peel is regulated for protective films by controlling coating weight. For the low coating weights generally used for removable products (e.g., special labels or protective films) situated in the critical coating weight domain where peel strongly depends on the coating weight, tolerances are more important. As can be seen from Table 5.6, the average coating weight tolerance values for protective films range from −16% to +14%. That means that such products also have a broader peel value distribution.

The main adhesive formulations used for protective films are crosslinked compositions. Therefore their cold flow is limited also. Thus it can be supposed that, in contrast to labels and tapes, the contact building of protective films has to be achieved at very low tack and peel values. Admitting that instantaneous peel and tack are interdependent, it is evident that low tack protective films need quite different application conditions to improve their initial peel value. Later, the buildup of adhesion increases the peel resistance up to the value required in practice.

Table 5.5 Coating Weights for Common Adhesive-Coated Protective Films[a]

Adhesive characteristics		Coating weight (g/m^2)	
Code	Type	Theoretical value	Practical avg. value
1	Acrylic	1.70	1.80
2	Acrylic	2.40	2.57
3	Acrylic	2.80	3.00
4	Acrylic	4.00	4.60
5	Acrylic	5.70	6.30
6	Acrylic	7.00	7.10
7	Acrylic	12.00	10.10
8	Rubber resin	1.90	2.26
9	Rubber resin	2.90	3.34

[a]Crosslinked, solvent-based acrylic and masticated, crosslinked, solvent-based rubber resin PSAs were used.

Table 5.6 Coating Weight Tolerances for Protective Films

Coating weight (g/m²)		Tolerances (g/m²)		
Theoretical	Average	Extreme values Minus	Plus	Average values
1.6	1.86	0.2	0.8	0.26
1.7	1.77	—	0.2	0.07
1.7	1.85	—	0.2	0.15
2.2	1.90	0.4	—	0.30

As seen from relations (5.4) and (5.6), both α and β depend on the adhesive, carrier, substrate, and time and temperature. For a given application of the same PSP (same adhesive, coating weight, and carrier) on a given substrate, α and β can be considered as parameters that depend only on time and temperature. Therefore in this case the peel value depends on the application conditions only, i.e., laminating pressure, time, and temperature:

$$P = f(p_1, t, \theta) \tag{5.7}$$

where p_1 is the laminating pressure, t is the laminating time, and θ is the laminating temperature. The comparison of correlations (5.1) and (5.7) shows that for the application of classic PSPs (tapes and labels) the intrinsic adhesive properties of the product (regulated by adhesive nature and coating weight) are more important for peel buildup. For protective films the application conditions are determinant. In some cases the SAF has to display good adhesion properties at very high temperatures and good blocking properties at elevated temperatures. For instance, a heat-sensitive EVAc film gives good adhesion when heated for 20 s at 120°C and good blocking resistance during 30 min at 50°C and 1 kg/cm² load [49].

Peel Buildup and Peel Gradient. For classic, PSA-coated pressure-sensitive products the buildup of adhesion is dependent on dwell time. For instance, an aluminum carrier and a tackified CSBR are used as insulating tape in climatechnics [50]. The peel value of this tape builds up to 37 N/25 mm after 24 h. As discussed in Chapter 4, acrylic HMPSAs with thermally reversible crosslinking (segregated structure) have been developed [51]. For such formulations the buildup of peel values can be observed also. The initial peel value of 38 oz/in. increases after storage for 1 week at 50°C to 68 oz/in. For extruded self-adhesive products also, time-dependent adhesion buildup can be observed. The adhesive properties of such self-adhesive films with a built-in adhesive component (e.g., polybutylene-tackified polyolefins) depend on the diffusion of the viscous com-

Adhesive Properties of PSPs

ponent (through the carrier) to the product surface [52]. As is known from plastic film processing, blocking of PVAc films depends on pressure, temperature, and time [53]. Table 5.7 illustrates the peel buildup over time for protective films. As can be seen from this table, peel buildup is a general phenomenon for products of different coating weights.

As shown by the data of Table 5.8, such peel buildup also occurs for highly crosslinked products with very low initial peel (e.g., separation films). Like instantaneous peel, peel buildup is also a function of substrate quality (Table 5.9) and is greater on glossy surfaces. Peel buildup also characterizes self-adhesive films (Table 5.10). Such films are warm laminated; therefore, peel buildup is also tested at elevated temperatures.

As illustrated by the data of Table 5.11, peel buildup is a general phenomenon for pressure-sensitive products with different adhesives, carriers, geometries, and buildup. Thus the control of removability is really the regulation of peel buildup.

Acrylics display the disadvantage of compliance failure, i.e., adhesion buildup [60]. As discussed above, in some cases adhesion buildup can be balanced by a decrease in adhesion due to a reduction in the chain mobility caused by crosslinking. For instance, adhesion buildup has been reduced by crosslinking of acrylic with dimethylaminoethylmethacrylate [61] or with polyisocyanate [62]. Gobran

Table 5.7 Peel Buildup for Protective Films

Code	Peel resistance (N/25 mm)	
	Instantaneous	After 24 h
1	2.50	2.61
2	3.50	3.57
3	4.30	4.33
4	7.50	7.69
5	14.00	14.31

Table 5.8 Peel Buildup for Separation Films

Product code	Laminate peel (N/10 mm) Storage time		
	24 h	1 wk	2 wk
1	0.08	0.21	0.47
2	0.03	0.10	0.21

Table 5.9 Dependence of Peel Buildup on Adherend Surface Quality

Product code	Protected surface	Nominal	\multicolumn{5}{c}{Peel (N/25 mm) Application time}
			24 h	1 wk	2 wk	3 wk	4 wk
1	Polished aluminum	0.50	0.40–0.65	—	—	—	0.45–0.65
2	Polished aluminum	0.40	0.20–0.40	—	—	—	0.30–0.45
3	Aluminum	1.20	0.50–1.50	0.70–1.80	0.70–1.80	—	0.75–1.80
4	Stainless steel	4.00	3.20–4.00	4.00–4.20	4.10–4.40	4.30–4.50	4.30–4.50

Table 5.10 Peel Buildup for SAF[a]

	\multicolumn{3}{c}{Peel (N/10 mm)}
Code	Instantaneous	2 days at RT	4 h at 80°C
1	0.15	0.10	0.22
2	0.70	0.75	0.40
3	0.80	0.85	0.30

[a]EVAc-based protective film on PC substrate was tested.

[63] used superimposed adhesive layers having different gradients of shear creep compliance to avoid adhesive buildup.

It should be mentioned that coating weight regulation is a problem for adhesiveless PSP constructions also. For such coextruded products the thickness of the self-adhesive layer is the regulating parameter having the role of the coating weight.

Static and Dynamic Dwell Time. According to Ref. 64, for protective films a difference exists between adhesion on the surface and bonding on it. Similar behavior has been observed for certain tapes where the dwell time from the moment of application of the tape to the buildup of the maximum peel adhesion has been called "time to conformation" [65]. Some time ago, for better wetting out and to increase the contact surface between adhesive and adherend, pressure and solvents were used for application of tapes. In such cases the "pure" (dwell) time dependence of the bond formation changed with the simultaneous change in stress rate, temperature, and chemical affinity. For special print transfer elements (like Letraset), the low level of application tack is essential for storage and applicability. Such products need very high application pressure [66]. Adhesive laminating under pressure is a common operation in other domains also. For instance, linear pressure in laminating by gravure printing attains 200 N/cm (pres-

Table 5.11 Peel Buildup for Selected PSPs

	\multicolumn{8}{c}{Peel resistance}		
	\multicolumn{8}{c}{Dwell time}		
Product	0	1 h	3 h	1 day	1 wk	2 wk	4 wk	8 wk	Unit	Ref.
Tape, repositionable	26.7	—	36.2	—	—	—	—	—	oz/in.	46
Tape, removable	5.65	7.15	—	6.15	6.50	—	5.25	—	N/25 mm	47
Label, water-based acrylic	13.00	—	—	—	—	—	—	12.50	N/20 mm	48
Label, removable	2.00	—	—	—	—	3.00	—	—	N/25 mm	49
Protective film	—	1.4	—	1.0	2.3	—	—	—	N/25 mm	50
Tape, paper	0.057	—	—	—	0.065	—	—	—	kg/cm	51

sure ϕ 160 mm); wet and dry laminating use pressures of 300 N/cm. Solventless laminating applies a pressure of 270 N/cm [67]. It is known from PSA characterization that for the FTM-4 high speed release test the samples are placed between two flat metal plates under a pressure of 6.87 kPa to ensure good contact. In such cases the static (forceless) and dynamic dwell times differ. After studying the effect of contact pressure on adhesive properties, Johnston [68] stated that PE film, unlike other films, does not attain a plateau (as a function of contact pressure) but continues to increase in tack value due to its high extensibility.

The built-in tack of a PSP and the application conditions (application pressure and temperature) allow the instantaneous adhesion of the product on a substrate. For high tack products (e.g., labels, tapes, and some protective films), a light application pressure if any and room or low application temperatures are used. For PSPs with very low tack and instantaneous peel (e.g., protective films) the bonding depends on the applied pressure and temperature also. As known, foamlike transfer tapes are also used under pressure to increase initial bond strength [69]. According to Kuik [70], the peel resistance of heat seal polar ethylene copolymers depends on the corona treatment, coating weight, and aging. The peel strength increases from 50 up to 150 g/25 mm if the surface tension varies between 0 and 32 dyn.cm. The peel value (on PET) increases with time. Here contact buildup is influenced by other than rheological parameters (e.g., functionalization degree of the surface) also. All these products can be considered as working below the critical domain of coating weight where the conformation factor is determinant for laminating. The importance of laminating conditions is illustrated by the data of Table 5.12 for a protective film used for PMMA plates. As can be seen, the importance of laminating conditions for the bonding characteristics is greater than that of the sample geometry.

Tack is one of the standard adhesive properties. The use of tack as a criterion for evaluating the instantaneous adhesion requires a measurable tack value. Although for protective films in some cases a measurable (low) tack value is given (Table 5.13), it cannot be used for characterizing the application performance of the product. This tack (T) does not work like normal tack. It does not ensure sufficient instantaneous adhesion on the product surface. Therefore the product is laminated under pressure. In such cases it would be better to use an application tack (T_a) with a known dependence on the lamination pressure p (and/or temperature θ):

Table 5.12 Peel Resistance Values as a Function of Test Conditions[a]

Sample width (mm)	Test method 1 Carrier 1	Test method 1 Carrier 2	Test method 2 Carrier 1	Test method 2 Carrier 2
25	0.90	1.34	0.75	1.35
25	1.05	1.43	0.71	1.28
100	0.78	1.35	0.65	1.16
100	0.99	1.45	0.71	1.26
100	1.13	1.39	0.58	0.87

[a]Water-based acrylic-coated clear (carrier 1) and pigmented (carrier 2) PE film was warm laminated on PMMA plate. Method 2 uses double-sided heating (plate and film).

Table 5.13 Rolling Ball Tack Values for Protective Films

Code	Coating weight (g/m^2)	Carrier thickness (μm)	Peel (N/10 mm)	Ball diameter (mm)	Mean value (mm)	Tolerance (mm)
1	12.6	154–156	0.46	4.5	212	14
				11.1	>400	—
				14.3	>400	—
2	7.6	100–108	0.29	4.5	267	33
				11.1	>500	—
				14.3	>500	—
3	3.4	50–55	0.20	4.5	290	45
				11.1	>500	—
				14.3	>500	—

$$T_a = f(T,p,\theta) \tag{5.8}$$

Johnston [68] measured tack as tear energy (E_t) according to the correlation

$$E_t = AC^{mx} \tag{5.9}$$

where A and m are constants including the effect of load and dwell time and x is the separation rate of the probe. If tack is considered (according to Bates [71]) as the ratio of the separation energies at optimum (E_{so}) and infinite ($E_{s\infty}$) dwell time,

$$T = E_{so}/E_{s\infty} \tag{5.10}$$

and the dependence of such energies on the application conditions is taken into account also, we actually have an application tack that is generally valid for different PSPs. It should be mentioned that the instantaneous tack of certain self-adhesive PSPs depends on the degree of surface treatment and age also. For such products the definition of separation energy at optimum time is more complex (see also Chapter 6).

Stress Rate Dependent Peel. As discussed earlier, crosslinking is used to modify the adhesion–cohesion balance. The use of crosslinked formulations may change the debonding mechanism also. According to Charbonneau and Groff [37], for UV-crosslinked formulations the characterization of adhesive performance needs a special peel measurement method. A so-called cleavage peel value is determined (Fisher Body Test Method TM 45-88). A breakaway cleavage peel value and then a continuing cleavage peel value are measured. The continuing cleavage peel value is about 50–60% of the initial value. The existence of two different peeling behaviors, i.e., a first high level starting peel value that supposes a high rate of peeling force and a second, normally steady-state peeling, denotes that the adhesive behaves like a multiphase mixture with a high strength (but low contact) component and a relatively soft adhesive matrix with common adhesive flow.

The dependence of the peel value on the peel rate is discussed by Benedek and Heymans [1]. The increase in peel resistance with stress rate generally can be considered a consequence of the time–temperature superposition principle (see also Chapter 3, Section 1). As is known from practice, for medical tapes the rate of removal is very low (15 cm/min) [72]. Such low speed delamination provides better removability, whereas high speed delamination may change the peel resistance. For instance, crosslinked silicone pressure-sensitive adhesives can be contacted with a release based on highly crosslinked silicones [27]. High speed (600 in./min) stripping of this release gives 25–30 g release force in comparison with low speed stripping (12 in./min), which gives 9–16 g. High speed stripping values do not depend on storage time. That means that at high speed both components (crosslinked and uncrosslinked) behave like elastic materials (no energy dissipation), but energy is stored. Soft (e.g., freezer tape) adhesives increase the release force with increasing separation speed more than permanent ones [73]. It should be mentioned that the dependence of the peel value and of the break nature on the peel rate is a common phenomenon observed in the delamination of PSPs during their application also. Such dependence causes the so-called inversion of the adhesive break when tapes are unwound at too low a temperature (lower than 10°C) or with too high a running speed [74] (see also Chapter 7).

Common PSA labels or tapes are usually designed to give a well-defined peel resistance under test conditions where delamination occurs at a standard debonding speed. For high quality special products (e.g., removable labels), uniform delamination over time, e.g., a uniform peel resistance value, is required. However, there are other applications, i.e., debonding conditions for several tapes and protective films, where peel force uniformity during debonding is not a quality criterion and variable peel resistance values are needed during debonding. It should be emphasized that in such applications peel should vary as a function of debonding speed. Moreover a nonlinear dependence is required.

Diaper closure tapes are used as refastenable closure systems for disposable diapers, incontinence garments, etc. Such systems allow reliable closure and refastening. The tapes have to exhibit a maximum peel force at a peel rate between 10 and 400 cm/min and a log peel rate between 1.0 and 2.6 cm/min. Tapes with the maximum values of these parameters have been found to be strongly preferred by consumers. It is supposed that the higher peel forces at low rates discourages the wearer from peeling open the closure, whereas the lower peel force needed at greater peel rates helps the user remove the closure without tearing the back sheet of the product. According to Miller and von Jakusch [75], the presence of a maximum peel force as a function of peeling rate strongly depends on the formulation. When no high T_g endblock reinforcing resin is used, no maximum appears in the peel force versus rate curve. The peel force increases monotonically with rate until the peel force exceeds the strength of the tape carrier material. The rate for maximum peel force depends on the concentration of the high T_g resin, varying between 120 and 225 cm/min. The reopenability increases with the level of high T_g resin also. Although the nature of the substrate (different polyethylene grades or polypropylene) influences the value of the peel force (10–20%), it does not influence the presence of a maximum as a function of peel rate, demonstrating that such an effect is due to the viscoelastic behavior of the adhesive or tape carrier material. As stated in Ref. 75, for the same adhesive composition the 180° peel test leads to much higher (300%) peel resistance. Figure 5.7 illustrates the form of the peel curves for a common label delamination, a diaper closure system, and a special zip-peel protective film.

As can be seen from Fig. 5.7, unlike label delamination, for special tapes and protective films peel resistance should present a barrier value (cleavage peel) followed by a rapid decrease in peel resistance. The time and temperature dependency of the pressure-sensitive macromolecular formulations usually allows an

Adhesive Properties of PSPs

PEEL

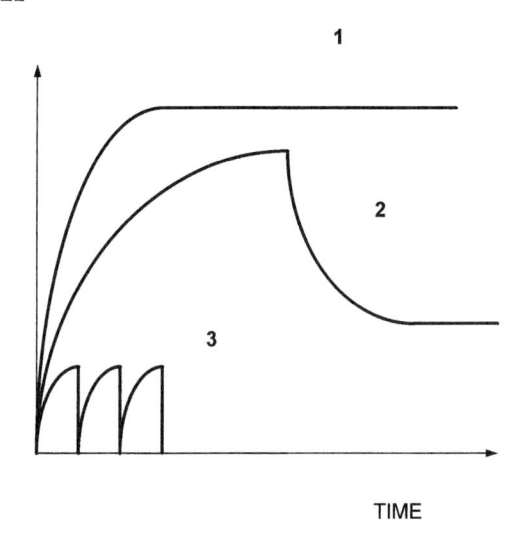

Figure 5.7 Typical peel curves for three types of PSPs. *1*, Label; *2*, diaper closure tape; *3*, special protective film (arbitrary scale).

increase in peel value as a function of delamination speed and leads to an increase in debonding resistance at higher speeds. A decrease in the debonding force at high speeds supposes a debonding mechanism quite different from common relaxation. According to Ref. 76, at increased rates of testing a slip-stick region appears that is characterized by a regular oscillation of the peel force. It may involve a regular oscillation between rubbery and glassy response of the adhesive. It may alternate storage and dissipation of the elastic energy in the stretched backing film.

The existence of a high starting peel value has been observed with other crosslinked adhesive formulations also. A peel test (90° peel) of such a (crosslinked) adhesive tape measures first the force required to start the breaking of the bond (initial breakaway peel) and then the force needed to continue the bond breaking, the "initial continuing peel" [77]. A pluck test (90° peel, with a slow peel rate of 2.5 cm/min) is carried out also. Initial continuing peel is lower (by about 30%) than breakaway peel. It is evident that such peel behavior is the result of a special adhesive or product buildup. Bonneau and Baumassy [28] found different peeling behaviors under the same peeling conditions for water-based formulations with acid resin or resin esters (Dermulsene, DRT). The acid resin used displays a narrower molecular weight distribution (M_w/M_n = 1.19) and higher T_g (10°C), and the ester resins have a broader MWD (1.19–2.48) and somewhat lower T_g (4–6°C). For the acid resin the debonding is jerky (stick-slip phenomenon); for the other resins the debonding is quite steady. According to Bonneau and Baumassy, this behavior is due to the "better wetting of the substrate" by the ester resins. Due to the greater modulus of the adhesive at low temperatures, a loss of wetting and a jerky debonding are observed for the acid resin. The resins were of similar softening points and compatibilities, and the molecular weight of the acid resin was about 400. The resin esters have a molecular weight of 1000–1300. For a given softening point and concentration of resin, the higher the increase in the tan δ peak temperature, the better the compatibility (see also Chapter 3, Section 1). Bonneau and Baumassy [28] found no difference by DMA study, between the compatibilities of the tested resins although in practice results have been different. It can be supposed that because of the tendency of the acid resin to crystallize, an apparently crosslinked incompatible system is built up that displays different peel behavior.

The use of classic peel measurement does not allow perfect characterization of practical adhesion behavior for the whole product range of PSPs. Debonding resistance characterized by peel is a force measured under well-defined conditions (speed, temperature, strain). For the majority of PSPs the application parameters (although different from those used for quality tests) are constant. That means that for these products peel resistance can be used as a standard measure of laminate bond strength. Good product quality means desired and uniform peel force level. A uniform peel force means a constant peel value during test independent of specimen length, i.e., testing time. On the other hand, it is obvious that there are some special products where the peeling force may change during debonding, and therefore the standard test method is not enough to characterize the real behavior of the product.

The adhesion force tested as peel is a so-called bonding/debonding peel, i.e., it is assumed that the debonding resistance is an index of the bonding force. For certain products it is more important to know not only the final value of the peel–stress function but also the form of this dependence (Fig. 5.8). As can be seen from Fig. 5.8, for the so-called zip peel, the peel resistance should not be uniform during delamination. The peel should show an increase up to a maximum value followed by an almost zero level, in order to allow slight and rapid delamination of large coated surfaces. As mentioned earlier, such behavior can be achieved by designing a special adhesive or carrier film rheology and composition. It can be supposed that the two-phase peel is due to energy storage in a highly elastic carrier (which together with the external force causes high speed debonding) or to the destruction of a crosslinked adhesive network in which hard crosslinked parts and soft viscous parts of the network differ in their debonding resistance. The first hypothesis accentuates the special role of plastic-based carrier materials.

Adhesive Properties of PSPs 281

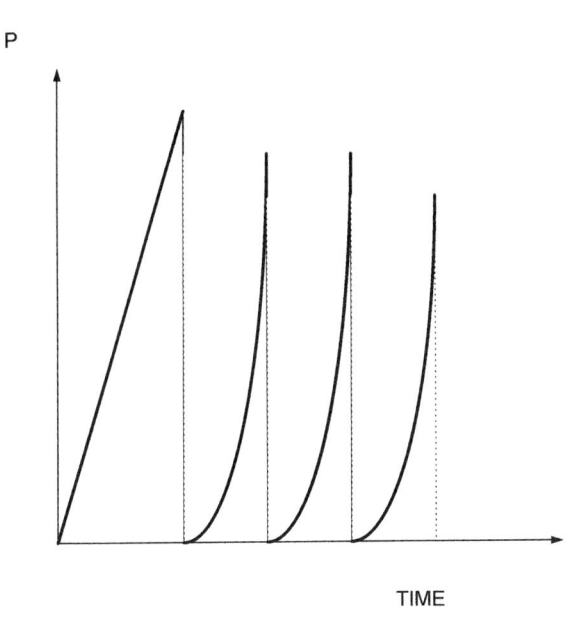

Figure 5.8 Typical zip-peel (sawtooth) plot, illustrating variation of peel force during debonding (arbitrary scale).

Carrier Deformation. As schematically presented in Fig. 5.9, the use of a very stiff adhesive or of a rigid carrier material (e.g., a foamed, crosslinked adhesive) can lead to changes in the peeling nature of delamination (peel is transformed into butt peel). On the other hand, as discussed earlier (see Chapter 3, Section 1), a deformable carrier can grade the value of the peel.

It can be supposed that in the first step of delamination the debonding force causes a deformation of the carrier material (ϵ_c), followed by the deformation of the adhesive layer (ϵ_a) and bond failure (B_f) (see Fig. 5.10).

It can be supposed that for a PSP with a nondeformable carrier material, the peel force acts on the adhesive bond instantaneously; for a product with a deformable carrier, first the film deforms. In reality, deformations of both carrier and adhesive occur simultaneously and lead to bond failure. The debonding time (t_{db}) is given by the time required for the deformation of the carrier (t_{Dc}), the time necessary for adhesive deformation (t_{Da}), and the time to bond failure (t_{Bf}):

$$t_{db} = t_{Dc} + t_{Da} + t_{Bf} \tag{5.10}$$

The debonding force is also the sum of the forces required for extension of the carrier material and of the adhesive layer and for bond failure. These processes

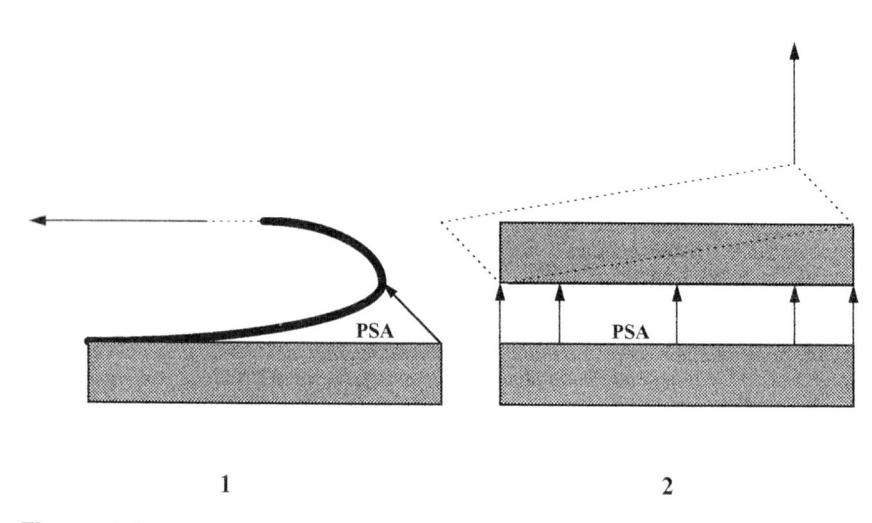

Figure 5.9 Peel degradation by nondeformability or excessive deformability of the carrier material. (1) Soft deformable carrier material; (2) rigid nondeformable carrier material.

occur over time, being partially superposed (see Fig. 5.10). The peel force (P) required to destroy the adhesive bond is the sum of the forces necessary to deform the carrier material (F_{Dc}) and the adhesive (F_{Da}) and to destroy the bond (F_{Bf}):

$$P = F_{Dc} + F_{Da} + F_{Bf} \tag{5.11}$$

Correlation (5.11) suggests that the peel force required for pressure-sensitive laminates based on deformable carrier material may be higher than that required for a nondeformable carrier. According to Dunckley [24], for common papers, paper strain appears at 1500 g/in. debonding force. The real forces in the paper are higher because they are partially absorbed by the deformation of the paper. As discussed in Ref. 1, soft carrier materials allow a low level of peel force. This is possible due to their energy absorption and the time and temperature dependence of the stress transfer. As is known from work with removable labels, high speed debonding leads to paper tear. Good removability requires that the adhesive have energy absorption properties (i.e., its flow, relaxation of its macromolecular components). This process needs time. The deformation time of the carrier material can be (at least partially) used for adhesive deformation too. Therefore it would be more correct to admit that in correlation (5.11) the time necessary for deformation is the sum of the time required for processes occurring somewhat simultaneously. Thus the time for debonding is given by the time for deformation (τ_D) and bond failure,

$$\tau_{db} = \tau_D + \tau_{Bf} \tag{5.12}$$

Adhesive Properties of PSPs 283

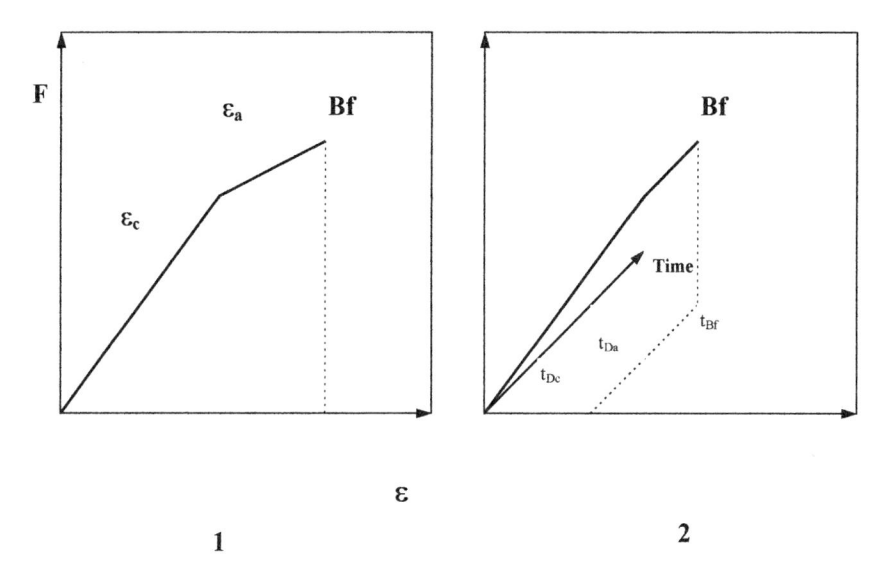

Figure 5.10 Schematic presentation of force transfer by debonding. (1) Force transfer via carrier to adhesive; (2) simultaneous carrier and adhesive deformation.

where the deformation time is a function of the mechanical characteristics and geometry of the plastic film carrier and the rheology of the adhesive layer. As is known from the testing of the mechanical properties of carrier materials, the mechanical resistance of the plastic carrier material is greater than the force necessary for adhesive bond failure. This force is higher than the forces required for adhesive flow. As discussed in Chapter 3, Section 2, in evaluating the carrier quality for labels its dimensional stability plays an important role. This is evaluated generally by measuring the forces necessary to achieve a given (low percentage) elongation, supposing that machining forces will not exceed this value. As can be seen from Table 3.6, the value of this force is generally much higher than the bond strength for removable pressure-sensitive laminates. That means that the deformable carrier material acts like a nondeformable one, transmitting the debonding force almost instantaneously and at a high level at the adhesive/substrate interface.

According Kim et al. [78], the measured peel value is an "engineering strength per unit width." For thin films it does not represent the true interface adhesion strength. The peel resistance may represent the product of the interface adhesion and other work expended in the plastic deformation of the film. The contribution of the plastic deformation of the film to the peel strength is found in some cases to be many times higher than the true adhesion. The major controlling factors in the peel strength are the thickness, modulus, yield strength, and strain hardening coefficient of the film, the compliance of the substrate, and the interface adhesion strength.

In reality, both the plastic carrier and the PSA are (at least partially) viscoelastic materials, and their mechanical characteristics depend on the material, geometry, and time and/or temperature. The peel force required to delaminate the pressure-sensitive construction includes a component required for elongation work. Variations in peeling energy due to deformation of the stripping member, which in turn changes with speed and other parameters, should be taken into account. The change in the stress rate acting on the bond by delaying the forces due to carrier deformation enhances debonding at lower peel forces. Thus the resultant peel is a complex function of carrier deformation (ε_c),

$$P = f(\varepsilon_c^\gamma / \varepsilon_c^\delta) \tag{5.13}$$

where γ and δ are exponents that account for the influence of the deformation work on the peel force. For thick, mechanically resistant or elastic materials, the value of δ is low, lower than γ. For plastically deformable thin films, δ increases and the carrier deformation can lead to a decrease in the peel force (Table 5.14). For a given material and given delaminating conditions, the geometry of the film is the main parameter influencing the carrier deformation. One can suppose that for each material used there is a critical thickness. Under this limit the deformation of the plastic film decreases the delaminating force. Therefore decreasing the carrier thickness for pressure-sensitive products, especially for protective films, presents a complex problem.

Table 5.14 Influence of Plastic Carrier Deformability on the Peel of PSPs

	Carrier characteristics			180° Peel resistance (N/25mm)	
Code	Thickness (μm)	Tensile strength (N/mm^2)	Elongation (%)	Nominal	Measured
1	23–31	27	419	1.2–1.5	0.15
2	32–37	23	415	0.4–0.6	0.02
3	40–48	31	532	2.5–3.7	2.0
4	42–46	33	558	1.2–2.2	1.1
5	42–47	28	494	2.5–3.0	2.2
6	58–67	18	259	2.5–3.0	2.4
7	65–75	18	348	2.50–3.0	2.7

The above considerations have a general character independent of the product buildup (construction of the adhesive layer). It has been shown by Kuik [70] for heat-sealable polymer films that the 90° peel is the result of adhesion and energy loss. Energy loss is due to the deformation of the plastic film during peeling, and this deformation is proportional to the resistance to dimensional change and material thickness. Therefore it is obvious that carrier deformation (depending on geometry and composition) will have a general influence on the bonding and debonding of PSPs. According to Boutillier [79], for an extensible polymer film the greater the work of adhesion the greater the dissipation energy (due to the polymer deformation) and the greater the peel energy. For a self-adhesive film (EVAc), the dissipated energy, deformation energy, and dissipated energy per unit volume increase with the thickness of the polymer. In the classic field of labels this is less evident, but elastic/plastic carrier deformation must always be taken into account by winding of tapes as a factor influencing bond strength. It must be taken into account by delamination of protective films, where carrier deformation modifies the peel force, i.e., removability. It is evident that carrier deformation is a function of the stress rate. Adhesive deformation is also. As discussed earlier, removability is a relaxation phenomenon, and relaxation needs time.

In the above examples (and according to the data of Table 5.14), the viscous and viscoelastic tensile deformation of the carrier (parallel to the direction of the peeling force) is a main factor influencing the transfer of the debonding force. Carrier deformation by flexural stress influences conformability by bonding and deformability (peeling angle) by debonding. The deformation of the carrier characterized by its tensile strength is more pronounced for self-adhesive films whose formulation is soft in comparison with adhesive-coated PSPs. The reduced conformability of a carrier during bonding can be balanced by the conformability of the adhesive for adhesive-coated PSPs only. The flexural deformability brought about by debonding is as common for adhesive-coated as for adhesiveless PSPs. Removability of SAFs is strongly influenced by their stiffness. As stated by Kuik [80] for polar terpolymers of ethylene with maleic anhydride, in blends of LDPE two different influences coexist. A terpolymer with a good hot tack (i.e., high polar component content) gives excellent sealability at low temperatures. The addition of LDPE to the terpolymer increases its peeling strength. This is due to modification of the stiffness of the film. The higher the density of PE, the greater the increase of adhesion.

For special PSPs the carrier conformability and deformability impose the use of special test methods for evaluating adhesion. For applications where combined stresses act on the adhesive joint, the test methods have to be adapted to the actual conditions of product application. One such case is that of the self-adhesive (tackified polyethylene) stretch films where so-called peel cling and lap cling are used to characterize the product (Fig. 5.11). Peel cling is evaluated by a T-peel

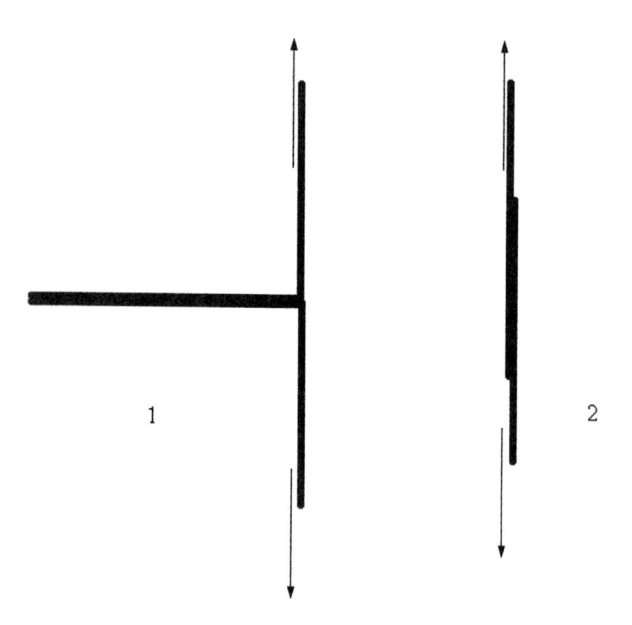

Figure 5.11 Schematic presentation of peel cling (1) and lap cling (2) measurement for self-adhesive films.

test where self-adhesion is tested. T-peel is used for flexible adherends [81,82]. According to Ref. 83, peel cling needs to be high enough to stop the loose end of the film from unwinding on the pallet. Its optimum value appears to be around 2.4 N/m. Lap cling needs to be sufficiently high to maintain the integrity of the wrapped load. Its optimum value is about 5 N/2.5 cm. Using adhesiveless soft PSPs for this test, the measured data will strongly depend on the stiffness, conformability, and elasticity of the film. Lap cling is measured like shear resistance, parallel to the joint. Mizumachi [84] studied the adhesion of PSAs by testing rolling ball tack and used for its characterization the coefficient of friction, considered as the sum of rolling friction caused by compressive deformation of the substrate material and friction caused by extensional deformation of the substrate. In the case of lap peel, slip friction appears and tensile stress combined with shear stress act on the adhesive joint. Therefore the value of lap cling is higher than the value of peel cling. Due to the slow buildup over time of an adhesive–substrate contact, the relative value of aged peel cling (in comparison with its initial value) is much higher than that of the aged lap cling [85]. Generally peel cling increases with the level and molecular weight of the tackifier (polybutene). As is known, the molecular weight of the tackifier resin and its level have a similar influence on the peel of adhesive-coated PSPs. On the other

Adhesive Properties of PSPs 287

hand, lap cling increases with the level of polybutene but decreases as the molecular weight of the polybutene increases due to decreases in its ability to migrate and its tackiness. Unlike the classic tackifying of elastomers, where the components must be compatible, the tackifying effect of polybutenes for polyethylene is due to the partial incompatibility of the two materials and the inherent tackiness of polybutenes. It is evident that tackifier diffusion also depends on the film thickness. This dependence also exists for slip migration [86].

A special method for evaluating the adhesive strength of adhesive coatings (e.g., SAF) exhibits similarities to the lap cling test [87]. In this case the adhesive strength (W) of a polymer coating on a metal substrate is determined by stretching the substrate and recording its elongation (ε) at the onset of debonding, followed by measuring the modulus of elasticity (E) of unsupported polymer film. The adhesive strength is calculated as εE. The results are comparable to the data obtained by peel test [87].

Removability

For certain applications PSPs are required that display a low peel force and give clean, deposit-free separation from the substrate. Generally, removable adhesive joints are those that can be destroyed without damaging the solid-state components of the joint. Conversely, as is known from work with pressure-sensitive labels and packaging tapes, permanent labels or tapes are those that can be peeled off only by destroying the face stock material (labels) or substrate (tapes). That is, for permanent laminates the mechanical resistance of the adhesive joint (R_{aj}) is higher than the resistance of the solid components of the joint, the resistance of the carrier (R_c) and that of the substrate (R_s):

$$R_c < R_{aj} > R_s \tag{5.14}$$

The resistance of the adhesive joint is characterized by the internal strength of the adhesive (A_c) and adhesive bond strengths toward the carrier (A_{ac}) and the substrate (A_{as}):

$$R_{aj} = A_c + A_{ac} + A_{as} \tag{5.15}$$

That means that both the internal cohesion of the adhesive layer and its adhesion (to the carrier and to the substrate) should be higher than the mechanical resistance of the solid-state components:

$$R_c < A_c + A_{ac} + A_{as} > R_s \tag{5.16}$$

As mentioned earlier, for a removable joint, during bond break the adhesive layer will be destroyed, leaving the solid-state components of the construction intact. In this case,

$$R_c \geq R_s > R_{aj} \tag{5.17}$$

Theoretically, this relationship ensures damage-free separation of the adhesive-bonded solid-state components. For practical use, the place of failure is important also. Generally, removability requires bond breaking at the adhesive/substrate interface (e.g., removable labels and protective films) or in some special cases at the adhesive/carrier interface (e.g., double-faced tapes). Thus,

$$R_c \geq R_s \geq A_{ac} > A_{as} \qquad (5.18)$$

Correlation (5.18) is valid for protective films also, where bond failure should occur at the adhesive/substrate interface in order to ensure a deposit-free peeloff (delamination) of the protective film after use.

Peel resistance and removability can be regulated according to Pasquali [11] by modifying the ratio between the adhesion surface (S_{ad}) and the application surface (S_{appl}). For a removable product this ratio R_{su} should be smaller than 1:

$$R_{su} = S_{ad}/S_{appl} < 1 \qquad (5.19)$$

This ratio can be modified by using partial coating of the PSA. Actually it is difficult to obtain pressure-sensitive products whose peel value does not change in direct proportion to the adhesive surface area. Therefore, as suggested in Ref. 88, for adhesive balance regulation the adhesion surface area should be diminished (see also later, Formulation). It should be emphasized that reducing the peel resistance to a substrate causes a reduction in peel toward the carrier also. Such behavior can be illustrated by a rolling ball test using adherent and nonadherent carrier material. Theoretically the rolling ball tack (RB_{pa}) of a permanent adhesive (with good anchorage on paper) measured with a PSA sample coated on paper must be better than its tack on siliconized paper (RB_{si}):

$$RB_{pa} \gg RB_{si} \qquad (5.20)$$

As shown by Fig. 5.12 for a PSA sample coated on release paper, the lack of adhesion between the adhesive layer and the carrier material causes separation of the adhesive layer during rolling of the ball. In this situation the adhesion of the PSA on the ball will be influenced by the flexural modulus of the adhesive layer and its increased contact surface and time. For a transfer-coated removable adhesive where the anchorage of the adhesive on the paper is lower, the rolling ball tack on paper or on release paper would have almost the same values:

$$RB_{pa} \cong RB_{si} \qquad (5.21)$$

As illustrated by the data of Table 5.15 for a tackified permanent acrylic PSA, the RB values on paper and release liner are very different. For a removable adhesive, the RB values are similar. This statement emphasizes the importance of primer coating for removability also.

Debonding (peeloff) requires a separation force. As shown earlier, for removable joints this separation force should be higher than the force of adhesion at the

Adhesive Properties of PSPs 289

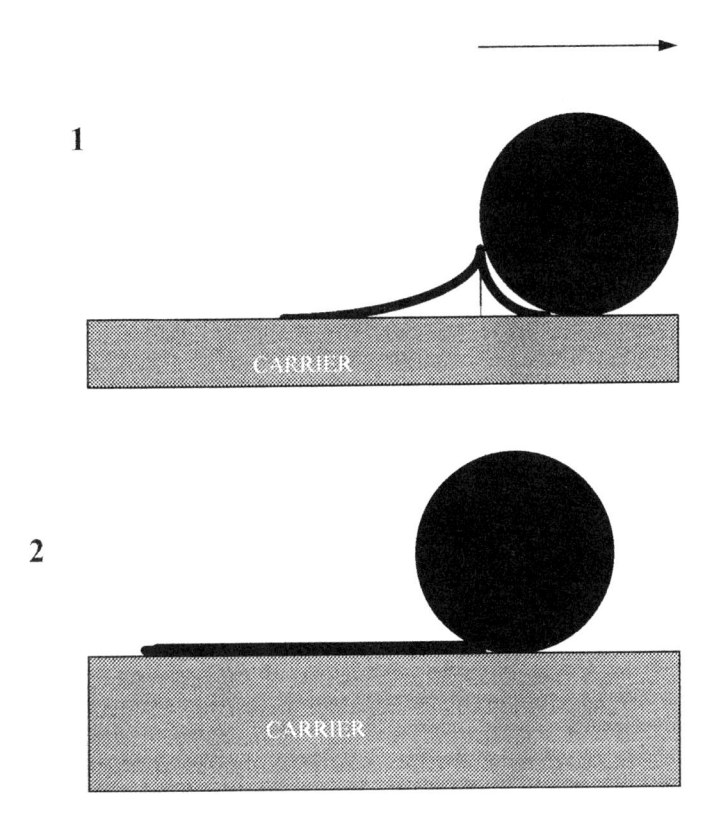

Figure 5.12 Delamination of the adhesive layer from the carrier during rolling ball tack test as a function of its anchorage. (1) PSA on release liner; (2) PSA on film carrier.

adhesive/substrate interface. The adhesion at that interface is a function of the chemical nature of the components, but it also depends on the nature of the applied force. That means that generally for removability, both the level of force and its time/temperature dependence are important. The buildup of adhesion is another time-dependent parameter of removability. As illustrated in Fig. 5.1, removable, repositionable, and permanent adhesives vary in the rate of adhesion buildup.

According to Gerace [89], the bond strength of a viscoelastic material varies as strain rates and temperatures vary. If this is true, removability is not an absolute performance characteristic but a relative one, i.e., regulating the debonding force and its rate also allows control of removability. Kuik [70] studied the dependence of the adhesion energy on coating weight and stress rate for polyethylene and polar ethylene copolymers and stated that the adhesion of self-adhesive

Table 5.15 Rolling Ball Tack on Adherent and Abhesive Carrier Material

PSA	Coated thickness (μm, wet)	Rolling ball tack (cm) Abhesive carrier	Paper carrier
Removable	60	4.0	3.5
		4.5	4.0
		4.0	3.5
	100	2.0	1.5
		2.0	1.5
		2.0	1.5
	200	1.5	1.5
		1.5	1.5
		1.5	1.5
Permanent, tackified	60	3.5	11.5
	100	3.5	13.0
	200	3.5	13.5
	60	2.5	3.5
	100	2.5	3.5
	200	2.5	3.5
	60	1.5	4.5
	100	1.5	5.5
	200	1.5	5.5
Permanent, nontackified	60	2.5	7.5
	70	3.0	9.0
	100	1.5	6.5

thermally sealed films increases with coating weight and decreases with the speed of debonding.

In industrial practice, debonding force acts via the carrier material, i.e., the carrier material will be tensioned and peeled off and debonding force is transmitted through this material. Therefore, as discussed earlier, the distribution of the debonding force as an absolute and time-dependent value will depend strongly on the plasticity and elasticity of the carrier material, i.e., on its stress-damping properties (see also Chapter 3, Section 1). That means that mutual interaction exists between the characteristics of the solid-state carrier material and adhesive properties. Both bond formation and debonding depend on the viscous flow of the solid-state carrier web.

Ideally, for permanent PSAs, failure is 100% cohesive, indicating that the maximum strength of the bond has been reached. In this case when failure occurs, a layer of adhesive remains on each adherend. However, specimens may exhibit adhesive failure initially and cohesive failure after aging, or vice versa. The type of carrier surface may also influence the test results. As an example, clay-coated papers give paper tear if the superficial strength of the layer is not enough [90].

Repositionability and re-adherability are performance characteristics dependent on peel buildup. Both allow the debonding of the PSP after its application on an adherend and its rebonding. Repositionability is an aging-related characteristic; it needs a slow buildup of the adhesion on the substrate. In this case peel buildup is delayed in time but is not limited as an absolute value. Repositionable labels may also be permanent. Re-adhering labels are removable after their application independently of aging time. That means that their peel buildup is of limited value. There are different ways to achieve removability (softening or crosslinking the adhesive, regulating its geometry, etc.). Re-adhering requires balanced adhesive properties and excellent bonding characteristics. Therefore, the only manufacturing techniques that can be used for re-adherable products are those that do not influence the instantaneous tack of the adhesive. Repositionable and re-adherable products are required in those PSP classes where instantaneous tack provides product applicability. For instance, abrasion-resistant automotive decorative films with resistance to the impact of stones must be repositionable [91]. In some cases crosslinked compositions having a reduced adhesive contact surface are used for re-adherable products [92]. In such formulations pressure-sensitive polymer filler particles having a diameter of 0.5–300 μm can be used also [93]. To improve the detackifying effect of the particles, the formulations may include an ionic low tack monomer [94]. The composition of the PSA polymer according to a Nichiban patent [93] includes a crosslinking agent also used for both the matrix polymer and the particles, and the matrix polymer covers the surface of the particles also. The particles are synthesized via suspension polymerization. The T_g of the particles should be lower than 10°C; these are elastic. Polyisocyanate derivatives, polyepoxides, or aziridines are suggested as crosslinking agents. A low coating weight (7 g/m^2) is used. The re-adherability of the product on paper has been tested by measuring the peel after repeated lamination and delamination (50–100 steps). Re-adhering and removable adhesive in a solid stick form (glue stick, applicator crayon) is used also [95]. Such products are removable unless the user applies a very heavy coat. A natural rubber latex, a tackifying agent, and a gel-forming component are blended. The adhesive-coated paper substrate can be readily lifted and removed from the contact surface and can be reapplied at least eight times to the paper adherend surface. The product must be coated along one edge with a strip of adhesive.

As mentioned earlier, removability is related to adhesion buildup, i.e., to the change in adhesive performance over time. Therefore, removability must always be examined with respect to aging also. A removable adhesive that ages poorly loses its cohesive strength and will not peel cleanly. In some cases the adhesive properties change after the application of the adhesive. As is known, aging stability characterizes the resistance of the adhesive to such undesired changes. However, in certain cases the changes in adhesive performance over time are inherent properties of the formulation used or they are designed by the formulator of the adhesive. Changes in the adhesive properties that are inherent in a given formulation are more pronounced for systems with an apparent character of pressure sensitivity, i.e., with a limited shelf life of tack or peel. Such systems may lose their tack or peel due to the loss of flow properties. Generally the flow properties disappear due to the buildup of a structural order (network) in the polymer. As shown in Ref. 96, 0.5% crosslinking agent is perfectly adequate to give elastic properties for polyalkyl methacrylates. Therefore, crosslinking, crystallization, plasticizer migration, and plasticizer volatilization can change the flow properties of a formulation. Postplasticizing due to absorption and migration of liquid laminate or atmospheric components also affects the flow properties adversely.

For instance, propylene copolymers with short-chain comonomers are used for so-called semi-pressure-sensitive adhesives [18]. Such products lose their tack after a well defined time because of crystallization. The nature of the comonomer influences the crystallization rate of the polymers. Their pressure sensitivity can be improved by using tackifier or plasticizer and SBCs. A loss of pressure sensitivity may be caused by crystallization of acid rosins also. Generally, a pronounced loss of pressure sensitivity is caused by crosslinking too (see later), although for some self-curing acrylics (according to the supplier [97]) the balance of adhesive characteristics is not altered by the degree of curing. As discussed in Chapter 3, the influence of crosslinking also depends on the molecular weight between the network points. For a low M_c, an increase in the degree of crosslinking is not accompanied by an increase in the elastic modulus [98]. The crosslinking method also affects the M_c and crosslinking density. The molecular weight between crosslinks may be larger than the M_n of the base polymer. As stated in Ref. 99, the crosslinking of polyurethane acrylates by electron beam restricted the mobility of the polymer chain less strongly than chemical crosslinking.

As discussed earlier, plasticizer migration from self-adhesive films based on plasticized PVC may cause those films to lose adhesion also. In a similar manner, volatilization of the solvent (plasticizer) from pressure-sensitive formulations with high T_g base polymers may produce a loss of adhesion. For instance, such formulations on acrylic basis have been used for self-adhesive letters (transfers). These are coated by means of printing on a relatively soft polyolefin carrier material and are applied on the substrate by pressing (embossing) the back side of the carrier. High T_g acrylates (known from printing ink technology) have been used for such formulations. Their pressure-sensitive adhesivity (flow) has been given by a very high boiling point (over 250°C) plasticizing solvent (e.g., bornyl

Adhesive Properties of PSPs

acetate). For such products, volatilization of the solvent imparts a semipressure-sensitive character to the product.

According to Czech [100], an inverse phenomenon appears in certain formulations used for splicing tapes. For water-dispersible, soft adjusted splicing tapes, the adhesion depends on humidity. The dependence of the adhesion (P) on the relative humidity (H_R) is given by the (empirical) formula

$$P = A \left(\frac{1 \times 12}{10 \, H_R}\right) \left(\frac{H_R^2}{200}\right) + 7 \tag{5.22}$$

where A is a parameter depending on the composition of the water-soluble PSA [100].

Adhesive Properties of Labels

Labels are a special class of pressure-sensitive products generally manufactured and supplied as laminates including at least two solid-state components. These solid-state components are bonded with a pressure-sensitive adhesive, and the surface quality of the release liner is regulated in a such a manner as to allow an easy separation of the components. The adhesion between the solid-state laminate components has to ensure the stability of the label. Therefore, the adhesive properties of label components are determined by the end use adherence of the product on the substrate and its temporary adherence on the release liner. Unlike tapes and protective films, for labels the adhesion between the release liner and the PSA has to be regulated exactly.

The adhesion between the pressure-sensitive layer and the release liner depends on both components. It influences the application speed of the labels (roll labels), and therefore it should be exactly controlled. Because of the low level of such adhesion it is sensitive to the test conditions. It is well known that peeling the label from the backing and peeling the backing from the label give different peel values. Generally, peeling the backing from the label gives the lower adhesion value. Depending on the mechanical characteristics of the solid-state components of the laminate and the type of silicone, a high speed release force test may or may not give a maximum of the peel force as a function of the debonding speed.

Labels need balanced adhesive properties. This balance refers mainly to sufficient tack and peel on surfaces that are difficult to bond. Cohesion is less important for labels. As stated by Chu [34], the application window for (permanent) labels is characterized by a G' value of 20 kPa at room temperature. In comparison, tapes with high cohesive strength have G' values between 50 and 200 kPa.

According to Bonneau and Baumassy [28], the main requirements for adhesives used for labels are (1) good adhesion to untreated LDPE and (2) retention of the adhesive properties over a wide temperature range. Labels exhibit only medium shear resistance. Shear values of 12–680 min (on steel, according to

PSTC) are acceptable for labels only. Packaging tapes need higher shear [24]. Table 5.16 shows some typical values of adhesive properties of labels.

Adhesive Properties of Tapes

Unlike labels, tapes are self-wound laminates. Their delamination is influenced by the release properties of the back side of the carrier material, by the adhesive, and by the delaminating conditions. Compared to labels, the adhesive properties of a tape on a given substrate are less influenced by the quality of the carrier material back, by the surface of the adhesive, and by application conditions (temperature, pressure, and speed). This is due to the different application conditions of labels and tapes. It should be taken into account that the adhesive on a label has to allow the label to instantaneously adhere and (sometimes) to instantaneously bond on a given surface. However, due to the special application conditions of a label (i.e., well-defined and relatively small dimensions, static adherend surface, and no or very low shear force due to its own weight only), its adhesive has sufficient time to build up the final bond force. Packaging tapes are applied on dynamic surfaces or on items with a tendency to move, where the adhesive has to instantaneously balance forces of motion during application of the

Table 5.16 Adhesive Properties of Labels

Grade	Adhesive[a]	Carrier	Tack Value	Tack Method, unit	Peel resistance Value	Peel resistance Unit	Shear resistance Value	Shear resistance Unit	Ref.
Permanent	WB AC	PET	40–45	LT,oz/in.	55	oz/in.	4–8	h	101
Permanent	AC, crosslinked	—	11	RB,cm	11	N/25 mm	—	—	102
Permanent	AC	PET	36–37	LT,oz/in.^{2}	37	oz/in.	3	h	103
	SBR	PET	54	LT,oz/in.^{2}	46	oz/in.	11	h	103
	EVAc	PET	54	LT,oz/in.^{2}	69	oz/in.	4	h	103
Permanent	SB AC	PET	—	—	11	kg/25 mm	—	—	104
	AC, crosslinked	PET	—	—	10	N/20 mm	>24	h	105
Permanent	AC	PET	1100	Polytack, g.in.^{2}	13	N/25 mm	8	min	26
Permanent	AC	PET	1350	Polytack, g.in.^{2}	25	N/25 mm	1	h	26
Permanent	AC, crosslinked	PET	—	—	12	N/25 mm	—	—	106
Removable	AC, crosslinked	PE	—	—	2	N/25 mm	—	—	107

[a]AC = acrylic; WB = water-based; SB = solvent-based.

Adhesive Properties of PSPs 295

tape—work shear and peel forces in the case of a PSA layer. Under such conditions the bonding force has to instantaneously attain almost its full final value. Such behavior is enhanced by using high coating weights. As shown by Gent and Kaang [108], the pulloff force acting on a packaging tape is a resultant of peel, shear, flexural, and tensile stress. It can be considered as a vector resulting from cling and lap peel debonding. Therefore the detachment energy is lower than the energy obtained from peeling. According to Schroeder [65], the adhesion of a tape depends on its tack, instantaneous peel, peel buildup, and wetting out of the substrate surface.

As mentioned earlier, in the first period of development of pressure-sensitive tape technology, dwell time was referred to as conformation time [65]. Different tapes require different lengths of time to conform. For certain products that time has been as long as 3 months. Generally tapes need high shear values, but the requirements for shear resistance differ according to the tape specialties. Because of the variety of carrier materials and substrates and the different requirements fulfilled by tapes, as illustrated by the data of Tables 5.17 and 5.18, their adhesive characteristics are different also.

Medical tapes require tack, peel, and good shear adhesion [129]. It is well known that for many adhesive formulations for tapes, in order to prepare high solids content solutions the elastomer is masticated. Therefore postcrosslinking is necessary. Advanced crosslinking causes low tack. Such high shear, low tack formulations are used for contact adhesives, mounting tapes, and double-faced foam tapes [25]. To provide high tack and conformability, high coating weights are used. For instance, for mounting tapes an 88 μm thick adhesive layer is coated, whereas for splicing tapes a 50 μm adhesive thickness is used. For mounting

Table 5.17 Adhesive and Mechanical Characteristics of Some Special Tapes

Application	Adhesive performance				Mechanical performance			Ref.
	Peel (N/10 mm)				Break strength (N/10 mm)	Elongation (%)	Tear strength (N)	
	Backing	PVC	PE	SST				
Marker tape	1.2	1.2	1.5	—	22.0	50–100	—	109
Thermally resistant tape	—	10.0	14.0	—	80.0	250	100.0	110
Easy tear tape	1.8	1.8	—	—	18.0	100	8.0–10.0	111
Flame-resistant tape	1.0	1.0	1.0	—	15.0	160–300	7.6–10.6	112
Harness wrap tape	1.0	1.2	1.2	—	15.0	1400	700.0	113
Building masking tape	1.0	—	—	5.7	11.0	400	—	—
Insulating tape	—	—	—	1.7	29.0	26	—	—
Insulating tape	—	—	2.0	6.0	—	—	—	106
Application tape	0.1	—	—	0.5	29.0	340	—	—
Closure tape	1.0	—	—	2.8	—	—	—	—
Carpet tape	—	4.1	—	—	—	—	—	—

Table 5.18 Adhesive Performance of Tapes

Product	Application	Adhesive[a]	Carrier[b]	Peel	Tack	Shear	Ref.
General	Packaging	—	PVC	230 g/10 mm	—	—	114
General	Packaging	SB, AC	—	27 lb/333 yd^2	—	—	115
General	Packaging	WB, AC	Paper	1450 g/in.	—	142 h	116
General	Packaging	HM, SIS	—	0.72 kN/m	0.65 kN/m	74°C	117
General	Packaging	HM, SEBS	—	800 g/25 mm	—	—	118
Special	Packaging heavy duty	—	PVCR	350 g/10 mm	—	—	114
Special	Packaging	Silicone	PET	4.6 N/10 mm	2.25 N/cm	>20 h	119
Special	Packaging	HM, CR, AC	—	5.7 N/10 mm	—	—	120
Special	Packaging	CR, AC	—	925 g/25 mm	—	—	121
Special	Bag closure	—	PE	50 oz/in.	—	—	122
Special	Closure	SB, SBR, NR	PE	2.8 N/10 mm	—	—	—
Special	Insulating	—	Alu	800 g/10 mm	—	—	123
Special	—	HM, SBC	—	6.0 kg/30 min	—	—	124
Special	Heat-resistant	Fluoro derivatives	—	22 g/10 mm	14 cm	—	125
Special	Mounting	WB, AC	PET	4.5 lb/in.	6.5 lb/in.	400 × 8.8 psi at 72°F	25
Special	Splicing	WB, AC	PET	2.2 lb/in.	1.5 lb/in.	250 × 8.8 psi at 72°F	25
Special	Removable	Rubber resin	PET	40 g/10 mm	17 N/cm	60°C/5 min	126
Special	Outdoor	—	Plastic woven	7.0 lb/in.	—	—	127
Special	Masking	Butyl rubber	—	0.73 kg/in.	—	—	83
Special	Electrical	AC	—	5.0 lb/in.	—	—	128

[a]CR, AC = crosslinked acrylic.
[b]PVCR = glass fiber reinforced PVC.

tapes, peel values on steel and on polycarbonate have been tested. For high temperature splicing tapes, shear at 250°C has been measured. For mounting tapes, the target values of 180° instantaneous peel/loop tack and 72°F shear are 34.5/6.5/400. For splicing tapes the corresponding values are 2.2/1.5/360. Here the shear resistance has been measured at 250°F. Aluminum carrier and a tackified CSBR coated with 40 g/m^2 are used as insulating tape in climatechnics [50]. Shear resistance after water immersion (storage) is tested for such tapes also. Shear values of about 600–1000 min (25 × 25 mm, 1000 g, steel) and peel values of 25–35 N/25 mm (20 min, steel) are required. Adhesives with excellent water resistance (humidity and condensed water resistance) are required for technical tapes and pressure-sensitive assembly parts in the automotive industry [50].

A special acrylic raw material for acrylic HMPSAs has been developed that can be cured photochemically. The uncrosslinked product does not have balanced adhesive properties. Because of its low molecular weight it is tacky but does not have sufficient shear resistance. The pressure-sensitive properties are achieved only by crosslinking [48]. Both the peel and shear resistance of this polymer are inadequate. The product needs to be tackified and crosslinked. A low level of tackifier resin (8.5%) significantly improves the peel and tack but reduces the shear resistance of the adhesive. For the unmodified polymer with a high coating weight of 50 g/m^2, an instantaneous peel adhesion of only 7.10 N/25 mm has been achieved. The shear value is about 7.3 h (23°C, 2 kg, 5 cm^2). The tackified formulation gives 10.5 N/25 mm as the peel value. A transfer-coated PSA (with 15% tackifier resin) provides after irradiation 25.6 N/25 mm peel resistance and 27.5 N/25 mm quick stick on steel and 18.75 N/25 mm peel and 44 min shear resistance (1.2 × 1 in.^{2}, 500 g) on polyethylene. The adhesive layer irradiated on the face stock material gives 20 N/25 mm peel on steel, 16.87 N/25 mm quick stick, 6.25 N/25 mm peel on polyethylene, and a shear value of 77.5 min. Unfortunately, the shear values given by the authors are not comparable. According to Ref. 41, such a formulation may be an alternative to solvent-based adhesives for tapes and labels.

As stated by Jacob [130], SIS block copolymers are used exclusively as raw materials for HMPSAs for tapes. As a less expensive alternative, SBSs may be formulated, but they not as good. As discussed by Jacob [130], the SBS formulations are significantly softer and their peel adhesion values are lower. A proposed formulation of SBS and tackifier for a PSA used for double-faced carpet tape having a high coating weight of 40 g/m^2 exhibits relatively low adhesive characteristics (peel strength of 16.76–21.25 N/25 mm, loop tack of 17–18 N/25 mm, rolling ball tack of 3–5 cm, and shear resistance on steel of more than 100 hs). According to Donker et al. [131], high cohesion is required for tapes. For packaging tapes a peel value of 18 N/25 mm (on steel), loop tack values of 20–30 N/25 mm, and rolling ball tack below 5 cm are preferred. For packaging tapes a flap test is also carried out. This gives a combination of shear and peel adhesion and tack, with the preferred value being 100 min. Shear adhesion to a carton at elevated temperature (40°C) is tested for packaging tapes. The preferred value for this test is 100 min.

As shown by Gerace [89], tapes may be debonded by cleavage and peel. Gerace [89] presents an empirical graphical method that depicts adhesive performance over a wide range of conditions (pull rate and temperature) that are typical for tapes. Lines of equivalent adhesion were drawn to describe the degree of adhesion of pressure-sensitive tapes used to bond exterior trim moldings to automobiles over all combinations of pull rate and temperature. Each adhesive tape was used to bond a 0.5 in. vinyl bar to a substrate painted with 27% dispersion lacquer. Since the vinyl bar was semirigid, the separation mode of the bar was a hybrid of cleavage (rigid member) and peel (flexible member). The peak load values for cleavage peel are shown as constant value curves called "isocleaves."

Urahama and Yamamoto [132] used a rolling adhesive moment tester to evaluate the adhesive properties of tapes. They correlated the rolling adhesive moment to the peel and proved the applicability of the time–temperature superposition principle. The effect of temperature on the 180° peel force can be predicted from the profile of the velocity spectrum of the rolling adhesive moment (in the temperature range of -20 to $+40$°C). The correlation coefficients for mechanical properties of the adhesive illustrated that the properties of acrylic adhesives and those of rubber-based adhesives had different effects on adhesion.

As discussed earlier, labels require a fine regulation of their adhesion to the release liner. In the case of tapes the unwinding resistance must be controlled. As known, the unwinding resistance depends on the adhesive and carrier characteristics (see Chapter 7). In the case of labels the surface characteristics of the top side of the face stock material have no effect on label delamination. For tapes this (printed or lacquered) surface must exhibit release properties. Its abhesive (release) and adhesive (printing ink anchorage) characteristics must be balanced [133]. As for labels, the dependence of peel resistance on the release (carrier back side) as a function of debonding speed plays an important role for tapes.

Adhesive Properties of Protective Films

In the application of protective films, adhesive properties should ensure a fast, low pressure, full surface adhesion, independent of the type of surface, laminating pressure, web thickness, and application temperature. While the laminate is functioning, its adhesive properties should ensure the monoblock character of the laminate (no debonding). At the time of separation of the protective film, its adhesive properties should allow easy, high speed debonding.

As discussed earlier, protective films must be removable. Removability can be achieved by appropriate formulation of the adhesive and design of the adhesive joint. The choice of an adequate raw material with a built-in removable character, the regulation of this character by way of formulating with viscous components

Adhesive Properties of PSPs

(improvement of the energy-absorbency), and the improvement of this character by crosslinking (hindering uncontrolled flow, and reducing tack) are common modalities for regulating the removability of the adhesive. On the other hand, improving the anchorage and stress distribution (by surface treatment, priming, and direct coating) are current modalities for controlling the removability via adhesive joint design. Other possibilities concern the regulation of the coating weight and coating geometry.

As can be seen from the above evaluation of the possible ways to achieve removability, both crosslinking and the reduction of coating weight involve strong reduction of tack. As mentioned earlier, the use of untackified formulations (acrylics) or tackifier formulations with low tackifier level (in comparison with that suggested for labels or tapes) provides low tack (see also Chapter 6). The use of high melting point resins (rubber resin formulations) and of hard comonomers (acrylics) diminishes tack also.

Protective films exhibit very low tack values (see Table 5.13). Therefore the adhesion–cohesion balance and the corresponding tack–peel–shear diagram can by simplified for protective films to a two-parameter plot (as shown in Fig. 3.11), where the adhesive properties are regulated via peel (i.e., the final strength of the joint) and shear; but shear itself is not a primary requirement, it is only a side effect of the high degree of crosslinking. Therefore because of the built-in character of the shear, the peel remains the sole adhesion parameter. Thus (at least theoretically) the adhesive properties of the protective films are controlled via peel, and PSAs for protective films are formulated for peel values only. As can be seen from the product literature [134,135], protective films are classified according to their adhesive strength, that is, peel. Table 5.19 presents the peel resistance values of common protective films.

As known from industrial practice, the peeling off of removable PSAs depends on peel force value and removability. Both a low force separation and an adhesive (residuum)-free debonding are required. The first is characterized by the peel value; the second depends on the subjective examination of the failure mode and place. The peel force is measured in a laboratory on a standard test surface and under standard conditions. The values measured are adequate to characterize a product, but are only indicative with respect to the use of this product for a given surface and the real peel value for this surface.

It is well known from work with PSAs that the value of the peel force differs according to the chemical nature of the substrate and its polarity. In the case of protective films this phenomenon is more complex. Substrate surfaces that are the same chemically can have quite different degrees of roughness. Moreover, although application (laminating) is carried out under pressure to compensate for the lack of tack, it influences the peel value too (see Table 5.12). Table 5.20 lists the main parameters affecting the peel value of protective films.

Table 5.19 Peel Values of Some Common Protective Films

Adhesive	Carrier thickness (μm)	Peel on SS	Tolerance	Peel on carrier back side
Rubber resin	50	0.12	0.03	—
Acrylic	50	0.20	0.07	0.01
Acrylic	50	0.50	0.20	—
Acrylic	50	0.60	0.20	0.10
Rubber resin	50	0.60	0.20	—
Rubber resin	80	0.90	0.30	—
Acrylic	80	1.10	0.30	0.30
Acrylic	50	1.10	0.30	—
Rubber resin	80	1.10	0.30	0.10
Rubber resin	90	1.10	0.30	—
Rubber resin	90	1.10	0.30	0.30
Rubber resin	90	1.20	0.30	—
Acrylic	50	1.40	0.30	0.40
Rubber resin	110	1.40	0.40	0.50
Rubber resin	80	1.40	0.40	0.50
Acrylic*	70	2.50	0.50	1.70

Product characteristics — Peel resistance (N/10 mm)

The protective films listed above are coated with solvent-based PSA except for the last (*) position; 180° peel on stainless steel (SS) was measured.

As can be seen from Table 5.20, the substrate has a primary influence on the bond strength of protective films. Both the substrate and the adhesive age. Therefore, along with the chemical nature of the surface to be protected, the surface roughness, and the application conditions, the age of the surface and that of the adhesive layer (and protective film) can act as supplemental parameters. Surface aging has an especially high influence with physically treated or lacquered items (Table 5.21).

It is known from tests on PSAs that adhesives of the same chemical nature exhibit different levels of peel on surfaces that are chemically different. Generally there is an univoqual correspondence between the surface quality (nature) and peel value, e.g., the different adhesives will always give higher peel values on steel or glass than on polyethylene. Because of the strong influence of surface roughness, carrier deformability, and laminating conditions, for protective films this univoqual correlation is not always valid.

In the application of labels and tapes, the peel on standard polar and unpolar surfaces serves as a sure application guide, that is, peel values determine the

Adhesive Properties of PSPs

Table 5.20 Parameters Influencing the Peel Resistance of Protective Films

Bonding parameters			Debonding parameters		
Product parameter	Substrate parameter	Application conditions	Product parameter	Substrate parameter	Delaminating conditions
Adhesive characteristics	Chemical nature	Temperature	Adhesive characteristics	Substrate stiffness	Temperature
Coating weight	Roughness	Pressure	Carrier stiffness		Rate of delamination
Adhesives age	Physical treatment	Time	Carrier deformability		Laminate age
Carrier conformability	Cleanness Surface age		Carrier elasticity		

Table 5.21 Influence of Aging on the Peel Resistance of Protective Films

			Peel resistance (N/10 mm)		
Code	Coating weight (g/m^2)	Product age (mo)	Stainless steel	Lacquer 1	Lacquer 2
1	4.0	1	0.90	1.05	1.10
	4.0	13	1.20	1.38	1.47
	4.3	16	1.19	1.50	1.65
	4.5	17	1.33	1.47	1.53
2	7.0	1	1.50	1.60	1.76
	6.7	15	1.63	1.71	1.80
	7.2	17	2.16	1.99	2.00
	5.1	22	1.83	1.79	1.89

choice of product for a given application. As can be seen from Tables 5.22 and 5.23, protective films having the same peel value are recommended for quite different applications, according to the roughness and chemical nature of the surface to be protected (see also Chapter 8).

It is known that rubber resin PSAs (having a much lower T_g) are softer than acrylics. Therefore, such products are recommended for rough surfaces, where intimate contact between adhesive and substrate is difficult to achieve. In this case products having the same standard peel are suggested for different applications. On the other hand, certain metals catalyze the aging of natural rubber, and in such cases rubber resin PSAs cannot be used. The above examples illustrate that parameters other than peel also enter into the choice of a protective film for a given application. Peel resistance, lamination, and working conditions influ-

Table 5.22 Application Domains of Protective Films as a Function of Their Adhesivity[a]

Product characteristics		Adhesivity				
Type of adhesive	Carrier thickness (μm)	Very low	Low	Medium	High	Very high
Acrylic	50	—	3	4	10	—
Acrylic	70	1	—	—	13	—
Acrylic	80	—	—	—	11	—
Acrylic	100	—	2	—	9	—
Acrylic	130	—	—	—	—	—
Rubber resin	50	—	—	8	16	—
Rubber resin	60	—	—	6	—	—
Rubber resin	80	—	—	7	—	—
Rubber resin	100	—	—	5	12	15
Rubber resin	130	—	—	—	—	14

[a]The application domains are described in Table 5.23.

ence the performance and thus the choice of protective films. In such a situation the manufacturer of protective films should have a broad range of PSAs on different chemical bases that have the same standard peel value in order to fulfill the different application requirements.

The peel value can be regulated through formulation and coating technology. "Formulation" refers to the choice of appropriate raw materials and the working out of a recipe. Regulating peel through coating technology means making the choice of laminate components, coating weight, coating technology, and drying conditions.

Working out a recipe involves testing it in both uncrosslinked and crosslinked formulations. The latter is affected more by adhesive manufacturing technology, primer coating, and drying conditions. The choice of laminate components has to do with the manufacture of a coating with or without a top coat (primer). The use of a primer improves the anchorage of the adhesive on the carrier film and reduces the free flow (uncrosslinked) zone of the adhesive, i.e., improves its removability. As illustrated by Table 5.24, the use of a primer allows the coating weight to be increased (giving more conformability), avoiding adhesive transfer.

Using the same raw materials (but different balances of the components), the same peel level can be achieved with a recipe with or without primer. However, generally for higher coating weights, primer-containing coatings are manufactured. As a common practice the adhesive and the primer may have the same formulation components, the primer being more crosslinked. The use of a

Table 5.23 Application Domains of Protective Films

Code	Application domain	Comments
1	Glossy plastics; moderate processing conditions	Glossy plastic (PMMA, PC, polystyrene) plates; easy peel
2	High glossy metal surfaces	Stainless steel, aluminum, copper, and copper alloys with high gloss
	Glass, moderate processing conditions	Glass surfaces
3	Polished metal surfaces	Stainless steel, aluminum, copper and copper alloys, polished
	Glass, moderate processing conditions	Glass surfaces
4	Lacquered surfaces with high gloss	Lacquered (polyester, PVDF, AC) surfaces with a gloss higher than 60%
	Glossy metal surfaces	Stainless steel, aluminum, copper and copper alloys, glossy
	Processing, transport, storage	
5	Glossy metal surfaces	Stainless steel, anodized aluminum
	Glossy PVC	Glossy PVC
	Processing, transport, storage	
6	Glossy metal surfaces	Stainless steel, anodized aluminum, glossy
	PVC	PVC
7	Glossy and semimatte metal surfaces	Stainless steel, glossy and semimatte; anodized aluminum, semimatte
	Matte plastics	Plastics (PMMA, PVC, polystyrene)
	Processing	
8	Structured rough surfaces	Structured rough metal surfaces
	Processing of profiles	Processing of profiles and thick coils
9	Lacquered surfaces	Lacquered (polyester, PVDF, AC) surfaces
	PVC profiles	PVC profiles and plates
	Matte structured laminates	Matte structured plastic laminates
	Processing	
10	Brushed, glossy metal surfaces	Stainless steel, aluminum, copper and copper alloys, brushed, with a gloss of 20–60%
	Lacquered surfaces	Lacquered (polyester, PVDF, AC) surfaces
	Matte laminates, moderate processing	Matte composite laminates
11	Brushed and matte metal surfaces	Stainless steel and aluminum, brushed; copper and copper alloys, matte
	Lacquered surfaces with medium gloss	Lacquered (polyester, PVDF, AC) surfaces with 20–60% gloss
	Matte laminates	Matte composite laminates
	Processing, transport, storage	
12	Brushed and matte metal surfaces	Stainless steel, burst; anodized aluminum, matte
	Matte structured PVC	Matte structured PVC plates and items
	Processing, transport, storage	
13	Lacquered matte surfaces	Lacquered matte (polyester, PVDF, AC) surfaces
	PVC profiles	PVC profiles and plates
	Matte structured laminates	Matte structured composite laminates
14	Thick coils	Thick metal coils
	Heavy building elements	Heavy building elements of stainless steel and aluminum
	Powder-coated surfaces	Epoxy and polyester powder-coated surfaces
	Difficult, complex processing	
15	Structured rough surfaces	Structured, rough stainless steel and aluminum surfaces
	Powder-coated surfaces	Epoxy and polyester powder-coated surfaces
	Processing	
16	Structured rough surfaces	Structured, rough stainless steel and aluminum surfaces
	Difficult, complex processing	

crosslinked contact cement as a primer and in the adhesive enhances adhesion, which in turn improves peelability [55]. For instance, the primer increases the bonding on the carrier from 0.5 to 2.4 kg [136].

As discussed earlier (see Fig. 5.4), the peel force increases as coating weight increases. That means that (at least theoretically) for the same adhesive formulation the same coating weight gives the same peel force value. This statement is generally valid for adhesives coated with different coating techniques in the domain above the critical coating weight (see Fig. 5.5). However, in industrial practice the situation is more complex because of the use of different coating devices (see Chapter 6) for very low coating weights of crosslinked adhesives. As can be seen from Table 5.25, different coating techniques can, for the same nominal coating weight but different practical coating weight values, lead to the same peel value as a result of differences in the geometry of the adhesive layer. In conclusion, unlike "normal" coating practice (average coating weight values of 15–20 g/m^2 of an uncrosslinked adhesive), where for a given adhesive there is a clear relationship between coating weight and peel value, the influence of the

Table 5.24 Coating Weight for Primed and Unprimed Protective Films

Product code	Adhesive coating weight (g/m^2)			
	Primed		Unprimed	
	Theoretical	Practical	Theoretical	Practical
1	4.00	4.60	3.70	3.75
2	—	—	4.50	4.28
3	4.5	6.60	4.50	3.97
4	4.5	4.30	4.50	4.20
5	4.5	4.80	4.50	4.15
6	5.5	4.96	4.50	4.10
7	5.5	7.30	4.50	3.95

Table 5.25 Influence of Coating Device on Adhesive Properties[a]

Coating device	Coating weight (g/m^2)		Peel resistance (180°; N/10 mm)
	Nominal	Measured	
Gravure cylinder, line 30	2.0	2.2–2.4	1.5
Gravure cylinder, line 40	2.0	2.2–2.4	1.5
Meyerbar	2.0	1.6–1.8	1.5

[a]Peel on stainless steel was measured.

coating device should also be taken into account in the manufacture of protective films.

2 REGULATING THE ADHESIVE PROPERTIES

Because of the various end use requirements for PSPs there is a need to regulate their adhesive characteristics. The main factors influencing the adhesion are

1. The properties of the adhesive
2. The properties of the carrier (face stock, liner) material
3. The construction and geometry of the laminate
4. The coating weight
5. The coating technology
6. The age of the laminate

It is evident that the relative importance of these parameters depends on the manufacturing technology used for the PSP, whether it is produced with or without adhesive, and the complexity of its buildup. The adhesive properties of a classic PSA-coated pressure-sensitive product are determined mainly by the adhesive properties of the PSA. Unfortunately, such characteristics cannot be defined independently from the composite structure of the product. As described in Ref. 1, there are numerous possible ways to regulate the adhesive properties by means of the chemical composition of the adhesive, the choice of carrier, and the manufacturing method. This chapter discusses the special features of the control of adhesive performance characteristics related to the main PSPs.

2.1 Regulating the Adhesive Properties with the Adhesive

The adhesive properties of a PSP depend on its components, its buildup, and the manufacturing method. Of the product components, the pressure-sensitive adhesive has the major influence on the adhesive properties of the product. Its composition and geometry are the most important parameters.

Regulating the Adhesive Properties of a PSP with the Chemical Composition of the Adhesive

The common way to regulate the adhesive performance characteristics of a PSP is to change its chemical composition. Although changes in the chemical formulation made to improve other than adhesive properties can influence the adhesive properties too (see later), formulation for the purpose of regulating adhesive properties is the most important modality used to tailor them. It should be mentioned that the manufacturer of the PSP has a limited influence on the chemical composition of the adhesive. As discussed in Chapter 4, except for the new (mainly radiation-based) technologies, the adhesive is synthesized by special chemical firms. Formulating offers more freedom to tailor the adhesive properties, although in some cases the number of available formulation additives is limited. For a classic elastomer-based PSA, formulation with a viscous component was at first the only way to achieve viscoelastic properties and the basic pressure-sensitive characteristics. Later, with the synthesis of viscoelastic raw materials, the need for such a formulation was eliminated. A PSA based on viscoelastic raw material is formulated to modify the adhesion–cohesion balance. For PSPs displaying permanent adhesivity, regulation of the adhesion–cohesion balance means attaining maximum peel and tack without the loss of (too much) cohesion. For removable PSPs, the adhesive properties are characterized by a (time-dependent or virtually time-independent) low level of adhesion. For such products formulation has to reduce the peel force without changing the bond break character, i.e., the joint should fail as adhesions break at a well-defined place (adherend surface). Taking into account the dependence of tack on peel and that of the peel on cohesion, it is difficult to formulate removable PSAs with a high tack or medium tack and high cohesion. Reduction of the peel force reduces the tack and the cohesion as well. As discussed earlier [see correlation (5.11)], the peel force is used for bond deformation and bond break, i.e., adhesive (and carrier) flow and failure of the adhesive–adherend contact. Adhesive flow is facilitated by using a high level of viscous components. Bond failure is enhanced by low cohesion and contact hindrance, i.e., soft or hard crosslinked adhesive formulations. That means that certain removable formulations are not tacky enough or their low cohesion can lead to cohesion failure and the deposition of adhesive on the adherend surface. Therefore adhesion and cohesion are generally well-balanced only for permanent labels. The most important special requirements, e.g., removability and adhesion on nonpolar surfaces, require modifying this balance to reduce the peel level (removability) or to improve the anchorage (PSA for nonpolar surfaces). Although having quite different formulations, label PSA recipes provide adhesion–cohesion balance for both classes of products, which ensures common conversion and application properties.

For tapes the adhesion–cohesion balance is shifted to higher cohesion values. However, because of their higher coating weight, such products can be characterized by using the common evaluation criteria, i.e., tack, peel, and shear resistance. In a quite different manner, protective films do not possess the usual adhesion–cohesion balance. The situation is more complex for adhesiveless products (see also Chapter 6).

Tackifying. The regulation of tack and peel by the use of viscous components (resins and plasticizers) is the main formulation method. It is used mostly for labels and tapes. In the case of protective films, rubber-based formulations are tackified only. Tackifying, i.e., modifying a raw material to achieve higher tack and peel or (rarely) better shear resistance, is a controversial technology where technical development and commercial considerations (base elastomer and tackifier price fluctuations) may impose global changes. As discussed in detail in Ref. 1, although basic assumptions have been accepted concerning the mechanism of tackification of elastomers and the characterization of formulation variants according to adhesive properties and rheological parameters (DMA) has made important advances, only a few experimental data have been published, and in many cases the results serve only as commercial arguments for the raw materials suppliers. The data given by different authors for an optimum tackifier level are astonishingly different. For instance, according to Pierson and Wilczynski [25], a special crosslinkable water-based carboxylated acrylic has been developed. A room temperature shear test at 8.8 psi (0.5 × 0.5 in., 1000 g) and an elevated temperature shear test at 4.4 psi (0.5 × 0.5 in., 500 g; 200 and 300°F) have been used for its characterization. Shear times of 83–1000 h at room temperature and a SAFT value of 240°C have been obtained, which, according to the authors, compares with the shear performance of high quality commercial solvent-based acrylics. Unfortunately, these values were measured without tackifier and the corresponding peel and tack values (of 1.1 lb/in., 180° peel, PSTC-1 on PET and quick stick of 0.7 lb/in.) were very low. To improve these values, tackifiers were used at a loading of 12–35% dry tackifier per total solids. The goal of tackifying was to increase peel and tack by 25%. At a 35% tackifier level, 72°F, and 4.4 psi, a shear value of 48 h was obtained. According to Ref. 23, higher softening point resins are less miscible with the base polymer. The highest resin level that contributes to the peel is about 50%, but for a high softening point resin this level is no more than 39%. The best overall balance is obtained with high softening point resins at a level of about 30%. This statement was confirmed by Green [26] for CSBR also. A higher softening point resin gives optimum tack at a lower loading. Natural rubber latex used at level of 10% or less improves the shear resistance of CSBR. It improves the aging characteristics also. The SBR latex crosslinks and hardens during oxidative aging, whereas the NR latex undergoes chain scission and softens; thus a balance is obtained. Bonneau and Baumassy [28] state that a 25–35% resin loading is optimum for a standard acrylic latex. Dunckley [24] suggests that a cohesive acrylic latex (Acronal 80 D) needs 60–70 phr tackifier for paper tear; Bamborough [32] says 80%. Such values are unrealistic. Dunckley [24] asserts (as discussed earlier by Benedek [137]) that carboxylated acrylic latex needs a higher tackifier level. He is speaking of soft loop tack of such tackified CSBR dispersions with very high tackifier loading and states that shear resistance increases with the resin loading. Surprisingly, the shear resistance of the studied formulations attains at 100 phr resin level only the shear value of HMPSAs (as is known, HMPSAs display low cohesion). Such behavior can be explained only by the use of special resins and experimental conditions. Generally, for CSBR, more than 50% resin loading is recommended [23]. It should be noted that the increase in shear resistance with the resin level was also demonstrated by Mancinelli [13], but in this case special block copolymers were used. Acrylic block copolymers having both soft and hard acrylic phases have been tackified. They possess a star-shaped radial structure. Peel values of 90–105 oz/in. have been found for such tackified acrylic block copolymers for hot melts. According to Ref. 13, optimal adhesive properties are characterized by a tack of more than 500 g (adhesive thickness of 25 μm), peel of more than 90 oz/in. (face stock polyester), and shear resistance of more than 500 min (½ × ½ in., 1 kg). Although shear resistance increases with tackifier loading (formulations with 75–80% resin were used), Mancinelli states that the crossover point from adhesive failure to adhesive delamination occurs in the 30–45 phr tackifier range. It should be mentioned that the higher tackifier level for such block copolymers that are harder (see Chapter 3, Section 1) is usual. Acrylic hot melt PSAs with thermally reversible crosslinking (like SBCs) are harder (Williams plasticity number 2.7) than the early common acrylics and exhibit better cold flow resistance [51]. The storage modulus of the acrylic block copolymer discussed by Mancinelli [13] has a plateau value of $10^{9.5}$ dyn/cm^2. The soft phase of the polymer has a T_g of −45°C; the hard phase displays a T_g of 105°C. It should be stressed that the tackifying of such multiphase polymers differs from the tackifying of formulations based on natural rubber or one-component viscoelastomers. Although in both cases the matrix–filler theory can be applied as a simplified explanation of the practical behavior of the tackified systems (see also Chapter 3, Section 1), the tackified block copolymer based system has multiple "filler components," the phase inversion of such systems produces different morphologies, and compatibility with the matrix and filler phase leads to quite different results, and except for some experimental resins the resin alone does not impart processibility. Natural rubber has a network structure, and in such a network the "hard phase" is not tackified. In segregated SBC it is. Therefore the tailoring of the adhesive properties of SBCs by tackifying is always associated with side effects that lead to a pronounced decrease in other parameter values. However, the tackifying of multiphase systems can lead to self-adhesive materials also.

Regulation of the adhesive properties of HMPSAs can be considered in some cases a by-product of viscosity control. According to Donker et al. [131], 80 Pa.s is the preferred maximum viscosity value for high speed coating. One possible way to reduce the processing viscosity of the HMPSAs based on SBC is to increase their processing temperature (upper limit 190°C) or decrease the melting temperature of the polystyrene domains. As discussed earlier (see Chapter 3, Section 1), the mobility of the midblock can be evaluated by means of dynamic mechanical analysis. It is related to the softness, viscous flow of the adhesive (i.e., energy loss modulus), the position of the rubbery plateau modulus,

Adhesive Properties of PSPs

and the ratio of the two moduli (storage and loss modulus), i.e., tan δ. The positions of tan δ related to temperature, the tan δ minimum temperature ($T_{\delta min}$), and the tan δ peak temperature ($T_{\delta max}$) characterize the rheological behavior. The melting of the polymer given by the disappearance of the endblock domains is indicated by $T_{\delta min}$ and the crossover temperature of the loss and storage moduli (T_{cross}). Between $T_{\delta min}$ and T_{cross} the end domains soften and disappear. The difference between $T_{\delta min}$ and T_{cross} gives information about the melt viscosity of the formulation. The contour lines run parallel to the melt viscosity and SAFT. There is a direct correlation between them. Most tackifiers used for HMPSAs are aliphatic and midblock-compatible and do not influence the styrenic domain. Oils act on the polystyrene domain (being compatible with these groups) and thus soften it. Unfortunately, they reduce cohesion. In practice a mixture of resins is used for HMPSAs formulated for tapes. One of the resins is midblock-compatible; the other is compatible with the styrene endblocks. Using a midblock- and endblock-compatible resin for SIS, the desired viscosity value can be obtained with a lower resin level (rubber/resin ratio of around 1/1). If resins are used that are partially compatible with the styrene domains (e.g., rosin esters, modified terpene resin, or a modified aliphatic resin), both an increase in midblock mobility and softening of the endblock are achieved. Such resins may have a higher molecular weight aromatically modified aliphatic hydrocarbon basis. In a standard HMPSA formulation with 100 parts SIS, 120 parts resin, and 20 parts oil, such resins give similar $T_{\delta max}$ and loss modulus values but a lower T_{cross}, i.e., a lower softening temperature for the endblock domains. As discussed earlier for packaging tapes, a peel value of 18 N/25 mm (on steel), loop tack values of 20–30 N/25 mm, and a rolling ball value below 5 cm are preferred. Such characteristics are achieved if the resin fraction does not exceed 0.53. An optimum value (100 min) for the flap test is measured if the oil fraction does not exceed 15 phr. Shear adhesion to cardboard at an elevated temperature (40°C) is tested for packaging tapes also. It is to be noted that although Donker et al. [131] stress the decrease in shear achieved by the formulating oil in common formulations, they state that shear adhesion to cardboard for a HMPSA formulated with a midblock- and endblock-compatible resin is not influenced significantly by the oil level. Midblock- and endblock-compatible resins allow formulation with less resin and oil. A formulation with 105 phr of resin and 12.5 phr of oil is suggested as an optimum. For an optimal PSA packaging tape formulation, $T_{\delta max}$ values of −6 to 0°C and loss modulus values of 72 and 92 kPa at the tan δ valley are measured. As a less expensive alternative to the use of SIS block copolymers for tapes, SBSs may be formulated. They are less suitable because of their greater rigidity (higher modulus) and their lower compatibility with tackifier. Aging problems make them inadequate also. The branched isoprene midblock is more flexible and more compatible than the butadiene (butene) one. In some cases the viscosity of SBS-based HMPSA formulations may change during storage. Jacob [130] states that the SBS formulations are, "at comparable resin loading," significantly softer and that the peel adhesion values are lower. The theories about tackifier practice and mechanisms discussed in Refs. 130 and 131 are quite different. Donker et al. [131] state the need for special tackifier resins acting on both (mid- and endblock) segments. Jacob [130] shows that most SIS tackifiers are compatible with both the aliphatic middle block and the styrene domains and thus decrease the cohesion. According to Jacob [130], to achieve the softness of polyisoprene segments without influencing the endblock polystyrene segments in SBS, it is necessary to use special tackifiers that are compatible with only the midblock and do not affect the styrene endblocks. A special hydrocarbon resin with a molecular weight of 700 (M_n) and a T_g of 42°C (ring and ball softening point of 85°C) has been developed. According to Ref. 130, increasing the oil level in an SBS formulation increases the peel and tack values but does not greatly decrease the SAFT value. According to Ref. 131, for SIS-based adhesives shear resistance is given by the block copolymer characteristics and the oil/resin ratio does not influence it. According to Jacob [130], for a tackifier resin that acts on the midblock of the SBS block copolymer, oil also influences the rigidity (modulus) of this midblock. Therefore, there is a need for oil. Oil-free or plasticizer-free SBSs have not been used commercially. Donker et al. [131] state that for SIS the oil does not affect the midblock. Standard formulation with 200 parts SBS, 125–150 parts resin, and 25–50 parts oil has been used. As is known, the plateau of the storage modulus gives an indication about the rigidity of the formulation. A big difference in the value of the storage modulus (2×10 Pa) was found by Jacob [130] for a formulation with 100 parts resin without oil and one with 175 parts resin and 50 parts oil. The formulation with oil shows a higher modulus value (4×10 Pa) although both formulations have the same T_g. Both formulations give adequate room temperature tack. Tests have been carried out with different stress frequencies at different temperatures (-50 to 160°C). An index of the energy absorption is the value of the loss modulus. High loss modulus (peak) at the practical frequency of bonding facilitates bonding. On the other hand, at the highest debonding frequency, the adhesive (storage modulus) must exhibit elasticity. The Dahlquist criterion states that the modulus value (plateau) should be lower than 3.3×10^6 Pa. With faster unwinding of tapes (increased frequency), the noise of the winding increases because of the increasing peel adhesion at higher frequencies. The tape tends to break more often. According to Krecenski et al. [76], a direct correspondence exists between the loop tack, peel, and loss modulus. The slope of the curve allows predictions of the peel and tack values depending on the formulation. A PSA formulation for double-faced carpet tape with 220 phr resin and 20 phr oil to 100 phr block (SBS) copolymer (although with a high coating weight of 40 g/m^2) displays relatively low adhesive characteristics (a peel strength of 16.7–21.2 N/25 mm), loop tack of 17–18 N/25 mm, and rolling ball tack of 3–5 cm. New midblock-reactive hydrocarbon resins improve the shear values.

According to Ref. 19, phase separation in the radial block copolymer is higher due to the increased organization of the molecule. Therefore physical crosslinking is higher also. The better connected pure polystyrene domains ensure higher elasticity of the product. The radial copolymer shows a longer linear portion of the rubbery plateau and lower values of tan δ. As is known, the loss tangent versus temperature plot characterizes cohesive strength. The lower the tan δ value, the greater the cohesive strength. Tack needs a low modulus for bonding but a high modulus at the strain rates and elongations that exist during bond breaking. Admitting that the storage modulus value at low frequencies (less than 10^{-2} rad/s) is related to wetting and creep (cold flow) while the value at high frequencies may be related to peel or quick stick properties (i.e., high speed debonding by tack/peel evaluation), both values are necessary for adhesive characterization. A parallelism (proportionality) has been found between the G' ratios at different frequencies and quick stick values (e.g., a $G'_{100/0.1}$ value of 60.9 corresponds to a quick stick value of 24.1, and a $G'_{100/0.1}$ of 5.5 gives a quick stick value of 15.3 N/15 mm). Increasing storage modulus ratios leads to higher quick stick values. In a similar manner, decreasing tan δ ratios (at different frequencies) increases the quick stick values. It is evident that the high elasticity of a radial polymer displayed in mechanical properties affects the adhesive properties, reducing cold flow (wettability) and tack. Cohesion is better for this polymer. On the other hand, on nonpolar surfaces where physical contact between PSA and surface is more important, peel is strongly affected (30% difference).

Softening of the Adhesive. As discussed earlier, removability can be achieved by softening the polymer with plasticizer and/or other viscous components. This possibility is used in Ref. 138 for acrylic, with paraffinic oil and polyisobutylene added to ensure repositionability. Internal crosslinking by special monomers (like *N*-methylolacrylamide) and the simultaneous use of plasticizers have also been suggested [139]. According to Ref. 21, removable price labels almost exclusively incorporate a paper carrier coated with rubber resin adhesive. Natural and synthetic rubber and polyisobutylene tackified with soft resins and plasticizer have been suggested. The rubber has been calendered and masticated. For these formulations the adhesion buildup on soft PVC and varnished substrates has been excessive after long storage times. No clear adhesion failure has been possible. Legging and migration is characteristic of such formulations. For certain removable (not pure PSA) adhesives, legging is required. As shown by Hintz [140], so-called combination adhesives made as blends of PSAs and non-pressure-sensitive EVAc dispersions (70/30) are used as one-side adhesives. They can be applied as flooring adhesives after a drying time of about 20 min. So called legging adhesives are prepared also. Here the addition of a resin allows regulation of the legging. Flooring adhesives of this type can be removed with or without water. For systems that can be removed without water, an EVAc-based primer is applied on the substrate to allow removability. Removable HMPSAs can contain up to 25% by weight of a plasticizing or extending oil to enable control of wetting and viscosity [14]. Butyl rubber (15–55 parts) untackified or tackified with PIB (7.5–40 parts) and crosslinked with zinc oxide (45 parts) is suggested for removable formulations [83]. Such formulations give (180°) peel values of 0.73–1.81 kg/25.4 mm. Their peel increases as the tackifier level increases, but at a tackifier/elastomer ratio of 40/15 legging appears. As shown in Table 5.26, the value of peel resistance required differs for different classes of removable PSPs. Therefore components for both softening and hardening the adhesive are used.

Crosslinking of the Adhesive. The chemistry and technology of crosslinking have developed very rapidly in the domain of PSAs also. Some years ago a monograph [141] stated that "the introduction of crosslinking increases cohesive strength . . . but reduces tack and solubility . . . and thus finds only limited use in pressure sensitive type adhesives." Actually it would not be possible to manufacture special tapes or protective films without crosslinking. A decrease in tack and increase in cohesion are the main consequences of crosslinking.

Some raw materials possess a built-in crosslinkable structure that works physically or chemically (see also Chapter 4). The best known is natural rubber, but synthetic elastomers possess crosslinked structures also (e.g., SBR or CSBR). It should be noted that crosslinking resulting from the synthesis of several elastomers influences their adhesive and processing performance. As discussed by Burroway and Feeney [142], the gel content of the SBR determines the processibility of the adhesive formulation. On the other hand, for the same tape formulation (100 parts rubber to 154.5 parts resin) it can change the value of aged tack between 1 and 3 cm, that of the peel between 48 and 53 g/cm, and that of the shear resistance between 10 and 93 min.

Table 5.26 Adhesive Characteristics of the Major Removable PSPs

Product class	Tack (RB, cm)	Peel resistance (180°, N/25 mm)	Shear resistance (HS, h)
Label	2.5–10	0.5–2.5	1–50
Tape	1.5–20	0.5–5.0	1–100
Protective film	10–70	0.05–1.5	10–200
Form	5–10	0.05–1.5	1–20
Separation film	>100	0.02–1.0	10–200

[a] Adhesive characteristics

[a] RB = rolling ball; HS = hot shear.

Physical crosslinking occurs in specially sequenced copolymers (mainly SBCs or other block copolymers). As discussed in Chapter 4, acrylic hot melt PSAs with thermally reversible crosslinking (such as SBC) have been developed [51]. Such formulations are harder than the early common acrylics and show better cold flow resistance. This type of crosslinking disappears at elevated temperatures, allowing the use of the formulation as a molten material. As discussed in Chapter 4, the cohesion given by physical crosslinking and its pronounced temperature dependence do not fulfill special end use requirements. In such cases a supplemental network is built up by using chemically or physically initiated (radiation-induced) crosslinking.

Special built-in polar monomers (e.g., nitriles, hydroxy derivatives) can strengthen the polymer also. Chemical crosslinking supposes the presence of unsaturation or of physically and/or chemically activatable multifunctional reactive groups in the polymer. Chemical crosslinking can be initiated by using external or built-in initiators or crosslinking agents. The use of external crosslinking agents is common for solvent-based acrylics, the synthesis of which presents opportunities for building in reactive groups, and the presence of a homogeneous reaction medium in the finished product (dissolved polymer solvent) without the interference of technological additives (surfactants, water, etc.) facilitates the easy regulation of postmodification reactions. Generally polymeric isocyanates (0.05–1.5%) are suggested as crosslinkers in order to achieve reproducible peel values [36]. As shown in Fig. 5.13, the use of crosslinking agents drastically (70–90%) reduces peel resistance.

Such pronounced peel reduction is acceptable for protective or separation films only, where the unbalanced adhesive characteristics are improved by film conformability and lamination under pressure. Therefore the theoretical possibilities for regulating the peel resistance are limited and vary in practical importance for each product class (Tables 5.27 and 5.28). Because of the predominance of paper as the face stock material for labels, adhesive anchorage is better; therefore in this case the use of a primer has a secondary role. Modification of the adhesive geometry is limited also because of the high requirements for smoothness of the PSA layer (e.g., "no-label look" products). As mentioned above a high degree of crosslinking is not possible for labels because of the required adhesion balance. Common tapes have a limited raw material basis; therefore in this case the PSA geometry, coating weight, and crosslinking (to achieve high shear and peel resistance) are more important. The adhesive properties of common protective films are controlled by varying coating weight and degree of crosslinking.

The composition of the PSA polymer according to a Nichiban patent [92], although applied as a patterned contact surface, includes a crosslinking agent that is used for both the matrix polymer and the particles, and the matrix covers the surface of the particles also. Mueller and Türk [21] proposed a water-based acrylic formulation having physically or chemically crosslinkable

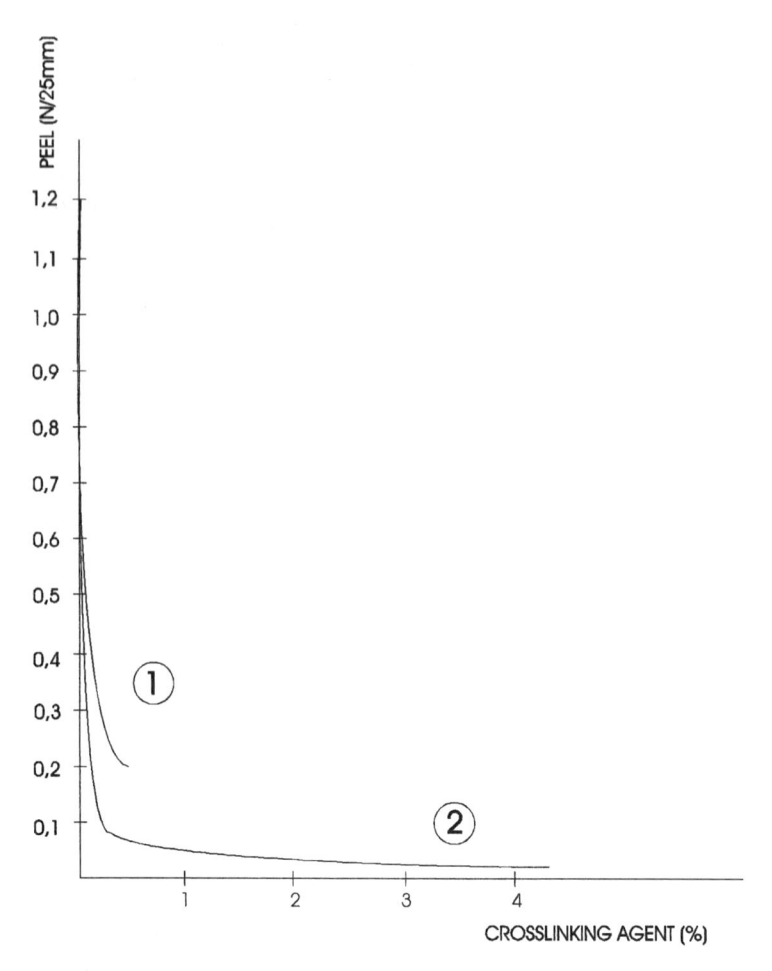

Figure 5.13 The influence of crosslinking on peel. (1) Aziridine crosslinking agent; (2) polyisocyanate crosslinking agent.

comonomers like acrylo- or methacrylonitrile, acrylic or methacrylic acid, N-methylolacrylamide, and plasticizer (10–30% w/w). Special problems appear in the manufacture of AC hot melts with respect to achieving enough cohesion and a low processing viscosity. Two approaches are known (see also Chapter 4). The first is to imitate the buildup of SBCs, i.e., to prepare physically crosslinkable segmented copolymers. The second approach is to synthesize low molecular weight linear random copolymers. In this case postpolymerization is needed to increase the molecular weight of the coated product. As stated by Auchter et al.

Adhesive Properties of PSPs

Table 5.27 The Use of Crosslinking for Regulation of Adhesive Characteristics

PSP	Type of adhesive	Nature of crosslinking	Characteristic improved
Protective film	Rubber resin	Chemical, using supplemental crosslinking agents	Removability; anchorage on the carrier
	AC	Chemical, using supplemental crosslinking agents	Removability; anchorage on the carrier
	AC	Chemical, self-curing	Removability
Tape	Rubber resin	Chemical, using supplemental crosslinking agents	Shear resistance; removability; peel resistance
	AC	Chemical, using supplemental crosslinking agents	Shear resistance; dimensional stability; removability; peel resistance
	AC	Chemical, self-curing	Peel resistance
	TPE resin	Radiation	Shear resistance; temperature resistance
		Radiation, using supplemental crosslinking agents	Shear resistance; dimensional stability; temperature resistance
	AC (oligomer) resin	Chemical, using supplemental crosslinking agents	Cohesion
		Radiation	Cohesion
Label	AC (oligomer) resin	Radiation	Cohesion
	Rubber resin	Chemical, using supplemental crosslinking agents	Removability
	AC	Chemical, using supplemental crosslinking agents	Removability
	AC	Chemical, self-curing	Converting properties

Table 5.28 The Importance of Specified Regulation Parameters of Peel Resistance for the Main PSPs[a]

	Control parameter of peel resistance				
Product class	Adhesive	Coating weight	Crosslinking	Primer	PSA geometry
Label	1	2	3	3	3
Tape	3	1	2	2	2
Protective film	3	1	1	2	2

[a]1, 2, and 3 denote the importance of the peel regulating parameter for a given PSP class, with 1 denoting the lowest importance.

[48], even a high molecular weight, highly viscous 100% polyacrylate processible as a hot melt is not of high enough molecular weight to display balanced pressure sensitivity. Therefore it should be crosslinked. The crosslinking possibilities of hot melts are limited, because thermal crosslinking cannot be used. Radiation curing is one of the new modalities used to crosslink such compositions. It is possible to synthesize a prepolymer that is polymerized by radiation. A patent [143] discloses the preparation of a low molecular weight spreadable composition to which may be added a small amount of catalyst or polyfunctional crosslinking monomer prior to the completion of the polymerization by heat curing. Unsaturated or saturated polymer chains are synthesized that can be polymerized by radiation [144]. A monoethylenically unsaturated aromatic ketone can be used for crosslinking PSA compositions by UV radiation [145,146]. A special acrylic raw material for AC HMPSAs has been developed that can be cured photochemically [48]. The photoinitiator is built into the polymer. In this case the uncrosslinked product does not have balanced adhesive properties. Because of its low molecular weight, it is very tacky, but it does not have sufficient cohesion. Balanced pressure-sensitive properties are achieved only if it is crosslinked. The polymer itself before crosslinking is a highly viscous fluid (at room temperature its viscosity is about 10–20 Pa.s) that can be processed at 120–140°C. The nonirradiated polymer has a storage modulus of about 10^3 Pa, which can be increased to 10^4 Pa by irradiation. In comparison, acrylic block copolymers with soft and hard acrylic phases and a star-shaped radial structure exhibit a storage modulus with a plateau value of $10^{9.5}$ dyn/cm^2 [13]. Their viscosity depends on the shear rate but is much higher. At 10 s^{-1} (177°C) a value of 248,000 cP and at 100 s^{-1} a value of 57,000 cP have been obtained. Melt viscosity of tackified compositions is less than 50,000 cP at 177°C.

For such polymers the storage modulus can be used as a measure of the crosslinking density. The copolymerized photoinitiator is an acrylic ester with benzophenone terminal groups [48]. As discussed in Chapter 4, the main crosslinking reaction is the addition of a hydrogen atom of the side alkyl group of an acrylic segment to the carbonyl group of the benzophenone side group of the vicinally polymer chain, i.e., the formation of a hydroxydiphenylpolyacrylomethane derivative. Although no unsaturation is present in the macromolecule, i.e., the system is inert to classic aging reactions, other intramolecular photoinitiated side reactions are possible due to the built-in photoinitiator. To avoid such side reactions, low radiation levels should be used. The use of low radiation levels supposes the use of a low concentration of formulating components that absorb UV light, i.e., the formulating freedom of such systems is limited by the classic tack–peel–shear interdependence and by the manufacturing procedure. For modifying UV-crosslinkable polyacrylates, esterified highly stabilized rosin is suggested. A low level (8.5%) significantly improves the peel and tack but reduces shear resistance. For the untackified adhesive with a high coating weight of 50 g/m^2, an instantaneous peel resistance of only 7.10 N/25 mm is achieved, i.e., such a polymer must be tackified. The shear value is about 7.3 h (23°C, 2 kg, 5 cm^2). The tackified formulation gives 10.5 N/25 mm as peel value. In comparison, tackified acrylic block copolymers (with much higher tackifier loading) show peel values of 25–29 N/in. [13]. Optimal properties require a tack of more than 500 g (adhesive thickness of 25 μm), peel of more than 90 oz/in. (face stock polyester), and shear values of more than 500 min (½ × ½ in., 1 kg). The SAFT value attains 80–86°C. Formulations with 75–80% resin were used. It should be noted that for radial SBCs, G' values at different strain frequencies give quick stick values in the following range: A $G'_{100/0.1}$ of 60.9 corresponds to a quick stick value of 24.1, and a $G'_{100/0.1}$ of 5.5 corresponds to a quick stick value of 15.3 [18]. For UV-curable formulations, according to Ref. 48, there is an upper limit to the amount of resin that can be added. It is supposed that the resin acts as a diluting agent in the polymer matrix, reducing the number or contact (crosslink) points. Therefore, for tackified formulations the radiation dosage must be increased. For instance, an 8.5% resin level requires doubling the radiation dosage! Polymers with photoinitiator properties have been developed for different coatings [147]. The photoinitiating polymer could be used as a base coat that could then catalyze the curing of a UV-hardenable top coat. Thus coatings with controlled thickness and very low molecular weight can be applied. Compounds containing aromatic or aliphatic ketone groups and silicone are bonded to alkylene side chains of oligomers to prepare polymeric photoinitiators [148]. UV-curable acrylic esters and polyvinylbenzophenone as initiator have been used as adhesives on plasticized (35%) PVC [149]. These acrylic oligomers can be cured by electron beam irradiation also [150]. In such crosslinked systems the flexibility of the crosslinking "bridges," i.e., the nature of the oligomers used, influences the adhesive properties. For instance, as stated in Ref. 151, with the use of polyisobornyl methacrylate instead of isooctyl acrylate the peel is increased from 53 to 88 N/100 mm. Peel buildup also depends on the oligomer used.

Styrene-butadiene rubbers (random or alternating copolymers) contain a residual double bond that enables further polymer chain growth as well as crosslinking to form three-dimensional network structures with improved mechanical properties without apparently affecting the T_g [25]. For such polymers, internal networks (gel) can be formed during polymerization. Gel content influences the adhesive properties and the ways of regulating them. Higher gel content is an indication of higher molecular weight. Above 30% gel content, CSBR tack decreases rapidly. Shear resistance increases with gel content. Pierson and Wilczynski [25] demonstrate that although the butadiene content decreased from 69.0% to 62.5% the increase in gel content from 37.0% to 45.0% reduces the loop tack from 100 to 800 gr (FTM). Deep freeze properties of such formulations are improved if the butadiene level of the polymer is increased above 80%. For such raw materials the built-in crosslinking density (gel) has to be taken into account by regulation of the adhesive properties. It is well known that supplemental crosslinking of rubber-based adhesives builds up new "gel structure." In certain cases the primer used for these adhesives has the same chemical composition as the adhesive; only its curing agent level and degree of crosslinking are different.

It can be concluded that except for permanent labels with balanced adhesive properties for a medium coating weight, advanced crosslinking is a current technology for manufacturing PSPs (Table 5.29).

Table 5.29 Crosslinked PSPs[a]

Product	Adhesive	Curing agent[b]	Uncrosslinked, peel	Crosslinked Peel	Crosslinked Tack	Crosslinked Shear	Ref.
Tape	AC	UV	440	52	—	—	149
Tape	AC	UV	1500	15	—	—	104
Tape	AC	BA	—	1000	—	—	152
Label	AC	PI	—	120	—	—	153
Label	AC	ME	—	1200	—	—	106
Label	AC	UV	—	1050	17.8 cm	>24 h	105
Tape	AC	PI	—	630	18 cm	<1 mm	118
Tape	AC	PI	—	900	—	—	121
Tape	Silicone	BPO	—	1000	5.6 N/25 mm	N/25 mm	119
Tape	Silicone	BPO	—	3500	24 cm	2 mm	154

[a]180° peel was measured on stainless steel in grams per 25 mm.
[b]BA = bifunctional acrylate; PI = polyisocyanate; ME = methylol functionalized monomer.

Regulating the Adhesive Properties of PSPs with the Adhesive Structure or Geometry

As discussed earlier, adhesive properties depend on the area of the contact surface between adhesive and substrate. This is a function of the adhesive geometry too. Adhesive geometry is characterized by the continuous or discontinuous character of the coated layer (due to the coating technique or its failure) and the coating weight. An increase in coating weight improves tack and peel and decreases removability. Therefore regulating the coating weight can help to ensure a balanced adhesion–cohesion level for compositions with chemically enhanced cohesion (tapes with higher coating weight) or removable formulations. As is known from the formulation of protective films, the coating weight depends on the degree of crosslinking used for the adhesive. It also depends on whether a primer is used (see Table 5.21). Regulating the adhesive properties by controlling the coating weight can be achieved by knowing the interdependence of coating weight and adhesive performance for a given formulation.

For regulating adhesive performance by controlling the contact surface there are various possibilities. In such cases the quantitative and shape-related characterization of the effective contact surface between adhesive and substrate is absolutely necessary to evaluate the adhesive properties. The contact surface can be reduced by using patternlike contact points or patternlike nonadhesive points (see Fig. 2.13). The patent of Pasquali [11] suggests the use of an impregnated textile network in which the adhesive-coated fibers act as the contact surface and so the adhesive contact surface can be regulated by modifying the dimensions of the network. The PSA surface can have a special discontinuous shape ensured by the PSA itself as a suspension polymer [94]. Thus spherical contact sites are formed. Pressure-sensitive polymer particles with a diameter of 0.5–300 μm can be used also as the filler [77]. Such pressure-sensitive filler particles are called for in Ref. 72. Partial coating of the PSA leads to the same results [145,146]. Striped or droplike coating of PSAs has been carried out for removability [155]. The adhesion buildup can be decreased according to Ref. 156 by using a powderlike surface coating of the PSA layer with particles having a diameter of less than 10 μm embedded in the PSA, with two-thirds of their diameter above the PSA level. The particles are pressed into the mass [156]. Contact surface can be reduced by patternlike crosslinking also [157]. The same effect is obtained if the carrier is coated with alternate adhesive and release stripes, according to Ref. 158.

These procedures have the following disadvantages:

1. The formulation contains plasticizer (migration), thickener, and surface-active agents (migration, hydrophilicity).

2. The adhesion increases over time. This increase depends on the surface nature and the quality of the adherend. Therefore a full surface coating does not ensure removability for all types of surfaces.
3. If the adhesive is coated in a discontinuous manner, the adhesion increases over time on the contact sites, and so there may be high peel resistance in the longest contact direction. Coating the PSA with spherical PSA particles (50–150 μm) does not display this behavior.
4. Unfortunately, the anchorage of the spherical PSA particles needs a continuous matrix and a high coating weight. On porous surfaces this coating weight is difficult to control. A supplemental primer is necessary. The spherical particles have to be coated as a monolayer.
5. Removable coatings having alternate adhesive and abhesive stripes have to be achieved in two different coating steps but with the same coating thickness.

Another procedure without such disadvantages uses pyramidal PSA-coated sites (30–600 μm base dimensions) coated by means of screen printing [144]. Rotary screen printing can be used with a coating weight of 1–20 g/m^2, using a screen geometry with 15–40 holes/cm, blade of 1.5–30 mm, blade thickness of 150–300 μm, and pressure of 2–6 mmHg. Regulation of the contact surface can be achieved inversely also, i.e., by crosslinking the whole adhesive surface and making it sufficiently fragile to allow crack migration of the uncrosslinked adhesive [145].

The contact surface can be reduced by using expandable fillers also. According to Ref. 104, a photocrosslinked acrylic-urethane acrylate formulation used for holding semiconductor wafers during cutting and releasing them after cutting exhibits a peel strength of 1700 g/25 mm without crosslinking and 35 g/25 mm with crosslinking. Adding an expandable filler to the formulation (6 parts to 100 parts adhesive) reduces the peel resistance. This composition exhibits 1500 g/25 mm peel resistance before crosslinking and 15–29 g/25 mm after crosslinking. Such a releasable foaming adhesive sheet can be prepared by coating a blowing agent (e.g., a microcapsular type) containing PSA (30%) as a 30 μm dry layer [104]. This type of adhesive is removable and improves cuttability.

2.2 Regulating the Adhesive Properties with the Carrier

In the classic manufacturing technology for a PSP using a nonadhesive carrier material that is coated with a PSA, the adhesive characteristics of the product are influenced by the surface and bulk properties of the carrier material. As discussed in Chapter 2, unlike labels that have balanced adhesion–cohesion given by the PSA coating and where product design and manufacture minimize the influence of the solid-state laminate components on the adhesive properties, in the manufacture of certain tapes and protective films, the conformability of the carrier and its mechanical resistance are taken into account as parameters of the adhesive properties.

The problems are more complex for extruded pressure-sensitive tapes or protective films where the diffusion characteristics of the carrier material can influence the migration of the pressure-sensitive components, i.e., the adhesive properties of the final product. Mechanical stresses acting on the finished product (and depending on the product's own mechanical characteristics) can also influence adhesive diffusion. In some cases the manufacturing method used for the polymer film (blown or cast) and its crystallinity influence the migration of the pressure-sensitive component and the adhesive properties (see also Chapter 6). Similar problems may appear for special products having a (radiation) polymerized carrier material. According to Ref. 72, using UV photopolymerization of acrylics, the PSA can be synthesized as an adhesive layer supported by a nonadhesive layer (carrier or rigid body) or as a double-faced system. Such a carrier may be nonadhesive, although for certain applications the carrier itself has some pressure sensitivity. Thick, triple-layered adhesive tape with filler material in the center layer can also be manufactured. Such a product is really a carrierless tape and thus has better conformability and lower flexural resistance, both of which influence the adhesive properties. For instance, a foamed pressure-sensitive adhesive sheet prepared from an agitated acrylic emulsion according to Ref. 160 exhibits a higher peel strength (4.3 kg/cm) than the same adhesive coated on a PUR foam carrier material (1.5 kg/cm).

Influence of Carrier Material Surface Properties on Adhesive Properties

Coatability is characterized by wetting out and anchorage of the coated adhesive on the carrier material. Both are influenced by surface polarity and porosity. Special problems are encountered with nonpolar plastics where no chemical affinity exists between the polar PSA and the nonpolar, nonporous carrier surface. For textured fiberlike porous surfaces, adhesive migration and breakthrough may improve anchorage but denaturate the top surface of the face stock and at the same time change the coating weight. Fillers or surface-active agents built into the carrier material of protective films or tapes can act as release-enhancing agents, but they disturb the adhesive contact. Fillers can modify the relaxation spectra of the polymers and increase the relaxation time as a result of the retardation of relaxation in the vicinity of the filler surface [161].

Influence of Carrier Material Bulk Properties on Adhesive Properties

Paper-based carrier materials are humidity-sensitive. Their mechanical characteristics and isotropy may change as a function of the climate. Plastics are more temperature-sensitive. Both types of carrier materials can change dimensionally and in mechanical resistance as a function of environmental conditions. As is known, the mechanical properties of plastics depend on the test/force rate also (see also Chapter 3). Therefore deformation of the carrier material during application (or test) may influence the adhesive properties also.

It can be seen that for PSPs coated with PSA a decrease in carrier film thickness increases their deformability and reduces their peel resistance. In practical use such films display sufficient adhesion to be processed together with the protected product. Difficulties appear during their conversion and application owing to their elongation by winding or laminating forces. In practice, the actual values of delamination force have to be taken into account. These result from the deformation of both the adhesive and the carrier. Laboratory tests using a standard carrier material or a reinforced carrier material do not allow the correct evaluation of practical adhesion values. For adhesive-coated PSPs the bulk properties of the carrier have to be taken into account only for thin or soft plastic films. For pressure-sensitive webs manufactured by coextrusion (e.g., EVAc-based or hot laminating films) or for carrierless transfer tapes manufactured by stiffening of the adhesive (using filling, crosslinking, foaming, etc.), the exact separation of adhesive and carrier components is more difficult, sometimes impossible, and therefore it is not possible to use a standard carrier for laboratory measurements. That means that tests must be carried out with the finished composite. The product has to be evaluated as a whole. The regulation of the adhesive properties by means of the carrier are discussed in detail in Chapter 6.

2.3 Regulating Adhesive Properties with Manufacturing Technology

The manufacturing technology used for PSPs also influences their adhesive properties. Adhesive manufacture and coating technology affect adhesive performance, carrier performance, and product buildup. As discussed above, the carrier manufacture influences its mechanical, chemical, and surface properties. In special cases adhesive manufacture may be the manufacture of the finished product (carrierless tapes, adhesive crayons, etc.). In certain cases the carrier manufacture may be the production of the finished PSP (e.g., SAF). Generally the manufacture of PSPs include both adhesive and carrier manufacture and their assembly into a finished product.

Regulating Adhesive Properties with Adhesive Manufacturing Technology

For PSPs having a carrier material with self-adhesive properties, the method of mixing the components influences the adhesive properties. This is evident in the use of tackifying raw materials like polybutenes, where the mixing technology (solid-state or molten, etc.) influences the adhesive properties of the film (see also Chapter 6). Water-based dispersions can be formulated with different levels of supplemental (formulation-related) surfactants. Molten or solvent-based additives can be used. Special techniques allow the use of water-based components without supplemental surfactant loading [50]. Hot melts can be manufactured via batch processing or continuously. The thermal oxidative influences on the base polymers are different for such procedures. Solvent-based rubber resin adhesives can be manufactured with masticated or unmasticated rubber. The degree of mastication can vary according to the equipment used. The tackifying components can be mixed in solution, molten, or solid-state form. As illustrated by the above examples, the various steps in adhesive manufacture influence the adhesive characteristics of the product (see also Chapter 6).

Regulating Adhesive Properties with Adhesive Coating Technology

The coating technology used for the liquid adhesive may influence the adhesive properties in various ways. The classic example is the regulation of adhesive properties via direct transfer coating or by the use of various coating devices. Different anchorage, depth of penetration, and geometry of the coated adhesive are achieved by the use of such technologies. As discussed above, the classic technology used for low coating weights and special adhesives leads to different peel resistance values for the "same" coating weight depending on the adhesive layer geometry (see Fig. 5.11). Such a phenomenon would be unacceptable for labels, where in certain cases the "roughness" of the release layer itself has to be improved. In the above example the changes in coating technology do not affect the adhesivity. In other cases, changes in the adhesive properties are due to chemical modification. For instance, using the same coating device and but other working parameters may alter the characteristics of HMPSAs. De Jager and Borthwick [162] studied the influence of coating speed on SAFT and rolling ball tack for SBC-based HMPSAs. No significant effect was found, but variation of the coating temperature within the interval 120–180°C caused changes in the rolling ball tack, especially in the MD/CD ratio. A more pronounced temperature effect has been observed in working with an open melt mill or closed die. Changes in temperature between 140 and 180°C produced a decrease in the rolling ball tack from 17 to 3 cm for the open melt mill. For the closed die, the rolling ball tack (5 cm) did not change in this interval. In a similar manner the holding power changed drastically from 30 h to 3 h for the open melt mill. For the closed system the change was less important (from 22 h to 17 h).

Nonclassic, special coating methods used for other products have a more pronounced effect on adhesive characteristics. As is known, for the manufacture of certain tapes, the solventless, masticated bulk adhesive is warm coated on a textured carrier material in order to achieve high coating weight or to avoid volatiles. In certain cases the carrier is impregnated with the adhesive. Some products are spray coated. Such coatings have a rough contact surface.

When radiant energy is used to improve the performance of PSAs or to manufacture them, the nature of the radiation energy, its level, and the technical details of the equipment and procedure influence the adhesive and other properties of the PSPs. For instance, according to Martens et al. [60], a very precise exposure of 0.1–7 mW/cm^2 optimizes the molecular weight of the resulting polymer when UV polymerization is used for tapes. As stated in Ref. 93, in the photopolymerization of acrylics with 280–350 nm UV light, the light intensity at the surface should be 4.0 mW/cm^2. According to Ref. 60, low radiation levels should be used for acrylic HMPSA prepolymers in order to avoid side reactions.

Regulating Adhesive Properties by Direct Transfer Coating. Because of the quite different rheology of the liquid adhesive coated directly on the porous surface of a paper carrier material in comparison with a dried, solidlike, transfer-coated PSA, the anchorage of the adhesive layer is different for each procedure. Therefore the adhesive properties of transfer- and direct-coated PSA are different too. Special examples of the change in adhesive properties as a function of the direct or transfer coating are given (as discussed above) for UV-cured formulations in Refs. 40 and 48. Because the radiation is partially absorbed, partially transmitted, and partially reflected, the maximum radiation level at the top and bottom of the adhesive layer may differ. The crosslinking degree is the maximum on one side. Therefore direct and transfer coating give different adhesive characteristics (see Chapter 6 also).

Regulating the Adhesive Properties with the Coating Device. Theoretically the use of a coating device with a smooth coating cylinder gives a smoothly coated adhesive layer. The use of a gravure cylinder can give a discontinuous paint-like coating; the use of a wire rod may lead to a coating having linear defects. A formulation required for a Meyer bar differs from one coated with a reverse roll coater [25]. As discussed earlier, removability and/or re-adherability may require patterned coating. In this case the choice of coating device is determinant (see Chapter 6 also).

2.4 Regulating the Adhesive Properties with Product Application Technology

As discussed earlier, for PSPs that have balanced adhesive properties (i.e., enough tack and peel) application conditions have less importance. It is mainly temperature that influences the applicability of the product (e.g., freezer labels). For low tack products the laminating conditions are determinant for the product application (see Chapter 8).

3 INTERDEPENDENCE: ADHESIVE PROPERTIES AND OTHER PERFORMANCE CHARACTERISTICS

Pressure sensitivity given by adhesive properties is the most important technical property of PSPs. As discussed in detail by Benedek and Heymans [1], adhesive properties also influence conversion and end use properties. For some pressure-sensitive products such as labels and tapes, where posttransformation of the adhesive-coated web is gaining more and more importance, the converting properties are important also. For other PSPs used as weblike, continuous products with or without less postconverting, the end use properties are more important. As discussed earlier, in some special product classes (tapes, protective films) pressure sensitivity is the result of adhesive and end use properties. Therefore it can be stated that generally the adhesive properties influence in a determinant manner the conversion or end use performance characteristics of PSPs. This influence is discussed in Chapters 7 and 8.

REFERENCES

1. I. Benedek and L. J. Heymans, *Pressure-Sensitive Adhesives Technology*, Marcel Dekker, New York, 1997, Chapter 6.
2. R. Köhler, *Adhäsion 6:*247 (1968).
3. J. H. S. Chang (Merck & Co. Inc., Rahway, NJ), EP 0179628 A2/24.10.1984.
4. Y. Aizawa (Cemedine Co. Ltd.), Jpn. Patent 6317945/25.01.1988; in *CAS Adhes 14:*4 (1988).
5. M. E. Fowler, J. W. Barlow, and D. R. Paul, *Polymer 28*(11):2145 (1987).
6. K. Dormann, *Coating 6:*150 (1984).
7. G. R. Hamed and C. H. Shieh, *J. Polym. Sci. 21:*1415 (1983).
8. D. R. Gehman, F. T. Sanderson, S. A. Ellis, and J. J. Miller, *Adhäsion 1:*19 (1978).
9. K. Chu and A. N. Gent, *J. Adhes.* 25(2):109 (1988).
10. M. Toyama and T. Ito, Pressure sensitive adhesives, in L. Naturman *Polymer Plastics Technology & Engineering,* Vol. 2, Marcel Dekker, New York, 1974, pp. 161–230.
11. J. C. Pasquali, EP 0122847/24.10.1984.
12. M.M. Mazurek (Minnesota Mining and Manuf. Co., St. Paul, MN) EP 4693935/15.03.1987.
13. P. A. Mancinelli, New developments in acrylic hot melt pressure sensitive adhesive technology, *TECH 12, Advances in Pressure Sensitive Tape Technology,* Technical Seminar Proceedings, Itasca, IL, May 1989, p. 165.
14. I.J. Davis (National Stock, Bridgewater), US Patent, 4728572/01.03.1988.
15. M. Schleinzer and G. Hoppe, Amorphe Polyalphaolefine als Basis Material zur Formulierung von HMPSA und semi pressure sensitive Adhesives, 19th Münchener Klebstoff und Veredelungsseminar, 1994, p. 120.

16. C. N. Clubb and B. W. Foster, *Adhes. Age 11:*18 (1988).
17. U.S. Patent 3321451; in S. E. Krampe and C. L. Moore (Minnesota Mining and Manuf. Co., St. Paul, MN), EP 0202831A2/26.11.1986.
18. A. Haas (Société Chimique des Charbonnage-CdF Chimie, France), U.S. Patent 4624991/25.11.1986; in *Adhes. Age 5:*26 (1987).
19. C. Parodi, S. Giordano, A. Riva, and L. Vitalini, Styrene-butadiene block copolymers in hot melt adhesives for sanitary application, 19th Münchener Klebstoff und Veredelungsseminar, 1994, p. 119.
20. R. G. Jahn, *Adhes. Age 12:*35 (1977).
21. H. Mueller and J. Türk (BASF A. G., Ludwigshafen, Germany), EP 0118726/19.09.1984.
22. G. W. H. Lehmann and H. A. J. Curts (Beiersdorf A. G., Hamburg, Germany), U.S. Patent 4038454/26.07.1977.
23. A. Midgley, *Adhes. Age 9:*17 (1986).
24. P. Dunckley, *Adhäsion 11:*19 (1989).
25. D. G. Pierson and J. J. Wilczynski, *Adhes. Age 8:*52 (1990).
26. P. Green, *Labels Label. 11/12:*38 (1985).
27. J. Pennace and G. E. Kersey (Flexcon Co. Inc., Spencer, MA), PCT, WO 87/035537/18.06.1987.
28. G. Bonneau and M. Baumassy (DRT), New tackifying dispersions for water based PSA for labels, 19th Münchener Klebstoff und Veredelungsseminar, 1994, p. 82.
29. A. T. DiBenedetto, *J. Polym. Sci., Part B., Polym. Phys. 25*(9):1949 (1987).
30. D. H. Kaelble, *Int. SAMPE Symp. Exhib. 33:*153 (1988); in *CAS Adhes. 15:*2 (1988).
31. H. Gramberg, *Adhäsion 10*(3):97 (1966).
32. D. W. Bamborough and P. M. Dunckley, *Adhes. Age 11:*20 (1990).
33. D. W. Bamborough, Water based adhesives dispersions for labels: Prediction of adhesive performance using dynamical mechanical analysis, 19th Münchener Klebstoff und Veredelungsseminar, 1994, p. 96.
34. S. G. Chu, Visco-elastic properties of pressure sensitive adhesives, in *Handbook of Pressure Sensitive Adhesive Technology,* 2nd ed. (D. Satas, Ed.), Van Nostrand Reinhold, New York, 1989.
35. S. Benedek, *Adhäsion 5:*16 (1986).
36. Specialty Adhesives, Product Data, *Duro-Tak,* Druckempfindliche Klebstoffe, National Starch & Chemical B. V., Adhesives Division, October 1987, p. 9, Lutphen, Netherlands.
37. R. R. Charbonneau and G. L. Groff (Minnesota Mining and Manuf. Co., St. Paul, MN), EP 0106559 B1/25.04.1984.
38. U.S. Patent 2925174; in R. R. Charbonneau and G. L. Groff (Minnesota Mining and Manuf. Co., St. Paul, MN), EP 0106559 B1/25.04.1984.
39. U.S. Patent 4286047; in R. R. Charbonneau and G. L. Groff (Minnesota Mining and Manuf. Co., St. Paul, MN), EP 0106559 B1/25.04.1984.
40. U. S. Patent 4181752; in R. R. Charbonneau and G. L. Groff (Minnesota Mining and Manuf. Co., St. Paul, MN), EP 0106559 B1/25.04.1984.

41. Can. Patent 747341; in R. R. Charbonneau and G. L. Groff (Minnesota Mining and Manuf. Co., St. Paul, MN), EP 0106559 B1/25.04.1984.
42. U.S. Patent 4223067; in D. K. Fisher and B. J. Briddell (Adco Product Inc., Michigan Center, MI), EP 0426198 A2/08.05.1991.
43. G. F. Vesley, A. H. Paulson, and E. C. Barber, EP 0202 938 A2/26.11.1986.
44. U.S. Patent 4330590; in G. F. Vesley, A. H. Paulson, and E. C. Barber, EP 0202 938 A2/26.11.1986.
45. U.S. Patent 4329384; in G. F. Vesley, A. H. Paulson, and E. C. Barber, EP 0202 938 A2/26.11.1986.
46. F. C. Larimore and R. A. Sinclair (Minnesota Mining and Manuf. Co., St. Paul, MN), EP 0197662A1/15.10.1986.
47. U.S. Patent 514950/18.07.1983; in F. C. Larimore and R. A. Sinclair (Minnesota Mining and Manuf. Co., St. Paul, MN), EP 0197662A1/15.10.1986.
48. G. Auchter, J. Barwich, G. Rehmer, and H. Jäger, *Adhes. Age* 7:20 (1994).
49. K. Nobuhiro and T. Hirokazu (Mitsubishi Paper Mills Ltd.), Jpn. Patent 63152686/25.06.1988; in *CAS Adhes.* 25:7 (1988).
50. A. Dobmann and A. G. Viehofer, 19th Münchener Klebstoff und Veredelungsseminar, 1994, p. 168.
51. A. H. Beaulieu, D. R. Gehman, and W. J. Sparks, Recent advances in acrylic hot melt pressure sensitive technology, Tappi Hot Melt Symposium, 1984; in *Coating* 11:310 (1984).
52. G. Panagopoulos, S. E. Pirtle, and W. A. Khan, *1991 Film Extrusion*, p. 103.
53. *Adhäsion* 3:83 (1974).
54. *Adhes. Age* 10:125 (1986).
55. D. J. Bebbington (Wiggins Tape Group Ltd.), EP 251672/07.01.1988; in *CAS Siloxanes* 14:4 (1988).
56. Y. Moroishi, T. Sugii, and K. Noda (Nitto Electric Ind. Co. Ltd.), Jpn. Patent 6381184/12.04.1988; in *CAS Colloids (Macromol. Aspects)* 21:6 (1988).
57. A. Osaku, S. Shinji, and M. Takeshi (Nippon Synth. Chem. Ind. Co. Ltd.), Jpn. Patent 62243669/24.10.87; in *CAS Colloids* 4:5 (1988).
58. L. K. Post, Development trends in high performance pressure sensitive acrylic adhesives, 19th Münchener Klebstoff und Veredelungsseminar, 1994, p. 191.
59. J. L. Walker and P. B. Foreman (National Starch and Chemical Corp.), EP 224795/10.06.1987; in *CAS Colloids (Macromol. Aspects)* 1:5 (1988).
60. Martens et al., U.S. Patent 4181752; in J. N. Kellen and C. W. Taylor (Minnesota Mining and Manuf. Co., St. Paul, MN), EP 0246352 A2/25.11.1987.
61. Gander, U.S. Patent 3475363; in J. N. Kellen and C. W. Taylor (Minnesota Mining and Manuf. Co., St. Paul, MN), EP 0246352 A2/25.11.1987.
62. Zang, U.S. Patent 3532652; in J. N. Kellen and C. W. Taylor (Minnesota Mining and Manuf. Co., St. Paul, MN), EP 0246352 A/25.11.1987.
63. Gobran, U.S. Patent 4260659; in J. N. Kellen and C. W. Taylor (Minnesota Mining and Manuf. Co., St. Paul, MN), EP 0246352 A2/25.11.1987.
64. BASF, T1-2.2-21d, November 1979.
65. K. F. Schroeder, *Adhäsion* 5:161 (1971).
66. Letraset Ltd., London, U.S. Patent 1545568; in *Coating* 6:71 (1970).

67. *Coating 9:*301 (1993).
68. J. Johnston, *Adhes. Age 12:*24 (1983).
69. F. Altenfeld and D. Breker, *Semi Structural Bonding with High Performance Pressure Sensitive Tapes,* 2nd ed., 3M Deutschland, Neuss, February 1993, p. 278.
70. John F. Kuik, *Papier Kunstst. Verarb. 10:*26 (1990).
71. R. Bates, *J. Appl. Polym. Sci. 20:*2941 (1976).
72. U.S. Patent 3321451; in S. E. Krampe and C. L. Moore (Minnesota Mining and Manuf. Co., St. Paul, MN), EP 0202831A2/26.11.1986.
73. *Coating 4:*88 (1984).
74. G. Meinel, *Papier Kunstst. Verarb. 19:*26 (1985).
75. J. A. Miller and E. von Jakusch (Minnesota Mining and Manuf. Co., St. Paul, MN), EP 0306232B1/07.04.1993.
76. M. A. Krecenski, J. F. Johnson, and S. C. Temin, *JMS—Rev. Macromol. Chem. Phys. C 26*(1):143 (1986).
77. D. K. Fisher and B. J. Briddell (Adco Product Inc., Michigan Center, MI), EP 0426198A2/08.05.1991.
78. J. Kim, K. S. Kim, and Y. H. Kim, *J. Adhes. Sci. Technol. 3*(3):175 (1989).
79. J. Boutillier, 11th Münchener Klebstoff und Veredelungsseminar, 1986, p. 4.
80. J. F. Kuik, Polyethylenes, copolymers and blends for extrusion and coextrusion coating markets, SP '89, Maacks Business Service, Zurich, Switzerland, 1989, p. 107.
81. ASTM D 1876-61T.
82. Eur. Standard NMP 458, No. 16-94D.
83. BP Chemicals, *Hyvis/Napvis Polybutenes,* PB 701, September 1983, London, England.
84. H. Mizumachi, *J. Appl. Polym. Sci. 30:*2675 (1985).
85. BP Chemicals, *Hyvis, Napvis, Ultravis, Cling Properties,* PB 104/1, January 1992, London, England.
86. J. H. Glover, *Tappi J. 71*(3):188 (1988).
87. O. V. Stoyanov, V. F. Mironov, V. P. Privalko, R. Ya. Deberdeev, R. M. Kuzakhanov, and S. M. Minisalikhova, *Lakokras. Mater. Ikh. Primen. 1:*51 (1988); in *CAS Coating, Inks Related Products 10:*2 (1988).
88. U.S. Patent 3.364063; in J. C. Pasquali, EP 0122847/24.10.1984.
89. M. Gerace, *Adhes. Age 8:*84 (1983).
90. *Coating 1:*35 (1988).
91. N. W. Malek (Beiersdorf A. G., Hamburg, Germany), EP 0095093/30.11.1983.
92. H. Miyasaka, Y. Kitazaki, T. Matsuda, and J. Kobayashi (Nichiban Co. Ltd., Tokyo, Japan), Offenlegungsschrift, DE 3544868A1/18.12.1985.
93. Jpn. Patent 2736/1975; in H. Miyasaka, Y. Kitazaki, T. Matsuda, and J. Kobayashi (Nichiban Co. Ltd., Tokyo, Japan), Offenlegungsschrift, DE 3544868 A1/18.12.1985.
94. U.S. Patent No., 3691140/12.09.1972; in H. Miyasaka, Y. Kitazaki, T. Matsuda, and J. Kobayashi (Nichiban Co. Ltd., Tokyo, Japan), Offenlegungsschrift, DE 3544868 A1/18.12.1985.
95. R. J. Shuman and B. D. Josephs (Dennison Manuf. Co., Framingham, MA) PCT, WO 88/01636.

96. K. W. M. Davy and M. Braden, *Biomaterials* 8(5):393 (1987).
97. Resins for Adhesives, UCB, Drogenbos, Belgium, 463-3, 1992.
98. U. Stirna, P. Tukums, N. P. Zhmud, and V. A. Yakushin, *Latv. PSR Zinat. Akad. Vestis Kim. Ser. 1:*69 (1988); in *CAS Crosslink. React. 13:*4 (1988).
99. M. Ando and T. Uryu, *Kobunshi Ronbunshu* 44(10):787 (1988); in *CAS Polyacrylates (J.) 5:*3 (1988).
100. Z. Czech, *Eur. Adhes. Sealants 6:*4 (1995).
101. F. T. Sanderson, *Adhes. Age 11:*26 (1988).
102. H. Nakagawa, A. Baba, S. Furukawa, and F. Shuzo (Nippon Shokubai Kagaku Kogyo Co. Ltd.), Jpn. Patent 62292877/19.12.1987; in *CAS Adhes. 11:*6 (1988).
103. P. Mudge, Ethylene-, vinylacetate based, waterbased PSA, TECH 12, *Advances in Pressure Sensitive Tape Technology,* Tech. Seminar Proc., Itasca, IL, May 1989.
104. E. Kazuyoshi, N. Hiroaki, T. Katsuhisa, K. Yoshita, and T. Saito (FSK Inc.), Jpn. Patent 6317981/25.01.1988; in *CAS Adhes. 12:*5 (1988).
105. M. A. Johnson, *J. Plast. Film Sheeting* 4(1):50 (1988).
106. T. Tsubakimoto, K. Minami, A. Baba, and M. Yoshida (Nippon Shokubai Kagaku Kogyo Co. Ltd.), Jpn. Patent 6335676/19.12.1987; in *CAS Adhes. 16:*1 (1988).
107. K. Akasawa, S. Sanuka, and T. Matsuyama (Nippon Synth. Chem. Ind. Co., Ltd.), Jpn. Patent 62243669/24.10.1987; in *CAS Colloids 4:*5 (1987).
108. A. N. Gent and S. Kaang, *J. Appl. Polym. Sci. 32:*4689 (1986).
109. Packard Electric, Engineering Specification ES M-2379.
110. Packard Electric, Engineering Specification ES M-1881.
111. Packard Electric, Engineering Specification ES M-2359.
112. Packard Electric, Engineering Specification ES M-2147.
113. Packard Electric, Engineering Specification ES M-4037.
114. Industria Nastri Adesivi, Data Sheet.
115. Ashland Chemicals, Aroset, EPO 32216-APS-102, Data Sheet.
116. R. G. Czerepinski and R. Gundermann (Dow Chem. Co.), U.S. Patent 4713412/15.12.1987.
117. *Coating 1:*8 (1986).
118. H. Satoh and M. Makino (Nippon Oil Co. Ltd.), Ger. Offen. DE 3740222/01.06.1988; in *CAS Adhes. 22:*4 (1988).
119. B. Copley and K. Melancon (Minnesota Mining and Manuf. Co.), U.S. Patent 878816/26.01.1986; in *CAS Siloxanes Silicones 14:*2 (1988).
120. R. M. Enanoza (Minnesota Mining and Manuf. Co.), EP 259968/16.03.1988; in *CAS Adhes. 22:*3 (1988).
121. T. Sugiyama, N. Miyaji, Y. Ito, and T. Tange (Nippon Carbide Industries Co. Inc.), Jpn. Patent 62199672/03.09.1987; in *CAS Adhes. 14:*4 (1988).
122. *Adhes. Age 12:*8 (1988).
123. J. P. Kealy and R. E. Zenk (Minnesota Mining and Manuf. Co., St. Paul, MN), Can. Patent 1224678/28.07.87.
124. M. Arakawa, S. Sakashita, H. Nagami, and T. Ono (Nitto Electric Ind. Co., Ltd.), Jpn. Patent 63142086/14.06.1988; in *CAS Adhes. 25:*6 (1988).
125. A. Saburo (Central Glass Co. Ltd.), Jpn. Patent 63117085/21.05.1988; in *CAS Adhes. 25:*7 (1988).

126. C. H. Hill, N. C. Memmo, and W. L. Phallen, Jr. (Hercules Inc.), EP 259697/16.03.1988; in *CAS Adhes. 14:*5 (1988).
127. *Adhes. Age 3:*8 (1987).
128. *Adhes. Age 9:*8 (1986).
129. *Coating 6:*184 (1969).
130. L. Jacob, New development of tackifiers for SBS copolymers, in 19th Munich Adhesive and Finishing Seminar, 1994, p. 107.
131. C. Donker, R. Ruth, and K. van Rijn, Hercules MBG 208 hydrocarbon resin: A new resin for hot melt pressure sensitive (HMPSA) tapes, 19th Münchener Klebstoff u. Veredelungsseminar, 1994, p. 64.
132. Y. Urahama and K. Yamamoto, *J. Adhes.* 25(1):45 (1988).
133. O. Kazuto, H. Kenjiro, N. Akio, and N. Kiyohiro (Nitto Electric. Ind. Co. Ltd.), Jpn. Patent 6348382/01.03.1988; in *CAS Colloids (Macromol. Aspects) 14:*5 (1988).
134. Poli-Film, America Inc., Cary, IL, Product Data Sheet, 1992.
135. Poli-Film Verwaltungsgesellschaft mBH, Wermelskirchen, Germany, *Poli-Film für metallene Oberflächen.*
136. T. Tatsuno, K. Matsui, M. Takahashi, and M. Wakimoto (Kansai Paint Co. Ltd.), Jpn. Patent 62285927/11.12.1987; in *CAS Adhes. 14:*3 (1988).
137. S. Benedek, *Adhäsion 12:*17 (1987).
138. DE-A-2407494; in P. Gleichenhagen, E. Behrend, and P. Jauchen (Beiersdorf A. G., Hamburg, Germany), EP 0149135B1/24.07.1987.
139. Jpn. Patent A 8231792; in P. Gleichenhagen, E. Behrend, and P. Jauchen (Beiersdorf A. G., Hamburg, Germany), EP 0149135B1/24.07.1987.
140. H. Hintz, Kunststoff Dispersionen für das Verkleben von PVC Bodenbelägen und Textiler Auslegeware, *Kunstharz Nachrichten (Hoechst) 19:*18 (1983).
141. J. J. Higgins, F. C. Jagisch, and N. E. Stucker, Butyl rubber and polyisobutylene, in *Handbook of Pressure Sensitive Adhesive Technology,* Ch. 14, D. Satas, Van Nostrand-Rheinhold, New York, 1982.
142. G. L. Burroway and G. W. Feeney, *Adhes. Age 7:*17 (1974).
143. Lehmann et al., U.S. Patent 3729338; in D. K. Fisher and B. J. Briddell (Adco Product Inc., Michigan Center, MI), EP 0426198A2/08.05.1991.
144. J. N. Kellen and C. W. Taylor (Minnesota Mining and Manuf. Co., St. Paul, MN), EP 0246352 A/25.11.1987.
145. U.S. Patent A 2510120; in P. Gleichenhagen, E. Behrend, and P. Jauchen (Beiersdorf A. G., Hamburg, Germany), EP 0149135 B1/24.07.1987.
146. Ger. Patent A 2535897; in P. Gleichenhagen, E. Behrend, and P. Jauchen (Beiersdorf A. G., Hamburg, Germany), EP 0149135B1/24.07.1987.
147. M. Koehler and J. Ohngemach, *Polym. Paint Colour J. 178:*203 (1988).
148. Loctite Co., Jpn. Patent 62179506/06.08.1986; in *CAS Siloxanes Silicones 13:*1 (1988).
149. H. Kuroda and M. Taniguchi (Bando Chem. Ind. Ltd.), Jpn. Patent 6343988/25.02.1988; in *CAS Adhes. 21:*4 (1988).
150. A. Dobashi, T. Ota, and T. Uehara (Hitachi Chem. Co. Ltd.), Jpn. Patent 6368683/28.03.1988; in *CAS Crosslink. React. 14:*7 (1988).

Adhesive Properties of PSPs

151. W. J. Traynor, C. R. Moore, M. K. Martin, and J. D. Moon (Minnesota Mining and Manuf. Co.), U.S. Patent 4726982/23.02.1988; in *CAS Adhes. 17:*4 (1988).
152. Y. Ikeda, Y. Watanabe, and H. Tadenuma (Japan Synthetic Rubber Co. Ltd.), Jpn. Patent 6386777/18.04.1988; in *CAS Adhes. 24:*3 (1988).
153. K. Hayashi and K. Shimobayashi (Nitto Electric. Ind. Co. Ltd.), Jpn Patent 63118383/23.05.1988; in *CAS Adhes. 24:*5 (1988).
154. I. Murakami, Y. Hamada, and S. Sasaki (Toray Silicone Co. Ltd.), EP 269454/01.06.1988; in *CAS Adhes. 19:*5 (1988).
155. T. Shibano, I. Kimura, H. Nomoto, and C. Maruchi (Sanyo Kokusaku Pulp Co., Ltd., Tokyo, Japan), U.S. Patent 4624893/25.11.1986; in *Adhes. Age 5:*24 (1987).
156. E. Pagendarm, Hamburg, Germany, Offenlegungsschrift, DE 3632816A1/26.09.1986.
157. U.S. Patent 3.865770; in F. C. Larimore and R. A. Sinclair (Minnesota Mining and Manuf. Co., St. Paul, MN), EP 0197662A1/15.10.1986.
158. Jpn. Patent A5695972; in P. Gleichenhagen, E. Behrend, and P. Jauchen (Beiersdorf A. G., Hamburg, Germany), EP 0149135B1/24.07.1987.
159. K. Tatsuo, O. Nozomi, and T. Naomitsu (Nitto Electric. Ind. Co. Ltd.), Jpn. Patent 6333487/13.02.1988; in *CAS Adhes. 12:*5 (1988).
160. A. Nakasuga and M. Kobari (Sekisui Chemical Co. Ltd.), Jpn. Patent 6389585/20.04.1988; in *CAS Colloids (Macromol. Aspects) 22:*5 (1988).
161. V. I. Pavlov, T. P. Muravskaya, R. A. Veselovskii, and M. T. Stadnikov, *Kompoz. Polym. Mater. 37:*14 (1988).
162. D. de Jager and J. B. Borthwick, *Thermoplastic Rubbers for Hot Melt Pressure Sensitive Adhesives—The Processing Factors,* Shell Elastomers, Thermoplastic Rubbers, Tech. Manual TR 8.11, p. 5, Shell Chemicals, Louvain La Neuve, Belgium, May, 1988.

6
Manufacture of Pressure-Sensitive Products

Considerable attention has been given to the general area of PSA manufacture [1], but a need now exists to review specific topics concerning the manufacture of pressure-sensitive products with and without pressure-sensitive adhesives. With the classic manufacturing procedure, PSPs are produced by coating a carrier web with a PSA. In this process, depending on the product class and application, multiple carrier and coating components can be used. Although paper and various special nonpaper, nonpolymeric materials have been developed, plastics are the main carrier materials for PSPs. In the early stage in the development of pressure-sensitive products, carrier films used in packaging materials were applied. Later, other, special plastic-based weblike materials were produced and recommended as carrier materials for PSPs.

In the manufacture of packaging films, weblike carrier materials were coated with adhesives to produce laminates, i.e., adhesive-bonded sheetlike composites. Later, similar coating and laminating technology was used in the production of PSPs. Actually, depending on the laminate components, type of adhesive, and lamination place, such adhesive-coated finished products are supplied as permanent or temporary laminates (e.g., laminated and overlaminating films, labels, tapes) or as a monoweb (tapes, protective films, etc.). For the latter products, laminating occurs during application (Table 6.1).

The manufacture of permanent laminates requires classic adhesives, which bond chemically. The coating of adhesives with permanent viscoelastic flow is the application domain of PSAs. Laminating is carried out because of the need

333

Table 6.1 Main Features of Laminated Products

Product	Buildup	Components	Type of bond between laminate components
Packaging film	Multiweb	Carrier$_1$, adhesive, carrier$_2$ (carrier$_{n+1}$, adhesive$_n$)	Permanent
Label	Multiweb	Carrier$_1$, carrier$_2$ (carrier$_{n+1}$)	Permanent
		Carrier$_1$, adhesive, carrier$_2$, release liner	Temporary
		Carrier$_{n+1}$, adhesive$_n$, release liner$_n$	
Tape	Monoweb	Carrier$_1$, adhesive	Temporary
	Multiweb	Adhesive$_1$, carrier$_1$, adhesive$_2$, release liner$_1$	Temporary
Protective film	Multiweb	Adhesive$_1$, release liner$_1$	Temporary
	Monoweb	Carrier$_1$, adhesive$_1$	Temporary
	Monoweb	Carrier$_1$	Temporary
	Multiweb	Carrier$_{n+1}$	Permanent

to protect the tacky adhesive layer. The back side of the carrier (e.g., in a tape or protective film) or a supplemental release liner (e.g., in a label or double-faced tape) protects the adhesive layer (see also Chapter 2).

The development of macromolecular chemistry allowed the synthesis of rubberlike or plastomerlike products that can be processed as film and (under well-defined conditions) exhibit self-adhesivity and pressure sensitivity (see Chapter 4). The performance of pressure-sensitive products is characterized by their laminating, bonding, and delaminating characteristics. Laminating and delaminating refer to the application and deapplication, respectively, of the product. According to classic manufacturing technology, such characteristics are coating-dependent properties. As illustrated in Fig. 6.1, laminating to produce weblike materials is used for the manufacture of PSP components (e.g., carrier), for the manufacture of the finished product (e.g., label), or for the end use application of PSPs (label, tape, or protective film). In the case of labels, delamination of the liner precedes end use lamination. In the case of protective films, the lamination of the carrier components may represent the final step in the manufacture of a self-adhesive film. Special materials having intrinsic adhesive (and abhesive) properties allow the same performance without coating. Such raw materials lead to the manufacture of pressure-sensitive products without pressure-sensitive adhesives and without the use of other than coating technology to build up the adhesive/carrier

Manufacture of PSPs 335

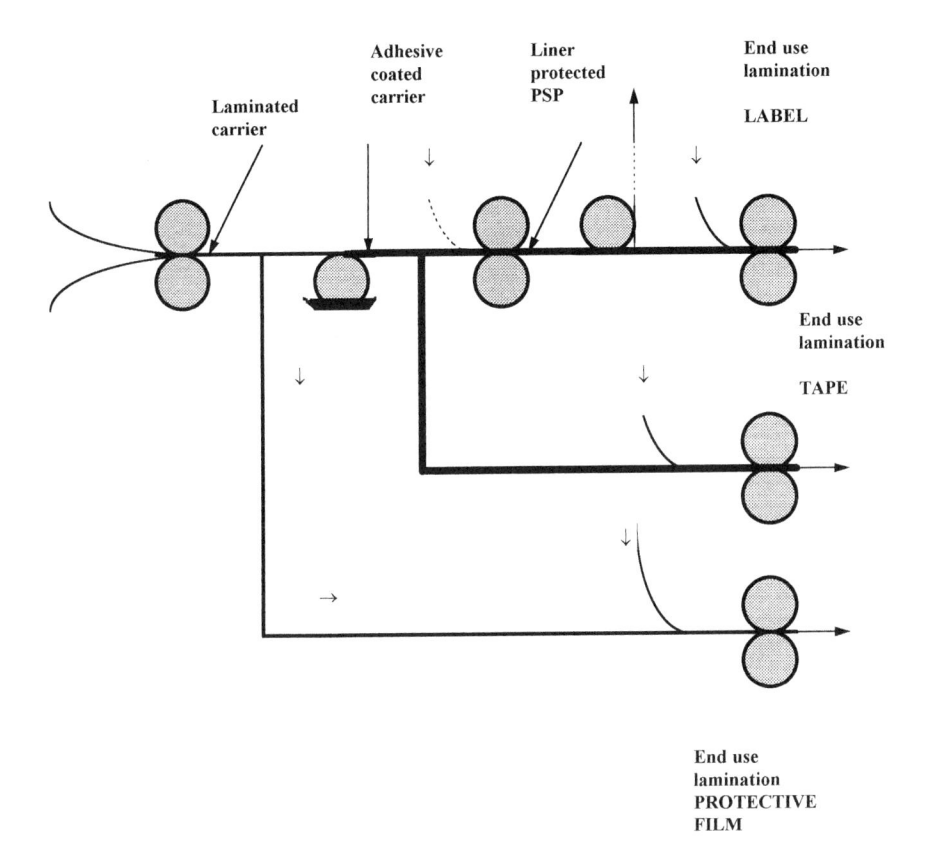

Figure 6.1 Lamination as a technological step in the manufacture of PSP components, of PSPs, and of their application.

assembly. Like common plastic films, such products are made by extrusion (Table 6.2).

As can be seen from Table 6.2, the manufacture of PSPs by coating with a liquid pressure-sensitive adhesive can be considered as a special case of the production of adhesive-coated PSPs. The manufacture of PSPs by adhesive coating itself is a special case of the manufacture of pressure-sensitive products. Generally the adhesive and end use characteristics of PSPs can be given by the carrier material also. It should be mentioned that the manufacture and application of such self-adhesive products are made possible by the special application conditions, which may include elevated temperature and pressure (Fig. 6.2). For PSPs with balanced adhesive properties, peel buildup is controlled mainly by the dwell

Table 6.2 Manufacturing Technology for PSPs

Pressure-sensitive product	Manufacturing technology		
Label	Coating	Coating of liquid state adhesive	Molten PSA Dispersed PSA
Tape	Coating	Coating of liquid adhesive Coating of solid adhesive	Molten PSA Dispersed PSA
	Extrusion Foaming-molding		
Protective film	Coating	Coating of liquid adhesive	Molten PSA Dispersed PSA
	Extrusion—physical treatment		

time, as illustrated by the top diagram in Fig. 6.2 (see also Chapter 5). For low tack self-adhesive films, peel buildup is also regulated by the application temperature and pressure.

The manufacture of PSAs and the general problems of PSA coating are discussed in Ref. 1. This technology can be considered a special part of PSP manufacture. The present chapter discusses the manufacture of PSPs as a whole, including the components that are necessary to build up a PSP by means of classic coating methods or by "plastics" processing.

In the classic manufacture of PSPs, a PSA is coated on a carrier web. The production of PSPs includes the manufacture and assembly of the individual components. Production of the components includes the manufacture of both the solid-state carrier material and the coating (adhesive, abhesive, primer, etc.).

1 MANUFACTURE OF COATING COMPONENTS

The main coating component is the PSA. Depending on its adhesive characteristics, on the type of carrier, and on product geometry, it should be protected by using different laminate constructions. Its delaminating is allowed by the abhesivity of the protective laminating component. Such abhesivity may be an intrinsic property of the solid-state material or may be given by a release layer. The nature of this release layer (coated on the same carrier or laminated as a separate solid-state component) depends on the product construction (class) and application conditions. The release effect is given in both cases by a special coated (or built-in) abhesive component. Other coating components are used to allow the anchor-

Manufacture of PSPs

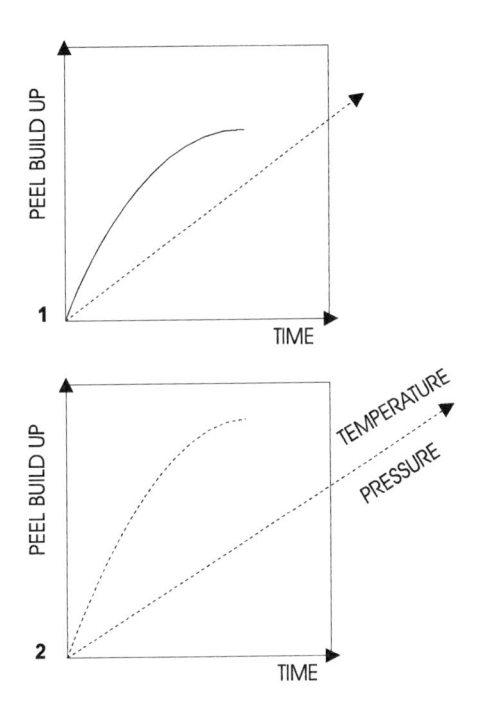

Figure 6.2 Application parameters that influence the bonding (peel buildup) of PSPs. (1) Adhesive-coated PSP; (2) adhesiveless PSP.

age of the adhesive, release agent, or printing inks. Additives for improving certain performance characteristics (e.g., electrical conductivity, flammability, recycling) can be coated on or built into the carrier material too.

1.1 Manufacture of the Adhesive Components

With the rapid growth in PSPs the need to develop new adhesive formulations is ongoing. In-house coating is also gaining increasing attention. The adhesives used as coating components for PSPs are mainly PSAs. Some of them are viscoelastic compounds with built-in PSA properties; others have to be formulated to achieve the required rheology and adhesive characteristics (see Chapter 5). Their processing (depending on the coating technology, equipment, carrier material, end use, and economic considerations) and their storage, application/ deapplication, and recycling require modification of the properties of the raw materials, i.e., formulation. Therefore the manufacture of the adhesive components includes their synthesis and formulation.

Synthesis of Adhesive

In their first stage of development, pressure-sensitive adhesives were formulated on the basis of natural macromolecular products. Natural rubber and natural resins were used. Some decades ago, synthetic elastic components (rubbers) and viscous raw materials (tackifiers, plasticizers, etc.) were developed and blended to produce (formulate) the adhesive. Advances in macromolecular chemistry allowed the synthesis of raw materials having built-in viscoelastic properties, i.e., pressure sensitivity (e.g., acrylics, ethylene-vinyl acetate copolymers, carboxylated rubber, polyvinyl ethers, polyurethanes, polyesters) (see also Chapter 4). Some of the polymers used for PSAs are raw materials for other adhesives or plastics also. Their synthesis is a special chemical/macromolecular technology. It is not the aim of this book to describe it.

It is to be noted that in certain cases it would be desirable for the adhesive converter to carry out its own polymerization. Such necessity appears for low volume special products or for adhesives where polymerization and converting technology have the same importance for the manufacture of the final PSP. For instance, to achieve a patterned contact surface (necessary for removability), the adhesive surface should have a special discontinuous shape ensured by the PSA itself if it is a suspension [2]. Thus spherical contact sites are formed. These are anchored on the carrier surface with the aid of a primer. The classic procedures concerning formulation for removability have some disadvantages (see later). The recipes contain plasticizer (migration), thickener, and surface-active agents (migration, bleeding, and softening). The adhesion increases over time. This increase depends on the nature of the surface and quality of the adherend. Therefore, full surface coating does not ensure complete removability. If the adhesive is coated in a discontinuous manner, adhesion increases over time at the contact sites, and thus there may be high peel resistance in the longest contact direction. The use of an adhesive with spherical PSA particles (50–150 μm) avoids such behavior [3]. Unfortunately, anchorage of the spherical PSA particles needs a continuous matrix. The matrix polymer has to cover the surface of the particles also. The particles have to be coated as a monolayer. On porous surfaces such a coating weight is difficult to achieve. The special PSA particles (with a T_g lower than 10°C) are synthesized via suspension polymerization. The composition of the pressure-sensitive polymer according to this patent includes a crosslinking agent also used for both the matrix polymer and the particles. Polyisocyanate derivatives, polyepoxides, and aziridines have been recommended as crosslinking agents. A low coating weight (7 g/m^2) has been applied. The product displays re-adherability. As can be seen from the above example, in this case the properties of the final product strongly depend on its synthesis.

Generally for the PSA converter the manufacture of the adhesive consists of its formulation. However, there are some special cases where the formulation is made to allow synthesis of the adhesive by postpolymerization. Polymerization or postpolymerization use special multifunctional monomers or macromers and classic (chemical) or physicochemical polymerization techniques. Polyaddition, polycondensation, or crosslinking are carried out (see Polymacromerization, Chapter 4, Section 2.1).

The manufacture of the adhesive can be carried out off-line or in-line (Table 6.3). Off-line manufacturing includes mixing the adhesive components followed by coating of the ready to use PSA. In-line manufacture of the adhesive consists of simultaneous coating and curing or postpolymerization. In this case a special "ready-to-coat" mixture of polymerizable and crosslinkable monomers, oligomers, or polymers (e.g., radiation-cured hot melts) is first applied on a carrier. Such a reaction mixture is transformed after coating into a ready-to-use adhesive. The postcoating synthesis of the PSA must be carried out by the converter.

In-Line Synthesis of the PSA. The simultaneous manufacture of the PSA and PSA laminate using radiation curing is discussed in Ref. 1. Based on experience with the radiation curing of thin coatings in the printing industry, in-line coating and polymerization of the base monomers (actually, of a blend of monomers, oligomers, and reactive diluting agents) have been carried out to manufacture pressure-sensitive adhesive-coated laminates. The main problems with this technology have had to do with chemistry, physiology, and rheology. The precise regulation of the polymerization (curing) as a function of the adhesive raw materials, layer thickness, and carrier features is difficult. In the first stage of development, side reaction products and effects (residues of unreacted monomers and initiators, carrier damage, etc.) created technical, environmental, and physiological problems. Additional technological aspects appeared also. The coating of such low viscosity fluids requires special devices. Therefore the use of prepoly-

Table 6.3 Manufacturing Possibilities for the Adhesive

Synthesis		Formulation	
In-line	Off-line	In-line	Off-line
Synthesis from monomers	Synthesis from monomers (polymerization)	Formulation for adhesion Formulation for other properties	Formulation for adhesion Formulation for other properties
Synthesis from oligomers (polymacromerization)	Synthesis from oligomers (polymacromerization)		
Synthesis from polymers (crosslinking)	Synthesis from polymers (grafting, crosslinking)		

Table 6.4 Macromer-Based PSPs

Product	Adhesive	Macromer	Ref.
Tape	Water-based acrylic	Acrylic-ethylene copolymer	6
Foam tape	Photocured acrylate	Isooctyl acrylate–N-vinyl-pyrrolidone copolymer, poly(t-butylstyrene)	7
Tape	UV-curable acrylic esters, polymeric photoinitiator	Polyvinylbenzophenone	8
	UV-curable acrylic esters, polymeric photoinitiator	Poly(2-(p-(2 hydroxy-2-methylpropanyl)-phenoxy)ethyl methacrylate)	9
	UV-cured butadiene rubber, polymeric photoinitiator	Aromatic-aliphatic ketones and silicone side chain substituted rubber	10
Tape	Ethylhexyl acrylate	Styrene-butadiene copolymer	11
Tape	UV-curable acrylic oligomer	Acrylic oligomers	12
Tape	Styrene-butadiene-acrylic	Tackified poly(styrene-butadiene) grafted with EHA	13
Tape	Acrylate	AC oligomer, EB cured with trimethylolpropane tris(thioglycolate) and 1,6-hexanediamine	14

merized or polymerized raw materials (reactive oligomers or polymers) and their postpolymerization have been preferred.

The full or partial postmanufacture of the PSA is the result of the chemical development induced by the trend toward solvent-free fabrication. Such formulations with 100% solids include hot melts and radiation-cured reaction mixtures. Technological reasons and (partially) insufficient progress in acrylic-based hot melts [4] forced the development of oligomer- and macromer-based curable formulations. It can be supposed that in the future polymacromerization will be used on a large scale (see also Chapter 4). The use of macromers is not new. It has been developed as graft copolymerization and crosslinking of plastomers. In water-based systems, adhesive modification via macromers is known also. For instance, an acrylic pressure-sensitive adhesive has been prepared by polymerization of an acrylic-vinyl monomer mixture in the presence of an alkali-soluble or dispersible support resin having a number average molecular weight of 21,000–150,000. The support resin is used as an emulsifier [5].

The raw materials for postpolymerization are common PSAs, PSAs with unbalanced adhesive properties, or nonadhesive oligomers (see Table 6.4). Such compounds are manufactured via off-line synthesis. They are macromolecular compounds having a low to medium molecular weight and residual functionality, i.e., they can be postpolymerized (or crosslinked). For this process the converter can choose to use either a classic polymerization technology (i.e., thermal, free radical initiated polymerization, polyaddition, or polycondensation) or a radiation-induced reaction. It is evident that in some cases this choice is limited by the chemical nature of the (pre)polymer. It is to be emphasized that in certain cases (i.e., acrylic hot melt development), postpolymerization of an oligomer is a necessity because of the lack of suitable macromolecular compounds.

Photopolymerization of the monomers neat, without any diluent that needs to be removed after the polymerization, provides processing advantages. Unfortunately, if UV radiation is used, postcuring is needed for high coating weights. In a special case of off-line adhesive synthesis, a hybrid procedure combines thermal postcrosslinking with radiation polymerization of the PSA. A patent to manufacture a tape describes the UV polymerization of acrylic monomers directly on the carrier [5]. The liquid monomers are applied to the carrier with a doctor blade or by roller coating or spraying. The polymerization occurs in a tunnel from which oxygen has been excluded. This patent uses a temporary carrier, an endless belt that does not become incorporated into the final tape. The off-line step of the polymerization is completed in an oven. Such tapes have PSA on both sides of the carrier.

It is possible to prepare a prepolymer that is polymerized by radiation to a coatable product that is postpolymerized in-line. A patent discloses the preparation of a low molecular weight spreadable composition to which a small amount of catalyst or polyfunctional crosslinking monomer can be added prior to the completion of the polymerization by heat curing [15]. The viscosity of the prepolymer makes it easy to apply to the support. This procedure was developed from an earlier monomer-based in-line procedure using UV photopolymerization (with a light intensity of 4.0 mW/cm^2 at the surface) of acrylics [16]. The PSA can be produced as an adhesive layer supported by a nonadhesive layer (carrier or rigid body) or as a double-faced system. As stated in Ref. 17, PSAs cured with UV or EB have not been a total success in performance or economics. In 1973, the costs of UV-crosslinkable inks and adhesives were twice as high as today [18]. In recent decades, technical and economic advances have been registered. For instance, a 1995 patent eliminates the requirement for photoinitiators for UV curing [19]. However, this method for synthesizing or modifying an adhesive should not be considered important for common pressure-sensitive products. Very thin or very thick adhesive layers (e.g., protective layers, transfer tapes etc.) are the suggested domain for this technology.

According to Ref. 20, postcuring is carried out to improve shear resistance. Copolymers of diesters of unsaturated dicarboxylic acids with acrylics (with a T_g of about 30–70°C below the use temperature) with a multifunctional (monomeric) crosslinking agent have been synthesized. This composition can be cured chemically or with any convenient radiation. The molecular weight of the first synthesized (pre)polymers should differ according to their molecular weight distribution (MWD). For a narrow MWD, a weight average molecular weight of 100,000, and for a broad molecular weight distribution a molecular weight of 140,000, are required to enable the desired response to EB curing. M_w/M_n ratios between 4.2 and 14.3 are achieved. Generally constituents having a molecular weight of less than 30,000 are nonresponsive to electron beam radiation [21]. The cohesive strength of the adhesive is proportional to the concentration of the hard monomer in the formulation. When a multifunctional monomer is used as crosslinking agent, its concentration should be preferably between about 1% and 5% by weight. Its presence enables a reduction of the dosage level for EB curing. A radiation dosage of 200 kilograms (kGy) is applied [20]. Thick PSA tapes (0.2–1.0 mm) have been manufactured by UV photopolymerization of acrylics [22]. Such an acrylic ester based PSA formulation is a combination of nontertiary acrylic acid esters of alkyl alcohols and ethylenically unsaturated monomers having at least one polar group, which may be substantially in monomer form or may be low molecular weight prepolymers or a mixture of prepolymer and additional monomers and may further contain a photoinitiator, fillers, and crosslinking agents (such as multifunctional monomers). The acrylic esters generally should be selected from those that possess PSA properties as homopolymers. Diacrylates are generally added as crosslinking agents (0.005–0.5% w/w) after synthesis of the prepolymer. As photoinitiator, 2,2-dimethoxy-2-phenylacetophenone can be used.

A special case of adhesive synthesis is the manufacture of PUR-based PSAs used for special tapes such as medical tapes. Unlike rubber-based formulations, which leave a deposit, special PUR adhesives can be cleanly removed from the skin [23]. Here a one-shot or prepolymer procedure can be chosen to synthesize the adhesive.

Postapplication crosslinking can be considered a special case of off-line synthesis (finishing) of an adhesive. It is proposed to improve shear or to ensure delamination (detachment) after use [24]. Such postcrosslinking is achieved by using thermal, free radical [25,26], or photoinitiated reactions [27]. A latent crosslinking agent has to be formulated in the recipe. A less alkylated amino formaldehyde condensate having C_{1-4} alkyl groups (e.g., hexamethoxymethylmelamine) has been proposed as crosslinking agent. First an intermediate product (having a viscosity of 0.3–20 Pa.s) is synthesized by UV photopolymerization; then a crosslinking agent is added, and the mixture is coated and polymerized by UV light to the final product. The product can also be used as a transfer tape [27]. By polymerization in mass, a thick coating can be manufactured in one step; the use of a solvent-based or aqueous adhesive requires more than one coat if a thickness of more than 0.2 mm is desired. As discussed in detail in Chapter 4, in-line synthesis by photopolymerization of macromers and macromeric photoinitiators is possible. This procedure is of increasing importance in the manufacture of carrier-free tapes. As illustrated by the data of Table 6.4, most systems are based on acrylates, but compounds containing aromatic or aliphatic ketone groups and silicone can be bonded to the alkylene side chains of polymers; e.g., butadiene rubber has also been hydrosilylated [10]. The in-line synthesis of the adhesive can be carried out in two steps to increase the pot life of curable formulations. For instance, adhesives with a long pot life are synthesized by coating the elastomers and the hardening agents on two different substrates, then putting them together and crosslinking them. Corona-treated 60 μm polyethylene has been coated with a mixture of 100 parts acrylic oligomer and 10 parts trimethylolpropane tristhioglycolate combined with a 20 μm film of PE coated with 0.5 g/m^2 1,6-hexadiamine, and irradiated with an electron beam of 5 Mrad to give an adhesive tape with a pot life greater than 24 hr and an adhesion of 410g/25 mm [14].

Off-Line Synthesis of the PSA. Off-line synthesis of PSAs is the common way to produce coatable adhesives. Such products can be macromolecular compounds having ready-to-use or ready-to-formulate adhesive properties or ready-to-postpolymerize reaction mixtures supplied for in-line adhesive synthesis.

Formulation of Adhesives

Formulation of PSAs is discussed in detail in Ref. 28. For most PSAs, formulation is the final manufacturing step, giving a product with the properties required for processing and end use. However, the same end use properties may differ strongly according to product class. Therefore the formulation of PSAs for PSPs displays numerous special features.

Formulation depends on the chemical basis of the adhesive. As discussed in Chapter 4, one-component viscoelastic raw materials (AC, EVAc, PUR, etc.) and multicomponent compositions having separate elastic and viscous components can be used to design a PSA. Their choice also depends on the coating technology (physical state, coating method, coating device, etc.). Formulating is also influenced by the PSP product class. The main raw materials used for PSPs have been discussed in Chapter 4.

Different classes of raw materials are used for various pressure-sensitive products. For different raw material classes, different formulating components and different technologies have been suggested. For PSPs based on natural rubber, formulation with resins has been the only way to achieve viscoelastic behavior and adequate pressure-sensitive properties. Hydrocarbon-based synthetic elastomers and polar synthetic elastomers (silicones, polyurethanes, polyesters, etc.) must also be formulated.

In principle, the scope of formulating is to develop a recipe with properties tailored to the product's end use. One of the most important end use criteria is the adhesive quality evaluated as the sum of the adhesive performance characteristics. On the other hand, formulation has to allow the use of the adhesive for the manufacture of PSPs (coating and converting properties) and to enhance the application of the final product (conversion and other end use properties). Ideally, the scope of the formulation is to fulfill the requirements of the entire spectrum of performance properties.

Formulation for the Adhesive Properties of PSPs. As discussed earlier (see Chapter 5), the simplest way to regulate adhesive properties is by formulation. The main purpose of formulation is to regulate the adhesion–cohesion balance. Practically, it is a trial to achieve the required tack, peel, and bond break characteristics, assuming that the formulated adhesive preserves the internal cohesion required for its conversion and end use. Therefore, in principle, the formulation of a common (label grade) PSA for adhesive properties consists of its tackification or detackification. Although the raw adhesive possesses sufficient cohesion before tackifying (Fig. 6.3a), improvement of its tack and peel by compounding with tackifier resin or plasticizer may cause a pronounced decrease of its shear resistance (Fig. 6.3b). The shear level required for a balanced adhesivity can be restored (Fig. 6.3c) by adding other formulation components to the recipe (e.g., high melting point resins, hard PSA, or crosslinkable components).

Tackification is required mainly for permanent adhesives; detackification is needed for removable compositions. Tackification is a general formulating method. Detackification is applied only in special cases. As discussed in Ref. 28, the adhesive properties can be improved by tackification by (1) formulation with other viscoelastic components, (2) formulation with resins and plasticizers, or (3) formulation with special additives. Formulation with the aid of viscous components allows tackification (regulation of tack and peel), removability, and easy processing.

Tailoring the chemical affinity of the adhesive for the substrate surface depends on its chemical composition. Detackifying agents have to work on the bonding surface. As is known from the production and use of packaging materials, detackifiers (debonding agents) are used for "easy peel" closures. These work in the bulk sealing layer. Their efficacy is based on their incompatibility with the main sealing component. Therefore in this case debonding occurs in the bulk adhesive layer. Such detackifiers cannot be applied for removable PSPs, where failure must occur at the adhesive/substrate interface. As a special case, the formulation of sealants should be mentioned. For sealants, cohesive failure is required [29]. According to Holden [29], in this case saturated hydrocarbon resins and saturated paraffinic oil (up to 400%) are suggested to achieve the desired behavior. Detackification uses special additives, resins, crosslinking agents, fillers, and abhesives (Table 6.5). For instance, a repositionable pressure-sensitive adhesive for mounting tapes includes a detackifying resin comprising a caprolactone polymer (1–30% w/w) [33]. To improve the detackifying effect of the elastic contact

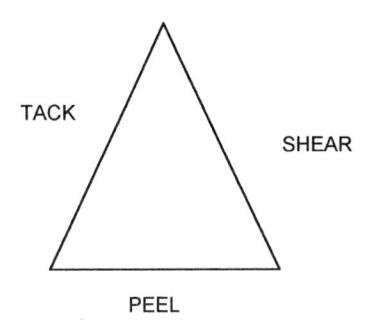

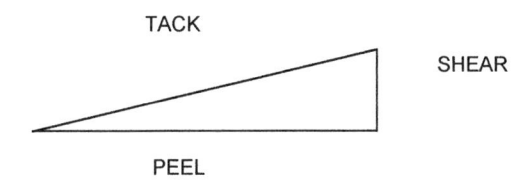

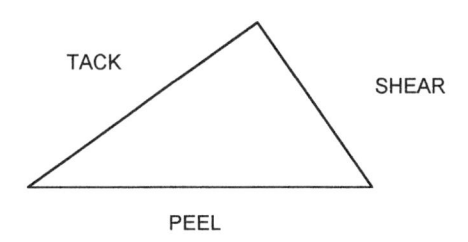

Figure 6.3 Schematic presentation of tackification steps. Untackified composition; tackified soft composition; tackified composition with balanced properties.

Table 6.5 Adhesive Formulation with Detackifier

PSP	Adhesive	Detackifyer	Ref.
Readherable PSA	Tackified polymer emulsion	Wax	30
Tape	Crosslinked acrylic	Tristearyl phosphite, 0.03–10%	31
Tape, sheet	EHA-VAc copolymer, crosslinked	Sorbitane trioleate	32
Tape, mounting	Acrylic	Caprolactone polymer	33
Sheet	Acrylic	Ionic, low tack PSA particles	34
Tape	HMPSA	Metallic salts of fatty acids	35

particles in a re-adherable product, the adhesive contains an ionic low tack monomer [34]. Detackification of an HMPSA using a metallic salt of a C_{14}–C_{19} fatty acid is described in Ref. 35. Sorbitane fatty acid esters (e.g., trioleate) have been used for ethylhexyl acrylate-vinyl acetate copolymers crosslinked with isocyanates to improve their removability [32]. Such formulations exhibit 250 g/25 mm peel on stainless steel after 3 days of storage at 60°C, in comparison with 950 g/25 mm for a common crosslinked PSA. The removability of acrylic-based PSAs is improved by small amounts of organofunctional silanes [36].

Tackification is described in Ref. 28. This chapter discusses the special aspects of tackification related to the product class (e.g., label, tape, or protective film). Tackification as discussed in Ref. 28 has been used to improve the adhesive properties of PSAs, i.e., the performance of adhesive-coated pressure-sensitive products. For the whole range of pressure-sensitive products, tackification consists of more than the modification of the adhesive; it also involves the carrier and other coating components (e.g., ink) (Fig. 6.4).

Tackification of the carrier is carried out for self-adhesive films. The mechanism can be quite different from that of the tackification of PSAs. As is known, PSA tackification assumes compatibility. Noncompatibility is taken into account only by the tackification of adhesives having composite segregated structures (i.e., SBCs). In the range of SAFs some products are tackified with compatible tackifiers, but there are also products where tackifier incompatibility is used to achieve the final performance. Vinyl acetate and ethylene-vinyl acetate copolymers are compatible with polyethylene, low molecular weight polyolefins, and some tackifier resins. Such compounds are used to prepare adhesive-free protective films. On the other hand, polybutylene is not compatible with certain polyolefins. Its incompatibility with and migration to the film surface are used to manufacture cling films. It is known that the addition of compatible polymers may depress the melting point of crystallizable polymers blended with an amorphous polymer, according to the Flory–Huggins theory [37]:

Manufacture of PSPs

```
                    ┌─────────────────────────┐
                    │  OTHER PSP COMPONENTS   │
              ┌────→│                         │
              │     └─────────────────────────┘
              │
              │     ┌─────────────────────────┐
              │     │       CARRIER           │
              │     │   ┌─────────────────┐   │
              │     │   │ Compatible      │   │
              │     │   ├─────────────────┤   │
              │     │   │ Noncompatible   │   │
TACKIFICATION │     │   └─────────────────┘   │
              │   ↗ └─────────────────────────┘
──────────────┤
                  ↘ ┌─────────────────────────┐
                    │       ADHESIVE          │
                    │   ┌─────────────────┐   │
                    │   │Partially compatible│ │
                    │   ├─────────────────┤   │
                    │   │ Compatible      │   │
                    │   └─────────────────┘   │
                    └─────────────────────────┘
```

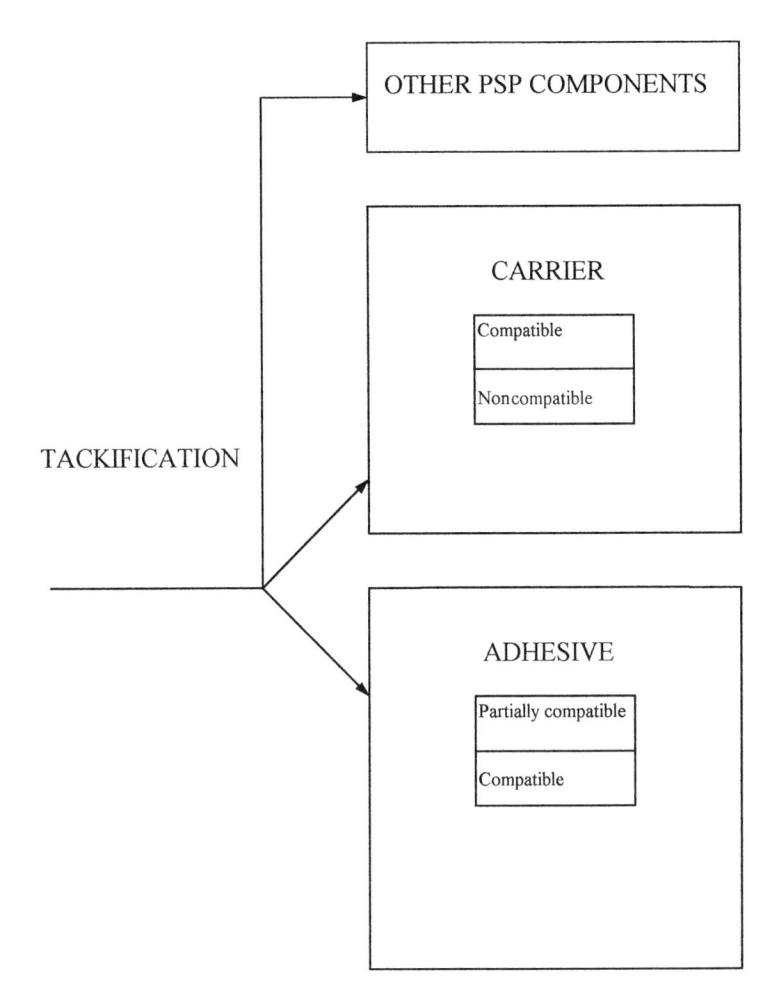

Figure 6.4 Schematic presentation of the general character of tackification for PSPs.

$$\frac{1}{T_m} - \frac{1}{T_m^\circ} = -\frac{RV_2}{\Delta H^0 V_1[(\ln \phi_2/m_2) + (1/m_1 - 1/m_2)\phi_1 + \chi_{12}\phi_1^2]} \quad (6.1)$$

where T_m and T_m^0 are the equilibrium melting points of the blend and homopolymer, respectively, ΔH^0 is the heat of fusion for the main polymer, V_1 and V_2 are the molar volumes of the repeat units; and m_1, ϕ_1 and m_2, ϕ_2 are the degrees of polymerization.

It should be taken into account that compatibility is temperature-dependent. Thermodynamic miscibility may exist in the whole composition range at ambient temperature, and phase-separated regions may appear at high temperatures for compatible mixtures. The time–temperature superposition principle can be applied. In such systems at high frequencies, neither component has relaxed and both contribute to G'. At mid-frequencies, one component relaxes. At low frequencies, this component does not contribute to the modulus, and the second component, which has still not relaxed, is responsible for the occurrence of a second plateau. This relaxation is similar to the relaxation observed in melts of mixtures of homopolymers having widely different molecular weights. At this frequency the low molecular weight polymer may act as a low molecular weight solvent. Such a binary blend exhibits a second plateau modulus GN_2^0 similar to that of polymer–solvent mixtures:

$$GN_2^0 = Gn^0 \phi^{2.5} \qquad (6.2)$$

where Gn^0 is the plateau modulus of the pure polymer and ϕ is its volume fraction. In the phase-separated region a $\omega^{0.5}$ dependence of the modulus can be observed for some polymers. A similar $\omega^{0.5}$ power low has been reported for chemically crosslinked systems at the gel point. According to Refs. 38 and 39, the rheology of such systems should be discussed in terms of fractal dimensionality, i.e., in terms of the similarity of supermolecular order over a large distance scale.

It should be taken into account that tack and peel do not improve (always) in parallel; the tackifier level required depends on the nature (chemical and physical status) of the tackifier, and the peel–shear interdependence influences the level of the debonding force and type of failure. It should be emphasized that the ideal adhesion–cohesion balance is different for labels, tapes, and protective films, therefore the scope of the formulation for the adhesive properties and the formulation possibilities for such products are different also. As illustrated by the data of Table 6.6, high tack and improved polyethylene peel are required (and achieved by tackification) for labels, whereas high shear and medium tack and peel are needed for packaging tapes. High tack and peel are obtained by using a low level (16–32%) of tackifier resin with soft acrylate (AC_1). A much higher resin level (45%) is necessary for SBR and NR–SBR mixtures for medium tack and peel. Therefore, such formulations are more suitable for tapes where high shear resistance is required. On the other hand, even mixtures containing cohesive acrylate (AC_2) and a low level (32%) of tackifier do not provide shear resistance competitive with that of formulations with SBR and SBR–NR mixtures with higher tackifier concentrations.

Generally the goal of tackification is to improve tack and peel and/or removability. This may be quantified by knowing the base performance of the adhesive and the chemical or macromolecular characteristics of the formulation components. For a high T_g ($-25°C$) crosslinkable carboxylated acrylic, Pierson and Wilczynsky [40] state the need to improve peel and tack by 25%. For such improvement the tackifiers have been tested at a loading of 12–35% dry tackifier per

Manufacture of PSPs 349

Table 6.6 Tackification of Selected PSPs

PSP	Formulation component	Concentration (parts, dry)	Peel (PE), 180° (N/25 mm)	Tack LT (N/25 mm)	Tack RB (cm)	Shear RT (h)
Label, permanent	AC 1 Tackifier 1	100 20	9.8	20.0	—	10
	AC 1 Tackifier 1	100 40	10.5	22.0	—	4
Label, permanent	SBR Tackifier 1	100 110	8.5	25.0	—	—
Label, permanent	NRL, SBR Tackifier 1	100 110	5.0	11.0	—	—
Tape, packaging	AC 2, AC 1 Tackifier 1	100 20	4.4	6.5	12	200
	AC 2, AC 1 Tackifier 1	100 40	6.0	11.0	12	144
	AC 2, AC 1 Tackifier 1	100 60	8.0	12.5	20	120
Tape, packaging	SBR Tackifier 1	100 40	8.5	9.0	8	>200
Tape, packaging	SBR Tackifier 1	100 60	10.0	12.0	20	>100
Tape, packaging	NRL, SBR Tackifier 2 + tackifier 3	100 120	6.5	10.0	20	>200
Tape, masking	NRL, SBR Tackifier 2 + tackifier 3	100 100	9.6	6.4	2	>200

Water-based formulations were used. AC 1 has a T_g of −42°C; AC 2 (self-crosslinkable) has a T_g of −58°C. Tackifier 1 is a dismuted rosin resin (ring and ball 50°C); tackifier 2 is a terpenephenol resin (R&B 85°C); tackifier 3 is a liquid rosin ester.

total solids. For a standard acrylic latex, 25–35% resin loading is used [41]. As stated in Ref. 42, the highest resin level that contributes to peel is about 50%; however, for a high softening point resin, a loading of less than 39% has been suggested. With high softening point resins the best overall balance is obtained at a concentration of around 30%. The T_g allows a forecast of the possible tackifier loading [43] (see also Chapter 3). A carboxylated acrylic latex needs a higher tackifier level. Water-releasable HMPSAs contain 15–40% tackifying agent [44,45]. Polyacrylate rubber based HMPSAs are tackified (40% resin) or plasticized formulations [46]. A common tacky PSA can be added to the formulation also (50 parts PSA to 100 parts acrylic rubber).

Typical resin dispersions contain 40–60% resin and 2–20% plasticizer [47]. It is evident that when such tackifiers are used, plasticizers are also added to the formulation. Internal crosslinking of the adhesive using special monomers such as N-methylolacrylamide, together with softening via plasticizers has also been suggested [48]. As discussed in Chapter 5, in HMPSAs the oil used in the formulation is a plasticizer and a viscosity-regulating agent. Formulating for shear resistance is a special domain, using crosslinking as its main technical modality. This technology is related mostly to the manufacture of tapes and protective film, so it will be discussed later.

As described in Chapter 5, formulating for removability achieves softening of the polymer with plasticizer and/or other viscous components. This is achieved for ACPSAs with paraffinic oil and polyisobutylene [34]. A repositionable transfer tape is manufactured by casting a thin carrier film that is coated with polybutyl acrylate plasticized with dibutylphthalate [49]. According to Ref. 50, removable price labels use almost exclusively paper carrier coated with rubber resin adhesive. Natural and synthetic rubber and polyisobutylene tackified with soft resins and plasticizer have been suggested for such compositions. A water-based acrylic formulation proposed by Mueller and Tuerk [50] includes physically or chemically crosslinkable comonomers like acrylo- or methacrylonitrile, acrylic or methacrylic acid, N-methylolacrylamide, and plasticizer (10–30% w/w) also. According to Jahn [51], a starting formulation for PSA for tape is based on a 70/17/10/10/0.45 part (dry) mixture of CSBR, coumarone indene resin, hydrocarbon resin, plasticizer, and thickener. A natural rubber latex based re-adhering, removable composition contains a tackifying agent also [52]. HMPSAs generally contain two different grades of resin, endblock-compatible and midblock-compatible [53].

Removability is related to energetic (rheological) and chemical adhesion. To ensure low peel resistance, the external force has to be balanced by the viscous and elastic (flow) deformation of the adhesive layer. Due to the self-deformation of the carrier material, reduced debonding energy is transferred at the adhesive/substrate surface (and adhesive face stock surface) where low adhesion should allow an adhesive (contact) break. In this situation formulation for removability has to allow energy absorbance (self-deformation) of the adhesive and low level adhesion on the adherend surface.

Adhesive deformability is achieved by increasing its viscous flow (e.g., using viscous raw materials or additives, see above). Adhesion break at the adherend surface is achieved by reducing (physically or chemically) the adhesive–substrate contact. This results from the changes in the adhesive and carrier rheology (e.g., by crosslinking) or in the properties of the carrier/adherend surface. Table 6.7 lists the main technical methods used to achieve removability.

Table 6.7 Technical Possibilities for the Manufacture of Removable PSPs

Modification of laminate components				Laminate manufacture	
Modification of the adhesive		Modification of the carrier			
Adhesive	Non-adhesive	Bulk modification	Surface modification	Coating	Post-lamination
Adhesive synthesis Crosslinking Plasticizing	Filling	Modification of carrier geometry Modification of mechanical properties	Tackification Physical surface treatment Mechanical surface treatment	Primer Coating method Modification of the PSA geometry	Crosslinking

As can be seen from Table 6.7, some of the technical ways to achieve removability depend on formulation, whereas other modalities depend on the geometry of the adhesive layer, i.e., they are (at least partially) a function of the coating technology.

According to Pasquali [54], contact surface reduction (a decrease in the ratio of contact surface to application surface of the adhesive) can be achieved by a formulation that regulates the rheology of the PSA and the chemical affinity between the adhesive and the substrate surface (see also Chapter 5). The control of adhesive flow to reduce the contact surface concerns the buildup of an internal structure in the bulk adhesive. Such a structure does not allow rapid adhesive flow and penetration of PSA in the "rough" substrate surface, i.e., it hinders the physical contact between adhesive and adherend. The same result is obtained by dividing the contact surface into small discrete areas.

The rheology of a highly viscous and partially elastic structure is regulated by means of crosslinking. The contact surface of the adhesive is divided by formulating an inhomogeneous adhesive mass with particles in the adhesive layer that do not build up contact because they are not adhesive or not adhesive enough or do not diffuse into the pores of the adherend. Such an adhesive layer can be produced by using inert filler particles that are placed on the contact surface. These particles are partially adhesive or elastic or "fully" adhesive but too voluminous.

The principles of crosslinking of PSAs are discussed in Ref. 28, which states that crosslinking can be chemically or physically (radiation) initiated. It is well known that chemical crosslinking supposes the presence of reactive groups (polar functional groups or unsaturation) in the polymer. Ultraviolet-induced crosslinking requires a chemical composition of this type; EB-induced crosslinking does not. Crosslinking due to built-in macromolecular reactive sites is generally the domain of solvent-based PSAs. Here long experience with the crosslinking of natural rubber can be called upon. The curing of NR uses external crosslinking agents (and heat). A slight mechanochemical destruction of the macromolecules of natural rubber (mastication) followed by postcrosslinking leads to products having balanced adhesive properties (labels). A more advanced mastication and crosslinking together with a higher coating weight allow the better shear and high tack and peel necessary for tapes. (Crosslinking may serve as a removability regulator also.) A much higher degree of crosslinking and low coating weight lead to low tack removable products, i.e., protective films (see Fig. 6.5).

As is known, the wet adhesive coating is dried in a drying channel. For uncrosslinked formulations, the drying channel serves for the volatilization of the liquid dispersing medium (solvents or water) used for the adhesive. For crosslinked formulations it serves for the crosslinking (via heat transfer) of the coated adhesive also.

As discussed in Ref. 2, acrylics display the disadvantage of compliance failure, i.e., adhesion buildup. This disadvantage has been reduced by internal crosslinking agents (e.g., multifunctional comonomers) like dimethylaminoethyl methacrylate or with polyisocyanates [3] (see Manufacture of Tapes in Chapter 6, Section 2.2).

Crosslinking as a means of regulating rheology has its limits. Such a high degree of crosslinking as that used for protective films is not practicable for labels. The application of external crosslinking agents for common functional water-based polymers (e.g., carboxylated acrylics or styrene-butadiene rubber) causes a pronounced loss of tack and peel. This can be understood by considering the higher molecular weight, high content of hard comonomers, and advanced crosslinking degree of the particle surface. As stated in Ref. 55, the crosslinking possibilities of hot melts are limited; thermal crosslinking cannot be used.

For peel regulation via nonadhesive contact sites, fillers have been suggested (see Chapter 4 also). For instance, according to Ref. 49, the PSA for repositionable transfer tape is softened and filled with glass powder. The adhesive surface can possess a special discontinuous shape ensured by the PSA itself if it is a suspension polymer. Spherical contact sites are formed [2] that are anchored on the carrier surface with the aid of a primer. PSA polymer particles having a diameter of 0.5–300 μm can be used as a filler also [3]. Such pressure-sensitive filler particles are crosslinked. The composition of the PSA according to Ref. 3 includes a crosslinking agent for both the matrix polymer and the particles. Polyisocyanate derivatives, polyepoxides, or aziridines are suggested as crosslinking agents. Removable pressure-sensitive tapes containing resilient polymeric microspheres (20–66.7%), hollow thermoplastic, expanded (acrylonitrile-vinylidene copolymer) spheres having a diameter of 10–125 μm, a density of 0.01–0.04 g/cm^3, and shell thickness of 0.02 μm, in an isooctyl acrylate-acrylic acid copolymer, have been prepared [56]. The particles are completely surrounded by the adhesive to

Manufacture of PSPs

1

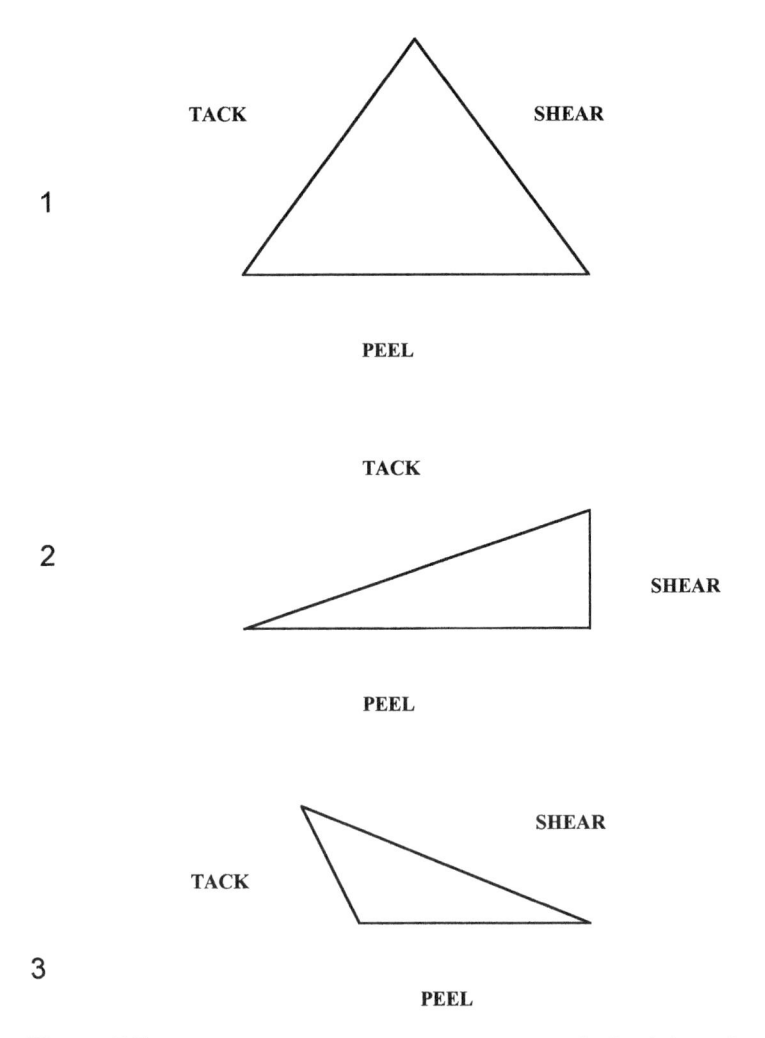

```
        TACK        SHEAR

              PEEL
```

2

```
           TACK

                         SHEAR

            PEEL
```

3

```
                   SHEAR
         TACK

                 PEEL
```

Figure 6.5 Schematic presentation of the adhesion–cohesion balance for masticated, tackified, and crosslinked adhesive formulations used for PSPs. (1) Label; (2) tape; (3) protective film.

a thickness of at least 20 μm. When the adhesive is permanently bonded to the backing and the exposed surface has an irregular contour, a removable and repositionable product is obtained, and when the pressure-sensitive adhesive forms a continuous matrix that is strippably bonded to the the backing (see Transfer Tape), the product has a thickness of more than 1 mm. This product is a foamlike transfer tape or foam tape. The (40 μm) cellulose acetate based tape could be removed from paper without delaminating the substrate. The contact surface can be

Table 6.8 Microspheres Used in Formulation of PSPs

PSP	Carrier	Adhesive	Filler	Ref.
Tape	—	AC	Adhesive copolymer microspheres, polymerized in suspension, 10–100 μm diam.	59
Tape, releasable from paper	—	—	PSA particles, 50–100 μm diam.	60
Tape, releasable from paper	—	—	Crosslinked, milled PSA, 50–100 μm diam.	61
Transfer tape	Cellulose acetate	Isooctyl acrylate-acrylic acid	Acrylonitrile-vinylidene copolymer, expanded spheres, 40 μm diam.	62
Sealing tape	—	AC, UV-cured	Silica microbubbles	63
Insulation tape	PET	UV-cured	Glass microbubbles	63
Sheet, releasable	—	AC	Microcapsule with blowing agent	57
Sheet, releasable	—	AC, UV-cured	Microcapsule with blowing agent	58

reduced by using expandable fillers also. According to Ref. 57, a photocrosslinked acrylic-urethane acrylate formulation suggested for holding semiconductor wafers during cutting and releasing them after cutting exhibits a peel strength of 1700 g/25 mm without crosslinking and 35 g/25 mm with crosslinking. Adding an expandable filler to the formulation (6 parts to 100 parts adhesive) reduces the peel resistance. This composition exhibits 1500 g/25 mm peel resistance before crosslinking and 15–29 g/25 mm after crosslinking. The releasable foaming adhesive sheet can be prepared by coating a blowing agent (e.g., microcapsules) containing PSA (30%) as a 30 μm dry layer [58]. This adhesive is removable and improves the cuttability. Table 6.8 lists some typical examples of the use of microspheres in the formulation of PSPs.

Raw Material Dependent Formulating. The aim of formulation and its modalities vary widely depending on the raw materials. For the classic way to obtain PSA (i.e., using an elastic component and a viscous component), formulation is the manufacture of the adhesive by mixing components. In the formulation of viscoelastic pressure-sensitive raw materials, formulating consists only of tailoring the PSA, i.e., modifying a given "finished" adhesive, to achieve other adhesive, conversion, or end use properties. Formulation should be designed according to the chemical and macromolecular characteristics of the base elastomers or viscoelastomers. There are certain raw materials that can be used for different formulations. However, the chemical and macromolecular characteristics of these materials limit their preferred use to well-defined application domains. For instance, tackifiers work best with base polymers whose glass transition temperature is below 0°C. Because VAc copolymers have a higher T_g, they do not work with tackifiers [64].

Formulations with natural rubber as base elastomer have to take into account NR's balanced adhesive properties (for labels), relatively low shear resistance (for tapes), poor solubility (need for milling), and softness in a highly crosslinked status (protective films). Current removable adhesives are rubber resin solvent-based and acrylic water-based recipes. Both of these formulations include high molecular weight polymers that reduce flow on a surface and prevent adhesion buildup [65].

HMPSAs contain lower molecular weight raw materials and a high level of micromolecular components. The adhesives may further contain up to 25% by weight of a plasticizing or extending oil to control wetting and viscosity. Hardness of formulations based on thermoplastic elastomers is given by physical crosslinking only (see Chapter 4 also). This is temperature-dependent. Therefore, in formulations using such raw materials, their low cohesivity at higher temperatures should be taken into account. In comparison with tackification of natural rubber, where adequate compatibility and chemical reactivity of the elastomer allow the use of numerous types of resins (including reactive ones), for SBCs the choice of tackifier and that of tackifier level are more difficult. Because of the higher coating weight needed in their application field, economic considerations are also important. As mentioned, HMPSA formulations usually incorporate a high level of plasticizer, usually a naphthenic oil or a liquid resin. The use of plasticizers results in a number of disadvantages including long-term debonding. According to Ref. 66, resins synthesized from a C_5 feedstock having a major portion of piperylene and 2-methylbutenes, with a defined diolefin/monoolefin ratio, M_w of 800–960, M_n of 500–600 (M_w/M_n ratio of at least 1.3), a softening point of about 57°C, and a T_g of about 20°C need no oils in formulations with saturated midblock thermoplastic elastomers. Saturated midblock thermoplastic elastomers include commercial ABA-type block copolymers where the polyisoprene or polybutadiene is hydrogenated. The block copolymers with a saturated midblock segment have an M_n of about 25,000–300,000. These thermoplastic elastomers are used in the formulation of adhesives for diaper closure tapes also [67]. It is desirable that the endblock resin have a softening point and glass transition temperature above those of the endblock and midblock of the block copolymer. But a liquid tackifier component is used also.

Polymers with built-in viscoelasticity have a broader chemical basis for the manufacture of all adhesive classes (i.e., hot melts and solvent-based and water-based formulations). Some of them can be used as raw materials for noncoated

PSP components also. As is known, acrylics have been developed as raw materials for PSAs in the form of elastomers (rubbers) and viscoelastomers also. Random and block copolymers have been synthesized as acrylic rubber. Both can be applied for solvent-based or hot melt adhesives. Viscoelastic acrylics are not suitable in their neat form as HMPSAs; they need postcrosslinking. Because of the similarities and differences in the properties of acrylic and styrene-diene block copolymers, their formulations differ. Acrylic block copolymers have both soft and hard acrylic phases [68] and a star-shaped radial structure. Their storage modulus has a plateau value of $10^{9.5}$ dyn/cm^2. Their soft phase has a T_g of −45°C, whereas that of the hard phase is 105°C. That means that their soft phase is harder and their hard phase is softer than those of corresponding SBCs. Their viscosity depends on the shear rate. Tackified acrylic block copolymers for hot melts exhibit peel values of 90–105 oz/in. The crossover point from adhesive failure to adhesive delamination is situated in the tackifier range of the 30–45 phr. Their shear resistance increases with tackifier level. This is unusual behavior. The SAFT value attains 80–86°C. Their melt viscosity is less than 50,000 cP at 177°C. For such polymers formulations with 75–80% resin loading have been suggested. In comparison, the highest resin level that contributes to the peel of water-based acrylics is about 50% and the best overall balance is obtained with high softening point resins at around the 30% level [42].

Formulation as a function of raw material basis (chemical composition of the raw materials) should also take into account the formulating additives. It is evident that solvents used as dispersing media differ according to the type of raw materials. As an example, isoparaffinic solvents are used for natural rubber, butyl rubber, polyisobutylene, EPR, and polyisoprene (see Chapter 4 also).

A special aspect of formulating concerns the adhesives polymerized or cured by radiation (EB or UV radiation or mixed techniques) where formulations based on monomers or oligomers (reactive monomers, oligomers) or polymers (curable HMPSAs) can be used. In this case the use of adequate radiation intensity is determinative. Unfortunately, the amount of radiation required also depends on its absorption by the components of the recipe. As stated in Ref. 55, a special raw material for acrylic HMPSAs developed for photochemical curing requires tackifying. For this polymer the photoinitiator is built into the polymer. Although the classic way is to use external photoinitiators (e.g., benzophenone), built-in monoethylenically unsaturated aromatic ketones can also be used for radiation crosslinking of polyacrylates [69] (see Chapter 4 also). The built-in crosslinker makes it unnecessary for the viscoelastomer to contain unsaturated polymerizable bonds (and therefore it gives improved aging properties), it does not produce skin irritation, and its UV polymerization is not affected by atmospheric oxygen [55]. In the classic technology using an external photoinitiator, the liquid monomers are applied to the carrier with a doctor blade or by roller coating or spraying and are cured in a polymerization tunnel from which oxygen has been excluded.

Manufacture of PSPs

Table 6.9 Technical Criteria for Usability of Tackifier Resins for Selected PSP Raw Materials

Raw material	Usability criteria
Natural rubber	Compatibility, reactivity, solubility, melting point
Styrene block copolymer	Compatibility, melting point
Radiation-curable raw materials	Compatibility, melting point, absorptivity, reactivity
Acrylics	Compatibility, melting point, solubility/dispersibility

As shown in Ref. 70, photopolymerization of acrylics can be carried out with UV light having a wavelength of 280–350 nm. According to Martens et al. [71], a very precise exposure of 0.1–7 mW/cm^2 optimizes the molecular weight of the resulting polymer. In the procedure given by Auchter et al. [55], the uncrosslinked product does not have balanced adhesive properties. Because of its low molecular weight it does not display enough cohesion, and its tack must be improved also. Its pressure-sensitive properties are achieved only by tackifying and crosslinking.

The copolymerized photoinitiator is an acrylic ester with benzophenone terminal groups. The main crosslinking reaction is the addition of a hydrogen atom of the alkyl side group of an acrylic segment to the carbonyl group of the benzophenone side group of the vicinal polymer chain, i.e., the formation of a hydroxy-diphenyl-polyacrylomethane derivative. Unfortunately, the building in of this reactive group promotes side reactions. To avoid such reactions, low radiation levels should be used. To modify UV-crosslinkable polacrylates, the use of an esterified highly stabilized resin is proposed. Because of the absorption of UV radiation by the resin (cyclical aromatized structure), only a low level of the resin is suggested [55]. For the tackified formulation, a resin loading of 8.5% improves the peel and tack but reduces shear resistance. Because the resin is a diluting agent in the polymer matrix and reduces the number of crosslink points, there is an upper limit to the amount of resin that can be added, delimited by shear reduction and radiation absorbance. Therefore, for tackified formulations the radiation dosage must be increased. Table 6.9 lists the main criteria for the choice of tackifier resins according to the raw materials used for the PSPs.

A similar technical method, the synthesis of a low molecular weight acrylic polymer (processible as a hot melt) and its postcrosslinking, have been proposed by Harder [72]. High concentration solution polymerization of acrylics has been carried out, and relatively low molecular weight, low viscous polymers have been synthesized and isolated, by evaporation and degassing, as solid-state raw materials for HMPSAs. According to Wabro et al. [73], such polymers also include restsolvents ($\approx$3%). The manufacture of highly concentrated acrylic polymer solutions is not new. High molecular weight polymers (used as fiberlike or filmlike materials) were patented some decades ago [74,75]. The polymers described in Ref. 72 are not of high enough molecular weight. Because of their low molecular weight such polymers have to be crosslinked after coating. They can be crosslinked with the use of EB and UV curing. Chemical crosslinking by NCO-containing systems can be carried out also. Their tackified formulation depends on the choice of crosslinking method.

Generally the choice of a raw material class determines the range of possible performances also. As discussed by Gerace [76], butyl tapes tested in comparison with acrylics for bonding exterior trim moldings to automobiles perform better at low temperature and pass cold shock tests. This behavior is due to the low T_g of the butyl system.

Coating Technology Dependent Formulation. Formulations may differ according to coating technology also. The pressure-sensitive layer can be coated on the carrier material as a dispersant-free, 100% solids component or as a liquid having a dispersed or dissolved adhesive in organic solvent or water. Hot melts or extrusion-coated thermoplastic adhesive layers are applied in the molten state. Thermoplastic elastomers combine the excellent processing characteristics of thermoplastic materials and a wide range of physical properties of elastomers. For tapes, cold pressing of masticated natural rubber based formulations is known also. Special compositions can also be dry and cold sprayed.

Even with a 100% solids content, radiation-curable PSAs can also be coated (at room or elevated temperature) in the liquid state. Because of the different abilities of the raw materials to be coated in the molten or dispersed states, their formulations also vary. Although the development of macromolecular chemistry allowed the synthesis of new elastomers and viscous components (resins), their number is limited, especially in the range of meltable compounds (i.e., raw materials for hot melts) and water-based materials.

As discussed in Chapter 4, coating of a prepolymer is preferred for radiation-curing PSAs also. For instance, the viscosity of a radiation-crosslinkable low molecular weight acrylic is about 10–20 Pa.s. It can be processed at 120–140°C [55]. Another patent [77] states that the monomers should be partially polymerized (prepolymerized) to a coating viscosity of 1–40 Pa.s before use. The data of Table 6.10 illustrate the viscosity domain of low viscosity oligomer-based, UV-curable HMPSAs. As can be seen from these data, the viscosity of these compounds is much higher than the usual viscosity of solvent-based adhesives but significantly lower than that of common HMPSAs.

Generally hot melts are polymer blends (elastomeric and viscous components). Styrene-butadiene block copolymers were the first rubbery raw materials to be coated in the molten state. Although later AC- and EVAc-based hot melts were developed, styrene-butadiene block copolymers remain the most important

Table 6.10 Viscosity of Some Oligomer-Based HMPSA Formulations

Product	Chemical composition	Viscosity	Temp. (°C)	Ref.
Tape, label	Acrylated polyester, UV- or EB-curable HMPSA	2000–10000 mPa.s.	100	78
Tape, label	Polyurethane-acrylate HMPSA	11,625 cP	120	79
Tape	Rubber resin, in heptane, solvent-based PSA	500 mPa.s.	RT	80
Label	Styrene block copolymer, HMPSA	5000–80,000 mPa.s	175	81
Label	Acrylic, UV-curable oligomer	10,000–20,000 mPa.s	120–140	55
Tape	Acrylic, UV-curable oligomer	40,000 mPa.s.	—	77

raw materials for HMPSAs. Like adhesive manufacture for natural rubber resin based PSAs, TPE based hot melt formulation is really the manufacture of the adhesive. This manufacturing process is absolutely necessary to transform an elastic compound into a viscoelastic material of coatable viscosity. Therefore it may be concluded that, depending on the physical status of the coatable adhesive, formulation refers to the manufacture or modification of the adhesive. According to Hunt [82], because of the complex equipment and know-how required for HMPSA formulation (in a different manner from WBPSA formulation), special suppliers of ready-to-use HMPSAs should be preferred.

Hot melts have to be manufactured by mixing the elastomeric and viscous components. The adhesive produced must also be tailored with respect to processibility. In reality, the choice of tackifier resins influences the viscosity of the molten polymer also, but plasticizer and oils are necessary to regulate it. Low molecular weight polyisoprenes (35,000–80,000) having a low T_g (−65 to −72°C) have been used as viscosity regulators for EVAc-based HMPSAs [83]. They replace common plasticizers and improve migration resistance and low temperature adhesion.

Hot melt formulations should take into account the temperature sensitivity of the carrier material. As is known, hot melts may be coated on plastic carrier materials. Fabric, polyvinyl chloride, cellulose acetate, and polyester films have been coated with hot melt for labels [84]. Because of the differences in monomer basis and manufacturing procedure (polymerization, polycondensation, or polyaddition) of these polymers and films (blown, cast, nonoriented or oriented, etc.), their temperature resistance is quite different (see later, Manufacture of the Carrier Material Section 1.2).

Formulation for hot melts should also take the coating equipment, i.e., the coating device, into account. Different coating devices require different viscosities. The importance of processing viscosity is illustrated by the following list of quality criteria for HMPSAs given by Shields [85]: adequate adhesion on kraft paper and undulating surfaces, better adhesion on PE at $-18°C$, better tack, no migration, and no processing problems (no odor and low viscosity). The addition of low melting point resins to a hot melt formulation with block copolymers (Kraton GX 1657) decreases the melt viscosity [86]. It should be mentioned that low viscosity resins may produce migration in hot melt formulations based on block copolymers. The formulation of styrene block copolymers uses resins that are compatible with one of the segments of the block copolymer. Such compatibility can be limited by special manufacture of the resin also. The insolubility of a rosin-based tackifier resin in the styrene blocks of a hydrogenated SBR and SIR copolymer can be improved by its transformation into a coordination complex having the formula $M(ORes)_2 \cdot n$ (HORes), where $0 < n < 6$, M is the metal, and Res is the resin [87].

Solvent-based PSAs have a broader raw material basis. They can be formulated with rubber resins, acrylics, polyurethanes, polyvinyl ethers, and other materials. In this case formulating can be either the manufacture of an adhesive (e.g., rubber resin PSA) or the modification of adhesive (e.g., AC, PUR, PVE). In both cases it is focused mostly on the design of the adhesive properties. According to Hunt [82], solvent-based acrylics are used for special products such as no-label look PVC labels and where optimum water resistance is required.

For water-based PSAs, formulation is mostly modification of the base adhesive (AC, EVAc, PVE, etc.), but it is also focused on the adhesive and conversion properties. In this case the modification of the base adhesive is more complex because of its built-in (bulk) chemical composition and its chemical composition as a dispersed system (having particles with a composite structure). In the first stage of their development, water-based PSAs were considered less aggressive than solvent-based compositions. As can be seen from an old test standard [88], the coating weight suggested for water-based PSAs was 25 g/m^2. In comparison, the value given for solvent-based products was only 20 g/m^2. The chemical nature of the main polymer chain, its water-activatable functional groups, and the chemical (or macromolecular) nature of the technological (dispersion) additives also influence the formulation. As discussed in Ref. 89, the molecular weight of the polymer and its T_g are the major variables that determine the PSA properties of latices based on SBR. These are characteristics of the main polymer backbone. Other factors that play a role in adhesive properties are the chemical structure of the additional monomers used for synthesis, the choice of alkali for neutralization of carboxylic end groups, and the chemical structure of the surfactant in the latex. For water-based dispersions, the chemical basis of the main polymer influences the technology used for its dispersion and the chemical and dispersion properties of the product. As a consequence the formulation is affected too. As shown in Ref. 90, vinyl acetate polymers and copolymers with an acid pH give acid dispersions that must be formulated in acid domain; therefore only nonionic tackifiers should be used. Chloroprene latices have to be neutralized. The choice of neutralizing agent requires a special know-how [91]. Formulation also depends on storage time. When water-based viscoelastic and viscous components are mixed (e.g., acrylics with tackifiers), the peel resistance increases with the length of storage of the mixture (up to a limit).

Coating technology and the particular coating device can require a certain viscosity and solids content and influence the formulation. As an example, coating with an air knife requires a viscosity of 100–300 mPa.s [92] and a solids content of about 30% w/w, and the practical coating weight (generally between 2 and 25 g/m^2) is 2–3 g/m^2 (dry). The maximum running speed (250 m/min) decreases for films to 90 m/min. Formulations required for a Meyer bar differ from recipes to be coated with a reverse roll coater [93]. According to Sanderson [94], acrylic tape adhesives can be designed (i.e., produced) that differ in viscosity and foam generation as a function of the coater used. An adhesive for rotogravure has a viscosity of 75–100 cP and foaming of 2–5% in comparison with adhesives for a Meyer rod (viscosity of 300–800 cP, foam generation of 10–20%) and an adhesive for knife over roll (viscosity of 4000–8000 cP, foam of 10–30%). Spray-dried emulsion polymer powders are also known; for these raw materials a special dispersing technology is required.

In Europe the large label stock suppliers formulate their own water-based adhesives, but ready-to-use formulations are also commercially available [82]. The types and grades of PSAs that are used vary according to geographical regions and countries. In Europe the proportions of solvent-based, water-based, and hot melt adhesives vary according to geographical area also. In 1989 the most water-based adhesives were marketed in the northern area (United Kingdom, Iceland, and Scandinavia). The southern region (France, Italy, and Spain) leads in the use of HMPSAs [93].

A special aspect of formulation is related to crosslinking. For common crosslinked formulations (e.g., PSAs for tapes and protective films), the formulation should ensure thermal crosslinking in a short period of time of drying at elevated temperatures in a drying channel (see later, Manufacture of Protective Films Section 2.2). The majority of crosslinked formulations give a crosslinked adhesive after coating and drying, crosslinking, and cooling of the pressure-sensitive web. However, there are certain formulations where crosslinkability is "stored" and crosslinking occurs later, during bonding or debonding of the PSP. For these products a special coating technology and a crosslinking technology dependent formulation are necessary. A superficial crosslinking or two-component formulation is suggested. Such technologies are described in Refs. 95 and 96. According to Ref. 95, tapes with adhesion-free surfaces can also be manufactured. The tack-free surface is achieved by superficial crosslinking of the adhesive using polyvalent cations (Lewis acid, polyvalent organometallic complex, or salt). This crosslinked portion of the adhesive has a greater tensile strength and less extensibility. In some cases the polymer should be crosslinked only along the edges of the adhesive layer. The tack-free edge prevents oozing and dirtiness. The crosslinking penetration depends on the type of polymer and on the solvent. Crosslinking can also be achieved by dipping. In this process, water or water-miscible solvents should be used.

Acrylate-based PSAs containing acid or acid anhydride groups and glycidyl methacrylate coated on a flexible carrier can be crosslinked between 60 and 100°C by using a zinc chloride catalyst [96]. A special formulation is carried out. Instead of acrylate copolymers with monomers containing epoxide groups and compounds containing an acid anhydride group, mixtures of two separately prepared copolymers may be used, for example an acrylic copolymer containing epoxide groups and an acrylic copolymer containing anhydride groups. Crosslinking catalysts (0.05–5% w/w) are added to the polymer. The polymer coated on a carrier is crosslinked after evaporation of the solvent (2–7 min at 60–100°C). As catalyst, zinc or magnesium chloride, monosalts of maleic acid, organic phosphoric or sulfonic acid derivatives, or oxalic or maleic acid can be used. Tertiary amines and Friedel–Crafts catalysts accelerate crosslinking. The crosslinked adhesive is storable, i.e., its adhesive properties are not changed by postcrosslinking during storage. Even heat-sensitive materials like plasticized PVC may be used as the carrier material for such adhesives. It is possible to make a prepolymer that is polymerized by radiation. A small amount of catalyst or polyfunctional crosslinking monomer may be added to the composition prior to completing the polymerization by heat curing. Such prepolymers are more viscous and are easily applied to the support [15]. Formulation for adhesives coated as a radiation-curable mixture of oligomers or polymers should take into account the inhibiting or accelerating effect of the tackifier resin during curing. It has been shown that saturated hydrocarbon resins are more effective than their unsaturated counterparts in their ability to enhance electron beam crosslinking of styrene-isoprene block copolymers [97].

Formulation Depending on Product Class. A variety of pressure-sensitive products are manufactured using various coating techniques, carrier materials, and different constructions. Therefore, formulation differs according to product class also. Economic considerations also limit the use of certain raw materials. Except for special requirements and products, PSAs for labels are formulated to achieve better peel and tack; those for tapes, to obtain better shear; and those for protective films, to realize better removability.

For general application labels, formulation criteria imposed by their end use depend on whether the labels will be produced on reels or sheets and on whether the product is to be permanent or removable. Other end use requirements have to be taken into account for special labels, e.g., increased or decreased water sensitivity or mechanical resistance.

Generally, cohesive formulations are used for tapes. Such formulations are crosslinked compositions. In adhesive recipes for tapes, natural rubber should be crosslinked [98]. For PE tapes, adhesives on natural rubber basis have been suggested [99]. In their formulation, polyisobutylene and rosin are added as tackifier. To afford better dissolution of the rubber and to reduce its molecular dispersity, the elastomer is masticated, i.e., depolymerized by calendering. After coating of the formulated adhesive, the molecular weight is "rebuilt" by crosslinking to achieve higher cohesion. According to Ref. 46, although acrylic rubbers can be dissolved without milling, mastication lowers their viscosity. Such adhesives can be cured by using isocyanates, UV curing agents (e.g., p-chlorobenzophenone). Starting from the same raw materials, formulations for labels, tapes, or protective films can be quite different. The degree of mastication of the rubber, the tackifier level, and the degree of crosslinking may vary widely.

In some cases mastication helps to realize a fully solventless mixing and coating technology. The adhesive is calendered on a continuous carrier material or in a porous one. Such "press rolling" technology ensures solventless formulation and manufacture (no need for drying, no migration, no skin irritation), which is preferred for medical tapes. Rubber for medical tapes is masticated also [100]. According to Schroeder [101], the adhesives used for tapes have to display color stability, transparency, and chemical, thermal, and electrical resistance.

As stated in Ref. 102, adhesives for protective films must display adhesion that permits removal of the film. The adhesive properties of protective films are characterized by removability as a major requirement. This performance is achieved by an extreme reduction of the coating weight, but other formulation-related technical means are also used to reduce peel resistance. The first formulating modality is the change of the adhesion–cohesion balance. Because of the requirement for deposit-free removability, reduction of peel values cannot be achieved via softening of the adhesive (see Chapter 5). Another possible means of detackifying is crosslinking.

Softening of the "adhesive" was attempted earlier with the use of PVC plastisols (plasticizer gelled PVC compositions) to manufacture oil- and solvent-resistant paper-backed adhesives for covering sensitive substrate surfaces. Softening of the plastic carrier materials leads to adhesive-free self-adhesive films (see Self-Adhesive Films Section 2.2).

Like recipes for tapes, most natural rubber based formulations for protective films use mastication. In this case calendering is used as one way to regulate viscosity and peel resistance. Because of the requirement for a low peel, a low coating weight is suggested for these products. Low coating weight can be exactly controlled by using low viscosity formulations. Because of the low peel requirement, a relatively low level of tackifier resins is recommended. A low level of tackified resins means a higher viscosity of the formulated adhesive. Therefore mastication plays a special role in the manufacture of adhesives for protective films. As a consequence, postcrosslinking is very important also.

Removability criteria influence the adhesive choice and formulation for tapes too. Because adhesion builds up due to the use of highly polar comonomers such as acrylic acid in acrylic PSAs, such components are not recommended for the synthesis of adhesives for masking tapes [103]. Decorative films need special formulation in order to avoid an interaction (migration) between the components of the carrier material and the adhesive [104].

Product Buildup Dependent Formulation. The nature of the carrier material influences formulation. Heat-sensitive carrier materials need "low" temperature warm melts. These are hot melts that can be used for temperature-sensitive applications. According to Ford [105], such products do not have a lower application temperature but the plastic-to-liquid transition temperature has been dramatically reduced while the softening point has been held constant.

The carrier influence is given by its surface and bulk characteristics. Because HMPSAs are based on relatively low MW polymers, plasticizer migration affects them more. On the other hand, many health care applications exist where it is desirable to coat a PSA on a porous web. A porous carrier is required, and the adhesive should also be porous [106]. In this case the use of HMPSA may reduce the penetration of adhesive into the web.

The surface quality of the adherend can require a special adhesive geometry or/and coating weight. This can be achieved by tailored formulations. Crosslinked, filled, and multilayer compositions are used. According to Czech [107], UV curing of monomer–prepolymer mixtures is suggested for medical tapes with a high coating weight. As stated in Ref. 108, EB-cured formulations can be used only for coating weights of less than 100 g/m^2.

The formulation of the adhesive depends on the printing method and also on the technology used for the face stock material. For instance, rating plates can be printed in sandwich printing, where on transparent materials are printed on the reverse side by using mirror lettering [109]. Where the adhesive is to be used to form laminates in which at least one surface is a printed surface, the presence of any residual surfactant can lead to discoloration or bleeding of the ink. This is a problem in applications such as overlaminating of books or printed labels, where the purpose of the outer surfacing film is to preserve the integrity of the printed surface.

End Use Dependent Formulation. End use requirements for PSPs are various; therefore, requirements for the PSAs also vary. According to a study conducted by the Skeist Laboratories [110], low temperature resistance for deep freeze price labels, balanced adhesion–cohesion for structural pressure-sensitive adhesives for holding and mounting, high shear and/or peel for packaging adhesive tapes, resistance to perspiration and body liquids for personal care products, and heat resistance for electrical and industrial uses are the main requirements for the adhesives needed in future label production. Temperature-resistant adhesives are required for thermally printed labels and copy labels [111]. Solvent-based acrylics are recommended for durable mounting tapes, outdoor PSPs, medical tapes, protective tapes, and high performance decals [112]. Liquid resins are recommended for plasters, and multifunctional anhydrides and multifunctional epoxies are suggested as crosslinking agents [113].

Quite unlike labels, where the adhesive properties are determinant for the end use, for other product classes the application and deapplication conditions have greater importance. There are special end use requirements for a given application or application field. The actual application conditions affect formulation within the same product class also. For instance, for bookbinding, nonyellowing carrier materials and adhesives are preferred [114]. According to Fuchs [115], for decorative films, a soft PVC carrier (coated with 18–28 g/m^2 HMPSA) is used. Some special pharmaceutical labels may have take-off (detachable) parts to allow multiple information transfer [116]. Parts of the label remain on the drug or on the patient. The detached part of label should have a medical adhesive so it can be applied to skin. Nonremovable and nontransferable constructions have been manufactured as antitheft labels (see Chapter 8 also). In an application of this type, a destructible, computer-imprintable vinyl film and a very aggressive PSA have been used. For such a product, the adhesive must bond in less than 24 h [117].

Water-resistant products are required for many applications. The water resistance of polymers depends on their polarity, crystallinity, and structure. The water resistance of dispersion-based coatings is primarily a function of their water-soluble additives [118] (see Chapter 4 also). Theoretically, for water-resistant applications, water-insoluble solvent-based adhesives should be used. Water-based acrylics that can survive immersion in water have been synthesized also [119]. For several applications PSAs have to resist atmospheric moisture but be removable with warm water [120]. Wine and champagne bottle labels have to be removable in water without supplementary addition of chemicals at 60°C [121,122].

Water-soluble formulations may have quite different raw material bases. They can be formulated by using common, commercially available polymers or by synthesis of special macromolecular compounds. Their formulation depends on the degree of solubility required and on the adhesive coating technology. Water-releasable labels and water-dispersible splicing tapes or medical tapes need quite different degrees of solubility. Building solubilizing functional monomers into a polymer or adding water-soluble agents to a formulation depends on the physical state of the adhesive also. Emulsion copolymerization of certain water-soluble monomers is difficult because of their tendency to homopolymerize in water. Therefore solution polymerization is used. On the other hand, water-soluble additives (surfactants) can be fed into water-based dispersions without a problem. The use of nontacky hydrophilic solubilizers in a hydrophobic molten HMPSA requires special knowledge. Ideally both the main elastomer or viscoelastomer and the tackifier or plasticizer should be water-soluble. In practice, such formulations are used only for difficult applications like splicing tapes or medical labels. For water-releasable labels, the use of solubilizing additives is suggested [121].

A water-soluble PSA hot melt is formulated on the basis of vinylpyrrolidone-vinyl acetate copolymers. The composition also includes a free monobasic saturated fatty acid with an acid number higher than 137 [123]. According to Ref. 124, water-soluble HMPSAs for labels are formulated with 35–60% vinylpyrrolidone or vinyl acetate-vinylpyrrolidone copolymer together with free fatty acids. Such labels can be peeled off in 5 min by subjecting the edge of the label to a stream of water at 35°C. Fatty acids with 8 to 24 carbon atoms act as solubilizer in a concentration of 3% or less [125]. As discussed in Refs. 44 and 126, water-soluble hot melt adhesives can be formulated with vinylpyrrolidone, pyrrolidone-vinyl acetate copolymers, and water-soluble polyesters. Polyvinylpyrrolidone-based formulations exhibit poor thermal stability. A hydroxy-substituted organic compound (alcohols, hydroxy-substituted waxes, polyalkylene oxide polymers, etc.) and a water-soluble *N*-acyl-substituted polyalkylenimine obtained by polymerizing alkyl-substituted 2-oxazolines can be used with better results. Such water-releasable HMPSAs contain 20–40% polyalkylenimine, 15–40% tackifying agent, and 25–40% hydroxy-substituted organic compound [44]. Water-soluble plasticizers whose molecular weight does not exceed 2000 are preferred (e.g., polyethylene glycols of molecular weight of about 200–800). Fillers with increased organophilicity (e.g., fatty acid ester coated mineral extenders like stearate/calcium carbonate compound) are suggested. As an antioxidant, distearyl pentaerythritol diphosphite is used. As a water-soluble base component, polyethyloxazoline has been proposed for water-soluble hot melts also [125]. Polyalkyloxazolines and water-insoluble polymers have been introduced for formulation of PSAs displaying (alkaline) water dispersibility, solubility, water activation, and solubility in alkalies [45]. The recipe comprises a rubbery polymer, a polyalkylenimine, a functional diluent based on acid functional polymeric compounds, hydroxy-substituted organic compounds, tackifiers (rosin acids) and plasticizers, waxes, and an acrylic acid copolymer to provide additional strength. Such a formulation for a repulpable general-purpose HMPSA contains a thermoplastic polymer (20–50%), polyalkylenimine (10–

30%), diluent (2–90%), wax (5–40%), tackifier (85–40%), plasticizer (0–40%), and filler. As thermoplastic components, vinyl polymers, polyesters, polyamides, polyethylene, polypropylene, and rubbery polymers prepared from monomers including ethylene, propylene, styrene, acrylonitrile, butadiene, isoprene, acrylates, vinyl acetate, functionalized acrylates, and polyurethanes can be used.

According to Ref. 127, water-dispersible hot melts are based on vinyl ester graft copolymers with water-soluble polyalkylene oxide polymers. Vinyl carboxy acid based polymers tackified with water-based resin dispersions are used also [128]. Water-dispersible HMPSAs can contain 25–40% SIS, 25–40% tackifier, 10–25% fatty acid, and 10–25% PVP [129]. A proposed formulation for a water-soluble HMPSA comprises 22, 39, 11, 14, and 14 parts of ethylene-acrylic acid copolymer, PVP, plasticizer, tackifier, and polyethylene glycol, respectively. A simplified formulation contains 60, 30, and 10 parts of PVP, polyethylene glycol, and phthalate plasticizer, respectively [129].

According to Czech [130], a water-soluble adhesive for splicing tapes can be manufactured using (1) a polymer-analogous reaction; (2) polymerization; or (3) polymerization and formulation.

By blending the PSA raw materials that have carboxyl groups with reactive hydroxyl or amine groups and neutralizing the mixture, it is possible to regulate the solubility of the polymer. The synthesis of a special PSA on the basis of aminoalkyl methacrylates or methacrylamide derivatives also leads to water-soluble polymers. Formulating the water-soluble viscoelastic components with water-soluble resins gives better adhesivity and better water solubility. Polyacrylates tackified with water-soluble PVE have also been proposed [131]. Amine-containing soluble polymers and soluble plasticizers have been proposed [132]. A water-activatable PSA containing a water-soluble tackifier (5–15% w/w) is described in Ref. 133.

Paper-based tape recyclable as paper is manufactured by using a water-soluble adhesive based on alkali-soluble acrylic acid esters and polyethylene or polypropylene waxes and alkali-dispersible plasticizer [134]. For splicing tapes that have to resist temperatures up to 220–240°C, carrier materials have to be water-dispersible, and the adhesive should be water-soluble but resistant to organic solvents. No migration is allowed. Their adhesive has to bond at a running speed of 160 m/min. Under such conditions, high instantaneous peel and shear resistance are required [135]. Water-soluble compositions used for splicing tapes include polyvinylpyrrolidone with polyols or polyalkylglycol ethers as plasticizer [135], vinyl ether copolymers [136], neutralized acrylic acid-alkoxyalkyl acrylate copolymers [137], and water-soluble waxes [138]. Such formulations must adhere to the paper substrate running at 1200 m/min and must be soluble in water at pH 3–9. For a water-soluble plasticizer, 20–70%

w/w ethoxylated alkylphenols, ethoxylated alkylamines, and ethoxylated alkylammonium derivatives have been proposed. For higher shear, short cellulose fibers (1–5%) are added as filler.

Water-based PSA formulations that are removable with warm water contain 7–30% hydrophilic monomers [139]. Polymeric oxazoline derivatives can also be used in water-soluble adhesive compositions. A reaction product of an acrylic interpolymer and 2-oxazoline has been suggested [140].

Polyacrylate-based water-soluble PSAs are recommended for medical PSPs (operating tapes, labels, and bioelectrodes) [132]. Operating tapes are used for fastening cover material in surgical operations. Cotton cloth and special hydrophobic textile materials coated with low energy polyfluorocarbon resins are used as carrier materials together with polyacrylate-based water-soluble PSAs. The adhesive has to be insensitive to moisture and insoluble in cold water. It has to display adequate skin adhesion and skin tolerance. The water solubility of the product is reached above 65°C at a pH higher than 9. Water solubility is given by a special composition containing vinyl carboxylic acid (10–80% w/w) neutralized with an alkanolamine [141]. Such tapes are used for biolelectrodes also because of their low electrical impedance and good conformability to skin. Other compositions contain acrylic acid polymerized in a water-soluble polyhydric alcohol and crosslinked with a multifunctional, unsaturated monomer. A prepolymer (precursor) can also be synthesized that is polymerized to the final product by using UV radiation [94]. This polymer contains water, which affords electrical conductivity. Electrical performance can be improved by adding electrolytes to the water. Such products can be used as biomedical electrodes applied to the skin.

The adhesive for medical tapes has to be resistant to light and heat, show skin tolerance, and transmit water vapor. The adhesive can contain acrylic, styrene copolymers, vinyl acetate copolymers, and polyvinyl ether [142]. For medical tapes applied to the skin, adhesives based on acrylic and natural rubber have been suggested [143]. To prepare high solids content solutions, the elastomer should be masticated [144]. According to Ref. 100, polyisobutylene (MW 80,000–100,000) is used for medical tapes because it does not adhere to the skin. Physiologically compatible adhesives can also be based on polyurethanes, silicones, acrylics, polyvinyl ether, and styrene copolymers. These should be breathable, cohesive, conformable, self-supporting, and nondegradable by body fluid [145]. Special polyurethanes have been suggested for medical tapes too. As shown in Ref. 23, unlike rubber formulations, which leave a deposit on the skin, and acrylic formulations, which may irritate the skin, PUR gels are inoffensive. As discussed in Ref. 73, generally the type of polyethylene glycol, its functionality (2–4), its structure (linear or branched), and its OH number influence its properties. For the polymer proposed in Ref. 23, the average functionality (F_1) of the polyisocyanates used being between 2 and 4 and the average functionality of the polyhydroxy compounds (F_P) being between 3 and 6, the isocyanate number (K) is given by the formula

$$K = \frac{300 \pm X}{F_I F_P - 1} + 7 \tag{6.3}$$

where $X \leq 120$ (preferably $X \leq 100$) and K is between 15 and 70.

As discussed in Ref. 146, a fluid-permeable adhesive useful for transdermal therapeutic devices is based on an acrylic, urethane, or elastomer PSA mixed with a crosslinked polysiloxane. The therapeutic agent passes through the adhesive into the skin. Uninterrupted liquid flow through the adhesive has to occur over a prolonged period (6–36 h) and at a constant rate [147]. Acrylic PSAs in medical tapes may display adhesion buildup over time or a weakening of cohesive strength due to the migration of oils. These disadvantages are avoided by crosslinking with built-in dimethylamino methacrylate or with isocyanate [147,148]. Table 6.11 lists some of the raw materials used for medical PSAs. As illustrated by the data in this table, a very broad range of raw materials are proposed for the manufacture of medical tapes. Such formulations are coated mainly without solvent (roll pressing, extrusion, 100% solids, etc.), but compositions soluble in organic solvents or water are suggested also.

Wet surface adhesives have to adhere on cold adherend surfaces coated with condensed water [154]. Water-resistant formulations are used for different products. Tackified formulations made with a special technology (using the tackifier's own surfactants only) give better performance. In this case the solids content is very high (20% more) in comparison with common formulations. Such adhesives with excellent water resistance (resistance to humidity and condensed water) are required for technical tapes and pressure-sensitive assembly parts in the automotive industry [155].

Low temperature resistant adhesives are required for deep freeze labels. As discussed in Chapters 3 and 4, sequence distribution strongly influences the rheology of block copolymers. Therefore, diblock polymers are preferred for such applications [156]. For instance, for a low temperature HMPSA, copolymers with A_1B_1 sequence distribution have been proposed, where A_1 is an aromatized vinyl polymer block (molecular weight 12,000) and B_1 is a diene block with a molecular weight greater than 150,000. Another patent [131] describes the use of natural rubber (50–75 parts), SBR (with 12% styrene), regenerated rubber (3–8 parts), and PVE (1–3%). As tackifier, polyterpene resin 1–4% (MP 115°C), rosin ester (1–3%), and carboxylated alkylphenol resin (2–15%) are suggested. Such a crosslinked (with isocyanate) formulation can be used between -29 and $+113°C$.

Formulations used for contact adhesion (PSA on PSA) generally contain natural rubber. For instance, Ref. 157 proposes a recipe based on NR and resin (150/50). Filled natural rubber–synthetic rubber mixtures can be used also [158].

Table 6.11 Raw Materials and Manufacture of Medical Tapes

Product	Manufacture	Components Adhesive	Carrier	Coating weight	Other characteristics	Ref.
Surgical tape	Coating	Acrylic HMPSA	Rayon	1.4 oz/yd^2	Application temperature maximum 350°F	141
Tape	Coating	AC prepolymer, UV-cured	—	—	Electrically conductive	95
Tape	Coating	Solvent-based AC	—	—	Solution in isopropyl acetate, 3000–110,000 cP	149
Tape	Coating	Solvent-based AC	Cotton cloth	—	Cold water insoluble, hot (65°C) water soluble	132
Tape	Coating	AC, crosslinked	—	—	—	147,148
Tape	Coating	Solvent-based carboxylated AC	—	—	—	141
Tape	Coating	AC, PVE, PVAc	—	—	K 80–130	141,142,144
Tape	Coating	SB, AC/RR	—	—	—	142
Surgical tape	Coating	Isobutyl acrylate-butylcarbamoyl methacrylate	—	—	—	150
Tape	Coating	EHA, PUR, PVE, styrene copolymer	—	—	—	151
Tape	Coating	RR, SB	—	17 lb/in^2	Solution in hexane (80%), toluene (3%), cyclohexane (17%)	152
Tape	Roll pressing	RR	—	—	Zinc resinates, zinc oxide filler	153
Tape	Roll pressing, extrusion	PIB	—	—	—	99
Tape	Coating	PUR	—	—	—	23

Manufacturing Technology of the Adhesive

Manufacturing technology includes both the procedure and the equipment used to manufacture the coating components of the PSPs. As discussed earlier, manufacture of the coating components generally consists of their formulation; therefore, the manufacturing equipment is formulating equipment.

Formulating Equipment. The raw materials determine both the coating technology and the formulating equipment. Formulating is the manufacture of coating material by mixing the product components in the solid, molten, or dispersed state. Certain raw materials are solid, and others are liquid or have to be transformed into a liquid system. Some raw materials have to be mixed; other compounds are supplied in ready-to-use form. Certain raw materials are supplied in a micronized state, whereas others are predispersed. Therefore, formulating equipment varies depending on the chemical basis of the adhesive, the coating technology, and the product classes.

Two-component formulations that include rubberlike products and tackifiers require that the components be micronized (cut, pelletized), mechanochemically destroyed (masticated, plasticized), and solubilized. To facilitate solvation, the surface area of the solid raw materials (e.g., elastomers) should be increased by physical size reduction. This can be accomplished by using a bale chopper or granulator; a sheet mill and slab chopper, or a Banbury mixer. For instance, for the manufacture of an adhesive for pipe isolation tapes based on butyl rubber, a crosslinked elastomer is required. A cutting extruder is applied to mix the pellets of rubber (100 parts) with naphthenic oil (120 parts) and resin (4 parts; Super Bekacite 2000) [159]. The construction and hydrodynamic characteristics of the equipment may influence the processing of HMPSAs. When polyacrylate rubber based HMPSAs are prepared, high shear mixing is required [46]. Mastication depends on the temperature, rotating speed, and clearance of the masticating cylinders [160]. According to Ref. 161, in the continuous processing of HMPSAs (based on EVAc) there is a relationship between plug diameter and melt viscosity. Melt viscosity decreases with plug diameter.

Formulating Problems. General aspects of raw materials such as storage conditions and warranty have to be taken into account. Special features concerning the compatibility of the components and their order of addition are important also. For instance, for natural rubber latices and adhesives based on natural rubber latex, a shelf life of 3 months is expected when they are stored in sealed containers at normal temperature [162]. Water-based acrylics have a minimum storage time of 6 months. Natural rubber latices are not compatible with other adhesives, and care should be taken to wash out the equipment [162]. Special formulating equipment is needed for crosslinked formulations. Curing of two-component adhesives (or primers) starts immediately after the components have been mixed together. Therefore only a limited time remains in which the adhesive can be applied. This period is referred as pot life.

Manufacture of Adhesives for Labels

According to Ref. 163, the adhesive for labels must have the following qualities: good adhesion on critical substrate surfaces; chemical, temperature, and UV resistance; adhesion on different surface structures; easy labeling; removability; repositionability; noncorrosivity; plasticizer resistance; mechanical resistance; adhesion on wet surfaces; dimensional stability; transparency; instantaneous/final adhesion; uniform adhesion toward the release liner; no emission of gaseous components; and FDA approval.

Evaluating these properties according to their priority, one can state that well-defined adhesive characteristics and the stability of these characteristics independently from the application and environmental conditions are needed. Solvent-based, water-based, hot melt, and radiation-cured adhesives can be used for labels. As illustrated by the statistical data, some years ago solvent-based adhesives were preferred in label manufacture. In 1984 mainly (60%) solvent-based adhesives were used for labels and tapes in the United States [164]. Now water-based formulations are the most common.

The choice of adhesive formulation depends on its end use. A preparation plant is necessary for the manufacture of the adhesive [165]. Its complexity depends on the nature of the adhesives to be manufactured. The manufacture of adhesives for labels includes their formulation, production, and quality assurance. The adhesive for labels can be synthesized in-line also. Radiation-induced polymerization or curing can be carried out to achieve a PSA-coated label material.

Because of the balanced adhesivity required for labels, label manufacture is based on coating with PSA. The formulation of a PSA has to ensure its coatability. The range of possible raw materials for PSAs for labels is very broad. Two-component rubber resin or one-component viscoelastic polymer-based formulations can be coated in the molten, dissolved, or dispersed state. Radiation curing has been recommended for liquid (dissolved) prepolymers or molten, high polymer based formulations. Because of the special role of the face stock material as information carrier and the high surface quality required for labels, only classic coating technologies using very fluid adhesive and giving a smooth adhesive surface are suggested for label manufacture. Started with rubber resin based PSA solutions, label coating now uses mostly acrylic water-based PSAs. Therefore the formulation of PSAs for labels is mainly a formulation of water-based compounds. In practice, it consists of blending liquid-state components. Water-based adhesives for labels have a 4 year shelf life and can be used in the temperature range of -30 to $+70°C$ [166]. Generally they conform with FDA Regulation 21 CRF 175 105 and BGA XIV.

Manufacture of PSPs 373

The adhesives for labels are tackified formulations having a resin level above 30% by weight (paper labels). This is required for their high speed, low pressure application on difficult adherend surfaces. For instance, products wrapped in plastic film (PE, PP, PVC, PA, ionomer/PA, coated cellophane) such as refrigerated meat need a PSA that can be applied at low temperature [167]. The service temperature of the film is -50 to $-150°F$. Tackified adhesives for labels are softer (more viscous) and therefore display a better wetting of nonpolar surfaces at low temperatures [163]. The dwell time, i.e., the time necessary to build a full surface contact, is shorter for such adhesives. According to Ref. 50, removable price labels use almost exclusively a paper carrier coated with rubber resin adhesive. Natural and synthetic rubber and polyisobutylene tackified with soft resins and plasticizer have been suggested for such formulations. Gasoline, toluene, acetone, and ethyl acetate have been used as solvents for rubber-based adhesives. The elastomer is calendered, or masticated. For this application Mueller and Tuerk [50] proposed a water-based acrylic formulation of crosslinkable comonomers such as acrylo- or methacrylonitrile, acrylic or methacrylic acid, N-methylol acrylamide, and plasticizer (10–30% w/w). For the plasticizer, common commercial products (e.g., alkyl esters of phthalic, adipic, sebacic, acelainic, or citric acid and propylene glycol or methyl vinyl ether) have been suggested.

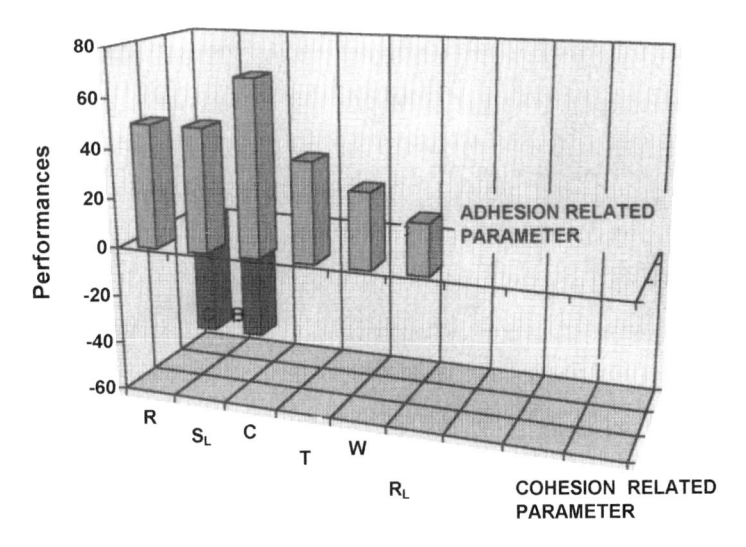

Figure 6.6 The influence of adhesion- and cohesion-related functional parameters on the whole performance of PSAs for labels. Adhesion-related parameters include reel application versatility (R), labeling speed (S_L), the use of nonpolar carrier materials (C) and ester-based tackifiers (T), low coating weight (W), and solvent-free release liner (R_L).

If a two-component crosslinking WBPSA is used, the mixing of components must be simple and the proportions tolerance must be high for operator convenience. Pot life at the coating station must exceed one working day. Foaming should be avoided.

Generally the development of adhesive raw materials and formulated adhesives for labels is affected by requirements for more adhesivity or/and more cohesion. Requirements for more adhesivity include versatility for roll application, increased labeling speed, adhesion on nonpolar carrier material, formulability with resin ester based tackifiers, reduced coating weight, and the use of solventless release. Requirements for more cohesion are related to cuttability and die cuttability and bleeding. Figure 6.6 illustrates the influence of these parameters on the performance characteristics of label PSA, on an arbitrary scale.

Manufacture of Adhesives for Tapes

The volume of PSAs used for coating tapes is higher than for labels (due to the higher coating weight). At the end of the 1980s the ratio of PSA manufacture for tapes compared to labels was 75/25 [168]. As can be seen from Table 6.12 there is a very broad raw material basis for various tapes. Rubber resin and acrylate based formulations compete with special raw materials (silicones, PVEs, polyesters, radiation-curable recipes, etc.). Monomer-, oligomer-, and polymer-based compositions are used. Tackifiers, plasticizers, and crosslinking agents are the usual components of such formulations. Because of the wide range of materials, tape manufacturing methods vary widely also.

As discussed in Chapters 2 and 8, the greatest volume of tapes are used in packaging. Such tapes are relatively simple products that have to fulfill mechanical assembly requirements like a packaging material. These are low cost products where raw material price and productivity play special roles.

The most common tape formulations are rubber-based and are related to end use. Special tapes can be manufactured on a rubber resin basis also. Rubber has to be masticated or plasticized for many tape applications. Hot melts have been used for labels, tapes, and decals since the early 1970s [183]. Natural and thermoplastic rubbers provide a relatively inexpensive raw material basis, and hot melt coating ensures the highest coating speed. According to Fries [184], hot melt technology consumes only 25% as much energy as a solution-based coating process. Therefore several tapes are formulated with HMPSAs. Radial and teleblock copolymers with isoprene-based branches have been suggested for PSA tapes [185]. A formulation with a tackified star block copolymer gives a higher shear resistance (at least 1000 min, according to ASTM D-3654, overlap shear to fiberboard at 49°C).

Water-based chloroprene dispersions exhibit improved shear performance and can be used for tapes. Silicone polymers have excellent electrical properties such as arc resistance, high electrical strength, low loss factors, and resistance to current leakage. These characteristics are very important for electrical insulating

Manufacture of PSPs

Tape 6.12 Chemical Basis of Tapes

Application	Base elastomer	Other components	Ref.
Packaging	NR, SBS	Terpenephenol resin	169
Packaging	NR, PIB	Rosin	99
Packaging	NR, polybutadiene	Rosin, crosslinker	98
Packaging	NR, SBS	Resin	170
Packaging	cis-Polybutadiene	Tackifier	98
Packaging	SIS, SBS, cis-polybutadiene	Resin, plasticizer	171
Packaging	AC	—	172
Packaging	EVAc, PE, HMPSA	Rosin, waxes	173
Packaging	SEBS, HMPSA	Hydrogenated hydrocarbon resin, polymethylpentene	174
Packaging	AC, WB	Epoxy-containing polyglycidyl compounds	175
Packaging	AC, HMPSA	Crosslinked with Zn carboxylate	176
Packaging	Acrylated polyester, HMPSA, UV- or EB-cured	—	78
Packaging	AC, SB	Polyisocyanate-cured	177
Packaging	NR	Aliphatic petroleum resin, oxidatively degraded EPR	178
Packaging, deep freeze	NR, SBR, PVE, regenerated rubber	Rosin ester, alkylphenol resin, crosslinker	130
Masking	AC	Terpenephenol	179
Masking	AC, oligomer, EB-cured	Trimethylolpropane tris(thioglycolate), hexanediamine	14
Insulating	Polybutadiene, SBC	Tackifier	180
Pipe wrapping	Polysiloxane	Tackifier	181
Pipe wrapping	Polysiloxane	Tackifier	182

tapes. These polymers have low surface energy and give excellent low and high temperature performance [186].

The formulation of the adhesive for tapes depends on the choice of carrier material also. For tapes with a PE carrier, PIB and NR have been proposed as elastomer [187]. Crosslinked butyl rubber based tapes for pipe insulation have been manufactured since 1963. A mixture of PIBs with molecular weights of 100,000 and 1500 is applied. Rosin is added as the tackifier. Tapes based on PE have been coated with adhesives based on natural rubber, polyisobutylene, tackifier resins (phenolformaldehyde, terpene, and coumarone resins), and fillers (SiO_2, ZnO, talcum, etc.). Air-seal tape is made with butyl rubber (on ethylene propylene foam) and solvents [187]. Such formulations contain natural rubber (100 parts) with polyisobutylene, hydrocarbon resin (50–70 parts), and low molecular weight polybutylene (4–50 parts) and mineral oil. For insulating tapes, a mixture of 100 parts of polyisobutylene having a molecular weight of 1000, 120 parts of low molecular weight polyisobutylene (MW 1500), and 150 parts of rosin has been proposed.

For the PVC carrier, natural rubber, chlorinated rubber, chloroprene, and perchlorovinyl have been suggested. Chlorinated rubber gives excellent adhesion on PVC, but primers can also be used. Acrylonitrile-butadiene-styrene and acrylonitrile-butadiene rubber blends are recommended as the primer [187]. Adhesive formulations based on chlorinated rubber (90 parts) include a plasticizer (29 parts) and a tackifier resin (35 parts). Such compositions can contain chlorinated rubber (60 parts), polyvinylbutyral (10 parts), phenolic resin (45 parts), and plasticizer (90 parts). Pipe insulation tapes with PVC carrier are coated with an adhesive based on natural rubber (crepe 27 parts), ZnO (20 parts), rosin (22 parts), or hydrated abietic acid (8 parts), polymerized trimethyldihydrochinoline (2 parts), and lanolin (4 parts).

Double-faced foam tapes are designed for industrial gasketing applications. They are based on closed-cell PE foams coated with a high tack, medium shear adhesive [188]. They can be applied as pipe insulation tape for gas pipe lines also [189]. Air seal tape is made with butyl rubber on ethylene propylene foam. Tapes applied as anticorrosion protection for steel pipes contain butyl rubber, crosslinked butyl rubber, regenerated rubber, tackifier (polybutene or resins), filler, and antioxidants. Low molecular weight polyisobutylenes (M_n less than 600,000) give better tack and peel [190]. Generally, blends of different molecular weights are used. Faktis improves the creep resistance for such tapes. Insulating tapes for gas and oil pipe lines are manufactured by coextrusion of butyl rubber and a PE carrier. According to Risser [191], butyl rubber, plasticizer, carbon black, and terpene phenol resins have been tested for sealing tapes. Polyacrylate rubbers can be used for HMPSAs, solvent-based PSAs, and reactive hot melts for pressure-sensitive sealant tapes and caulks [46]. Polyacrylate rubbers for these HMPSAs are tackified (40% resin) or plasticized. A common tacky PSA can be added to the formulation also (50 parts to 100 parts acrylic rubber). The composition can be cured by using isocyanates or UV curing agents (e.g., p-chlorobenzophenone). The recipe can include a high loading of fillers. Such formulations can contain 200 parts of hydrated alumina as filler for each 100 parts of rubber, 50 parts of carbon black, and 50 parts of plasticizer. Polyisobutylene can be added also.

Deep freeze tapes are coated with an adhesive based on natural rubber, styrene-butadiene rubber, regenerated rubber, rosin ester, or polyether [192]. Generally, SBR polymers used in the adhesive tape industry contain a gel, either a microgel or a macrogel [193]. A microgel is built into the polymer during synthesis by crosslinking between polymer chains within the latex particle. A macrogel results from crosslinking between particles, which builds up large masses of polymer, generally due to instability in the polymer. A microgel can be broken up mechanically to give a microgel. When a macrogel is broken, the active radicals can react internally to produce a more tightly crosslinked microgel or they may react externally to give an unstable viscosity; therefore the dissolution of SBR is a difficult process. A microgel yields an easily dipersed, low viscosity, smooth, shiny mastic; a macrogel gives a poorly dispersed, high viscosity, and grainy mastic. Macrogel formation can be prevented by improving the rubber stability. Most of the viscosity increase occurs during the first few days of aging and depends on the solids level. The addition of aromatic solvents results in greater viscosity on aging. Although a stable viscosity could be achieved through proper selection of solvent/resin/rubber levels and total solids, the characteristics of the elastomer are the major determining factor in viscosity stability.

Masking tapes are used in the varnishing of cars [194] (see Chapter 8 also). They have to adhere to automotive paint, enamel, steel, aluminum, chrome, rubber, and glass and resist the high temperatures used for drying and curing varnishes. Natural rubber latex tackified with PIB is recommended for such tapes. Polyvinyl ether can be used also. A mixture of polyvinyl-n-butyl ethers of different molecular weights (1000–200,000 and 200,000–2,000,000) is suggested for such tapes [195]. Butyl rubber (15–55 parts) untackified or tackified with PIB (7.5 to 40 parts) and crosslinked with zinc oxide (45 parts) has been proposed also [196]. Such formulations give (180°) peel values of 0.73–1.81 kg/25.4 mm. Peel increases with increases in the tackifier level, but legging appears at a tackifier/elastomer ratio of 40/15.

Water-based acrylics are recommended as adhesives for tapes also. Such PSAs are used for packaging tapes with OPP carrier [197]. CSBR has been proposed for high shear tapes [51]. The direct feed-in of the resin in such formulations is suggested in Ref. 198. Therefore a tackifier resin with MP < 95°C is recommended.

A special chapter of adhesive tape formulation concerns medical tapes. In this application field, old rubber-based or polyvinyl ether based formulations compete with special acrylics, TPEs, and polyurethanes (see Table 6.10).

According to Ref. 199, pressure-sensitive adhesives for medical tapes are based on natural rubber or polyvinyl ether. The formulation contains natural rubber (100 parts), a terpenephenol resin (80 parts), and solvent (600 parts). Polyisobutylene is used for medical tapes [200]. This type of tackified polyisobutylene has been suggested for other tapes also [201]. It has been plasticized at 80°C. Butyl rubber (MW 6000–20,000) does not need mastication to be dissolved, unlike many natural and synthetic elastomers. It is clear and tacky, it is not fragile after aging, and it can be used together with PIB as a tackifier for medical tapes [202]. Air- and moisture-permeable nonwovens (PET nonwoven, embossed nonwoven) or air-permeable tissue (breathable face stock) are used for medical tapes coated with a porous, nonsensitizing acrylic adhesive for prolonged application of a medical device on the human skin. They minimize skin irritation due to their ability to transmit air and moisture through the adhesive system. The air porosity rate has to attain a given value (50/100 cm^3 per square inch) [203].

The first adhesives for medical tapes were based on the reaction product of olive oil and lead oxide (lead oleate) plus tackifier resins and waxes [204]. Lead oleate was used as an adhesive component for medical tapes later also [205,206]. Natural rubber, tackified with rosin, is also recommended for adhesives for medical tapes [67]. The rubber for such tapes is masticated [207]. Polyvinyl ether is suggested for medical tapes alone or with polybutylene and titanium dioxide as filler. These compounds are warm coated without solvent (via coating by calendering). By cooling of the roll pressed mass, a porous adhesive is achieved [208]. Polyvinyl acetate, polyvinyl chloride, polyethylene, and ethylene copolymers can be used as adhesives for medical tapes too [209]. Such compounds are hot pressed in the fibrous carrier material.

According to Ref. 73, the PUR-based PSAs used for medical tapes include polyethylene polyol, polyisocyanate, resin, catalyst, and wetting agent. The type of polyethylene polyol, its functionality, and its structure and hydroxy number are the primary parameters. As the isocyanate component, isophorone diisocyanate, trimethylhexamethylene diisocyanate, hexamethylene diisocyanate, naphthylene-1,5-diisocyanate, etc., are recommended. Hydrocarbon resins, modified rosin, polyterpene, terpenephenol, coumarone-indene, aldehyde- and ketone-based resins, and β-pinene resins are suggested as tackifiers. As catalyst, Sn(II) octoate is used. According to Ref. 73, the components of the formulation are mixed as follows: 100 parts polyethylene polyol, 50 parts resin, 1 part catalyst, 7 parts diisocyanate, and 1 part wetting agent. The resin is dissolved in the polyethylene polyol. This solution is blended with the isocyanate in an extruder, coated (closed blade), and crosslinked (120°C, 1.5–2.5 min). Two-component polyurethane PSAs are suggested for products of higher coating weights and thermally nonsensitive carrier materials.

Chemical-resistant tapes are based on silicones [184]. Linear polyamides are used only for special tapes [73]. Polymers based on isophthalic or adipic acid and 1,6-hexanediol with a molecular weight of less than 30,000 are preferred for hygienic hot melts.

For adhesive-coated PSPs, the cold flow of the adhesive depends on its viscosity and coating weight. Therefore, a low coating weight is recommended for tropically resistant tapes. For such tapes, silica (particle diameter 0.01–0.03 μm) is proposed as filler in order to increase the viscosity of the adhesive. Starch is suggested also as a water absorbent [209] (see also Chapter 4).

A special crosslinkable carboxylated, high T_g ($-25°C$) acrylate has been developed for mounting tapes and double-coated foam tapes. This compound contains tackifiers at a loading of 12–35% dry wt [40]. For transfer tapes used at elevated temperatures, a solvent-based crosslinked acrylic is suggested [210]. Formulations of transfer tapes that are conformable to uneven surfaces use fillers. Glass microbubbles can be incorporated also as filler to improve immediate adhesion to rough and uneven surfaces [22]. Such tapes are produced by polymerizing in situ with UV radiation. Thick PSA tapes (0.2–1.0 mm) have been prepared by the photopolymerization (UV) of acrylics (see later, Manufacture of Tapes Section 3.2). The filled layer can be laminated together with an unfilled layer. The thinner layers are from about 1 to 5 mils thick.

Foamlike pressure-sensitive tape is manufactured according to Vesley et al. [211] by using glass microbubbles as the filler. Dark glass microbubbles are embedded in a pigmented adhesive matrix. The glass microbubbles have an ultraviolet window that allows UV polymerization of the adhesive composition. Foam-backed adhesive tapes are commonly used to adhere an article to a substrate. Pressure-sensitive tapes filled with glass microbubbles have a foamlike appearance and character and are useful for purposes that previously required a foam-backed PSA tape. The adhesive matrix may include 0.1–1.15% w/w of carbon black also without affecting the UV polymerization. Glass microbubbles as fillers display the advantage of having higher distortion temperatures. The glass microbubbles (at least 5% by volume) are embedded in a polymeric matrix with a coating thickness of 0.2 mm. The thickness of the pressure-sensitive layer should exceed three times the average diameter of the microbubbles. This enables the buildup of intimate contact with rough and uneven surfaces while retaining the foamlike character. Optimum performance is attained if the thickness of the PSA layer exceeds seven times the average diameter of the bubbles. The bubbles may be colored with metallic oxides. The average diameter of the stained glass microbubbles should be between 5 and 200 μm. Bubbles having a diameter above 200 μm would make UV polymerization difficult. The PSA matrix may also have a cellular structure. The die embedded in the PSA matrix should have a UV window also. In order to enhance the cohesion of the adhesive layer, polar monoethylenically unsaturated monomers are included in the acrylic polymerization recipe (less than 20%) and photoactive or heat-activatable crosslinking agents are also added. The monomers may be partially polymerized (prepolymerized) to a coating viscosity of 1–40 Pa.s before the microbubbles are added.

Nonflammable tape formulations must also include a carrier material and adhesive. Nonflammable tapes are coated with an adhesive with 0–40% polychloroprene rubber [212]. An adhesive for PVC-based tapes includes cold masticated nitrile rubber (100 parts), calcium silicate (40 parts), titanium dioxide (10 parts), oil (25 parts), coumarone indene resin (MP 45–55°C; 25 parts), chlorinated rubber (100 parts), and solvents (methyl, ethyl ketone and toluene) [213].

Special end uses require special adhesive formulations. For tear tapes, crosslinked acrylic formulations with *N*-methylolmethacrylamide as comonomer are suggested [214]. Tapes used for automobile interiors have to resist high temperatures [215]. Common rubber resin based adhesives do not meet these requirements, so crosslinked acrylics are used. In some cases the composition contains glass microbubbles [22,216].

Adhesives for diaper tapes have to display high shear resistance [217]. Crosslinking monomers (e.g., *N*-methylolacrylamide) are added to increase shear resistance. Unsaturated dicarboxylic acids can be included also as crosslinking agents, and polyfunctional unsaturated allyl monomers and di-, tri-, and tetra-functional vinyl crosslinking agents have been suggested [218,219]. The polyolefinically unsaturated crosslinking monomer used to improve shear resistance should have a T_g of -10°C or below. Shear resistance is improved by water-based polymerization using a special stabilizer system comprising a hydroxypropyl methyl cellulose and an ethoxylated acetylenic glycol.

Adhesives with excellent water (humidity and condensed water) resistance are required for technical tapes and pressure-sensitive assembly parts in the automotive industry [155]. This can be achieved with water-based raw materials using the tackifier's own surfactants.

According to Ref. 220, PSAs can be manufactured as powders also. A PSA based on an ABA thermoplastic elastomer with a softening temperature higher than about 23°C is cooled below -20°C and pulverized; then a tackifier resin with a softening point of about 85°C is pulverized also. A dry blend of the elastomer and tackifier is prepared. This powder can be dry coated and sinterized by heating the adherend surface to 177°C.

Pressure-sensitive adhesive tapes whose adhesion can be decreased by UV irradiation are prepared from an elastic polymer, a UV-crosslinkable acrylate, a polymerization initiator, and a methacrylate photosensitizer containing an NH_2 group [22]. An ethylene glycol-sebacic acid-terephthalic acid copolymer (having a T_g of about 10°C), a tackifier resin, dipentaerythritol hexaacrylate as crosslinking monomer, benzyldimethylketal as photoinitiator, and dimethylamino acrylate as photosensitizer are mixed. This adhesive has been coated on a 100 μm PVC carrier primed with a modified acrylate. It displays a peel resistance of 140 g/25 mm before irradiation and 52 g/25 mm after UV irradiation.

Contact hindrance for tape adhesives can be achieved by using special fillers. For instance, a removable tape for paper contains PSA microparticles obtained from a crosslinked milled PSA. The 50–100 μm particles are coated with 5 g/m^2 [221].

Manufacture of Adhesives for Protective Films

The adhesive for protective films is a composition without tack; therefore there is no need to achieve an adhesion–cohesion balance [222]. For PSA labels the nature of the adhesive is more important than its coating weight and crosslinking degree. As mentioned earlier, the adhesive properties of labels are tested and are stated for a standard coating weight. For tapes, the coating weight is an important parameter used to regulate adhesive properties. For protective films, both the coating weight and the degree of crosslinking in the adhesive play important roles in the control of adhesive properties.

The choice of adhesive for labels (except for economic and environmental arguments) is determined mainly by adhesive properties (by the peel value required for a given substrate). For tapes the same criterion is valid. Adhesives for protective films (rubber-based or acrylic) are suggested for certain applications according to their adhesivity and the softness of the composition. That means that quite different adhesives may exhibit the same peel value as a function of the type of substrate and application conditions, and generally the specifications classify the products by using such empirical, inexactly formulated performance indices as "light, medium, and strong" adhesion. Therefore, unlike the label suppliers, manufacturers of protective films offer the "same" products (with respect to standard peel value) with adhesives that are chemically or physically different.

Generally the adhesives for protective films are crosslinked formulations. Therefore it is evident that the coating machines for labels, tapes, and protective films exhibit different characteristics (see Manufacture of the Finished Product Section 3.2). Rubber resin solvent-based adhesives were first manufactured for protective films, both crosslinked and uncrosslinked. Later, solvent-based crosslinked and water-based crosslinked and uncrosslinked acrylics were tested. The recipes include masticated natural rubber and cyclized rubber tackified with a low level of resins and crosslinked with isocyanates. Isocyanates are recommended as curing agents for solvent-based acrylic adhesives, and polyaziridine or isocyanates (in a concentration of 0.3–2.0%) for water-based acrylics. Table 6.13 lists the main raw materials and manufacture technology for adhesives for protective films. Table 6.14 illustrates the use of crosslinking agents for different PSPs.

As can be seen from Table 6.14, the most commonly used external crosslinking agents are isocyanate derivatives, applied in a concentration comparable to the level of internal multifunctional crosslinking monomers. The most common solvent-based acrylates for protective films have a glass transition temperature of about −20°C. Such adhesives need crosslinking to improve their internal cohesion. In the last decade or so, new products have been developed that do not require external crosslinking. Such formulations generally contain at least two components. The base polymer is harder (with a T_g of about +7°C). Such an ad-

Table 6.13 The Main Raw Materials Used for Coated Adhesives for Protective Films

Base polymer	Additive	Characteristics
Natural rubber	Resin, plasticizer, curing agent, solvent	Crosslinked
Styrene-butadiene rubber	Resin, plasticizer, curing agent, solvent	Crosslinked
Cyclized rubber	Resin, solvent	—
Polyurethane	—	Crosslinked
Polyacrylate	Crosslinker, solvent/dispersant	Crosslinked/uncrosslinked

Table 6.14 Crosslinking Agents for PSPs

Product	Adhesive	Crosslinking agent Type	Concentration (%)	Ref.
Protective sheet	Acrylate	Polyisocyanate	5	222
Tape	Acrylate with OH groups	Diisocyanate	0.5	177
Tape	Tackified AC	Isocyanate	1.0	223
Tape, low temperature	NR, SBR, PVE, carboxylated alkylphenol resin	Isocyanate	0.1–5.0	165
Tape	Water-based acrylic	Polyethylene glycol-diglycidyl-ether	0.1–0.5	224
Tape, protective film	EHA-VAc copolymer	Isocyanate	2.0	225
Tape, label	EHA-oligomer-VAc	Isocyanate	0.5	226
Label	AC, carboxylated	N-Methylolacrylamide	0.1–10	227
Tape	AC, carboxylated	Aziridine	0.1–10	228

hesive is self-curing and exhibits better UV resistance. The common base polymer for solvent-based acrylates is so soft that it is not possible to use it for the Williams plasticity test. The new self-crosslinking products exhibit a Williams plasticity of 2–2.8 mm (PN).

Manufacture of Adhesives for Forms

Business forms are very complex, multilayered constructions that are manufactured like multilayered labels. The formulation of the adhesive includes a range of permanent or removable adhesives of different, exactly controlled, adhesivities. The special adhesives have very low adhesion (allowing the detachment of die-cut label parts, e.g., cards). They are generally formulated from very hard, high T_g acrylates, high T_g vinyl acetate-acrylate copolymers, or low T_g crosslinked acrylates.

Formulation of the Primer

As discussed in Chapters 2 and 5, for certain applications top coats are required to improve the adhesion of the postcoated layers (adhesive, supplemental carrier, printing ink, release, etc.) to the carrier material. Primers do not have a universal chemical composition (see Chapter 4 also). They differ with respect to coating technology, the nature of the solid state surfaces to be coated, and the postcoated layer (Table 6.15).

Primers for Coated Adhesives. Adhesion promoters are used to improve the adhesion, the anchorage of a liquid postcoat (mainly adhesive, release layer, or printing ink) on a solid-state carrier material. Chemical primers provide a bond between substrate and adhesive. They are supplied as solvent-borne, water-borne, or solid-state products. Available equipment can influence the choice of primer. It determines whether the primer is to be applied in-line or off-line. Common application systems include rotogravure, transfer gravure, and smooth roll. Water-based primers are advantageous from the standpoint of compliance with environmental requirements.

Primers are recommended mainly for plastic film carriers [239]. On untreated PE surfaces, common acrylic PSAs exhibit a peel force of only 4 N/cm [240]. To improve anchorage, primers have to be applied [101]. They are used for labels, tapes, and protective films. The manufacture of tapes commonly includes in-line primer, release layer, and adhesive coating [232]. As a classic primer, solutions (in toluene) of butadiene acrylonitrile-styrene copolymers have been preferred. For solvent-sensitive SPVC films, primer dispersions based on Hycar latex and Acronal 500 D (1/1) have been suggested by Reipp [232]. According to Wechsung [187], for PVC tapes coated with a chlorinated rubber adhesive, primers based on butadiene-acrylonitrile rubber and butadiene-styrene rubber (1 : 1) with a butadiene/styrene ratio of 1/8 have been proposed. For such products the primer layer has a thickness of 2–6 g/m^2 [233]. Tapes for deep freeze applications use a soft PVC carrier, which needs a primer coating for the rubber-based PSA. According to Malek [234], for such tapes the carrier is primed with rosin. Abrasion-resistant automotive decorative films are made on a PUR or EVAc/PVC basis. Their carrier is coated with a PUR primer [234].

The chemical nature of the primers varies. ABS, isocyanates, and silane derivatives can be coated as primers [233]. In some cases (e.g., protective films, tapes) primers are crosslinked versions of the base adhesive used to coat the carrier material. Thus, a primer for a polyethylene tape and with adhesive based on butyl rubber consists essentially of a butyl rubber, a polyisocyanate, and an or-

Table 6.15 Primers for PSPs

PSP	Carrier	Primer	Adhesive	Coating Method	Coating Weight (g/m^2)	Ref.
Tape	Cellophane	Butyl rubber, grafted	SBC, polybutadiene	Coating, extrusion	—	229
Tape	—	NR, SBC, nitrile rubber, tackifier, epoxy resin	NR, crosslinked	Coating	—	230
Tape	PP	AC, PP, maleated	—	—	—	231
Tape	PP	Chlorinated PP	—	Coating	0.2–0.5	232
Tape	PVC	AC, WB	AC	Coating	—	232
Tape	PVC	ABS	RR	Coating	—	232
Tape	PVC	SBR	Chloroprene	Coating	2.0–6.0	187,233
Tape	PVC	Rosin resin	RR	Coating	—	165
Tape	PVC	PUR	RR	Coating	—	234
Tape	PVC	Butyl rubber, isocyanate	Butyl rubber	Coating	—	170
Tape	PVC	SBR	RR	Coating	—	235
Tape	PVC	CSBR, WB	RR	Coating	—	236
Tape	PP	Chlorinated rubber, chlorinated PP, EVAc	RR	Coating	—	237
Tape	PP, PET	Silicone, NR, polyterpene resin	Silicone	Coating	—	238
Protective film	PE, PP	NR, resin, isocyanate	NR, resin	Coating	0.2–1.5	—

ganic solvent [170]. On tapes with impregnated paper carrier, the butadiene-styrene rubber may function as an abhesion promoter also [235]. A primer based on SBR latex is coated on the carrier for a removable PSA [236]. As a primer for coating polypropylene, a mixture of chlorinated rubber, EVAc copolymer, and chlorinated polypropylene can be used [237]. Polar ethylene copolymers are suggested as adhesion promoters also [241]. The effect of primers based on chlorinated polyolefins depends on the coating weight and coating conditions [242]. For silicone tapes with PE or PET carriers, a primer containing organosilicones, nitrile rubber, and polyterpene resin is recommended [238].

Generally, water-based primers are used for paper, and solvent-based primers are used for films. Water-penetrable carrier materials are primed with water-based solutions of resins or latex. Hydrophobic carriers are primed with solutions of nitrile rubber or SBR [233].

A discontinuous adhesive layer can be preprimed to provide a continuous matrix. As discussed earlier, the PSA surface can have a special discontinuous shape ensured by the PSA itself, if this is a suspension polymer [2]. Thus spherical contact sites are formed. These are anchored on the carrier surface with the aid of a primer. Theoretically primer thicknesses approaching a single molecular layer are required, but in reality the primer coating is 0.5–1.0 μm thick [243]. A release coating may also need a primer. According to Ref. 95, a barrier layer based on vinyl chloride, vinyl acetate, and vinyl alcohol is applied to paper before siliconizing.

Primers for Extruded Layers. Extrusion primers are used to improve adhesion between carrier layers manufactured via extrusion (see Section 2). Various raw materials and formulations have been proposed for primers. Some of them act as a primer for postcoating of the liquid adhesive also. The know-how for their manufacture and use comes from the manufacture of multilayer packaging films. As a tie coat or primer for PE and EVAc, extrusion coating carboxylated ethylene polymers can be used. Such top coats exhibit excellent anchorage on many substrates and are not sensitive to water, grease, oil, and many chemicals.

Coextrudable primers are known from the manufacture of laminated plastic plates or metal/plastic laminates [165]. Generally they are ethylene copolymers with a melt flow index of 6–25. Higher MFIs allow better deep drawing properties. These compounds improve anchorage on PET, PVC, PP, LDPE or HDPE, EVAc, PC, and styrene copolymers (HIPS, SAN, and ABS). Extrusion primers can be used as plasticizer barriers for soft PVC also. For such applications, polymethyl methacrylate or PUR (3–10 g/m^3) can also be used. Ethylene-acrylic acid copolymers (EAA) with randomly distributed carboxyl groups provide excellent adhesion characteristics for bonding with polar surfaces [244]. In EAA copolymers, the carboxyl groups disrupt crystallinity and also provide a site for hydrogen bonding between molecules [245]. They can be used on paper too. EAA polymers are applied mainly to prime foil (0.5–1.0 lb/3000 ft^2). Such products are supplied as waterborne dispersions also [246]. Ethylene-vinyl acetate copolymers have been introduced as primers for extrusion coating [247]. They can be used as solvent-based or water-based primers. According to Ref. 248, such formulation allows anchorage of postcoated layers on cast polypropylene. EVAc copolymers ensure sealing temperatures situated about 40°C lower than those needed for polyethylene [247].

The use of primers can be avoided if the adhesion of the carrier material is improved by the inclusion of adhesive components. Biaxially oriented multilayer PP films for adhesive coating have been manufactured to improve adhesion to the adhesive coating by mixing the PP with particular resins. In this case in order to prepare an adhesive tape that can easily be drawn from a roll without requiring an additional coating on the reverse side, at least two different layers having different compositions are coextruded and the thin back layer (1/3) contains an antiadhesive component [249].

1.2 Manufacture of the Abhesive Components

As discussed earlier (see Chapter 2), most PSPs need a release layer or a separate release liner. Different grades of release liners for labels, tapes, hygienic items (sanitary towels, sanitary pads, diapers), envelopes, etc. are manufactured [250]. In order to design their formulation and coating method, first their function has to be clarified. The release liner for a label serves multiple functions [163]: protection of the adhesive layer, handling of the label, ensurance of lay flat during converting, and other functions during converting (e.g., optical guide).

According to Ref. 73, for tapes the following properties are important for a release liner: unwinding performance, exactly regulated release force toward the PSA, dimensional stability, tear resistance, environmental stability, and cuttability and die cuttability.

Generally the liner has to have trouble-free release and dimensional stability, and perform well during die cutting, perforating, prepunching, and fan folding. Some PSPs (e.g., transfer tapes or label sandwich constructions) require release layers that exhibit different release forces.

Generally the manufacture of the release liner includes the manufacture of the abhesive (formulation and technology), the manufacture of the release carrier material, and the coating technology. In some cases the release agent is embedded in the carrier material or the carrier material itself has release properties given by its chemical or physical characteristics or its geometry. In such cases the manufacture of the release liner consists of the manufacture of the carrier or its modification. As discussed in Chapter 4, special chemicals are used as abhesive components. Some are macromolecular compounds having a very complex production technology. Other products are common commercial polymers or micromolecular substances. It is not the aim of this book to discuss the synthesis of these components, but their characteristics and their coating technology are briefly described.

Abhesive agents are used in various technical domains. The processing of plastics needs such materials for the mold. In the packaging industry, release lacquers are used for cold sealing adhesives. For instance, calcium behenate exhibits a strong lubricating action due to its long fatty acid chain. It is used for bottle blowing, film blown extrusion, and food packaging. It is applied as a main stabilizer, and lubricant (1–2 phr) in PVC processing. Silicone mold release agents are used in the injection molding of plastics. Polyamide-based release layers have been developed for cold sealing. Various types of release liners using silicones, CMC, alginates, and shellac have been produced [251]. A chemical release agent is applied together with a mechanical release agent. For instance, paper coated with profilated natural latex as a release agent [252]. Release coatings are also formulated with melamine resins. These formulations have limited pot life. Self-crosslinking release coatings are used together with precrosslinked adhesives. The release layer is formulated and manufactured according to the end use properties of the PSP, mainly according to the degree of abhesivity required during application of the PSP. Generally abhesive agents that have different chemical bases display different degrees of abhesion. Silicones are the sole polymers with a broad range of abhesion performance, i.e., with general usability for all classes of PSPs.

Formulation of Abhesive According to Its Chemical Basis

Silicones are the most used abhesive compounds. Their main application field as solid-state release liners is in the manufacture of labels, but separate silicone-based abhesive components have been used for medical tapes also [253,254]. Double-side coated mounting tapes and transfer tapes must have a solid-state liner with multiple release layers (see Chapter 2 also). For such products the release force generally depends on the following parameters [163,255]: quality of paper carrier, liner stiffness, silicone nature, silicone formulation, silicone coating weight, silicone affinity toward the adhesive, temperature, adhesive coating weight, and age of the PSA laminate.

Formulation of Silicones. The most used release coatings having general applicability are those based on silicone polymers. They were first manufactured for greaseproof and release papers. As release coatings, silicones and silicone-modified elastomers were developed. Solvent-based, water-based, and solventless conventional silicones and radiation-curable silicones are manufactured [256]. Such compounds are coated on paper and plastic carrier materials. Silicone release papers have been coated with 0.3–0.8 g/m^2 coating weight [257].

The technical problems concerning the use of silicones have been related to their formulation for controlled release, their dosage during coating as a very thin abrasion-sensitive layer, and their use for different adhesives and PSPs having different delaminating speeds. As is known, the release force depends on the adhesive. For instance, the same release gives 350 mN peel force with SBR based PSA and about 150 mN peel force with acrylics [258]. Classic silicone coatings were not adequate for silicone PSAs. Silicone pressure-sensitive adhesives can be laminated together with a highly crosslinked silicone release layer [259,260]. Fluorosilicone coatings improve readherability [261]. Silicone acrylates have been developed as release coatings for radiation-curable film and paper [262]. The photoinitiated curing of silicones was discussed more than a decade ago by Timpe et al. [263].

Formulation of Other Compounds. For some PSPs there is no need for the high level of abhesive performance given by silicones. On the other hand, for adhesive coatings with very low tack and peel, the use of silicones would give laminates with insufficient bond strength. In such cases the abrasion sensitivity and poor anchorage of silicones would influence the peel level also. Another problem is silicone transfer, which can occur from any release liner to any adhesive and render the adhesive coating completely tackless. Trace amounts of migratable silicone can cause detackification. A rough silicone surface or silicone transfer can cause a loss in subsequent adhesion, which is the adhesion measured after an adhesive has been in contact with a silicone release liner.

For such applications release layers other than silicone are recommended (see Chapter 4 also). According to Ref. 264, the release performance of a carrier can be improved by using 10–15% polyacrylonitrile latex and 2.5–5% melamine formaldehyde resin. Copolymers of octadecyl acrylate-acrylic acid or vinyl stearate-maleic anhydride have been proposed as release layers for tapes. Tristearyltetraethylenepentenamine and polyvinyloctadecyl carbamate have been suggested also [73]. Modified starch or starch and a water-soluble fluorine compound (water-soluble salts of perfluoroalkyl phosphates, perfluoroalkyl sulfonamide phosphates, perfluoroalkoxyalkyl carbamates, perfluoroalkyl monocarboxylic acid derivatives, perfluoroalkylamines, and polymers having as skeleton epoxy, acrylic, methacrylic, fumaric acid, vinyl alcohol, each of which contains a C_6–C_{12} fluorocarbon group on a repeating unit of the polymer) can be used as the release layer on paper for an adhesive containing a minute spherical elastomeric polymer [265].

Tapes with an olefinic carrier material are noisy. Modification of the adhesive by addition of a mineral oil substantially reduces the noise level during unwinding of the tape, according to Galli [266]. Further improvement can be achieved by using a release coat and/or special treatment of the carrier back. Polyvinyl carbamate or polyvinylbehenate can be used as the release layer [267]. The common manufacturing process used for tapes includes the in-line primer coating, release coating, and adhesive coating [232]. As an example, the back side of the tape carrier can be coated by rotogravure with a 0.15% solution of Hoechst Wachs V. Gasoline (boiling range 60–100°C) has been used as the solvent. Blade coating and drying at 30–80°C have been proposed.

For special applications, vinyl acetate polymers can also be coated as the release layer. An abhesive layer based on PVAc has been proposed for tapes [268]. PVA is formulated with silicones as a release agent for certain tapes. Maleic copolymers with polar vinyl and acrylic monomers such as styrene, vinylpyrrolidone, or acrylamide can be used also [269]. Other water-based or solvent-based silicone-free release materials have also been developed. These silicone-free release coatings need no postdrying curing [270]. Solvent coatings are free of migratable species, hydrophilic components that can cause humidity-induced variations in release performance, and they can be dried faster. Water-based formulations have the tendency to foam and contain migratable species.

The polyvinyl carbamates are the reaction products of C_{12-18} aliphatic isocyanates and polyvinyl alcohol. They adhere well to plastic films as well as cellulosic substrates and are normally used for tackified rubber formulations. The carbamates exhibit the lowest initial release values but show the greatest peel buildup. These compounds are recommended for tapes [271]; less than 20% alcohol can be added to their solution in toluene to improve their solubility. Although the re-adhesion (subsequent adhesion) values are unchanged after aging on carbamate and polyvinyl ether release coatings, the rolling ball values are worse. This suggests the presence of low molecular weight migratable species in the release coatings. The DSC plot presented in Fig. 6.7 shows the presence of low molecular weight fraction with a lower melting point in the release agent. Experimental results demonstrate that the higher its concentration the lower the degree of release. The presence of low molecular weight products in the carbamate-based release agent is the result of the multiple polymer-analogous reactions used for the synthesis of such compounds. The polyvinyl alcohol itself is the result of a hydrolysis reaction of polyvinyl acetate that has a broad molecular weight dispersion. The reaction of the macromolecular alcohol with isocyanates leads to side products also.

Behenates and carbamates are applied as release layers for protective films. Polyvinyl carbamate is the main product used for non-silicone-based release coatings, mostly for film. Other carbamate derivatives have been proposed also. Polyethyleneimine octadecyl carbamate has been suggested for tapes with good rewindability [272]. Such kraft paper based tapes are coated with PE. For instance, a 10 μm layer of PE with a density of 0.918 g/cm^3 and a 10 μm layer of PE with a density 0.944 g/cm^3 are laminated on PE and coated with the carbamate.

Acrylic copolymers based on a fatty alcohol acrylate such as stearyl methacrylate are dissolved in hydrocarbon solvent and diluted to 1–4% solids for coat-

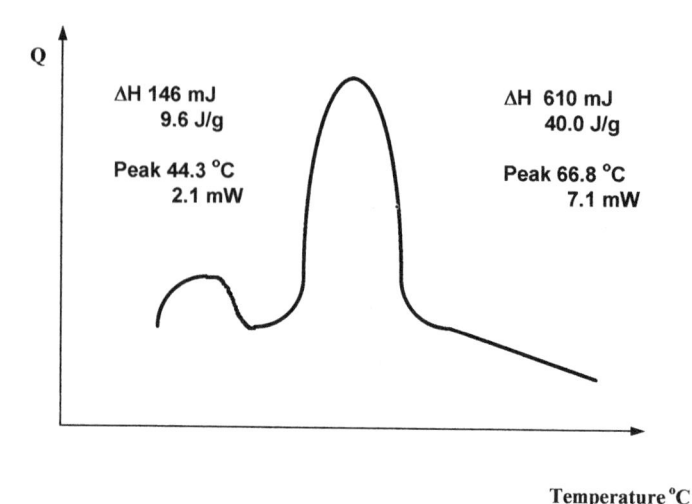

```
Q

       ΔH 146 mJ                    ΔH  610 mJ
          9.6 J/g                      40.0 J/g

       Peak 44.3 °C                 Peak 66.8 °C
          2.1 mW                       7.1 mW
```

Temperature °C

Figure 6.7 Differential scanning calorimetric plot of a carbamate-based release agent.

ing. Such release agents are used with rubber resin adhesives as well as with certain acrylic adhesives where a higher release value is acceptable, such as in self-wound tapes. Such compounds are not suitable for labels.

Polyamide resins are used for paper tape applications (and cold seal coatings). They are less expensive than carbamate but their aged release values are high. The polyamide release values are not significantly different from those of the uncoated polyester film. Octadecyl vinyl ether copolymers can be used with rubber resin or acrylic adhesives but are expensive. Polytetrafluoroethylene (40–80%) can be embedded as an abhesive filler in a binder based on cellulose derivatives (e.g., nitrocellulose) [273].

The water-based release agents are crosslinkable or self-crosslinking compounds. Vinyl acetate and acrylic-based emulsions are used. These compounds have been suggested for tapes, but their adhesion is inadequate for use in plastic films. As mentioned earlier with respect to silicones, the performance of any given release coating is dependent on the adhesive composition against which it is measured. This statement is valid for nonsilicone release coatings also. Table 6.16 presents some nonsilicone release materials used for PSPs and their application.

Formulation According to Coating Technology

A special domain of release coating is the manufacture of radiation-cured abhesive materials. In this case the polymer displaying release properties is cured

Table 6.16 Nonsilicone Release Coatings

PSP	Carrier	Chemical basis	Coating weight (g/m^2)	Release force	Ref.
Tape	PVC, 120 mm	VAc-VC copolymer-PVA octadecylisocyanate adduct	0.2	150 g/50 mm	274
Tape	—	Octadecyl acrylate-acrylic acid copolymer	—	—	73
Tape	—	Polyvinyl acetate copolymer	—	17.0 oz/in.	268
Flexible PSP	—	Methyl methacrylate-styrene copolymer	—	—	275–277
Tape	—	Hexafluoropropylene-tetrafluoroethylene copolymer-polyamic acid	—	—	3
Tape	Paper	Natural rubber	1.5–2.0	70–100 g/25 mm	252
Tape	—	Modified starch	—	—	73
Tape	—	Vinyl stearate-maleic anhydride copolymer	—	—	73
Tape	—	Tristearyltetraethylenepenteneamine	—	—	73
Tape	Paper/PE laminate	Polyethyleneimine octadecyl carbamate	—	—	272
Tape	—	Polyvinyloctadecyl carbamate	—	—	73

(polymerized) with radiation. Such polymers are silicone-based compounds. The first experiences with EB-cured silicones were with solvent-based silicones (e.g., for a coating weight of 1 g/m^2, with a dose of 1 Mrad). For such abhesive layers the release values increased with time [209]. As stated in Ref. 278, no carrier damage was produced by EB curing. However, economic considerations preclude its general use.

The development of UV curing in printing brought about its application as a common method for siliconizing also. In principle, cold and hot UV curing systems are known [279]. Hot UV curing systems use IR and UV radiation together. Cold UV systems use only UV radiation to cure silicone coatings. UV-IR systems were first commercialized in Europe for solventless silicones in 1984. UV-cured silicones have been proposed for narrow webs, and EB curing for broad webs [280]. A special manufacturing technology for a radiation-cured release layer is described by Brack [281]. In this case the release agent is incorporated as an incompatibile compound in the release binder. The film is cured to a flexible solid layer by irradiation. The abhesive substance migrates to its surface.

Diorganopolysiloxanes have been suggested as water-based release dispersions [280]. These release layers have to adhere to polyethylene-coated paper and must display abrasion resistance. The formulation contains 10–70% w/w polyvinyl acetate and 2–20% polyvinyl alcohol [280].

Abhesive substances can also be embedded in the carrier material. Such a carrier is manufactured for insulating tapes. Special insulating tapes have to adhere to refrigerated surfaces. A self-adhesive tape for thermal insulation used to secure a tight connection between heat exchange members (metal foils) and thermal insulation materials based on polyurethane foam is prepared by applying an adhesive (which is able to adhere to refrigerated surfaces) on an olefin copolymer carrier material that contains 0.2–2% fatty acid amides or silicone oils as the release agent and is bonded to the polyurethane [282]. Silicone rubber can also be included in the carrier material. According to Babayants et al. [283], 1 phr silicone rubber embedded in a 100 μm LDPE film reduces the release force from 350 to 70 g/25 mm.

Formulation According to Product Classes

As discussed earlier, different application technologies require different degrees of abhesion. Therefore abhesive choice and formulation depend on the product class. Labels and certain tapes need low release force abhesives based on silicones. These abhesives can be modified to suit the product end use. Generally siliconized paper has been suggested as release liner for labels, but special compounds can be used also [284]. According to Omote et al. [285], a release layer for a paper carrier is based on modified starch or starch and a water-soluble organic fluorine compound, like salts of perfluoroalkyl phosphates, fluoroalkylsulfonamide phosphates, perfluoroalkoxy carbamates, perfluoroalkyl monocarboxylic acid derivatives, perfluoroalkylamines, and polymers having a skeleton of acrylic acid, methacrylic acid, vinyl alcohol, etc. each of which contains C_6–C_{12} fluorocarbon groups on a repeating unit of the polymer.

Copolymers of maleic acid with acrylic and vinyl monomers, styrene, vinylpyrrolidone, and acrylamide are used as release coatings for tapes [286]. Nitrocellulose, polyethylene, rubber, proteins, and silicones have been proposed as release layers for medical tapes [287]. Aluminum film coated with a tackified polyisobutylene based adhesive is applied as street marking tape. The release layer for this adhesive is PVA [288]. Water-based release coatings suitable for paper tapes have been developed for rubber-based adhesives. Such products are

Table 6.17 Release Agents for Tapes and Protective Films

Product	Carrier for liner	Adhesive	Release liner buildup	Release agent	Release force	Ref.
Tape	LDPE	Polyester acrylate	Embedded	Silicone	70 g/25 mm	289
Tape	LDPE	AC	Embedded	Methyl hdrogen siloxane + α-diolefin	95 g/25 mm	290
Tape	—	AC	Coated	PVA + synthetic microballoons	—	291
Tape for porous substrate	PE laminated kraft paper	—	Coated	Polyethylenimine-octadecyl carbamate	—	272
Tape, double-side coated foam	Paper, 60 lb	—	Coated	Silicone	—	292
Protective film	PVC/PE laminate	—	Coated	Silicone	—	222
Masking tape	Crepe paper	—	Coated	Silicone	—	293
Tape	Plastic film	Silicone	Coated	Fluoroalkylsilicone	—	261

precrosslinked vinyl acetate copolymer emulsions containing variable levels of a functionalized additive for adjusting the degree of release (e.g., ethylene glycol diacrylate, or pentaerythritol triacrylate). Release values of 13.9–17.0 oz/in. are obtained. Silicone-modified release coatings are also used that give an initial release value of 8.7 oz/in. [289].

As mentioned above, polyvinyl carbamate is used as the main release agent for protective films. It can be applied alone or in mixtures with chlorinated polyolefins. The melting point and MWD of the polymer have to be well defined in order to ensure the required solubility and anchorage of the product. As is known, polyvinyl carbamate exhibits low solubility; therefore generally solutions having less than 1% solids are coated. Dissolution of the polyvinyl carbamate and the transport and storage of the release solution have to be carried out above room temperature. Toluene is the best industrial solvent for this release agent.

Polyamide-based release agents are used for cold seal coatings. Blends of such agents with Teflon derivatives have been proposed as release agent for protective films also. Unfortunately, their solubility, anchorage, and adhesion are too low. Table 6.17 presents data on some of the release agents used for tapes and protective films.

Technology of Abhesive Formulation

Quite unlike adhesive formulation technology, the formulation of abhesive materials is mainly a simple dissolution or dilution of special, almost ready-to-use, formulated chemicals. Except for some micromolecular compounds and the vinyl carbamate homo- and copolymers, which have to be used in solutions of low viscosity, raw materials applied as abhesives are supplied as liquids. The equipment used for their formulation depends on the nature of the raw materials. Polyvinyl carbamate-based release materials are used as solutions with low solids content (0.6–1.5%) in toluene. Such solutions are coated warm with a ceramic cylinder. Because of their low solubility they are stored warm and under continuous stirring.

2 MANUFACTURE OF THE CARRIER MATERIAL FOR PSPs

The carrier materials for PSPs can be classified as paper-based or non-paper-based. Of the non-paper-based carrier materials, synthetic films are the most important. Generally the carrier is the most expensive component of the laminate. According to Hunt [82], a decade ago laminate costs typically accounted for about 40% of selling price of labels. They now account for 70%. Table 6.18 lists the main carrier materials for selected PSPs.

Manufacture of PSPs

Table 6.18 Major Carrier Materials for PSPs

Label		Tape		Protective film	
Material	Treatment	Material	Treatment	Material	Treatment
Paper	Uncoated Coated Laminated	Plastic film	Oriented Nonoriented Laminated Embossed Foamed	Plastic film	Nonoriented Oriented Laminated
Plastic film	Oriented Nonoriented Laminated	Paper	Coated Laminated Embossed Pleated Impregnated	Paper	Coated
Textile	Nonwoven Woven Laminate	Textile	Woven Nonwoven Impregnated Laminated	Textile	—
		Foam	—		
Special materials	Metal Ceramics	Special materials	—	Special materials	—

2.1 Carrier on Paper Basis

Paper was the first carrier material used for PSPs. Various products such as labels, tapes, and protective films were manufactured on paper basis. Later synthetic films replaced paper partially (e.g., in the manufacture of labels) or almost totally (e.g., in the manufacture of tapes and protective films) as the adhesive carrier material. In 1993–1994, films were used for 9% of the European self-adhesive products and paper for 38% [294].

Manufacture of Paper

Paper is produced by a specialized industry. Quite differently from plastics manufacturing, where some of the film producers are also converters (PSP producers), paper is supplied for converters as a separate raw material. It is not the aim of this book to discuss paper manufacturing. For paper converters its processing characteristics and surface and bulk properties are more important. The surface characteristics of paper influence its processing and end use also.

General Requirements for Paper as Carrier Material

There are some general requirements with respect to paper quality as carrier material. These include its surface and mechanical characteristics. Surface characteristics influence mainly its coatability. Its bulk properties (lay flat, dimensional stability, and flexibility) influence its coating and conversion properties (printability, cuttability, and labeling). Printability is also a function of paper surface quality. Non-impact printing procedures require a relatively porous surface [295]. Different paper qualities such as uncoated, coated, cast coated, and colored are used. Synthetic papers have also been introduced. These are polymer (mainly HDPE) based materials [296].

The surface quality of paper varies. In Europe, coated papers have higher gloss and density than in the United States [297]. The quality of coated papers depends on the coating procedure used, the coating weight, the drying process, and compatibility between coated layer and paper [298].

The nature of the pigments used for surface coating of paper is very important for its coating performance. Various compounds such as kaolin, clay, talcum, calcium sulfoaluminate, titanium dioxide, calcium silicate, sodium aluminum silicate, and barium sulfate are used [298]. These pigments possess different surface energies and pH stabilities. In top-coated papers the surface layer is based on about 90% white pigment and 10% binder (vehicle). The binder influences the adhesion or anchorage of the top coat, its stiffness, porosity, and ink absorbance. This layer has a porous surface, with a pore diameter between 0 and 40 μm [299]. The average diameter of the pores for fine quality papers is 2–4 μm. Kaolin is the most used filler for paper. There are gravure printing and offset grade kaolins (having different particle diameters). Calcium carbonate is less expensive and allows fast ink penetration. Calcium carbonate increases the porosity of the paper. This is very important for heat set-rotary printing [300]. The mean coating weight of the top-coat layer is about 15 g/m^2, i.e., the layer has an average thickness of 12 μm and a pore volume of 3 mL/m^2. The printing and adhesion properties of the top coat depend on its chemical composition. Caseine, starch, PVA, acrylic, or styrene-butadiene latex are used as binder. Polymer dispersions have been used since 1949 [299]. It should be mentioned that the uniformity of the surface distribution of binders and pigment, i.e., the coating structure, influences the printing defects ("print mottle") [301]. As stated by Topliss [302], the main face stock materials have been white uncoated paper, white coated paper, non-white paper, paper-backed foils, vinyls, and polyacetate. Above average growth has been foreseen, especially for vinyls, polyester, and cellulose acetate. Growth in demand for coated paper has been higher than for uncoated material.

Papers for offset printing with 70 g/m^2 can be coated on both sides [300]. Calcium carbonate, clay, and satin white are used as the top layer. Low weight coated papers for gravure and roll offset printing, with at least 45 g/m^2, can be double-side coated too. Gravure printing requires high gloss and no "missing dots."

The bulk properties of paper influence its dimensional stability. These also depend on environmental conditions. Saturated and coated fourdrinier paper can exhibit up to a 10% weight gain and greater than 1% width gain over the range of 0–97% relative humidity [303]. The humidity content of the paper (the ideal dry status of paper) depends on its quality. Newspaper materials have a humidity content of 9%, whereas a quality paper contains 6% water for the same external relative humidity of 9% [304]. Dimensional stability is a function of the geometrics and dimension of the carrier material. For labels, thin films (15–20 μm) and cardboard of 300 g/m^2 can both be used [305]. Narrow web (420–620 mm) flexo printing machines can use webs with a thickness of 20–300 μm [306]. Printing methods influence the choice of carrier material requiring tailored surface quality and thermal resistance for the carrier. As shown by Strahler [307], most label printers prefer a working temperature of at least 35°C for UV offset printing.

Special Requirements for Paper Used as Carrier Material for Different PSPs

Various PSPs have various end use functions. Market requirements lead to well-defined characteristics of the label material. For instance, labels for lubricants have to display durability, visual impact, and flexibility [308]. Toiletries and cosmetics require a no-label look, high consumer image, and stain resistance. The hardware and electrical markets need visibility and heat resistance. For toys, non-toxic labels with a long life and high visual impact are necessary. The chemical industry requires carrier materials that are resistant to chemical attack and have good visibility and durability. Mechanical stresses during use can be quite different. Therefore mechanical requirements for carrier materials are different also. A paper carrier material has to have certain special properties for use in each PSP class. For instance, fragile pressure-sensitive materials either are very thin or have low internal strength. Such materials based on paper or film are used for labels and tapes also. Different face stock materials are suggested for envelopes, warning labels, sterile needle cartridges, etc. [309]. A porous carrier can be required for medical or mounting labels and tapes. Paper for release liners has to meet various mechanical and surface-related requirements and should not contain inhibitors of silicone curing agents (e.g., heavy metals, sulfo and carboxy groups) [310]. Fan folders are used for table labels; the machines used need a low paper weight of 67–68 g/m^2 [311]. Copy labels must allow go-through of the image to the release.

Paper as Carrier for Labels. Paper is the most used carrier material for labels. For most labels it is used as both the face stock and the liner material. A 70–80 g/m^2 paper has been suggested as a generally versatile carrier material. Common label papers have a weight of 70–80 g/m^2; pressure-sensitive labels,

83–90 g/m² [312]. In 1965 paper was considered the sole carrier material for labels [313]. Now 80% of labels have paper as the carrier material [314]. The number of paper qualities used for the same application is very large. For instance, for wine bottle labeling a single supplier of labels uses 25 different paper qualities coated with various water-based PSAs [315]. For labels, different types of materials, thermopapers, vellum, and clay-coated papers are suggested [316]. Uncoated, coated, cast coated, and colored papers are applied. First cast coated papers were introduced. Now machine-coated paper qualities are preferred. The proportion of satinated kraft paper increases [317]. Low weight, high rigidity papers are required [318]. For table labels printed via non-impact printing, open porous papers are preferred [319]. Metallized paper and film/paper laminates are used also. Metallized paper displays low curl in comparison with metallized film. According to Ref. 320, paper, fabric, vinyl, acetate, and polyester films are suggested as carrier materials for HMPSA-coated labels. Fluorescent paper labels were introduced in 1958 [321]. Latex-impregnated paper is proposed for weather-resistant labels, calendered (90 g/m²) paper for opaque labels, paper having high absorptivity for labels printed with so-called cartridge inks, oil- and fat-resistant paper for thermal printing [322]. Kraft paper is reinforced with a mid polypropylene film for tapes [323]. According to Senn [324], 55–100 g/m² papers should be used for paper laminates and 12–180 g/m² papers as release liner for film laminates. Because of the roughness of the paper it is difficult to achieve release values less than 450 g; therefore in this case polyethylene-coated papers are recommended [325]. Such papers have release forces that can be regulated between 20 and 400 g.

According to Ref. 319, the main requirements for a paper carrier material for labels can be listed as high gloss, high surface strength, good wettability, and good printing ink anchorage. Paper as a face stock material for labels should display special mechanical and surface characteristics. Adequate mechanical characteristics are required to ensure its dimensional stability and special behavior during cutting and labeling. Surface characteristics should allow its coatability (printing and varnishing). Pressure-sensitive labels displaying security peel-off must have either exceptional strength to prevent removal or be so fragile that they cannot be removed in one piece [326]. On the other hand, high speed converting machines solicit the label material mechanically. Electronic confectioning machines for labels cut, die-cut, punch, and perforate the paper at a running speed of 200 m/min [304]. Electronic data systems use primarily a paper engineered for strength and with high absorbency to capture ink [327]. Special papers like radiation luminescent papers have also been manufactured [328].

The mechanical and surface characteristics of paper can be improved by impregnating, coating, and laminating. Latex-impregnated paper is a lower cost alternative to PVC [329]. HDPE, usually blended with LDPE, is used on photographic paper to avoid curl [330]. Paper-backed aluminum foils and metallized face material have the highest opacity (100%). Latex-impregnated and plastic-coated papers possess an opacity degree of 84–88% [331]. Plastic-coated paper attains a gloss degree of 96%.

Special paper is required for direct thermal coating. Common organic direct thermally printable papers include a colorless leuco dye and an acidic color developer [332]. These are coated and held onto the surface of the paper with a water-soluble binder. During printing the two components melt together and react chemically to form the color. In order to limit this image to the heated area, inorganic fillers—calcium carbonate, clay, etc.—are used. The problem of image stability (protection again chemicals) is solved by applying a top coat as a transparent film-forming layer. It is possible to also apply a barrier coat to the underside of the paper to prevent adhesive or plasticizer from migrating from the opposite side. Therefore these papers are classified as non-topcoated (non-smudgeproof) and topcoated (smudgeproof) papers. Non-topcoated papers are applied for price weight labels of perishables and quick sale items, for tags, and tickets. Topcoated papers are suggested for frozen food and long-life labels. Latex-impregnated paper is used for data system applications that require moisture resistance and good flexibility. They provide smudge-resistant ink absorption [333].

Paper as Carrier for Tapes. At the beginning of tape manufacture, paper was the only material used as carrier. According to Stott [334], paper masking tapes were used in the 1920s for car varnishing. Crepe paper coated with crosslinkable silicone adhesive has been applied up to 180–200°C. Different types of paper, dimensionally stable or deformable, tear-resistant or fragile, have been used. Some paper carriers are water-resistant, and others are water-soluble. Paper for splicing tapes has to resist temperatures up to 220–240°C and be water-dispersible but resistant to organic solvents. It must also be migration-resistant. The tear resistance of the tape has to be as high as that of the paper to be bonded [16,130]. Extensible, deformable paper is required for medical tapes [335]. Glossy and crepe papers are used for tapes. For instance, a pressure-sensitive adhesive tape for protection of printed circuits, based on acrylics and phenolic terpene resin as adhesive (with a 100–200 g/m^2 siliconized crepe paper as temporary carrier), is manufactured by coating the adhesive mass first on a siliconized polyurethane release liner and then transferring it under pressure (10–30 N/cm^2) to the final carrier material [293].

Paper as Carrier for Protective Films. Actually there are only a few special protective papers. Anticorrosion protective papers [336] and weathering-resistant papers are used [337]. Self-adhesive wall covering consists of a layer of fabric having a visible surface, a barrier of paper that has one surface fixed to another fabric layer, a pressure-sensitive adhesive coated on the barrier paper, and a release paper [338].

Paper as Carrier for Release Liner. According to Hufendiek [163], the paper used for release liners can be normal, densified, glassine, plastic-coated, or clay-coated.

The weight of the release liner depends upon the method of label conversion, and the type of die-cutting and tooling as well as the strength of the sheet. According to Reinhardt et al. [339], silicone penetration in paper decreases as the density of the paper increases. Low weight (65 g/m^2) open papers allow silicone to penetrate, and the level of silicone penetration and its distribution in the paper influence the release force. Base materials for release liners in 1987 were calendered paper (50.5%), clay-coated paper (21.3%), polymer-coated paper (6.7%), other kraft papers (12.9%), and plastic films (2.0%) [340]. Different degrees of release are achieved for a double-side coated release liner by using the same release component but one side with polyethylene-coated release paper [341].

Polycoated kraft, clay-coated kraft, and densified kraft papers have been proposed for use in the manufacture of release liners [342]. For siliconizing (release liner), various paper qualities with weights of 40–220 g/m^2 have been suggested. Satinated (glassine), clay-coated, single- and double-side polyethylene-coated papers, and normal kraft papers are used [343]. Conventional glassine paper liners account for about 60% of the market [82]. Clay-coated papers are widely used in the computer sector because of their good lay flat and handling. Polyethylene extrusion-coated papers combine the strength of long fiber papers with the smooth nonabsorbent surface of film layers.

The liner plays an important role in the functionality and cost of most PSPs. Since the release liner is discarded after the label is applied, the development of more environmentally friendly materials is important. As stated by Allen [344], there is a trend to use the same paper qualities for face stock material and release liner. According to DeFife [345], the most important features of the release liner are tear strength, dimensional stability, lay flat, and surface characteristics.

The tear strength of the carrier material is important for label conversion and dispensing. Good caliper control and liner hardness (lack of compressibility) are the main parameters. Dimensional stability, the ability to maintain the original dimensions when exposed to high temperature and stresses, is required for print-to-print and print-to-die registration (see also Chapter 7). If the release liner stretches under heat and tension, graphics will be distorted. Liner stretch can affect the dispensing of the labels also. Special attention should be given to the influence of supercalendered base papers with different densities and closeness.

The smoothness of the release liner affects several performance features [345]. The adhesive surface is a replica of the release surface. If it is rough, the level of initial adhesion will be reduced. If a rough label is laminated with a clear face stock material, label clarity will be negatively affected. If the liner is too smooth, air entrapment during dispensing can become a problem. The roughness of the back side of the label can affect web tracking on the printing press. Smooth liners can weave, causing cross-web print registration. Also, in laser printer applications, the roughness of the liner plays a critical role in sheet feed ability and tracking through the printer. Single-side, double-coated label papers have been developed for special surface quality requirements [346].

High speed table labels require lay flat; the weight of the release paper should be between 67 and 85 g/m^3 [347]. An 80 lb kraft liner is used for repositionable labels in order to avoid curl; a 78 lb kraft release liner is used for thick (18 mil) embossed vinyl [348]. Lay flat is very important for sheet labels or pin perforated and folded labels. Labels that are butt cut, laser printed, and fan folded must lie flat to ensure proper feeding and stacking. Paper liner lay flat is affected by the liner material's resistance to humidity, dimensional stability at different temperatures, and stiffness (thickness). The use of radiation-cured or low temperature heat-cured silicones produces less changes in the humidity balance of the paper. Glassine liners are harder, (densified) and are used for high speed dispensing [349]. To reduce silicone consumption, silicone coaters require high and uniform gloss and low porosity of the paper surface [350].

Satinating reduces paper porosity (the specific gravity of paper increases from 0.90 to 1.20 kg/dm^3), and its permeability to air decreases to one-tenth of the initial value. Fillers increase the specific gravity of paper to 1.25 kg/dm^3. The components of the pigments and binder for topcoated papers must be free of substances that can poison the silicone catalysts [350]. According to Moser [349], the most important characteristics of glassine silicone papers are (1) their ability to be siliconized, (2) machining properties, (3) dimensional stability and lay flat, (4) die cuttability, and (5) storage performance. Theoretically, these characteristics are influenced by the porosity, surface structure, and surface energy of the material. Measurable characteristics are the length of penetration, porosity, smoothness, gloss, solvent holdout, and surface tension. As stated by Moser [349], porosity is the most important parameter for solvent-based siliconizing, and surface structure is the principal factor for solventless siliconizing of such papers. Surface tension does not influence the versatility of the paper to be siliconized.

Generally, paper thickness and stiffness affect the processing of the laminate. As discussed in Ref. 351, these are the main parameters for cuttability. According to Wabro et al. [73], low weight (70 g/m^2), low thickness (60 μm) siliconized papers are recommended for laminates that are not confectioned (e.g., decals and foam-based products). The standard paper grade for die-cut products has a weight of 95 g/m^2 and a thickness of 80 μm. Polyethylene-coated papers have a weight of 120 g/m^2 and a thickness of 100 μm. Tear-resistant release liners with improved cuttability are manufactured by using paper having a weight of 110–145 g/m^2 and thickness of 120–155 μm [73].

For one-sided splicing tapes, the siliconized paper liner remains on the adhesive and has to be recyclable also. Carrier-free splicing tapes must also be water-soluble. The liner for transfer tapes has to allow detachment of the adhesive, i.e., it has to display different degrees of release on either side [130]. Siliconized paper is also used as a pleated liner for tapes.

2.2 Synthetic Films as Carrier

According to Schroeder [101], fiberlike materials and textiles (water-resistant and water-soluble) based on cellulose, plastic films (soft and hard PVC, cellulose derivatives, PET, etc.), foams (open-pore PUR, closed-pore PVC, closed-pore polychloroprene, etc.) are used as carrier materials for tapes. The paper and nonwovens can be impregnated with water-soluble or water-insoluble resins to achieve a reinforced or filmlike character; the foam carriers are coated with adhesive on one or both sides.

A decade ago labels and tapes were the main market areas for film carrier materials. Oriented PVC (60%) with a thickness of 25–70 μm, OPP (40%) with a thickness of 25–40 μm, and PET (2%) with a thickness of 23 or 36 μm were used [352]. According to Ref. 353, the growth areas for film carrier materials for labels are beverages (45%), personal care products (20%), pharmaceuticals (15%), and on-demand digital applications. Among the non-paper materials, synthetic films are the most important carrier materials. PVC, polystyrene, PET, metallized materials, and synthetic papers are used. PVC, PP, PE, PET, PUR, Zellglas, and PMMA can be coated with HMPSA. Soft PVC is used as a carrier material for decorative films [115]. PTFE is used as a carrier for splicing transfer tapes [101]. Special carrier materials such as polyester, anodized aluminum, stainless steel, and ceramics resist temperatures up to 1400°C [354].

Synthetic films were introduced as carrier materials for PSPs because plastic films have certain technical and commercial advantages over paper. There are certain application fields like toiletries and pharmaceuticals where film labels are used almost exclusively. Some special pharmaceutical labels may have take-off (detachable) parts also to allow multiple information transfer. Parts of the label remain on the drug or on the patient [355]. These labels use a film carrier. Labels for textiles are manufactured on PE basis [356].

General Requirements for Synthetic Films as Carrier Material

Independently of their use for a given product class (label, tape, protective film, etc.) or their raw material basis, synthetic film carrier materials have to satisfy certain requirements. The general requirements for synthetic films as carrier materials concern their weblike character and surface quality. Related to surface quality, there are requirements to ensure coatability (adhesive coating and printing) and machinability (during conversion or application and in the applied state). Blocking, smoothness, hardness, and electrical conductivity must be regulated.

Good sliding properties, optical properties, static resistance, scratch resistance, and shock resistance are also required, as are lay flat, minimal thickness tolerances, no wrinkles, no gel, and surface tension higher than 40 dyn/cm.

Depending on their use for certain product classes and the actual application of a product, synthetic films used as carrier materials have to meet certain special requirements. When selecting a suitable raw material for carrier manufacture, consideration must be given to the desired film properties, the processing method to be applied, and, not least, economic factors. As an example, LLDPE is not foldable or stiff. It is difficult to cut or die cut and to punch because of its high elasticity. Its extensibility is a disadvantage for the converter [357]. In choosing the paper for the release liner it should be taken into account that the stiffness of the laminate is the sum of the stiffnesses of the components (taking into account the reinforcing effects; see also Chapter 3, Section 2). Therefore for thin face stock materials a heavy release liner should be used, and for thin release materials, a heavy face stock. There are some applications where the thickness of the liner is determined by technical functions, e.g., copy labels with a self-copyable release liner. White, thin (50 g/m^2) release paper is recommended for self-copyable products [325]. For instance, for name plates made with a 100 μm stiff face stock film, a thin release paper is suggested. Table 6.19 summarizes the main requirements for synthetic films used as carriers for various PSPs. As can be seen from this table, labels need aesthetically high quality face stock material because their primary use is as carriers of information. Tapes for fastening are weblike products that require excellent mechanical properties. Protective films have to be able to conform to and deform with the product to be protected.

The development of carrier materials has been influenced by requirements for more film or more paper carrier material. Factors in film carrier development take into account its better surface quality (S), excellent aesthetics (E) and chemical resistance (Ch), its versatility as release liner (R), and the general development of film-based packaging materials (Pa), which need the same label material as that used for packaging film. The improved cuttability (Cu), better processibility

Table 6.19 Main Technical Requirements for Plastic Film Carrier Materials

Label	Tape	Protective film
Coatability	Coatability	Coatability
Printability	Mechanical resistance	Conformability
Cuttability and die cuttability	Tearability	No elasticity
Chemical and environmental resistance	Foldability	Processibility
	Chemical and environmental resistance	Chemical and environmental resistance

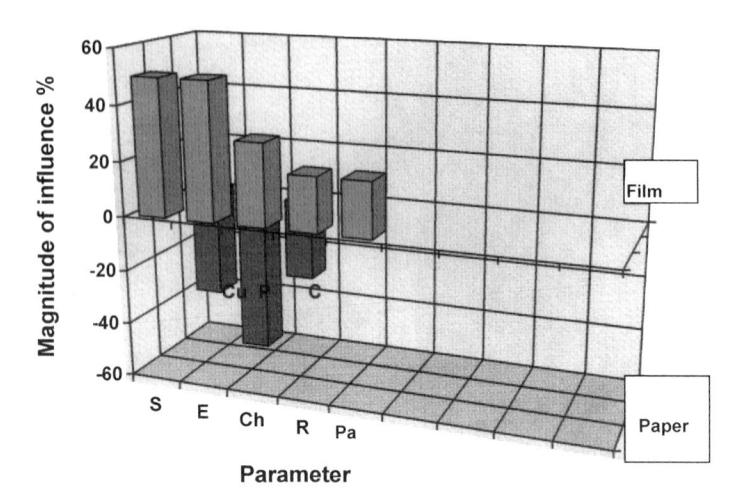

Figure 6.8 Technical and economic parameters influencing the face stock development.

(P), and lower costs (C) enhance paper carrier development. Figure 6.8 illustrates (on an arbitrary scale) the estimated magnitude of influence of such parameters on the development of label face stock.

Special Requirements for Synthetic Films as Carrier Materials

As discussed in Chapters 2, 7, and 8, carrier materials used for different product classes need different special properties. Their performance characteristics are determined primarily by the manufacture and application if the PSP.

Requirements for Plastic Films for Labels. Plastic films used as face stock materials for labels are coated as webs and laminated and converted as webs. Reel labels are dispensed (labeled) in web form. Therefore they have to satisfy the general quality requirements for the machining of weblike materials. On the other hand, labels are used for information transfer, i.e., they have an aesthetic character. Their message is transferred onto their surface by using various printing and lacquering techniques. It is to be supposed that the requirements concerning the mechanical properties for dispensing are higher than those for coating and converting, and dimensional stability related requirements for printing are more severe than those arising during conversion. Therefore the main requirements for carrier materials used for labels concern their dimensional stability and surface quality. The film has to be stiff enough for dispensing and flexible enough to conform to various container shapes and to resist repeated deformation [358]. For label converters, the stock must be able to meet printing and die-cutting specifications for computer imprintables. For end users, the label has to withstand harsh environments and rough handling. That means that plastic films having a pronounced rigidity and surface quality are recommended as carriers for labels. Such films must allow good register control, easy die-cutting, and reliable high speed dispensing. The specialist for label design and the manufacturer of the labels should know which polymers and plastic film manufacturing methods produce plastic film suitable for label applications. Caliper, gloss degree, opacity, and corona treatment are the main performance characteristics of synthetic face stock materials [331].

Polyvinyl chloride was the first filmlike synthetic material used as a carrier for labels. Thin PVC is used as the carrier for tamper-evident labels [359]. Polyvinyl chloride exhibits the desired tear and temperature resistance for computer applications. It has excellent UV stability [360]. The main disadvantages of PVC are its caliper variation, shrinkage, and plasticizer migration. In the last decade, PE and PP have replaced PVC.

Filled HDPE can be used as face stock material for water-resistant dimensionally stable labels that can be printed by different methods [361]. They have to possess good chemical resistance also. For battery labels, BOPP has been suggested [362]. PE carriers used for labels require one year treatment stability [363]. Common HDPE has densities of 0.952–0.960 g/cm^3. Higher densities are recommended when an optimum modulus (rigidity) is required [364]. Faster cooling gives less stiffness due to the formation of smaller crystallites. Like polyethylene-coated paper, polyethylene laminates have good humidity resistance [365]. Mixtures of polystyrene and polyethylene are used as face stock materials for labels also. As a face stock material, polystyrene displays adequate printing properties, die-cuttability, and labeling ability [366].

A typical example of special requirements for a label carrier film is name plates. A film that can be printed via flexo, letterpress, and computer printing methods must have a matte metallic appearance like that of anodized aluminum. It has to display the handling characteristics of hard aluminum. No curl should occur after removal. A heat resistance of 150°C is needed. The matte surface has to absorb ink, maintaining sharp graphic contrast, good color intensity, and excellent legibility. It has to be printable with a range of common or special computer printers. Abrasion, chemical, and environmental resistance are needed. Opportunities for development of new films are given in the fields of security, computer, battery, specialty, and promotional labels [308].

Requirements for Plastic Films for Tapes. The carrier for tapes has to support and absorb the stresses during application in order to allow bonding of the adhesive onto the surface. The carrier material is chosen according to the nature, magnitude, and direction of the forces. It is important to know the yield, tear resistance, and elongation of the material in the machine direction and in the cross direction. It is imperative to know if the carrier splits parallel to the machine direction. Dart drop resistance and foldability are required also. Elmendorf tear resistance also plays an important role (see Chapter 3, Section 2). The ability of a film to retain a crease is an important property in packaging and is also required for some tapes [367].

In the mid-1960s, cellulose hydrate, cellulose acetate, PVC, crepe paper, and fabric were the main carrier materials for tapes coated with solvent-based adhesives [313]. In the 1980s PVC, PET, and OPP were the most common carrier materials for tapes [101,368]. Now, nonoriented, oriented, mono- or multilayer films, and film/film and film/fabric reinforced laminates are recommended for tapes. Fiber-reinforced films (applied for wet adhesive tapes [369]) are used for PSA tapes also. Electrically conductive polyolefins with a high loading of carbon black (30% w/w) have been proposed for insulation tapes. Chemically pretreated polyethylene (octene-based LLDPE) is recommended for tapes [187,370]. Biaxially oriented polypropylene with a thickness of 30–35 μm is suggested also [187]. It costs 20–30% less than PET [362]. Shrinkable adhesive tapes based on polyolefins are manufactured by extrusion. As central section polypropylene and ethylene-propylene copolymers are coextruded [371].

Requirements for Plastic Protective Films. Protective films are conformable, laminated covers that have to pass through the same processing and/or storage cycle as the protected item. Such films are only temporary and do not contribute to the value of the finished product. Therefore, protective film carriers are generally common materials although they have to be soft, mechanically resistant, thin, and deformable, with a balanced plasticity/elasticity, medium temperature resistance, and a low manufacturing cost.

Film Manufacture

Most raw materials for films are produced from synthetic monomers via polymerization (polyaddition or polycondensation). The polymers are processed as thermoplasts to produce a film. The physical (bulk or surface) transformation of the film is the next step in product manufacture. Table 6.20 summarizes the main manufacturing procedures for carrier films.

Most plastic films are produced by extrusion. Another method is casting from solution or dispersion. In this procedure the dissolved and/or dispersed film raw material is cast on a conveyor belt and dried. A high quality film free of tension is produced by casting. This method is used to manufacture PVC (4–100 μm) film with defined, low tensile elongation [372]. Cellulose acetate, butyrate, propionate, and ethyl cellulose can be processed by casting [373]. Such cast or support based coating lines manufacture a continuous strip from a liquefied polymer mass. In comparison with extrusion coating, the liquefied material to be coated generally contains liquid components and the coating is only temporary. Upon solidification the film layer becomes a self-supporting material, and it is stripped

Table 6.20 Main Manufacturing and Postprocessing Procedures for Plastic Carrier Films for PSPs

Process	Main procedure	Variants
Manufacturing	Extrusion	Blowing: Mono, coex
		Flat die: Mono, coex
	Calendering	
	Casting	Solvent-based
		Radiation-cured
	Sinterizing	
	Impregnation	
	Polymerization in bulk	
	Lamination	Homogeneous: Adhesiveless, with adhesive
		Heterogeneous: Adhesiveless, with adhesive
Postprocessing	Orientation	Monoaxial
		Biaxial
	Physical treatment	Corona: atmospheric, special
		Plasma: Inert, reactive
		UV
	Chemical treatment	
	Laquering	

off the support base before being wound to form a roll of cast film material. There are two types of support bases, reusable materials and endless belts. Roll coaters and curtain or flow box coaters are used, depending on the dispersing medium, the viscosity of the material to be coated, and the thickness of the end product. A special polymer film having the characteristics of paper can be manufactured using the technology of plastisols, by precipitation of polypropylene dissolved in dichloromethane [374].

Calendering has been developed also. It has been applied specially for manufacture of PVC films. Generally thermoplasts having a thermoplastic domain with a viscosity of 10^2–10^3 Pa.s and a broad softening range of 10–30°C can be processed by calendering [374,375]. Copolymers of vinyl chloride, vinyl acetate, cellulose acetate, polyethylene and its copolymers, polypropylene, and polybutylene can be transformed into film via calendering. Films with a minimum thickness of 70–100 μm can be manufactured using this procedure. Polypropylene films with a thickness of 100–800 μm are manufactured by calendering. The raw materials for calendered PP film have to meet special requirements; they have to be processed at about 200°C ± 10°C. ZN polypropylene has a MWD of 6–13, but it can be degraded to achieve a narrower MWD ($M_w/M_n < 6$), which enhances calendering. Slowly crystallizing polypropylene is suggested for calendering. Such films exhibit higher stiffness (crystallinity of the calendered film is about

62%, in comparison with the crystallinity of 82% for the extruded films) and lower tensile strength and are recommended for furniture [376].

The principle of extrusion is well known. An endless screw rotates in a heated cylinder, which leads the molding materials forward to be compressed, plasticized, and homogenized. In front of the cylinder there is a tool to form the molding material. The molten plastic is molded through a die. In the main manufacturing procedures a flat die is used to mold a flat film ("cast" film) or a circular die is used to mold a tubular film. The flat film is calendered (via cooled chill rolls) to achieve the final dimensions at normal temperature. The tubular film is blown up with air to stretch it to the final dimensions and to cool it.

In choosing a polymeric raw material for a plastic film carrier, the most important criteria are the following [377]: mechanical properties (tensile resistance, elongation, and shrinkage); chemical resistance; permeability (to oil, gas, water); sealability; surface quality; stiffness; thermal resistance; adhesion; coefficient of friction; deep drawability; and melting temperature. From the above listed characteristics the resultant mechanical properties (resistance to different stresses, elongation, and stiffness), the properties related to dimensional stability (shrinkage and deep drawability), the properties related to surface quality (coefficient of friction and sealability) and the performances related to thermal resistance (melting temperature, sealability and warm deep drawability) have a special importance for carrier materials used for PSPs also.

Roll-down properties of tapes are very important (see also Chapter 7). Biaxially oriented, multilayer PP films for adhesive-coated tapes have been modified to improve their adhesion to the coating by mixing the PP with particular resins. In this case in order to prepare an adhesive tape that can easily be drawn from a roll without requiring an additional coating on the reverse side, at least two different layers having different compositions are coextruded, and the thin back layer (1/3) contains a release component [249].

Blow Procedure. In the blow procedure the molten polymer mass is brought to the blowing head and is constricted to a thickness of about 2 mm between the exterior and interior walls. If necessary, several layers may be put together in the blowing head to build up a multilayered coextruded film. When the molten material leaves the blowing head, the film is blown with air impinging at a certain angle under pressure. The diameter of the bubble is mechanically controlled with an adjustable aperture [378]. The turbulence of the air produces a tube of a special shape. At a certain height (distance from the head) there is a freezing zone, where the molten material is cooled. Bubble cooling may use either conventional external air rings with blowers or internal cooling. In the processing of polyethylene, molten PE (120°C) is cooled to 60–80°C. The wall tolerances are balanced within ±5% by (1) rotating the entire blowing area or (2) turning the tube around with a rotating lay flat.

Haul-off rates are dependent on extruder output, gauge, and bubble diameter. These factors, together with melt temperature, cooling rate, die gap, and blowup ratio (BUR), control the degree of molecular orientation in the film. This in turn determines the shrinkage characteristics of the film in both the machine and cross directions. Of the machine parameters, the neck-in influences orientation. The material is oriented in the machine direction. The neck-in is the ratio of the constriction of the bubble to the diameter (BUR). The greater the constriction, the higher the blowup ratio and the better the isotropy of the film [379].

Narrowing the die leads to a closer balance of transverse and machine direction tear. The advantages of narrower die gaps—that they provide a thinner melt cross section and allow for improvement of various film properties, gauge control, cooling, and output—are well known. On the other hand, a large die gap means thicker polymer layer, poor heat transfer, longer hot temperature time, better relaxation, and lower anisotropy [378]. Processing with a long neck (for HDPE) is associated with a high blowup ratio with more freezing height. A calibration unit for better cooling, a large die gap, and a lower operating temperature for better bubble stability are required. A high melt index film with low temperature processing conditions blown with a very high degree of orientation may give excellent stiffness [380].

Low density polyethylene is considered the easiest material to process into blown film [360]. LDPE is highly branched, gives much entanglement, has a broad molecular weight distribution, and exhibits reduced mechanical properties. The high degree of branching of LDPE promotes easy extrusion and bubble stability. The narrow MWD of LDPE reduces the ease of extrusion and bubble stability.

Special internal bubble cooling systems (IBC) are required to ensure a trouble-free start-up in processing HDPE [381]. High density polyethylene requires a high blowup ratio and high frost line heights, i.e., bimodal grades. The frost line height at which the melt is normally considered to be frozen is important. Lowering the frost line by faster cooling gives smaller crystallites (like chill roll cooling) and normally better optical properties but does not allow relaxation before freezing [364]. Faster cooling gives less stiffness due to the smaller crystallites. HDPE has a long melt memory, so the die must be designed to allow time for any melt strain to disappear. The stiff film has a tendency to crease and wrinkle, so that nip height, collapsing geometry, and tension are critical [364].

In the early 1970s, a linear low density PE with no long branches, little entanglement, a narrow MWD, and improved mechanical properties was introduced [382]. There are fundamental rheological differences between LDPE and LLDPE. Linear low density polyethylene is less viscous at low shear rates and more viscous at higher shear rates than LDPE. Extrusion equipment is quite different for different PE types. Straight grooved barrel extruders have been developed for HDPE. Helically grooved barrels developed originally for high molecular weight HDPE have been used for LLDPE also [382]. Barrier screws may be used for different types of PE [383]. Grooved barrels were developed in the 1970s. Such devices are not affected by the high back pressures produced by the die. With conventional screws, HDPE and LLDPE produce high pressures and high abrasion. To reduce abrasion by the material the barrier screw was introduced, which has a homogenization section where the molten and solid-state materials are separated. Any solid is left in the plasticizing unit because the channel has a decreasing volume. With LDPE and LLDPE a BUR of 1.5–2.5 is used; for HDPE, a higher BUR is needed. Films with thicknesses of 20–300 μm can be processed. For low friction or tacky film applications, roller framers with an aluminum or nonmetallic roll are needed [381].

The most important polymer variables to be taken into account by PE processing are the melt index, density, MWD, and long-chain branching [384]. Optimized operating conditions include polymer temperature, output rate, BUR/bubble stability, thickness, and frost line height.

Flat-Die (Cast) Procedure. This procedure, also called the chill roll method, includes one or more chill roll units (cooling drums or rollers) as well as a chill roll water bath process. Polishing and smoothing rollers do not remove enough heat from the coating; therefore additional reverse side chilling of the web or water bath is required. Generally for the cast procedure the extrusion temperature is higher. For low density polyethylene, conventional blowing techniques operate at 160–220°C; cast techniques use melt temperatures of 200–260°C [385]. A temperature profile for cast PET film should be situated about 10°F higher than the blown profile [367]. The cast procedure allows better cooling of the molten material by using a cooling cylinder with controlled temperature ($\pm 10°C$) [386]. The exact control of the cylinder speed, the regulation of the rotation speed of the casting and cooling cylinder, gives a better film quality. The finish of the film depends on the smoothness of the roll and its relative velocity. Films with a thickness of 0.01–0.3 mm can be manufactured (3000 mm width, 1000 kg/h) using the flat die procedure. The cooling determines the morphology of the film and the mechanical and other properties (e.g., transparency, gloss, tensile strength, internal tensions, and dimensional tolerances) [387]. Productivity also depends on cooling and product thickness. For instance, for a film with a thickness of 0.1 mm, a cooling temperature of about 36°C is achieved with a running speed of 120 m/min. For a 0.5 mm film, the speed decreases to 40 m/min; and for a 1.0 mm film the running speed is only about 20 m/min. For high speed lines one or more additional rolls are often used to increase heat removal capacity [388].

When casting film the melt is brought into close contact with the chill roll by the use of an air knife, vacuum box, or electrostatic pinning [364]. Film polished on both sides and less than 200 μm thick is difficult to manufacture by cast procedure; large roll gap forces are required [389]. The running (rotation) tolerances of the glossy cylinder are at least 5 μm. That is, a change of 10 μm in the clearance should be taken into account, which means a tolerance of at least 10 μm for a 100 μm film. Cast film can be oriented monoaxially; a 1/7 stretching ratio is suggested [390]. Oriented PP film (30–50 μm) and oriented LDPE film (80–110 μm) are recommended for insulation and packaging tapes [388] (see later).

Coextrusion. Generally the properties of plastic films can be improved by means of (1) coextrusion, (2) extrusion of polymer blends, and/or (3) orientation of the film. Downgauging and improving the film tolerances are supplemental technical possibilities [369]. Coextrusion consists of directly combining several plastics in the die during extrusion. For semicompatible or almost compatible polymers (e.g., LDPE/LLDPE), the mechanical properties of coextruded films are close to those of films made from blends [391].

Coextrusion has economic and technical advantages. As shown by Hensen [377], the increase in the number of layers (e.g., splitting a layer) provides economic benefits. A reduction of up to 40% of raw material costs is achieved by using this procedure. Coextrusion also allows some mechanical properties to be improved independently from the nature of the layer. Generally coextrusion increases the material stiffness. For instance, for HMHDPE a 20% improvement in stiffness has resulted from coextrusion [377]. Thus the stiffness/raw material costs ratio increases by about 25%.

In coextrusion at least two different plastics are plasticized separately and the melts are extruded through multiple-layered dies (slot or annular). The plastics are brought together in the die orifice itself or shortly after. Coextrusion casting and coextrusion coating are possible [388].

The main possible coextruded film constructions based on polyolefins and used for packaging films have been listed by Verse [392] as LDPE/LDPE, LDPE/EVAc, LDPE/HDPE, LDPE/HDPE/LDPE, LLDPE/HDPE/LLDPE, and LDPE/EVAc/PP. The combination of two LDPE layers offers the advantage of different colors and/or foamed/nonfoamed layers. The use of EVAc can impart special mechanical or adhesive characteristics. A symmetrical buildup of the layers ensures lay flat (no curl) and the build up of a medium layer containing recyclate. In PE/PP coextruded films EVAc is the adhesion primer. Generally a coextruded film containing three different layers is curl-free. Four-component films may display curl if the components have different rheological or mechanical properties. Films produced by three-layer coextrusion have HDPE as the outside layer, HDPE or PP as the middle layer, and EMAc as the internal or heat seal layer [393].

The release layer can be coextruded also [249]. For instance, polydiorganosiloxanes are suggested as fillers in coextruded polypropylene carrier materials with tackified top layer. A dimethylpolysiloxane is preferred (0.3–2.0% w/w) as additive. The release layer of such a coextrudate has a thickness of

0.5–10 μm. The total thickness of the tape carrier film is about 15–50 μm. Generally polyethylene embodied in a polypropylene carrier acts as a release agent.

Blown Coextrusion. The high branching of LDPE promotes easy extrusion and bubble stability. The narrow molecular weight LLDPE is less stable than LDPE and tends to form gels if the material stagnates in the die. In coextrusion, residence time increases in the die, especially in the outer layer. When tubular coextrusion is used, problems arise with lay flat and a small rolling up effect of the composite films due to the different behavior of the melts during drawing, the solidification ranges, and the great differences in heat expansion coefficients. The typical multilayer film is built up with three layers referred to as the print skin, core layer, and inner skin; the core can contain recyclate and may have a soft texture top surface with a paperlike back [394].

Cast Coextrusion. Melt fracture and interfacial stability are the main phenomena delimiting the possibilities of cast extrusion. Increasing layer thickness and decreasing the extrusion rate or increasing the die gap opening and decreasing the side layer melt viscosity may improve the performance characteristics of the film [388]. Coextruded outer skin layers of HDPE are used to afford excellent chill roll release for LDPE at high speeds [330].

Comparison of Blown Film and Cast Film. It can be concluded that the choice of a polymer and film grade for the manufacture of carrier material depends on many factors. It is evident that most extruders are specialized equipment with a barrel designed for a given PE type (LDPE, HDPE, or LLDPE). The use of another grade is in some cases possible, but problems concerning the productivity and quality of the product may arise. Assuming that the extruder allows the use of all raw materials, the next problem is the choice of production parameters. The first problem is the choice of a die in order to determine the final BUR ratio. As discussed earlier, a large BUR allows balanced MD/CD properties but lower productivity. The choice of the die gap is also complex. It influences residence time in the material, and thus gels building up and haze may occur. The choice of the cooling place and rate are also important. In order to allow relaxation of the internal tensions in the (HDPE, LLDPE) melt for HDPE for such materials, "long neck" extrusion (with a delayed cooling of the film) is preferred. The higher sensibility of LLDPE related to shear and residence time is taken into account by the choice of composition of the outer layer. Narrow gaps generally result in better machine direction tear resistance, downgauging potential, and orientation balance in the film. However, these benefits are achieved at the cost of adding processing aids to the polymer to prevent melt fracture [382]. According to Feistkorn [395], the advantages of cast films are (1) high transparency, (2) high stiffness, (3) different (textured) surface qualities, and (4) adhesivity on one or both sides.

Thickness control is more difficult for blown film than for cast film because of the lack of control during cooling. Control is better when the melt strength is higher, so using lower temperatures, a lower melt index, and LDPE rather than LLDPE improves thickness control. As mentioned earlier, lower temperatures are recommended to avoid gel buildup also [364]. According to Djordjevic [388], the physical properties of a blown coex film are better than those of a cast coex film, but the optical properties of the cast coex film are better than those of the blown film. Blown films never look as brilliant as the corresponding flat film produced by the slot die process. This is due to the always present streaks caused by the annular die and the uneven surface appearance. To achieve optimum optical properties for a PE film, the frost line should be kept low [396]. Wrinkles may be caused by bubble collapse due to high friction on the slats [382].

The minimum thickness of blown films is about 6 μm, that of cast films about 10 μm, and that of biaxially oriented films (depending on the raw material) about 2–5 μm [377]. The maximum winding speed is 20–140 m/min for blown films, 120–400 m/min for cast films, and 280–350 m/min for biaxially oriented film.

Blown film equipment designed optimally for processing one grade of polyethylene is generally not suitable for processing other types of polymers. The higher viscosity of LLDPE at typical working shear rates makes its processing more difficult [364]. The ratio of investment costs for blown vs. cast film equipment is 1:5–1:10 [397].

Main Synthetic Film Materials

Various synthetic (polymer) materials are used as carriers for PSPs. Films are the best known carrier materials, but foams, fabrics, and plastic profiles are used also. According to Fust [398], in 1984 about 14% of label carrier materials were manufactured on plastic basis, and in 1990 the proportion of plastics attained 40%. In 1996 the main film label carrier materials [399] were polyethylene (41%), polyvinyl chloride (30%), polyethylene terephthalate (13%), polypropylene (8%), and polyimides (4%). Reactivity, ecology, handling, and printability were the main evaluation criteria for their use [392]. Most synthetic carrier films are thermoplastic materials. Table 6.21 lists the main synthetic films used as carrier materials for PSPs and their manufacturing technology.

Polyethylene. Polyethylene is produced by specialized chemical firms using different process technologies (high pressure tubular process, high pressure autoclave process, gas-phase process, slurry process, solution process, etc.) and different catalysts (oxygen, peroxides, heterogeneous or homogeneous organometallic compounds). The properties of the polymers are different. Different grades of polyethylenes are used for films and as carrier materials for coated or laminated webs. For PSPs, low density polyethylene (LDPE), medium density polyethylene (MDPE), high density polyethylene (HDPE), and linear low density polyethylene (LLDPE) have been suggested.

Low density polyethylene gives a tough, conformable film with good thermal characteristics. It is the most common film used as a carrier material for protec-

Table 6.21 Main Plastic Films Used as Carrier Materials for PSPs

Plastic film	Manufacturing procedure	Pressure-sensitive product application
LDPE	Extrusion, blowing	Protective film, label, tape
	Extrusion, chill roll	Label
	Oriented	Label, tape
	Coextruded with HDPE, PP, LLDPE	Label, tape, protective film
HDPE	Extrusion, blowing	Label, tape
VLDPE	Extrusion, blowing	Tape
	Extrusion, chill roll	Protective film
PP	Extrusion, blowing	—
	Extrusion, chill roll	Label
	Oriented monoaxially	Label, tape
	Oriented biaxially	Label, tape
	Coextruded with PE	Label, tape, protective film
LDPE-HDPE blends	Extrusion, blowing	Label, tape, protective film
	Extrusion, chill roll	Label, protective film
EVAc-PE blends	Extrusion, blowing	Label, tape, protective film
	Extrusion, chill roll	Label, tape, protective film
Polystyrene-PE blends	Extrusion, chill roll	Label
Polystyrene	Extrusion, nonoriented	Label
	Extrusion, oriented	Label
Polyester	Extrusion, oriented	Label, tape, protective film
PVC	Extrusion	Label, tape, protective film
	Extrusion, oriented	Tape, label, protective film
	Casting	Label
	Calendering	Label, tape
Polyamide	Extrusion, blowing	Label, tape
	Extrusion, chill roll	Label, tape
PUR	Extrusion, blowing	Label, tape
	Extrusion, chill roll	Label, tape
Cellulose acetate	Extrusion	Label, tape

tive films and meets about 75% of the market requirements. Medium density polyethylene is slightly stiffer than LDPE. It exhibits medium elongation characteristics.

High density polyethylene is the stiffest polyethylene. It has more rigidity and better barrier properties than the other grades. It is manufactured by running the process at a high BUR (more than 3:1) using a high stalk bubble profile. This is necessary to cope with the intrinsic tendency of HDPE films to develop excessive machine direction orientation leading to anisotropy in the film (mechanical properties) [400]. HDPE blown films can exhibit very low machine direction tear strength. Such films can split very easily. The high stalk blowing of the bubble and large BUR ratios are intended to impart some biaxial orientation to the film and balance the mechanical properties. Transverse direction orientation is achieved by delaying the blowup until the melt has cooled somewhat. On the other hand, the delayed blowup of the bubble makes it unstable. To solve this problem, special air rings and mechanical bubble stabilizers (iris technology) are used. The best results are obtained with blown HDPE using high blowup ratios and high frost line heights; to achieve these, very broad, preferably bimodal, grades are needed [364]. This can be realized by using faster cooling or nucleation with LDPE blends. Because film blocking is not a problem with HDPE, the bubble collapsing frame sits closer to the die than in LDPE or LLDPE extrusion. Another problem that arises from the high BUR is that the diameters are kept on the low side. A small diameter means a low specific die rate (output per unit time per unit die circumference). A special PE with a high melt flow index, low processing temperature, and high degree of orientation gives very stiff films [401]. A modulus of 400 N/mm can be achieved for this polymer at a density of 0.930 g/cm^3.

Linear low density polyethylene exhibits better elasticity than other grades and a medium stiffness. Therefore for carrier materials used for protective films, mixtures of LLDPE with LDPE are recommended. Their ratio affects the properties of the final product. LLDPE melt fracture must be avoided to obtain the best optical properties. The higher melt temperatures needed for LLDPE are dangerous because of gel building. Higher temperatures promote oxidation, and oxidation is a source of gel particles [364]. The physical properties of LLPDE compared to LDPE may be listed as follows: higher tensile strength and elongation, better puncture resistance, superior toughness, better machine direction–cross direction balance, and deep drawability. LLDPE was introduced to the U.S. market in 1977 [402]. Copolymers of ethylene with butene (gas-phase polymerization), hexene, or octene (liquid-phase polymerization) have been manufactured. Copolymers with hexene give an LLDPE with a low MFI that displays better tear resistance. Such films may be used as self-adhesive materials also [403].

Very low density polyethylene (VLDPE) is defined as a linear low density polyethylene with a density of less than 0.915 g/cm^3 [380]. Common VLDPE has a density of 0.905–0.915 g/cm^3. It may be oriented also. The orienting ratio used decreases with increases in the density (from 1/8 to 1/6). VLDPE displays excellent mechanical and optical characteristics (less than 2% haze). It is proposed for films (carrier for tapes) and tissues. VLDPE and ULDPE with densities below 0.900 g/cm^3 are used where superhigh flexibility and autoadhesion are required [364].

The main market segments in the European polyethylene film market are [404] food packaging, non-food packaging, consumer goods, industrial, agricultural, hygienic and medical, and special film markets. In the industrial films segment, which includes construction and protective films, antistatic films, surface protection films and adhesive films, labels and sleeves are the most important applications. In the range of special films, lamination films, oriented films and tapes, foiled films, and perforated films are the main applications. For tapes, mainly oriented films are recommended.

Polypropylene. Homopolymers with high crystallinity and block copolymers exhibiting a melting range of 160–165°C have been prepared; the melting range of random copolymers is situated at 135–160°C [405]. The temperature resistance and elasticity of PP are higher than those of PE; therefore, PP is used for higher temperature applications. It can be sterilized, but it becomes brittle at +5°C [373]. Polypropylene offers better clarity, better barrier properties, higher impact strength, and a weaker memory effect. Polypropylene has a very low solvent retention value (from printing inks). It is more expensive than PE, and because of its built-in elasticity it is more difficult to handle after manufacture. Compared to PE, polypropylene is more grease- and chemical-resistant. It is difficult to compound polypropylene with PE. There are some patents that recommend the use of a compatibilizer such as ethylene-propylene elastomer to improve blend homogeneity.

Polyolefins can be used as nonoriented or mono- or biaxially oriented films. Oriented films are applied mainly for tapes. Polyvinyl chloride and BOPP have been the main plastic carrier materials for tapes [406]. Rigid calendered PVC has been used as mono- or biaxially oriented material with a plain or embossed surface and a thickness of 28–70 μm. BOPP has been applied as a blown or cast film, biaxially oriented, with a gauge of 25–38 μm [406]. BOPP films are also now produced with a thickness of 40–50 μm. With the common stretching process, thick BOPP films cannot be produced economically. For these purposes it has been suggested that several thin films be combined to form a multilayer film (via heat laminating). According to Ref. 407, the idea of putting a series of BOPP films one on top of the other to achieve an increased film gauge is not a new one. In 1985 a patent was granted that covers the production of plates made of heat-sealable BOPP films. The performance characteristics of thick BOPP films can be compared with those of PVC, amorphous PET, or cellulose acetate film. Because of its low density, polypropylene offers the advantage of increased yields per unit area. The yield per unit area with PVC is about 53% lower than for PP; for PET it is 47% lower, and for cellulose acetate 43% lower than for polypropylene [407]. Normally the extruded film is stretched to six times its initial length, and its thickness decreases. Such warm laminating is affected by temperature, pressure, and time. With an increase in speed, many air bubbles are entrapped; therefore more pressure (maximum 250 N/cm) is needed. A heated, rubber-coated press and heated laminating steel cylinder are used.

Manufacture of PSPs

Table 6.22 DSC Phase Transformation Characteristics of Some Olefinic Raw Materials Used as Carriers for PSPs

Product	Material	ΔH mJ	ΔH J/g	Peak °C	Peak mW
Warm laminated protective film for plastic plates	LDPE, 50 μm	934	114.5	111.5	7.7
Warm laminated protective film for 0.3 mm coils, BA	C_8-LLDPE, 30 μm	679	165.6	127.7	7.9
Label carrier film	MDPE, LLDPE, 80 μm	751, 348	97.2, 45.1	117.9, 127.0	9.7
Deep drawable protective film	LDPE, PP, 70 μm	104, 443	17.3, 73.4	109.9, 154.7	0.9, 2.9
Hot laminated protective film	LDPE, 120 μm	930	111.8	112.3	10.9

As discussed in Ref. 408, oriented polypropylene film is thermofixed, i.e., shrinkage (recovery of original dimensions) is avoided up to about 110°C. However, certain solvents (e.g., toluene) can cause "deblocking" of the shrinkage mechanism at lower temperatures, i.e., the loss of dimensional stability. Polypropylene can be combined with nonolefinic films too. For instance, for a pressure-sensitive label having a wrinkle-resistant lustrous, opaque facing layer for application to a collapsible wall type of container (squeeze bottle), the carrier has a thermoplastic core layer that has upper and lower surfaces and voids, with a void-free thermoplastic skin layer fixed to the upper surface and optionally to the lower surface of the core layer and discrete areas of pressure-sensitive adhesive. As the core layer, a blend of isotactic polypropylene and polybutylene terephthalate is used, as the skin layer, polypropylene, and as the adhesive, circular dots of HMPSA [409]. With appropriate formulations, the range of olefin-based raw materials can be used for the manufacture of carrier films for different pressure-sensitive products (Table 6.22).

Polyvinyl Chloride. PVC is polymerized in emulsion or suspension. Emulsion PVC contains stabilizers (surfactants, protective colloids); therefore, its electrical properties are not as good, but its resistance against static electricity is better than those of suspension PVC [410]. Hard PVC plates having a thickness of less than 1 mm are called films [411]. According to DIN 4102, PVC is nonflammable. Films of PVC have been manufactured via extrusion, calendering, and casting.

During calendering the particles of EPVC (having a diameter of about 0.1 μm) are pressed together in a calender and later sinterized at higher temperatures [411]. Casting is based on precipitation of the polymer from a solvent. Because of the lower tensions induced in the film during manufacturing, cast and calendered PVC have better performance characteristics as carrier materials for PSPs than extruded PVC.

Cast vinyl marking films are used for producing labels, emblems, stripes, and decorative markings for trucks, automobiles, and other equipment. They are durable, conformable, and dimensionally stable and withstand weather and handling conditions. They can be processed by screen printing, roll coating, steel rule die cutting, thermal die cutting, and premasking.

Nonoriented and biaxially oriented, embossed, printable PVC (25 μm) has been used as carrier material for tapes [368,406]. Rigid calendered PVC has been applied as mono- or biaxially oriented material with a plain or embossed surface and a thickness of 28–70 μm [412]. Calendered PVC is used for plotter film also.

Polyvinyl chloride formulations include plasticizers, fillers, antiblocking agents, and lubricants. Micromolecular and polymeric plasticizers are used (e.g., saturated polyesters) [413]. Fillers for imparting opacity and antiblocking agents are also included [414]. The level of opacity additives (0.5–5.0%) depends on whether the PVC is hard or soft. For antiblocking purposes, only 0.5–1.0% of such an agent is added. Calcium behenate exhibits a strong lubricating action due to its long fatty acid chain. It is used as the main stabilizer lubricant for PVC. For rigid PVC film, the raw material may be either emulsion or suspension PVC. Suspension PVC is suggested primarily for small rolls, for office and household use, and for long rolls for automated packaging [406]. At the end of the 1980s suspension PVC made up less than 5% of the total market. Some years ago, rigid PVC exhibited the advantage (in comparison with PP) that it could undergo direct dyeing during film production.

Polyethylene Terephthalate. Biaxially oriented PET for the packaging sector has good dimensional stability and thermal and mechanical resistance as well as good transparency and luster. This film possesses excellent tear resistance and good resistance against oil and chemicals [410]. It can be heat-sterilized. A range of films are offered: transparent film with thicknesses of 15–50 μm, in standard, chemically treated, corona-treated (52 dyn), and opaque grades; white film having a thickness of 12–36 μm; and other colors [415]. PET has to be oriented or crystallized to obtain tough usable articles. Amorphous nonoriented PET is too brittle. Chemically toughened PET (copolymer) films have been developed also [367]. Elongation for the chemically toughened PET is higher than for oriented PET. Polyethylene can be coextruded with PET with a coextrudable adhesive [367]. Thin PET film (2–20 μm) can be manufactured at a productivity rate of less than 720 kg/h [377]. Cast PET film having a thickness of 20–300 μm is manufactured at a rate of less than 5500 kg/h [370]. Polyester is used as carrier for labels for cosmetics, toiletries, pharmaceuticals, chemical products, and shrink sleeves [416]. It is applied as a release liner for tapes also [73]. PET is available in both thermosetting and thermoplastic forms. It is unsuitable for heat sealing and thermal die cutting. Oriented polyester can be thermoformed. Raw, topcoated, and pretreated grades are available.

Polystyrene. Films of polystyrene can be used as carrier materials for PSPs. They can be either nonoriented or oriented. Nonoriented polystyrene films display clarity and excellent printability. Such films are notch-sensitive and susceptible to web breaks [418]. Unoriented polystyrene films have relatively little elongation, i.e., little deformation during printing. Such films can be be used between −40 and +150°F. Unfortunately, they can build up a static charge. Polystyrene possesses excellent printability but low chemical and weathering resistance [412].

In the last decade, biaxially oriented polystyrene films (OPS) have been developed. These films have thicknesses of 0.2–0.6 mm [418]. They are made from cast polystyrene film oriented in a two-step procedure. The orientation degree is 2.5–6.5. The advantages of OPS films are the following:

1. They can be high gloss or matte with excellent transparency.
2. They have good antistatic properties.
3. They have good rigidity even with low thickness.
4. They have a low specific gravity (1.05 g/cm^3), i.e., good yield per kilogram.
5. They are nontoxic and nonhygroscopic.
6. They are stable at extreme temperatures (−60 to +70°C).

Oriented polystyrene is suitable as an antistatic support film for transfer letters and symbols and adequate for printing by litho, letterpress, screen, and flexo printing. Polystyrene has also been used as a carrier for labels [419]. It complies with food industry regulations, including FDA and BGA requirements. No-label look labels have been manufactured from OPS for lubricant containers. Labels for toiletries and cosmetics as well as motoring, gardening, cleaning, and other consumer products use this material [420]. These labels can be recycled together with the labeled product. The availability of transparent and opaque white, matte, or glossy films gives greater flexibility for graphics.

Polyamide Films. Carrier materials can also be based on polyamide. Polyamide films display abrasion resistance and resistance to oil [410]. Such films possess good printability. Polyamide 6 was considered the best of all polyamide grades. 6,6-Polyamide is also used, because of its higher melting point and greater stiffness [421]. Blown and cast PA films are produced. For the manufacture of PA blown film, medium to high viscosity polyamides are used. For cast film, a low viscosity PA is processed.

Cellulose Acetate Films. Extruded transparent cellulose acetate film was developed by Eastman in 1967 [422]. Cellulose acetate films exhibit reduced shrinkage (0.6% at 71°C, 24 h) and low triboelectricity and have good printability. They can be used as the carrier material for self-adhesive tapes and tamper-evident products. For instance, a 50 μm cellulose acetate film has been modified to be a brittle film with a low tear strength. The film has to be die cut and skeleton stripped. For cellulose acetate films, no pretreatment is required for normal printing. They can be used as overlaminate, seal or label, as clear white and computer-imprintable white films. Such films are an alternative to PVC also [423]. Cellulose acetate has been introduced as a carrier for labels [420].

Cellulose Hydrate. Cellulose hydrate (Zellglas, cellophane) can be used as the carrier material for PSPs also. Cellophane is a natural material with excellent deadfold and antistatic properties and a wide range of colors but is not moisture-proof [424]. There are different grades characterized by sealability (S), flexibility (F), lacquering (P,D, and X), moisture resistance (M), etc. [425].

Ethylene-Vinyl Acetate Copolymers. EVAc copolymers with a density of 0.930 g/cm^3 have been suggested as high gloss, clear, nonadhesive carrier films [426]. The ethylene copolymers such as EVAc, EBA, and EMA are miscible with LDPE [327]. The use of VAc copolymers for carrier films is enhanced by their flame-retardant properties also. Halogen-free flame-retardant noncorrosive materials (FRNCs) are preferred for insulating wires. Such materials have to resist 20–180 min. Flame resistance increases with VAc content [427].

Polyurethanes. Linear polyurethanes can be processed as cast or blown films. Such films with a melting range of 75–170°C can be hot laminated without adhesive [428]. For instance, a pressure-sensitive adhesive tape for the protection of printed circuits, based on acrylics and phenolic terpene resin, with a 100–200 g/m^2 siliconized crepe paper, is made by coating the adhesive mass first on a siliconized polyurethane release liner and then transferring it under pressure to the final carrier material [179].

Polycarbonate. Polycarbonate is supplied as films and sheets and is used for membrane switch graphics, decals, product identification, name plates, etc. It can transmit more light than PET and retains its transparency in thicker films and sheets.

Laminates. A quite different technology is used in the manufacture of film laminates (homogeneous or heterogeneous, concerning raw materials, component geometry, or structure). The range of such products is very broad. Thirty years ago, flexible packaging used over 600 different types of laminations [429]. In recent decades, their diversity and manufacturing methods have been strongly developed.

Manufacture of Foams

Generally weblike foams or foamed films are manufactured by firms that specialize in foam manufacturing or the production of special films. Foams with good tensile strength and elongation are manufactured by mixing LDPE, blowing agents, additives, peroxides, and other polyolefins, and extruding, grinding, and foaming [430]. Iononomer foams with high tensile strength have also been developed [431]. Fire-resistant composites comprise plastic films (e.g., polyimide) and in situ foamed silicones [432] or polyamide-polyimide foams [433]. The most used foams are PUR-based. A special industry has developed for their manufacture.

Formulation and Manufacture of Synthetic Carrier Films for Labels

The requirements for chemically and aesthetically improved labels have forced the introduction of plastic films as face stock materials for labels. As discussed earlier, plastic films used for labels have to display stiffness and dimensional stability. Such properties can be achieved through the choice of an appropriate raw material, the choice of an appropriate film processing technology, or both. The choice of an appropriate raw material means formulation on the basis of a high molecular weight polymer that is processible as a thermoplast to give a balance of stiffness and flexibility and of plasticity and elasticity in order to allow cuttability (plasticity), labeling (flexibility and plasticity), and printing (dimensional stability).

The choice of raw material is limited by the range of available polymers and the compatibility of the mixture components. The choice of processing technology refers to the extrusion technology and postextrusion processing technology (orienting). A combination of such variants, i.e., the use of a raw material compound (e.g., polystyrene/polyethylene), processing via multilayer coextrusion, and orientation of the film, allows the desired properties to be achieved. According to Ref. 163, the choice of carrier for labels is influenced by several factors: the nature and surface of the substrate, the end use environment (weathering and applications climate), mechanical requirements, printing methods, processing conditions, and special requirements (e.g., FDA, BGA approval).

The main carrier materials used for labels are [163,434] PET, OPP, LDPE, HDPE, polystyrene, PC, PA, cellulose acetate, PVC, polyacrylate, and laminates. According to Hufendiek [163], the choice of a plastic carrier material for labels depends on the following parameters: internal or external use; postprintability and writability; temperature resistance; chemical resistance; weatherability; abrasion resistance; labeling ability, and stiffness.

It is known that stiffness influences labeling ability also. The choice of label materials is a function of the application field requirements and economic considerations too. Multilayer materials are being used with filled medium layers for economic reasons [435,436]. According to Ref. 437, the carrier materials for labels can be classified as monolayer (paper, films), coated monolayer (coated, printed, metallized paper and foils), and composite. The broad range of quite different carrier material grades required for labels having the same end use is illustrated by wine bottle labels. Such labels use glossy, matte, antique finish, and aluminum laminated papers with permanent and washable adhesives [438].

Polyvinyl Chloride for Labels. Some years ago PVC was the most commonly used carrier material for labels. For special products (e.g., sheet labels for outdoor application), cast PVC was used [439]. According to a forecast concerning label production and usage to the year 2000, the annual growth of film carrier material consumption is estimated as 12–14% for PET, 13–17% for PE, 18–22% for OPP, and 3–5% for PVC [440]. Soft PVC 80–120 μm thick is used for self-adhesive labels. Films that are less than 60 μm thick are made from HPVC; and for large surface area products, calendered PVC is used [441].

Polyolefin Films for Labels. Polyolefin films are the most common plastic carrier materials for labels. They are used as the carrier for the face stock and for the release liner also. Some years ago, nonoriented blown polyethylene films were used only as label carrier materials. Later biaxially oriented, cast, and coextruded PE films were developed. Cast LDPE, oriented LDPE, LDPE–polystyrene blends, LDPE–LDPE laminates, and coextrudates can be used also. The main disadvantage of LDPE as the carrier material for labels is its limited stiffness. Different processing methods, conditions, and material combinations have been suggested to eliminate this disadvantage. For instance, cast PE is used in cosmetics and toiletries [442]. Polyethylene/polystyrene film (clay-coated for ink receptivity) is used as a carrier for labels also. From the range of polyethylenes, HDPE is the most useful material for labels. Where a combination of high stiffness and puncture resistance is required, i.e., as a substitute for cardboard and paper (synthetic paper), bimodal HDPE is used [443]. Bimodal HDPEs are tailor-made products for coextrusion and for the production of labels [444]. Crosslinking of PE improves its dimensional stability. A crosslinked PE exhibits an elongation of 175% at a tension of 20 N/cm^2 [445]. Cross direction mono-oriented films display low cross directional shrinkage and are used for labels [446]. Polyethylene lamination films are usually produced by the blown film process [447].

The high stiffness required for labels can be achieved by orienting the PE film. Films made using a machine direction orientation exhibit outstanding stiffness in the machine direction and flexibility in the cross direction. Lay flat by printing is completed by on-pack wrinkle-free squeezability [82]. For most applications an isotropic material is necessary, so biaxially oriented films should be used. A special product is manufactured that has different lamination angles of the oriented film. This is produced as 75–215 μm film [448,449]. Cross laminating of two films in which the orientation of the molecules runs at an angle of 45° with the machine direction leads to outstanding physical and mechanical properties surpassing those of traditional materials [449]. For instance, the ultimate tensile load of such a 50 g/m^2 film can attain 30 N/10 mm (DIN 53455), and its Elmendorf tear resistance is about 17 N (ASTM D 1922). Biaxially oriented and cross laminated films have also been manufactured.

According to Waeyenbergh [434], PP films for labels have to fulfill the following requirements: usability for transparent and opaque face stock, high mechanical strength (yield and stiffness), moisture and chemical resistance, good weathering characteristics, good printability (matrix printability also), good embossing and die-cutting characteristics, ability to be combined with other films, ability to be metallized, and recyclability. Chemical resistance is related to the end use of the labels and to the environment [437]. For application on containers, resistance to organic solvents, lubricants, and oils and to humidity, steam, and water are required.

Nonoriented and mono- and bioriented thermofixed PP films are available. Oriented films give a higher yield at lower gauges. For instance, a PP film having a thickness of 20 μm can be manufactured to produce a 56.2 m^2/kg film; a biaxially oriented product having a thickness of only 12.5 μm can be used for a 89 m^2/kg film [450]. Coated pearlized, white opaque, and metallized PP films have been manufactured as carriers for labels [451]. Pearlized white film may give a plastic-coated paper effect. Such (coextruded) films can incorporate white pigment in the core or voids in the common "cavitated" films. Transparent films having a thickness of 40 or 50 μm and opaque or white films with a gauge of 28, 35, 40, 50, or 60 μm have been suggested for labels. Coextruded white, metallized, nonlacquered and acrylic lacquered oriented polypropylene films for use as carriers for labels are manufactured with a thickness of 21–60 μm [452]. Polypropylene films provide good machinability, and when coated they give a glossy surface for a high quality print. Polypropylene is used as a carrier for labels for plastic bottles or containers as a matte, white, pretreated material [453].

Very stiff polypropylene (with a modulus of 2400 N/mm^2) and low modulus polypropylene (with a modulus of 100 N/mm^2) can be used as carrier materials for labels and tapes [454]. The stiffness of common materials used for tapes (e.g., OPS, PET, and PVC) is situated at 2500–3500 N/mm (E modulus). The higher stiffness required for labels can be achieved with special PP homopolymers. Filled polypropylene or heterophasic copolymers exhibit a modulus of 3000 N/mm^2. Biaxially oriented polystyrene replaced PVC, but has since been replaced by PET and PP [423]. BOPP and nonoriented transparent PP with a thickness of 60 μm are recommended for no-label look labels [453,454]. Biaxially oriented PP films are used for cosmetics and other bottle labels [455]. Generally films are made with a thickness of 30–80 μm. Films having a thickness of 50 μm are suggested for high speed labeling [455]. Such films possess a surface tension of about 52 dyn/cm. Because of their higher stiffness, lacquered films may have a lower thickness. For instance, a nonlacquered film with a thickness of 100 μm corresponds to a lacquered material with a thickness of 90 μm. Topcoated superwhite opaque PP films with thicknesses of 50 and 60 μm have been developed for labels. The top coating of such films can be printed by letterpress, UV silkscreen, litho, UV flexo, and thermal transfer processes. Topcoated PE and PP films display the advantage of better printability and environmental resistance. About 15–25% of the polyolefin films used in Europe as label carriers are topcoated. Such films have to be lacquered via rotogravure to achieve a high quality coating.

To improve the mechanical characteristics (stiffness and tensile strength) of label carrier film, coextrudates can be used also. Such films are built up with a middle layer of LLDPE (e.g., with a thickness of 36 μm) and two external LDPE layers (e.g., with a thickness of 22 μm).

Label films have to display excellent cuttability. That supposes that the cutting forces in both direction MD/CD are the same (see Chapter 7 also).

Polyester Films for Labels. Computer-printable topcoated PET films are used for stiff labels such as name plates. For the production of high quality labels, metallized and printed lacquered polyester films are suggested [456]. Due to its thermal resistance and dimension stability, PET is the most important carrier material used in hot stamping and for holograms. As known, the hologram on a special photoresist is coated with a nickel layer; this is the master embossing tool. As a transfer film for hot stamping, 19–23 μm PET is used. This is a metallized film coated with a special lacquer that can be embossed. During hot stamping the transfer temperature is about 120–160°C. Diffraction films are made using the same procedure (they contain two-dimensional images only). Pressure-sensitive holograms can be laminated, cut, and overprinted. For pressure-sensitive holograms a 50 μm PET film has been proposed.

Common, commercially available polyester films are manufactured with a thickness of 12–100 μm [457]. Such film may be topcoated. According to Ref. 390, the bulky colored white film allows more application flexibility because white pigments may change the adhesive performance. Generally PET is chemically pretreated. Common PET exhibits a contact angle of 54° with water (in comparison with 66° for PP) [458]. For writability, special coatings are used [459]. For transparent films a solution of precondensed melamine resins and cellulose ether or ester containing a dispersed inorganic powder has been patented as a writeable coating [453]. Polyester laminated with polypropylene can be used in medical applications where hot steam (134°C) sterilization is applied [461].

Polystyrene Films for Labels. Polyethylene–polystyrene blends have been developed as recyclable carriers for labels, which complies with European food contact regulations. Such films display squeezability, printability, roll conformity, cuttability, and dispensability [443,462]. Polystyrene films were the first "no-label look" materials.

Laminates for Labels. Film/film and film/paper laminates are suggested as carrier materials for labels. Humidity-resistant papers are made with a PE coating (extrusion coating or laminating). Polyethylene (LDPE or HDPE) is coated on paper by extrusion (with a coating weight of 20–30 g/m^2). By laminating, LDPE is applied on paper with a coating weight of 10–20 g/m^2 [455]. For extrusion coating a PE grade with a density of 0.915–0.925 g/cm^3 and MFI of 190/2.16 is recommended [463]. For special end uses, where the transparency of a partially delaminated multilayer label is required, paper/film laminates with the PE film as the middle layer are suggested [464]. A white opaque polypropylene film over-laminated with PET is used for labeling soft drinks. It has a very low coefficient of friction with respect to metal, which allows trouble-free high speed application [465].

Other Face Stock Materials for Labels. Metallized paper and laminated metallic films are used for labels [466]. Since 1981 there has been a trend to replace aluminum/paper laminates with metallized papers. Such papers are less humidity-sensitive and their demetallizing offers new decoration possibilities [467]. Aluminum base layers are used for special laminates for security film technology [468].

Conformable and porous carrier materials are suggested for medical labels and tapes. Such materials include nonwoven fabric, woven fabric, and medium to low tensile modulus plastic films (PE, PVC, PUR, low modulus PET, and ethyl cellulose). For conformability, the films should have a tensile modulus of less than about 400,000 psi (in accordance to ASTM D-638 and D-882) [469]. Preferred carrier materials are those that permit transpiration and perspiration and/or tissue or wound exudate to pass through them. They should have a moisture vapor transmission of at least 500 g/m^2 over 24 h at 38°C (according to ASTM E 96-80), with a humidity differential of at least about 1000 g/m^2. A conventional polyethylene terephthalate film has an approximate value of 50 g/m^2 of moisture vapour transmission. Cellulose acetate can be used as a porous carrier material [470]. The coating technology for textiles and the manufacture of impregnated carrier materials are described by Witke [471–473]. The manufacturing process includes coating with dispersed or dissolved plastics and resins or with solid-state coating components.

Carrier manufacture for labels may include coating also. For certain label manufacturers the surface quality (gloss, adhesive anchorage) should be improved chemically. Therefore the film has to be coated with a lacquer. The top-coated film is converted into a pressure-sensitive laminate and later processed and printed on the lacquered side as a narrow web. For such a carrier, a high coating quality is required. For instance, such quality is given by gravure printing using a line number of 44.

Formulation and Manufacture of Synthetic Carrier Films for Tapes

As discussed earlier, the first plastic films for tapes were introduced from the range of common packaging films. Their development occurred in parallel with the general development of raw materials, film manufacturing, and transformation methods. The development of other weblike polymer-based products (fabric, nonwoven, foam, etc.) influenced the development of the packaging industry and of carrier materials also. Hard and soft PVC, cellulose hydrate, polyethylene, cellulose acetate, and polypropylene are the most used carrier materials for tapes [474]. Generally oriented films have been applied.

The procedure used for postforming the carrier is more important for tapes than for labels. For instance, a carrier for tapes for low temperature application may have transverse cuts or holes [475]. Easy tear, breakable tapes allow easy tear or splitting of the tape during handling. Easy tear breakable hank tapes are used in the same finished product as easy tear paper tapes. These tapes are applied by hand. They are used for hanking and spot tapes when fast, easy removal is required. They are highly filled, having a low elongation and low break force. PVC/PVAc has been proposed also [476].

A large variety of homogeneous and composite materials are suggested as carriers for tapes depending on the end use requirements. HPVC, SPVC, PP, PE, PET, PETP, polyimide, PET/nonwoven, PET/glass fiber, paper/PET/glass cloth, PE/EVAc foam, PE foam, PU foam, aluminum, paper, cloth, and nonwovens are used as carriers for tapes. For instance, electrically insulating tapes are based on PET, polyimide, PET/non woven, glass cloth, paper, PET/glass fiber, and other carriers. High temperature (300°C) resistant special labels and tapes for printed circuit boards use polyimide films as carriers also [477]. Such polyimide materials have been recommended for transparent pressure-sensitive sheets coated with silicone-based adhesives. The pressure-sensitive laminate resists 8 h at 200°C [478]. High temperature nonsilicone-type adhesive systems based on high temperature resistant rubber on PTFE or polyester backing are used as aircraft PSAs. Such products are suggested for holding composite layup pieces in place during bonding [479]. Polyvinyl isobutyl ether and polyvinyl alkyl ether on cellophane are used for medical tapes [480]. Biaxially oriented cast films are recommended as carriers for packaging tapes [481].

Polyolefins for Tapes. Narrow MWD HDPE is a raw material for flat film for monofilaments and tapes [442]. Such products require excellent mechanical properties during and after stretching. Therefore a medium MWD polymer is used for blown film manufacture and also for tapes. Mechanical strength and elongation are required for these products. Therefore, the blown film extrusion of

HDPE is carried out at high blowup ratios to reduce the imbalance of the mechanical properties of the film. Transverse direction orientation is achieved by delaying the expansion of the bubble (special air rings and bubble stabilizers are necessary, and the collapsing frame is situated closer) until the melt has cooled somewhat. Polyethylene for tapes is chemically pretreated [184]. Polyolefins are suggested for medical tapes also [482,483]. Using a special method, a monoaxially oriented PE film may be sealed (laminated under pressure, without adhesive) on another PE or PP film in order to obtain improved mechanical characteristics [384]. Oriented LDPE (80–110 μm) is proposed for insulation tapes, and oriented PP (30–50 μm) is recommended for packaging tapes [390].

LLDPE and mixtures of LDPE and LLDPE are recommended for tapes also. Such formulations make use of the deformability of LLDPE and processibility of LDPE. The difference between the molecular structures of LLDPE and LDPE and the narrow MWD are responsible for the differences in rheological behavior during processing. LLDPE has a higher viscosity under shear. This viscosity is due to the friction of the polymer with the metal and to shear on the surface of the metal [393]. Extensibility and deformability are general requirements for certain mounting tapes also (see Chapter 8). The elasticity of LLDPE is used to prevent blocking during unrolling. Such tapes have a coextruded backing layer of LLDPE [485]. Ethylene acrylic copolymers are amorphous, low modulus compounds. Such products are recommended as blown or cast film carriers for special elastic tapes [389].

Polypropylene can be synthesized that have quite different mechanical properties. As mentioned before, very stiff polypropylene (modulus 2400 N/mm^2) and low modulus polypropylene (100 N/mm^2) are suggested as carrier materials for labels and tapes. Flameproof grades are manufactured also. Diaper closure tapes are used as refastenable closure systems for disposable diapers, incontinence garments, etc. [486]. These products include two- and three-tape systems. The two-tape systems include a release tape and a fastening tape. The fastening tape comprises a carrier material such as paper, polyester, or polypropylene. The preferred material is polypropylene (50–150 μm) with a finely embossed pattern on each side. Such tapes allow reliable closure and refastenability from an embossed, corona-treated polyethylene surface used for the diaper cover sheet.

Coextrusion of PP is carried out to give products of different colours and different surface properties (adhesion and sealability) and to improve their mechanical properties. Biaxially oriented multilayer PP films for adhesive coatings have been manufactured to improve their adhesion to the adhesive coating by mixing the PP with particular resins [487]. The resins add up to 25% by weight. Coextruded ethylene/propylene can be stretched for use as a tape carrier also [488].

For tapes and protective films the back side of the self-adhesive (coated or uncoated) products can work with a supplemental release coating [489] or, by with an appropriate carrier material with low level of adhesivity (e.g., polyolefins), as an uncoated liner [490]. The abhesive agent can be coated on the back side of the carrier material as a separate layer or embedded in the carrier material [249]. If coextruded structures are used, the total thickness of the support film is about 15–50 μm, whereas the thickness of the abhesive layer is about 1.0–5.0 μm. For instance, a polyorganodisiloxane, particularly a dimethylsiloxane with a well-defined viscosity, can be used as the abhesive component. Such an abhesive substance is added to the layer in an amount of 0.2–3.0% w/w. The second layer of a biaxially oriented multilayer polypropylene film that faces the adhesive has a thickness of less than one-third of the total thickness of the adhesive tape and contains the antiadhesive substance [491].

Masking tapes can have special carrier constructions to allow conformability. According to Lipson [492], a masking tape of polypropylene has a stiffened, wedge-shaped, adhesiveless longitudinal section of PET extending from one edge, with an accordion-pleated structure to conform to small radii. Some masking tapes have to be flexible, stretchable, and contracting, capable of maintaining the contour and curvature in the position in which they are applied [493]. They are suitable for protecting a surface on a substrate that has irregular contours. Such tape carrier materials can contain VAc copolymers. For instance, smooth white crosslinked polyolefin foams having an average cell diameter of 0.50 mm and a yellowness index of 20.4 have been prepared [494]. To produce a smooth carrier material, polymer compositions with high water absorption can be extruded also. Such a film contains a mixture of EVAc (80 parts), absorbent resin (10 parts), and calcium oxide (10 parts) and displays a water absorption of 50–100 g/g [495].

Polyvinyl Chloride for Tapes. Hard (rigid) and soft (plasticized) PVC have been suggested as carriers for tapes. Such films are the most used carrier materials for packaging tapes. About 80% of all tapes are packaging tapes. PVC does not need a release and is not noisy [434].

Monoaxially or biaxially oriented films are applied with a thickness of 3–90 μm, transparent or clear, glossy or embossed [496]. Embossed films have a thickness of 28 μm in comparison with 40 μm for glossy films. In recent years, the thickness of the PVC film has been reduced from 40 to 25 μm for packaging tapes; for thicknesses of less than 30 μm an embossed film is suggested, which is used as a mechanical release agent also.

Soft PVC may contain up to 50% plasticizer [184]. It is applied for electrical tapes and decor films [497]. Copolymers of vinyl chloride serve as clear tape for packaging and household use. Polyvinyl chloride films with thicknesses of 30 and 50 μm are carrier materials for packaging tapes [165]. Freezer tapes are based on primed soft PVC [498]. Such tapes for deep freeze applications are manufactured with a soft PVC carrier that needs a primer for the rubber-based

PSA that is used [165]. Generally PVC used as a carrier for tapes has been primed [411]. Sealing tapes based on PVC carriers are printable. Sandwich printing is possible also (flexo printing for the top side and gravure printing for the back side between film and adhesive). Such tapes are used for closing [499]. Polyvinyl chloride with silicon carbide as filler is recommended for electrically conductive tapes [500]. PVC has been introduced as carrier for medical tapes too [501–505]. Cover tapes for bathroom applications are made on special PVC basis, with a bacteriostatic agent incorporated in the vinyl formulation. These tapes are embossed to produce a secure nonslip surface. They are coated with a repositionable adhesive that builds up adhesion from 26.7 oz/in. to 36.2 oz/in. [506]. For such tapes a 78 lb kraft release liner is used.

The development of PVC substitution by OPP as carrier material for tapes is illustrated by the following data. In 1979 about 80% of all tapes were manufactured with PVC carrier material. In 1995 only about 37% were PVC-based. The polyvinyl chloride carrier material represented 60% of the manufacturing costs of a packaging tape [507]. According to Ref. 508, in 1989 the costs for the OPP carrier represented 43% of the global costs. The coated components (primer, release, adhesive) accounted for 27%.

Other Films for Tapes. Double-sided mounting tapes of PET with silicone adhesive on one side and a silicone release on the other are used in electronics [509]. Cellulose acetate and hydrate have been introduced as carrier for clear tapes in household and office [510]. Electrically conductive adhesive tapes have been manufactured by coating an acrylic emulsion containing 3–20% nickel particles onto a rough flexible material (e.g., Ni foil) [511]. High temperature resistant polyimide films have been proposed for insulating tapes. Such films are flame- and temperature-resistant between -269 and $+400°C$.

Fiberlike Carrier Materials for Tapes. Fiberlike fabric (textile) materials were the first carrier webs used for PSPs [510]. Actually such products are used for applications with high mechanical requirements. Fiberlike materials (water-resistant and water-soluble) based on cellulose, combined with plastic films (soft and hard polyvinyl chloride, cellulose derivatives, polyester), foams (open-pore polyurethane, closed-pore polyvinyl chloride, closed-pore polychloroprene, etc.) are suggested as carriers for tapes [101]. Paper and nonwovens can be impregnated with water-soluble and water-insoluble resins; foamlike carriers are coated with adhesive on one or both sides. As separating solid-state components, glossy or crepe papers and nonwovens coated or impregnated on one or both sides are used. Fiber-based plastic carrier materials, cellulose-based nonwovens, polyethylene-coated cloth, polyamide cloth, and siliconized cloth are used for various tapes. A special two-layer nonwoven laminate is applied for tapes [512,513]. Combinations of materials such as chemically bonded cellulose with cotton and synthetic fibers or thermally bonded polypropylene (50%) with cellulose (50%) have been proposed for special tapes. Impregnation is a general method for strengthening the carrier for tapes [514].

Woven and nonwoven materials have been suggested as carriers for medical tapes [515,516]. Air- and moisture-permeable nonwovens (PET nonwoven, embossed nonwoven) or air-permeable tissue (breathable face stock) are used for medical tapes. The carrier has to meet air transmission criteria. For such tapes cotton cloth has also been proposed as the carrier material [142]. A special conformable carrier material contains a layer of nonwoven web of randomly interfaced fibers bonded to each other with a rewettable binder dispersed throughout and at least one additional layer of the same composition. The fiber of the additional layer is laid directly on the first layer before the second layer is added [512]. Cotton cloth and hydrophobic special textile materials coated with low energy polyfluorocarbon resins are used as carrier materials for surgical tapes. Reusable textile materials such as Goretex, polyamide, polyethersulfone/cotton, 100% cotton, or other hydrophobized carrier materials can also be applied. Porous PE can also be used as a carrier for medical tapes [517]. PVAc, PVC, PE, and copolymers are also recommended as adhesive for medical tapes. The adhesive is hot pressed in the fibrous carrier. Such materials possess the advantage of preventing contact between small fibers and the wound.

Electrical tapes have to possess high dielectric strength and good thermal dissipation properties. They are used for taping generator motors and coils and transformor applications, where the tape serves as overwrap, layer insulation or connection, and lead-in tape [518]. The carrier is woven glass cloth impregnated with a high temperature resistant polyester resin or a combination of impregnated woven polyester and glass cloth. This last variant is recommended where conformability is required. For insulating tapes, resistance to delamination and tear resistance of impregnated fabric or nonwoven based tapes are required. Heat crosslinkable, pressure-sensitive insulating tapes are manufactured by using an adhesive based on an aromatic polyester coated on a textile carrier [519]. Mounting tapes have been manufactured with woven plastic carrier materials also. Such tapes can have a woven plastic backing [520]. As a nonwoven material, Tyvek can be used. This is an HDPE fiber based carrier material [521,522]. It possesses the opacity of paper and the tear resistance of fabric. Generally there are two different types of nonwovens. Certain materials are manufactured with a special resin as an adhesive; others (e.g., DuPont's Reemay polyester or Monsanto's Cerex nylon) do not have an adhesive matrix. Thermoplastic acrylic films can be laminated with nonwovens also [523]. Binders and binding techniques for nonwovens are discussed in Ref. 524. Filament-reinforced tape is used for fiber-optic cables and is useful for binding optical cable components [525]. Such tapes have a carrier of PET and a PUR low adhesive component on the back [526]. Cloth-based tapes with good dimensional stability and easy tear property are manufactured by laminating synthetic resin layer on one side of cloth with a warp of 10–25 yarns and weft of 10–20 yarns and coating adhesive on the other side [527].

Plastic Foams as Carrier for Tapes. Foamed polypropylene films can be used to replace expensive satin and acetate films for tapes [406]. These products are actually films that have a relatively low degree of foaming and the geometrics of a common film carrier material. Common foam carrier materials are thicker, of lower density, softer, and more elastic. Rubber, neoprene, PUR, PP, PE, and soft PVC foam are applied for double-sided tapes [528]. Double-faced foam tapes are designed for industrial gaskets. They are based on closed-cell PE foams coated with a high tack, medium shear adhesive [529]. Foam as carrier (PUR) alone or with film is proposed for medical tapes for fractures; the adhesive is sprayed on the web [530]. Laminated structures of PET foam are used also. Masking tape based on foam is discussed in Ref. 531. According to Ref. 73, PUR and PE foams are the most used foamlike carrier materials for tapes. Polyethylene foams are aging-resistant and can be applied up to 90°C. PUR foams resist temperatures up to 150°C but are destroyed by UV light. Tensile strength, stiffness, density, flexural modulus, flammability, and impact resistance are the main characteristics of the foams used as carriers for tapes [532]. Generally MDI-based PUR foams from the United States are softer and have a lower specific gravity than European products [533]. Foams of high mechanical strength are manufactured from water-based polybutadiene dispersions [534].

Other Carrier Materials for Tapes. Film/film laminates, film/plastic laminates, film/paper laminates, and metallized film are applied as carrier materials for tapes also. Such materials are manufactured by specialized firms. Metallic foils are used for mounting tapes [535]. For instance, a special mounting tape for an instrument housing assembly is designed like a tear type with transfer adhesive. The tape consists of a metallic foil strip of predetermined thickness and an acrylic transfer adhesive. Metallic foils are suggested for tamper-evident products also. Aluminum film is proposed as street marking tape [235]. Aluminum carrier and a tackified CSBR coated with 40 g/m^2 is used as an insulating tape in air conditioning [155]. The converting characteristics of aluminum films are described in detail in Ref. 536. Paper containing asbestos fibers and impregnated with an elastomer has been suggested for double-faced tapes [537]. Silicone-impregnated paper can also be used as a carrier material. For instance, a pressure-sensitive adhesive tape for protection of printed circuits based on acrylics and phenolic terpene resin with a 100–200 g/m^2 siliconized crepe paper is made by coating the adhesive mass first on a siliconized polyurethane release liner and then transferring it under pressure onto the final carrier material [171].

Formulation and Manufacture of Plastic Films for Protective Films

The formulation and manufacture of protective films differ according to film buildup, adhesive-coated or self-adhesive. General quality requirements such as lay flat, narrow thickness tolerances, and the absence of wrinkles and gel particles are valid for both. As can be seen from Table 6.23, the carrier materials for protective films have lower mechanical performance characteristics than the films used for labels or tapes.

Polyolefins as Nonadhesive Carrier for Protective Films. Polyethylene is the most used carrier material for protective films [101]. The first PE-based polyolefin films with UV protection (black color) for outdoor use were introduced onto the market in 1986 [538]. Polyolefin carrier films for protective films are manufactured as blown or cast films. Monolayer or coextruded, nonoriented and oriented films have been developed. As raw materials PE and PP are generally used, but other compositions may be formulated also. Economic and environmental considerations forced the use polyolefins as carrier material for protective films and the downgauging of such films. It must be emphasized (as discussed in Chapter 5) that a minimum mechanical strength (expressed as maximum admitted deformability) is required for such films. Excessive deformation of the films degrades their adhesion. As can be seen from Table 6.24, tensile strengths of less than 8–10 N do not provide enough adhesion for these films. That means that common monofilms cannot have a thickness of less than 40 μm and coex (polyolefin) film has a minimum thickness of about 25 mm.

Special protective films with improved deformability such as deep drawing films and security (mirror tape) films need sophisticated formulations that give a balance of plastic deformability and dimensional stability (Table 6.25; see also Chapter 8).

Table 6.23 Typical Mechanical Properties of the Main PSP Carrier Materials[a]

Property	Label	Tape	Protective film
Tensile strength, N/mm^2	22/19	220	18/11
Elongation by break, %	350/420	25	450/650
Force at break, N/cm	20/17	120	17/10
Elongation by maximum force, %	280/300	25	300/550
Force by 5% elongation, N/cm	19/17	80	11/11

[a]MD/CD values were measured for an 80 μm polyolefin film.

Manufacture of PSPs 433

Table 6.24 Dependence of Adhesion on the Mechanical Strength of the Carrier Film[a]

Tensile strength F_{max} (N)	Peel resistance on SST (N/10 mm)
8.0	No adhesion
10.0	0.20
12.0	0.43
14.0	0.65
16.0	0.80
18.0	0.95
20.0	1.10

[a]Polyolefin film.

Table 6.25 Screening Formulations for the Carrier of Some Special Protective Films

Protective film	Formulation components (%)				
	LDPE	MDPE	LLDPE	EBA	EVAc
Deep drawable	20	—	80	—	—
	—	—	70	—	30
	—	—	90	—	10
Mirror tape	—	40	20	40 (3%)	—
	—	40	20	40 (8%)	—

Polyvinyl Chloride as Nonadhesive Carrier for Protective Films. Polyvinyl chloride was used as the first carrier material for common protective films [538]. Later it was successfully tested for special applications such as deep drawable films. Problems related to its recycling and environmental impact have recently arisen that forced its replacement with more acceptable polymers.

Polyolefins as Adhesive Carrier Material for Protective Films. Stretch films were introduced for wrap-around packaging two decades ago. One-side adhesive oriented (400%) thin films were used [546]. The low level of crystallinity of VLDPE provides extruded films with an intrinsic cling that are used in coextruded structures for industrial wrapping stretch film. The stretch film is usually a multilayered structure in which at least one of the external layers is constituted by VLDPE or by a blend of ULDPE and LLDPE. For a standard VLDPE the difference between melting and softening range is about 60°C; for ULDPE it is about 80–90°C. That means that such films are very conformable at 20–40°C

[379]. While the tackiness of stretch wrapping film manufactured by the chill roll casting process can also be achieved through the use of special raw materials in conjunction with the very fast rate of cooling on the chill roll, in the case of blown film extrusion it is necessary either to use compounded raw materials that already incorporate the necessary additives or to feed the additives into the system during extrusion.

Cast (chill roll) films are applied as self-adhesive protective films (without PSA coating). Such films have a thickness of 50–70 μm. These products have to adhere onto extruded and cast plastic (PMMA, PC, etc.) plates. Very low density polyethylene and common LDPE are used as raw materials. Generally LDPE films are corona-treated [540] (see Manufacture of the Finished Product).

As known for cast plates the application is carried out by room temperature, for extruded plates by 60°C. In some cases the film has to support thermal forming where the protective film coated plastic plate is heated (ca. 5 minutes at 180°C). Polyolefin films containing EVAc, EBAc or PB have been manufactured and tested (see Section 3). For blown films the incorporation of liquid polyisobutylene offers additional technical superiority over EVAc films, with PIB slowly migrating to the film surface (see Chapters 4 and 8 also).

Polar Films as Self-Adhesive Protective Material. As discussed in Chapter 4, filmlike peelable protective coatings have various special compositions. For certain compounds, peelability is wet debonding. For instance, an *N*-vinylpyrrolidone-based protective formulation becomes emulsified in 10–20 min [541]. Resin dispersions have also been proposed for removable protective coatings [542]. Common formulations may contain methyl methacrylate-ethyl acrylate copolymer (1/1–1/5), styrene-butadiene latex, melamine and carbamide resins, plasticizer, anticorrosion protection additives, and pigments [543].

Removable protective coatings for metals have been manufactured from PVC also [544]. They are applied by spraying, bursting, or dipping and are removed as film. Self-adhesive PVC coatings have been used as plastisol too [545,546]. Gelled PVC coatings have been used for protecting metal [547]. In this way a soft PVC film-like coating was manufactured.

Ethylene copolymers with polar monomers are used as self-adhesive carrier materials (see Section 3). Ethylene-vinyl acetate copolymers were the first of these materials from this product range. The sealing temperature of an EVAc copolymer film with 12% VAc is 130–150°C; an increase in the VAc content lowers it. Such raw materials are recommended for cold sealing also. Copolymers with 15–18% VAc have been suggested for packaging films to modify their sealing properties. Commercial EVAc grades contain 10–40% VAc. It should be noted that their properties are strongly influenced by the molecular weight and side chains of the polymer. Only high molecular weight EVAc polymers exhibit adequate adhesion to plastics [548]. A content of 32% VAc in EVAc leads to a partially crystalline polymer; at the 40% vinyl acetate level, a completely amorphous polymer is achieved. Polymers with 15% VAc display PE-like properties. Polymers having 15–30% VAc give PVC-like performance, whereas polymers with more than 30% VAc are elastomer-like. For self-adhesive protective films, both the mechanical and blocking properties play important roles. The elongation of VAc copolymers increases with VAc concentration. The main increase is given by up to 15% VAc content. The optimum tensile strength is achieved for a content of 20–30% VAc. The clarity of EVAc films increases with the VAc content; however, many films with less than 15% VAc are manufactured because of their blocking. Blocking is reduced by slip and antiblocking agents and by cooling during manufacturing. In the manufacture of such films the suggested BUR is 4/1. For this ratio the MD/CD shrinkage values are of about 50%. During manufacture, tensionless film transfer and air flow parallel to the film are recommended. EVAc foams can be manufactured also.

Polymer blends having graft copolymerized PE with carboxylic groups have been produced for SAFs also [549]. For instance, a cold-stretchable self-adhesive film is based on an ethylene–α-olefin copolymer (88–97% w/w) and with polyisobutylene, atactic polypropylene, *cis*-polybutadiene, and bromobutyl rubber [550]. This copolymer has a density of less than 0.940 g/cm^3 and exhibits an adhesive force of at least 65 g (ASTM 3354-74). Polyvinyl acetate films coextruded with LDPE have been recommended for three-layered SAFs that have a core layer of polyethylene mixed with polybutylene. For instance, such a film may have the following buildup: LDPE (9 μm)/LDPE-polybutylene (37 μm)/EVAc (4 μm). Partially hydrolyzed PVAc (ester number 50–80) has been proposed for use as a peelable protective film for rough surfaces [551]. EVAc with chlorinated PP has been suggested for use as a removable protective film [552]. Ethylene-ethyl acrylate copolymers have been proposed for removable clear protective films on metals [553].

Formulation and Manufacture of Plastic Films for Release Liner

A separate release liner is used for labels, tapes, bituminous building products, and a wide variety of industrial applications [554]. Plastic films can be used as separate release carriers or as a supplemental layer coated on a main carrier material (paper, nonwoven, or film) to ensure a smooth surface. As is known from the manufacture of carriers for tapes, generally such a carrier functions as a release agent also. It possesses release properties, or it is transformed into an adhesive-repellent material by coating it with or embedding release substances. The use of plastic films as carriers for tapes allows their high speed automatic application due to their flexibility and high mechanical strength. Like paper, plastic liners for tapes can be pleated. With its excellent mechanical resistance, such a liner can be delaminated very quickly [555]. The siliconized HDPE liner displays high stiffness and good die-cuttability. For PSA-coated fabric and nonwovens having a high degree of extensibility during processing, extensible film based release materials should prevent wrinkle buildup.

Embossed plastic liners have also been developed. Such liners are used with or without supplemental siliconizing. Plastic-based release liners work as release agents due to their low adhesivity or reduced contact surface. Low adhesivity is given by the nature of the film raw material (nonpolar polymers) or by the supplemental release coating of the film. Reduced contact surface is ensured by a special, folded, embossed surface of the plastic film. Using such surfaces, polar plastics may function as an abhesive component also. As stated in Ref. 496, a contoured special PVC film displays only one-tenth of its full surface as the contact area. The shape of the embossing plays an important role also. Release liners with folded, contoured film are used for tapes that are applied on round items, i.e., where conformability is very important. Embossed PVC film is recommended as the liner for plasters and adhesive films.

The abhesive layer can be coated on the back side of the carrier material, embedded in the carrier material, or applied on a separate carrier material as liner [247]. If coextruded structures are used, the total thickness of the support film is about 15–50 μm and the thickness of the abhesive layer is about 1.0–5.0 μm. Polyorganodisiloxanes, particularly a dimethylsiloxane with a well-defined viscosity, can be applied as the abhesive component. The abhesive substance is added to the layer in an amount of 0.2–3.0%.

Paper as a carrier material was discussed earlier. Its composition can also include plastics. Glassine, polyethylene-coated kraft, and clay-coated kraft papers are the most used carrier materials for release liners [554]. Paperboard laminated with PE can be used as release liner for "stickies" to give a smooth surface for siliconizing.

According to Ref. 73, the main plastic films used as release liners for tapes are polyester, polypropylene, polyethylene, and polystyrene. The polyolefins are used for 60% of plastic release liners. From the range of plastic films, HDPE, LDPE, PP, and PET are used as carrier materials for release liners. Low density polyethylene and HDPE are adequate mainly for double-faced tapes and for foam tapes [73]. High density polyethylene possesses good tear resistance and stiffness. It is suitable for low cost release liners for industrial applications, e.g., bitumen films, tapes, and insulation panels [554]. LDPE offers good tear resistance also and is inexpensive. It is suggested for industrial applications and for graphic products, where high clarity is not a priority or colored products are required.

Biaxially oriented polypropylene can be used as a release liner, being siliconized by heat-, UV-, and EB-curable systems [556]. Oriented PP possesses high clarity, better gauge control, a glossier surface, and improved stiffness than nonoriented PP [554]. It is suggested mostly for labels. For instance, a silicone-treated release liner of OPP having a thickness of 50 μm has been proposed for labels [434]. Sandwich printing is usual for plastic labels coated with hot melts.

Manufacture of PSPs

Such products have a transparent polypropylene or PET carrier and an OPP release liner [557]. Oriented polypropylene is used for closure tapes for diapers [73].

Polyester displays the same performance characteristics as OPP and better temperature resistance. PET was first introduced as a plastic release liner. Later, after its use as face stock material, oriented blown PP films (with a maximum thickness of 75 μm) were tested. Double- and single-sided release can be produced with different release levels. The main advantages of polyester are its smoothness and transparency. As is known, only plastic-based liners provide a really glossy "no label" look for the adhesive layer [558]. Paper carriers give textured surfaces due to the paper fibers. Generally plastic films used as release liners display the following advantages:

1. They are transparent.
2. They can be produced in different colors.
3. They can be made in specific thicknesses.
4. They are dimensionally stable.
5. They are sealable.
6. They are smooth.
7. They exhibit fiberless tear.
8. They need no blocking after die cutting.
9. They are recycleable.

The transparency of the films allows light-sensitive web control and regulation. Different colors allow the use of various release liner colors for various product groups. The ratio between stiffness and thickness (specific thickness) is higher for plastic films than for paper. This allows higher processing speeds. Plastic films do not change dimensionally when the atmospheric humidity changes. No elongation and wrinkles appear. No remoisturization is necessary after coating. For bag closure tapes, sealability is one of the main requirements. Smooth, glossy release surfaces give smooth and glossy adhesive surfaces also. A glossy barrier surface does not need smoothing primers before siliconizing. A clean, fiberless tear or cutting section is required for medical or electronic tapes because of contamination danger. Blocking after guillotine cutting of paper-based PSPs is a disadvantage given partially by their compressibility and the porosity of the cut section. Plastic films (polyolefins) can be recycled by extrusion. LLDPE causes less grit [559].

Postextrusion Modification of the Plastic Carrier Material

Stretching and orientation have different effects on the molecular structure of polymers. In comparison with a nonoriented film the stretched film displays better mechanical properties (tensile strength, stiffness), optical properties, barrier properties, and shrink properties [446].

Monoaxial or biaxial stretching is used. Cross direction mono-oriented films exhibit low cross direction shrinkage and are recommended for labels [446]. Biaxially oriented films are manufactured by bubble stretching or by the cast film tenter frame process. Bubble stretching allows a good balance of the properties in the machine and cross directions, and the cast film tenter frame process leads to certain anisotropy due to the two-step character of the procedure. The oriented films used are polyethylene, polypropylene, polyester, and cellulose acetate [560].

Surface Modification of Synthetic Carrier Films

Bond strength and bond permanence are greatly dependent upon the type of surface that is in contact with the adhesive. The main purpose of surface preparation is to ensure that adhesion develops in the joint between two substrates to the extent that the weakest link is in the adhesive itself and not at the interface with the adherend (for removable PSA the weakest link apparently has to develop between the adhesive and the substrate). The anchorage of a coating on the carrier surface depends on the polarity and geometry of the surface. Its polarity affects its antistatic characteristics also. The modification of surface polarity is of special importance for certain self-adhesive plastic films where a functionalized surface has to ensure instantaneous bonding.

Modification of the Surface Polarity. Surface polarity influences the adhesion of the coated layers on the carrier material. Paper, certain metals, and plastics are sufficiently polar to bond with PSA. Other plastics (e.g., nonpolar polyolefins), special papers, and modified metal surfaces do not possess the required bonding affinity. For instance, a common acrylic PSA on an untreated PE surface exhibits a peel force of 4 N/cm; on polypropylene its adhesion is only 0.45 N/cm [243,540]. For adhesiveless PSPs, increased surface polarity provides self-adhesion in the first phase of adhesion buildup. In a special case of cover films, i.e., semi-self-adhesive films, which adhere to the protected surface without pressure and without following buildup of the adhesion, the increased polarity of the surface provides the only adhesiveness of the product [561]. The increase in surface polarity may have negative side effects also. As is known, the increase in polarity improves the surface tension. The unwinding noise of tapes depends on their surface tension.

The surface of nonpolar carrier materials has to be modified. A broad range of methods have been developed to carry out the modification. Coating and treatment methods compete. Flame, chemical, and physical or electrical treatment of the web surface have been proposed. As is known, in some cases primers are used to improve the adhesive anchorage. Bulk additives (or additives in a separate coextruded layer) can work as adhesion-improving agents (tackifiers) also. For instance, biaxially oriented multilayer PP films for adhesive coating have been modified to improve their adhesion to the adhesive coating by mixing the PP with particular resins [249]. For such films, the preferred tackifiers (15–25% w/w) are nonhydrogenated styrene polymer, methylstyrene copolymer, pentadiene polymers, α-pinene or β-pinene polymers, rosin or rosin derivatives, terpene resins, and α-methylstyrene-vinyltoluene copolymers. To improve their adhesion even further, corona discharge treatment has been suggested.

A broad range of materials can be physically treated. Generally PE and PP carrier materials have to be treated before use. PET, PA, Zellglas, and aluminum may be treated in some cases too. Polyethylene has to be pretreated or precoated with a primer [366]. According to Prinz, primers used for UV printing of plastics can be replaced by corona treatment [562] and transfer-coated PSA can be corona treated also [563]. The final carrier material should also be treated, and both treated materials should be laminated together [564]. Special treaters have been developed for narrow web printing [565].

The energy level needed for treating various plastic films varies in the order [566]

$$\text{OPP} > \text{PVC} \gg \text{HDPE} > \text{LDPE} > \text{PS} > \text{PET} \tag{6.4}$$

Among the polypropylene films, the required dosage of treatment energy increases as follows [496]:

$$\text{CPP}_h < \text{CPP}_c < \text{MOPP}_c \ll \text{BOPP}_c < \text{C}_a\text{PP}_c < \text{BOPP}_h \tag{6.5}$$

Cast polypropylene homopolymer (CPP_h) needs less treatment energy than cast copolymer (CPP_c), mono-oriented copolymer (MOPP_c), bioriented copolymer (BOPP_c), calendered copolymer film (C_aPP_c), or bioriented homopolymer (BOPP_h). Corona treatment is recommended for PP, PS, LDPE, LLDPE, HDPE, EVA, EVOH, PA [301] and paper [302].

Corona treatment generates polar sites, increasing the free energy of the surface and bonding [567]. Depolymerization, oxidation, and crosslinking of the treated surface are the main effects of such treatment [568]. Nitrogen can also be fixed on the surface of corona-treated polyolefins [540]. According to Wilson et al. [303], a treatment energy of 500 J/m^2 brings about the formation of about 2×10^{14}–3×10^{14} CO groups on the polymer surface. These functional groups may react with the postcoated layer. For instance, the anchorage of acrylates crosslinked with trifunctional polyisocyanate is improved by the reaction with the OH groups of the corona-treated surface [569].

The effect of corona treatment is a function of the type of film and its age, slip, and length of storage. Such effects are more pronounced for a film treated during manufacture [570]. As stated in Ref. 571, the main methods for testing the effectiveness of corona treatment are based on measurement of the contact angle or the adhesion. The decrease in the polarity given by corona surface treatment (measured as contact angle) depends on the type of film and its age. It is about 3–5 mN/m after the first two days for an aged film and 1–3 mN/m for a "fresh" film.

It is abrasion-sensitive also [571]. Reference 572 discusses the interdependence between the effectivity of the corona treatment, storage time, and concentration of the slip agents. As stated, the shelf life of the corona-treated PP is longer than that of similarly treated PE [364]. Orientation also influences the corona treatment. The treatment of OPP and BOPP is more difficult, requiring high dosage. The energy level needed for the corona treatment of BOPP may be 10 times as high as that needed for PE [373,573]. Generally, corona treatment of PP during film manufacture (before migration of slip) gives better results [373]. Corona discharge activation of elastomer surfaces leads to the new polar surface having a very short shelf life. Reconstruction of the elastomer surface may occur within some minutes after treatment (e.g., 15 min for ethylene-propylene or styrene-butadiene copolymers) [574]. Electrically conductive printing inks make corona treatment difficult [575].

To improve the self-adhesion of polyolefin films, their functionalization and high pressure/temperature lamination have been developed. Functionalization means corona treatment. High pressure/temperature lamination means lamination in a laminating machine. Such effects are used for other technologies also. As shown in Ref. 576, a polyolefin laminate can be manufactured by laminating together two corona-treated surfaces at a temperature that is lower than the softening temperature of the films. High temperature and pressure ensure better conformability and contact. Full surface contact and full surface treatment are absolutely necessary. Therefore, the corona treatment of very thin films of PE and PP requires the use of a pressure cylinder to eliminate air bubbles [577–579].

Corona treatment requires working with high voltage (10–25 kV), high frequency (30–50 Hz) electric fields [580]. Values of 12 kV [581] or 12–20 kV [540] and frequencies of 1–3.8 MHz [580], 20 kHz [540], 15–25 kHz [582,583], and 20–40 kHz [584] have also been tested. For polypropylene, treatment energies of 0.15–40 J/cm^2 have been used [540]. According to Prinz [584,585], a greater frequency improves the corona treatment effect. Blocking increases at low frequencies. The so-called streamers (discharge channels) appear above a frequency of 10 kHz.

As stated by Markgraf [586], over the years additive levels have steadily increased. High slip polyethylene contains 2000 ppm slip now in comparison with 800 ppm a decade ago. Therefore corona treatment of such films became difficult. Corona treatment can be improved by using a hot air jet to blow the slip agents from the film surface [587]. Slip additives can be volatilized by using a high temperature air jet, and better treatment stability is achieved. Indirect corona treatment without a counter electrode has been developed also. With this system it is possible to treat flat materials of any thickness, and the working distance can be increased up to 20 mm [588]. Electrically inhomogeneous materials (containing metals) and contoured items can be treated with "spray corona" systems, which use a gaseous agent [589]. According to Gerstenberg [588], the shelf life of PP films treated by indirect corona is longer than that of untreated films.

Another method of corona treatment that prevents the fast fading of surface energy was developed in the 1970s [583]. This process is based on the deposition of an ultrathin chemically active layer, a so-called corona deposition. Corona deposition is a process that deposits an ultrathin SiO_x layer made from silanes and oxidizing agents. Therefore the corona discharge is carried out in a controlled (nitrogen) atmosphere with controlled amounts of silanes and oxidizing agents. The surface tensions achieved with this procedure are higher (50–52) than the 40–43 dyn/cm achieved by common corona treatment.

Polyethylene for tapes must be chemically pretreated [187]. Etching produces a hydrophilic surface and creates sites for adhesion. Most commercial etching compositions are solutions of chromic acid or chromic acid plus sulfuric acid. Potassium bichromate, sodium hypochlorite, potassium permanganate, zinc chloride, and aluminum, barium, and magnesium chlorides can be used also. For PE, various chemical treatment methods have been tested [591]. For instance, treatment with water-based potassium bichromate and sulfuric acid solutions, sodium hypochlorite, potassium permanganate, etc. can be used. According to a patent [592], solutions of metal salts (e.g., Zn, Ba, Al, Sn, and Mg chlorides) improve the anchorage on a chemically oxidized PE surface. For tapes, potassium bichromate, sulfuric acid, and zinc chloride treatments have been tested [593]. Polyethylene terephthalate can be chemically treated with acid–potassium bichromate mixtures (5–30 s at 75–80°C) or with sodium hydroxide and acid–potassium bichromate [594].

Flame treatment is preferred for packaging tapes [595,596]. Flame treatment can offer advantages where moisture must be driven out [597]. The temperature of the flame is about 1200°C [598]. According to Cada and Peremsky [599], chemical and flame treatments give better adhesion values. According to Dorn and Bischoff [600], corona treatment ensures better (PP/steel) bond strength than flame treatment. For extrusion coating the best adhesion values have been obtained by using a primer [601].

Poly(ethylene-vinyl acetate) foam can be treated with fluorine [602]. For instance, a weblike foam material used as carrier for double-faced tapes with a thickness of 2 mm has been pretreated at a running speed of 5 m/min. In this case the improvement of the surface tension is due to the fluorination of the end methylene groups of vinyl acetate [580]. Unlike PE foam, the poly(ethylene-vinyl acetate) foam partially loses its treatment rapidly (5 days). The method allows the treatment of complex profiles and forms (e.g., bubble packs to be coated with PSA, or EPDM profiles) and of foams more than 2 mm thick. This method does not cause pinholes. It is used for the manufacture of barrier layers also [603]. First chlorine was applied as a gaseous reactive agent together with UV radiation

[580]; later a mixture of fluorine and nitrogen was tested. Recently the use of sulfonation (treatment with an SO_3 atmosphere) was studied.

In corona treatment, the air contains 100 ppm ozone [604]. Most rigid pollution control regulations call for the elimination of ozone. Because ozone forms during corona treatment, ozone-eliminating systems are required. On the other hand, ozonization is also used for surface pretreatment. In this case the the ozone concentration is 12,500 ppm. A combined treatment with ozone and UV light has been tested also [580]. The use of benzophenone with UV radiation helps to remove a hydrogen atom from the polymer and generates radicals that can change the surface polarity [605]. The range of increasing difficulty in removing hydrogen is given as

$$\text{SBR} > \text{Kraton} \gg \text{EPDM} > \text{PP} > \text{PE} \tag{6.6}$$

Plasma treatment can be used too. As shown by Kaplan and Rose [606], a 4–12-fold improvement of the adhesion on plastic parts was attained (depending on the types of adhesive and plastic) by oxygen and ammoniacal plasma treatment. Sheets or rolls can be plasma treated [607]. The advantages of the procedure are that there are no pinholes, no side effects, low energy, and no static charges. The charge is maintained for a longer time. There are many variants of plasma used for treatment [608]: chemically inert plasma, reactive nonpolymerizable plasma, reactive graft polymerization plasma, and reactive polymerization plasma.

Oxygen plasma treatment allows higher adhesion for medium molecular weight PE than for high molecular weight LDPE [609]. The effect of oxygen plasma is twofold. Plasma treatment can be applied as pretreatment for better adhesion, degreasing, roughening, activating, and priming. Active surface groups are generated that allow good adhesion properties, and reduction of the defects gives a smoother polymer surface [610]. Plasma treatment affects deeper molecular layers than corona treatment and produces crosslinking [611]. Plasma treatments are subatmospheric processes, working at low pressure (60–150 Pa), between room temperature and 250°C [612], at 0.1–1.5 mbar, 2.45 GHz [613]. According to Ref. 614, the use of oxygen leads to a pronounced decrease of the contact angle with water. For improvement of the surface affinity and adhesive anchorage, chemically reactive plasma can be used. If a reactive nonpolymerization plasma treatment is applied (fluorine, oxygen, or nitrogen), multiple effects appear. Low molecular weight substances are eliminated from the polymer surface, and polar groups are formed. According to Dorn and Wahono [589], a polypropylene surface having a contact angle of 90–92° with water before treatment displays contact angles of 63–85° after plasma treatment, in comparison with corona treatment, where contact angle values of 60–62° have been obtained. According to these authors [589], PP can be plasma treated with oxidative gaseous mixtures only, whereas PE can be plasma treated with nonoxidative gases also. Dorn and Bischoff [600] stated that by treating polypropylene with oxygen plasma by 1 HPa and 27.12 MHz, the same dependence of bonding strength on time was observed as with oxidative chemical treatment with chromesulfuric acid.

The formation of double bonds, crosslinking, and redox reactions are initiated by plasma treatment. Dorn and Bischoff [600] admit that polypropylene tacticity is also modified. The superficial weak boundary layer is eliminated. This effect was observed by Rasche [615] for polypropylene too, but it has not been found for polyethylene. Plasma treatment with graft polymerization leads to an increase in surface adhesion also. First an inert plasma treatment is carried out. Then the activated surface, which has long-living radicals, is reacted with polymerizable monomers. This procedure is really a plasma treatment followed by coating at low temperatures to modify the polymer surface. Plasma treatment has been used to simultaneously improve the wetting characteristics of PET and reduce static charge accumulation via triboelectricity [616]. For instance, the wetting angle with water has been reduced from 75° to 35–60° by using acrylic acid as the reagent, which is coated to a thickness of 15–20 Å. Unfortunately, although the electrical conductivity has also been improved by an order of magnitude, it does not fulfill the practical requirements. Actually plasma treatment for film is mainly a batch process [617]. Plasma-treated surfaces lose their surface treatment also. However, according to Ref. 611, the storageability of the treatment effect is better for the plasma process. For instance, a loss of 15–20% of the treatment effect has been found in about a month [618].

Surface oxidation by UV light or ozone has an overall effect similar to that of certain plasma and flame treatments. Treatment with benzophenone under UV light produces crosslinking of the surface without increasing its hydrophilicity [619].

Perforation using corona discharge is possible too. Special electrodes allow the manufacture of holes of 2–5 μm [620] (see also Chapter 7). Unlike glow discharge treatments, this process could be carried out under ambient conditions as a continuous method [621].

Different treatment methods give different surface tension values for various plastics and a different rate of decay of these values during storage. As stated by Armbruster and Osterhold [622], sulfonation provides the highest surface tension values for PP/EPDM, but surface recovery is the highest also. Plasma treatment leads to stable surface tension values. Direct (on-line) measurement of surface tension (adhesion improvement) during corona treatment has been developed. The method is based on the continuous evaluation of the friction between the treated web and a special transducer surface [623].

Modification of Antistatic Properties. Static electricity accumulates during web processing and can achieve a level of 40,000 V [624]. Agents improving the electrical and antistatic properties of plastic films can be added to the formulated raw materials during film manufacture, or they can be coated on the film. Electrically conductive PE (HDPE) is used for extrusion of tubes and for film [625]. Antistatic agents like polyethoxythiophene can be coated on the film surface [626] (see also Chapter 4). The electrical properties of plastic films can be modified by using radiation-induced crosslinking also. For instance, a polyolefin film having an electrical resistance of 1600×10^{-2} ohm has been manufactured via radiation crosslinking [627]. Conductive fillers improve the electrical conductivity and antistatic characteristics. The most commonly used filler to achieve enhanced electrical conductivity is carbon black. Nentwig [628] suggests a level of 15–45% carbon black as filler. There is a logarithmic dependence between electrical conductivity and carbon black concentration. From the technological point of view, triboelectricity can be reduced by avoiding separation operations during web transport and processing [629]. Rotary machine parts, forced transport, and the use of ionizing bars are the suggested methods. Passive and active charge-eliminating devices have also been developed. Ion spray bars used to eliminate static electricity during the conversion of weblike products work with a tension of 4.5–8 kV [630]. Web cleaning devices use electrically charged air jets and vacuum [631].

Modification of the Blocking Properties. Blocking occurs in both sheetlike and reel-form products. To avoid the bonding of adjacent (contacting) plastic surfaces, antiblocking agents are used (see Chapter 4). Blocking causes tapes or solid-state tape components to resist unwinding. Tapes with an olefinic carrier material are noisy. If the unwinding is carried out at high speed as in the operation of cutting the spool for rolls with semiautomatic and automatic machines, a high noise level is encountered. It is supposed that the noise level is related to the mechanical properties and surface characteristics of the carrier material, to adhesive/abhesive properties, and to unrolling conditions. Modification of the adhesive by the addition of a mineral oil results in a substantial reduction of the noise level when the tape is unwinding, according to Galli [266]. Further improvement can be achieved by using a release coat and/or special treatment of the carrier back. Polyvinyl carbamate, polyvinyl behenate, and other polymers can be used as the release layer (see Section 2.4). It should be mentioned that in some cases blocking and sealability are required properties for a carrier material. For special applications, blocking may be required. For instance, the use of heat-sealable face stock allows the manufacture of single material type pocket labels [434].

Modification of the Surface Geometry. Embossed carrier materials are used for diaper tapes, medical tapes, wall coves, decor films, etc. Embossing can be carried out at either room temperature or high temperature. For instance, PVC wall coverings have been embossed by high temperature (120–140°C) embossing [632] (see also Chapter 7).

2.3 Other Carrier Materials

Various other materials are used as carriers for PSPs. Weblike irregular surface items or discontinuous items having various shapes and a complex buildup have also been applied as carriers. Cotton fabric (cloth) [633,634], glass fiber reinforced materials [635], and other materials have been developed as carrier materials. Special materials have been suggested for medical, insulating, and mounting tapes (see Chapter 8).

2.4 Manufacture of the Release Liner

General Considerations

In principle, an adhesive-coated PSP is built up from a carrier material and an adhesive layer. In some cases another carrier material with a dehesive layer is also required. This component is the release liner. The manufacture of the release liner includes the production of the carrier material and its coating with a release layer. Certain plastic films do not exhibit chemical affinity or contain antiblocking agents (see Chapter 4). Therefore in such cases the uncoated (nonpolar) plastic films can act as release liners. According to Uffner and Weitz [636], antiblocking agents are materials added to an adhesive (or carrier) formulation to prevent the adhesive coating made therefrom from adhering to its backing when the adhesive-coated carrier material is rolled or stacked at ambient or elevated temperatures and relative humidities. (Blocking can be caused by heat or moisture, which may activate latent tack properties of the adhesive composition.) Certain technological additives (e.g., surface-active agents in PVC, or fillers) can work as embedded release agents also.

In other cases the abhesive properties of the liner are given by a special coating. As discussed earlier (see Section 1.2), various release formulations are used. Silicones are the most versatile release compounds. Different silicone systems, 100% solids, solventless and water-based silicones, low temperature radiation curing, and hot air curing options are known [262]. Solvent-free silicones have been on the market since the 1960s. Such materials have been coated using offset gravure [637]. Low temperature cure solventless systems are two- to three-component formulations coated with a coating weight of about 1 g/m^2. Their cure profile shows that at 110°C (where normal systems require several hours of curing) such materials need a curing time of only 15 s. At 66°C their cure time is less than 1 s [638]. In comparison, silanol-terminated polydimethysiloxane-based silicone release coating (viscosity 80 cP) applied with a coating weight of 0.3 lb/ream with a gravure roll and rubber transfer roll is cured in 7 s at 202°F [259]. Because of their high curing rate, UV curing systems allow a 20% increase in productivity [639]. Such siliconizing lines may run at rates of 1000 m/min. Coating weights are 1.0–2.0 g/m^2 for paper and 0.5–1.0 g/m^2 for film. In contrast, UV curing needs initiators, and such additives may affect the quality of the product [640]. The subsequent adhesivity of PSA on UV- or EB-cured release papers is about 95%.

Diorganopolysiloxanes have been suggested as water-based release dispersions [276]. Such release coatings have to adhere to polyethylene-coated paper and display abrasion resistance. The forecast of Fries [184] concerning the forced use of emulsion-based siliconizing procedures (i.e., a 40% market share for 1985) has not been met by actual technical development. In 1990 about 40% of release liner production used solvent-based silicones, 40% solventless silicones, and only 20% water-based silicone emulsions [641]. Various carrier materials (paper, film, fabric, and laminates) are release-coated.

Manufacturing Technology for Release Liner

Almost all label stocks and double-faced pressure-sensitive tapes use a separate release web. Both release papers and release films are manufactured. Mechanical and/or chemical release has been suggested for various products. Coated and uncoated release liners can be used. Coated or embedded release agents or both can be applied. Static and dynamic (postworking) release have been developed. The coating devices suggested for abhesive components depend on their chemical composition and physical state. Silicone coatings are generally used, but wax or other materials can be applied. Solventless, low solvent, solvent-based, and water-based silicones are used. These systems are discussed by Weitemeyer et al. [640]. The release formulation is designed to provide variable release levels depending on the method of conversion and on whether the adhesive is continuously coated or pattern (zone) coated.

Release papers are generally high density paper (glassine type) materials (30–60 g/m^2) coated with a release layer. The weight of the release liner depends on the method of label conversion and the type of die-cutting and tooling as well as on the mechanical properties of the laminate. The general manufacturing scheme for the production of such liners involves compounding the release formulation, coating, curing, and remoisturizing when appropriate. Siliconizing of a film may require the use of a primer (top coat). For instance, carboxylated ethylene polymers crosslinked with zirconium derivatives are used as the top coat for siliconized films. Silica is suggested as antislip. According to Larrimore and Sinclair [77], a barrier layer based on polyvinyl chloride, polyvinyl acetate, and polyvinyl alcohol is coated before siliconizing on paper. Vinyltriacetoxysilane can be used as a component in formulating primers for siliconizing paper [642].

Various coating devices have been developed and experimented with for siliconizing. For instance, a 100% solid silicone coating system that is a combination of rolled and extruded coating was developed in 1986 [643]. Solvent-free silicones are coated using a six-cylinder coating head with a tolerance of 0.7–4.0

g/m². According to Moser [644], a Meyer bar can be used for solvent-based siliconizing of glassine paper. For solvent-free siliconizing, a four-cylinder gravure coating has also been proposed. In both cases a coating of about 0.5 g/m² was applied. A conventional (42 lb/ream) supercalendered, densified kraft paper with silicone release that is highly crosslinked has been suggested for silicone pressure-sensitive adhesives [259]. A gravure roll with 150 lines per inch (lpi) with a rubber transfer roll is recommended for silicone coating of about 0.3 lb/ream.

Radiation curing offers numerous advantages. For instance, for high temperature siliconizing, the remoisturizing of overdried papers and methods for handling papers after overdrying are important [259,638]. The complete im mediate curing of radiation-curable release coating eliminates face-to-back blocking of rolls of the finished product and provides greater assurance of uniform release characteristics from the outside of a roll to its core [645]. Electron beam curing allows simultaneous crosslinking of both sides of a two-side coated release liner (transfer tapes). Common release coating formulations have relatively short pot lives. Formulations for EB curing typically have a shelf life of 6 months to 1 year. Silicone acrylates allow room temperature curing using UV or EB radiation [646]. Such systems do not need a catalyst, and no postcuring occurs; therefore, in-line siliconizing, adhesive coating, and lamination can be carried out [647,648]. UV curing allows a running speed of 400 m/min; a coating weight of 1 g/m² is applied [649].

Glassine, polyethylene-coated kraft paper and clay-coated kraft paper, high and low density polyethylene, and polyester are the most common liner carrier materials. Various carrier materials are used for release liners for labels and tapes. They differ according to the end use of the PSP also. Common labels and tamper-evident products need different release liners. The liners for single-side coated or double-side coated tapes differ also. For instance, lightweight machine-coated paper face stock combined with strong adhesives can be used, and security edge cuts and perforations applied on the press for tamper-evident labels. These products require special liners. Double- and single-sided release liners can be coated [554]. For instance, a dual strippable release liner is manufactured by coating the release layer on a kraft paper/polypropylene laminate [650]. This release liner exhibits a release force of 123.5–170 N/m. A pressure-sensitive removable adhesive tape useful for paper products consists of two outer discontinuous layers (based on tackified SIS block copolymer) and a discontinuous middle release layer [651]. This tape can be manufactured with less adhesive than conventional tapes, and its top side can be printed and perforated. For medical adhesive tapes a special release coating has been proposed [652]. The formulation contains silicone rubber, butoxytitanate-stanniumoctoate complex, oligomeric dimethylphenylpolysiloxane rubber, and triacetoxymethylsilane. For food packaging, water-based emulsions have been preferred [641].

The extrusion technology for the manufacture of the release liner can use silicone derivatives also. For instance, according to Ref. 286, release agents that do not migrate in the PSA were prepared by reaction of hydrogen siloxanes with unsaturated hydrocarbons. Methyl siloxane has been reacted with α-diolefins. Polyethylene mixed with 3% of the product and used with a layer thickness of 15 μm gives values of adhesion to PSA tape of 50, 90, and 110 g/25 mm at peel rates of 0.3, 3, and 20 m/min and adhesion retention of the tape of 98.3%.

3 MANUFACTURE OF THE FINISHED PRODUCT

Generally PSPs exhibit pressure sensitivity due to their coated or built-in pressure-sensitive layer. The classic manufacture of the finished product means buildup of a pressure-sensitive product from its components, i.e., the carrier material and the pressure-sensitive adhesive. In practice, other product components such as the release liner and cover film can be built in in one or more layers, giving a laminate with a complex structure (see also Chapter 2). As discussed in Chapter 2, the pressure-sensitive product can be manufactured in one step also if a carrier material with built-in pressure sensitivity is used. Such a procedure offers primarily economic advantages. Therefore economic considerations forced the development of extrusion-made products. Table 6.26 summarizes the main characteristics of the manufacture of adhesiveless and adhesive-coated PSPs.

From the technological point of view, PSPs can be manufactured by using coating, extrusion coating, or coextrusion. Each of these methods has its advantages and disadvantages (see Table 6.27). Therefore their applicability for a given product has to be rigorously examined.

Table 6.26 Major Raw Materials and Manufacturing Procedures for Adhesiveless PSPs

Raw materials	Manufacturing procedure	PSP
Polyethylene	Extrusion	Protective film
Ethylene-vinyl acetate copolymers	Extrusion, coextrusion	Protective film, business forms
Ethylene-olefin copolymers	Extrusion	Protective film, decorative film
Ethylene-acrylate copolymers	Coextrusion	Protective film, tape
Butene-isobutene copolymers	Coextrusion	Protective film, tape
Vinyl acetate copolymers	Extrusion	Business forms
Vinyl chloride copolymers	Extrusion, casting	Tape

Manufacture of PSPs

Table 6.27 Manufacturing Possibilities of PSPs

	Advantages of the manufacturing procedure					
	Free choice of product components		Production parameter			
			Running speed			
Manufacturing procedure	Carrier	Adhesive	Very good	Good	Fair	On-line
Coating	+	+	+			±
Extrusion coating	−	−		+		±
Coextrusion	−	−			+	+

Table 6.28 Characteristics of the Manufacturing Procedure for Major PSPs

		Product class		
Product component	Manufacturing versatility	Label	Tape	Protective film
Adhesive	Limited use of certain grades possible	+	+	−
	Limited use of certain coating weight values possible	+	+	−
	Crosslinking-free formulation possible	+	+	±
	Primer-free formulation possible	+	±	±
Carrier	Limited carrier grades possible	−	+	+
	Release-free carrier possible	−	−	+
Equipment	Lamination-free production possible	−	±	+
	Limiting of the coating devices types used possible	+	+	−

In manufacturing PSPs the specific features of the product class and trends in production technology must also be taken into account. The importance of the product components in the buildup of a PSP depends on the product class. Each product class has its own limits concerning the choice of laminate components and manufacturing equipment. For instance, label manufacture is strongly related to printing technology. The choice of label components and manufacturing method is determined by the requirements of printing technology. The manufacture of labels should start with their design and should include origination and pre-press electronic design and artwork, digital artwork and reproduction, plate-making, and plate types. Table 6.28 lists the parameters describing the manufacturing versatility of a production technology for the main pressure-sensitive products.

Technology change refers to the new technologies or material components used in production. Solvent-free, water-based, and 100% adhesives are being used more and more to meet the toxic products emission standards. The printing industry has also undergone changes to meet these standards. Food laws are getting tougher in controlling the effect of the migration of various constituents of the package and their interaction with the packaged products. These changes have increased the use of solvent-free and adhesive-free products.

3.1 Manufacture of the Finished Product by Extrusion

As discussed earlier, for special applications it is sometimes possible to produce a virtually adhesive-free (noncoated) carrier film based on a common plastomeric polymer. The pressure sensitivity of such materials is achieved via physical treatment or/and a special chemical composition using viscoelastic raw materials. Their manufacture is a one-step procedure (extrusion), which may include on-line physical treatment also. Principally plastomer, tackified plastomer, and viscoelastomer-based films can be manufactured (Fig. 6.9). Physically treated extruded plastomer films are used as hot laminating films. The application tem-

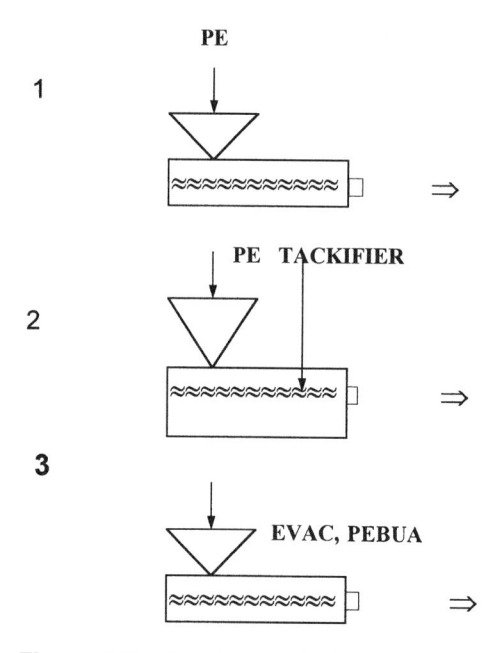

Figure 6.9 Manufacture of PSPs by extrusion: (1) Hot laminating films; (2) tackified plastomer-based films; (3) self-adhesive films with intrinsic adhesivity.

Table 6.29 Main Raw Materials and Manufacturing Procedures Used for Self-Adhesive PSPs

Raw material	Manufacture technology	PSP
Polyethylene	Extrusion	Protective films
Ethylene-olefin copolymers	Extrusion	Protective films
Ethylene-vinyl acetate copolymers	Extrusion, coextrusion	Protective films, decors
Ethylene-acrylate copolymers	Coextrusion	Protective films, tapes, labels
Polyisobutylene copolymers	Coextrusion	Protective films, tapes
Vinyl acetate copolymers	Extrusion	Forms, tapes

perature can be reduced by tackifying the plastomer with incompatible tackifiers. Room temperature laminated, with adhesive-coated PSPs competitive films can be manufactured with viscoelastic polymers such as ethylene-vinyl acetate or butyl acrylate copolymers.

Nonpolar olefin polymers are used as corona-treated pressure-sensitive hot laminating films. If coextruded they have an adhesive layer designed to adhere to the surface to be protected, while the top surface layer, which can be printed, provides slip characteristics and protection. Functionalized polar olefins and other polymers have also been developed as self-adhesive films. The main raw materials and manufacturing procedures used for self-adhesive PSPs are listed in Table 6.29.

Nonpolar Hot Laminating Films

Nonpolar hot laminating films do not have a special formulation for viscoelasticity. In actuality, they are plastomers and do not exhibit pressure-sensitive adhesivity. Their instantaneous adhesivity is provided by the increased contact built up at high temperatures, under pressure, and with physical polarization treatment. However, they have to be discussed together with self-adhesives films that display room temperature self-adhesivity because their applications and test methods are similar and because a hot laminated formulation may undergo "soft" transformation into a pressure-sensitive one.

LDPE-Based Hot Laminating Films. These films are manufactured from common PE. Their self-adhesivity is due to a special physical treatment. Hot laminating films are used for coating coils and for protecting unvarnished and varnished metallic webs and plastic plates. Because of their one-component formulation and their manufacture as a monolayer blown film, they are the most inexpensive protective films. However, the mechanism of their bonding and the technology of their manufacture are not completely understood. Their production is based mainly on empirical knowledge.

The use of electric charges to ensure instantaneous adhesion is a general technical procedure. It is suggested for sheet handling and temporary adhesion also. A high tension (100,000 V) electric field is applied to charge the film [653]. Generally, adhesion on a metal surface depends on the energetic state of the metal (wetting out, adsorption), the electric potential (electrical double layer), the morphology and geometry of the metal surface (roughness), and the chemical structure of the metal surface [654]. The same is valid for the plastic film surface to be applied, where other viscoelastic parameters also interfere. Physical treatments modify some of these surface characteristics.

Adhesion of corona-treated surfaces is well known. As stated in Ref. 571, films given a surface treatment at more than 41 mN/m show autoadhesion. Generally the treatment quality is tested by measuring surface tension. In practice, films that have a certain surface tension are used to obtain a particular debonding resistance (peel). Therefore for practical use the peel resistance (P) and treatment quality (Q_s) of the film surface are assumed to be related,

$$P = f(Q_s) \tag{6.7}$$

and the surface quality is also evaluated in terms of surface tension (ζ),

$$P = f(\zeta) \tag{6.8}$$

As can be seen from the data of Table 6.30, for a given polymer (plastic film) and a given adherend surface there is a correlation between the debonding peel resistance and the surface tension of a corona-treated polyethylene film. The adhesion of the hot laminated film increases with the surface tension (treatment degree) of the film.

In practice, the surface tension of a polymer film depends on its chemistry, formulation, manufacture, storage, and treatment. Regulation of the treatment has its limits; therefore the peel level is controlled by regulating both the surface

Table 6.30 Adhesion Characteristics of Hot Laminated Film[a]

Carrier thickness (μm)	Surface tension (dyn)	Peel resistance (90°) (g/30 mm)
120	52	410
120	52	360
150	48	260
150	42	140

[a] An LDPE film was laminated on a lacquered coil (241°C) with a pressure of 1.5 bar and running speed of 30 m/min.

characteristics of the film (i.e., surface tension) and the intrinsic properties of the web to be coated. In this case the resultant peel for a hot laminated, corona-treated polyolefin film can be written as a function of the surface tension of the film and the surface quality of the adherend (Q_a):

$$P = f(\zeta, Q_a) \tag{6.9}$$

That means that in practice the desired peel is achieved by regulating the surface tension of the protective film and by modifying the formulation of the lacquer for the coated coil. Unfortunately, surface tension alone is not sufficient to characterize the adhesion properties of the treated plastic film. As shown by the data of Table 6.31, quite different peel values can be obtained for the same initial surface tension of samples of polyolefin protective film on the same adherend after a period of time. Therefore in correlation (6.9) one should (for a given adherend quality) include another supplemental parameter (β) that takes into account those surface characteristics that are not characterized by the surface tension:

$$P = f(\zeta, \beta, Q_a) \tag{6.10}$$

Bonding depends on the laminating conditions (C_l) also:

$$P = f(\zeta, \beta, Q_a, C_l) \tag{6.11}$$

where C_l takes into account the influence of laminating temperature, pressure, and time. For a given laminating pressure and temperature, the adhesion of a given film is a function of the treatment degree and the characteristics of the surface to be protected. Similar adhesion behavior is known in the packaging industry where self-adhesive heat-sealable films are used. According to Kuik [655], the peel of heat-sealable polar ethylene copolymers depends on the corona treatment, coating weight (i.e., layer thickness), and aging. It has been shown that the adhesion of LDPE on steel (measured as the coefficient of friction) increases with temperature (above 150°F) and attains a maximum at 220–250°F [656]. The weldline strength of LDPE measured in newtons per 15 mm (time 0.5 s, pressure

Table 6.31 Peel Buildup of Hot Laminated Films

Treatment degree (dyn/cm)	Peel resistance (90°, g/30 mm)	
	Instantaneous	After 1 month
42	230	435
42	235	370
44	270	295
44	280	520

of 0.5 N/mm^2) has a value of about 10 at 110°C, but this value increases to 13 at 130°C.

It is known that corona treatment depends on the crystallinity of the surface and the relative humidity of the air. As a result of the treatment, various functional groups (hydroxy, ketone, carboxy, epoxy, ether, ester, etc.) are formed. The energy of positive ions acting in the discharge "streamers" attains about 100 eV, which is sufficient to destroy bonds with energies of about 4 eV [585]. Each of the new polar groups ensures a different surface energy level, i.e., adhesion. Theoretically the same "apparent" surface tension can be achieved with different induced functional groups (depending on their concentration). It is not possible to tailor their synthesis. It is evident that peel buildup may be different for surfaces of different chemical compositions. According to Markgraf [586], the buildup of a laminar flow of ozone is determinant for corona treatment. It is possible to corona treat a substrate and have no indication of the change in surface tension even though the adhesion is improved. The effects of treatment depend on the nature and age of the film also. HDPE or PP is more difficult to treat than LDPE [615]. According to Ref. 585, LDPE can be processed with the same equipment and parameters at a treatment energy level of 7.5 W·min/m compared to the 25 W · min/m required for PP.

Slip agents decrease the effect of surface treatment. For instance, a level of 0.12% of erucaamide as slip requires a 300% increase in treatment energy to obtain the same surface tension [596]. Surface tension has both polar and nonpolar components. Their increase can be quite different, depending on the treatment energy [513].

Another problem is the dependence of the degree of treatment, measured as surface tension, on the level of treatment energy. According to Prinz [562,563], the treatment of polyethylene is a surface phenomenon up to a treatment energy of 10^4 J/m^2. The autoadhesion of PE attains a maximum at 500 J/m^2. The surface tension for this energy level is 60×10^{-3} J/m^2. That means that for plastic films there is a limit to the treatment energy of about 500 J/m^2; for paper, board, and metallic films this limit is 2000 J/m^2. The required corona energy level (D) depends on the generator power (G_p), web width (W_w), and web speed (s_w) [585]:

$$D = G_p/W_w s_w \qquad (6.12)$$

The bonding strength increases continuously with treatment time for PP/steel adhesive joints [600]. As can be seen from Table 6.32, surface tension is dependent on the treatment energy level (generator output). As shown in Table 6.32 and stated by Menges et al. [540], this dependence is almost linear up to the saturation level. The dependence of adhesion on treatment energy level varies with the

Table 6.32 Dependence of Surface Tension on Treatment Energy Level[a]

Treatment energy level (kW)	Surface tension of film (dyn/cm^2)
0.5	38
0.7	43
0.9	46
1.1	48
1.3	50
1.5	52
1.7	52
1.9	52

[a]The surface tension of a PE film was measured after 4 days storage.

type of adhesive. For instance, for a classic epoxy-based adhesive, a maximum adhesion on polypropylene has been found to be a function of the treatment energy level [433]. There is no such maximum for pressure-sensitive tapes. The adhesion between adherend and treated film increases over several months time. The delay in adhesion buildup increases with treatment level. The saturation level of surface tension depends on the type of polymer; for polypropylene it is achieved at an energy level of 2 J/cm^2. It is astonishing that the surface tension attains its maximum at relatively low levels of treatment energy. The oxygen concentration attains its maximum at 20 J/cm^2 (depending on the experimental conditions). For the same surface energy level per square centimeter under different experimental conditions, the same surface tension can be obtained at quite a different oxygen concentration.

It is evident that the degree of treatment depends on both the constructional characteristics of the machine and treatment conditions. Generally the discharge channels ("streamers") have diameters of several micrometers and a service life of about 10 ms. The required processing time of the material to be treated depends on the kind of material. For instance, for fabric, it is 0.05–1 s [657]. According to Gerstenberg [657], the discharge channels (the active treatment zones) have a diameter of 30 μm and a shelf life of only 6–20 ns.

The manufacturer must take into account the fact that the exact nature of the dependence of adhesion on the degree of treatment is not yet fully understood. To attain the desired final bond strength, the laminating conditions (temperature and pressure) and the substrate surface are more important than the chemistry of the film. Therefore hot laminating films are products that need perfect cooperation between the lacquer manufacturer, film manufacturer, and laminator (coil coater). For the coil coater, the parameters influencing the adhesion of the protective film are the following:

Electrostatic treatment (polarization) of the film
Storage time of the film (stability of the treatment)
Type of lacquer
Color and gloss of the lacquer
Lamination temperature
Period of time between the lacquering oven and lamination (running speed dependent)
Period of time between lamination and cooling (running speed and equipment buildup dependent)
Temperature of the coil after cooling
Evolution of adhesion during storage

Taking the above parameters into account, hot laminating is a complex technology. For quality assurance of such products, special apparatus and test conditions are required. The suitability of a given corona-treated hot laminated film (tested on a given surface) for use on another slightly different adherend is minimum. Application requires long-term preliminary tests of the adhesion buildup because the adhesion increases over time. It should be mentioned that warm lamination under pressure is used for laminating a PE film on another PE (or PP) film. Stretched and unstretched films can also be warm laminated.

Warm Laminating Films Based on VLDPE. The use of polyolefins as self-adhesive pressure-sensitive products (without adhesive) is based on the flow properties of low modulus polymers. As discussed earlier (see Chapter 3, Section 1), such behavior is the result of special chain buildup and amorphous structure. Linear low density and very low density polyethylene exhibit self-adhesivity. This property is used in their application as shrink film. As mentioned in Ref. 462, if they are used on LDPE, their self-adhesion can lead to excessive bonding. Therefore new products have been developed that have an antiblocking Surlyn layer. VLDPE with a density of 0.885 g/cm^3 is self-adhesive, and it is difficult to process for blown film with hot seal adhesivity. Very low density PE (0.905 g/cm^3) has a low modulus, 87.5 N/mm^2, in comparison with a value of 170 N/mm^2 for common LDPE (ASTM-D-882), which gives it a low seal initiation temperature [658]. Such very low density, low modulus films can be used as warm-laminated protective films also.

Butene-based LLDPE can be used as a cling-modifying component in slot-cast stretch films for wrapping pallets and paper reels and for bundling [659]. The polymer should be used in blends at a level of 2–30% LLDPE for 20–25 μm monofilms or as a 3–5 μm layer for coextruded films of the same gauge. The formulation with special raw materials in conjunction with the very fast rate of cooling on the chill roll provides enough tackiness for stretch wrapping film. In the case of blown film extrusion it is necessary either to use compounded tacky raw materials that already incorporate the necessary additives or to feed the additives into the system during extrusion [660]. The adhesive strength of LLDPE films on stainless steel can be improved by grafting with vinyltrimethoxysilane [661].

Common hot laminating films used since the early 1970s are LDPE-based. Such films are thermoplastic and do not display autoadhesion under room temperature and low pressure. Their application is possible due to their conformability and the self-diffusion of a thin polarized molten surface layer during their application under high temperature and pressure. As discussed earlier (Chapter 3, Section 1), such polymer deformation is a viscous flow. The bonding deformation of elastomers or viscoelastomers is a viscoelastic flow. That means that viscoelastically bonded surfaces ensure instantaneous debonding resistance given by the elastic characteristics and viscosity of the material. Plastically bonded macromolecular compounds exhibit an instantaneous debonding resistance depending only on the (temperature-dependent) viscosity of the material. Mutual interpenetration of the contact surfaces, regulated by the diffusion of the macromolecular compound, remains for "true" PSAs a mostly hydrodynamically and rheologically controlled phenomenon. For plastomers used as adhesives, the chemistry of the contacting surfaces is more important. The surface treatment has to produce an attraction between the contacting surfaces of hot laminated films (based on a double layer of electric charges) and ensure chemical affinity between the polymer and the contacted surface due to the induced polar functional groups.

Nonpolar Self-Adhesive Films

In some cases the finished product is a carrier with "true" built-in pressure sensitivity. As discussed earlier, plastic films may undergo cold flow, and therefore plastic film based carrier materials of special chemical compositions may work as one-component pressure-sensitive products, where pressure sensitivity is to be understood as bonding at high temperature under increased pressure. In this case the manufacture of the finished product by extrusion is actually the processing of plastics. As is known, the main plastic films are manufactured via extrusion. In a similar manner PSPs that have a built-in adhesivity are produced by extrusion. Their manufacturing technology is extrusion technology, characterized by the formulation of the raw materials and the equipment used for their manufacture.

The formulation of one-component (carrier only) pressure-sensitive products is a problem of plastics processing with some aspects in common with the formulation of carrier films (see Section 2). The raw materials and recipes suggested for such a formulation include the manufacture of a solid-state carrier material with mechanical characteristics and pressure sensitivity designed to allow its application and deapplication. In some cases the adhesive characteristics of the product are given by the thermoplasts used for the manufacture of the carrier film (together with physical surface activation); in other cases adhesive components have to be built in. In certain cases the product has a monolayer construction; in other cases it consists of a multilayered sandwich. In such cases the reciprocal adhesion of the layers can sometimes be improved by using primers, or barrier layers have to be built in to prevent the migration of the adhesive component to the other side. For many applications no release layer is required, but some products have a built-in release layer.

It is known that butylene or halogenated butylene derivatives are recommended for use as tackifiers for elastomers. For instance, the tack and green strength of blends of bromobutyl and EPDM have been improved by increasing the level of bromobutyl [662]. In a similar manner butylene derivatives work as tacky additives and lead to adhesive films if embedded in a plastomeric carrier material. For instance, flexible adhesive tapes for temporary protection of fragile surfaces have been manufactured by coextruding a mixture of polyethylene and synthetic resin adhesive (50 μm) and a mixture of butyl rubber and synthetic resin adhesive (15 μm) [663].

Self-Adhesive Protective Films Based on Polybutylene and Polyisobutylene. Polybutene has been used as a tackifier of plastomers in the manufacture of silage wrap films and for stretch and cling films. The pressure sensitivity of the products is achieved by using polybutylene as a viscous tackifying component. Because of its incompatibility with the main polymer, the viscous component migrates to the surface of the film and imparts the needed adhesion [664]. Polybutene (3–6%) is used together with LDPE or LLDPE [665]. For LLDPE a level of 3–5% tackifier is generally used [664]. Its concentration depends on the adhesion required, the grade of polybutylene, and processing conditions.

Polybutylenes are not plastomers; they are viscoelastic compounds. Their processing consists of mixing a plastomer with a viscoelastic or viscous compound. As discussed earlier, a normal feed-in of such components is not possible. Polybutylene (mainly an isobutylene-butylene copolymer) can be added to the recipe as a masterbatch (10–20%), as tackified pellets (4–6%), or directly (2–8%).

Blending can be accomplished by direct compounding during the manufacture of the pellets or by mixing with PIB masterbatch. The use of a masterbatch is less expensive than compounding. It has the advantage that the masterbatch can be blended with polymers of choice. Unfortunately, a high masterbatch concentration is required. The masterbatch is sticky, and its dosage is difficult to control.

When tackified pellets are used there is no need for processing know how. This is an advantage. On the other hand, like the masterbatch, the granules are tacky and difficult to handle. The extrusion of such raw materials is difficult.

Block building may occur during storage. There is an upper limit to the PIB level also. Masterbatch and pellets are expensive special products.

There are many variants of direct feed-in. Direct feed-in is the direct addition of a solid-state or molten component with the aid of a cavity mixer or other injection device. As for material costs, direct injection into a hopper is the least expensive. Unfortunately, screw slip limits injection. Injection at the hopper throat uses a mixture of PB and solid PE. In this case a reduced throughput screw is required.

Direct injection into the polymer melt avoids problems with screw slip, but high pressure injection, which has high capital costs, is necessary. For the addition of components in the molten state, extruder injection is proposed. The equipment must allow the heating and dosage of PB. A heatable storage tank, a precision speed-controlled pump, a connecting hose, and a lance are needed. A mixture of PB and molten PE is injected into the extruder body. Direct injection into the polymer via the center of the screw is also possible [660] by means of a lance that injects the additives through the screw shank into the screw channel [666]. A special technical modality has been developed that uses a cavity mixer to inject the molten materials between the extruder barrel and screen changer die [667]. A low cost process for injecting PIB uses a specially designed screw and feed sections [668].

Stretch films with a clinging effect on one side have been manufactured for packaging by using PB and PIB formulations and technologies [669]. Polyisobutylene is blended with LLDPE. The adhesive is on the inside, and the outer side has slip properties. Such a product can be manufactured as a one-layer film, but a three-layer coex is better. In this case an LLDPE layer provides the stretch properties and another layer ensures the adhesive characteristics. Generally the addition of polybutene to polyethylene film depends on the film manufacturing process, the type of polyethylene, the buildup of the film, and the application of the film. The parameters that affect the quality of the product include the type of PB, type of PE, processing conditions, concentration of the components, and storage of the finished film.

Peel cling (see Chapter 5) increases with the molecular weight of the polybutylene and its addition level. Lap cling decreases as the molecular weight of the polybutylene increases and increases with addition level. The density of the PE used influences the migration of PB; therefore the choice and dosage of PE are very important. The suggested PB level also depends on the process used to manufacture the film. Generally, for blown films, 2–8% PB, and for cast film only 2–4% PB, are suggested. According to Ref. 669, 3–5% tackifier is recommended for blown film and 1–3% for cast film.

Die gap, frost line height, melt temperature, and winding tension influence the adhesivity of film [664]. Among these parameters, winding tension and frost line height are the most important. The storage time and temperature influence the migration of the PB and therefore the self-adhesive properties also. It is suggested that the finished product be stored under well-defined conditions (heated room) for at least 5 days after manufacture. It should be taken into account that (because of the more amorphous nature of the film) PB migrates rapidly in cast film. Curable isobutylene polymers with crosslinkable silicone groups have been synthesized also [670].

Self-adhesive protective films on polybutylene basis are applied for the protection of plastic plates, in coil coating, and in the automotive industry. For the customer, SAFs on PB basis have the advantage of better adhesion than EVAc-based films. Therefore the laminating conditions are less important. On the other hand, because of their higher adhesion, their application and handling are more difficult and blocking may appear. For the manufacturer this is a relatively new product class, more expensive than PE-based hot laminating films. The manufacturer of such films needs special machines and know-how. Therefore the production of such SAFs is not recommended for small firms or newcomers.

Polar Self-Adhesive Films

Increasing the polarity of polyolefins improves their adhesion. For instance, ethylene-propylene block copolymers are grafted with maleic anhydride [671]. If hot pressed for 10 min at 200°C, these films give a peel adhesion of 8.5 kg on stainless steel and 3.8 kg on aluminum. Vinyl acetate and acrylate copolymers of ethylene are used commonly as raw materials for the manufacture of polar self-adhesive films. Their use has been suggested for removable coatings also [560]. Polyvinyl alcohol with an ester number of 50–80 has been used as solution. Special compositions have been suggested also (see Chapter 4).

Self-Adhesive Protective Films Based on EVAc Copolymers. It is known that EVAc copolymers display polyethylene-like or adhesive-like properties depending on their vinyl acetate content. In practice, their processing for SAFs is the processing of a plastomer, which itself can build up a carrier material. Unlike PIB-based self-adhesive films, where the pressure sensitivity properties are given by a low molecular weight tackifier component that is incompatible with the carrier polymer, EVAc copolymers have pressure sensitivity properties themselves and are compatible with the carrier material.

The main application of EVAc-based polar SAF films is for the protection of plastic plates and films (PC, PMMA), laminates, profiles, furniture, sanitary items, etc. Such films possess the advantages of deposit-free removability, resistance to environmental stress cracking, and cuttability. Their main disadvantages are that (1) it is difficult to regulate their adhesion (peel), (2) their adhesion depends on the processing (application) conditions, and (3) peel may build up. Such films replaced paper-based protective films for PC plates (750 μm to 13.3 mm). They have a thickness of 50–100 μm and are applied at temperatures of 60–

135°C. They have to exhibit a peel resistance of 140–290 g/25 mm. Films having a thickness of 50 μm are recommended for the protection of plastic films with a thickness of 175 μm to 1.5 mm. Such films are coextruded, having an adhesive layer of 12.5 μm with 7.0% EVAc. The adhesion buildup for such films is a complex phenomenon, and the formulation alone does not provide a controllable product quality.

Their main advantage for the manufacturer of such films is the possibility of a one-step production procedure. On the other hand, the manufacturer depends strongly on the raw material suppliers, who have the manufacturing know-how for producing the films. Special extrusion, winding, and confectioning conditions are required. To avoid blocking, high speed machines with finely regulated winding characteristics are necessary. Such EVAc-based film is a mass and end product; no supplemental working steps (added value) are possible.

For protection of PMMA plates, ethylene-vinyl acetate copolymers with less than 7% VAc have been suggested. Generally, protective films based on EVAc are manufactured by using polymers with a VAc content of 3–28%. An MFI of 1.5–3 is recommended. The VAc content of the polymer required depends on the application temperature of the substrate to be protected and decreases as that temperature increases. The BUR is also a function of the application temperature, decreasing as that temperature increases. To achieve a homogeneous mixture, the output of the extruder should be limited. Polypropylene can be used as carrier material for EVAc-based SAFs also. The self-adhesive layer is built up by coextrusion of an EVAc layer (EVAc with 18% VAc, layer thickness of about 10 μm) or with an LLDPE layer.

Bilayer heat-shrinkable insulating tapes for anticorrosion protection of petroleum and gas pipelines have been manufactured from photochemically cured low density polyethylene with an EVAc copolymer as adhesive sublayer [672]. The linear dimensions of the two-layer insulating tape decrease by 10–50% during curing depending on the degree of curing (application temperature 180°C). The adhesive strength of such tapes decreased by 10–45% after 1 year. The use of EVAc copolymers has been proposed for carrierless self-adhesive films (tapes) applied for bonding roofing insulation and laminating dissimilar materials [173]. The formulations contain EVAc–polyolefin blends, rosin, waxes, and antioxidant. They are processed as hot melt and cast as 1–4 mm films.

SAFs Based on Other Compounds. The autoadhesion of soft PVC films depends on the choice of polymer, plasticizer, slip agent, and filler [673]. Self-adhesive PVC coatings were first developed as plastisols. They replaced common plastisols, which have the disadvantage that they require a prime coat for anchorage [674]. Self-adhesive lamination of PVC is also practiced in the manufacture of films. For instance, a 70 μm PVC film was laminated with a 10 μm LDPE film [274].

Self-adhesive, sealable polypropylene films with up to 20% terpene resins have been tested [675]. Chlorinated polyethylene, acrylonitrile-butadiene rubber, liquid chorinated paraffin, and fire-retardant fillers (e.g., hydrated alumina, $CaCO_3$, zinc borate, Sb_2O_3) have been compounded and molded into oil-resistant tapes (0.7 mm thick) for electric wires and cables [676]. Such a laminate can be obtained by coextruding LDPE and a thermoplastic elastomer [677].

Peelable protective films (30 µm) comprising methacrylate (e.g., ethylene-methyl methacrylate) copolymers, an organic filler, and slip have been blow molded. They are useful for the protection of rubber articles and show a peel strength of 165 g/25 mm in comparison with 500 g/25 mm for ethylene-ethyl acrylate copolymers [678]. The production of a protective film from polymethyl methacrylate solution has been described [679]. A water-removable protective varnish has been manufactured from a vinyl chloride-vinyl acetate copolymer [680]. The water-washable film (70 µm, 2 m^2) has been peeled off or removed with a stream of water in 3 min.

Generally the end use properties of protective films manufactured with different technologies such as plastomer extrusion/corona treatment, adhesive-filled plastomer extrusion, modified plastomer extrusion/corona treatment, or the coating of nonadhesive carrier material are different. The application conditions necessary to achieve adhesion, the value of the peel resistance, and its buildup over time strongly differ. The regulation of peel buildup is more difficult for adhesiveless PSPs, because it depends strongly on manufacturing and application parameters.

Manufacturing Technology and Equipment

The manufacturing technology used for PSPs produced by extrusion is the same as that used for the extrusion of carrier materials (see Section 2). It must be emphasized that for self-adhesive products the flat die (chill roll) procedure is preferable to others.

3.2 Manufacture of the Finished Product by Carrier Coating

This chapter is designed to enable PSP producers to coat and laminate PSA and silicone release liners in-house rather than having to purchase these expensive constituents of their finished products. The manufacturing technology for producing PSPs via coating or lamination is part of classic lamination technology. Lamination is the bonding of two different layers by pressure with an adhesive used as glue or as molten polymer. Laminating procedures include thermal, extrusion, and adhesive lamination.

The adhesive lamination process is influenced by the choice of lamination technology, laminating components (adhesive, substrate), and laminating machines. Dry lamination is a procedure in which wet adhesive is coated onto a carrier material, dried, and then laminated, generally on-line, with another carrier material. PSAs can be used as adhesives for dry laminating also [681]. The criteria for choosing one or the other technique should be based on (1) technical performance, (2) cost, and (3) environmental impact [682].

The classic way to manufacture a PSP is to coat a carrier material with a PSA. Printers can siliconize an adhesive coat in their own plant and save as much as one-third on their manufacturing costs [683]. Manufacture of the finished product by carrier coating is a coating technology completed by other conversion steps that are common for weblike products. For some products, laminating is also required. Coating a carrier material supposes an affinity between it and the adhesive. For polar carrier materials this affinity is a given. For nonpolar carrier materials it must be realized by means of a primer coating or surface treatment. Although surface treatment is carried out during carrier manufacture, generally the surface treatment must be refreshed before the carrier is coated.

Water-based PSAs require stand-alone equipment, space for a drying channel, and remoisturizing. Equipment and energy costs are much higher for these adhesives than for HMPSA coating, although production capacity is higher [82]. Hot melt coating does not need drying ovens; it can be easily adapted to narrow web production. Such coating technology is more flexible. Pattern and strip coating, high finish for no-label look coating, and coating in specific shapes are possible [82]. The manufacture of PSPs via adhesive coating and laminating has some of the general advantages of adhesive lamination [682] such as (1) simplicity of the line; (2) the ability to laminate any preformed web; (3) lay flat, when the web tensions are under control; (4) minimum make-ready times; and (5) no extra material needed for caliper adjustment, edge bead, etc.

Coating Technology

Coating and printing technology are used for the manufacture of the finished product by coating. Coating is less complex technologically than printing. According to Frecska [684], the main quality criteria in the printing process are (1) accuracy in the reproduction of all image details; (2) accuracy in the placement of the image elements, and (3) uniformity of ink deposit, both within the image and from image to image. Generally this last quality criterion is valid only for adhesive coating.

Coating methods are discussed in Ref. 1. Their special product-related features are also described here. The main methods used for coating liquid materials are based on cast, spraying, roller, blade, and screen printing techniques [685], but calendering has been used also. Screen printing is a special procedure that can be used for adhesive coating, and ink printing also presents the following advantages for PSA coating: (1) The contour (shape) of the adhesive layer can be exactly controlled, even for complex designs; (2) various face stock materials such as metals, glass, or plastics can be used; (3) there are no adhesive losses; and (4) coating weight can be exactly controlled within a range of 7–60 μm [686]. For screen printing, solvent-based or water-based adhesives are used. Acrylics or rubber resin PSA formulations have been introduced. At the beginning of the development of this technology (1970s), only slightly adhesive PSAs were used.

The main aspects of coating technology refer to direct or transfer coating, monolayer or multilayer coating, in-line or off-line coating, and coating technology as a function of the type of adhesive and the adhesive end use properties of the laminate. For special products requiring a very smooth adhesive layer, direct coating has the advantage that air bubbles are avoided.

Processing lines are designed for the individual processor and combine individual units manufactured by various mechanical engineering companies. Various coating devices such as direct gravure, offset gravure, reverse roll, and knife over roll can be used [687]. Advances in the development of raw materials caused important changes in coating technology and forced improvements in the coating devices. Gear-in-die coating devices designed specially for hot melts and highly viscous materials can coat 20–900 g/m^2 with a coating tolerance of 2 g/m^2 [688]. The coating weight range broadened in recent years. Thirty years ago, pressure-sensitive products were generally coated with 30–100 μm of adhesive. Some time ago the coating weight of PSA-coated webs was between 20 and 60 g/m^2 (dry) and the running speed was 3–12 m/min [689]. The average coating weight for permanent labels is now about 20–25 g/m^2, for protective films 2–7 g/m^2, and for tapes 20–40 g/m^2. A decade ago, higher (2–40%) prices for finished products in the United States allowed the use of greater release coating weights [690]. Electron beam curing was considered a rapidly growing sector with possible industrial applications in the mid-1980s [690]. The choice of manufacturing technology and equipment became important. Past trends called for multipurpose machines with moderate speeds; now specialized production requires high speeds and large web widths [691].

Coating Technology as a Function of Chemical Composition. As discussed earlier, the chemical composition of the adhesive and its formulation determine both the type of coating and the coating technology required. Thermoplastic elastomers are coated as hot melts. Acrylics can be coated as solvent-based or water-based formulations, but acrylic hot melts and radiation-cured compositions are also known.

The main advantage of thermoplastic elastomers is their thermoplasticity, their ability to be processed as plastomers. The coating of HMPSAs has to take into account their high specific heat and low thermal conductivity [692]. Styrene block copolymers require three times as much energy to melting as EVAc copolymers.

Manufacture of PSPs

Table 6.33 Primer for Selected PSP Manufacture Procedures

	Primer coating			
			Optional	
Manufacture procedure	Required (more than 70%)	Not required	Different formulations possible	Different technologies possible
Coating	−	+	+	+
Extrusion coating	−	+	+	−
Coextrusion	+	−	±	−

Radiation-curable systems can have very different compositions and viscosities. As discussed earlier, the coating of UV-curable formulations was developed from the coating of monomer mixtures. Generally a (UV and EB) radiation-curable monomer-based formulation includes a hard monomer system, a soft monomer system, and multifunctional monomers [7]. Such low viscosity formulations require quite a different coating technology than the coating of oligomer- or macromer-based formulations developed later. Macromer formulations start from monomer mixtures containing a hard monomer ($T_g > -25°C$), a soft monomer ($T_g < -25°C$), a tackifying monomer ($T_g < 25°C$), and multifunctional monomers (e.g., pentaerithrytol triacrylate). For instance, a mixture of acrylic acid, butyl acrylate, and diethylfumarate has been polymerized to a macromer that can be coated as hot melt and postcrosslinked via EB curing [693].

According to Klein [694], primers are coated with gravure, flexo, or reverse roll coating (see also Chapter 4). Coating weights of 0.01–0.5 g/m² are used. The problem of the primer coating for extruded or adhesive-coated pressure-sensitive products is related to the need for a primer coating and to the technical means available for coating a top layer (Table 6.33). Such a coating is generally required for coextruded products with a complex buildup, but is also needed for the main polyolefin-based adhesive-coated carrier materials. Advances in carrier formulation, carrier treatment, and adhesive coating allow primer-free procedures also, but generally such compositions have to be more crosslinked or they are coated with a lower coating weight (see also Chapter 5).

Coating Technology as a Function of the Physical State of the Adhesive. The replacement of solvent-based adhesives with water-based adhesives has been a continuing trend in industry for reasons of air quality, safety, and economics. For instance, acrylic adhesives for packaging tapes offer the following advantages: (1) They are waterborne; (2) they exhibit superior instantaneous adhesion to cardboard, good shear resistance, good tack, clarity, UV stability, and temperature re-

Table 6.34 Manufacturing Versatility of Selected Coating Methods for Various PSPs

Coating method	PSP	Adhesive Type	Coating weight (g/m²)	Viscosity	Coating speed (m/min)	Carrier	Ref.
Slot die	—	HMPSA	10	>120,000 mPa.s	—	Paper, film, fabric	702
Slot die	—	AC HMPSA	20–200	50–500 Pa.s, 130–160°C	—	—	703
Slot die	—	—	8–300	—	250	—	704
Slot die	—	HMPSA	7–10	<120,000 cP	200	—	705
Slot die	—	HMPSA	1–2500	—	—	—	706
Slot die	—	—	—	35000–600,000 cP	—	—	707
GID	Labels, decor film, carpet tape	—	—	5000–80,000 cP, 175°C	—	Paper, film	708
GID	—	—	30	200–1,000,000 mPa.s	—	—	703
Roll coater	—	—	20–1000	200–2000 Pa.s	—	—	703
Roll coater	—	—	—	500,000–2,000,000 cP	—	—	707
Roll coater	—	HMPSA	—	<30,000 cP	—	—	709
Roll coater	—	HMPSA	—	<100,000 cP	—	—	709

466

Method	Product	Col3	Adhesive	Viscosity	Col6	Substrate	Ref
2 Rolls	—	—	—	—	—	—	702
3 Rolls	—	—	—	—	—	—	710
3–4 Rolls	—	—	HMPSA	15,000 cP	—	—	710
Reverse roll	—	—	AC WB PSA	<30,000 cP	—	—	710
Reverse roll, nip feed	—	—	—	1000–2000 cP	—	—	94
Reverse roll	—	20	—	—	110–120	—	711
Reverse roll	Tape/label, medical	17 lb/3000 ft^2	RR, SB	3000–5000 cP	—	PET	712
Three rolls, reverse roll	Label	12–200	—	300–100,000 cP	200	—	713
Four rolls, reverse roll	Tape	60–100	—	10–4000 cP	300	—	713
Roll blade	Tape	20–200	—	2000–10,000 cP	60	—	713
Roll blade	—	30–50	—	—	150	—	711
Knife over roll	—	—	AC WB PSA	4000–8000 cP	—	—	94
Reverse roll	Tape/label, medical	17 lb/3000 ft^2	RR, SB	3000–5000 cP	—	PET	712
Rotogravure	—	—	AC WB PSA	75–100 cP	—	—	94
Reverse gravure	—	18–25	WB PSA	—	300	—	711
Wire rod	—	10–20	—	—	—	—	711
Wire rod	—	14–30	—	—	—	—	711
Wire rod	—	—	AC WB PSA	300–900 cP	—	—	94
Screen	—	6–7	WB and SB PSA	—	—	—	685
Screen	Mounting	17 lb/3000 ft^2	RR, SB	1500–2000 cP	—	PET	714
Rotary screen	—	30	—	5000–10,000 mPa.s	—	—	703
Calendering	Tape	200–1000	—	200–2000 Pa.s	—	—	703
Extruder	—	—	—	500–5000 Pa.s	—	—	703
Spray	—	3	HMPSA	—	—	—	704

sistance; and (3) they are ready to coat [695]. However, the transition from solvent-based to water-based adhesives has not been simple and is by no means complete. Aqueous adhesives have found limited acceptance, being regarded as generally inferior in either performance or coating rheology or both. An additional deficiency of latex adhesives is their tendency to build up coagulum on the metering roll when the dispersion is applied with a reverse roll coater. The coagulum may arise either from poor mechanical stability under shear or from imperfect doctoring, leading to a thin coating on the roll that dries and cannot be redispersed. These particles transfer to the applicator roll and mark the coating [696]. Water-based systems are dried more efficiently with large volumes of air rather than with air of higher temperature [697]. It should be noted that the energy consumption for coating and drying increases exponentially with coating weight [698].

Solvent-based crosslinked formulations have been developed for each product class due to their relatively broad raw material basis and the ease with which their degree of crosslinking can be regulated through the choice of raw materials or manufacturing conditions. The crosslinking of solvent-based formulations has been used since the beginning of PSA technology. Natural rubber can be used together with polar elastomers and reactive resins (phenolic resins). The resulting crosslinked adhesive has both temperature and solvent resistance [699]. The reaction is carried out by mastication of the components. Crosslinked solvent-based formulations require special drying conditions to achieve the simultaneous drying and crosslinking of the adhesive.

According to Ref. 700, HMPSAs are used for medical and hygienic products, masking tapes, plotter films, office tapes, labels, and nameplates. Hot melt coating machines can coat paper, fabric, vinyl, acetate, and polyester films [701]. Hot melt coatings can be defined as thermoplastic coatings that have a minimum viscosity of 9000 cP. Less viscous coatings are generally applied with roll coaters that have two to four metering and application rollers. Coatings of higher viscosities (over 50,000 cP) are processed with a slot orifice coating device that acts like an extruder. It should be mentioned that data in the literature are very contradictory concerning the limits (viscosity) of the usability of each coating method (Table 6.34).

For hot melts, slot die coating is used below 7 and above 10 g/m^2 [715]. For tapes, generally coatings of 5–35 g/m^2 are applied [715]. The introduction of offset rollers between the die or application roller and the substrate is a useful technique to lower the coating weight [716]. Slot die application is normally offered by machinery manufacturers for use up to 120,000 cP (at 200°C) and roller coating below 100,000 cP. Hot melts can be coated without problems up to a viscosity of 100,000 mPa.s on a roller coating device [717]. According to Fries [184], coating viscosities for hot melts can increase to 17,000–20,000 mPa.s (at 180°C). Rotary screen hot melt applicators for zoned and patterned coating are available also [82].

Hot melt coaters can use a carrier preheating drum to improve the adhesion of the HMPSA and its surface profile. Most HMPSA coating machines for labels are intended for continuous coating of 18–24 g/m^2. Such machines have an in-line filter with a filter cartridge and replaceable coating heads [82]. By inserting various shim plates and Teflon profiles, operators can change application width and pattern. Patterned coating can be carried out by using special flexible tools [718]. Full-width, striped, beaded, spot, pattern, or zone coating is possible. For the coating of high gloss adhesive layers on film carrier materials, rotating bars can be used to eliminate streaks. Intermittent or continuous operation is possible. On modern machines the coating head automatically swivels away from the substrate if the unit stops. Laminating rolls mounted to the side of the coating roller are either cooled or heated to maintain an accurate finish and coating weight. A drum melt system ("melt on demand") is usually used for label coating.

Spraying has also been developed for hot melts. At first, only stripes, dots, or full surface coating of HMPSAs was possible. By spraying larger surfaces, temperature-sensitive materials can be coated with at least 50% economy [719]. This method is most used with nonwovens [82]. Contactless coating methods for HMPSAs have been developed to allow the use of hot spray technology in the nonwovens market. In this case the adhesive and air are premixed in the nozzle. Controlled fiberization was introduced in 1986 [720]. The fiberization system uses a set of air jets to orient the bead of adhesive, drawing it down into a fine fiber applied in a helical pattern. Spiraling the adhesive maintains good edge control, even in intermittent applications up to 33 m/min. The spiraling also cools the adhesive, permitting the application of hot melt on heat-sensitive substrates. Another coating process uses a head that combines elements of slot coating and spray technology. The adhesive is distributed to the desired pattern width through internal slot channeling. As the adhesive is extruded from the slot nozzle, heated air impinges on it from both sides, stretching the adhesive and breaking it into very fine fibers. Hot melt coating systems are described in Refs. 704, 710, and 721–723.

In-line siliconizing and HMPSA coating were developed in 1985 [724]. HMPSAs have some major deficiencies: excessive thermoplasticity, poor UV resistance, and limited plasticizer tolerance [184]. However, McIntire [725] predicts savings of 30% by combining off-line siliconizing and hot melt processing. In 1988 the proportion of HMPSAs in Europe in the hot melts reached about 25% [726]. In the last decade about 10% growth has been observed.

As discussed earlier, the physical state of the adhesive can vary for the same raw material class or adhesive synthesis method. For instance, 100% acrylics can be coated as low viscosity monomers, medium viscosity oligomers, or high viscosity hot melts using postcuring technology. In 1988 there was no industrial use of postcuring of HMPSAs [727]. Now there are numerous machines designed especially for the manufacture of tapes that use this technology. Formulations radiation-cured with UV light [728] and EB [729] have been developed. As is known, EB curing displays the advantage of high coating weights (50–80 g/m^2) [730]. Radiation processing of acrylic HMPSAs has also been developed [731] (see also Chapter 4). The development of new low cohesion acrylic warm melts requires the use of a postcrosslinking agent [55,732]. Therefore the large-scale application of UV postcuring is enhanced by this raw material development. Before crosslinking the new polymer is a highly viscous fluid (at room temperature) and can be processed at 120–140°C (its viscosity is about 10–20 Pa.s). The UV lamps used in the curing channel either use electrodes or are microwave powered. These adhesives are cured as tackified formulations. For tackifying UV-crosslinkable polyacrylates, esterified, a highly stabilized rosin is used at a low level (8.5%) and high coating weight (50 g/m^2). For tackified formulations the radiation dosage must be increased. For instance, an 8.5% resin level requires doubling of the radiation dosage.

A special feature of this radiation crosslinking is its anisotropy. The radiation is partially absorbed, partially transmitted, and partially reflected. The maximum radiation is given at the top of the adhesive layer. That means that the degree of crosslinking is greatest on this side. Therefore direct and transfer coating give different adhesive characteristics. A uniform, isotropic adhesive layer can be achieved by using a UV-transparent face stock (backing) and irradiation from both sides. The manufacture of such UV-crosslinkable acrylates is more expensive than that of acrylic dispersions. However, they are an alternative to solvent-based adhesives, where, without too much capital investment, a common hot melt processing line can be equipped with UV curing lamps.

As mentioned earlier [219], the PSA can be coated as a powder also. A dry blend of elastomer and tackifier is obtained as a powder and can be coated, then sinterized by heating the adherend surface (at 177°C). Coating of soft, tension-sensitive PVC carrier materials requires a relaxation station on the coating machine [715].

Coating Technology as a Function of the Product Class. The product class determines the type of carrier and its geometry, the type of adhesive and its geometry, the type of release agent and its geometry, and how they are built up. Adhesive geometry includes the coating weight also. Analyzing the changes that occurred between 1980 and 1994 in the domain of pressure-sensitive technology, Bedoni and Caprioglio [508] state that line speed and web width increased, adhesive thickness measurement and feedback became the standard, multitechnology combined machines became available, and changeover time was reduced. The advances listed above should be evaluated separately for each PSP class. The special buildup of pressure-sensitive products in each class needs particular coating techniques. These techniques have been developed in various ways.

Rigid, thick, and uneven surfaces are usually coated on sheet-fed curtain coaters or spray coaters. In curtain coating a continuous stream of liquefied coating

Manufacture of PSPs

Table 6.35 Coating Methods Used for the Main PSPs

| Coating method | Pressure-sensitive product ||||
	Label	Tape	Protective film	Other products
Cast coating	+	+	−	+
Roller coating	++	++	++	+
Blade coating	+	+	++	+
Slot die	+	++	−	+
Extrusion	−	++	−	−
Screen printing	++	−	−	++
Calendering	−	++	−	−
Spray coating	−	+	−	++

++ Main coating method

falls onto the moving material. Spray coating deposits an atomized stream of material.

Impregnation can also be used for the manufacture of carrier and finished tape. In this process the coating permeates the fibers or the spaces between the fibers. At saturation the impregnation level reaches the point where the fibers cannot hold any more liquids or the spaces between the fibers are completely filled. In the dip process, impregnation and saturation machines apply simultaneous coatings. Impregnation or saturation can be used to achieve removability (see Chapter 4), dosage of a liquid component (see Medical Tapes), or reinforcement of an adhesive or carrier material. Roll press coating of a (temperature-) softened adhesive mass is also used for tapes. Foaming (simultaneous or postcoating) has been developed for the manufacture of special tapes. These methods are not used for common labels. The release technologies suggested for labels and other PSPs differ also. The release technology suggested may differ for different labels. The manufacture of a release liner via coating involves a complex chemistry and great precision in the coating of low coating weights using gravure or flexo anilox rolls. An in-house choice is made between solventless thermal processing and radiation curing [82]. The choice of a manufacturing technology is strongly influenced by the product class (Table 6.35).

Economic considerations forced the in-line manufacture of PSPs. Production equipment has been developed for simultaneous adhesive/abhesive coating and lamination. In-line extrusion and coating of film-based pressure-sensitive products is possible also. Table 6.36 summarizes the advantages and disadvantages of in-line coating. The relaxation processes in the extruded film are time-dependent. Such processes may lead to changes in the geometry or surface quality of the films. In-line production of coated films has to take such phenomena into account.

Table 6.36 Advantages and Disadvantages of In-Line Manufacture of PSPs

Advantages		Disadvantages	
Technical	Economic	Technical	Economic
Treatment exists	No supplementary technical (handling and maintenance) operations are required Own carrier Global know-how	Time-dependent, postmanufacture processes work in the film and change its surface quality and geometry	Coating speed is lower Mixed technology

Coating Technology as a Function of Product Buildup. Product buildup includes the nature, number of layers, and construction of the PSP laminate. Details concerning the laminate's components and buildup were discussed in Chapter 2. For products with a multiweb structure, multiple coatings are required. For such products the different coating steps can be carried out using different coating methods. For instance, for a label, the PSA, the release agent, the primer, and the lacquer can be applied by using different coating methods. However, if possible, the same aqueous or hot melt technology should be used for the different coating operations (e.g., primer and adhesive coating or release and adhesive coating).

Porous or heat-sensitive face stocks require the use of transfer coating [733,794]. Fabric is coated via a transfer procedure [115]. Undulated (crepe) paper is coated with 60–80 g/m^2 HMPSA. Generally, plastics are difficult to coat because of their lack of porosity and their nonpolar surface. Cratering, adhesive-free places, and coating weight differences are the main defects [734]. The wetting out of plastic films depends on temperature; vapor saturation of the atmosphere, and electric charges on the surface [735].

The common manufacturing procedure for tapes includes in-line primer coating, release coating, and adhesive coating [232]. As a classic primer, solutions (in toluene) of butadiene-acrylonitrile-styrene have been preferred. For instance, for solvent-sensitive SPVC films, primer dispersions based on Hycar latex and Acronal 500 D (1/1) have been suggested by Reipp [232]. The back side of the tape carrier has been coated via rotogravure with a 0.15% solution of Hoechst Wachs V in gasoline. Gasoline (boiling range 60–100°C) has been proposed for use as a solvent. Blade coating and drying at 30–80°C have been recommended.

As mentioned earlier, thick PSA tapes (0.2–1.0 mm) have been prepared by photopolymerization of acrylics [22]. For such polymerization a minimum UV light intensity of 0.1 mW/cm^2 is often a practical limit. Multilayered composites having different adhesion and mechanical properties can be manufactured with this technology. As discussed above with respect to photopolymerizable acrylic warm melts [56], light-initiated polymerization allows the buildup of anisotropic products. According to the procedure disclosed in Ref. 45, the acrylic ester PSA formulation is a combination of non-tertiary acrylic acid esters of alkyl alcohols and ethylenically unsaturated monomers with at least one polar group that may be substantially in monomer form or may be a low molecular weight prepolymer or a mixture of prepolymers. The formulation can be polymerized as a thick layer (up to about 60 mils). The thick layer may be composed of a plurality of layers, each separately photopolymerized. The acrylic esters generally should be selected in major proportion from those that as homopolymers have some PSA properties. As crosslinking agents, diacrylates in a concentration of about 0.005–0.5% by weight are preferred. Typically the crosslinking agent is added after the formulation of the prepolymer. The recommended photoinitiator is 2,2-dimethoxy-2-phenylacetophenone. The polymerization chamber has an atmosphere of nitrogen. The length of polymerization zones and the density of lamps in these zones affect the manufacture. The thickness of the layer is a determining parameter with respect to productivity, a thicker layer requiring a greater degree of exposure. Thick multilayered PSPs are made using this type of manufacturing process. When the thick layer is sandwiched between two thinner layers, the thick layer can be used as a carrier. This support carrier layer can be 25–45 mils thick [22]. Glass microbubbles can be incorporated to enhance immediate adhesion to rough and uneven surfaces. Such carrierless foamlike tapes are prepared by polymerizing in situ with UV radiation. According to Ref. 73, weblike products, i.e, tapes, were first manufactured by using UV-initiated polymerization as disclosed in Ref. 736, by thickening of the monomers, coating, and UV polymerization. Using this procedure a coating weight of 1000 g/m^2 can be achieved. The curing time is 60 s.

According to Ref. 20, thick, triple-layered adhesive tape with filler material in the center layer can also be manufactured by UV photopolymerization. In this way a carrierless tape can be produced. As the filler, fumed silica is suggested. The first polymerization zone for an adhesive layer is shorter than the second zone for the polymerization of the carrier. The first prepolymer layer contains 1.0 part photoinitiator and 0.6 part of 1,6-hexanedioldiacrylate (crosslinker). The second prepolymer layer includes 1.0% by weight photoinitiator, 0.3 part of a crosslinking, multifunctional monomer, and a polyvinyl acetate polymer used as filler. The second adhesive layer has the same composition as the first one. A line speed of 5 ft/min is suggested. Cooling is provided by the use of cooled nitrogen gas. The adhesive is dried for 1 h at 350°F.

As stated by Schmalz and Neumann [737], for common PSPs, UV curing displays the following disadvantages in comparison with EB (and thermal) curing: Rough paper surfaces are unsuitable; curing is insufficient and cannot be used for laminates. Because of absorption of the radiation by the adhesive layer, UV-curable products must have a limited coating weight.

A multilayered structure can be obtained by postcrosslinking also. This is carried out after coating, by using immersion or printing techniques. According to Larimore and Sinclair [77], a carrierless PSP (the adhesive layer provides adequate strength to permit it to be used without a carrier material) with good conformability can be manufactured by applying a special procedure. The product has an adhesive surface and an adhesion-free surface. The tack-free surface is achieved by superficial crosslinking of the adhesive with polyvalent cations (Lewis acid, polyvalent organometallic complex, or salt). The crosslinking penetration depends on the polymer and solvent used. Crosslinking can be achieved by dipping also. Water or a water-miscible solvent should be used. For such tapes without backing, strength can be provided by a tissue-like scrim (cellulose or polyamide) embedded in and coextensive with the adhesive layer. The superficially crosslinked adhesive layer may be stretched to fracture the crosslinked portions, exposing the tacky core in order to bond it. Crosslinking in tiny separated areas can be achieved by printing techniques (halftone printing). Thus microscopically small tack-free areas can be created to separate narrow tacky portions. A coating weight of 81.5 g/m^2 (thickness 0.2 mm) is applied [738].

Decalcomania (transfers) are carrierless products similar to transfer tapes. For them the solid-state component is a technological (application) aid only. Such products can be manufactured by using screen printing. First a clear carrier lacquer is printed. This layer provides the mechanical resistance of the product. The image (in mirror print) is coated on this layer by using screen (or offset) printing. The next layer is a transparent layer followed by a screen-printed PSA [739]. In another procedure, the PSA is coated on release paper. The PSA-coated liner is laminated together with a printed film [740].

According to Ref. 741, multilayered self-adhesive labels (such as business forms) comprise a carrier label containing silicones, an adhesive layer, a printed message, another carrier layer (e.g., polyester), a release layer (e.g., silicone), a second adhesive layer, a third carrier layer (e.g., polyethylene), and a printed message. Labels of this type may have more than two adhesive–carrier layer–message layer units.

Manufacture of Labels

In 1982 in Europe as much as 30–40% of the HMPSAs produced were used for labels [569]. However, taking into account the balanced adhesive properties required for labels, except for some low cost products the majority of labels are manufactured with water-based or solvent-based adhesives. Therefore their coating technology uses coating devices specialized for such adhesives. Their choice depends on the coating weight and economic considerations. Label manufacture generally consists of coating a medium thick (15–25 μm) dry adhesive layer hav-

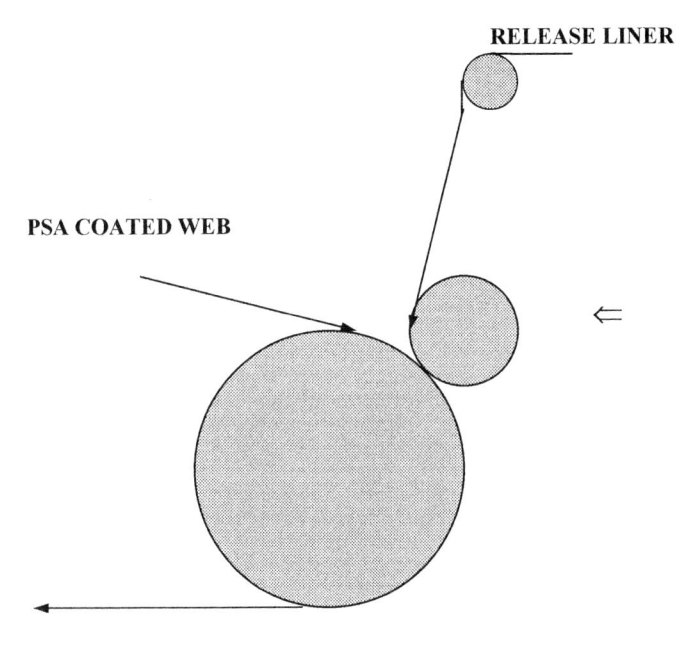

Manufacture of PSPs

RELEASE LINER

PSA COATED WEB

LABEL LAMINATE

Figure 6.10 Schematic representation of the laminating operation in label manufacture.

ing a very smooth image on a nontextured, smooth carrier surface and laminating the adhesive-coated web with the release liner or the final face stock material. According to Ref. 441, the usual width of the coating machines (of 70, 100, 140, and 200 cm) is determined by the 70 $\times$ 100 cm sheet format.

Labels use a separate release liner sheet that is normally applied in a single pass. Label stock manufactured on a one-station coater might require two or three passes to complete the process of applying a primer and adhesive and laminating. Two- or three-station coaters would perform all of these operations in a single pass. Such multiple-station coaters deposit three or more coatings on one or more carrier materials in a single machine pass. Coaters can be roll-fed coaters or sheet-fed. Sheet feed printing presses have the ability to print images of various sizes on any size sheet. Sheet feed coaters can apply a top or bottom coating. From the technological point of view, the main characteristic of label production machines is the existence of a laminating unit, which is necessary to build up the label laminate from the face stock material and the separate solid-state release liner (see Fig. 6.10). Common tapes and protective films do not need such equipment.

Forms (which, because of their construction, can be considered complex labels) are manufactured by coating and laminating "prefabricated" webs (Fig. 6.11). For instance, in the first manufacturing step the paper carrier is coated with release agent and/or adhesive. The web, coated with adhesive 1, can be laminated with a film, and the release liner, coated with adhesive 2, can be laminated with another release liner to give a temporary laminate. The sandwich structures thus manufactured can be combined to give a pressure-sensitive product. In coating practice, adhesives with different bonding strengths and release liners with different abhesive characteristics are combined to achieve a partially detachable product.

Manufacture of Tapes

Tapes are used for a very broad range of applications. Their performance and buildup vary also (see Chapter 8). Their raw material basis and coating technology vary widely according to their general or special character. Mainly rubber resin and acrylic adhesives are used for tapes. Both can be applied as hot melts, solutions, or water-based dispersions with the appropriate equipment.

The coating technology for tapes generally includes multiple coating of the carrier (face and back) with different (primer, adhesive, and abhesive) layers. Crosslinking of the adhesive coated with higher coating weights is a general feature of tape manufacture. The major coating systems include systems for [742]

1. In-line production (e.g., release–corona–primer–PSA; corona–primer–PSA).
2. Very thin coatings (e.g., solventless silicones). These systems have problems with broad webs and high running speeds.
3. Medical products (e.g., in-line and tandem production; solvent-based and water-based PSAs; soft PVC; nonwoven, fabric, and paper carriers).
4. Crepe paper coating. (Such systems are strongly influenced by the amount of pleating.)
5. High coating weight on a textile carrier.
6. Wetting out of difficult carrier materials.
7. Labels (e.g., for smooth coated layer, rough carrier surface, nonuniform carrier profile).
8. Protective films (with a broad carrier web).
9. Rubber solutions with high solids content.

The solvent-based technique is the most proven and most popular technology for producing tapes [743]. The development of tapes started with rubber resin based formulations. Tapes include a release coating, primer, and adhesive coating [744]. Primer and release coatings can be dried in either vertical or horizontal ovens [743]. Tape manufacturing can require the in-line double-sided coating of the carrier. For instance, a release coating (0.2 g/m^2) is applied on the bottom side and a primer (1 g/m^2) is coated on the top of the carrier. In this case gravure coat-

Manufacture of PSPs 477

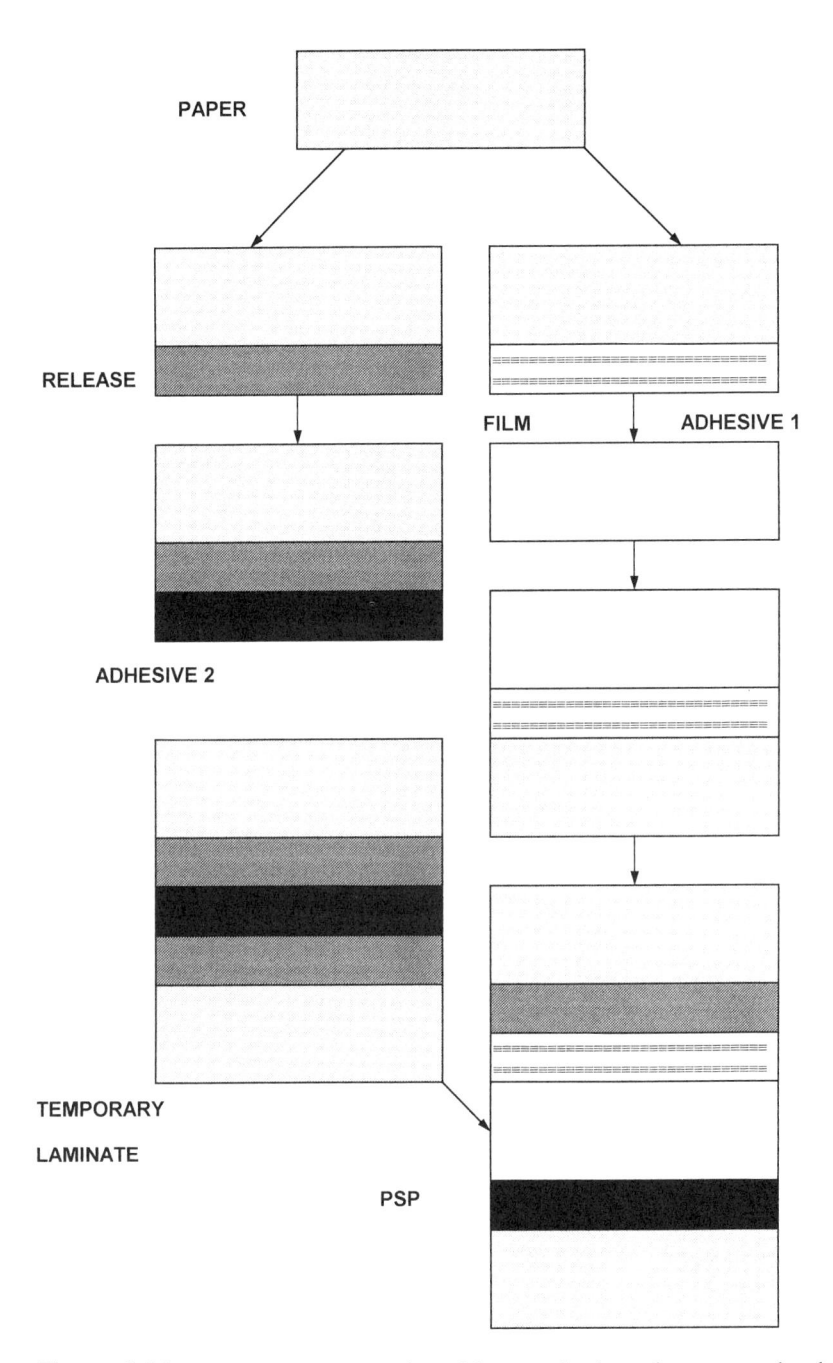

Figure 6.11 Schematic representation of form production using precoated webs.

ing is recommended, but alternatively two smooth rollers or reverse gravure can be used [745].

Packaging tapes are manufactured with water-based formulations too. A medium coating weight (20–30 g/m^2) is suggested. The coating technology is similar to that of common labels. A large proportion of tapes of general use (packaging, carpet, office, etc.) are formulated with hot melts. Such formulations can be coated by slot die coating. The main advantages of slot die coating for HMPSAs are that it can be used for processing high viscosity HMPSAs; the coating device is simply constructed; the coating weight is easily regulated; and the device is a closed melting system with low oxidation. The disadvantages of the slot die system are that the die can be cleaned only by changing the formulation and problems are caused by dirt. The problem of cleaning is a determining factor for HMPSA die coating; it depends on the construction of the die [747]. The extent to which the coating device is abraded depends mainly on the running speed. Generally a highly viscous melt (up to 400,000 mPa.s) can be coated [748]. Hot melts for tapes can be coated by using a porous coating cylinder and a knife that forces the adhesive through the cylinder as in screen printing (the rotary screen printer is a hollow perforated applicator roll) by so-called rotation extrusion [749]. Spray coating of hot melts requires less adhesive (up to 50% less) [750].

The solvent content can be reduced by adding high solids or using roll press calendering. High solids content (50%) solvent-based (viscosity of 500 Pa.s) concentrated rubber resin adhesives have been coated on PVC carrier for tapes [80]. Tapes can also be coated by calendering. In this case if multicomponent crosslinking agents are used, one crosslinking component can be used as the primer coat. The crosslinking agent used for a rubber-based PSA for tapes can be incorporated in an oil and mixed with the adhesive components in a cavity transfer mixer [728]. Such a crosslinking agent can be used in the partially crosslinked adhesive together with a crosslinking activator in the primer. A precrosslinked butyl rubber can be used also instead of partial crosslinking in the Banbury mixer. The degree of partial crosslinking prior to mixing with other components can vary between 35% and 75%. The proportion of partially crosslinked rubber in the composition varies inversely to the percentage of crosslinking. *p*-Quinone dioxime, *p*-dinitrosobenzene, phenolic resins, etc., can be added as the crosslinking agent. A precrosslinking during mixing of the components in a Banbury mixer is followed by a supplemental crosslinking (additional crosslinker) in a second mixer (a two-roll mill), where a quantity of *p*-quinone dioxime is added. For the initial crosslinking, slower acting phenolic resins are used to obviate the lumping problem. The amount of crosslinker used for premixing is about 3 to about 10 phr. After the second mixing the resulting adhesive is coated on a carrier, typically by calendering, to form a tape. This method has been used for the manufac-

Table 6.37 Coating Weights Used for Tapes

Type of tape	Carrier	Adhesive	Coating weight	Ref.
Packaging	—	—	5–35 g/m^2	754
Insulating	Aluminum	Silicone	50 µm	182
Packaging	PP	NR	30 µm	178
Special	Paper	EPR	20 µm	755
Packaging	PET	AC	35 g/m^2	756
Insulating	Aluminum	Silicone	50 µm	181
Insulating	PET	Silicone	42 µm	757
Insulating	Aluminum, paper	—	40 g/m^2	—
Removable from paper	—	AC	5 g/m^2	61
Removable	PE	AC	10 µm	651

ture of medical tapes, where the components of the adhesive were calendered and the hot adhesive was coated under pressure on the carrier material [751].

Tapes made with special extensible or pleated carrier materials are manufactured via transfer coating. For instance, a pressure-sensitive adhesive tape for the protection of printed circuits, based on acrylics and phenolic terpene resin with a 100–200 g/m^2 siliconized crepe paper, is made by coating the adhesive mass first on a siliconized polyurethane release liner and transferring it under pressure (10–30 N/cm^2) onto the final carrier material [334]. Coextruded ethylene/propylene can be stretched for tape carrier also [751]. It is press rolled with the adhesive. Microencapsulated adhesive can be coated on tapes also [752]. In Europe, EB crosslinking is used for tapes with a 1300 mm width machine using the scanning principle. This equipment uses normal adhesive formulation [753]. It must be emphasized that the majority of tapes are coated with a medium coating weight. This is illustrated by the data of Table 6.37.

Manufacture of Packaging Tapes. As stated in Ref. 743, the majority of tapes are packaging tapes. Packaging tapes are manufactured by coating. Recent productivity developments in tape manufacture are related mostly to these products. A decade ago coating machines for tapes ran at 60–220 m/min for solvent-based PSAs and 100–250 m/min for water-based PSAs [507]. Now they run at about 250–500 m/min depending on the type of adhesive. According to Ref. 743, the production speed of such tapes increased from about 250 m/min in 1985 to about 600 m/min in 1995 and the coating machine width increased from about 1400 mm to about 2000 mm in the same period. The average coating speed increased from about 100 m/min to about 300 m/min. The average coating machine width increased from 1300 mm to 1500 mm. A new coating machine for tapes having a width of 1200–2400 mm is run with HMPSA on a 25 µm BOPP film as carrier material at a speed of 400–600 m/min. Hot melt PSAs, acrylics (mostly water-based), and rubber resin formulations have been used for such tapes. Advances have also been made in coating devices. An attempt has been made to combine the forced adhesive feed of gear-in-die systems with the web transport and adhesive coating of roller-based systems. Generally, with gravure systems there is the problem that the gravure cylinder is not filled with adhesive because of its viscosity, rheology, speed, foam, etc. In Accugravure systems, high pressure is used to fill the cylinder; in the Vario coater a knife is pressed on the cylinder to allow better filling [758]. The Accugravure system ensures the complete filling of the gravure lines under hydraulic pressure.

In the United States, biaxially oriented PP tapes have become the most important packaging tapes [759]. These products require high tack toward corrugated paperboard at low temperatures and high shear. There is a trend on the market to provide customers with printed packaging tapes. The text or graphics is imprinted on the nonadhesive side of the carrier material before the tape is made. The problem with hot melts is that a release coating has already been applied to the backing material before it is imprinted. The coating is necessary because hot melts do not release easily from the nonadhesive side of the carrier material.

The most important developments in machine design are unwinders with fully automatic reel changeovers at top line speed, with a very short and constant overlapping tail; preconditioning rolls with tension nips; in-line flame treater; release coating unit with total solvent capture; air flotation dryer; coating device with slot orifice die with automatic adhesive thickness control; and rewinders with independently indexing arms, allowing changeover at full speed [743]. Except for the combination of BOPP with water-based adhesive and that of filament-reinforced BOPP with hot melt adhesive (with higher stiffness and lower contact surface), strapping, packaging, and masking tapes usually require a primer [743]. Extensible reinforced packaging tapes have been developed that use a fiberlike reinforcing material laminated on or embedded in the tape carrier material [760]. A closure tape based on crosslinked acrylate terpolymer uses a reinforced web and a reinforcing filler. Such an adhesive is useful for bonding the edges of heat-recoverable sheets to one another to secure closure. An amine-formaldehyde condensate and a substituted trihalomethyltriazine are used as additional crosslinking agents [761].

Manufacture of Insulating Tapes. Insulating tapes are generally carrier-based adhesive-coated products, but they can also be carrierless (extruded) or adhesive-less (extruded). Special (temperature-, chemical-, and humidity-resistant) adhesives with high coating weight are coated on soft or metallic carrier materials. For instance, silicone pressure-sensitive adhesives that can be cured at low temperatures have been developed from gumlike silicones containing an alkenyl group, a tackifying silicone resin, a curing agent, and a platinum catalyst. Such adhesives for heat-resistant aluminum tapes are coated to a thickness of 50 μm and cure in 5 min at 80°C. Heat-resistant insulating tape formulations can include a fluoropolymer [762]. Such a tape displays a rolling ball tack of 14 cm and peel value of 45 g/20 mm. An aluminum carrier and a tackified CSBR coated with 40 g/m^2 are used as insulating tape in air conditioning.

Manufacture of Mounting Tapes. Mounting tapes are generally products coated on both sides. Such tapes may or may not include a carrier. Generally their manufacture consists of the coating (on both sides) of a carrier material that is postlaminated with a solid release liner. Mounting tapes with carrier can also be "adhesiveless," i.e., self-adhesive. Such products are extruded. Transfer tapes are carrierless mounting tapes that are manufactured by coating, coating and foaming, or extrusion (Table 6.38).

The coater of double-faced tapes performs both transfer and direct coating. The first coating station deposits the adhesive on the release liner (direct coating). In a combining nip, the adhesive-coated release liner is sandwiched with the carrier material (transfer coating). The duplex structure then proceeds to the second coating station, where a second adhesive layer is applied on the carrier material (direct coating). The manufacturer of foam mounting tapes or other double-faced tapes must often use relatively thick films (6–8 dry mils or higher) of PSA. This is done to ensure that the adhesive film will be able to conform to a surface that is irregular or unpredictable [763]. Generally, high coating weights are suggested. Polished cylinders and wire-wound rods are used for higher coating weights, particularly for paper. They have a coating weight tolerance of 15%. For instance, for mounting tapes a 88 μm thick adhesive layer is coated; for splicing tapes, a 50 μm adhesive thickness is used [40]. For double-faced tapes, different degrees

Table 6.38 Manufacture of Carrierless Tapes

Type of tape	Chemical basis	Manufacturing process	Ref.
Special	Acrylic emulsion, ammonium stearate, tackifier	Mechanically foamed, coated on siliconized PET, impregnated with AC adhesive	766
Sealing	AC, UV-cured, silica-filled	Filled, coated, cured	765
Releasable sheet	AC, filled with expandable microbubbles	Filled, coated, cured, blown	58
Releasable sheet	AC urethane, filled with expandable microbubbles	Filled, coated, cured, blown	
Transfer	AC copolymer	Filled, coated, cured, blown	56
Insulating	Chlorinated polyethylene, acrylonitrile-butadiene rubber, liquid chlorinated paraffin	Filled, extruded	676

of release can be achieved on the two faces by using the same release agent but with polyethylene-coated release paper on one face. Splicing tapes are manufactured by transfer coating [130].

A Belgian patent [71] describes the UV polymerization of acrylic monomers directly on the carrier to manufacture a tape. The liquid monomers are applied to the carrier with a doctor blade, roller, or sprayer in a polymerization tunnel from which oxygen has been excluded. This patent calls for a temporary carrier that is an endless belt that does not become incorporated in the final product, and the polymerization is completed thermally in an oven. The tapes have PSA on both sides of the carrier.

Curing by UV light is a preferred method in the manufacture of transfer tapes. A prepolymer having a viscosity of 0.3–20 Pa.s can be synthesized by UV radiation. The crosslinking agent is added, and the mixture is coated [27] and polymerized by UV to the final product. A foam carrier can be applied too. Both surfaces of the carrier may have low adhesion. For a crosslinkable formulation, highly volatile vehicles must be used as solvents in order to avoid excessive heating. Such a tape has a coating weight of 0.5–1.5 mm. A thick coating can be laid down in one pass according to Ref. 27; if solvents or aqueous adhesive are used, multiple coating is necessary to achieve a thickness of more than 0.2 mm. Foam carriers coated on one or both sides can have an overall thickness of 0.1–2.0 mm. A knife coater is recommended for coating a photopolymerizable acrylic prepolymer filled with glass microbubbles and used as transfer tape [22,209]. Sealant tapes are made by extrusion on a release film [46].

According to Harder [72], the main raw materials used for PSA tapes have been acrylonitrile copolymers, butyl rubber, natural rubber, polyacrylates, polyester, polyurethane, SBS and SIS block copolymers, and silicones. The number of suppliers for acrylic HMPSA is very small. Such materials are generally low viscosity prepolymers that must be postcured or polymers having excessive viscosity (see also Chapter 4). For instance, a HMPSA on acrylic basis obtained by solution polymerization can be coated at 90–140°C and is crosslinked by using EB and UV radiation or chemical crosslinking (with NCO-containing systems).

Coating lines can be divided into equipment for adhesives with viscosities of less than 500, 200–2000, and 500–5000 Pa.s. [72]. Most acrylic HMPSAs belong to the lowest viscosity range and can be coated with die systems. Coating weights of 20–200 g/m^2 can be applied. Adhesives belonging to the medium viscosity range have higher molecular weight and demand lower coating temperatures and shear stresses. A combination of roll coater and calender coater (with two heated rolls) is used. Coating weights between 20 and 100 g/m^2 are recommended. For high coating weights, high viscosity acrylic HMPSAs are coated by using an extruder.

A foamlike pressure-sensitive transfer tape with good shear strength and weather resistance that is useful for bonding uneven surfaces has been prepared by forming foamed sheets from agitated acrylic emulsions, impregnating or laminating them with adhesives, and drying. A similar film is coated on a PET release film as a 1.2 mm adhesive sheet with a density of 0.75 g/cm^3. It displays (in the bonding of acrylic polymer panels) an instantaneous shear strength of 4.5 kg/cm and shows, after weatherometer exposure, twice as much resistance as a common polyurethane film [764]. Tapes with microspheres useful for sealing windows and bonding side moldings onto automobiles contain microspheres and silica [765]. Such tapes are UV-cured on PET. For such products the hydrophobic silica improves the adhesive properties. Similar carrierless UV-curable tapes are filled with glass microbubbles and silica.

Manufacture of Medical Tapes. Medical tapes are carrier-based products that are generally manufactured by adhesive coating. Because of the absorption, dosage, and storage function of such tapes, they are coated on special porous carrier materials with a high coating weight of special adhesives. These adhesives do not contain volatiles. Therefore medical tapes were first manufactured by roll pressing or with high solids, later by extrusion and radiation curing. According to Ref. 759, commercial pressure-sensitive adhesives used for skin application are based on acrylic adhesives, which are easier to remove and cause less skin irritation than rubber-zinc oxide adhesives, which adhere well to the skin but may cause skin irritation because of their higher adhesion level. Acrylic solutions and PUR solutions are both used. Both adhesives exhibit a continuous decrease of G' as a function of temperature. Acrylics used for finger bandages show a more abrupt decrease; their tan δ peak is shifted to higher temperatures (see also Chapter 3).

The first compositions for medical tapes contained natural rubber or polyisobutylene. Later polyvinyl ethers were used. Polyvinyl ethers can also be coated by extrusion. As discussed in Ref. 766, polyisobutylene or a blend of polyisobutylene (5–30% w/w) and butyl rubber, a radial SBC block copolymer (3.0–20% w/w), mineral oil (8.0–40% w/w), water-soluble hydrocolloids (15% w/w), tackifier (7.5–15% w/w), and a swellable cohesive strengthening agent are used to formulate a medical adhesive. A mixture of *cis*-polybutadiene, an ABA block copolymer, and tackifier resins have also been proposed for medical adhesive. Such an adhesive mixed with an anti-inflammatory paste is spread on a film carrier material for medical poultices [767].

Polyvinyl ethers do not produce skin irritation; therefore their use is suggested for medical applications [768]. Polyvinyl ether processed as a hot melt can be used for double-faced tapes with a textile carrier [769]. The water vapor porosity of PVE is about the same as that of the skin, 259 g/m^2 over 24 h.

Acrylic adhesive compositions for medical tapes that do not leave adhesive residues on the skin contain *tert*-butyl acrylate-maleic anhydride-vinyl alcohol copolymers having a molecular weight of 2500–3000 and acrylic acid-butyl acrylate copolymers [770]. The adhesives for medical tapes have to meet the following requirements [759]: They should not irritate the skin, they should adhere well but be easily removable, and they should be water-resistant. For medical uses the direction of extensibility is very important. Cross-direction elasticity may be given by a special nonwoven material [771]. The wound plaster placed in the direction of elasticity on the carrier is narrower than the carrier. Soft PVC is used for plasters (70–200 µm) with a silicone liner. These films are dimensionally stable up to 200°C [496].

Cotton cloth and hydrophobic textile materials coated with low energy polyfluorocarbon resins are used as carrier materials for surgical tapes and are coated with a polyacrylate-based water-soluble PSA. Reusable textile materials such as Goretex, polyamide, polyether sulfone/cotton, 100% cotton, or other hydrophobic carrier materials are suggested for medical PSPs (operating tapes, labels, and bioelectrodes). Special tapes are used for biolelectrodes because of their low electrical impedance and their conformability to the body. These products are manufactured by photopolymerization [77]. Compositions include acrylic acid polymerized in a water-soluble polyhydric alcohol and crosslinked with a multifunctional unsaturated monomer. A prepolymer is synthesized that is polymerized to the final product with UV radiation [92]. This polymer contains water, which affords electrical conductivity. Electrical performance can be improved by adding electrolytes to the water. Electrically conductive PSAs can be manufactured as gels containing water, NaCl, and NaOH [772]. Such gels are useful in detecting voltages in living bodies. Such polymers based on maleic acid derivatives have a specific resistivity of 5 kohm (1 Hz).

Radiation curing has been proposed for medical tapes also. This method is limited only by the coating weight. UV coating is practicable up to 70 g/m^2 [773]. UV-curable formulations can use difunctional or polyfunctional vinyl ethers for better adhesive bond strengths after curing [774].

Manufacture of Application Tapes. Application tapes are mounting tapes used for placing signs, letters, decorative elements, etc., die cut from plotter films. Such products are special removable tapes with a carrier base. Paper and film carriers are used for application tapes. Special adhesive formulations and special coating methods ensure removability.

In some cases the coating technology used depends on the adhesive surface geometry. As discussed earlier (see Chapter 4), a reduction in the adhesive contact surface enhances removability. This reduction can be achieved by choosing an appropriate coating device. As is known, film-based application tapes have to be conformable but removable. Therefore for such products gravure printing is favored. The patternlike, striped surface structure of the adhesive ensures less contact surface and better removability. A decrease in the contact surface decreases the buildup of static electricity also. Such a coating can be unwound with less buildup of static charge [775].

As mentioned earlier, application tapes are produced as paper- or film-based PSPs. Although used for the same application, products based on paper are quite different in buildup from those based on plastic film. Products with a film carrier are more conformable than those with paper. For excellent lay flat, the carrier contains HDPE and is manufactured via coextrusion. To avoid curl from unwinding during application, adhesion should be minimum on the back side. Because of the different surface qualities of plotter films (calendered or cast PVC, cast PE, etc.), which function as the adherend surface for application tapes, a range of application films have been developed that have different coating weights. The choice of hard acrylic dispersions as base formulations for application films allows the design of an uncrosslinked formulation for the film-based products. Paper-based products are formulated with NR latex as the main component. Both grades of tape can be produced with or without a supplemental release layer.

Manufacture of Masking Tapes. Generally the manufacturing technology used for masking tapes is the same as that used for protective films. However, there are some special tapes that are produced with a complex technology. For instance, a laminated pressure-sensitive adhesive tape having good chemical, heat, and water resistance that is applied for soldering portions of a printed circuit board is manufactured using a kraft paper carrier material impregnated with latex, treated with a primer or corona discharge, and coated on one side with a rubber resin adhesive and on the other side with a polyethylene film and release agent [776].

Thick masking tapes with "bubblepack" characteristics and good cuttability can be manufactured with classic coating technology, including an expandable filler in the adhesive. According to Ref. 57, a photocrosslinked acrylic-urethane acrylate formulation used for holding semiconductor wafers during cutting and releasing the wafers after cutting exhibits a peel strength of 1700 g/25 mm without crosslinking and 35 g/25 mm with crosslinking. Adding an expandable filler to the formulation (6 parts to 100 parts adhesive) reduces the peel resistance. The resulting composition exhibits a peel strength of 1500 g/25 mm before crosslinking and 15–20 g/m^2 after crosslinking. A releasable foaming adhesive sheet can be prepared by coating blowing agent containing PSA (30%) for a 30 μm dry

Table 6.39 Adhesive Characteristics of Selected PSPs

Adhesive formulation	Peel resistance (N/25 mm)	Rolling ball tack (cm)
Unfilled AC PSA	18.0	2.8
Filled (10%) AC PSA, not expanded	17.0	3.5
Filled (10%) AC PSA, expanded (140°C)	8.0	3.2

Table 6.40 Peel Buildup for Expanded PSP[a]

Dwell time	Peel resistance (N/25 mm)
0	10
1 day	15
2 days	19
5 days	PT[b]

[a] A 30 μm (wet) PSA layer was coated on paper; peel on glass was measured.
[b] PT = paper tear

layer [58]. Blowing up of the adhesive layer reduces the contact surface. Such products exhibit lower bond strength (Table 6.39). However, they are only temporarily removable. Adhesive flow leads to rapid buildup of the peel strength (Table 6.40). Therefore for adequate removability, both blowing up of the adhesive layer and crosslinking are needed.

Manufacture of Protective Films

The choice between adhesiveless and adhesive-coated protective films is determined by the main characteristics of the products. The manufacture of adhesiveless protective films or protective films with embedded adhesive was discussed earlier. The procedures yield removable products with a maximum adhesion of 1.5 N/10 mm. Such peel resistance values can be achieved with special adhesives or/and crosslinked formulations and low coating weight. Therefore, adhesive-coated protective films generally need a crosslinked adhesive with low coating weight.

Similar low coating weights of a crosslinkable adhesive are used in laminating also. According to Klein [777], coating weights of 0.8–1.5 g/m^2 are used for laminating smooth surfaces, 1.2–1.8 g/m^2 for printed films, and 3.5–4.5 g/m^2 for paper laminates. Therefore, at least theoretically, the coating technology for protective films is related to the well-known procedure used for laminating films. Its choice depends on the type of adhesive and its coating weight and geometry, on the type of carrier and its geometry, and on the end use of the product.

Hot melts are limited by their low heat resistance, low penetration into porous substrates, and relatively high application temperatures of 150–200°C. Even at these application temperatures, some hot melts are highly viscous (which affects their ability to be coated at very low coating weights) and can cause damage to sensitive substrates. The applicability of hot melts is limited by the restricted formulating freedom for removable recipes too.

Table 6.41 Coating Weight Tolerances for Protective Films

Product code	Coating weight (g/m^2) Theoretical	Coating weight (g/m^2) Measured	Extreme tolerance values Minus	Extreme tolerance values Plus	Average tolerance (%)
1	1.6	1.86	0.2	0.8	16
2	1.7	1.77	—	0.2	4
3	1.7	1.85	—	0.2	9
4	2.2	1.90	0.4	—	7
5	2.4	2.57	0.3	0.4	7
6	2.8	3.00	0.1	0.9	7
7	2.9	3.34	—	0.7	15
8	4.0	4.6	—	0.5	15
9	4.5	4.28	0.1	—	5
10	5.7	6.30	—	0.4	11
11	7.0	7.10	—	—	1.5

Aqueous crosslinked adhesives for protective films can be coated using a gravure cylinder (quad system) of 65, 96, or 120 quads/in. For instance, 65 quads/in. gives a coating weight of 8 g/m^2 and 120 quads/in. gives a coating weight of 4.4 g/m^2. In this case a doctor blade is used to smooth the adhesive surface. An intermediate cylinder and a pressure cylinder are used also. According to Fust [391], flexo, letterpress, and gravure printing provide a coating weight of 1.0–3.0 g/m^2. Therefore it would be possible to use them for the manufacture of protective films also. Screen printing works with a coating thickness of 5–10 μm; therefore, it is suitable for label, tape, or protective film production. Depending on the roughness of the product surface and on the laminating, delaminating, and processing conditions, chemically different adhesives with different coating weights are applied with various coating devices. The mean nominal coating weights are between 1.5 and 7.0 g/m^2. For low coating weights, positive tolerances are preferred (see Table 6.41).

As discussed in Chapter 4, the peel resistance values required for the removable protection of different surfaces vary widely. Generally peel values of 0.3–1.3 N/25 mm are recommended. In practice there are much lower values also. For such products the coating weight is between 0.5 and 1.5 g/m^2. As can be seen from Fig. 6.12, for such low coating weights, the range of coating tolerances is very broad and may be as much as ±0.5 g/m^2.

As is known, the ratio of the viscosities of adhesive compositions based on unmilled or milled rubber is about 5–10, which means that at least theoretically low viscosity rubber resin solutions may be coated for protective films, although

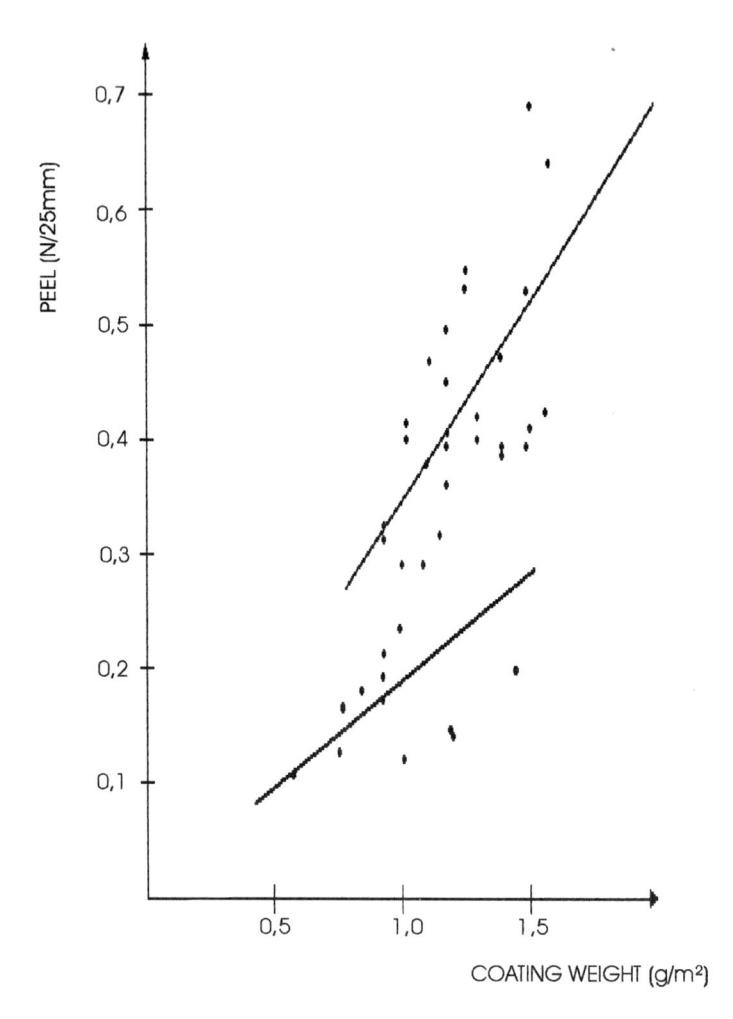

Figure 6.12 Coating weight tolerances for protective films with low adhesion. Two crosslinked rubber resin adhesives with different degrees of crosslinking gave different peel values.

the resin concentration of such recipes is much lower than for common label or tape formulations (Table 6.42). Unfortunately, crosslinking leads to a decided increase in viscosity.

Release coating for protective films and tapes presents a complex problem. It depends on the nature of the adhesive coat and on its coating weight. As can be seen from Table 6.43, different release coatings exhibit different debonding

Manufacture of PSPs 489

Table 6.42 Elastomer/Tackifier Concentration in Removable Products

PSP	Elastomer	Non-resin tackifier	Plasticizer	Tackifier	Ref.
Removable sheet	NR (65–75)	PIB (0–10), PB (0–10)	—	Hydrocarbon resin (25–35)	—
Masking tape	NR (70)	—	—	Terpene resin (50)	154
Masking tape	AC (100)	—	—	Terpenephenol resin	293
Protective sheet	NR (100)	—	DBP (5–10)	Rosin (100)	—
Protective sheet	NR (100)	—	—	Rosin (55)	—
Protective sheet	NR (100)	—	—	Rosin (60)	—

Table 6.43 Dependence of Peel Resistance on the Back of Protective Film on the Release Agent Used

Coating weight (g/m²)	Peel resistance (N/10 mm)			
	Without release coating	Release coating 1	Release coating 2	Release coating 3
2.5	0.20	—	0.15	0.05
4.0	0.25	—	—	—
5.0	0.30	—	0.05	0.05
6.0	0.33	0.20	0.05	—
7.0	0.43	0.15	—	—
9.0	0.65	—	—	0.50
10.0	0.65	0.20	0.27	—
11.0	0.68	0.20	—	—
12.0	0.72	—	—	—
13.0	0.77	0.20	—	0.80

forces and different increases in release force per unit coating weight. The best formulations display a "constant" release force for different coating weights. The carbamate-based release coatings (1 and 2 in Table 6.43) provide low level release forces for different coating weights. A polyamide-fluoropolymer base formulation gives very low release forces at low coating weight but high release forces (as high as without coating) at high adhesive coating weight.

Main Equipment

Generally the following parameters are determinant concerning the versatility of a coating equipment to produce PSPs belonging to different product classes: the ability of the coating device to process PSAs of different physical states, with different coating geometries, the on-line slitting and cutting possibilities, the existence of lamination equipment, the possibility of in-line flexo printing, and the possibility of in-line release coating. According to Massa [691], the main processes in self-adhesive coating and lamination are web handling, coating, drying and/or curing, and moisturizing.

In recent decades specially designed new machines have been built, but "second-hand" modernized older machines still exist [778]. For such machines modular design is important for flexibility and for the possibility of upgrading or rebuilding in the future [691].

Coating Device. Generally an apparatus for applying low and medium viscosity liquid adhesive includes a rotating applicator cylinder, an adhesive tank to supply the adhesive to the roller, and at least one doctor blade operationally associated with the applicator roller. The liquid coating components may interact with the coating cylinder; therefore it is recommended to take the adequate materials [779]. The applicator cylinder may be polished or engraved.

Generally gravure coating is used for laminating (solvent-based, solventless, and water-based systems) and for coating solvent-based, water-based, or hot melt PSAs [780]. The characteristics of such cylinders are described in Ref. 781. The main parameter affecting the choice of coating device is the rheological behavior of the adhesive. It is well known from the practice of laminating that adhesives with different flow characteristics require different coating devices. For instance, for low viscosity (14–15 s, DIN cup No. 4) solvent-based adhesives, smooth reverse rolls are recommended; for the same adhesives with 20 s viscosity, gravure rolls are suggested. Gravure and reverse gravure require low viscosities, 200–500 cP [782]. Such coaters are well suited for water-based PSAs. Reverse gravure ensures narrow tolerances for the coating weight. On the other hand, for such a coating device it is necessary to change the coating cylinder for different coating weights. The maximum viscosity is limited to about 1000 cP [691]. At high speeds air may be entrained with the cylinder, i.e., foaming may appear. Air entrainment may be avoided by using an enclosed supply chamber (Fig. 6.13). For highly viscous crosslinked adhesives, gravure coating may lead to a discontinuous or contained coating layer. Such a coating is preferred for certain removable products (e.g., application tapes). The main coating device water-based for PSAs has been the one-roll reverse coater with metering bars. This is an inexpensive, easily operated, flexible device. Its main disadvantage appears at high speeds, where foaming may be a problem. The maximum working speed is about 200 m/min [691].

Manufacture of PSPs 491

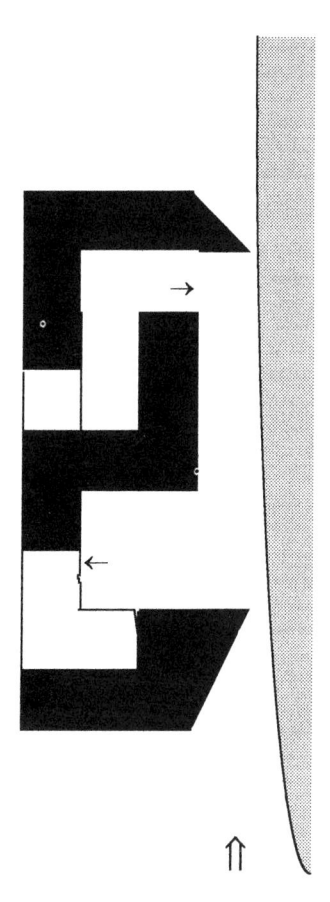

Figure 6.13 Enclosed supply chamber.

A gear-in-die slot orifice die has been developed also. This device has a liquid pumping section built into the die. The pump is designed to cover the complete die length. Advantages include higher solids, accuracy, no foaming, and low shear. An air knife can also be used as a coating device [783].

Common application systems for hot melts include rotogravure, transfer gravure, and smooth roll [691]. The parameters that affect the processing of HMPSAs are viscosity, coating speed, and coating weight [737]. Rubber-based conventional hot melts exhibit a viscosity range of 10,000–60,000 mPa.s [737]. Rubber-based EB curing hot melts have a viscosity range of 20,000–100,000 mPa.s; acrylate-based EB- or UV-cured systems have lower viscosities (10,000–30,000 mPa.s at 120–170°C). For low viscosities (10,000 cP), roll coating is suggested, and for higher viscosities (35,000–500,000 cP), slot die coating [706,707]. Slot die is used for 10 g/m² and 120,000 mPa.s or more [702]. Slot die coating systems can coat 1–2500 g/m². According to Ref. 635, for hot melts, slot die application is suggested for use up to 120,000 cP (200°C) and roller coating for 100,000 cP. Coating weights of 1–80 g/m² (in special cases down to 7 g/m² and above 80 g/m²) are obtained [784]. Coating trials with acrylic block copolymers allowed a coating speed of 500 ft/min using a die coater [785]. The slot die is recommended for labels, plasters, pressure-sensitive foams, and pressure-sensitive textile materials [786]. Slot die coating devices may run faster (exceeding 400 m/min) than roller systems, have a broader working width (3000 mm or more), produce less oxidation of the PSA, and coat viscosities up to 1×10^6 mPa.s more exactly (±0.5%), but changing the coating material or coating width, startup, and cleaning are more complex. Therefore roller systems are suggested for frequent product changes or short production runs. Slot die coating is suitable for one-product, high speed, 24 h operation.

Roll coaters can coat hot melts with viscosities between 500,000 and 1×10^6 or 2×10^6 cP [706]. Roll coating systems for HMPSAs are suggested for 100–1500 g/m² [786]. According to Drechsler [787], roller coating systems for hot melts are adequate up to viscosities of 15,000 cP. For such viscosities a three-cylinder roller coating system is suggested. For lower viscosities (7000 cP) a central drive system is used. For viscosities over 5000 cP, forced feed of the hot melt onto the gravure cylinder is recommended [788]. Heated coating cylinders with 20 lines/cm and halfmoonlike cells of 80–100 μm depth are suggested for a coating weight of 12 g/m², and cylinders with 18–34 lines/cm and cells of 40–50 μm depth for a coating weight of 7–8 g/m². The temperature of the gravure cylinder should be kept 10–15°C higher than that of the molten hot melt adhesive [789]. According to Ref. 790, gravure cylinders are suggested for hot melt coating for viscosities less than 5000 cP. Reverse roll coaters and a kiss coater are used up to 20,000 and 30,000 cP, respectively.

Spray coating technology for hot melts has been developed from first-generation melt blowing to systems with controlled fiberization. Their main functional criteria are fiber size, fiber density, absorbency control, edge control, product aesthetics, minimal operator involvement, and minimal air consumption [791]. As substrates are bonded at thousands of points of contact, fiberization allows the manufacture of light weight products and can be applied for thermally sensitive carrier materials also. Coating weights as low as 2 g/m² can be achieved [792].

Coating of HMPSA supposes the existence of premelters, melt tanks, melt extruders, filters, feed lines, pumps, and heating systems. Such systems are described in detail manner in Refs. 793 and 794.

Quite different coating weights are required for permanent and removable labels, tapes, or protective films. Different adhesive layer buildups may be neces-

Manufacture of PSPs

Table 6.44 Pattern Coating for PSPs

Product	Application	Adhesive	Coating method	Ref.
Tape	Removable from paper	HMPSA	Pattern	651
Label	Wrinkle resistance for squeeze bottle	HMPSA	Circular dots	409
Tape	Diaper closure	HMPSA	Pattern	677
Protective paper	Plastic plate protection	NRL	Profilated coating	252

sary for special products. Therefore other, special coating devices are used also. For instance, one special coating device works according to the principle of rotary screen printing [703]; it is recommended for tapes and labels. It can also coat double-faced materials. Screen printing has been recommended for adhesive layers of 7–60 μm [681]. In this case the coating device includes a melt supply, coating slot die, gear pump, temperature controllers, metering bar, coating of the backup roller, chill roll, and tension control.

As illustrated by Table 6.44, very different products need pattern coating. Special coating devices have been developed for pattern coating, i.e., the coating of discrete portions of the product. A reverse gravure system does not usually allow pattern coating. A special patterned reverse gravure coating is described in Ref. 795 in which a tampon cylinder (which contacts the gravure roll) is profilated. The printing cylinder is used for the pattern coating of hot melts; dots in excess of 24,000/in.^{2} are coated with a speed of 500 ft/min, and specific zone coat patterns—squares, circles, rectangles, or stripes—can be applied. Pattern transfer and laydown are affected by the cylinder mesh size, adhesive viscosity, doctor blade pressure and tangent point, temperature, and operating speed [796]. Patterned coatings can be applied with a perforated cylinder also [797]. Tools of the same thickness as the coating weight and with an elastic cylinder are also used for patterned coating [798].

A removable display poster is manufactured by coating distinct adhesive and nonadhesive strips on the carrier [799]. The adhesive strips lie in the same plane, the plane being elevated with respect to the product surface. Another procedure, described by Gleichenhagen [2], uses a pyramidal PSA coating site (30 × 600 μm base dimensions) coated via screen printing. Rotary screen printing can be used with a coating weight of 1–20 g/m^2, a running speed of 10–100 m/min, screen geometry of 15–40 holes/cm, blade 1.5–30 mm; blade thickness 150–300 μm, and pressure of 2–6 N/mm. Patterned hot melt coating may reduce adhesive costs by 50% [800]. According to a patent for removable PSPs [629], the carrier may be coated alternately with adhesive and release stripes. A patterned coating can be replaced by special postcoating reduction of the adhesive surface as discussed in Ref. 801. The regulation of the adhesive properties can be achieved by

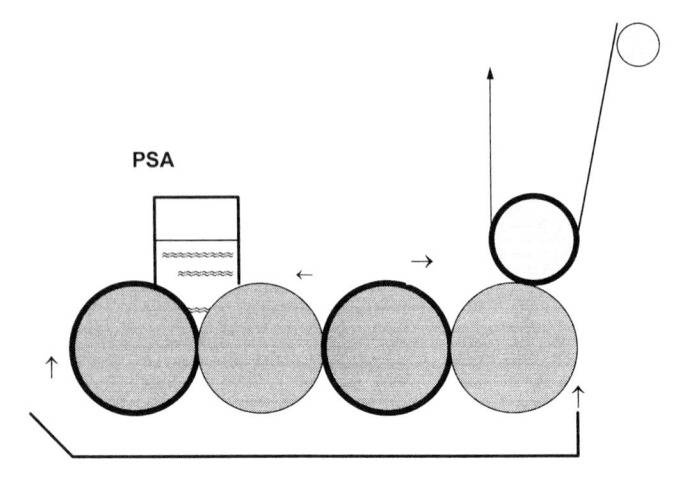

Figure 6.14 Nip feed, four-cylinder coating device for low coating weights.

partial detackifying after the adhesive is coated on a temporary release liner and transferred onto the final support.

Special coating machines allow sheetlike materials to be coated also. The procedure uses transfer coating of the adhesive from a transfer cylinder or tape onto a primed and release-coated sheetlike face stock material [639].

Primers are supplied either as solvent-borne, water-borne, or solid-state products. Equipment availability can influence the choice of primer. It determines whether the primer is to be applied in-line or off-line [640]. The need for a prime coat also influences the complexity and economics of PSP manufacture.

Different adhesives and coating procedures give different coating images. For crosslinked adhesives whose fluidity is limited in time (because of their gelling) this phenomenon can lead to quite different product build-up with the same application characteristics. As discussed earlier, the same grade of protective film manufactured with different coating devices will have different coating weights (but exhibit the same application characteristics). On the other hand, the same type of coating device but with different characteristics (e.g., number of Meyer bars) may lead to higher coating weights (as usual) and much higher buildup of the coating weight. The very low coating weights that are usual for protective films can be coated with a four-cylinder nip-feed coating device (Fig. 6.14).

Drying. The drying capacity has to be increased for water-based coatings in order to provide the extra energy required to evaporate water compared to other solvents. This is difficult to achieve without extending the length of the dryers, reducing the drying speed, and affecting the printing register, a critical point when coating plastic webs. Drying influences surfactant migration also. According to Pitzler [734], in the transfer coating of acrylic dispersions, the migration to the film surface is increased by drying.

Drying speed is influenced by the type and formulation of the adhesive. Because of the use of multiple coated layers with different dispersing media, the buildup of the drying equipment may be complex. For instance, for drying in-line coated double-faced tape (with release and primer), because of the different solvents and coating weights applied, the use of two different drying units is suggested. For drying the release layer, a tangential air jet dryer, and for the primer, an air jet with a longer channel, are recommended. The use of solvent-based adhesives requires the removal of organic solvents during drying and a special solvent recovery system.

On the other hand, the same drying channel is frequently used to handle both solvent-based and water-based adhesives. For solvent-based coatings, the air volume necessary for drying (V_{min}) is given by the formula [794]

$$V_{min} = \frac{G_{max}(273° + \vartheta_{max})}{C_r K_s f \times 293} \tag{6.13}$$

where G_{max} is the maximum amount of solvent, ϑ_{max} the maximum temperature of recirculated air, C_r the lower explosion limit, K_s a security parameter, and f a parameter that takes into account the aerodynamic conditions in the channel. In (6.13) G_{max} is given as a function of the coating weight C_w, web width W_w, and web speed W_s.

For water-based adhesives the relative humidity of the air is the most important parameter of drying. As stated in Ref. 802, different adhesives need different drying temperatures. For instance, the drying temperature recommended for a thermoplastic adhesive is about 200°C, whereas thermosetting adhesives need 250°C, self-crosslinking adhesives require 200°C, and emulsion-based adhesives are dried at 212°C.

Infrared drying offers the possibility of regulating the drying intensity in the cross direction, and is characterized by low space requirements [803]. IR drying can be used as intermediate drying between coating and printing devices [804]. A combination of IR and air drying gives better results [691]. Drying consumes 70% of the energy of a printing machine [805]. Exhaust and recycling have to be adjusted according to the solvent concentrations [691]. If multiple layers are coated, adequate drying time should be allowed for each layer.

A special system is described in Ref. 806. For this system the thin water- or solvent-based wet coat is transferred onto a special heat-storing endless release belt. This system can work with high temperatures without the risk of skin formation. This dryer is considerably shorter than standard machines.

A classic machine for OPP adhesive tapes for cardboard sealing, designed for a production speed of 350 ft/min, uses a solvent-based PSA with a coating weight of 20 g/m² (solids content 28%), a solvent mixture of hexane (70%) and toluene (30%), and an OPP carrier layer about 40 μm thick [806].

Electrostatic control is fundamental for safe conversion. The formation of an electrostatic charge is associated with low relative humidity also. Therefore humidifiers (humidification systems) are used. For paper processing, a temperature of 20–22°C and air at 55–60% relative humidity are suggested [807]. Deionizing bars or systems (static eliminator bars, which neutralize the charge with a vacuum hood to remove contaminating particles) are recommended [808].

Laminating. Laminating is required for multiweb constructions. These include PSPs that have a separate release liner (e.g., labels and tapes). According to Bulian [682], the most important part of a laminating machine is a nip between a rigid roller and an elastic roller, so that enough pressure is achieved to bring the two materials into contact with one another.

Auxiliary Equipment

Various kinds of special equipment ensure the weblike processing of PSPs. Winding and printing devices are the most important special parts of the PSP manufacturing equipment. Winders are constituents of web-handling equipment in the paper, film, and metal processing industries, among others. Printing is a complex industry itself. It is not the aim of this book to discuss in detail the problems related to the construction and use of such equipment. Therefore only some special features are discussed here.

Winding Equipment. Unwinder web tension control is a major parameter. The winding tension affects the pressure in the roll and the diffusion (migration) of viscous components in the carrier for self-adhesive film (SAF). It depends on the hardness of the film and has to be adjusted on the winder [665].

Many versions of unwinders and rewinders are known [691]: simplex, duplex, nonstop, fully automatic, with disc brakes or with dc motors, with different methods of reel core chucking and locking, and with constant or variable tension. Flying splice unwind/rewind stands with electronic tension control units situated between the unwind reel stand and the first coating station and between the last coating station and the rewind stand are used. Tension can be controlled by fully regenerative dc motor driven draw rollers operating through a low friction dancer roller with potentiometer and tachometer feedback reference [809].

Controlled, constant web tensions are very important for thin films. Generally a two-component regulating system is used. The first component regulates the torque, and the second acts as feedback. Unwind reel diameter is controlled, and torque is regulated proportionally. A sensitive dancer roller measures the tension. Fine control of web tension was first required in the coating of soft PVC films [810]. Similar machines were developed later for thin protective films.

Winding machines with a web tension control of 50–200 g/mm² and web tensions of 0.1–10 kN are common [202,419]. The maximum winding speed is

20–140 m/min for blown films, 120–400 m/min for cast film, and 280–350 m/min for biaxially oriented film [163] (see also Chapter 7). Web cooling before winding may be required also [811]. Winding is described in detail manner in Refs. 812 and 813. The lowest web tension is about 20 N [814]. Generally winding should be tight enough that the film does not shrink later due to crystallization [815]. Web tension should not exceed 10 N for thin plastic films. Common winding machines used for blown film manufacture do not meet these requirements; because of the eccentricity of the roll, forces of 200–300 N may appear. A center rewinder allows production of hard or soft rolls with laminates up to 200 g/m².

Web Control Equipment. On-line thickness is measured with sensors that are not in contact with the web. Optoelectronic (laser triangulation, laser scanner, light transmittance), capacitive (condensator), inductive roll sensor, contact, pneumatic (pressure die), radiometric (β radiation), or acoustical (ultrasound) principles are used. According to Ref. 816, the sensors most often used for measuring film thickness usually fall into one of three categories: caliper, nuclear, or infrared. Caliper gauges make a direct physical measurement of total thickness. Such gauges can contact the product on both sides or may have a relatively thin contact layer of air on both sides. Such devices are not capable of measuring the individual layers of the laminate. Nuclear sensors pass β or γ radiation through the material. They cannot measure the thickness of individual layers either. Infrared sensors use the 1.30–2.70 μm portion of the near-infrared spectrum and work like a spectrometer. The limited number of filters in a sensor restricts the number of polymer layers in a laminate that can be measured simultaneously. Web side control by ultrasound is possible at running speeds as high as 1200 m/min [817].

Printing Equipment. The majority of PSPs are printed during their manufacture. The printing can be carried out on one side or on both sides in different qualities. Printing during manufacture of the pressure-sensitive laminate may provide a technical aid for postcoating (e.g., labels) or postprocessing of the PSP (e.g., protective films) or serve for advertising. A solvent-based or water-based printing technique may be used. Printing with water-based inks is more difficult because of the different properties of the carrier liquid and ink components (see also Chapters 4 and 7). Drying and cleaning of the gravure cylinder require special operations. The risk of scumming problems is increased, and scum is more difficult to remove. With solvent-based formulations, when scum appears, it is common practice to blow compressed air on the printing cylinder between the doctor blade and the impression roller.

For labels, generally both printing and overprinting (see Narrow Web Printing Chapter 7, Section 1.3) are carried out. Narrow web printing is done after confectioning and is more and more often carried out by the end user. Labels are printed with various printing methods, including letterpress, flat-bed, semirotary,

rotary, flexography; lay and rotary offset, lithogravure, hot foil, and in-line finishing. Reduced setup times, good production speeds, and the ability to change sizes without changing cylinders are important parameters [396]. The introduction of UV-cured inks allows flexographic printing on nonpaper substrates that used to be printed with letterpress machines (see Chapter 7).

REFERENCES

1. I. Benedek and L. J. Heymans, *Pressure-Sensitive Adhesives Technology,* Marcel Dekker, New York, 1997, Chapter 5.
2. U.S. Patent 3691140; in P. Gleichenhagen, E. Behrend, and P. Jauchen (Beiersdorf AG, Hamburg, Germany), EP 0149135B1/24.07.1987.
3. H. Miyasaka, Y. Kitazaki, T. Matsuda, and J. Kobayashi (Nichiban Co., Ltd., Tokyo, Japan), Offenlegungsschrift, DE 3544868A1/18.12.1985.
4. *Druck Print 4:*18 (1988).
5. G. R. Frazee (S. C. Johnson and Son. Inc.), EP 258753/22.08.1986; in *CAS Emulsion Polym. 19:*2 (1988).
6. G. R. Frazee (S. C. Johnson and Son. Inc.), EP 259842/16.03.1988; in *CAS Colloids (Macromol. Aspects) 19:*2 (1988).
7. W. J. Traynor, C. L. Moore, M. K. Martin, and J. D. Moon (Minnesota Mining and Manuf. Co.), U.S. Patent 4726882/23.02.1988; in *CAS Adhes. 17:*4 (1988).
8. H. Kuroda and M. Taniguchi (Bando Chem. Ind. Ltd.), Jpn. Patent 6343988/25.02.1988; in *CAS Adhes. 21:*4 (1988).
9. M. Koehler and J. Ohngemach, *Polym. Paint Colour J. 178:*203 (1988).
10. Loctite Co., Jpn. Patent 621795006/06.08.1986; in *CAS Siloxanes Silanes 13:*1 (1988).
11. F. Buehler and W. Gronski, *Makromol Chem. 188:*2995 (1988).
12. M. A. Johnson, *J. Plast. Film Sheeting* 4(1):50 (1988).
13. M. Dai, L. Zhang, R. Zhu, and K. Zang, Chin. Patent 8505449/14.01.1987; in *CAS Adhes. 17:*4 (1988).
14. D. Akihito, O. Tomohisa, and U. Toshishige (Hitachi Chem. Co. Ltd), Jpn. Patent 6368683/28.03.1988; in CAS *Crosslink. React. 14:*7 (1988).
15. Belg. Patent 675420; in D. K. Fisher and B. J. Briddell (Adco Product Inc., Michigan Center, MI), EP 0426198 A2/08.05.1991.
16. Lehmann et al., U.S. Patent 3729338; in D. K. Fisher and B. J. Briddell (Adco Product Inc., Michigan Center, MI), EP 0426198 A2/08.05.1991.
17. *Adhes. Age 7:*53 (1994).
18. E. Fink, *Paper Film Foil Convert. 8:*65 (1973).
19. *Adhes. Age 11:*44 (1995).
20. D. K. Fisher and B. J. Briddell (Adco Product Inc., Michigan Center, MI), EP 0426198 A2/08.05.1991.
21. Y. Sasaaki, D. L. Holguin, and R. Van Ham (Avery Int. Co., Pasadena, CA), EP 0252 717 A2/13.01.1988.
22. U.S. Patent 4223067; in D. K. Fisher and B. J. Briddell (Adco Product Inc., Michigan Center, MI), EP 0426198 A2/08.05.1991.

23. M. von Bittera, D. Schäpel, U. von Gizycki, and R. Rupp (Bayer AG, Leverkusen, Germany), EP 0147588 B1/10.07.1985.
24. R. R. Charbonneau and G. L. Groff (Minnesota Mining and Manuf. Co., St. Paul, MN), EP 0106559 B1/25.04.1984.
25. U.S. Patent 2925174; in R. R. Charbonneau and G. L. Groff (Minnesota Mining and Manuf. Co., St. Paul, MN), EP 0106559 B1/25.04.1984.
26. U.S. Patent 4286047; in R. R. Charbonneau and G. L. Groff (Minnesota Mining and Manuf. Co., St. Paul, MN), EP 0106559 B1/25.04.1984.
27. U.S. Patent 4181752; in R. R. Charbonneau and G. L. Groff (Minnesota Mining and Manuf. Co., St. Paul. MN), EP 0106559 B1/25.04.1984.
28. I. Benedek and L. J. Heymans, *Pressure Sensitive Adhesives Technology,* Marcel Dekker, New York, 1997, Chapter 8.
29. G. Holden and S. Chin, *Adhes. Age 5:*22 (1987).
30. R. Higginson (DRG, UK), PCT WO 8803477/19.05.1988; in *CAS Colloids (Macromol. Aspects) 22:*6 (1988).
31. K. Akasoka, S. Sanuki, and T. Matsuyama (Nippon Synth. Chem. Ind.), Jpn. Patent 62243669/24.10.1987.
32. S. Shinji and Y. Yoshiuki (Nippon Synth. Chem Co.), Jpn. Patent 62243670/24.10.1987; in *CAS Adhes. 13:*4 (1988).
33. J. W. Otter and G. R. Watts (Avery Int., Pasadena, CA), U.S. Patent 5346766/13.09.1994.
34. DE-A-2407494; in P. Gleichenhagen, E. Behrend, and P. Jauchen (Beiersdorf AG, Hamburg, Germany), EP 0149135 B1/24.07.1987.
35. EP 4728572.
36. J. L. Walker and P. B. Foreman (National Starch Bridgewater, NJ), EP 224795/10.06.87; in *CAS Silicones Siloxanes 2:*3 (1988).
37. M. Avella and E. Martuscelli, *Polymer 10:*1734 (1988).
38. R. Stadler, L. de Lucca Freitas, V. Krieger, and S. Klotz, *Polymer 9:*1643 (1988).
39. F. Chambon and H. H. Winter, *J. Rheol. 31:*683 (1987).
40. D. G. Pierson and J. J. Wilczynski, *Adhes. Age 8:*52 (1990).
41. G. Bonneau and M. Baumassy, New tackifying dispersions for water based PSA for labels, 19th Münchener Klebstoff u. Veredelungsseminar, 1994, p. 82.
42. T. G. Wood, *Adhes. Age 7:*19 (1987).
43. P. Dunckley, *Adhäsion 11:*19 (1989).
44. Colon et al., U.S. Patent 4331576; in W. L. Bunelle, K. C. Knutson, and R. M. Hume (H. B. Fuller Co., St. Paul, MN), EP 0199468 A2/29.10.1986.
45. S. L. Scholl, K. C. Knutson, and W. L. Bunelle (H. B. Fuller Licensing and Financing Inc., St. Paul, MN), EP 0193427 A2/07.08.1986.
46. M. Vipin, Application of acrylic rubbers in PSA, in *TECH 12,* Advances in Pressure Sensitive Tape Technology, Tech. Proc. Ithasca, IL, May 1989, p. 191.
47. U.S. Patent 4052368, Morrison; in W. L. Bunelle, K. C. Knutson, and R. M. Hume (H. B. Fuller Co., St. Paul, MN), EP 0199468A2/29.10.1986.
48. Jpn. Patent A 8287481; in P. Gleichenhagen, E. Behrend, and P. Jauchen (Beiersdorf AG, Hamburg, Germany), EP 0149135 B1/24.07.1987.
49. Vitta Corporation, U.S. Patent 3598679; in *Coating 6:*338 (1969).

50. H. Mueller and J. Tuerk (BASF AG, Ludwigshafen, Germany), EP 0118726/ 19.09.1984.
51. R. G. Jahn, *Adhes. Age 12:*35 (1977).
52. R. J. Shuman and B. D. Josephs (Dennison Manuf. Co., Framingham, MA), PCT, WO 88/01636.
53. *Adhäsion 4:*172 (1985).
54. J. C. Pasquali, EP 0122847/24.10.1984.
55. G. Auchter, J. Barwich, G. Rehmer, and H. Jäger, *Adhes. Age 7:*20 (1994).
56. W. K. Darwell, P. R. Konsti, J. Klingen, and K. W. Kreckel (Minnesota Mining and Manuf. Co., St. Paul, MN), EP 257984/21.08.1986.
57. E. Kazuyoshi, N. Hiroaki, T. Katsuhisa, K. Yoshita, and T. Saito (FSK Inc.), Jpn. Patent 6317981/25.01.1988; in *CAS Adhes. 12:*5 (1988).
58. T. Kurono, N. Okashi, and N. Tanaka (Nitto Electric Ind. Co. Ltd.), Jpn Patent 6333487/13.02.1988; in *CAS Adhes. 12:*5 (1988).
59. T. Kinoshita (Sanyo Kokkusaku Pulp Co. and Saiden Ind., Tokyo), U.S. Patent 4645783/24.02.1987.
60. Sherwin Williams Company, Cleveland, OH, U.S. Patent 1519362; *Coating 1:*24 (1971).
61. Y. Kitazaki, T. Matsuda, and Y. Kobayashi (Nichiban Co.) Jpn Patent 62263273/ 16.11.1987; in *CAS Colloids 7:*7 (1988).
62. W. K. Darwell, P. R. Konsti, J. Klingel, and K. W. Kreckel (Minnesota Mining and Manuf. Co.), EP257984/02.03.1988; in *CAS Adhes. 20:*3 (1988).
63. Minnesota Mining and Manuf. Co., St. Paul, MN, Jpn. Patent 6317091/19.07.1988; in *CAS Adhes. 25:*7 (1988).
64. M. C. Bricker and S. T. Gentry, *Adhes. Age 7:*30 (1994).
65. Fr. Patent 2331607; in F. F. Lau and S. F. Silver (Minnesota Mining and Manuf. Co., St. Paul, MN), EP 0130087B1/02.0.1985.
66. V. L. Hughes and R. W. Looney (Exxon Research and Engineering Co., Florham Park, NJ), EP 0131460/16.01.1985.
67. C. Parodi, S. Giordano, A. Riva, and L. Vitalini, Styrene-butadiene block copolymers in hot melt adhesives for sanitary application, 19th Münchener Klebstoff und Veredelungsseminar, 1994, p. 119.
68. P. A. Mancinelli, New developments in acrylic hot melt pressure sensitive adhesive technology, *TECH 12,* Advances in Pressure Sensitive Tape Technology, Tech. Proc., Ithasca, IL, May 1989, p. 161.
69. E. L. Scheinbart and J. E. Callan, *Adhes. Age 3:*17 (1973).
70. J. N. Kellen and C. W. Taylor (Minnesota Mining and Manuf. Co., St. Paul, MN), EP 0246482 A2/25.11.1987.
71. Martens et al., U.S. Patent 4181752; in J. N. Kellen and C. W. Taylor (Minnesota Mining and Manuf. Co., St. Paul, MN), EP 0246 A2/25.11.1987.
72. C. Harder, Acrylic hotmelts—Recent chemical and technological developments for an ecologically beneficial production of adhesive tapes—State and prospects, European Tape and Label Conference, Brussels, April 28–30, 1993.
73. K. Wabro, R. Milker, and G. Krüger, *Haftklebstoffe und Haftklebebänder,* Astorplast GmbH, Alfdorf, Germany, 1994, p. 48.
74. C. I. Simionescu, N. Asandei, and I. Benedek, *Rev. Roum. Chim. 7:*1081 (1971).

75. C. I. Simionescu, N. Asandei, and I. Benedek, Rom. Patent 54213/26.05.1969.
76. M. Gerace, *Adhes. Age 8:*15 (1983).
77. F. C. Larimore and R. A. Sinclair (Minnesota Mining and Manuf. Co., St. Paul, MN), EP 0197662A1/15.10.1986.
78. H. F. Huber and H. Müller, Radcure '86, Conf. Proc. 10th, 12/1–12/12 (1986).
79. National Starch and Chemical Corp., Jpn. Patent 6306076/12.01.1988; in *CAS Adhes. 19:*2 (1988).
80. C. Cervelatti, G. Capaldi, and L. E. Jacob (Exxon Chemical Patents Inc.), EP 273585/06.07.1988.
81. *Coating 1:*12 (1988).
82. B. Hunt, *Labels Label. Int. 5/6:*34 (1997).
83. T. R. Mecker, Low molecular weight isoprene based polymers—Modifiers for hot melts, Tappi Hot Melt Symposium 1984; in *Coating 11:*310 (1984).
84. Esso Research and Eng. Co., U.S. Patent 3351572; in *Coating 7:*210 (1969).
85. D. A. Shields, Hot melt pressure sensitive adhesives. A case history, Tappi Hot Melt Adhesives and Coating Short Course, May 2–5, 1982, Hilton Head, SC.
86. *Coating 7:*187 (1984).
87. C. H. Hill, N. C. Memmo, W. L. Phalen, Jr., and R. R. Suchanec (Hercules Inc.), EP 259697/16.03.1988; in *CAS Adhes. 14:*5 (1988).
88. Klebstoffvorprodukte, *Prüfung von Haftklebstoffen,* D-EDE/K, Juni/Juli 81,2, BASF AG, Ludwigshafen, Germany.
89. *Adhes. Age 7:*36 (1986).
90. *Adhäsion 2:*123 (1974).
91. J. C. Fitsch and A. M. Snow, *Adhes. Age 10:*23 (1977).
92. *Coating 6:*198 (1996).
93. *Coating 7:*232 (1989).
94. F. T. Sanderson, *Adhes. Age 11:*26 (1983).
95. Engel, U.S. Patent 514950/18.07.1983; in F. C. Larimore and R. A. Sinclair (Minnesota Mining and Manuf. Co., St. Paul, MN), EP 0197662A1/15.10.1986.
96. G. W. H. Lehmann and H. A. J. Curts (Beiersdorf AG, Hamburg, Germany), U.S. Patent 4038454/26.07.1977.
97. E. E. Ewins, Jr. and J. R. Ericson, *Tappi J. 71*(6):155 (1988).
98. Minnesota Mining and Manuf Co., U.S. Patent 1594178; in *Coating 8:*240 (1972).
99. E. Djagarowa, W. Rainow, and W. Dimitrow, *Plaste Kautsch. 1:*28 (1970).
100. Adhesive Tapes Ltd., Br. Patent 861358; in *Coating 6:*185 (1969).
101. K. F. Schroeder, *Adhäsion 5:*161 (1971).
102. BASF, T1-2.2-21, November 1979, Teil 3, Blatt 5.
103. EP 0213860.
104. BASF, T1-2.2-21, November 1979, Teil 3, B.3, *Selbstklebende Dekorationsfolien.*
105. P. Ford, *Eur. Adhes. Sealants 6:*22 (1995).
106. Johnson and Johnson, New Brunswick NJ, U.S. Patent 3161554; in *Adhäsion 6:*277 (1966).
107. Z. Czech (Lohmann GmbH, Neuwied, Germany), DE 43 03 183 C1/04.02.1993.
108. *Coating 12:*344 (1984).
109. *Avery Rating Plates and Type Plates Last a Fairly Long Time,* Prospectus, Avery Etiketter-Logistik GmbH, Eching b. München, Germany.

110. *Adhes. Age* 7:36 (1983).
111. *Finat Label. News* 3:29 (1994).
112. L. M. Schrijver, *Coating* 3:70 (1991).
113. R. Milker and Z. Czech, 16th Münchener Klebstoff und Veredelungsseminar, 1991, p. 136.
114. *Coating* 1:29 (1978).
115. G. Fuchs, *Adhäsion* 3:24 (1982).
116. *Neue Verpack.* 1:156 (1991).
117. *Coating* 1:24 (1969).
118. *Coating* 8:248 (1969).
119. *Coating* 3:68 (1985).
120. A. Kenneth, J. R. Stickwell, and J. Walker (Allied Colloids), EP 0147/03.07.1985.
121. S. Benedek, *Adhäsion* 3:22 (1987).
122. S. Benedek, *Adhäsion* 4:25 (1987).
123. H. Monsey and A. Maletsky, U.S. Patent 4,331,576/25.05.1982; in *Adhes. Age* 12:53 (1983).
124. H. Colon and A. Maletsky, EP 0057421 A1/11.08.1982.
125. GAF, Frechen, Germany, *Preliminary Technical Information,* PTI No. 61/01.1989.
126. Morrison, U.S. Patent 4052368; in W. L. Bunelle, K. C. Knutson, and R. M. Hume (H. B. Fuller Co., St. Paul, MN), EP 0199468 A2/29.10.1986.
127. T. P. Flanagan (National Starch and Chemical Investment Holding Co., Wilmington, DE), EP 0212135B1/04.03.1987.
128. D. K. Ray-Chaudhuri, T. P. Flanagan, and J. E. Schoenberg, (National Starch and Chem. Co., Bridgewater, NJ), Ger. Patent 2507683/09.10.1975.
129. GAF, *Rahmenrezeptur,* GAF, Frechen, Germany, 1983.
130. Z. Czech, *Adhäsion* 11:26 (1994).
131. Z. Czech (Lohmann GmbH, Neuwid, Germany), DE 44 31 053/01.09.1994.
132. U.S. Patent 3661874; in F. D. Blake (Minnesota Mining and Manuf. Co., St. Paul, MN), EP 0141504 A1/15.05.1985.
133. D. J. Terriault and M. J. Zajaczkowski (Adhesives Research Inc., Glen Rock, PA), U.S. Patent 5352516/04.10.1994.
134. Eastman Kodak Co, Rochester, NY, U.S. Patent 3152940; in *Adhäsion* 2:47 (1966).
135. U.S. Patent 3096202; in P. Gleichenhagen and I. Wesselkamp (Beiersdorf AG, Hamburg, Germany), EP 0058382 B1/25.08.1982.
136. Br. Patent 941 276; in P. Gleichenhagen and I. Wesselkamp (Beiersdorf AG, Hamburg, Germany), EP 0058382 B1/25.08.1982.
137. U.S. Patent 3441430; in P. Gleichenhagen and I. Wesselkamp (Beiersdorf AG, Hamburg, Germany), EP 0058382 B1/25.08.1982.
138. U.S. Patent 3152940; in P. Gleichenhagen and I. Wesselkamp (Beiersdorf AG, Hamburg, Germany), EP 0058382 B1/25.08.1982.
139. K. Allan, J. R. Stockwell, and J. Walker (Allied Colloids Ltd., Bradford, UK), EP 0147067A1/03.07.1985.
140. A. Goel (Ashland Oil Inc., Ashland, KY), U.S. Patent 4626575/02.12.1986; in *Adhes. Age* 5:26 (1987).
141. U.S. Patent 3.865770; in F. C. Larimore and R. A. Sinclair (Minnesota Mining and Manuf. Co., St. Paul, MN), EP 0197662 A1/15.10.1986.

142. DDR Patent 64111; in *Coating 1:*24 (1969).
143. E. H. Andrews, T. A. Khan, and H. A. Majid, *J. Mater. Sci. 20:*3121 (1985).
144. *Coating 6:*184 (1969).
145. D. Allen, Jr. and E. Flam (Bard Inc., Murray Hill, NJ), U.S. Patent 4650817/ 17.03.1987.
146. J. R. Pennace, C. Ciuchta, D. Constantin, and T. Loftus, (Flexcon Co. Inc., Spencer, MA) WO 8703477 A/18.06.1987; in *Adhes. Age 5:*24 (1987).
147. U.S. Patent 475373; in S. E. Krampe and C. L. Moore (Minnesota Mining and Manuf. Co., St. Paul, MN), EP 0202831 A2/26.11.1986.
148. U.S. Patent 3532652; in S. E. Krampe and C. L. Moore (Minnesota Mining and Manuf. Co., St. Paul, MN), EP 0202831 A2/26.11.1986.
149. Pittsburgh Plate Glass Co., U.S. Patent 3355412; in *Coating 7:*274 (1969).
150. R. L. Sun and J. F. Kenney (Johnson and Johnson Products Inc.), U.S. Patent 4762688/09.08.1988; in *CAS Hot Melt Adhes.* 26:1 (1988).
151. D. Allen, Jr., E. Flam, and C. R. Bard, U.S. Patent 4650817/17.03.1987; in *Adhes. Age 5:*24 (1988).
152. Morton Thiokol Inc., Adhesives and Coatings, Morstik 103 Adhesive, MTD-MS103-12/1988.
153. *Coating 1:*23 (1969).
154. *Finat Label. News 3:*29 (1994).
155. A. Dobmann and A. G. Viehofer, 19th Münchener Klebstoff und Veredelungsseminar, 1994, p. 168.
156. S. Toshiki, T. Yasuo, I. Hidaharu, and M. Takumi (Japan Synthetic Rubber Co. Ltd.,), Jpn. Patent 6366254/24.03.1988.
157. *Coating 6:*175 (1972).
158. Bakelite Xylonite Ltd., Br. Patent 1081291; in *Coating 4:*114 (1969).
159. E. L. Scheinbart and J. E. Callan, *Adhes. Age 3:*17 (1973).
160. T. K. Bhaumik, A. K. Bhowmick, and B. R. Gupta, *Plast. Rubber Process. Appl.* 7:43 (1987).
161. R. J. Nichols and F. Kheradi, The interrelationship of machine design and processing of permanent hot melt adhesives, Tappi Hot Melt Adhesives and Coating Short Course, May 2–5, 1982, Hilton Head, SC.
162. Sealock, *L1233 Adhesive,* Product Data, Andover, Hampshire, UK.
163. F. Hufendiek, *Etiketten-Labels 1:*9 (1995).
164. *Adhäsion 3:*19 (1984).
165. M. Fairley, *Labels Label. 7/8:*34 (1995).
166. *Convert. Today 11:*13 (1990).
167. *Adhes. Age 1:*6 (1985).
168. *Adhäsion 11:*9 (1988).
169. Minnesota Mining and Manuf. Co., St. Paul, MN, U.S. Patent 3129816/1987.
170. Kimberley Clark, U.S. Patent 799429; in *Coating 1:*9 (1970).
171. G. Grzywinski and E. J. Foley (Scott Paper Co.), *Adhes. Age 12:*281 (1988).
172. K. Nakamura, Y. Miki, and Y. Nanzaki (Nitto Electric Industrial Co. Ltd.), Jpn. Patent 6386787/18.04.1988; in *CAS Adhes.* 19:5 (1988).
173. H. Suchy, J. Hezina, and J. Matejka, Czech CS 247802/15.04.1987; in *CAS Adhes.* 19:4 (1988).

174. K. Mitsui (Mitsui Petrochemical Ind. Ltd.), PCT, WO 8802767/21.04.1988; in *CAS Adhes. 22:*4 (1988).
175. I. Yorinobu, W. Yasuhhisa, and T. Hiroshi (Japan Synthetic Rubber Co.), Jpn. Patent 6386777/18.04.1988; in *CAS Adhes. 24:*3 (1988).
176. R. M. Enanoza (Minnesota Mining and Manuf. Co.), EP 259968/11.08.1986; in *CAS Adhes. 22:*3 (1988).
177. T. Sugiyama, N. Miyaji, I. Yoshihide, and T. Tange (Nippon Carbide Ind. Co. Inc.), Jpn. Patent 62199672/03.09.1987; in *CAS Adhes. 14:*3 (1988).
178. T. Hiroyoshi, H. Kuribayashi, and E. Usuda (Sumitomo Chem. Co. Ltd.), EP 254002/27.01.1988; in *CAS Adhes. 14:*3 (1988).
179. T. Moldvai and N. Piatkowski (Inst. Cerc. Pielarie si Incaltaminte), Rom. RO 93124/30.12.1987; in *CAS Adhes. 19:*3 (1988).
180. T. Kishi (Sekisui Chem. Co. Ltd.), Jpn. Patent 6369879/29.03.1988; in *CAS Adhes. 19:*6 (1988).
181. H. Yaguchi, H. Fukuda, L. Masayuki, and T. Ohashi (Bridgestone Corp.), Jpn. Patent 6327583/05.02.1988; in *CAS Adhes. 19:*6 (1988).
182. I. Murakami, Y. Hamada, and O. Takuman (Toray Silicon Co. Ltd.), EP 253601/20.01.1988; in *CAS Adhes. 14:*3 (1988).
183. *Coating 11:*341 (1972).
184. J. A. Fries, Hot melt pressure sensitive paper label application: An overview of the adhesives market, *Hot Melts—The Future is Now,* Tappi Symposium, June 2–4, 1980, Toronto, Canada.
185. U.S. Patent 4163077; in F. F. Lau and S. F. Silver (Minnesota Mining and Manuf. Co., St. Paul, MN), EP 0130087B1/02.01.1985.
186. L. A. Sobieski and T. J. Tagney, *Adhes. Age 12:*23 (1988).
187. A. B. Wechsung, *Coating 9:*268 (1972).
188. *Adhes. Age 1:*6 (1985).
189. Mobil Plastics, Virton, Belgium, *Mobil-Bicor 5:*4 (1984).
190. P. Penczek and B. Kujawa-Penczek, *Coating 6:*232 (1991).
191. A. J. Risser (Cities Cervice Co.), U.S. Patent 3759780/18.09.1973.
192. G. L. Burroway and G. W. Feeney, *Adhes. Age 7:*17 (1974).
193. A. N. Anisimov et al., UdSSR Patent 249525; in *Coating 1:*24 (1969).
194. BP Chemicals, *Hyvis/Napvis Polybutenes,* PB 301, September, 1983, London, UK.
195. *Coating 3:*12 (1985).
196. J. Verseau, *Coating 11:*309 (1971).
197. Beiersdorf AG, Hamburg, U.S. Patent 1569888; in *Coating 8:*20 (1972).
198. *Eur. Adhes. Sealants 6:*36 (1995).
199. E. Djagarowa, *Plaste Kautsch. 19:*748 (1969).
200. *Coating 6:*185 (1969).
201. *Adhes. Age 4:*6 (1983).
202. *Coating 6:*184 (1969).
203. G. Chand, Br. Patent 723226; in *Coating 6:*184 (1969).
204. G. Chand, Br. Patent 790087; in *Coating 6:*184 (1969).
205. Johnson & Johnson, U.S. Patent 2882179; in *Coating 6:*184 (1969).
206. Adhesive Tapes Ltd., Br. Patent 861358; in *Coating 6:*185 (1969).
207. Johnson and Johnson, Br. Patent 798471; in *Coating 6:*185 (1969).

208. B. B. Blackford, Br. Patent 8864365; in *Coating* 6:185 (1969).
209. EPA 0100146.
210. G. F. Vesley, A. H. Paulson, and E. C. Barber, EP 0202 938 A2/26.11.1986.
211. Johns Manville Corp., U.S. Patent 3356635; in *Coating* 7:210 (1969).
212. A. F. Carr, *Coating* 11:334 (1973).
213. J. Grabemann and R. Hauber (Hans Neschen GmbH & Co KG, Bückeburg, Germany), DE 42 31 607/01.09.1994.
214. U.S. Patent 2.925174; in EP 0120708, p. 2.
215. W. E. Lenney (Air Products and Chemical Inc., USA), Can. Patent 1225176/04.08.1987.
216. U.S. Patent 3257478; in W. E. Lenney (Air Products and Chemical Inc., USA), Can. Patent 1225176/04.08.1987.
217. U.S. Patent 3697618; in W. E. Lenney (Air Products and Chemical Inc., USA), Can. Patent 1225176/04.08.1987.
218. U.S. Patent 3998997; in W. E. Lenney (Air Products and Chemical Inc., USA), Can. Patent 1225176/04.08.1987.
219. F. Korpmann and G. Perry, U.S. Patent 4325770/20.04.1982.
220. Y. Kitazaki, T. Matsuda, and Y. Kobayashi (Nichiban Co., Ltd.), Jpn. Patent 62263273/16.11.1987; in *CAS Colloids* 7:7 (1988).
221. I. Benedek, *Eur. Adhes. Sealants* 2:25 (1996).
222. H. Kenjiro and S. Kotaro (Nitto Electric Ind. Co. Ltd.), Jpn. Patent 63118383/23.05.1988; in *CAS Adhes.* 24:5 (1988).
223. H. Satoh and M. Makino (Nippon Oil Co. Ltd.), Ger. Offen. DE 3740222/01.06.1988; in *CAS Adhes.* 22:4 (1988).
224. I. Yorinobu, W. Yasuhisa, and T. Hiroshi (Japan Synthetic Rubber Co.), Jpn. Patent 6386777/18.04.1988; in *CAS Adhes.* 24:3 (1988).
225. K. Akasaka, S. Sanuki, and T. Matsuyama (Nippon Synthetic Chem. Ind. Co. Ltd.), Jpn. Patent 62243669/17.04.1986; in *CAS Adhes.* 11:5 (1988).
226. H. Nakagawa, A. Baba, S. Furukawa, and S. Fukuchi (Nippon Shokubai Kagaku Kogyo Co. Ltd.), Jpn. Patent 62292877/19.1.1987; in *CAS Adhes.* 11:6 (1988).
227. V. Stanislawczyk (BF Goodrich), EP 264903/27.04.1988; in *CAS Colloids (Macromol. Aspects)* 17:7 (1988).
228. H. Miyasaka, Y. Kitazaki, T. Matsusa, and J. Kobayashi (Nichiban Co. Ltd.), Jpn. Patent 62263273/16.11.1987; in *CAS Colloids (Macromol. Aspects)* 7:7 (1988).
229. T. Kishi (Sekisui Chem. Co. Ltd.), Jpn. Patent 6369879/29.03.2988; in *CAS Adhes.* 19:6 (1988).
230. K. Maeda and Y. Kitazaki (Nichiban Co. Ltd.), Jpn. Patent 6348380/01.03.1988; in *CAS Adhes.* 19:3 (1988).
231. T. Tatsuno, K. Matsui, M. Takahashi, and M. Wakimoto (Kansai Paint Co. Ltd.), Jpn. Patent 6228597/11.12.1987; in *CAS Adhes.* 14:3 (1988).
232. H. Reipp, *Adhes. Age* 3:17 (1972); in *Coating* 6:187 (1972).
233. *Adhäsion* 1/2:44 (1987).
234. N. Wasfi Malek (Beiersdorf AG, Hamburg, Germany), EP 0095093/30.11.1983.
235. TNII Bumagi, SSSR Patent 349782; in *Coating* 5:122 (1974).
236. EP 0251672.
237. Kjin Co., Jpn. Patent 28520; in *Coating* 2:71 (1972).

238. Mystik Adhesives Prod. Inc., U.S. Patent 2878142/1976.
239. *Coating* 6:185 (1969).
240. P. R. Mudge (National Starch and Chem. Co., Bridgewater, NJ), EP 022541 A2/ 16.06.1987.
241. H. H. Hub, Composition, characteristics and application of polar ethylene copolymers, Polyethylene 93, Oct. 4, 1993, Session III, Maack Business Service, Zurich, Switzerland.
242. *Coating* 12:450 (1986).
243. EP 0120708.
244. Dow, *Primacor Polymers,* Specification CH 272-047-E-991, Dow Chemicals, Horgen, Switzerland.
245. A. Ridgeway and L. K. Mengenhagen, EAA/polybutylene blend for packaging applications, Tappi Proceedings, Polymers, Laminations & Coatings Conference, 1992, p. 52.
246. *Tappi J.* 5:103 (1988).
247. R. N. Henkel, *Paper Film Foil Convert.* 12:68 (1968).
248. I. Benedek, Ebert Folien AG, Wiesbaden, Germany, Deutsches Gebrauchsmuster, G 9113755.1/05.11.1991.
249. EP 4673611.
250. *Verpack. Rundsch.* 5:556 (1988).
251. D. H. Teesdale, *Coating* 6:246 (1983).
252. Rohm & Haas Co., Philadelphia, PA, U.S. Patent 3152921; in *Adhäsion* 2:80 (1966).
253. Vorwerk & Sohn, DBP 1078285; in *Coating* 6:184 (1969).
254. Johnson & Johnson, U.S. Patent 2973859; in *Coating* 6:184 (1969).
255. *April* 48:1360 (1985).
256. M. J. Owen, *J. Coat. Technol.* 53(679):49 (1981).
257. H. Toepsch, *Wochenbl. Papierfabrik.* 11/12:320 (1971).
258. *Coating* 1:12 (1984).
259. J. Pennace and S. Borasso (Flexcon Co., Inc.), PCT Inc., Int. Appl., WO 87 03537/ 18.06.87; in *CAS Silicones* 2:3 (1988).
260. J. Pennace and G. E. Kersey (Flexcon Co. Inc., Spencer, MA), PCT, WO 87/035537/18.06.1987.
261. P. L. Brown and D. L. Stickles (Dow Corning Co.), EP 251483/07.01.1988; in *CAS Siloxanes Silicones* 13:5 (1988).
262. G. A. H. Kupfer, Siliconizing with radiation curing systems, Tappi Hot Melt Adhesives and Coatings, Short Course, Hilton Head, SC, June 1982.
263. H. J. Timpe, R. Wagner, and U. Müller, *Adhäsion* 2:28 (1985).
264. V. F. Andreev et al., SSSR Patent 263409; in *Coating* 5:130 (1971).
265. T. Shibano, I. Kimura, H. Nomoto, and C. Maruchi (Sanyo Kokusaku Pulp Co., Ltd., Tokyo, Japan), U.S. Patent 4624893/25.11,86; in *Adhes. Age* 5:24 (1987).
266. G. Galli (Manuli Autoadesivi Spa., Cologno Monzese, Italy), EP 0191191A1/ 20.08.1986.
267. Ital. Patent 21842 A/82; in G. Galli (Manuli Autoadesivi Spa., Cologno Monzese, Italy), EP 0191191A1/20.08.1986.

268. *Adhes. Age 8:*42 (1987).
269. Norton Co., U.S. Patent 1123014.
270. R. A. Bafford and G. E. Faircloth, *Adhes. Age 12:*353 (1987).
271. Kalle Folien, *Release Coat K,* Data Sheet, Hoechst AG, Wiesbaden, Germany, 1992.
272. N. Yataba and Y. Miki (Nitto Electric Co. Ltd.), Jpn. Patent 6333485/13.02.1988; in *CAS Hot Melt Adhes. 13:*1 (1988).
273. Acheson Ind. Inc., Port Huron, MI, DBP 1278652; in *Coating 7:*274 (1969).
274. H. Kenjiro, O. Kazuto, A. Nagai, and K. Kiyohiro (Nitto Electric Ind. Co. Ltd.,), Jpn. Patent 63137841/09.06.1988; in *CAS Adhes. 24:*5 (1988).
275. T. Kikuta and H. Hori (Nippon Shokubai Kagaku, Kogyo Co. Ltd.), Jpn. Patent 6351478/04.03.1988; in *CAS Coatings Inks Related Products 17:*11 (1988).
276. T. Kikuta and H. Hori (Nippon Shokubai Kagaku, Kogyo Co. Ltd.), Jpn. Patent 6333482/13.02.1988; in *CAS Coating Inks Related Products 17:*11 (1988).
277. T. Kikuta and H. Hori (Nippon Shokubai Kagaku, Kogyo Co. Ltd.), Jpn. Patent 6322812/13.02.1988; in *CAS Emulsion Polym. 18:*2 (1988).
278. K. Nitzl and L. Birk, *Coating 12:*344 (1984).
279. *Coating 12:*345 (1984).
280. A. Fau (Rhone Poulenc Chimie, Courbevoie, France), EP 0169098 B1/22.01.1986.
281. K. Brack (Design Coat Co.), U.S. Patent 4288479/08.09.1981; in *Adhes. Age 12:*58 (1981).
282. G. Camerini (Coverplast Italiana SpA), EP 248771/20.05.1986.
283. V. D. Babayants, V. V. Kolesnitschenko, and S. G. Sannikov, *Lakokras. Mater. Ikh. Primen. 3:*45 (1988); in *CAS Coatings, Inks Related Products 17:*9 (1988).
284. L. Bothorel, *Emballages 278:*372 (1972).
285. M. Omote, I. Sakai, and T. Matsumoto (Nitto Electric. Ind. Co.), Jpn. Patent 6386788/18.04.1988; in *CAS 19:*4 (1988).
286. Br. Patent 1123014; in *Coating 7:*274 (1969).
287. *Coating 6:*184 (1969).
288. TNII Bumagi, SSSR Patent 300561; in *Coating 7:*368 (1969).
289. Y. Naoto (Toyo Ink Mfg. Co. Ltd.), Jpn. Patent 62218467/25.09.87; in *CAS Siloxanes Silicones 2:*4 (1988).
290. S. Ohara and R. Kitamura (Goyo Paper Working Co. Ltd.), EP 254050/27.01.1988; in *CAS Adhes. 13:*11 (1988).
291. H. Komatsu, Jpn. Patent 62246973/28.10.1987; in *CAS Colloids (Macromol. Aspects) 4:*10 (1988).
292. *Adhes. Age 1:*6 (1985).
293. T. Moldvai and N. Piatkowski (Inst. Cerc. Pielarie, Incaltaminte, Bucharest), Rom. Patent 93124/30.12.1987; in *CAS Adhes. 19:*3 (1988).
294. *Etiketten-Labels 1:*49 (1996).
295. *Papier Kunstst. Verarb. 9:*57 (1988).
296. S. Heimlich, *Papier Kunstst. Verarb. 11:*26 (1973).
297. *APR 5:*100 (1988).
298. O. Huber, *Wochenbl. Papierfabr. 17:*657 (1973).
299. R. Stockmeyer, *Deut. Drucker 13:*42 (1988).

300. A. T. Franklin, *Printing Trades J. 1044:*46 (1974).
301. T. Arai, T. Yamasaki, K. Suzuki, T. Ogura, and Y. Salai, *Tappi J. 5:*47 (1988).
302. V. B. Topliss, *Finat News 2:*45 (1989).
303. J. E. Wilson, R. A. Ravier, D. A. Ewaniuk, and R. S. McDaniel, PTSC XVII Technical Seminar, May 4, Woodfield Shaumburg, IL.
304. E. Chicherio, *Coating 10:*285 (1982).
305. *Etiketten-Labels 5:*91 (1995).
306. *Etiketten-Labels 5:*22 (1995).
307. E. B. Strahler, *Coating 5:*163 (1996).
308. M. Bateson, *Finat News 3:*29 (1989).
309. D. Lacave, *Labels Label. 3/4:*54 (1994).
310. R. H. Feldkamp, *APR 14:*367 (1986).
311. *Papier Kunststoffverarb. 9:*57 (1988)
312. *Coating 5:*212 (1990).
313. C. Bayer, *Adhäsion 9:*349 (1965).
314. *Etiketten-Labels 3:*9 (1995).
315. *Etiketten-Labels 5:*24 (1995).
316. *Etiketten-Labels 3:*32 (1995).
317. *Druck Print 10:*32 (1987).
318. *Etiketten-Labels 3:*10 (1995).
319. J. Paris, *Papier Kunststoffverarb. 9:*57 (1988).
320. *Paper, Film Foil Converter 8:*28 (1973).
321. M. Fairley, *Labels Label. Int. 5/6:*28 (1997).
322. Mactac, *Spitzentechnologie für Selbstklebende Materialien,* MACtac Europe S.A., Soignies, Belgium.
323. Domtar Ltd., Can. Patent 857847; in *Coating 2:*38 (1972).
324. H. Senn, *APR 16:*170 (1986).
325. L. C. Fehrmann, *Adhes. Age 6:*48 (1970).
326. *Labels Label. 3/4:*44 (1994).
327. J. Young, *Tappi J. 5:*78 (1988).
328. *Coating 8:*242 (1972).
329. *APR 18:*405 (1987).
330. B. H. Gregory, Extrusion coating advances—Resins, processing, applications, markets, Polyethylene '93, The Global Challenge for Polyethylene in Film Lamination, Extrusion, Coating Markets, Oct. 4, 1993, Maack Business Services, Zurich, Switzerland.
331. *Raflatac Synthetic Materials Available as Rollstock,* Booklet, Raflatac OY, Tampere, Finnland 1983.
332. *Paper, Film Foil Converter 9:*32 (1989).
333. E. Park, *Paper Technol. 8:*15 (1989).
334. D. M. Stott, *Surf. Coatings 11:*296 (1969).
335. Johnson and Johnson, U.S. Patent 3403018; in *Coating 1:*24 (1969).
336. Ludlow Corporation, U.S. Patent 1157154; in *Coating 12:*353 (1970).
337. K. Hamada, U. Uchiyama, and S. Takemura, Jpn. Patent 13765/69; in *Coating 12:*353 (1970).

338. I. P. Rothernberg (Stik-Trim Industries Inc., New York), U.S. Patent 4650704/ 17.03.1987.
339. B. Reinhardt, M. Hottenträger, B. Gather, B. Hartmann, and M. D. Lecher, 19th Münchener Klebstoff und Veredelungsseminar, 1994, p. 48.
340. C. M. Brooke, *Finat News* 3:34 (1987).
341. Arhoco Inc., U.S. Patent 350 9991; in *Coating* 5:130 (1971).
342. *Adhes. Age* 1:6 (1985).
343. *Papier Kunstst. Verarb.* 1:20 (1996).
344. J. R. Allen, New silicone coating base papers, Tappi Hot Melt Adhesives and Coating Short Course, May 2–5, 1982, Hilton Head, SC.
345. J. R. DeFife, *Labels Label.* 3/4:14 (1994).
346. *Packag. Today* 12:9 (1995).
347. *Papier Kunstst. Verarb.* 9:57 (1988).
348. *Adhes. Age* 10:125 (1986).
349. S. Moser, *APR* 16:452 (1986).
350. L. Placzek, *Coating* 1:2 (1986).
351. I. Benedek and L. J. Heymans, *Pressure Sensitive Adhesives Technology,* Marcel Dekker, New York, 1997, Chapter 7.
352. R. Hinterwaldner, *Adhäsion* 6:11 (1984).
353. *Finat News* 4:10 (1996).
354. *Label Buyer Int.* Spring 1997, p. 26.
355. *Neue Verpack.* 1:156 (1991).
356. *Etiketten-Labels* 5:136 (1995).
357. J. Michel, *Verpack. Rundsch.* 12:1425 (1985).
358. *Convert. Today* 11:8 (1990).
359. *Labels Label.* 3/4:82 (1994).
360. R. Kasoff, *Paper, Film Foil Converter* 9:85 (1989).
361. *Coating* 3:65 (1974).
362. M. H. Mishne, *Screen Printing* 5:60 (1986).
363. *Etiketten-Labels* 3:9 (1995).
364. W. J. Busby, Processing problems with PE films, Polyethylene 93, Maack Business Service, Zurich, Switzerland, Session VI, p. 3.
365. *APR* 40:1102 (1988).
366. *Druckwelt* 14/15:27 (1988).
367. T. McCauley, *Tappi J.* 6:159 (1985).
368. H. Roder, *Adhäsion* 6:16 (1983).
369. *Handling* 5/6:12 (1994).
370. Dowlex, *Polyethylene Resin,* Data Sheet 2740E.
371. K. Suenaga (Nitto Electric Industrial Co.), Jpn. Patent 6386784/18.04.1988; in *CAS* 21:5 (1988).
372. Lonza, *Cast Plastic Films for Technical Applications,* Lonza Werke GmbH, Weil am Rhein, Germany, 1993.
373. P. Keiston, *Müanyag Fóliák,* Müszaki Könyvkiadó, Budapest, 1976, p. 45.
374. Mitsubishi Rayon Co., Fr. Patent 2020683; in *Coating* 8:240 (1972).
375. *Kunststoffe* 75(10):xxiv (1985).

376. F. Altendorfer and A. Wolfsberger, *Kunststoffe* 6:691 (1990).
377. F. Hensen, *Papier Kuntsst. Verarb. 11:*32 (1988).
378. BASF Kunststoffe, *Lupolen* B 581, d/12.92, BASF AG, Ludwigshafen, Germany, p. 38.
379. M. A. Barbero and A. Amico, A new performance ULDPE/VLDPE from high pressure technology—Potential applications, Polyethylene 93, Oct. 4, 1993, Maack Business Service, Zurich, Switzerland, Session III, p. 6.
380. A. Stroeks, New developments in polyolefins (LDPE, LLDPE, VLDPE) for flexible packaging), Specialty Plastics Conference '87, Polyethylene and Copolymer Resin and Packaging Market, Dec. 1, 1987, Maack Business Service, Zurich, Switzerland.
381. *Convert. Today 1:*16 (1992).
382. W. W. Bode, *Tappi J.* 6:133 (1988).
383. *Neue Verpack.* 5:56 (1991).
384. *Trouble Shooting Guide for Processing Films,* Conversion Industry Reference Report MBS No. 903, Maack Business Service, Zurich, Switzerland, 1994.
385. Dow, *LDPE,* Data Sheet 2583-F-586.
386. *Neue Verpack.* 8:66 (1993).
387. W. Michael and R. Harms, *Papier Kunstst. Verarb. 10:*7 (1990).
388. D. Djordjevic, Tailoring films by the coextrusion casting and coating process, Specialty Plastics Conference '87, Polyethylene and Copolymer Resin and Packaging Markets, Dec. 1, 1987, Maack Business Service, Zurich, Switzerland.
389. H. Gross, *Kunststoffe 11:*1548 (1994).
390. B. Kunze, S. Sommer, and G. Düsdorf, *Kunststoffe* 84(10):1337 (1994).
391. A. La Mantia and T. Manlio, *Acta Polym. 11/12:*696 (1986).
392. N. Verse, *Papier Kunstst. Verarb. 10:*23 (1990).
393. E. B. Parker, Polyethylene polar copolymers and their application, Specialty Plastics Conference '87, Dec. 1, 1987, Maack Business Service, Zürich, Switzerland, p. 259.
394. *Finat News 4:*27 (1995).
395. W. Feistkorn, *Coating 9:*310 (1995).
396. L. Kovács, Müanyke Zsebkönyv, Mūszaki Könyvkiadó, Budapest, 1979, Ch. 7.2.4.4.1.
397. *Coating 2:*35 (1984).
398. K. Fust, *Coating 2:*66 (1988).
399. *Finat Label. News 3:*12 (1996).
400. W. A. Fraser, Novel processing aid technology for extrusion grade polyolefins, Specialty Plastics Conference '87, Polyethylene and Copolymer Resin and Packaging Markets, Dec. 1, 1987, Maack Business Service, Zurich, Switzerland.
401. *HamLet,* Polyethylene 93, Oct. 4, 1993, Maack Business Service, Zurich, Switzerland, Session III, p. 56.
402. *Kaut. Gummi Kunstst.* 6:564 (1985).
403. *APR 42:*1499 (1986).
404. *Polyethylene Film Market Applications,* Western Europe Conversion Industry Reference Report, MBS 904, Maack Business Service, Zurich, Switzerland, 1994.
405. E. Beier, *Technica 21:*35 (1995).

406. *Klebeband Forum No. 27,* Hoechst Films, Hoechst AG, October 1989.
407. *Papier Kunstst. Verarb. 2:*18 (1996).
408. *Coating 3:*64 (1974).
409. G. L. Duncan (Mobil Oil Co.), U.S. Patent 4720416/19.01.1988.
410. K. Taubert, *Adhäsion 10:*379 (1970).
411. G. Meinel, *Papier Kunstst. Verarb. 19:*26 (1985).
412. P. Dippel, *Adhäsion 4:*44 (1988).
413. *Kaut. Gummi Kunstst. 39*(9):778 (1986).
414. *Coating 11:*272 (1985).
415. *VR Interpack 96 Special,* E25.
416. *Etiketten-Labels 35:*21 (1995).
417. Dow, *Printing Trycite,* Dow Chemical USA, Designed Products Department, Midland, MI.
418. U. Reichert, *Kunststoffe 80*(10):1092 (1990).
419. G. M. Miles, *Papier Kunstst. Verarb. 2:*64 (1988).
420. Labels Label. 7/8:70 (1988).
421. C. D. Weiske, *Kunststoffe 8:*518 (1971).
422. B. Wright, *Adhes. Age 12:*25 (1971).
423. *Convert. Today 11:*9 (1991).
424. *Packag. Today 12:*31 (1995).
425. *Adhäsion 8:*296 (1973).
426. *Adhäsion 1:*15 (1974).
427. F. Haag and E. Rohde, *Kaut. Gummi Kunstst. 39*(12):1216 (1986).
428. *Coating 6:*154 (1974).
429. S. Sacharow, *Adhäsion 6:*268 (1966).
430. J. S. Cheng Shiang (Dow Chemical Co.), U.S. Patent 4,738,810,/19.04.88; in *CAS Adhes. 17:*6 (1988).
431. Gilman Brothers Co., *Plastics Technol. 7:*72 (1968).
432. J. S. Razzano and R. B. Bush (General Electric Co.), U.S. Patent 4728567/ 01.03.1988; in *CAS Siloxanes Silicones 17:*4 (1988).
433. R. G. Nelb and K. G. Saunders (Dow Chem. Co.), U.S. Patent 4738990/ 19.04.1988; in *CAS Siloxanes Silicones 17:*5 (1988).
434. L. Waeyenbergh, 19th Münchener Klebstoff und Veredelungsseminar, 1994, p. 139.
435. *Packlabel News,* Packlabel Europe 97, Ausgabe 2.
436. *Packag. Today 12:*28 (1995).
437. R. Hummel, Basismaterial für Haftetiketten, 19th Münchener Klebstoff und Veredelungsseminar, 1994, p. 58.
438. *Coating 4:*139 (1997).
439. *Druck Print 9:*65 (1986).
440. M. Fairley, *Labels Label. Int. 5/6:*76 (1997).
441. P. Hammerschmidt, *APR 7:*190 (1986).
442. R. Nurse, HDPE applications, PE developments, Polyethylene '93, Oct. 4, 1993, Maack Business Service, Zurich, Switzerland, Session 3, p. 1.
443. Bimodale HDPE, Polyethylene 93, Oct. 4, 1993, Maack Business Service, Zurich, Switzerland, Session III, p. 77.
444. G. Bolder and M. Meier, *Kaut. Gummi Kuntsst. 8:*715 (1986).

445. *Packag. Today* 12:28 (1995).
446. *Dowlex for Oriented Films,* CH-254-032-E-787, Data Sheet, Dow Chemicals, Horgen, Switzerland.
447. *Dowlex for Lamination Films,* CH-254-052-E-288, Data Sheet, Dow Chemicals, Horgen, Switzerland.
448. *Etiketten-Labels* 5:30 (1995).
449. Van Leer, *Valeron Film,* Van Leer Flexible Packaging, Van Leer Flexibles, Essen, Belgium, 1985.
450. *Adhäsion* 1:14 (1974).
451. *Etiketten-Labels* 5:6 (1995).
452. *Convert. Today* 11:23 (1990).
453. *Mobil-OPP Art* 3:1 (1993).
454. *Etiketten-Labels* 5:16 (1995).
455. *Kunststoffe* 83(10):737 (1993).
456. E. Pilipponen, *APR* 40:1100 (1988).
457. R. Hummel, *Adhäsion* 1/2:49 (1972).
458. *Adhäsion* 9:19 (1984).
459. B. Martens, *Coating* 6:187 (1992).
460. Y. Yuzo, K. Kyotaka, N. Kanji, and K. Hironori (Nippon Foil Manuf. Co. Ltd.), Jpn. Patent 62180780/08.08.1987; in *CAS Colloids* 3:13 (1988).
461. *Sengewald Rep.* 1:2 (1984), (Halle, Germany).
462. *Coating* 5:173 (1996).
463. *Kunststoffe* 83(10):725 (1993).
464. *Verpack. Rundsch.* 9:994 (1983).
465. *Paper, Films Foil Converter* 9:28 (1989).
466. *Finat Label. News* 3:29 (1994).
467. *Coating* 1:22 (1985).
468. D. Boettger, *Labels Label.* 3/4:58 (1994).
469. U.S. Patent 3321451; in S. E. Krampe and C. L. Moore (Minnesota Mining and Manuf. Co., St. Paul, MN), EP 0202831 A2/26.11.1986.
470. J. Verseau, *Coating* 7:181 (1971).
471. W. Witke, *Coating* 12:321 (1985).
472. W. Witke, *Coating* 8:278 (1988).
473. W. Witke, *Coating* 9:340 (1988).
474. *Coating* 6:187 (1972).
475. Mystik Tape Inc., IL, U.S. Patent 3161533; in *Adhäsion* 6:277 (1966).
476. H. Becker, *Adhäsion* 3:79 (1971).
477. *Labels Label. Int.* (5/6):18 (1997).
478. T. Nakajima, K. Oda, K. Azuma, and K. Fujita (Nitto Electric Ind. Co.), Jpn. Patent 6327579/05.02.1988; in *CAS Siloxanes Silicones* 17:5 (1988).
479. *Adhes. Age* 9:82 (1984).
480. DBP 1079252; in *Coating* 6:185 (1969).
481. U. Füssel, *Kunststoffe* 81(10):915 (1991).
482. Scholl Manuf. Co., Br. Patent 925810; in *Coating* 6:184 (1969).
483. Bunyan, Br. Patent 815121; in *Coating* 6:184 (1969).

484. *KI 1083*:4 (1991).
485. R. Bulet and J. Michel (Stamicarbon B. V.), Neth. Patent 86 01984/01.03.1988; in *CAS Adhes. 20*:3 (1988).
486. Minnesota Mining and Manuf. Co., St. Paul, MN, EP 0306232B1/07.04.1993.
487. German Offenlegungsschrift, 3144911; in EP 4673611.
488. S. Aoyanagi, H. Suzuki, and S. Takeda (Kanzaki Paper Manuf. Co. Ltd.), Jpn. Patent 6315872/22.01.1988; in *CAS Adhes. 14*:4 (1988).
489. German Offenlegungsschrift 3.216.603; in EP 4.673.611.
490. Vorwerk & Sohn GmbH & Co. KG, Wuppertal, DE 4040 917/02.07.1992.
491. G. Crass and A. Bursch (Hoechst AG, Frankfurt am Main, Germany), U.S. Patent 4673611/16.07.1987.
492. R. B. Lipson (Kwik Paint Products), U.S. Patent 5468533; in *Adhes. Age 5*:12 (1996).
493. M. J. Huber (Quality Manuf. Inc.), U.S. Patent 546692/07.11.1995; in *Adhes. Age 5*:12 (1996).
494. T. Horie, T. Kino, and T. Nagaresugi (Japan Styrene Paper Co.), Jpn. Patent 6823834/11.12.1988; in *CAS Siloxanes Silicones 11*:3 (1988).
495. K. Shiraishi, Jpn. Patent 63117068/21.05.1988; in *CAS Colloids (Macromol. Aspects) 22*:4 (1988).
496. Hoechst AG, *Klebeband Träger u. Abdeckfolie,* Datenblatt, Ausgabe 07/92, Hoechst AG, Wiesbaden, Germany.
497. *APR 18*:572 (1987).
498. Allmänna Svenska Elektriska AB, DBP 1276771; in *Coating 6*:184 (1969).
499. Kalle Folien, *Siegel Band,* Hoechst., Mi 1984, 38T 5.84 LVI, Wiesbaden, Germany.
500. Pritchett and Gold, Br. Patent 821959; in *Coating 6*:184 (1969).
501. Blackford-Gross, Br. Patent 829715; in *Coating 6*:184 (1969).
502. Scholl Manuf. Co., Br. Patent 871504; in *Coating 6*:184 (1969).
503. Scholl Manuf. Co., U.S. Patent 2953130; in *Coating 6*:184 (1969).
504. American White Cross Laboratories Inc., Can. Patent 647454; in *Coating 6*:184 (1969).
505. Johnson & Johnson, U.S. Patent 2882179; in *Coating 6*:184 (1969).
506. *Adhes. Age 10*:125 (1986).
507. P. Hammerschmidt, *Coating 4*:194 (1986).
508. D. Bedoni and G. Caprioglio, Modern equipment for label and tape converting, 19th Münchener Klebstoff und Veredelungsseminar, 1994, p. 37.
509. *Adhäsion 11*:37 (1994).
510. *Adhäsion 1/2*:27 (1987).
511. R. Shibata and H. Miyagawa (Hitachi Condenser Co. Ltd), Jpn. Patent 63 86785/18.04.1988; in *CAS Adhes. 20*:4 (1988).
512. J. E. Riedel and P. G. Cheney (Minnesota Mining and Manuf. Co., St. Paul, MN), U.S. Patent 4292360/29.09.1981; in *Adhes. Age 12*:58 (1981).
513. R. Milker, *Coating 3*:60 (1984).
514. BFGoodrich, U.S. Patent 3788878; in *Coating 10*:47 (1974).
515. Scholl Manuf. Co., U.S. Patent 3039459; in *Coating 6*:184 (1969).
516. Johnson and Johnson, U.S. Patent 3077882; in *Coating 6*:184 (1969).

517. Lohmann KG, U.S. Patent 3086531; in *Coating* 6:185 (1969).
518. *Adhes. Age* 12:67 (1984).
519. Westinghouse Electric Co., U.S. Patent 3772064; in *Coating* 19:47 (1974).
520. *Adhes. Age* 3:8 (1987).
521. Monsanto Co., U.S. Patent 3752733; in *Coating* 19:47 (1974).
522. Kendall Co., U.S. Patent 3723236; in *Coating* 6:154 (1974).
523. *Adhäsion* 9:254 (1976).
524. J. R. Wagner, *Tappi J.* 4:115 (1988).
525. J. M. Oelkers and E. J. Sweeney, *Tappi J.* 8:69 (1988).
526. L. E. Grunewald and D. J. Classen (Minnesota Mining and Manuf. Co., St. Paul, MN), EP 256662/24.02.1986; in *CAS Adhes.* 15:3 (1988).
527. K. Sakai and K. Inoue (Marubeni Co.), Jpn. Patent 6348379/01.03.1988; in *CAS Adhes.* 15:3 (1988).
528. *Scotchmount, Doppelseitige Klebebänder mit Schaumstoff Träger*, Brochure, 3M Deutschland GmbH, Neuss, Germany.
529. *Adhes. Age* 1:6 (1985).
530. A. D. Little, Inc., U.S. Patent 3039893; in *Coating* 6:185 (1969).
531. H. Meichner, Rehau, Germany, Gebrauchsmuster, GM 8607368/18.03.1986.
532. S. B. Driscoll, L. N. Venkateshwaran, C. J. Rosis, and L. C. Whitney, *Soc. Plast. Eng. Annu. Tech. Conf., Tech. Pap.* 1:450 (1985); in *Kaut. Gummi Kunstst.* 39(2):161 (1986).
533. A. S. Wood, *Mod. Plast.* 14(3):36 (1984).
534. Y. Tanaka, K. Mai, H. Takegawa, S. Watanabe, and A. Midorikawa (Dainippon Ink and Chemical Inc.), Jpn. Patent 62177043/03.08.1987; in *CAS Emulsion Polym.* 10:2 (1988).
535. G. D. Bennett (Simmonds Precision, New York), U.S. Patent 925735/20.01.1987; in *Adhes. Age* 5:28 (1988).
536. W. Geier, *APR* 20:618 (1987).
537. W. R. Grace and Co., Fr. Patent 1391908; in *Coating* 11:336 (1972).
538. *APR* 41:138 (1986).
539. *Verpack. Rundsch.* 7:33 (1986).
540. G. Menges, W. Michaeli, R. Ludwig, and K. Scholl, *Kunststoffe* 80(11):1245 (1990).
541. R. L. Francisco (Texo Co.), U.S. Patent 4732695/22.03.1988; in *CAS Coatings, Ink Related Products* 13:9 (1988).
542. *Adhäsion* 2:45 (1974).
543. U. Zorll, *Adhäsion* 90:236 (1975).
544. Takdust Products Co., *Rubber Plast. Age (Lond.)* 6:559 (1968).
545. *Coating* 8:244 (1969).
546. H. Tsukamoto, *Jpn. Plast. Age* 6:56 (1969); in *Coating* 1:18 (1971).
547. M. Michel, *Adhäsion* 4:154 (1966).
548. *Adhäsion* 3:83 (1974).
549. Chemplex Co., Rolling Meadows, IL, DE-PS 33 13 607/14.04.1983; in *Coating* 5:191 (1988).
550. A. Haas (Société Chimique des Charbonnage-CdF Chimie, France), U.S. Patent 4624991/25.11.1986; in *Adhes. Age* 5:26 (1987).

Manufacture of PSPs 515

551. Les Colloides Industriels Francais, DBP 1289219; in *Coating 9:*272 (1969).
552. Nitto Electrical Ind. Co. Ltd, Jpn. Patent 24229/70; in *Coating 2:*38 (1972).
553. R. J. Litz, *Adhes. Age 8:*38 (1973).
554. *Labels Label. 3/4:*64 (1994).
555. *Adhäsion 9:*19 (1984).
556. *Etiketten-Labels 5:*30 (1995).
557. *Etiketten-Labels 5:*25 (1995).
558. *Etiketten-Labels 3:*53 (1995).
559. *Neue Verpack. 2:*61 (1991).
560. Eastman Kodak Co., U.S. Patent 3718728; in *Coating 6:*154 (1974).
561. I. Benedek, E. Frank, and G. Nicolaus (Poli-Film Verwaltungs GmbH, Wipperfürth, Germany), DE 4433626A1/21.09.1994.
562. E. Prinz, *Coating 2:*56 (1978).
563. E. Prinz, *Coating 10:*269 (1979).
564. *Coating 10:*270 (1979).
565. *Paper, Film Foil Converter, 9:*32 (1989).
566. Softal Electronic, *Erhöhung der Oberflächeneenergie von Kunststoffolien durch Softalisierung,* Softal Electronic GmbH, Hamburg, Germany.
567. R. M. Podhajny, *Convert. Packag. 3:*21 (1986).
568. T. J. Blong and D. F. Klein, The influence of processing additives on optical, surface and mechanical properties of LLDPE blown film, Polyethylene 93, Oct. 4, 1993, Maack Business Service, Zurich, Switzerland.
569. H. J. Fricke and L. Maempel, Kleben & Dichten, *Adhäsion 11:*14 (1994).
570. K. W. Gerstenberg, Corona treatment for wetting and adhesion on printed materials, European Tape and Label Conference, Brussels, Apr. 28, 1993.
571. *Deut. Drucker 12:*26 (1988).
572. *Adhäsion 23:*136 (1979).
573. *APR 29:*798 (1988).
574. D. F. Lawson, *Rubber Chem. Technol.* 60(1):102 (1987).
575. *Coating 2:*54 (1987).
576. Milprint Overseas Corporation, Milwaukee, WI, U.S. Patent 1504556; in *Coating 8:*240 (1972).
577. *Coating 9:*340 (1995).
578. *Coating 7:*33 (1984).
579. V. Eisby, *APR 29:*794 (1988).
580. *Adhäsion 6:*255 (1966).
581. *Papier Kunstst. Verarb. 11:*44 (1986).
582. R. Milker and A. Koch, *Coating 1:*8 (1988).
583. *Coating 1:*20 (1978).
584. E. Prinz, *Coating 10:*360 (1979).
585. Softal Electronic, Report 102, Softal Electronic GmbH, Hamburg, Germany.
586. D. A. Markgraf, *Convert. Packag. 3:*18 (1986).
587. K. W. Gerstenberg, *Coating 8:*260 (1983).
588. K. W. Gerstenberg, *Coating 5:*172 (1991).
589. L. Dorn and W. Wahono, *Kunststoffe* 81(9):764 (1991).
590. J. Nentwig, *Papier Kunstst. Verarb. 2:*50 (1996).

591. E. Djagarowa, *Plaste Kaut. 9:*678 (1969).
592. Jpn. Patent 3571/1964; in *Coating 1:*19 (1969).
593. *Coating 6:*174 (1972).
594. A. M. Slaff, *Coating 7:*198 (1973).
595. Midland Silicones Ltd., London, DBP 1293374; in *Coating 1:*6 (1970).
596. *Adhäsion 7:*198 (1974).
597. *Convert. Today 1:*13 (1991).
598. W. Möhl, *Kunststoffe 81*(7):576 (1991).
599. O. Cada and P. Peremsky, *Adhäsion 5:*19 (1986).
600. L. Dorn and R. Bischoff, *Maschinenmarkt 43:*64 (1987).
601. *Das Papier 10A:* 1985; in H. Klein, *Coating 12:*431 (1986).
602. R. Milker, *Coating 11:*294 (1985).
603. *Neue Verpack. 2:*60 (1991).
604. *Papier Kunstst. Verarb. 7:*49 (1986).
605. R. A. Bragole, *Adhes. Age 4:*24 (1974).
606. S. L. Kaplan and P. W. Rose, *Plast Eng. 44*(5):77 (1988).
607. *Adhäsion 7/8:*39 (1996).
608. *Papier Kunstst. Verarb. 6:*10 (1988).
609. *Adhäsion 1:*16 (1987).
610. C. Bichler, M. Bischoff, H. C. Langowski, and U. Moosheimer, *VR Interpack 96,* Special, E37.
611. *Deut. Papierwirtsch. 2:*IV (1988).
612. R. Mannel, *Papier Kunstst. Verarb. 10:*48 (1988).
613. W. Möhl, *Kunststoffe 81*(7):576 (1991).
614. *Caoutch. Plast. 64*(674):101 (1987).
615. M. Rasche, *Adhäsion 3:*25 (1986).
616. Y. Nishiyama, S. Y. Mo, and K. S. Bae, *Nippon Kagaku Kaishi* 1118 (1985); in *Adhäsion 3:*32 (1986).
617. *Convert. Today 1:*13 (1991).
618. A. Kruse, K. D. Vissing, A. Baalmann, and M. Hennecek, *Kunststoffe 83*(7):522 (1993).
619. U. Zorll, *Adhäsion 7:*222 (1978).
620. *Coating 9:*279 (1969).
621. *Eur. Adhes. 6:*22 (1995).
622. K. Armbruster and M. Osterhold, *Kunststoffe 80*(11):1241 (1990).
623. *Papier Kunstst. Verarb. 6:*37 (1995).
624. *Coating 6:*163 (1974).
625. *Kunststoffe 84*(11):1569 (1994).
626. K. H. Kochem, *Kunststoffe 82*(7):578 (1992).
627. *Adhäsion 5:*211 (1967).
628. J. Nentwig, *Kunstst. J. 19:*8 (1991).
629. *Trespaphan Information 2,* September 1991), Hoechst Folien, Business Unit Trespaphan, Hoechst AG, Wiesbaden, Germany.
630. *APR 18:*686 (1986).
631. H. Klein, *Coating 12:*431 (1986).
632. *Coating 1:*28 (1978).

633. *Coating 1:*29 (1978).
634. Minnesota Mining and Manuf. Co., St. Paul, MN, U.S. Patent 2814601, in Coating 1:24 (1987).
635. Minnesota Mining and Manuf. Co., St. Paul, MN, U.S. Patent 2882183, in Coating 1:24 (1987).
636. M. W. Uffner and P. Weitz (General Aniline and Film Co., New York, U.S. Patent 3345320/03.10.1967.
637. G. L. Booth, 100% Solids silicone coating, Tappi Hot Melt Adhesives and Coating Short Course, May 2–5, 1982, Hilton Head, SC.
638. M. D. Fey, Low temperature cure solventless silicone paper coatings, Tappi Hot Melt Adhesives and Coating Short Course, May 2–5, 1982, Hilton Head, SC.
639. *Etiketten-Labels 35:*18 (1995).
640. C. Weitemeyer, J. Jachmann, D. Allstadt, and H. Brus, TEGO silicone acrylates RC for release coatings, Tappi Hot Melt Symposium 85, June 16–19, 1985, Hilton Head, SC.
641. *Papier Kunstst. Verarb. 9:*50 (1990).
642. H. J. Northrup, U.S. Patent 3691206; in *Coating 6:*154 (1974).
643. T. Bald, A 100% solid silicon coating system, PSTC Technical Seminar, May 7–9, 1986, Ithasca, IL.
644. S. Moser, *APR 16:*452 (1986).
645. *Tappi J. 9:*277 (1987).
646. *Adhäsion 4:*23 (1983).
647. P. Lersch, T. Ebbrecht, and D. Wewers, *Coating 2:*44 (1993).
648. *Kunststoffe 2:*85 (1992).
649. B. Hunt, *Labels Label. 2:*29 (1997).
650. K. B. Kasper and D. R. Williams (Schoeller Technical Papers Inc.), EP 252712/13.01.1988; in *CAS Siloxanes Silicones 17:*7 (1988).
651. K. Palli, M. Tirkkonnen, and T. Valonen (Yhtyneet Paperitehtaat Oy), Finn. Patent 74723/30.11.198; in *CAS Adhes. 14:*4 (1988).
652. T. N. Skuratovaskaya, O. V. Annikov, N. Z. Kvasko, V. I. Stolyarov, M. D. Podvolotskaya, Yu. A. Yushelevski, and A. S. Filenko, USSR Patent 1395721/15.05.1988; in *CAS Crosslinking React. 22:*13 (1988).
653. *Coating 1:*25 (1978).
654. *Adhäsion 11:*15 (1986).
655. J. F. Kuik, *Papier Kunstst. Verarb. 10:*26 (1990).
656. *Polyethylene 89,* p. 53, Maack Business Service, Zurich, Switzerland.
657. W. Gerstenberg, *Coating 9:*304 (1992).
658. Dow, *XZ87131.32,* Experimental Data Sheet, Dow Chemicals, Horgen, Switzerland, 1994.
659. Neste Chemicals, *Technical Information, Polyethylene F0133,* Oct. 2, 1991.
660. Battenfeld, *Press-Info,* 2713-GB, February 1988, p. 2.
661. C. Momose, K. Nakakawara, and M. Matsui (Mitsubishi Densen Kogyo K. K.), Jpn. Patent 6351488/04.03.1988; in *CAS Adhes. 20:*3 (1988).
662. T. K. Bhaumik, B. R. Gupta, and A. K. Bhowmik, *J. Adhes.* 24(2/4):183 (1987).
663. J. P. Croquelois and P. Phandard (Papeteries Elce, S.A.), Fr. Patent 22600981/08.01.1988.

664. G. Panagopoulos, S. E. Pirtle, and W. A. Khan, *1991 Film Extrusion*, p. 103, Tappi.
665. *Papier Kunstst. Verarb. 9:*23 (1987).
666. *Br. Plast. Rubber 5:*24 (1988).
667. Rapra Technology Limited, The Extrusion of Tacky Polyethylene Film, RTL/1705 6.12.90/PAA.
668. *Convert. Today 1:*15 (1992).
669. *Deut. Papierwirtsch. 3:*133 (1987).
670. K. Noda and K. Isayama (Kanegafuchi Chem. Ind. Co. Ltd.), EP 252372/ 13.01.1988; in *CAS Siloxanes Silicones 12:*4 (1988).
671. I. Hitoshi, T. Tokihiro, and K. Kiyonori (Tokuyama Soda Co. Ltd.), Jpn. Patent 62270642/25.11.1987; in *CAS Adhes. 13:*1 (1988).
672. V. M. Ryabov, O. I. Chernikov, and M. F. Nosova, *Plast. Massy 7:*58 (1988).
673. M. W. Janczak, *Polymery 9:*381 (1964).
674. *Coating 8:*4 (1969).
675. *Neue Verpack. 2:*61 (1991).
676. M. Toshio (Furukawa Electric Co Ltd.), Jpn. Patent 23 05.1988; in *CAS Adhes. 24:*6 (1988).
677. Y. Torimae (Kao Corp.), Jpn. Patent 63165474/08.07.1988; in *CAS Adhes. 24:*5 (1988),
678. K. Yamada, K. Miyazaki, Y. Owatari, Y. Egami, and T. Honma (Sumitomo Chem. Co., Ltd.), EP 257803/02.03.1988; in *CAS Coatings Inks Related Compounds 17:*6 (1988).
679. A. E. Khoklovkin, N. B. Vladimirokaya, and E. S. Bushkova, *Sovrem. Lakokrasokh. Mater. Tekhnol. Primeneniya, Mater. Semin. N.;* in *CAS Coatings Inks Related Products 18:*2 (1988).
680. M. Chladek, J. Jilek, and F. Benc, Czech CS 246986/15.10.87; in *CAS Coatings, Inks Related Products 18:*13 (1988).
681. W. E. Havercroft, *Paper, Film Foil Converter 10:*52 (1973).
682. GF. Bulian, Extrusion coating and adhesive laminating: Two techniques for the converter, Polyethylene 93, 4/6, 1993, Maack Business Service, Zurich, Switzerland.
683. Coating and Laminating In House, Cowise Management and Training Conference, Amsterdam, Mar. 20, 1997.
684. T. Frecska, *Screen Print. 2:*64 (1987).
685. G. Perner, *Coating 6:*237 (1991).
686. *Quleques produits dont vous ne pouvez vous passer,* Etilux, Booklet, S. A. Etilux N. V, Brussels, Belgium.
687. *Adhes. Age 8:*34 (1983).
688. *Coating 4:*139 (1997).
689. *Adhäsion 9:*351 (1965).
690. R. Hinterwaldner, *Adhäsion 6:*11 (1984).
691. C. Massa, Pressure sensitive tapes and laminates marketing and production techniques, 19th Münchener Klebstoff und Veredelungsseminar, 1994, p. 35.
692. R. Schieber, *Adhäsion 5:*21 (1982).
693. B. K. Bordoloy, Y. Ozari, S. S. Plamthottam, and R. Van Ham (Avery Int.), EP 263686/13.04.1988; in *CAS Rad. Curing 19:*1 (1988).

694. H. Klein, *Coating* 12:430 (1986).
695. Rohm & Haas Co., Philadelphia, Polymers, resins and monomers, *Adhesives, Rhoplex PS-83D,* 1982, p. 2.
696. U.S. Patent 0212358, p. 2.
697. *Adhes. Age 3:*41 (1985).
698. E. Bradatch, *Papier Kunstst. Verarb.* 6:33 (1987).
699. P. Beiersdorf & Co. AG, Hamburg, U.S. Patent 1569882; in *Coating 11:*336 (1972).
700. M. Guder and J. Auber, Die Verarbeitung von HMPSA im industriellen Bereich unter Technischen und Ökonomischen Aspekten, 19th Münchenener Klebstoff und Veredelungsseminar, 1994, p. 232.
701. *Paper, Film Foil Converter 8:*28 (1973).
702. W. Schaezle, Schmelzkleberbeshichtung, *Adhäsion 9:*5 (1983).
703. *Druck Print 8:*24 (1986).
704. H. J. Claasen, *APR 50–51:*1741 (1986).
705. *Coating 11:*291 (1982).
706. H. J. Einfeldt and H. J. Meissner, Schmelzklebstoff Auftragssysteme, 16th Münchener Klebstoff und Veredelungsseminar, 1991.
707. H. Klein, *Coating 6:*210 (1986).
708. *Coating 1:*12 (1988).
709. *Coating 12:*344 (1984).
710. G. Drechsler, *Coating 6:*153 (1971).
711. J. Türk, *Papier Kuntsst. Verarb. 10:*22 (1985).
712. Morton Thiokol Inc., Adhesives and Coatings, *Morstik 103 Adhesive,* MTD-MS103-12/1988.
713. Tappi, Hot Melt Symposium, June 16, 1988, Hilton Head, SC; in *Adhäsion 4:*234 (1986).
714. Morton Thiokol Inc., Adhesives and Coatings, *Morstik 16 Adhesive,* MTD-MS106-12/88.
715. *Coating 11:*394 (1990).
716. *Coating 11:*291 (1982).
717. G. W. Drechsler, 11th Münchener Klebstoff u. Veredelungsseminar, Oct. 20–22, 1986; in *Coating 3:*97 (1987).
718. Planatolwerk W. Hesselmann, Rosenheim, DBP 1288070; in *Coating 9:*272 (1969).
719. *Adhäsion 1/2:*36 (1985).
720. J. Weidauer, *Eur. Adhes. Sealants 5:*26 (1995).
721. H. Klein, *Adhäsion 7:*248 (1973).
722. F. M. Fischer, *Coating 3:*90 (1973).
723. V. Bohlmann, *Papier Kunstst. Verarb. 2:*12 (1979).
724. W. Grebe, *Papier Kunstst. Verarb. 2:*46 (1985).
725. F. S. McIntire, The future of inline UV silicone and hot melt pressure sensitive adhesive coatings for label and tape products, Hot Melt Adhesives and Coating Short Course, Technical Association of the Pulp and Paper Industry, Hilton Head, SC, June 1982; in *Adhes. Age 8:*35 (1983).
726. *Coating 5:*176 (1988).
727. E. G. Huddleston (The Kendall Co., Boston, MA), U.S. Patent 4692352/08.09.1987.

728. M. Dollinger, UV Strahlungstechnik für die Verarbeitung, von UV härtenden Klebstoffen, Möglichkeiten, neue Entwicklungen, 11th Klebetechnik Seminar, Jan. 25, 1989, Rosenheim.
729. P. Holl, Elektronenstrahlvernetzung/Vulkanisation von Klebstoffen für flexible Materialien, 11th Klebetechnik Seminar, Jan. 25, 1989, Rosenheim.
730. U. Schwab, *Coating 5:*171 (1996).
731. E. Smit, Pressure Sensitive Adhesives and Adhesive Coating, Apr. 24, 1996, Cowise Management and Training Service, Amsterdam.
732. H. Braun, UV-curable acrylic based hot melt pressure sensitive adhesives; in *Coating 12:*477 (1995).
733. A. Dobmann and J. Planje, *Papier Kunstst. Verarb. 1:*38 (1986).
734. G. Pitzler, *Coating 6:*218 (1996).
735. H. Kamusewitz, Die thermodynamische Interpretation der Adhäsion unter besonderer Beachtung der Folgerungen aus der Theorie von Girifalco und Good, Dissertation, Halle, 1988; in G. Pitzler, *Coating 6:*218 (1996).
736. Novacel, France, DE 1594193; in K. Wabro, R. Milker, and G. Krüger, Haftklebstoffe und Haftklebebänder, Astorplast GmbH, Alfdorf, Germany, 1994, p. 48.
737. H. Schmalz and W. Neumann, Coating lines for the production of self adhesive tapes and labels using radiation curable hot melt systems, European Tape and Label Conference, Brussels, Apr. 28, 1993.
738. H. J. Voss, *Coating 5:*150 (1987).
739. H. Hadert, *Coating 1:*11 (1969).
740. Johnson & Johnson, Can. Patent 583367; in H. Hadert, *Coating 1:*11 (1969).
741. U. P. Seidl (Schreiner Etiketten und Selbstklebetechnik GmbH), Ger. Offen. DE 3625904/04.02.1988; in *CAS Siloxanes Silicones 16:*3 (1988).
742. H. Klein, *Coating 10:*372 (1988).
743. D. Percivalle, High speed machines for the production of self adhesive tapes and labels, European Tape and Label Conference, Brussels, Apr. 28, 1993, p. 133.
744. H. Klein, *Adhäsion 5:*190 (1973).
745. H. Klein, *Coating 3:*4 (1993).
746. C. S. Watson, *Coating 11:*399 (1987).
747. H. Klein, *Coating 11:*390 (1986).
748. H. G. Reinhardt, Application methods for highly viscous melts, 16th Münchener Klebstoff und Veredelungsseminar, 1991, p. 15.
749. B. C. van Oosten, *APR 16:*382 (1988).
750. H. W. Jakobs, *APR 16:*380 (1988).
751. *Coating 6:*184 (1969).
752. S. Aoyanagi, H. Suzuki, and S. Takeda (Kanzaki Paper Manuf. Co. Ltd.), Jpn. Patent 6315872/22.01.1988; in *CAS Adhes. 14:*4 (1988).
753. *Adhäsion 4:*4 (1985).
754. *Coating 11:*393 (1990).
755. Y. Mizutani, T. Noguchi, H. Kuroki, and T. Imahama (Tosoh Corp.), Jpn. Patent 63265369/18.11.1987; in *CAS Hot-Melt Adhes. 11:*1 (1988).
756. G. Wouters (Exxon Chemical Patents Inc.), EP 251726/07.01.1988; in *CAS Emulsion Polym. 18:*2 (1988).

757. B. Copley and K. Melancon (Minnesota Mining and Manuf. Co.), U.S. Patent 878816/26.01.1986; in *CAS Siloxanes Silicones 14:*2 (1988).
758. 14th Münchener Klebstoff und Veredelungsseminar, 1989, p. 9.
759. D. Satas, *Adhes. Age 8:*30 (1988).
760. Universum Verpackung GmbH, Rodenkirchen, U.S. Patent 1297790; in *Coating 1:*274 (1969).
761. T. J. Bonk, T. I. Cheng, P. M. Olsom, and D. E. Weiss, PCT, WO 87/00189/ 15.01.1987.
762. A. Saburo (Central Glass Co. Ltd.), Jpn. Patent 63117085/21.05.1988.
763. Y. Hamada and O. Takuman (Toray Silicone Co., Ltd.), EP 253601/20.01.1988.
764. *Adhäsion 6:*14 (1985).
765. Minnesota Mining and Manuf. Co., Jpn. Patent 63175091/19.07.1988; in *CAS Adhes.* 25:7 (1988).
766. A. Nagasuka and M. Kobari (Sekisui Chem. Co., Ltd.), Jpn. Patent 6389585/ 20.04.1988; in *CAS Adhes. 19:*6 (1988).
767. E. R. Squibb and Sons, Princeton, NJ, U.S. Patent 4551490.
768. T. Kishi (Sekisui Chem. Co. Ltd.), Jpn. Patent 6369 879/29.03.1988; in *CAS Adhes. 19:*6 (1988).
769. H. W. J. Müller, *Adhäsion 5:*208 (1981).
770. I. A. Gritskova, L. P. Raskina, S. A. Voronov, G. K. Channova, D. N. Avdeev, E. B. Malyukova, T. K. Vydrina, N. D. Sandomirskaya, and O. M. Solozhentseva, *Otkrytiya Izobret. 42:*87 (1987); in *CAS Coating Inks Related Products 11:*11 (1988).
771. Beiersdorf A. G., Hamburg, Germany, DBP 1667940; in *Coating 12:*363 (1973).
772. K. Takashimizu and A. Suzuki (Advance Co. Ltd.), Jpn. Patent 63 92683/ 23.04.1988; in *CAS Adhes. 19:*5 (1988).
773. *Coating 6:*198 (1993).
774. *Druck Print 4:*19 (1988).
775. Application Tape Folie, *Technische Anforderungen,* Tacfol Klebfolien GmbH, Germany.
776. R. Higginson (DRG, UK), PCT/WO 88 03477/19.05.1988; in *CAS Colloids (Macromol. Aspects) 22:*6 (1988).
777. H. Klein, *Coating 7:*270 (1988).
778. G. W. Drechsler, *Coating 5:*167 (1996).
779. *Kaut. Gummi Kunstst.* 37(3):252 (1984).
780. G. W. Drechsler, *Coating 3:*62 (1984).
781. W. K. Behrendt and R. Schock, *Maschinenmarkt* 88(81):1648 (1982).
782. B. W. McMinn, W. S. Snow, and D. T. Bowman, *Adhes. Age 11:*37 (1995).
783. H. Klein, *Coating 6:*198 (1996).
784. A. H. Beaulieu, D. R. Gehman, and W. J. Sparks, *Tappi J. 9:*102 (1984).
785. D. H. Treesdale, *Coating 11:*290 (1982).
786. M. Guder and J. Auber, Die Verarbeitung von HMPSA im Industriellen Bereich unter Technischen und Ökonomischen Aspekten, 19th Münchener Klebatoff und Veredelungsseminar, 1994, p. 232.
787. G. Drechsler, *Coating 6:*153 (1971).
788. F. M. Fischer, *Coating 3:*90 (1973).
789. L. W. Fuller GmbH, Lüneburg, *Wachse, Hotmelts, Klebstoffe* 1977.

790. H. Hadert, *Coating* 7:203 (1970).
791. J. Raterman, Evolution of pressure sensitive adhesive spray technology, Tappi Hot Melt Symposium 85, June 16–19, 1985, Hilton Head, SC.
792. *Eur. Adhes. Sealants* 6:24 (1966).
793. W. Schaezle, *Coating* 7:450 (1981).
794. H. Klein, *Coating* 3:74 (1993).
795. K. H. Honsel, Bielefeld, Germany, DBP 2110491, in Coating 12:363 (1973).
796. *Adhes. Age* 5:34 (1987).
797. *Coating* 3:172 (1969).
798. K. Ochi, S. Ouzo and K. Toykma (Nippon Corkick Repyo K.K., Tokyo, Japan) EP 0070524/16.07.1981.
799. M. Hasegawa, Tokyo, U.S. Patent 4460634/17.06.1984.
800. G. E. Davis, Precision controlled application of continuous patterned hot melt adhesive, Tappi Hot Melt Symposium, June 15–19, 1985, Hilton Head, SC.
801. Jpn. Patent A 5695972; in P. Gleichenhagen, E. Behrend, and P. Jauchen (Beiersdorf AG, Hamburg, Germany), EP 0149135 B1/24.07.1987.
802. Ashland Chemicals, *Test Methods,* Technical Booklet.
803. *Siebdruck* 33(3):43 (1987).
804. *Druckwelt* 6:50 (1988).
805. H. Mürmann, *Papier Kunstst. Verarb.* 11:26 (1987).
806. 14th Münchener Klebstoff und Veredelungsseminar, 1989, p. 9.
807. E. Mandershausen, *Coating* 10:271 (1974).
808. *Convert. Today* 4:22 (1991).
809. *Convert. Today* 11:43 (1990).
810. *Adhäsion* 1/2:18 (1987).
811. H. Klein, *Coating* 19:346 (1992).
812. U. Temper, *APR* 42:1156 (1988).
813. H. Klein, *Coating* 5:177 (1996).
814. *Coating* 1:15 (1978).
815. A. Erdmann, *Papier Kunstst. Verarb.* 3:46 (1997).
816. S. I. Shapiro, *Tappi J.* 5:97 (1988).
817. *Kaut. Gummi Kunstst.* 37(1):57 (1984).

7
Converting Properties of PSPs

The uses of PSPs are made possible by their adhesive and mechanical characteristics. As discussed throughout the preceding chapters, PSPs are manufactured generally as weblike products. They are applied as a weblike or finite products whose dimensions and/or surface characteristics differ from those with which they are manufactured. Therefore, before application they have to be finished. In this case finishing means the transformation of the continuous weblike product that has the optimal geometry for manufacture into a product that has the optimal characteristics for use. This production step is called conversion. Benedek and Heymans [1] have discussed convertibility as the sum of the convertibility of the adhesive and that of the laminate. Convertibility of the adhesive has been described as its coatability. For some PSPs, coating is included in their conversion. Certain PSPs are not coated. For some products converting includes cutting (slitting), laminating or delaminating, die cutting, and other operations (called confectioning). Other PSPs are not laminated and not cut but torn. Numerous pressure-sensitive products are used for applications where the adhesive and mechanical properties have to be complemented with other properties. Therefore these products have certain special performance characteristics called conversion properties.

For PSPs manufactured with an adhesive coating, converting properties are also influenced by the adhesive. Although the characteristics of the solid-state components of the laminate (face stock and release liner) and the manufacturing technology needed for the laminate influence the conversion properties also, in this process the requirements of coaters and laminators primarily concern the adhesive. Good die cutting and stripping characteristics and good convertibility of the adhesive are required [2]. The adhesive has to exhibit good resistance to gum balls, edge flow, and face bleed and to meet the composition requirements of FDA 21 CFR 175.105 for indirect contact with food products.

1 COATING PROPERTIES OF PSPs

In manufacturing a PSP by way of coating, the solid-state components of the laminate must be coated with adhesive and in some cases with a release layer (e.g., labels, tapes, and certain protective films). The release layer can be manufactured as a separate solid-state component also. In some cases the adhesive is coated on the release liner (transfer coating). Therefore, the coating properties of the carrier material for the liner as well as those of the abhesive-coated release liner are important. For both PSP classes (adhesive-coated and extrusion-manufactured products) there is a need for postcoating (e.g., lacquering) also. Therefore coating properties generally play a special role in the manufacture of PSPs.

It should be emphasized that coating properties are the sum of the coating performance characteristics of the solid-state component to be coated (face stock or release liner), the coated mass (adhesive, abhesive, lacquer, and printing ink), and the coating technology. Unfortunately, the coater can regulate only some of these parameters.

1.1 Adhesive Coating

The coating of the adhesive on a carrier material depends on the rheology of the liquid adhesive (dispersed, dissolved, or molten), on the characteristics of the carrier material, and on the coating technology. For a given coating technology the coating properties depend on the rheology of the adhesive and the surface characteristics of the carrier material. For a dispersed or dissolved adhesive the coating properties depend on the components of the dispersed system and its wetting characteristics. The surface of the carrier material influences the coating through its texture (structure), porosity, surface tension, and chemical affinity.

As is known, there are certain tape manufacturing procedures in which a soft, solvent-free adhesive is calendered onto or into a fiberlike textile material (roll press coating). In such cases the penetration of the adhesive ensures its anchorage. Penetration as an adhesion-improving factor should be taken into account in the manufacture of removable products also.

Chemical affinity is a parameter that ensures the anchorage of many solvent-based adhesives on plastic films or of functionalized (carboxylated) latices on pa-

Conversion Properties 525

per. It is the reason for precoating a carrier with a reactive (e.g., polyisocyanates, reactive resin) primer also. Chemical affinity is also a factor in the regulation of the bonding force of polar self-adhesive films.

Wetting out depends on the surface tension of the carrier material and the surface tension and rheology of the adhesive. Surface treatment of the carrier and appropriate choice of the surface-active agents for a formulation allow control of dynamic wetting out. In adhesive coating the manufacturer has the ability to regulate the coating properties by proper formulation of the adhesive, by the choice of the carrier material, by the use of a primer, and by the choice of coating device or coating technology.

Solvent-Based Coating

In coating solvent-based adhesives, highly viscous, low surface tension liquids are used; therefore wetting-out is easily controlled. In this case the main problem is related to the chemical sensitivity of many plastic carrier materials to organic solvents. Another special problem is the change in viscosity of crosslinked systems.

Water-Based Coating

A variety of water-based products such as cold sealing adhesives, adhesives for lamination, primers, and finishing coats are routinely used in film conversion. Blisters, grit, ribbing, haze, and foam are always found with such pressure-sensitive adhesives. Ideally, water-based coatings should be smooth and dry quickly, stick under stress and heat, and resist humidity.

1.2 Abhesive Coating

Siliconizing is a well-known technology that uses solvent-based and water-based or solventless thermally cured silicones or radiation-cured 100% solids. Therefore the silicone coating specialist must have the skill to operate in each of the related fields. Surfaces that are difficult to to coat (such as plastics or coated papers) appeared only in the last decade. On the other hand, there are many mixed systems, where "solventless" formulations contain a certain level of solvent to regulate the viscosity. Adhesive formulation know-how and coating technology know-how can be accumulated by the manufacturer; the skills related to silicone formulation and coating belong more to the suppliers of such materials.

Special problems appear in the coating of carbamate-based release liners. There are only a limited number of solvents that can be used. Carbamates have very low solubility even in hot solvents, and the release solution has to be maintained and coated at elevated temperatures on chemically and thermally sensitive carrier material. The anchorage of such release layers is low.

1.3 Printing

Adhesives are coated on the carrier material with the use of special coating devices. Other components (inks, antistatic agents, primers, etc.) are coated with common printing techniques. In some cases the adhesive or the release liner can be coated with these printing techniques also.

Most PSPs such as labels or certain tapes are information carriers. The information has to be printed on them before they are used. Generally printing occurs during the manufacture of the weblike, nonconfectioned product. For some products (mainly labels and tapes), a postprinting of the confectioned (cut, die-cut product) can be carried out also, and in some cases printing (writing, stamping, etc.) can be done during their end use. Therefore generally one can speak of printing and postprinting. Some time ago the basic printing methods for printing PSPs during the manufacture of the weblike product and for postprinting (in the prelaminated, confectioned, or postlaminated state) were the same. Differences were given by the web width and the combination of the special materials and printing/confectioning technologies. Some years ago, special non-impact printing methods and computerized design and printing of PSPs were developed. Thus the last step in conversion or printing is transferred to the end user (or "almost" end user). Therefore, now (and in the future) printing and postprinting technologies may not be the same. As a consequence the surface quality (printability) required for a certain printing method used in the manufacture of the weblike product (e.g., flexo printing of a plastic film) could differ from the surface quality required by postprinting the confectioned product (e.g., screen printing or laser printing).

General Printing Considerations

Printability includes quite different performance characteristics such as lay flat, dimensional stability (under various mechanical and environmental stresses), good anchorage, and adequate machinability [3]. When choosing a printing base certain parameters have to be taken into account [4]: required quality standards, performance ratio, and subjective assessment of the finished product. Quality includes printability, roughness on front and reverse side, gloss, opacity, density, thickness, tensile strength, and bending stiffness.

The importance of printability is quite different for various PSP product classes. For labels it is determinant. Printing of labels is most important for extended text labels. Aesthetic products such as labels or certain tapes and those carrying information have to be printed with high quality images or text. Resolution is one of the main characteristics of a printing procedure used as a criterion in the choice of printing method. According to Frecska [5], resolution is the minimum distance between two objects that gives a clearly separated image at a

Conversion Properties

given magnification for a given optical system. The resolutions of the main printing methods are listed in Table 7.1.

Generally there are reciprocal influences between the printing ink and the paper or film [6]. Printing properties depend on the substrate to be printed. Printing quality depends on the surface finish, roughness, and wettability [7]. For direct thermally printed papers the choice of paper (nontopcoated or topcoated) can be dictated by the method of preprinting [8]. Nontopcoated grades have less protection against solvents or monomers. Therefore waterbased or low-solvent flexo printing are best suited for this grade. Topcoated papers can absorb out the monomers in UV-curable inks. The nonabsorbancy of films has to be compensated for by the proper choice of ink and drying conditions. Inks for normal and thermal papers and for nonabsorbent carrier materials such as aluminum, PVC, and PP have been developed. Good printability [9] requires

Sufficient smoothness (low surface roughness)
More or less surface absorbancy, depending on the printing ink
Sufficient surface energy
Sufficient resistance to solvents in printing inks

Table 7.1 Performance Characteristics of the Main Printing Methods Used for PSPs

Printing method	Resolution (dpi)	Other characteristics	Ref.
Offset	2400 × 2400	—	57
Mask	400 × 400	—	65
Thermal transfer	300 × 300	Low chemical and	86, 87, 91
	300 × 400	abrasion resistance of the image	94
DCI	1200 × 1200	—	98
Laser	600 × 600	Low chemical resistance of the image, temperature-resistant materials required	92
Ink jet	150 × 150	Porous face stock and	90
	360 × 360	long drying time required	107
Dot matrix	150 × 150	Good chemical, weathering, and abrasion resistance, good anchorage on plastics, noisy	22, 92

Sufficient resistance to heat produced during ink drying or hot foil stamping
Absence of separating agents or other impurities from the imprintable side

The chemical composition of the film substrate affects the bond. Additives (compatible or migratory) in the polymer can also influence the bond. Slip agents can contaminate the surface. Surface-active agents create an adhesively weak layer. Lower molecular weight oligomeric fractions from the polymer can cause problems. Excessive additives on the film surface (e.g., slip and processing aids) can cause printing problems. For instance, linear low density polyethylene sometimes contains more low molecular weight species and will require more antiblock additive to prevent blocking [10]. Residual oil on the film surface is dangerous. The smoothness of the surface can also affect the bond. The rough and porous surface of paper provides some mechanical adhesion, while the smooth surface of film does not. Surface energy determines wetting and also influences bonding.

As is known from printing practice, different plastic films display different degrees of printability. Acetate films display good printability; they can be used as label or tape carrier materials. A 50 μm cellulose acetate film needs no pretreatment for normal printing. It is computer-printable also [11]. For polystyrene films, flexography, roll feed lithography, and gravure printing are recommended. Sealing tapes are based on PVC. They can be printed by using sandwich printing (flexo for the top side and gravure printing for the back side between the film and the adhesive) [12]. For PVC and vinyl chloride copolymers, solvent-based gravure and flexo printing inks have been suggested [13]. Predecoration of packages involves the application of graphics before the container is filled with product; this is used by bottle manufacturers and contract packagers [14]. Predecoration methods include application of the label in the package mold, direct screen printing on a container, hot stamping, heat transfer decoration, shrink sleeve system, and pressure-sensitive systems. Decoration can also be applied in-line with filling. In choosing a carrier material and a printing technology, the composition of the printing ink should be taken into account also. Printing inks generally consist of two basic components: pigment and vehicle.

Gravure, flexo, and screen processes use low viscosity inks and simple inking systems. Lithographic and letterpress systems apply printing inks of higher viscosity and use a complex metering system. Vehicle systems contain film-forming parts (resins and oils), volatile materials (solvents, water, viscosity regulators), and additives (plasticizers, dispersing agents, drying agents). Different printing processes need different inks of various compositions and rheology. Flexographic inks are low viscosity recipes based on water or alcohol. Lithographic inks are water-soluble high solids compositions. Gravure inks are solvent-based high viscosity compositions.

Different printing procedures use different solvents and drying temperatures and cause different stresses in the web. Therefore printability is a function of the carrier thickness also. Gravure printing is suggested for polyethylene films that are at least 25 μm thick and for polypropylene films with a minimum thickness of 12 μm.

Printability depends on the surface wettability, i.e., degree of treatment. As is known, the type of film also influences pretreatability. The loss of treatment effect also depends on the nature of the film (see also Chapter 6). Blown films maintain their degree of treatment for a longer period of time than cast films. Triboelectricity affects printability also. As is known from printing and web handling, polyester displays a high level of triboelectricity.

Special Printing Considerations

Special printing parameters should be taken into account depending on what carrier material is used, the product class, and the type of ink. Most PSPs are plastic-based. Plastics printing is a special domain. The printing of nonpolar plastics with water-based inks, as a consequence of technological trends, causes additional problems. Curling and shrinkage are some of the problems that appear during printing, partially as a consequence of the printing process (see later, Printing-Related Performance Characteristics of the Carrier Material).

Printing on Plastics. The printability of plastics is described by Verseau [13]. Electrostatic charges, lack of porosity, high surface tension, thermoplasticity, sensitivity to solvents, and environmental stress cracking are some of the main disadvantages of printing on plastic films [15]. Before being coated with varnishes or lacquers, a plastic surface has to be pretreated with antistatic agents to deionize the air [16]. It is recommended that static charge elimination systems be placed after the printing cylinders to help dissipate charges in the atmosphere.

Nonpolar carrier materials are difficult to coat. In some cases a top coating is applied before printing. As is known, polypropylene is an alternative for PET for "no-label look" labels [17]. For such products a 60 μm polypropylene film is used. This is coated with a lacquer to allow printing via screen printing (solvent-based and UV-cured), flexo printing (water-based, solvent-based, and UV-cured), and UV-cured letterpress printing. According to Waeyenbergh [18], good ink anchorage is achieved for OPP by the choice of suitable ink and lacquer, adequate drying capacity, film surface finish (pretreatment level or primer), and corona treatment on the printing press. The product class should be taken into account also because of the special requirements for different products (see later).

Technological Trend. There is growing interest in printing technologies such as gravure and flexo that use water-based inks. There are differences between the characteristics of water-based inks and those of solvent-based inks. Apart from the fact that different resins and pigments are used in their formulation, major

530   *Chapter 7*

differences arise from the physical and chemical properties of water and organic solvents (see also Chapter 4). These differences concern the boiling point, evaporation rate, evaporation heat, and surface tension. Water has a higher boiling point than ethyl acetate or ethyl alcohol. Its evaporation rate is higher also. On the other hand, water consumes more heat energy during evaporation.

It is well known that the surface tension of water is much higher than that of most other solvents, and therefore it displays lower wettability. Water-based gravure ink requires cylinders with small cell volumes. The same approach has been confirmed for flexo printing by the use of ceramic anilox with line screens [19]. According to one forecast [20] the estimated market share of new printing machines in the period 1984–2000 is about 30% for flexo printing, 30% for UV flexo, 28% for rotary screen/combination offset, and 12% for rotary letterpress printing.

Printing Methods

There are a number of different printing methods. The main procedures are impact (flexo, gravure, screen, etc.) and non-impact (ink jet, laser, thermotransfer, etc.) printing. The requirements concerning the carrier material for these procedures are quite different. For instance, non-impact printing procedures need paper that has a porous surface [21]. The use of such printing methods is also related to their versatility with respect to broad or narrow web printing. Table 7.2 lists the main printing methods used for labels, tapes, and protective films.

Table 7.2 Main Printing Methods Used for PSPs

Pressure-sensitive product	Printing method	
	Broad web	Narrow web
Label	Gravure	Flexo
	Flexography	Thermal
	Offset	Dot matrix
	Letterpress	Ink jet
	Screen	Laser
	Tampon	Digital offset
	Mask	
	Hot stamping	
	Cold foil	
Tape	Flexo	Flexo
		Thermal
		Dot matrix
		Ink jet
Protective film	Flexo	—

Printing of Labels

As discussed earlier, label manufacturing includes printing and label overprinting. Label overprinting is really a converting step required to fix supplementary variable data on the label [22]. For label converters the stock must meet printing and die cutting specifications. Because the printing of labels is the most demanding domain of PSP printing, a short description of the main features of the different printing techniques used for labels follows. Generally the choice of the best printing method depends on run length, quality of printing required, type of equipment available, and on whether the material is in sheet or web form. The type of adhesive and the end use of the printed item influence the choice of printing method also. As an example, sandwich printing is usual for plastic labels coated with hot melts. These have a transparent polypropylene or PET carrier and an OPP release liner [23]. Rating plates can be sandwich printed also, with the reverse side of a transparent material printed with mirror lettering [24]. According to Fust [25], the main printing methods used for labels are flexography, letterpress, offset, gravure, screen, dry/hot stamping, and magnetography.

The most important method is letterpress printing, and the second is flexographic printing. In Europe, gravure and offset printing are too expensive; therefore, letterpress, flexographic, and screen printing are considered by Fust [25] to be the printing methods of the future. Screen printing should be used for more expensive (quality) labels than flexo or letterpress printing. The choice of printing procedure is influenced by printing quality, product range, dimensions and form, and costs [26].

Gravure Printing. In gravure printing the image is etched into the metal cylinder. The wells are filled with the ink, and the excess ink is wiped off with a doctor blade. In this procedure the pressure of the knife on the cylinder attains 12 N/mm^2, i.e., 2100 atm [27]. Gravure printing presses for labels have to secure excellent printing quality and productivity as well as the possibility to perform all the other conversion operations required for labels, such as lamination and die cutting, in line with the printing. According to Schibalski [28], the hot sealing adhesive of in-mold labels has to be printed on varnishing or gravure printing machines. In-mold labels have gravure printing, sheet-set offset gravure, narrow web UV, special inks, and special die cutting. There is a general trend in the packaging industry to replace gravure printing by flexo and offset printing [29]. The printing cylinders are cleaned with special burst cylinders.

Flexographic Printing. Flexography is a form of web-fed (letterpress) printing. This technology uses flexible (rubberlike) printing plates mounted on a printing cylinder. The print pattern is raised, and the ink is applied only to these raised portions. In-line, stack, or central impression flexo are the main machine configurations [30]. Flexo printing of films with one-cylinder printing machines requires high cylinder precision and web regulation [31]. Flexo for label printing has had a growth rate of 6% per year [32,33].

Classic flexo printing uses four cylinders in the printing device. An ink transport cylinder transfers the ink to the gravure cylinder. The gravure cylinder transports the ink to a plate cylinder that has a soft printing plate (stereotype). A pressure cylinder ensures contact with the web [34]. The first generation of flexo printing machines used solvent-based low viscosity printing inks, giving low quality printing [35]. For instance, flexo printing ink with a viscosity of 21–23 s (Zahncup No. 2) was suggested for polypropylene [36]. In the 1980s a new generation of rotary flexo printing machines was developed. Water-based and UV-cured printing inks were used. UV-cured printing gives a seal-resistant (over 200°C) image.

For labels a common flexo rotary press has one to three printing devices, a die-cutting device, a winder for labels and waste material, and a laminator [37]. Common flexo printing inks for labels are based on alcohol or water with a maximum of 3% solvent. Flexo presses for overprinting labels can print in eight colors, have a 420–620 mm web width, and run at a speed of 150–200 m/min, depending on the carrier material [38]. Narrow web (420–620 mm) flexo printing machines can use webs with a thickness of 20–300 μm [39]. In the United States, flexo printing is the main label-printing procedure [40]. Flexo postprinting machines are suitable for producing screen rulings of up to 60 lines/cm [41].

Screen Printing. Screen printing is the least technologically advanced printing method, but it can be used for the greatest number of types of substrates [42]. It can be combined with other printing methods such as letterpress and hot foil stamping [25], and it is recommended for plastics [43]. The main market segment for screen printing include food, pharmaceuticals, cosmetics, toiletries, and household items.

Rotational screen printing machines have been in use since 1986; they run at 50 m/min [44]. According to Ref. 45, rotary screen printing was developed earlier, in the 1970s. Today it allows speeds of 30 m/min; tolerances lower than 50 μm, and printed areas of 5 m^2 [45]. Special machines have been developed for printing labels [46]. For this procedure the drying time is short and energy costs are low, but there may be registry problems [47]. Polyvinyl chloride films having a minimum thickness of 0.1 mm and hard foams with a maximum thickness of 20 mm can be printed via screen printing [48]. Reel-to-reel screen printing lines allow labels to be printed with a production capacity of 10–20 m^2/min and web width of 150–750 mm. Posters, labels, membrane switches, and transfers can be manufactured. Laminators, embossing units, die cutters, and sheet-fed dryers (sheet feed from roll) can be included. The image can be transferred onto surfaces with different levels [49]. Weather-resistant decals are printed via screen printing [50]. Screen-printed polycarbonate labels are used for in-mold labeling [48].

Polyester with a maximum thickness of 350 μm is printed via screen printing. The printing screen should be cleaned shortly before use with special chemicals [51]. Special chemical pastes are used for cleaning screens [52].

Offset Lithography. In offset lithography the image and non-image areas are in the same plane. The image area is oil-receptive and hydrophobic. The ink adheres to this portion. The non-image area is hydrophilic. Special offset printing methods use a water-free offset printing plate. The printed image is built up on the plate by erosion [53]. There are two basic types of lithography, direct and offset. In offset lithography (used commonly) the image is transferred with a blanketed rubber offset cylinder. In roll offset printing, during drying the humidity content of the paper may decrease from 5–7% to 1–2%, which causes dimensional changes [54]. Excessive running rates during roll offset printing can cause the cylinders to overheat and deform (bomb) due to the high dilatation coefficient of plastic sleeves. This deformation produces printing defects [55]. Offset printing machines (sheet) can be used for lacquering also [56]. Offset printing allows a resolution of 2400 × 2400 dpi [57]. In the mid-1990s in Europe, about 50 roll offset machines were improved with EB curing systems [58]. Narrow web label postprinting machines use impact printing methods: offset, flexo, or dry offset [40].

Letterpress Printing. Letterpress printing is largely used for labels. Label printing machines run at 100 m/min [59]. Rotational letterpress machines have been developed. They are complex pieces of equipment with numerous cylinders and stable construction [35], so they are more expensive than flexo printing machines. Letterpress-printed labels are printed with 60 lines/cm at a running speed of 60 m/min [60].

Ultraviolet drying is applied also [42]. For UV-curing letterpress machines, 15–20 cylinders are used to transport the highly viscous UV curing printing ink to the hard printing plate. Web contact is ensured by a soft, flexible pressure cylinder. So called short roller train coating devices have also been developed; these constructions have only three cylinders. A special gravure cylinder transfers the ink to the soft plated cliché cylinder [60].

Letterpress printing is the most common method for printing labels in Japan [61]. In Europe in the 1980s, 70% of labels were printed by letterpress, 20% by flexo, 5% by offset, and about 3% by screen printing. Flexographic printing is growing also, but the main printing method is actually letterpress printing [40]. Gravure, flexographic, offset, screen, non-impact, and digital printing are the printing methods most used for labels.

Tampon Printing. Tampon printing is an indirect gravure (deep) printing method. The image is transferred from the (steel) gravure plate with the aid of an oval or circular rubber pad [45]. Tampon printing has been in use since the mid-1970s. It allows on-the-spot printing of nonuniform nongeometrical surfaces with small details at high running speed. It is less expensive than screen printing or hot stamping [62,63].

Mask Printing. Mask printing can be used for direct imaging also. Here the printing tool is a perforated carrier. The ink penetrates into the perforations [64]. The image is given by built-up perforations. The mask, or master, is manufactured by using a thermal procedure. A PET film is perforated with the aid of a computer-controlled thermal head, giving a resolution of 400 dpi [65]. The master is used for a scanner and an image processor for printing or for modifying the image before printing.

Hot Stamping. Hot stamping is a lithographic printing procedure that uses predried ink. It may be considered a lamination method also [66]. Hot stamping is a dry lamination. The printing tool is electrically heated. The transfer temperature is about 120–160°C. During hot stamping the image is transferred with a pressure of 100 kN onto a surface of 150 × 250 mm [67]. The dry transfer of the image has the advantages [68] that (1) no drying is needed and (2) the image can be modified as it is being transferred.

Hot stamping allows protective lamination too and is often used for labels. It has the supplemental advantages that (1) there are no problems with pretreatment; (2) since no solvent is used there is no problem with solvent sensitivity; and (3) a metallic effect can be achieved; normally such an effects is achieved only by metallizing [69]. Hot stamping does not pollute [70]. No registry problems appear. This is an one-shot operation; no intermediate drying steps are necessary; common inks can be combined with partially metallized inks in the same image; different surface structures can be combined, and fine color nuances are obtained. The substrate must not have a geometric form, but various substrates can be used. The ink for this procedure is less expensive than the transfer film [45].

There are two types of hot stamping printing machines: one with a stamping cylinder (for circular or flat printing) and the other with a flat device. The circular/flat principle ensures continuous lamination on a large surface without air bubbles and overheating [70]. There are different types of hot stamping: plan, with structure, and with relief [71]. Hot stamping can be considered as a special case of embossing also. Embossing is a conversion operation used for self-adhesive wall coverings too [72].

The first German patent concerning hot stamping appeared in 1892 (granted to Ernst Oeser). Today materials of this type are used in many variants of color and processing methods [70]. Hot stamping die-cutting machines have been developed. These systems offer the possibility to produce decorative labels, resistant type plates, and informational labels. Depending on specified requirements, the machine can be equipped with various work units (stamping station, numbering station, laminating station, and die-cutting station). Longitudinal and cross-cutting devices are used. The machine has a head for printing and a head for die cutting [72]. Generally material is processed in reels [69]. Hot foil labels are printed a a rate of more than 10,000 impressions per hour.

Hot stamping can be used for in-mold labeling also [70]. Because of the high temperature and pressure used in this procedure (in-mold labeling is carried out with a meltable adhesive), it may be considered as a case of hot stamping.

Hot stamping of holograms is a very complex procedure (having a running speed of 70–90 steps/min) that requires images to be exactly positioned. Holographic hot stamping was developed in 1986 [69]. Hot stamping of holograms is adequate for the transfer of holograms, having a replica of the image on the stamp. It needs a very exact placement of the image. Single-pass holograms have been developed that allow holograms to be created in-line on any narrow-web press. The process uses a patented plastic imaging foil with a metal layer 20–100 μm thick that adheres to the label stock during the stamping process through a lacquer coating and a heat-activated adhesive. Hot stamping and embossing the foil directly on the machine eliminate security risks [73].

Holography was discovered by Dénes Gábor in 1948, but the first holograms that could be seen in white light were not developed until 1963. The retrievable storage of holograms was achieved in 1980 by embossing, and the transfer of multiple embossed copies of holograms was developed in 1984 [74]. The main steps in the printing of a hologram are (1) the manufacture (mastering) of the hologram; (2) replication of the master hologram; and (3) transfer of the replica. The master hologram is used to prepare the shims, the embossing tools for hologram transfer. With a special rotary embossing system the holograms can be transferred by using a nickel embossing tool on a PET film for hot stamping [74]. The carrier film should display tear resistance. Therefore a 19–23 μm PET is recommended as the transfer film for hot stamping. This is a metallized film coated with a special lacquer that can be embossed. The hologram on a special photoresist is coated with a nickel layer; this is the master embossing tool. Two methods are used to transfer the holograms: the hot stamp method and PSA lamination. Thus PSA labels can be "printed" with holograms by using hot stamping or a PSA hologram. Diffraction films are made using the same procedure. They contain two-dimensional images only. Pressure-sensitive holograms can be laminated, cut, and overprinted. For pressure-sensitive holograms, a 50 μm PET film has been used. In 1984 a special method was developed to permanently bond holograms onto a substrate for use as identification [74]. Hot foil hologram presses can apply up to four holograms in one pass on a web up to 250 mm wide [75]. Label manufacturing machines are coupled with hologram-dispensing units, allowing the application of 70–90 pieces per minute [66].

Transfer films are hot stamping films applied to large substrate surfaces. They have PET or PP (coex or oriented) carriers and a flexible and conformable ink composition [76]. Heat transfer decoration uses a preprinted paper carrier web that transfers the graphic image onto a treated container via a thermal applicator [77].

Cold Foil Transfer. The cold foil transfer method may replace hot stamping in certain applications. In these cases a special adhesive is used to fix the image on the carrier. The adhesive is cured with a UV curing system having a time window for lamination. A special cold laminating head transfers the image onto the substrate [78]. Photos can also be laminated onto labels [79].

Label Overprinting Methods. These procedures include mechanical printing; dot matrix, electronic, and computerized systems; direct thermal and thermal transfer; ink jet and laser printing; ion deposition; magnetography; and digital color printing. Direct thermal printing is the largest segment, followed by methods that use a toner (laser, ion deposition, electron beam, and magnetography) [80]. The non-impact techniques of direct thermal and thermal transfer printing are the most reliable [81]. Ion deposition has been used in the label industry since 1983; it prints at a speed of 90 m/min [81]. Ion deposition works like a copier, but uses an increased pressure for toner absorbance [82].

Printing methods can be classified as impact and non-impact printing methods. Non-impact printing methods allow low noise level, high speed, and high printing quality. The computer label business is one of the fastest growing segments of the label market [83]. Variable data bar-coded labels (routing labels for mailing, inventory control labels, document labels, individual part and product labels, supermarket shelf labels, health sector labels) are the most important sector of computerized non-impact printing. The main non-impact printing methods are [84] electrostatic and electrosensitive procedures, magnetography, ink jet printing, and thermography.

Universal product codes (bar codes) are a major reason for the industry's growth, as their use has increased at the rate of 25–50% per year [81]. Digitized variable image printing (VIP) allows customer data supplied on various media to be downloaded, manipulated electronically, and printed.

Thermal printing, which uses heat to color the face stock, comprises direct thermal printing and transfer thermal printing. Direct thermal (thermosensitive or thermochemical) printing applies a heat-sensitive printing material. This undergoes a color change when heated. This method of printing was developed in Japan for facsimile paper in the early 1970s. In thermal transfer printing there is a thermal print head (which is rapidly heated and cooled again) and an ink transfer ribbon. This method does not use a heat-sensitive label stock. Ink ribbons are based on a wax/color composition. During printing the ink, melted on impact and for a very short time (milliseconds), is transferred onto the substrate [85].

Thermotransfer printing for labels is used for different substrates (paper up to 300 g/m^2, PET, PVC, fabric, etc.). According to Hufendiek [22], transfer thermal printing allows "high density" bar codes. It was developed for paper labels. Because the ink does not penetrate into the paper, its chemical resistance and abrasion resistance are relatively low. Thermal transfer printers with a near edge type printing head allow a resolution of 300 dpi (12 dots/mm), a running speed of 300 mm/s with a 10 mm print-free area, and a running speed of 250 mm/s [86,87]. Materials weighing up to 350 g/m^2 can be printed. Fanfold material can be processed also. Single labels, strip labels, and label strips can be printed. Labels having a width of 30.2–164 mm cover almost all areas of application. Within recent years the technical performance characteristics of thermotransfer printing machines have been improved substantially. Common thermotransfer printers have running speeds of 128 mm/s [88]; high speed, high resolution thermal printers are capable of printing at speeds of up to 50 mm/s, with excellent abrasion resistance, and bar codes in either picket or ladder form are produced [89,90]. According to Ref. 91, thermotransfer printing presses work at 150–600 mm/s and give a resolution of 300 dpi [91,92]. Depending on the need for smudge or scuff resistance, wax-based, resin/solvent, or multicoated ribbons are used [89]. Compatibility guides for thermal transfer ribbons and pressure-sensitive films summarize the test results for ribbon and substrate combinations such as speed, burn setting, print quality, and smudge, scratch, and chemical resistance [93].

Presses for printing small series of labels (20,000 pieces) are recommended for color thermal transfer printing. They are run at a speed of 100 mm/s, have print dimensions of 240 × 374 mm, and can cut and perforate. Such presses produce tractor punching and transverse cuts. A resolution of 300 × 400 dpi is achieved [94]. Thermal transfer printing is used for price and weight labels and for computer-printed tamper-evident labels [95]. For instance, a direct thermally printable label described in Ref. 96 has a five-layered construction. The paper carrier is coated with a top layer that includes a heat-sensitive ink embedded in a lacquer. Some labels also have a protective overlayer. Thermal transfer printers are used for rating plates. Their resolution is 7.6–11.4 dots/mm and their printing speed can reach 175 mm/s [24].

Electrostatic printing is based on the same principle as xerography and electrography. Both give an invisible latent image. Xerography uses an intermediate image carrier. In electrography the image built up by electric charges is formed directly on the surface to be printed [97]. The main electrostatic procedures are electrophotograpy and electrography. Electrophotography builds up the image using light and an electrically charged substrate. The image is made visible with the aid of a toner. In electrography the image is built up directly on the substrate with charges. Laser, LED magnetographic, and electron beam devices generate latent images onto a photoconductive drum. Direct charge imaging (DCI) is a belt type of imaging system. It is a toner-based printing method, where imaging is given by a continuous and seamless dielectric belt rather than a photoconductive drum [98]. This belt carries the latent image. Fusing is accomplished with heat. This method theoretically allows speeds of 300 m/min with 1200 × 1200 dpi resolution. Variable information printing allows the manufacture of personalized labels in very small quantities (100–500, for example) [99].

Electrophotography is a type of laser printing and uses toner to fix the image. With laser beam printing the image is built up by a raster image processor and thermally fixed by laser beam [100]. A laser beam writes the image from the computer onto a drum [92]. The laser beam can be applied to burn out the image in a two-layered laminate also. With laser printing based on electrical conductivity, the image is given by selective discharge of the image carrier material via laser beam and thermal fixing of the image. The use of temperature for fixing limits the choice of carrier materials. High temperature resistant carrier materials such as PET can be used [22]. The chemical resistance of the printed image is low. Laser printing is used for printing overlaminated labels also [101]. Laser-printable labels (sheet) have a warranty of 4 years and may be used between −20 and +120°C [102]. Laser printing allows a resolution of 600 × 600 dpi [92] and requires temperature-resistant laminate components. Temperature-resistant, high melting point tackifiers should be used for the adhesive formulation. The printing equipment should ensure adequate fixing of the printed image (at a minimum possible temperature) using a special fixing device (Fig. 7.1).

An electron beam printing system is used for high speed in-line application of variable data [103]. With an electron beam printing system (variable image printing), the numbers, text, graphics, etc. can be printed by using a computer program. They can be oriented in any direction. The resolution of the method is 300 × 300 dpi. The machine is based on a cold pressure fusing technique that presses the dry toner into the substrate. A fastening is recommended to ensure adhesion without smudging or flaking. Varnishing, UV-cured lacquering, or lamination is suggested. The electrosensitive method uses an electrolyte as base substrate. (There are electro-erosion printers also; they work on a metallized substrate.)

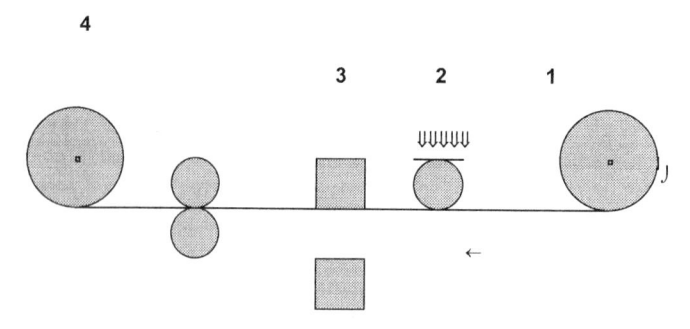

Figure 7.1 Schematic buildup of a laser printing line. *1*, Winder; *2*, laser printing unit; *3*, fixing unit; *4*, rewinder.

Magnetography applies magnetic fields for image conservation. Electron beam printers run at a speed of 100 m/min [40]. Modern laser printers print 2500 characters per second [104]. A laser cartridge generates a charged printing "form" onto a rotary light-sensitive cylinder, which attracts the toner [105]. The toner develops the image, which is fixed by fusing or pressure. Such methods completely re-image the cylinder after every impression [105]. Desktop laser printers allow a printing speed of 4–20 pages/min; printers for large computers achieve 75 pages/min [106].

An ink jet (or bubble jet) printer has small (0.03–0.06 mm) ink-filled channels that are heated, with the heating controlled by computer. The molten ink bubbles are directed via electric charges (and an electric field) to build up the image. Because of the fluid state of the ink, this procedure can be used for rough surfaces also. Unfortunately it needs a porous substrate surface (or top coat) and a long drying period. The resolution of the image is low [22]. Bubble jet printers give digital printing with a speed of 150 mm/s and a resolution of 150 × 150 dpi [90] to 360 dpi [107]. Ultrahigh speed ink jet printing systems have been designed to print at speeds of up to 300 m/min [108]. An ink jet allows printing of 1800 characters per second [109].

Dot matrix printing is an older printing method that may be used for "low density" bar codes only. Its resolution is limited to 150 × 150 dpi, and it can use a colored ribbon [92]. It is an impact method that mechanically "embosses" the ink layer onto the carrier. Because of its liquid state, the ink can rapidly bond chemically with low energy plastic surfaces also [22]. Very good chemical, weathering, and abrasion resistance is given. Unfortunately this procedure is very noisy and the resolution of the image is low.

Digital printing is used in many cases where parts of the printing equipment are computer controlled [110]. Digital printing is actually the digital buildup of the image onto a printing tool with the aid of a computer. This is the computer-to-machine or direct imaging technique. Currently direct imaging is applied in electrophotography or offset printing.

Until recently digital printing was carried out primarily on paper and was available only in black. Now the options range from black to as many as six colors on paper and film. Available technologies include dot matrix, color laser, and thermal digital printers and color copiers. LED systems, magnetography, EB, electrography, digital offset, and ink jet processes will affect the label market. Computerized digital label converting equipment can print and die cut labels [111].

Digital offset technology was developed to satisfy the demand to increase the number of colors used for printing labels. Digital offset printing is a combination of offset printing and xerography. The print image is digitally built up and transferred to an image cylinder. The electrostatically charged printing ink is transferred from this cylinder to the rubber image transfer blanket and from there to the carrier material. Recently a one-shot color technology was developed. The six colors (images) are transferred to an intermediate image transfer blanket and from this blanket in one step to the carrier. The advantages of the system are the following:

1. There are no registry problems.
2. The printing quality is regulated in the preprinting step (therefore it is possible to use less dimensionally stable materials).
3. There is no need for a drying system.
4. No change-over in printing is necessary if new products are needed.
5. No printing plate (form) manufacture is required.
6. There are no machine parts that depend on product geometry.

This procedure is recommended for production runs of less than 100,000 [112]. This is a flexible system that allows (at least theoretically) a new image for every cylinder rotation and a printing speed of 100 sheets/h. The face stock material used for digital printing has to have an electrostatically receptive surface, heat-resistant adhesive, and resistance to humidity changes. Running speeds of more than 10,000 impressions per hour are claimed. The common construction is a machine with three heads, two for printing and one for cutting [113].

Combined Printing. For printing during product manufacture (e.g., protective films or tapes), in-line presses are used, where the inking and impression cylinders are in line with the web. Printing presses with UV flexographic and rotary screen or digital offset printing are used [114]. UV flexography allows printing on non-paper substrates that was originally accomplished with rotary letterpress machines. The printing press has a flat bed and rotating die cutting and UV lamination equipment. For label printing by flexography, letterpress, or screen printing, UV-curable inks have been developed [115]. Special so-called cold UV systems have been introduced that work with only a low level of thermal (IR) radiation [116]. For PVC decorative films, water-based gravure printing is used [117]. Films used for nameplates should be printable via flexo, letterpress, and computer printing methods. Such films must have a matte, metallic appearance similar to that of anodized aluminum. The matte surface has to absorb ink and maintain sharp graphic contrast, good color intensity, and excellent legibility. It has to be printable with a range of ordinary or special computer printers. Cationic printing inks have been tested industrially in Europe since 1993 [118]. Free radical UV systems contain acrylics as the base macromolecular compound. Cationic systems include low viscosity epoxies. Both are 100% solids in comparison with common solvent-based inks, which have 30% solids.

Important growth in the label industry has come from blank labels or nearly blank labels [119], for which more printing is done at the point of use. Therefore combined printing methods have been among the main processes used in Europe (letterpress, flexo, offset, and digital) [120].

Lacquering. The standard pretreatments carried out by foil manufacturers are not sufficient to meet the extremely high requirements for printability in the field of computer printing. Although PVC foils can be quite easily printed, OPP foils may require considerable effort to satisfy the special requirements. The possibilities for chemical print treatment are often limited if the result is to be highly transparent. Certain pretreatments can be achieved only with opacity. Therefore lacquering of the face stock material is a necessity in many cases. Synthetic carrier materials with a surface coating can be printed with the usual paper-printing inks in all normal printing processes; they can be used in copiers, laser printers, and color ink jet printers [121].

There are a number of ways to produce a white label based on a clear colorless carrier material. A white pigment can be added to the adhesive, or a white top coat can be applied to the clear film. High opacity ensures accurate scanning [122]. High reflectivity given by a glossy surface offers high print contrast for bar codes, which is necessary for scanning accuracy and high first-read rates. Dimensional stability and abrasion resistance are important in maintaining bar code scanner and visual readability. A vapor-deposited aluminum layer is sensitive to mechanical stress (wear) and oxidation and must not remain unprotected on the front of the label. Therefore it is covered with a top coat [123]. Such a primer offers excellent conditions for standard printing processes. Full surface lacquering of paper labels followed by UV curing improves their gloss [35]. Linerless labels are supplied as a continuous tape-like monoweb material. A special coating on the top surface of the label prevents blocking of the adhesive layer [124].

Varnishing can be done on-line or off-line by dispersion or electron beam curing. There are a wide variety of dispersion, ink unit, oilprint, functional, and NC varnishes. UV, cationic, and calendered varnishes can be added. Varnishes can be applied by using water pans, varnishing units, varnishing modules, or coating units (towers). When varnishing is done within the offset press, the varnish is transferred to the substrate directly from the printing plate. Direct varnishing ensures that thicker varnish and ink coatings can be deposited, improving the mechanical properties and gloss of the product. According to Hummel [9], varnishability is a quality parameter for semifinished products for pressure-sensitive labels. Common lacquering methods are described by Hadert [125].

According to Ref. 126, gravure printing gives the best results; three- or four-cylinder coating devices with a dosage based on cylinder rotation speed allow the coating of up to 4 g/m^2 (wet) lacquer. Lacquering is carried out by using 60-line gravure printing to a gravure depth of 45–50 µm [127]. UV-cured lacquer for labels has a coating weight of 2–4 g/m^2. The coating device is a gravure cylinder (80–120 lines/cm, 12–20 µm deep) [128].

Printing of Tapes

Meinel [129] states that requirements for printable tapes include very different performance characteristics: the carrier surface must be abhesive to ensure release properties (i.e., unwind resistance) but at the same time it must exhibit good affinity to the printing inks and good anchorage.

Tapes can be printed via gravure, flexo, offset, or screen printing procedures, like labels [130,131]. Tapes with OPP as carrier have been printed with solvent-based inks. Water-based technology was developed toward the end of the 1980s [132]. Sealing tapes based on PVC carrier are printed by sandwich printing (flexo for the top side and gravure printing for the back side between the film and the adhesive) [12]. Vinyl marker tapes must be capable of being printed on a standard printer with vinyl ink [133]. Unfortunately, whereas the printability of polyethylene tapes can be evaluated by testing the surface tension (wettability), the printability of PVC tapes cannot [129].

The most common method for printing tapes is flexographic printing [134]. Chromed steel gravure cylinders having 60–100 engraved pyramidal per square centimeter points are used together with steel or plastic (polyamide) blades. A decade ago the majority of tapes (70%) were only one color. Now the majority of tapes are printed with multiple colors. Common three-color printing machines for tapes have a width of 100–300 mm and run at speeds of up to 200 m/min. Tapes are generally flexo printed [135].

Seng [118,136] discussed the printing of polypropylene tapes using UV-cured systems (flexo printing inks). As is known, radiation drying of printing inks is based on microwaves and IR radiation. Radiation curing uses UV light, X-ray radiation, and electron beams that are capable of breaking the C—C, C—H, or C—O links (i.e., have an energy of more than 3.6–4.3 eV). Ionic (cationic) and free radical polymerization can be used. Cationic initiators are onium salts that give an acid by photolysis (see also Chapter 4). The UV light absorption domain for most proposed cationic systems is 275–450 nm. Good results are obtained with inks based on cycloaliphatic epoxides (chain-opening polymerization via oxirane rings). Ionic UV systems possess the advantages that they are not inhibited by oxygen and they result in less shrinkage and less toxicity. Unfortunately, such systems are less reactive, react more slowly (curing is not finished after radiation; postcuring is necessary), and exhibit lower penetration. Therefore double-cure (dual-cure) systems are proposed. A peroxidic initiation is followed by UV curing.

Printing of Protective Films

Generally, protective films do not carry information. They are printed for opacification, as a technological aid, or for publicity. Generally flexo printing is used. In special cases gravure or screen printing is applied. Taking into account that certain protective films (e.g., films for plastic plates) are postprocessed, printing

Conversion Properties

inks used for protective films have to support the elevated processing temperatures. For instance, inks used for protective films for PC plates have to be stable up to 160°C. The ink must be compatible with the protective lacquer on the mask and should not penetrate through the mask during cold or hot line bending of the sheet or during drape forming.

Printing-Related Performance Characteristics of the Carrier Material

Some special performance characteristics of the carrier material allow evaluation of its printability. These are related to the dimensional stability of the material. Shrinkage, lay flat, smoothness, and stiffness affect printability. It is evident that direct printability tests are also necessary in order to evaluate printability. Label printability can be evaluated by evaluation of the print quality (visual appearance); the color tone of the material, and the curling of printed material [54]. Actually, print quality is evaluated automatically at productivity rates as high as 100,000 labels per hour [86].

Shrinkage. Dimensional stability includes the stability of the geometrical dimensions and shape (form) of a PSP. Shrinkage is a phenomenon that changes the original dimensions of the PSP with or without changing its original shape. This is the result of built-in and "processed-in" tensions (Fig. 7.2).

The carrier material (e.g., plastic film) is tensioned during manufacture and conversion. The material suffers detensioning (relaxation) as a function of time and temperature. This is the manufacturing-induced component (S_M) of the shrinkage. For instance, the shrinkage of extruded PVC is partially the result of tensions arising during extrusion of the film (see also Chapter 6). Cast or calen-

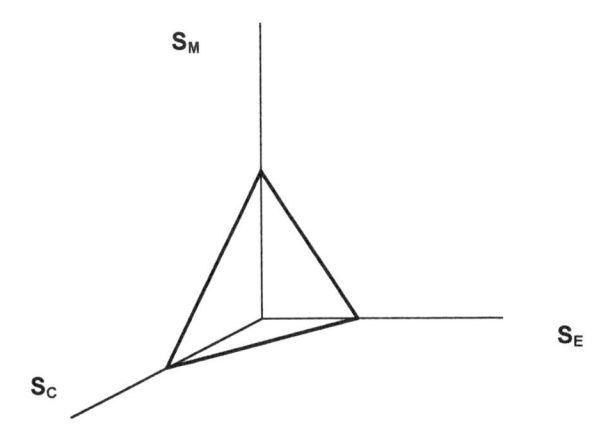

Figure 7.2 The main components of shrinkage. S_M = manufacturing-induced component; S_E, environmental component; S_c = coating-related component.

dered films are more dimensionally stable. On the other hand (independently of the manufacturing procedure), each material is influenced by environmental factors: the temperature and substances that migrate from the atmosphere into the carrier or from the carrier into the environment. In the case of paper, there is an equilibrium with the humidity of the air. In the case of plastics, solvents can interact with the film and in special cases (e.g., cellulose derivatives, polyamide, polycarbonate) with water. This is the environmental component (S_E) of the shrinkage. Other influences result from coating the carrier with adhesives or printing inks (S_C). For instance, the film can be printed by means of screen printing, which uses solvents. The solvents can migrate into the adhesive and dilute it. Certain solvents can penetrate into the film also. Therefore screen printing of PVC increases its shrinkage. In a similar manner, the elastic modulus of paper (E_M*) is related to the humidity elongation coefficient (β) in the machine (M) and cross (C) directions according to the correlation [137]

$$E_M \beta_M = E_c \beta_C = \text{constant} \tag{7.1}$$

where the humidity elongation coefficient is a function of the relative humidity of the air (H_a), its variation (ΔH_a), and the variation in the length of the sample (Δ_l):

$$\beta = (\Delta_l \times 100)/(l \times \Delta H_a) \tag{7.2}$$

It should be mentioned that shrinkage, i.e., dimensional stability, is also a function of temperature resistance. Generally the processing temperature range is given by the film supplier for common carrier materials. Shrinkage is due to tensions in the film, but shrinkage-resistant adhesives (e.g., crosslinked acrylics), solvents used in the printing process, and plasticizers affect it [138]. The importance of shrinkage differs for different PSP classes. Acceptable shrinkage values vary depending on the postconversion steps or end use of the product (Table 7.3). Because of the high print quality of labels and the mutual interaction between la-

Table 7.3 Shrinkage Test Conditions and Values for Selected PSPs

Product	Carrier	Comments	Temp. (°C)	Time (min)	Shrinkage value (%)	Ref.
Closure tape	OPP	15 × 15 cm sample	125	10	1.0	—
Protective film	Polyolefin	—	120	5	1.1	—
Plotter film	PVC	Bonded on aluminum	70	2880	0.25–2.0	139
Medical plaster	PVC	DIN 53374	—	15	<7	140
	Polyolefin	—	125	15	26	
Label	—	DIN 40634	—	—	1.1	9

bel and printing, the control of shrinkage plays a weighty role in their quality. In the domain of tapes there are applications where shrinkage is required.

Shrinkage of the carrier can be used for special PSPs like heat-shrinkable insulating tapes; during application, the dimensions of these tapes decrease by up to 10–50%. As an example, bilayered heat-shrinkable insulating tapes for anticorrosion protection of petroleum and gas pipelines have been manufactured from photochemically cured low density polyethylene with an EVAc copolymer as adhesive sublayer [141]. The linear dimensions of the two-layered insulating tape decrease depending on the degree of curing (application temperature 180°C). The application conditions and physicomechanical properties of the coating were determined for heat-shrinkable tapes with 5% shrinkage at a curing degree of 30% and shrinkage test force of 0.07 MPa [142]. Shrinkage of self-adhesive EVAc-based protective films is used as a quality criterion.

Lay Flat. Lay flat is the stability of the shape of a PSP. For laminated products it is the sum of the geometrical stabilities of the components. Inadequate lay flat is generally manifested as curl of the solid-state components. Like shrinkage and most other dimensional changes during processing and application, curl is a result of tensions or their relaxation in a carrier material. Such tensions may appear as the result of the composite structure of the laminate or laminate components. Paper itself is a composite built up from fibers. Its humidity balance produces changes in the fiber diameter (across the original machine direction) [143]. These changes may be different in the middle and external paper layers. Therefore paper curl can be caused by humidity changes also. In this case the stress σ in the paper layer is the result of deformation by mechanical forces ϵ, and by humidity e [137]:

$$\sigma = E(\epsilon - e) \tag{7.3}$$

where E is the modulus. It is well known that to attain a better humidity balance, carbamide, glycerine, or carbamide-calcium nitrate is added to paper [143].

Some films have an anisotropic buildup. Edge curl (corner drag) of printed films depends on the film thickness, the thickness of the printing ink layer, differences in the elasticity of the printed and carrier layers, and the diffusion of low molecular weight substances in the carrier film [144]. In particular the printing ink film is hardened more at a high drying speed and is less extensible at room temperature than PVC film. This can lead to curl. "Frozen" processing tensions in the film may lead to curl also. When plastic is melted, forced through a slit in a die, subjected to pressure by finish rolls, and cooled rapidly, it develops strains. When the sheet is later subjected to heat and/or solvents, it relaxes and warps, buckles, bows, or "dishes." The larger the piece and the heavier the gauge, the greater is the potential for problems. The shear rate during the processing of raw materials is quite different for calendering (e.g., $10-10^2$ s^{-1}) and extrusion (10^2-10^3 s^{-1}); therefore, the tensions that develop in the materials are different also [145]. Curl of the film edges may be due to, or amplified by, screen printing. Drying of labels printed by screen printing may produce tensions in the material depending on the printing ink used. These tensions cause curling. Ideally, printing inks have to allow rapid drying and optimum screen opening. Very elastic acrylic inks with alcohol as solvent may be processed with low curling. If a printing ink on PVC basis (copolymer) is used for PVC, low curling occurs. High gloss printing inks on an acrylic basis give low curling also. Unfortunately, mixed compositions that allow high printing speed and open screen produce high curling for PVC. Curling depends on coating weight and machining conditions also. Nip pressure and web tension should have low values [24].

Curling depends on the type of adhesive also. Because of the lack of sensitivity to atmospheric humidity, polyvinyl methyl ether has been added to formulations for paper tapes to avoid curl. As proposed by Sigii et al. [146], a water-based acrylic PSA with good cohesive strength that prevents adherends from recovering from bending is based on ethyl hexyl acrylate, ethyl acrylate, and acrylic acid.

Curling is a function of the laminate buildup also. The use of a stiff release liner (e.g., 80 lb kraft liner) for labels helps to avoid curling [147]. The liner plays an important role in the functionality and cost of most PSPs. Tear strength, dimensional stability, lay flatness, and surface characteristics are the most important features of the release liner [148]. Tear strength is important for label conversion and dispensing. Good gauge control and liner hardness (lack of compressibility) are the key parameters. Dimensional stability, the ability to maintain the original dimensions when exposed to high temperature and stresses, is important for print-to-print and print-to-die registration. If the liner stretches under heat and tension, the labels may be distorted. Liner stretch can affect label dispensing also. Lay flat is very important for sheet labels or pin-perforated and folded labels. Labels that are butt-cut, laser-printed, and fanfolded must lie flat to allow them to be properly fed and stacked. Liner lay flat is affected by resistance to humidity, dimensional stability at different temperatures, and the stiffness (thickness) of the liner. The use of radiation-cured or low temperature thermally cured silicones allows a better humidity balance of carrier paper and better lay flat.

When unwound from the roll, tapes should have no tendency to curl [149]. It should be mentioned that curl of tapes may be due to inadequate carrier design (two-layered coex films with different mechanical properties), processing, and transport tensions or printing. Shrinkage and lay flat are very important quality characteristics for heat-laminated protective films also. Lay flat influences the control of delamination, the separation of the protective film from the protected item.

Films made with a machine direction orientation display outstanding stiffness in the machine direction and flexibility in the cross direction. Because an isotropic material is necessary for labels, ideally the films should be biaxially oriented. Lay flat in printing is complemented by on-pack wrinkle-free squezeability [119]. In other words, lay flat and conformability are contradictory requirements. Migration of plasticizer from the substrate, insufficient adhesion due to slip and release material layers on leather and plastics, insufficient adhesion on round adherends, and damage due to UV light have been the main problems with labels for many decades [150]. Paper labels should be laid out in the long grain direction to provide maximum conformability [151].

Smoothness. Liner and face stock smoothness affect several performance features. The adhesive surface is a replica of the face stock or release surface. If it is rough, the level of initial adhesion will be reduced. If a rough label is laminated with a clear face stock material, its clarity will be negatively affected. If the liner is too smooth, air entrapment during dispensing can become a problem. The roughness of the label back can affect web tracking on the printing press. Smooth liners can weave, causing cross-web print registration. Also in laser printing the roughness of the liner plays a critical role in sheet feed ability and tracking through the printer. Smoothness may influence telescoping also. Mounting tapes used in the fashion and textile industries are protected against telescoping during storage with a woven or nonwoven supplementary layer [152].

Stiffness. Stiffness influences the conformability of the PSP in the printing press and wrinkle buildup. It affects die cutting and dispensing also. For instance, glassine liners are hard and densified; they are used for high speed dispensing (see Chapter 3 also).

Elongation During Printing. During printing, tensions may build up in the material and cause dimensional changes. They depend on the forces applied in the machine and on the temperature in the machine, and they are also a function of the material characteristics. There is a correlation between machining conditions and dimensional changes in films. There is a correlation between tensile strength during printing, tensile strength during print drying, and elongation during printing for various films. For instance, oriented polypropylene exhibits a higher elongation at 80°C than polyamide at the same temperature [153]. There is a correlation between the gravure roll inserting tensile strength and elongation during printing. The tensile strength of the material influences its elongation during lamination also. Lamination temperature also affects elongation.

Wrinkle Buildup. Wrinklebuild up during printing is due mainly to overdrying of the paper carrier. The humidity content of paper may decrease to 1–2%, changing its dimensions. This phenomenon depends on the stiffness of the paper. For papers of lower weights (40–80 g/m^2), the buildup of wrinkles is more accentuated. Such papers wrinkle with shorter wavelengths than papers of 100 g/m^2. Wrinkle buildup depends on machine construction also. As stated in Ref. 154, tension fluctuations can cause wrinkles. Rewinders with independent indexing arms allow reel changeovers at full speed without any tension flutters.

Conversion of Labels

Conversion of labels is the production of narrow rolls (or sheets) of finished labels, printed, die-cut, slit, and rewound. A revarnishing may be necessary also, requiring a separate drying unit (e.g., UV dryer) [155]. The conversion of label material includes the operations necessary to transform the coated laminated web into labels by cutting, die cutting, printing, perforating, postprinting, and lacquering. MD or CD cutting (slitting) of the web are common operations for tapes and protective films. Die cutting is generally the operation that transforms a PS web into a label (or business form) laminate, but it is used for special tapes also. Printing of the full width web is a common operation for tapes and protective films also. Postprinting is specific for labels. Postlacquering is required for special label applications only. Narrow web printing offers the advantage that it can be used for other operations also [105]. Users can modify or attach a variety of nonstandard devices to manufacture different products. An in-line sheeter nonimpact variable information printer, or web tinter can be attached. Flexographic varnishing units, plough folders, spot and pattern gluing modules, self-sealing adhesive modules, and card and label applicators can be associated with the main printer. Such equipment can be used to produce forms also. Electronic web tension devices, web guide sensors, and reinserters of preprinted web are used also. In-line sheeters allow form producers to compete by extending the cut sheet laser and ink jet market. Batcher stackers (with chipboard inserters) and wrapping machines complete the equipment. Certain form-printing machines offer custom-built commercial jaw folders to produce complete printed parts. Certain security labels have a complex construction. Their conversion includes a range of operations. Such labels are used for ware hanging, advertising or product information text, bar codes, and electronic security elements [82].

2 CONFECTIONING PROPERTIES

Confectioning requires roll and material handling equipment; slitter, sheeter, and surface treating apparatus; die cutting and embossing equipment; and windowing and tag machines. Electronic confectioning machines for labels cut, die cut, punch, and perforate the paper at running speeds of up to 200 m/min. The dimensions of the processed materials and their characteristics can vary widely. Thin films of 15–20 μm and cardboard of 300 g/m^2 can be processed [156].

Special preprocessing and postprocessing operations are related to printing [84]. Thus, humidification/dehumidification, ionization/deionization, decurling, coating/varnishing, cleaning, line/hole punching, and preprinting can be carried out before printing. Varnishing, overprinting, laminating, die cutting/waste re-

Conversion Properties

moval, cross perforating, slitting, and quality control should be carried out after printing.

2.1 Cutting

The general aspects of cuttability are discussed in a detailed manner in Ref. 1. Cutting is the operation carried out to transform the full width web into end use width reel material by slitting or into sheet material by transverse cutting. Simultaneous cutting in both directions is required for plotter film (pattern paper) cutting. Die cutting is a partial cutting of the laminate into laminate sections. The importance of cutting and die cutting differ for various PSP classes (Table 7.4).

Products supplied and used as continuous weblike materials have to be cut into narrow webs. Here slitting is the main cutting operation. These products include protective films and the main tapes. In-line slitters remove the edges of (trim) the material and also divide (slit) the web laterally into two or more narrow widths (ribbons) that are formed into individual rolls on the rewinder. Off-line slitters receive full-width rolls and form rolls with small diameters. Off-line slitters are normally used when the finished rolls have smaller diameters than the unwinding roll and when there are many narrow ribbons. The jumbo size coated rolls are transferred from the rewinder to a special slitter-rewinder that produces small retail or consumer size rolls of tape. Narrow-width rolls range from 12 mm to 25 mm in diameter and contain 3–30 m of tape per roll. Sealing tapes for laminated glass are supplied preslit [157]. Common slitting machines (600/800/100) have a changeover time of 40 s for finished rolls [158].

Other products are finite elements cut out from the continuous web. Here cutting is carried out throughout the whole section of the web (e.g., transformation

Table 7.4 Main Characteristics of the Cutting Process

PSP	Product form	Cutting operation	Cutting line	Cut material
Label	Reel	Slitting	Linear	Laminate
		Guillotining	Linear	Laminate
		Die cutting	In plane	Laminate component
	Sheet	Guillotining	Linear	Laminate
		Die cutting	In plane	Laminate component
Tape	Reel	Slitting	Linear	Laminate component
		Die cutting	In plane	Laminate component
Protective film	Reel	Die cutting	In plane	Laminate component
Plotter film	Reel	Slitting	Linear	Laminate component
		Die cutting	In plane	Laminate

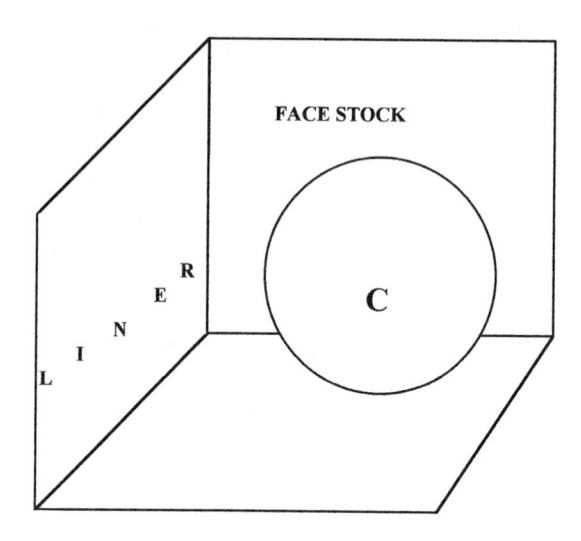

Figure 7.3 Product buildup parameters influencing cuttability (C).

Table 7.5 Dependence of Cuttability on the Flexural Resistance of the Laminate[a]

Product code	Flexural resistance (mN/10 mm)	Cuttability	Comments
1	226	Inadequate	Permanent, RR, HMPSA
2	331	Fair	Permanent, WBPSA
3	336	Inadequate	Permanent, WBPSA
4	384	Inadequate	Removable, HMPSA
5	483	Fair	Permanent, WBPSA
6	607	Very good	Permanent, RR, crosslinked HMPSA

[a]Paper-based laminates were tested by guillotine cutting.

of reel material into sheet material) or through the face stock material only (die cutting) to preserve the weblike character of the product, i.e., to ensure automatic handling, conversion, and end use.

Cuttability depends on the plasticity and elasticity of the material. Both the solid-state components of the PSP and the adhesive influence cuttability (Fig. 7.3). For instance, linear low density polyethylene is neither foldable nor stiff. It

Conversion Properties 551

is difficult to cut or die-cut it and to punch it because of its high elasticity. Its extensibility is a disadvantage for the converter [159]. In this case special cutting tools with a higher pressure should be used.

The influence of the solid-state components and that of the adhesive on cuttability is discussed in detail in Ref. 1. As stated there, the mechanical performance characteristics (e.g., flexural resistance) and geometry of the solid-state components are decisive concerning the cuttability of pressure-sensitive laminates (Table 7.5).

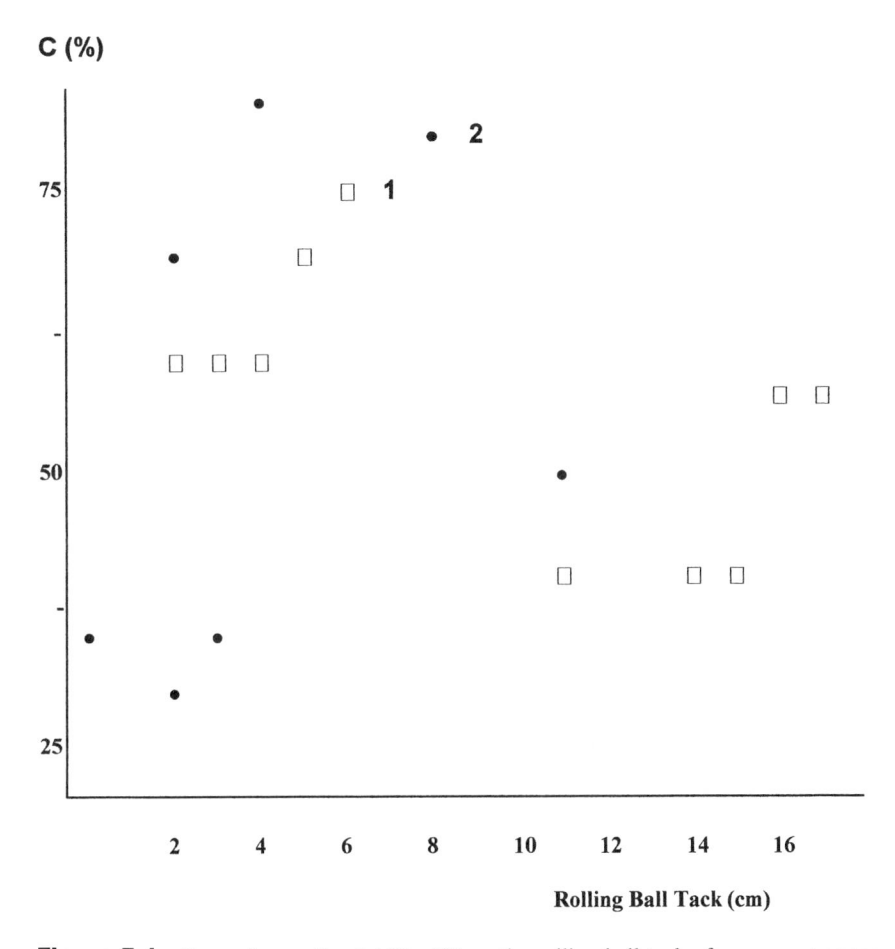

Figure 7.4 Dependence of cuttability (C) on the rolling ball tack of permanent paper label laminates. ([B]) Solvent-based rubber resin adhesive; (●) water-based acrylic adhesive.

As discussed by Medina and DiStefano [160], formulating modalities that improve the high frequency modulus of the adhesive will facilitate conversion (see also Chapter 3, Section 1). This is confirmed by the data of Fig. 7.4 also, which illustrate the dependence of the guillotine cuttability on the tack of the adhesive. Soft, high tack adhesives exhibit low cuttability. Waste stripping properties of an adhesive are very important. Such characteristics depend on the adhesive properties too. The slitting properties of the tapes strongly depend on their rewindability. Rewindability (considered a function of the peel resistance on the carrier back side of tapes) is a function of adhesive performance also. However, in the case of die cuttability, the deformability and tear characteristics of the solid-state components and the global rigidity of the laminate play a major role.

A special case of cutting is the processing of plotter films. For such films cuttability is really an end use performance characteristic. Texts and logos are cut with a speed of up to 400 mm/min. Roll and sheet cutters have been developed that not only have a cutting mode but can also be used for drawing [161]. Plotter cutting is very high speed cutting. For instance, tabletop plotter cutting machines work at a speed of 800 mm/s and can process films with widths of 50–634 mm [162].

2.2 Die Cutting

Conversion involves processes such as die cutting roll stock and guillotine cutting sheet stock. Die cutting is the technical operation allowing the transformation of the weblike face stock material into discrete items such as labels. Therefore die cutting has great importance for labels. For special medical tapes, where discontinuous constructions are used to allow breathing and diffusion, die cutting is important also [96]. The various types of cutting include cutting through, cutting out of the top layer, perforation, engraving, and marking [163].

For labels or label-like products, the cutting equipment is generally integrated in the converting/printing line. For instance, type plate printing units possess an automatic cutter [24]. Operating on-line finishing equipment supposes the existence of a die cutter and matrix stripper, re-reeler, sheeter, etc.

Die cutting requires that the label be designed and its shape and the distribution of label area on the web be calculated to allow web (and waste matrix) transport and removal after die cutting. The unusable areas surrounding the cut labels are stripped from the release paper and wound into a roll. The cut labels remain attached to the release paper and are wound into rolls also.

Flat bed cutters, rotary cutters, and wraparound dies are recommended for labels. Die cutting machines can be either rotary or non-rotary. Nonrotary mechanical die cutting machines have a maximum speed of 350 tacts/min [164]. Flat bed dies from 7 to 12 mm in line height using laser technology or grids, rotary dies with a die plate 0.41–1 mm thick, and magnetic or nonmagnetic, milled or etched versions have been developed. The development of form–label combinations where a die-cut siliconized patch transports one or more butt- or kiss-cut paper labels forced the development of rotary die cutting technology coupled with matrix waste removal.

Stanton Avery developed the rotating die cutting machine [165]. Such machines include a geared die cylinder with bearers and an anvil roller assembly. Flexible dies that wrap around a stainless steel cylinder containing magnetic inserts are less expensive. Flexible steel strip cutting knifes have been known since the 1930s. Rotary die cutting machines can cut cardboard as thick as 200–500 μm [166]. The clearance and tolerances of rotary die cutting machines depend on the nature of the material to be cut [167]. The critical cutting range, i.e., usable clearance, for common materials is 0.010–0.015 mm. For cutting through PE films or nonwovens it is about 0.005 mm. Cast PE exhibits better die cuttability and precision registry [168].

Steel strips have been used as cutting tools in the cardboard industry since the 1930s [169]. Magnetic cutting plates are used for a cutting height of 0.42–1.00 mm. A cutting angle of 60–100° is used [170]. A dual-knife sheeter for high speed work is a rotary dual-knife cutter with a delivery and stacking system. Metal-to-metal contact occurs between the upper and lower knives. For such machines sheet length and squareness accuracy of ±0.2 in. can be maintained continuously [171].

As discussed in Ref. 1, cutting and die cutting depend on the solid-state components of the laminate (face stock and release liner). Their quality and combination are very important. A good die cutting is provided by a uniform caliper densified paper liner [168]. Cuttability tests have been carried out by Hartmann et al. [172] using PE, PE/PP, PET, and PET/SiO_2 and paper (80 g/m^2) as liner and face stock material to test cutting knife wear. The influence of the face stock material and release liner was studied with rotary die cutting [173] also. The investigations show the importance of an appropriate release liner for each face stock material, depending on its cutting and tear mechanisms. The versatility of the release liner as a cutting basis depends on its chemical nature, mechanical properties, and dimensional tolerances.

The thickness and density of the paper used as face stock material are less important [172,173]. Paper can be compressed and split. Therefore, in paper cutting the deformation of the cutting tool is minimum and cutting forces are minimal. Adjustment of the cutting tool is easier and stable. Cutting of polyethylene includes splitting and separating the split polymer into two separate phases. Because of the plastic flow of the material there is a need for a hard bottom surface with low dimensional tolerances and no compressibility. In this case, cuttability (its quality and speed) depends on the density of the release paper used. Kraft paper is not recommended as a release backing for PE labels. The best results are obtained with high density glassine paper and PET. It is evident that material combinations for face stock and release liner lead to the best die cuttability results. The cutting of PET is similar to that of paper. It is a crack- or slitting-dependent process, initiated at the first contact points between the knife and the PET face stock materials (zip effect). Tear resistance influences cuttability. Therefore release papers with broad tolerances give the best cuttability results. Polyester needs a hard but low quality glassine paper as a release (backing) material. Therefore one can state that the cuttability of different face stock materials can be estimated as

$$\text{Paper} > \text{PET} > \text{PE} \tag{7.4}$$

and the versatility of different release liner materials may be estimated as

$$\text{PET} > \text{glassine paper} > \text{kraft paper} \tag{7.5}$$

Voluminous and fragile (inelastic) solid-state laminate components generally improve cuttability. Therefore postlamination blowup of the laminate improves cuttability. According to Ref. 174, a photocrosslinked acrylic-urethane acrylate formulation used for holding semiconductor wafers during cutting and releasing the wafers after cutting, with enhanced cuttability, can be prepared by crosslinking and blowing up the adhesive. Such a releasable foaming adhesive sheet can be prepared by coating a blowing agent (e.g., a microcapsular type) containing PSA (30%) as a 30 μm dry layer [175]. This adhesive is removable and improves cuttability.

Cuttability of label films can be evaluated by measuring the cutting forces in both the machine and cross directions. For optimum die cuttability, the MD/CD ratio of the measured values has to be about 1. As illustrated by the data of Fig. 7.5, the manufacturing parameters of the plastic film strongly influence its die cuttability. The choice of BUR together with the thickness of the film (for a given formulation) allows regulation of the die cuttability.

As stated in Ref. 176, current laser technology offers a limited power/cost ratio; therefore in-line laser cutting is not viable. Several laminates show undesirable side effects (discoloration of the carrier and differential cutting of paper and liner) because of the thermal effects of the laser beam. More perspectives are given to laser marking, where special foils accept laser marking. In this case the minimum width that can be engraved is about 1 mm. Thermal die cutting was developed for pressure-sensitive decals. According to Ref. 177, a 0.004 in. cutting line heated at 300°C will cut vinyl but not paper. Blank self-adhesive labels precut on silicone backing paper are also manufactured.

Conversion Properties 555

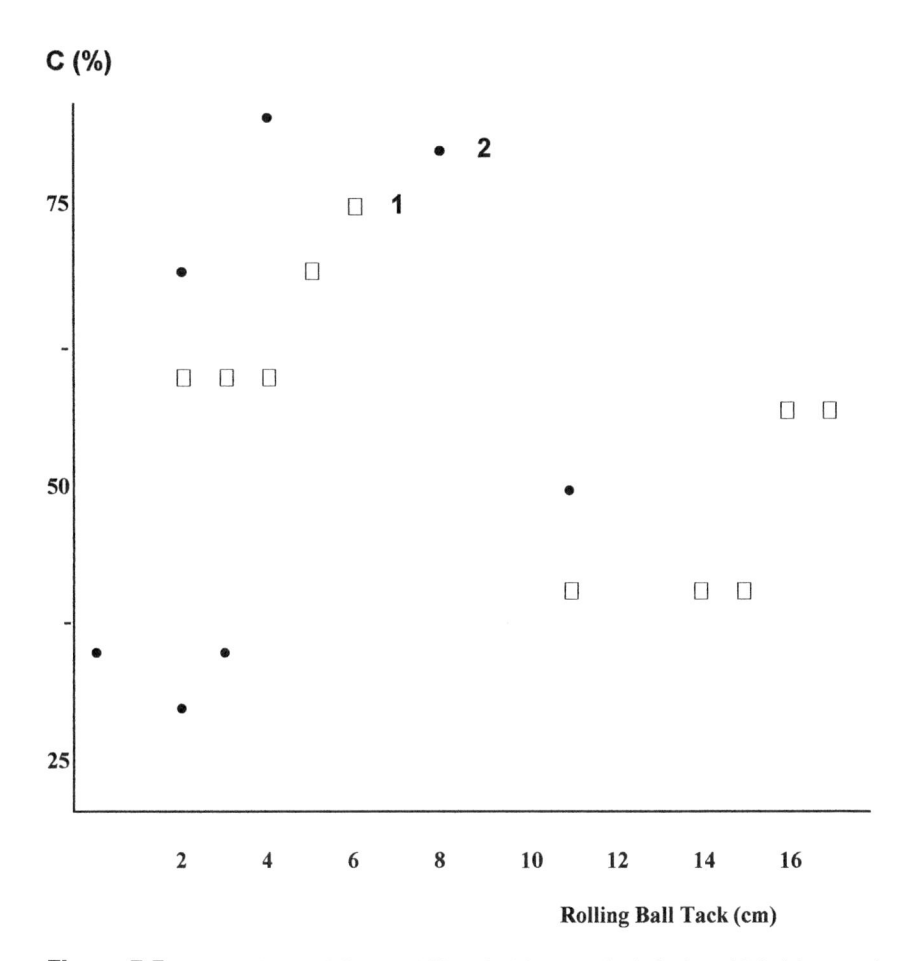

Figure 7.5 Dependence of die cuttability of a blown, polyolefin-based label face stock material on the film thickness and BUR. Symbols as in Fig. 7.4.

2.3 Perforating, Embossing, and Folding

Perforating and microperforating are common operations in the conversion of labels and tapes. Perforating is used mainly to weaken the carrier material to achieve detachability. Microperforating is performed to achieve porosity. Label die cutting, liner perforation, and pin-feed hole cutting (for labels printed on a dot matrix printer) are necessary also. Slits or perforations in the face stock improve the tamper-evident properties of labels [178]. For computer use, pin-feed hole punching is necessary. It can be accomplished through the use of male/female dies. Commonly a standard rotary die is used to die cut through the liner. The labels can be either rolled or fanfolded. Special acrylic adhesives with excellent adhesion at high temperature are suggested for PSPs subjected to the punching process [179]. Some price labels require perforation for transport [180]. Special pharmaceutical labels can have take-off (detachable) parts also to allow multiple information transfer. Parts of the label remain on the drug or on the patient [181]. The laminated structure is perforated to allow multiple detachment (i.e., multiplication) of the labels [182]. An adhesive tape used for holding together printer paper is perforated along the centerline in the long direction [183]. Electrical insulating tapes used for taping may be perforated also [184]. Common mechanical perforation machines have 25, 40, 50, or 70 teeth per inch [185].

Microperforating can be used to transform some carrier materials into a porous web [186]. Microporous webs are needed for medical tapes, condensed water free packages, etc. Various methods are used for perforation: hot needles, liquid or gas jets, ultrasound, high frequency lasers, and electrostatic microperforation. Hot needles are used for PE and PP films and nonwovens. The hole diameters are 200–500 μm. Webs having a width of 1500 mm can be processed at a speed of 10–30 m/min. Gas and liquid jets are used for perforating soft webs (foam, nonwovens, board). Thermoplast-like polyolefins and PVC can be cut with ultrasound. High frequency cutting is limited to thicknesses of 5–30 μm. Laser perforating is used for narrow webs. It allows running speeds of 600 m/min and is used mainly for fine paper. The hole diameters are 50–200 μm.

Electrostatic microperforation can be used only for electrically nonconductive materials. Webs with a width of 50–1000 mm and weights of 5–100 g/m^2 are processed at a running speed of 300 m/min. Paper, plastics (LDPE, PET, PP, and PVC), laminates (PET/PE), nonwovens, and other materials can be processed. The diameter of the holes is 2–70 μm, the pore number of 1.6 Mio/m^2. The interpore distance is 1.0 mm. The reciprocal distance of the stripes is 0.25 mm. The procedure uses electrodes with high voltage and high frequency (500–5000 Hz). Microperforations (diameters of 2–70 μm) and macroperforations (diameters of 50–500 μm) can be realized [186]. Special electrodes allow the manufacture of perforations having a diameter of 2–5 μm by using corona treatment [187]. Flame perforation is used also; this procedure allows the manufacture of hundreds of holes per square inch with a hole diameter of 0.4 mm [188].

Embossing is generally carried out during the manufacture of plastic carrier films (see also Chapters 6 and 8). Embossing of metal films, metallized PE, and paper is carried out more easily and runs at much higher speeds (up to 400 m/min) than that of PE [189]. Microembossed films are used for image holograms, two- and three-dimensional background designs, decorative laminates, and labels [190]. Embossed paper or film provides mechanical release.

Conversion Properties

2.4 Winding Properties

Winding properties influence the manufacture of carrier material (see Manufacture of Plastic Carrier), coating, laminating, and converting (see Chapter 6, Section 3), and the end use of PSPs. Winding properties play a special role for PSPs applied as webs (tapes and protective films). Such characteristics are primary conversion parameters for tapes because of their high degree of confectioning (slitting). Table 7.6 illustrates the common values of unwinding resistance for tapes. As can be seen from Table 7.6, the unwinding force of common tapes is situated as order of magnitude in the debonding force (peel resistance) domain of removable labels or protective films.

The stresses that appear during winding depend on the running speed of the web. Their order of magnitude and their degree of influence differ during film manufacture, coating, and confectioning. In the case of film manufacture the maximum winding speed is 20–140 m/min for blown films, 120–400 m/min for cast film, and 280–350 m/min for biaxially oriented film [196]. Common coating speeds are 100–200 m/min [81]. The rewinding speed for a narrow web is generally lower. For instance, one tape specification [197] states that when unwound at a rate of 7.3 m/min the adhesive of the tape shall not transfer to the adjacent layer. During unwinding, a tape may tear or split [198]. As shown in Ref. 129, the unwinding resistance (R_w) is a function of temperature (T_w) and unwinding speed (v_w):

$$R_w = f(T_w, v_w) \tag{7.6}$$

Table 7.6 Unwind Resistance of Different Tapes

Grade	Carrier material	Unwind resistance Value	Method	Ref.
Flame resistant	PVC	11.1–18.1 N/32 mm	OCT-906	192
Closure	PE	1.4 N/10 mm (30 m/min)	BDF A 07	—
Special	PP	1.5 N/50 mm	JIS/Z 0237	193
Special	PVC	0.4 N/20 mm	—	194
Packaging	PE	0.02 N/10 mm	—	195

In a similar manner, tape tear number (N_t) depends on the unwinding speed and on the temperature according to the correlation [198]

$$N_t = f(v_w, T_w) \tag{7.7}$$

It is well known from the unwinding and slitting of tapes that the reel should be tempered before confectioning. The adhesion on the back of the tape (blocking) depends on the temperature. The resistance to unwinding can be considered as equivalent to peel resistance. Therefore the unwinding force is a function of unwinding speed and temperature. Generally the unwinding force depends on a number of parameters:

$$F_w = f(F_a, E_a, E_f, h_a, h_f, w_t) \tag{7.8}$$

where F_a is the adhesion force; E_a and E_f are the moduli of elasticity of the adhesive and film, h_a and h_f are the thicknesses of the adhesive and film, and w_t is the width of the tape. As is known, the adhesive is a viscoelastic substance; therefore the modulus of elasticity of the adhesive depends on time and temperature, i.e., on the unwinding speed and temperature. The modulus of elasticity of the film carrier material also depends on the temperature. In this case the adhesion force is the peel resistance of the PSA from the back of the tape. The peel resistance from the abhesive layer protecting the active face of the pressure-sensitive product is also measured for labels. For labels the release layer is a separate component, and delamination occurs during end use; it is an end use performance characteristic. For tapes and protective films, delamination occurs during rewinding. It is both a converting and end use performance characteristic.

Maschke [198] studied the dependence of the number of tear cases on the unwinding temperature and unwinding speed. Unwinding temperatures of 22 and 28°C were chosen, and the number of cases of tape tearing per 10,000 m² of processed tape surface was registered. As shown by Maschke [194] increasing the unwinding speed from 120 m/min to 220 m/min (at 22°C) increases the tear number almost 100 times. Increasing the unwinding temperature from 22 to 28°C decreases the tear number by only 10 times. Therefore in this case in order to balance the negative influence of the speed increase, a temperature increase of 12–15° K would be necessary. As found by examination of the tear surface, tape tear is due mainly to bubbles and bubble "chains" and small (diameter less than 2 mm) holes. Bubbles appear on the tape carrier at places where the film is too thin.

Telescoping of a slit roll is related to the unwinding quality. For narrow webs it is more critical. Its maximum acceptable value is generally given as a product specification. For instance, according to Ref. 197, for a 32.0 mm PVC tape, telescoping shall be limited to 6.35 mm.

3 DISPENSING AND LABELING

The labeling process is influenced by the label and by the product to be labeled [199]. The main label-dependent parameters of labeling are adhesive type, coating weight, degree of release, labeling system, labeling speed, label geometry, labeling tolerance, label material, label printing, and converting. The product to be labeled influences the instantaneous and final bond strength by its surface temperature, surface structure, and geometry and by its content.

Many pressure-sensitive labels are still affixed by hand, but a growing percentage are applied by automated equipment. Adhesives used for products with automatic labeling should have a level of quick stick that will provide for secure adhesion with minimal application pressure [149]. Conventional dispensers for paper-based pressure-sensitive materials do not require a special applicator system because of the inherent stickiness of paper. For conformable and squeezeable labels, special systems are needed. Tight label placement tolerances of 0.015 in. are usual for application systems.

The label parameters required in label application differ for wet and pressure-sensitive labels. According to Ref. 200, the main labeling parameters of wet labels are (1) the quality of the carrier paper, (2) the nature of the adhesive, (3) the surface to be bonded, and (4) the labeling equipment.

The main features of labels with good labeling performance are (1) lay flat, (2) narrow die cut tolerances, (3) good mechanical properties (tear resistance in wet and dry states), and (4) low stiffness. Although some of these statements are valid for PSA labels also, other requirements are quite different. Because of the differences in the labeling systems used for wet and PSA labels, different carrier characteristics are required for each class. Labeling speed for common pressure-sensitive labels is lower than for wet labels. For instance, labeling speeds of up to 150 packs/min can be achieved with a print-and-apply system for food packaging [201].

Label application machines (labelers) of quite different constructions and capabilities have been developed. Suppliers of these machines offer hardware and supplies. A hand labeling device contains a storage roll, a pressure roll, and a cutting device [202]. Generally a hand-held labeler used to print and apply PSA labels releasably laminated on a carrier web comprises a housing having a handle, a label roll printing device, delaminating device, applicator, and some means for advancing the carrier web. A hand-operated printing and labeling machine may possess a carrier band on which the labels are temporarily positioned. The machine has a conveyor system that draws the label carrier head around a label-detaching edge. Pneumatic (touch blow) label transport and application has been known for many years [203].

Labeling speed depends on the dimensions of the labels. A common labeling system allows the application of 50 labels/min with label dimensions of 120 ×

1450 mm. The same equipment ensures a labeling speed of 30 labels/min for labels of 13 × 350 mm. Labels having dimensions of 230 × 999 mm can be dispensed a a rate of 600 labels/min [204]. Labeling speed is a function of label stiffness also. Biaxially oriented PP films are applied in cosmetics and bottle labeling [205]. For this use films with a thickness of 30–80 μm are recommended. For high speed labeling, films having a thickness of 50 μm are suggested [206]. Tamper-evident constructions have to stay put as the waste matrix is removed yet break up or tear when removed from the substrate [207].

Labeling speed depends on the labeled item also. For instance, a hand-held labeling device is used for marking wires and cables [208]. Hang tabs can be laminated manually or by an automatic dispenser or automatic machine applicator. Special machines for labeling plastic containers and cups have a productivity of 300 cups/min [209]. Nonoriented transparent PP is used in a thickness of 60 μm for no-label look labels [206]. No-label look labeling of glass bottles is carried out with a productivity of 450 pieces/min. Labeling speed of 80 m/min is attained [210]. Labeling machines with tandem labeling provide better productivity. These machines can simultaneously apply two labels at a running speed of 250 containers/min [211].

So called touch blow labels use no mechanical contact in label application. The label is blown onto the substrate. This method is used for labeling mechanically sensitive items and deformable surfaces [212]. Labeling machines for modern industrial production lines dispense 800 labels/min [213]. Microprocessor-controlled non-contact pressure-sensitive labelers achieve a labeling speed of 30,000 labels/h and a label placement accuracy of 1 mm [214]. The new labeling systems are complete packages incorporating their own label material and label-handling modules and are complementary to the relevant production lines [215]. Line scan photosensor technology checks for label placement and defects [216]. Technical specifications for labeling machines include label dimensions, label unwind/outside diameter, label core diameter, backing paper rewind, label stop accuracy (e.g., +0.05 mm), and dispensing speed.

Tape application machines are known also. These devices were first developed for hot laminated tapes [217]. They are used mainly for heavy insulating tapes. Transfer tapes are also often die-cut. Such tapes cannot be cleanly dispensed from a common adhesive transfer gun, which has no cutting blade. They tend to elongate during dispensing and leave excess adhesive both at the broken edge of the transferred strip and at the orifice of the gun [218].

REFERENCES

1. I. Benedek and L. J. Heymans, *Pressure Sensitive Adhesives Technology*, Marcel Dekker, New York, 1997, Chapter 7.
2. *Adhesives Age* 3:8 (1987).

3. H. Haas, *Coating* 3:102 (1987).
4. A. K. Das Gupta, A label paper satisfying the highest standard, 2nd International Cham Tenero Meeting for Pressure Sensitive Materials, Mar. 29, 1990, Locarno, Switzerland.
5. T. Frecska, *Screen Printing* 2:120 (1988).
6. *Adhäsion* 10:6 (1985).
7. J. C. Pommice, J. Poustis, and F. Lalanne, *Paper Technol.* 8:22 (1989).
8. E. Park, *Paper Technol.* 8:15 (1989).
9. R. Hummel, Basic material for self adhesive labels, 19th Münchener Klebstoff und Veredelungssemnar, München, Germany, 1994, p. 58.
10. W. J. Busby, Processing problems with polyethylene films, PE '93, Session VI, p. 3, Maack Business Service, Zurich, Switzerland.
11. *Convert. Today* 11:9 (1991).
12. Kalle Folien, *Siegel Band*, Hoechst, Mi 1984, 38T 5.84 LVI, Hoechst A. G., Wiesbaden, Germany.
13. J. Verseau, *Coating* 7:189 (1971).
14. J. M. Casey, *Tappi J* 6:151 (1988).
15. E. C. Schütze, Kunststoffe als Bedruckstoffe, Seminar "Bedrucken von Kunststoffen," Fachhochschule Stuttgart, Sept. 25, 1985; in *Coating* 11:311 (1985).
16. *APR* 42:1498 (1986).
17. *Etiketten-Labels* 5:16 (1995).
18. L. Waeyenbergh, 19th Münchener Klebstoff u. Veredelungsseminar, 1994, p. 138.
19. *Mobil-OPP Art* 2:2 (1993).
20. M. Fairley, *Labels Label. Int.* 5/6:76 (1997).
21. *Papier Kunstst. Verarb.* 9:57 (1988).
22. F. Hufendiek, *Etiketten-Labels* 1:9 (1995).
23. *Etiketten-Labels* 5:25 (1995).
24. Avery, *Avery Rating Plates and Type Plates Last a Fairly Long Time*, Booklet, Avery Etikettier-Logistik GmbH, Eching b. München, Germany.
25. K. Fust, *Coating* 2:65 (1988).
26. C. Hars, *Verpack. Rundsch.* 5:526 (1988).
27. G. M. Milles, *Papier Kunstst. Verarb.* 11:41 (1986).
28. W. Schibalski, *Coating* 3:87 (1985).
29. *Coating* 7:245 (1993).
30. Dow Chemical USA, *Printing Trycite Plastic Films*, Designed Products Department, Midland, MI.
31. B. Hoffmann, Qualitätssteigerung im Flexodruck durch moderne Maschinen Konzeptionen, Seminar "Bedrucken von Kunststoffen," Fachhochschule Stuttgart, Sept. 25, 1985; in *Coating* 11:311 (1985).
32. *Etiketten-Labels* 3:10 (1995).
33. R. Neumann, *Coating* 6:194 (1987).
34. *Etiketten-Labels* 5:95 (1997).
35. B. Ellegard, *Etiketten-Labels* 3:47 (1995).
36. R. J. Ridgeway, *Paper Film Foil Converter* 12:41 (1968).
37. *Etiketten-Labels* 5:91 (1995).
38. K. W. Holstein, *Neue Verpack.* 4:59 (1991).

39. *Etiketten-Labels* 5:22 (1995).
40. K. Ehrlitzer, *Etiketten-Labels* 3:34 (1995).
41. *Papier Kunstst. Verarb.* 10:54 (1996).
42. *Druckwelt* 17(9):31 (1986).
43. *Labels Label.* 7/8:67 (1988).
44. *Etiketten-Labels* 5:91 (1995).
45. *Siebdruck 40*, Drupa 94,5,Sonderausgabe, Halle 3, Düsseldorf.
46. *Pack Rep.* 11:48 (1983).
47. M. Emrich, *Druck Print* 1:34 (1986).
48. P. Dippel, *Coating* 4:44 (1988).
49. R. Davis, *Fasson Facts Int.* 1:2 (1969).
50. *Polygraf* 23:2260 (1987).
51. *Coating* 4:115 (1987).
52. *Druckwelt* 17(9):30 (1986).
53. *Coating* 12:494 (1995).
54. *Druckwelt* 12:34 (1988).
55. K. Hanke, *Deut. Drucker* 8:16 (1987).
56. A. Jentzsch, *Druckwelt* 18:44 (1988).
57. K. Hanser, *Druck Print* 1:15 (1989).
58. *Papier Kunstst. Verarb.* 6:37 (1995).
59. H. Roman, Bedrucken von Folien im Buchdruckverfahren auf Etikettendruckmaschinen, *Coating* 11:312 (1985).
60. *Pack Rep.* 11:48 (1983).
61. H. Mathes, *Etiketten-Labels* 1:21 (1997).
62. W. Kaiser, *Siebdruck* 8:46 (1988).
63. W. Kaiser, *Siebdruck* 9:48 (1988).
64. *Coating* 12:494 (1995).
65. R. Hummel, Prägefoliendruck auf Formkörpern, Prägefoliendruck auf Kunststoff Folien und Laminaten, Seminar "Bedrucken von Kunststoffen," Fachhochschule Stuttgart, Sept. 25, 1985; in *Coating* 11:311 (1985).
66. J. Teichmann, *Papier Kunstst. Verarb.* 6:21 (1995).
67. *Druckspiegel* 6:160 (1996).
68. Seminar "Bedrucken von Kunststoffen," Fachhochschule Stuttgart, Sept. 25, 1985; in *Coating* 11:311 (1985).
69. R. Hummel, *Coating* 19:358 (1992).
70. H. J. Teichmann, *Papier Kunstst. Verarb.* 11:10 (1994).
71. H. Klein, *Coating* 4:102 (1986).
72. *Convert. Today* 10:41 (1991).
73. B. Hunt, *Labels Label.* 2:47 (1997).
74. K. Unbehaun, *Deut. Drucker* 6:43 (1988).
75. *Labels Label.* 3/4:11 (1994).
76. *Coating* 12:452 (1986).
77. J. M. Casey, *Tappi J.* 6:151 (1988).
78. Arpeco, *Cold Foil Transfer*, Booklet, Arpeco Engineering, Mississauga, ON, Canada.
79. *Etiketten-Labels* 5:26 (1995).

80. *Etiketten-Labels 1*:38 (1996).
81. J. Young, *Tappi J. 5*:78 (1988).
82. *Etiketten-Labels 1*:30 (1995).
83. A. Prittie, *Finat News 3*:35 (1988).
84. Folex Symposium, Kommunikationstechnik in der näheren Zukunft, Nov. 27, Gravenbruch; in *Offset-praxis 2*:50 (1988).
85. *Etiketten-Labels 5*:142 (1995).
86. *Packlabel News*, Packlabel Europe 97, Ausgabe 2.
87. TDI, Label Printer, Avery Dennison Deutschland GmbH, Eching.
88. *Etiketten-Labels 5*:6 (1995).
89. *Labels Label. 3/4*:11 (1994).
90. Sato Europe GmbH, Hilden, General Printer Specification M8485S.
91. *Verpackungs-Berater 5*:26 (1996).
92. K. Hanser, *Druck Print 1*:15 (1989).
93. *Labels Label. 2*:29 (1997).
94. *Etiketten-Labels 5*:133 (1995).
95. *Etiketten-Labels 5*:136 (1995).
96. *APR 48*:1360 (1985).
97. *Coating 5*:117 (1974).
98. *Int. Forms 3*:16 (1997).
99. *Labels Label. 2*:38 (1997).
100. *Polygraf 23*:2237 (1987).
101. *Aplitape*, October 1995, South Plainfield, NJ.
102. *Verpackung 3*:10 (1995).
103. *Etiketten-Labels 5*:6 (1995).
104. *Verpackungs-Berater 5*:26 (1996).
105. B. Hunt, *Int. Forms 3*:22 (1997).
106. Taktik, Samuel Jones and Co. Ltd., *Laserdruckern*, Herzogenrath, Germany.
107. *Etiketten-Labels 5*:9 (1995).
108. *Labels Label. 3/4*:34 (1994).
109. *Papier Kunstst. Verarb. 10*:62 (1987).
110. *Coating 12*:494 (1995).
111. *Convert. Today 10*:41 (1991).
112. *Etiketten-Labels 3*:40 (1995).
113. FNAT *Label. News 1*:15 (1995).
114. *Etiketten-Labels 3*:30 (1995).
115. *Etiketten-Labels 35*:28 (1995).
116. *Etiketten-Labels 5*:22 (1995).
117. *APR 18*:686 (1986).
118. H. P. Seng, *Coating 9*:231 (1993).
119. B. Hunt, *Labels Label. Int. 5/6*:34 (1997).
120. *Etiketten-Labels 1*:49 (1996).
121. M. Fairley, *Labels Label. Int. 5/6*:28 (1997).
122. R. Kasoff, *Paper, Film Foil Convert. 9*:85 (1989).
123. Folien, Gesamtkatalog '96, Geonit Siebdrucktechnik GmbH, Erfurt, Germany, p. 23.

124. United Barcode Industries, Denmark, *Drucker Zubehör*, Technical Booklet.
125. H. Hadert, *Coating* 6:175 (1969).
126. *Papier Kunstst. Verarb.* 5:98 (1986).
127. R. Mannel, *Papier Kunstst. Verarb.* 2:17 (1996).
128. J. Wahl, *Etiketten-Labels* 3:44 (1995).
129. G. Meinel, *Papier Kunstst. Verarb.* 10:28 (1985).
130. J. Dietz, *Coating* 11:296 (1982).
131. *Coating* 11:22 (1969).
132. R. M. Podhajny, *Convert. Packag.* 3:21 (1986).
133. Packard Electric, Engineering Specification ES-M-2379.
134. P. M. Guella, *Papier Kunstst. Verarb.* 10:39 (1985).
135. *Verpack. Rundsch.* 3:334 (1987).
136. H. P. Seng, *Coating* 9:324 (1993).
137. C. Fellers, L. Salmen, and M. Htun, *APR* 35:1124 (1986).
138. R. R. Lowman, *Finat News* 3:24 (1987).
139. Grafityp, *Selbstklebefolien massgeschnitten*, Grafityp, Houthalen, Belgium, October 1992.
140. Hoechst, *Klebeband Träger u. Abdeckfolie*, Datenblatt, Ausgabe 07/92, Hoechst AG, Wiesbaden, Germany.
141. V. M. Ryabov, O. I. Chernikov, and M. F. Nosova, *Plast. Massy* 7:58 (1988).
142. P. Zimmermann, *Verpack. Rundsch.* 3:298 (1972).
143. Solvay, *Beschichtung von Kunststoffen mit Ixan WA*, Solvay & Cie, Brussels, Belgium, p. 29.
144. *Siebdruck* 1:14 (1988).
145. *J. Metall, Kaut. Gummi Kunstst.* 3:228 (1987).
146. T. Sigii, Y. Moroishi, and K. Noda (Nitto Electric Ind. Co. Ltd.), Jpn. Patent 6386 778/18.04.1988; in *CAS Adhes.* 19:2 (1988).
147. *Adhes. Age* 10:125 (1986).
148. J. R. DeFife, *Labels Label.* 3/4:14 (1994).
149. Packard Electric, Engineering Specification ES-M-2147.
150. C. Bayer, *Adhäsion* 9:349 (1965).
151. A. W. Norman, *Adhes. Age* 4:35 (1974).
152. Kendall Co., Can. Patent 853145/1987.
153. Technical Information 2003, *Printing on EVAL Films*, Kuraray Co. Ltd, 1992.
154. D. Bedoni and G. Caprioglio, Modern equipment for label and tape converting, 19th Münchener Klebstoff und Veredelungsseminar, 1994, p. 37.
155. *Convert. Today* 10:41 (1995).
156. *Etiketten-Labels* 5:91 (1995).
157. *Eur. Adhes. J.* 6:23 (1995).
158. *Papier Kunstst. Verarb.* 1:25 (1996).
159. J. Michel, *Verpack. Rundsch.* 12:1425 (1985).
160. S. W. Medina and F. V. DiStefano, *Adhes. Age* 2:18 (1989).
161. Grafityp, *CSR, Turbo, Computerized Signmaking Robot*, N. V. Grafityp, Houthalen, Belgium, 1997.
162. *Graphitec, RFC3100-60, Tischschneideplotter*, Multiplot Grafiksysteme, Bad Emstal, Germany, 1997.

163. G. Strasser, *Etiketten-Labels* 3:24 (1996).
164. K. W. Holstein, *Neue Verpack.* 4:59 (1991).
165. *Etiketten-Labels* 3:8 (1995).
166. *Etiketten Labels* 1:44 (1996).
167. M. Bringmann, *Etiketten-Labels* 5:72 (1995).
168. Packlabel News, Apr. 8–10, 1997, Frankfurt, Germany.
169. *APR 31*:1014 (1986).
170. *Etiketten-Labels* 5:32 (104).
171. *Paper Film Foil Converter* 9:30 (1989).
172. B. Hartmann, F. Zörgiebel, and R. Wilken, *Etiketten-Labels* 3:4 (1996).
173. *Etiketten-Labels* 3:14 (1996).
174. E. Kazuyoshi, N. Hiroaki, T. Katsuhisa, K. Yoshita, and T. Saito (FSK Inc.), Jpn. Patent 6317981/25.01.1988; in *CAS Adhes.* 12:5 (1988).
175. T. Kurono, N. Okashi, and N. Tanaka (Nitto Electric Ind. Co. Ltd.), Jpn. Patent 6333487/13.02.1988; in *CAS Adhes.* 12:5 (1988).
176. *Finat News* 4:10 (1996).
177. *Screen Print.* 22:204 (1987).
178. D. Lacave, *Labels Label.* 3/4:55 (1994).
179. Sekisui Chemical Ind. Co. Ltd., Jpn. Patent 072785513/24.10. 1995; in *Adhes. Age* 5:12 (1996).
180. *Papier Kunstst. Verarb.* 12:44 (1986).
181. *Neue Verpack.* 1:156 (1991).
182. LOS Lager Organisations System GmbH, Mannheim, Germany, Gebrauchsmuster, G 87 06 322.0/02.05.1987.
183. H. Nakahata, Jpn. Patent 07316510/05.12.1995; in *Adhes. Age* 5:11 (1996).
184. Packard Electric, Engineering Specification ES-M-1881.
185. *Etiketten-Labels* 1:50 (1996).
186. W. Grosse, *Coating* 6:201 (1993).
187. *Coating* 9:279 (1969).
188. *Coating* 12:492 (1995).
189. *Coating* 5:185 (1988).
190. *Convert. Today* 8:13 (1994).
191. Better Packages Inc., U.S. Patent 3510037; in *Coating* 5:130 (1971).
192. Packard Electric, Engineering Specification ES M-2147.
193. K. Nakamura, Y. Miki, and Y. Nanzaki (Nitto Electric Ind. Co. Ltd.), Jpn. Patent 6386787/18.04.1988; in *CAS Adhes.* 19:5 (1988).
194. H. Kenjiro and O. Kazuto (Nitto Electric Ind. Co. Ltd.), Jpn. Patent 63137841/09.01.1988; in *CAS Adhes.* 24:5 (1988).
195. Four Pillars, Tapes, Data Sheet.
196. F. Hensen, *Kunststoffe* 84(10):1325 (1994).
197. Packard Electric, Engineering Specification EM-S-2147.
198. K. Maschke (Hoechst Folien), 4th Klebeband Forum, Frankfurt am Main, November 1990.
199. *Pack Rep.* 11:48 (1983).
200. *Packung Transport* 6:20 (1983).
201. *Labels Label.* 2:24 (1997).

202. Alois Stöckerl GmbH, München, DBP 1183007; in *Adhäsion* 2:82 (1966).
203. Thatcher Glass Manufacturing Co., Inc., New York, U.S. Patent 3152940; in *Adhäsion* 2:80 (1966).
204. *Folio, Avery Dennison Info J.* 1997. p. 2.
205. *Etiketten-Labels* 5:30 (1995).
206. *Etiketten-Labels* 5:16 (1995).
207. B. Hunt, *Labels Label.* 2:29 (1997).
208. *Labels Label.* 3/4:11 (1994).
209. *Pack J.* May 14, 1996, p. 3.
210. *Nahrungsmittel Genussmittel Verpacken* 3:2 (1977).
211. *Pack Rep.* 6:40 (1991).
212. *Etiketten-Labels* 5:91 (1995).
213. *Verpackungs-Berater* 3:21 (1997).
214. *Willett News*, Willett Sarl, Roissy, France, 1997.
215. B. Topliss, *Finat News* 2:45 (1989).
216. *Labels Label. Int.* 5/6:5 (1997).
217. *Adhäsion* 10(2):82 (1966).
218. C. L. Vernon and E. A. Stanek (Minnesota Mining and Manuf. Co., St. Paul, MN), EP 0147093A2/03.07.1985.

8
End Uses of Pressure-Sensitive Products

The end use properties together with the adhesive properties are the most important performance characteristics of PSPs [1]. The number of applications for PSPs is growing. Semifinished products (such as sheets, tubes, films, and profiles) and finished ones (bottles, containers, etc.), natural and industrial products (fruits, machine parts, etc.) are made pressure-sensitive. Security products, envelope sealing strips, and materials such as Post-it notes are examples [2]. The proportions of the global volume represented by the various PSPs is continuously changing. In 1983 in Europe the main pressure-sensitive product applications were tapes, labels, and postage [3]. In 1987, labels, tapes, hygienic products, decorative tapes, construction products, protective items, envelopes, medical products, and tiles were considered the most important fields of application [2]. According to Jordan [4], the main application fields for pressure-sensitive products are medical and hygienic uses (plasters, tapes, dosage systems), masking tapes, plotter films, common tapes (office tapes, marking tapes, packaging tapes), labels, protective films, insulation tapes, decorative foils, and carpet and other double-faced tapes. A study carried out by the Skeist Laboratories [3] shows that in 1983 market expansion was expected in graphics, mounting, cushioning, and insulating products. Now low temperature resistance for deep freeze price labeling, structural pressure-sensitive adhesives for holding and mounting, high shear/peel adhesives for packaging tapes, adhesives resistant to perspiration and body liquids for personal care products, and heat-resistant PSPs for electrical and industrial uses are considered the most important. According to Ref. 5, in 1996

Table 8.1 Main Conversion and End Use Properties of PSPs

Pressure-sensitive product	Converting properties	End use properties
Label	Slitting ability Cutting ability Die-cutting ability Printability	Printability Labeling ability Delabeling ability
Tape	Slitting ability Die-cutting ability Printability	Tearability Cuttability Special electrical characteristics Special thermal characteristics Special dosage characteristics
Protective film	Slitting ability Printability	Laminating ability Mechanical and thermal processibility Delaminating ability

foodstuffs (15%), cosmetics (13%), pharmaceuticals (7%), industrials (10%), and variable image printed products (40%) were the main application domains for labels. Table 8.1 lists the main conversion and end use properties of PSPs.

As illustrated by Table 8.1, the end use properties of PSPs have to fulfill the different requirements of a number of application fields. Some of them are general requirements related to the application technology of a product class or adherend or to the processing of labeled, taped, protected, etc. products. These are special criteria related to the application field. There are products manufactured with different technologies that are proposed for the same application (e.g., certain adhesive-coated and adhesiveless protective films). On the other hand, some classic products (see Chapter 2) have the same buildup but quite different applications.

1 GENERAL CONSIDERATIONS

Generally end use properties must be evaluated as "instantaneous" application characteristics and storage- or aging-dependent characteristics. Both depend on the product buildup and application conditions. It is evident that labels, tapes, or protective films applied from rolls have to meet some general requirements related to their weblike character (e.g., ease of unwinding). Labels, tapes, and protective films that are marked by hand during or after application (i.e., writable items) have to have certain similar surface characteristics. Labels and tapes ap-

End Uses of PSPs

Table 8.2 Application and Deapplication Conditions for Major PSPs

Pressure-sensitive product	Application conditions	Deapplication conditions
Label	Impact or non-impact labeling High speed application Room temperature or low temperature application Very low application pressure Repeated application possible Manual or automatic application Conversion during application Conversion after application	Mechanical or chemical or thermal deapplication Repeated deapplication possible Partial deapplication possible
Tape	Impact lamination Room temperature, low temperature, or high temperature application Low or high pressure application High stress application Low or high speed application Converting during application Hand or automatic application	Mechanical deapplication
Protective film	Impact lamination High temperature and high pressure lamination	Mechanical delamination High stresses during delamination Delamination of more importance than lamination High temperature, high pressure processibility after lamination

plied on the same type of packaging material or item (e.g., bags) have to be recyclable with the item.

On the other hand there are special PSA applications, e.g., in the automotive and electronics industries, for which specific bond strength or environmental or physiological behavior, etc., may be required. The achievement of specific bond strengths in application to any substrate depends on the adhesive formulation used, the substrate itself, and the environment circumstances of the application. Two areas of prime concern involve adhesion to nonpolar surfaces and to flexible packaging that can change form [6]. Pressure-sensitive laminating can also be used for overlaminating. Some years ago thermal laminating was the main encapsulation laminating method. Pressure-sensitive laminating equipment costs considerably less than thermal laminating equipment [7].

As can be seen from Table 8.2, generally the application/deapplication conditions for different types of PSPs are quite different. Labels are laminated under low pressure at room temperature. Their bonding speed is high. Tapes and protective films are laminated under high pressure (tension) at room or elevated temperatures. The application speed is lowest for tapes. Delamination of labels is a low speed process. Delamination of tapes and protective films occurs at high speed. Removable, repositionable, and re-adherable products require different application conditions and bonding characteristics.

2 LABELS AND THEIR APPLICATION

Pressure-sensitive adhesives have been used since the late nineteenth century mainly for medical tapes and dressings [8]. Industrial tapes were introduced onto the market in the 1920s and 1930s, and self-adhesive labels in 1935. Various end uses are known for labels, but a label is always an information medium. Some of the end uses are general, classic application fields where PSA labels replaced wet adhesive labels or other products designed to place information or aesthetic, decorative elements on a product surface. Others are new domains where only a pressure-sensitive product can meet practical requirements.

2.1 Application Conditions

The conditions under which PSPs are applied differ according to the PSP classes and special uses. The application conditions for labels depend on the product to be labeled, the labeling method, and the type of label. Among the product's characteristics, its surface has special importance in application. Application conditions include the application temperature, environment, and speed of application. As mentioned earlier, they may also include re-adherability.

Influence of the Product Surface on Label End Use

Substrate surfaces used in laminating (labeling) vary widely. Their diversity is due to the raw materials used for their manufacture and the processing technology used for them. Special application fields need special application (adherend) surfaces. Therefore the label designer has to have information about the adherend surface in order to tailor the product to serve its intended purpose.

According to Fust [9], at the end of the 1980s the main labels markets were food, cosmetics, computers, pharmaceuticals, sanitary products, and beverages. Food, pharmaceuticals, and cosmetics accounted for 36% of the label market at that time [9]. Heger [10] stated that in 1965 less than 5% of packaging materials were plastics but by 1987 more than 40% were plastic-based. In 1988 in Germany, 44% of packaging materials were paper-based [11]. Now less than 35% are.

The variety of adherend surfaces in a given application field is illustrated by the automotive industry, which uses various unmodified and modified plastics. Composite materials and alloys, PCMA, and PUR foam are used as interior coverings [12]. PVC is no longer used at all. Thus in the United States in 1988 about 80% of interior automotive PVC parts had been replaced with other materials [13]. Most electroplated plastics in the United States are used for decorative products for automobiles and appliances. The main manufacturing processes for metallized finished plastics are electroplating and electroless plating, evaporative vacuum metallizing, vacuum sputter coating, and hot stamping. The surface quality of metallized items depends on the metallizing procedure used.

Because of the increased use of post-improved surface qualities, i.e., items whose final surface is quite different from that of the bulk material, it is difficult to find the correct adhesion level for a PSP without preliminary testing. Injection molded, blown, or thermoformed plastic items may have an outer layer of release material such as silicone [14]. Polystyrene sheets or thermoformed polystyrene parts (e.g., refrigerator shelves) are covered with clear OPS. In households, glass and glass-like surfaces are the most common labeling substrates. In some cases the glass surface is coated with TiO_2 or SnO and PVA [15]. Epoxy coatings and acrylics can be used also [16]. Colored bottles may have a greater amount of mold release agent applied in their manufacture. Lightweight single-use bottles are chemically hardened and smoothed with a surfactant, polyethylene, or wax [17]. Low friction coatings for glass bottles are based on PVA and plasticizer [18]. In other cases the surface quality of the glass is changed by condensed humidity. Therefore wet tack is necessary to ensure the bonding of labels to condensation-coated surfaces. Injection-molded containers have mold release agents on their surface.

Composite materials also replace other common surfaces. Such composite structures are manufactured by laminating, varnishing, or common processing of plastics. To improve the water resistance of paper or cardboard, chlorinated rubber and paraffin are used [19]. The water resistance of paper can also be improved by using ethylene-acrylic acid copolymers [20]. Paper can be coated with PE dispersions also. Dispersed polyethylene powder (particles of 5–12 μm diameter) are used to achieve a thin coating [21]. PE laminates dominate the market of plastic-coated papers [22]. It is well known that such laminates (milk and juice boxes) are difficult to label [23]. Industrial containers and drums made of HDPE or impregnated cardboard are also difficult to label. UV-cured epoxides are used for coating cans [24]. Polystyrene is used for the surface coating of imitation woods [25]. Various materials such as paper, paperboard, aluminum, glassine, and plastic films are extrusion-coated with PE [26]. Papers with extrusion-coated PE display different surface properties in the machine and cross directions. Extrusion coating orients the PE material [27]. When blow-molded items are rapidly cooled during injection, their surface may have a different structure than that of the bulk material [28].

The flexible packaging industry represents a difficult domain of label application because of the various combinations of packaging materials, their different thicknesses and shapes, and their end use conditions. According to Bulian [29], the main types of materials used by the flexible packaging industry in the form of thin sheets suitable to "wrap" a product are paper, aluminum foil, polyethylene, polyester, polypropylene, PVC, PA, polystyrene, PC, and PAN. For the packaging of fatty or moist products, 6–15 μm thick HDPE film is used [30]. In the packaging industry there is a trend to replace PS with PP [31]. Blow-molded containers have been manufactured from from PAN and PC/PET/PC, PP/EVOH, PET/EVOH/PET, PC/EVOH/PC combinations [32].

In-mold labeling allows a plastic item to be simultaneously manufactured by injection molding and labeled with a special label placed in the mold. No supplemental postlabeling or postprinting of the finished item is necessary. The label is made from the same polymer as the labeled item, and both are recycled together. However, the material structure may change due to the cooling and/or crystallizing conditions. In-mold labeling is used for injection-molded items, but it can also be used with thermoformed plastics [33]. It is evident that in this case the surface properties of the adherend are quite different.

According to Maack [34], the complexity of packaging materials and the variety of packaging surfaces have been increased by recycling and downgauging. Downgauged or recycled plastic materials require special adhesive properties. Recycling of post-consumer PET containers is well established. Wide-mouthed glass jars for coffee or milk powder, for example, have been substituted by coinjected PET barrier containers. In many cases glass bottles have been replaced with HDPE bottles. Puncture-, impact-, and tear-resistant shrink and stretch films (based on LLDPE, VLDPE, and LLDPE/LDPE blends and coextrudates) have replaced kraft paper wrapping. Cling films (13–15 μm thick) have replaced PVC as wrap, household, and catering film (LLDPE blends with EVAc or PP). Thinnest films (15–20 μm) are manufactured for rack and counterbags. Heavy duty bags produced from coextruded blown LLDPE/HDPE/LLDPE or similar coextrudates have been introduced to replace multi-ply paper bags. Such materials may have an antislip melt-fractured surface. The presence of certain fillers, pigments, and lubricants in these recycled materials may cause gel formation. Many hygienic plastic applications (diapers, garments, drapes, sanitary towels, etc.) use highly filled materials. Some companies develop products with degradable layers to facilitate accelerated decay if disposal is not by incineration. In medical packaging, injection- and coinjection-molded PET, PP/elastomer, and HDPE/elastomer blends find increasing use as well as coextruded PP/HDPE, HDPE/LMDPE, HIPS/HDPE, and other compositions. Labels for electronic parts have to adhere to paper, cardboard, wood, steel, and glass [35]. Coextruded and coinjected packaging materials can have different surface qualities for the same main carrier. New packaging techniques (aseptic, blow-fill-seal, boil-in bag, retort plastics, barrier containers, etc.) need new materials. Labels have to adhere (or re-adhere) to such surfaces and to be recyclable.

Influence of Product Form on Label End Use

The form and flexibility or rigidity of the product influence the application and labeling conditions also. As is known, application on flexible surfaces is difficult. The flexibility of packaging materials is increasing as a consequence of downgauging. New packaging films may be manufactured with a reduced wall thickness. As an example, biaxially oriented EVOH film with a wall thickness less than 10 μm can be manufactured with sufficient strength for many packaging applications [36]. In some cases the label has to be used for unicates having various forms. The adhesive for nameplates has to bond to difficult surfaces, to high or low surface energy plastics, textured or curved surfaces, and cold surfaces. No-label look transparent labels need a plasticizer-resistant PSA. As shown by Fries [37], such PSAs have to display improved holding power with a plasticizer content of 15%. A common label grade PSA shows a holding power of less than 1 h (with a 2 kg/in.^{2} load), and a plasticizer-resistant grade, more than 3 h.

Other Product Characteristics

Pressure-sensitive labels are also applied directly to foods such as fruits and vegetables. For these, FDA 175.125, Part (b) applies (Pressure Sensitives, Raw Fruits and Vegetables). The regulation does not distinguish between produce whose skin is normally eaten and produce whose skin is discarded. For adhesives with indirect contact with food, IPBC (3-iod-2-propenylbutyl carbamate) has been suggested as stabilizer. This product satisfies 21CFR 171.1 (b) and is on the list of adhesive additives [38].

Application Climate/Aging

Roll storage aging is the resistance to adhesive properties changes upon long-term contact with the face material and release liner. Migration of plasticizers, surface-active agents, and antioxidants may affect adhesives, printing inks, face stock, etc. Environmental resistance may be evaluated by retention of bond strength and other properties following exposure to water, humidity, temperature changes, sunlight, etc. The surface temperature of the adherend is another factor influencing its adhesive properties. Most general-purpose adhesives are formulated for tack at room temperature. If the adherend temperature is lower, a higher degree of adhesive cold flow is required to provide adequate wetout. Some products are labeled at room temperature and subjected to lower temperatures later in their life cycle. For tackified formulations, peel reduction at 0°C may attain 300% [39]. Due to the increase in the modulus of the adhesive at low temperatures, a loss of wetting and jerky debonding may be observed for certain adhesive formulations. According to Dunckley [40], freezer labels should display the same viscoelastic properties at $-40°C$ as at $+20°C$. At 20°C a storage modulus value of 10^5 has been suggested for common labels. For freezer labels also, a value of 10^4 is recommended. Cable tape with good temperature resistance is made with butadiene rubber (more than 90% *cis*-butadiene content), an SBC (SIS or SBS), and tackifier. This blend exhibits a T_g of $-104°C$. The peel resistance at $-10°C$ is 790 g/15 mm and at $+40°C$ it is 370 g/15 mm. The creep is measured at -5 and $+40°C$ also. The rolling ball tack decreases from 28 to 18 between 3 and $-5°C$ [41].

Generally there is a difference between labeling and service temperatures. Suggested labeling temperatures are between -10 and $+10°C$ [42]. Temperature-resistant labels are used between 40 and 126°C [43]. Computer labels are designed for a service temperature of -20 to $+90°C$ [44]. Solventless rubber-based adhesive for filled HDPE milk and fruit juice containers have to operate between -25 and $+70°C$ with a minimum surface application temperature of $-5°C$ [44]. According to Dobmann and Planje [45], office labels have to function at $+23°C$, labels for refrigerated meals at $+40°C$, and labels that are applied to frozen foods, at $-18°C$. Bag closure labels must be able to withstand outdoor temperatures and moisture [46]. Labels for butt splicing have to adhere to PP, PE, PA, and coated papers [32]. Such labels have high temperature stability up to 117°C. For high temperature splicing tapes, shear at 250°C has been tested [47]. For ice-proof gums, water resistance for more than 72 h is required [48]. Blood bag labels have to be water-resistant and able to withstand very low temperatures, so they are overlaminated [49]. Logging tags have to withstand long exposure to salt water; therefore, they are printed on a special synthetic face stock material

[50]. Surgical tapes are used for fastening cover materials during surgery; the adhesive has to be insensitive to moisture and insoluble in cold water and display adequate skin adhesion and skin tolerance. The water solubility of the product is reached at 65°C and at a pH value of more than 9 [51]. Aging of the end use substrate is a very important practical evaluation criterion. Therefore shrinkage should be tested on different surfaces such as 2 mil cast vinyl and 4 mil calendered vinyl [47]. Plasticizer resistance is important for exterior graphics on vinyl, semipermanent vinyl labels, and automotive decals; the target values are between 0.3 and 0.7%.

2.2 Application Methods for Labels

Labels are applied manually or by labeling machines. Generally, labeling quality depends on the machine, the label components, and the substrate. According to Grebe [51], labeling is influenced by the labeling system, substrate surface, label material, adhesive, and transport and storage conditions (see also Chapter 7).

The application of pressure-sensitive labels is a cold lamination. It involves the use of an adhesive-backed carrier on a release liner. Equipment for this method varies from hand-fed units to sophisticated equipment with automatic feeders and cutoffs. Sometimes the labeling device is a component of bag-making equipment, corrugated and folding carton converting machinery, envelope machinery, or engraving equipment. The release from the liner should be adequate to prevent the predispensing of labels but not so tight as to cause the web to break or the liner to tear. Tear-sensitive labels with low mechanical resistance are dispensed by machine also. Pressure-sensitive labels for tamper-evident application have to be applied automatically [12].

In the label industry a difference exists between roll and sheet laminates. Requirements are generally less critical for roll applications. Sheet application needs PSA with sufficient anchorage to prevent pulling out but sufficiently low tack level to prevent gumming of the guillotine. Labeling machines for roll and sheet labels have been developed [52]. Labeling speeds as high as 240 labels/min are achieved. Application of small roll labels needs almost no mechanical lamination. As is known from PSA testing practice, a loop tack test simulates the real application conditions of labels. They are blown onto a surface using air pressure [53].

A special case of label application is in-mold labeling where the item to be labeled is applied (melt around) to the solid-state static label. According to Teichmann [54], hot stamping can be used for in-mold labeling also. Microprocessor-controlled PSA label application offers variable speed settings to match product requirements and may have a special dispensing edge and a collapsible backing paper rewind. Where a high degree of conformability is desired, flexible label stock such as saturated paper should be used rather than a high gloss paper. The layout should be in the long grain direction to provide maximum conformability [55].

2.3 Postmodification of Labels

As discussed earlier (see Chapter 7), some conversion operations are transferred from the manufacture in the end use phase of labels. Postdesign, postprinting (writing), and postcutting are carried out by the end user. Therefore end use properties include label versatility for these operations also. Writability, printability, computer printability, laser printability, and cuttability are special characteristics of labels in development. For instance, promotional form labels can have many pages, and their front pages can have a different print quality [56]. Laser beams can be used for cutting, punching, sealing, and printing labels. The label material (adhesive and carrier) has to fulfill the related requirements. The use of lasers in paper processing is examined in Ref. 57.

2.4 Main Types of Labels

The diversity of labels with respect to construction and end use is increasing. Label markets and applications include food, toiletries, cosmetics, pharmaceuticals, computer and industrial goods, and others. The computer label business is one of the fastest growing segments of the label market [49]. Electronic data processing (simple and high quality) labels were the main market segment in the United States in 1986. Variable data bar-coded labels were introduced. Routing labels (for mail and goods transport), inventory control labels, document labels (for libraries and documentation centers), individual part and product labels (e.g., for the automotive, aerospace, and electrical and electronic industries), supermarket shelf labels, health sector labels (blood bags, drugs, medical equipment, and surgical utensils) are the most important sector of non-impact printing. Labels for variable data and short-run jobs like price labels, short-run product labels, labels and tags for bulk packaging, tickets, baggage tags, and other transport documents have been introduced. The film label market has grown at a rate of more than 30% per year [58]. Film labels provide a no-label look. Such labels are needed for cosmetics and office use also [59], i.e., in applications that require that the label itself virtually disappear when applied on the background. New labeling methods like sleeve labeling and in-mold labeling have been developed. Extended text labels were introduced. Magnetically encoded mailing labels are used by the German postal service for the automatic identification and tracking of packages. These contain a bar code and magnetic stripe to allow optical and magnetic processing [49]. It is not the aim of this book to list or describe all types of labels or their end use. However, certain products are discussed to clarify some technical aspects related to their manufacture.

According to a statistical evaluation by Thorne [60] in 1988, PSA labels were used mainly for on-pack information (83%), on-case information (71%), design (32%), off-pack promotions (20%), bar codes (20%), and security (15%). According to Linder [61], the market segments that are the most important for labels are food, EDP/computer, cosmetics/toiletry, and price/weight.

Labels can be classified according to their main performance characteristics related to application conditions (temperature-resistant labels, water-resistant/water-soluble labels, etc.) or special application field (price/weight labeling, bottle labeling, etc.). In their application field they may have different end uses— automotive labels, temperature labels in the category of technological labels, tamper-evident labels or nondestructible labels in the category of antitheft labels, etc. Some products (e.g., temperature-resistant, temperature display, freezer labels) have to support special climatic or temperature conditions.

Temperature-Resistant Labels. In almost every label application field there are products that have to be applied on substrate surfaces at extreme temperatures. According to Dunckley [40], temperature-resistant labels are used generally between 40 and 260°C.

Temperature Display Labels. Reel labels are used between 40 and 260°C and display the measured (substrate) temperature values values with an exactness of ±1% [62,63].

Freezer Labels. As discussed earlier, low temperature labeling needs labels with a special PSA composition that ensures adhesive flow for bonding at low temperatures. Plastic film (PE, PP, PVC, PA, ionomer/PA, coated cellophane) wrapped products (refrigerated meat), require a special, low temperature bonding PSA that has a low modulus in the temperature range between −50 and +150°F [64]. As discussed in Chapters 3, 5, and 6, the formulation of such adhesives needs special know-how. According to Bonneau and Baumassy [39], one of the main requirements for PSAs used for labels is the retention of the adhesive properties over a wide temperature range. According to Dunckley [40], freezer labels should display the same viscoelastic properties between −40 and +20°C. For common labels at 20°C, a storage modulus value of 10^5 is suggested, and for freezer labels, a value of 10^4. At low temperatures tack and peel may be too low. It was found by practical tests that at 0°C the peel of an untackified acrylic is almost 25% less than at 23°C [39]. Therefore by formulation screening, 180° peel at 0°C should be tested also. For rubber resin based hot melt formulations, low molecular weight polyisoprenes with a low T_g (−65 to −72°C) have been used as viscosity regulators [65]. They replace common plasticizers and improve both migration resistance and low temperature adhesion.

Water-Resistant Labels. Textile, automotive, and bottle labels (see later) need water resistance. Labels used for marking textiles or clothing should resist 50 laundry cycles and 10 drycleaning cycles [66].

Price–Weight Labels. Many applications for labels such as bottles, flexible packaging, and textiles are special cases of price–weight labeling. Price–weight labeling is one of the most dynamic application fields. Price marking and price–weight applications represent about 15% of European label consumption [67]. In the United States the 1988 forecast for the volume increase of thermal printed UPC (Universal Product Code) labels by 1990 was 20% [68]. The application requirements for products from this range vary. Simple and complex labels have been developed for different classes of products. Even the simplest products have to meet various criteria. In principle, the "same" price–weight labels should be usable on unpacked items (i.e., banana label), items packed in film [69], or items with a plastic plate label substrate (hang tags) [70]. Some price labels require perforation for transport [71]. According to Ref. 72, removable price labels use almost exclusively a paper carrier coated with rubber resin adhesive. Natural and synthetic rubber and polyisobutylene have been the main formulation components, with soft resins or plasticizers added as tackifiers. For these solvent-based adhesives, gasoline, toluene, acetone, or ethyl acetate have been used as solvent. The rubber has been calendered and masticated. Legging and migration are characteristic of these formulations. The adhesion buildup on soft PVC and varnished substrates attains too high a level after long storage times. By debonding no clear adhesions failure has been possible.

Bottle Labels. Although glass has a polar surface, the labeling of glass bottles with pressure-sensitive labels has developed only in recent decades. Economic and technical problems made this development difficult. As mentioned before, a chemically treated glass surface, condensed water on the bottle, water-resistant labels, and water-soluble adhesives call for specially tailored adhesive properties. The literature data concerning bottle labeling are summarized by PTS [73]. Pressure-sensitive labels are also used for wine and champagne bottles [74]. In 1977 PET was introduced for carbonated soft drinks [75] and glass bottle labeling was replaced at least partially with plastic bottle labeling. In-mold labeling is a special case of plastic bottle labeling where first the label is manufactured, then the plastic bottle. A combination label and folded information booklet (prospectus) is used for bottle labeling of pharmaceutical products [76]. Thus no supplementary instructions are needed. Organizational and logistic problems for food and beverages are solved by using bottle labeling [77].

A special case of application is wine bottle labeling where different water-based adhesives are used on a large variety of (mainly) paper face stock materials. Such labels have been developed since 1985 [78]. Their design has to meet certain requirements (see Chapter 6 also). These products include permanent labels with cold water resistance (ice water) and permanent labels that are removable with hot water. Items in all these categories have to possess good adhesion on moist, condensation-covered surfaces [79]. Special water-soluble compositions are required for bottle labeling. The adhesive must display good adhesion to both wet and dry (polar and nonpolar) surfaces and removability with hot or cold water or with commercial detergents and alkalies [80].

Bag Closure Labels. These products are applied to polyethylene bags that contain stacks and buckets that are stored outdoors. The labels are applied to packages at the end of the production process. They must be able to withstand outdoor temperatures and moisture [46]. Acrylic adhesives are proposed for such labels. Bag reclosure labels (tapes) are used also, mainly for tobacco, coffee, textiles, and dry and wet tissues [81] (see also End Use of Tapes, p. 581).

Nameplates. Equipment, apparatus, electric appliances, and other durable goods have always been given serial numbers. Traditionally this has been done by using nameplates (rating plates, type plates, inventory labels) to identify the manufacturer and state conditions of use (voltage, amperage, etc.). They have to provide information over a period of years on serial numbers, date of manufacture, article number, and performance data. Such plates are used in the most varied industrial and commercial domains. They have to be light, chemically resistant, and weatherproof and be able to withstand abrasion. Generally they have to be printed on the spot.

These data once had to be stamped by hand. But stamping was expensive and time-consuming. Up to the mid-1960s, these plates were mounted with screws or rivets. The application of adhesives allowed (required) the use of thinner metal plates and faster mounting. The data could be processed with high impact typewriters. At first only paper and cellulose acetate could be computer-printed, and they were not suitable for nameplates. The development of synthetic films, especially computer-imprintable PET, provided a solution to this problem. Metallized 90 μm PET has been used.

To be printed with flexographic letterpress, thermal, laser, or computer printing methods, a film must have a matte metallic appearance similar to that of anodized aluminum. It has to display the handling characteristics of hard aluminum. No curl should occur after removal. A heat resistance of 150°C is needed. The adhesives used must have adequate aging resistance. The adhesive for nameplates has to bond on difficult surfaces, plastics with high or low surface energy, textured or curved surfaces, and cold surfaces. The liner has to exhibit trouble-free release, dimensional stability, and good die cutting, perforating, prepunching, and fanfolding. The matte surface has to absorb ink and maintain sharp graphic contrast, good color intensity, and excellent legibility. It has to be printable with a range of common or special computer printers and demonstrate abrasion, chemical, and environmental resistance. Polyester and vinyl film and special heat-resistant materials have been suggested for this purpose.

Laser-printable labels (sheet) are warranted for 4 years and may be used between -20 and $+120$°C [82]. Laser-engraved acrylic films for nameplates are tamperproof. Such labels are ideal for applications where either contact-free data recording is needed or direct laser inscription of workpieces is not possible [83]. It must be emphasized that the formulation of "laser-resistant" adhesives, i.e., products that exhibit enhanced temperature resistance, requires special tackifier resins (see Chapter 6 also).

For printing such products, letterpress or screen printed presetting is used also. The printer should have a comprehensive library of types and standard bar codes, electrical symbols, etc. If thermal transfer printing is used, the resolution is 7.6–11.4 dots/mm and the printing speed reaches 175 mm/s. Rating plates can be printed in sandwich printing also, where the reverse side of a transparent material is printed with mirror lettering [84].

Instruction Labels. Instruction labels are removable. They have a PET or PVC carrier and are used in clean rooms and on delicate electronic parts [85]. Such labels for technological use are used in various fields such as the automotive [86] and textile industries. Iron-on labels used in the textile industry have to resist temperatures up to 180°C (10–20 s) during application and be chemical- and water-resistant [87]. Tamper-evident polyester label stocks provide an alternative to PVC and acetate films [88] (see Anti-theft Labels, p. 582). Various carrier materials can be used: acetate for envelopes, PVC for warning labels for electronic equipment, paper for sterile needle cartridges (see also Medical Labels), etc. [89]. Topcoating or treatment of PET makes it printable via different methods including impact, thermal transfer, and flexo printing. Applications include rating plates for use at high temperatures and exposure to cleaning solutions. Automotive parts labels have to withstand extreme abrasion, extreme temperatures, and exposure to chemicals. They must carry bar codes that will remain readable for long periods. Pharmaceutical labels can have a sandwich structure, i.e., a secondary removable label attached to the main label. The secondary label is printable via thermal printing. The laminated structure possesses perforations also to allow multiple dettachment (i.e., multiplication) of the labels and instructions [90].

Medical Labels. As discussed earlier, medical tapes were the first pressure-sensitive products introduced on the market (see Chapters 1 and 2). These products have special requirements imposed by their application under sterile conditions and on difficult substrates (see Medical Tapes also). The main requirements for medical tapes are valid for medical labels also. Table 8.3 lists the main functional requirements for medical PSPs.

Medical labels are used for wound care or in operating rooms. Surgical labels have to adhere well on the skin and not lose adhesiveness in hot or humid environments. No adhesive transfer should occur during removal of the adhesive from the skin. When the label is removed, no residue is left on the product to attract microorganisms [85]. For certain health care applications, pressure-sensitive labels are subjected to sterilization or autoclaving. Gas-sterilizable packages have to be resistant to ethylene oxide. Polyacrylate-based water-soluble PSAs are sug-

Table 8.3 End Use Requirements for Medical PSPs

Requirement	Product	Ref.
Physiologically compatible	Label, tape, electrode	91
Body fluid nondegradable	Label, tape, electrode	91
No skin irritation	Label, tape, electrode	92,93
Skin adhesion, 95%	Label, tape, electrode	92,93
Low adhesive transfer to skin (cohesion)	Label, tape, electrode	85,91–93
Conformability	Label, tape, electrode	91,94
Moisture insensitivity	Label, tape, electrode	95,96
Cold water insoluble, hot water (65°C) soluble adhesive	Label, tape, electrode	91,95
Breathability (50 s/100 cm^3 per square inch)	Label, tape	91,97
Vapor transmittance (500 g/m^2, 24 h, 38°C)	Label, tape, electrode	98
Liquid transmittance (6–36 h)	Label, tape	99
Light resistance	Label, tape, electrode	98
Heat resistance (350°F)	Label, tape, electrode	96,98
Detachable	Label	100
Ethylene oxide resistance	Label	50,101
Gamma radiation resistance	Label, tape	101
Electrical conductivity	Bioelectrode	50

gested for medical PSPs (operating tapes, labels, and bioelectrodes) [50]. A fluid-permeable adhesive useful for transdermal therapeutic devices applied to human skin for periods of up to 24 h is based on an acrylic, urethane, or elastomer PSA mixed with a crosslinked polysiloxane. The therapeutic agent passes through the adhesive into the skin. Uninterrupted liquid flow through the adhesive has to occur over a prolonged period (6–36 h) and at a constant rate [99]. The skin adhesive has to present an enhanced level of initial adhesion when applied to the skin but resist adhesion buildup over time. An acrylate-based skin adhesive should have a creep compliance value of at least 1.2×10^{-3} [94]. The earliest medical tapes were based on mixtures of natural rubber plasticized and tackified with wood resin derivatives and turpentine and pigmented with zinc oxide [94]. Later, acrylics were suggested for medical tapes [102]. The irritation caused by the removal of the tape was overcome by including certain amine salts in the adhesive. Adhesive balance, stretchiness, and elasticity are required for medical PSAs. Typical examples of conformable carrier materials used for medical labels and tapes include nonwoven fabric, woven fabric, and medium to low tensile modulus plastic films (PE, PVC, PUR, PET, and ethyl cellulose). The preferred carrier materials are those that permit transpiration and perspiration and/or the passage of tissue or wound exudates. They should have a moisture vapor transmission of at least 500 g/m^2 over 24 h at 38°C (according to ASTM E 96–80) with a humidity differential of at least about 1000 g/m^2.

Antitheft Labels. Pressure-sensitive tamper-evident labels are widely used for tamperproof packaging and sealing [103,104]. This market includes labels for traceability and identification, indicators of shock damage or temperature change, and security [105]. Generally permanent labels are tamper-proof. They tear when any attempt is made to remove them. However, for special requirements tamper-evident seals, security tags, tactile labels, special backgrounds, special (e.g., magnetic) inks, controlled front or reverse delamination, security threads and numbering, holograms, etc. have been used. Papers with watermarks, papers that react to UV light or heat, and iridescent papers are used. The choice of an adequate technical solution for tamper evidence is a complex problem.

Security labels for the pharmaceutical industry are used to protect product integrity. Changes in the product could involve indicators of time/temperature, light sensitivity, moisture, impact sensitivity, contamination, loss of vacuum, etc. Time/temperature indicators change color to reveal an underlying graphic when exposure to a specific temperature exceeds a specified time limit [105]. A special ink can be printed flexographically on the label surface. Using a simple testing kit, it can be validated within 2–3 min [106]. Security face stock materials can work by using color change in a peel-away, reverse peel, or peel-apart closure version [107]. The peel-away label is a standard one, placed across the opening to be closed. Removing it activates the color loss and exposes a hidden message that indicates tampering. Unlike many fragile security materials, it leaves no deposit on the adherend. Reverse-peel labels are used on transparent surfaces as a uniformly colored overlayer. When removed, the label splits apart, losing its color to reveal the hidden message. With peel-apart closures, the hidden message is incorporated in a laminated pouch or envelope with a transparent overlay. Opening activates the color loss and exposes the hidden message. A pharmaceutical label that cannot be removed with steam or water from a container once the label has been applied is built up as a combination of an image-producing self-contained carbonless label overlaminated with an opaque carrier material [108]. According to FDA recommendations, tamper-proof packaging can be achieved by using a packaged product that is wrapped in a transparent, boldly labeled printed foil, packagings sealed with foil or adhesive tape, or shrink film [109].

In the healthcare field, the goal is a zero defect product, i.e., zero-defect packaging and documentation. In such products resins sensitive to visible light (e.g., laser beams) may be used [110]. Safeguards against theft or improper use, childproofing, and user friendliness are necessary also. Product integrity and safeguards against mishandling and abuse have to be ensured [111]. As a technical solution, nonremovable and nontransferable constructions have been manufactured as antitheft labels. Destructible, computer-imprintable vinyl film has been used for this purpose. The vinyl film fractures when removal is attempted [112]. Such a product gives tamper evidence performance on chipboard cartons, glass, painted metal, polyethylene, and polypropylene. Security cuts enhance performance. Excellent conformability, good printability, and destructibility are required. Bonding in less than 24 h is necessary [113].

Fragile pressure-sensitive materials either are very thin or have low internal strength. They are coated with a very aggressive PSA [88]. The construction includes a matte topcoated transparent film. This film is coated on its back with a release layer that has moderate adhesion to the film. This release should be printable. A primer is used to achieve good anchorage of the printing ink (graphics in mirror image) to ensure effective splitting during removal. The adhesive film should be slightly narrower than the web width to prevent possible adhesive contamination of the laminating roll. During manufacture, nip pressure should be light and web tension should be adjusted to avoid label curling. Adequate die cutting is provided by a uniformly densified paper liner. Making slits or perforations in the face stock materials improves tamper-evident properties [88]. A decade ago a special method was developed to permanently bond holograms onto a substrate, and such products can also be used as identification [114].

Generally these tamper-evident products are made by ordinary direct or transfer coating of a face stock material such as film or paper. To achieve tamper-evident properties, special face stock materials are used; currently a polymer film is used, in several cases with a sandwich structure. Unfortunately, the manufacture and coating of special carrier materials involve technical and economic problems because of the low tear resistance of these materials. A quite different construction can be manufactured also. One tamper-proof label is based on chrome polyester [112]. On removal the word "Void" appears on both the marked item and the label, preventing their reuse. Another tamper-evident construction includes a release liner and a composite layer coated on the release liner. The upper part of the composite layer is nontacky. It is a filmlike layer with low tear resistance. The bottom layer anchored on the face stock layer is a common PSA. The film-forming polymer should be manufactured (cast) by the same coating procedure that was used for the PSA layer, but first the liner should be coated with an aqueous dispersion. This dispersion-made face stock layer of the tamper-evident PSA layer is made by using a low flexibility, low tack, medium cohesion polymer on a basis of acrylate, styrene acrylate, or ethylene vinyl acetate, with or without filler. To prepare labels that can be dispensed with a labeling gun, 6–250 g/m^2 coating weights are used.

In-Mold Labels. Labels in this special category are applied on injection-molded, thermoformed, or blow-molded containers or soft spreads. These labels are used for toiletries and chemicals. As carrier material, paper and synthetic paper, PP, and metallized films and papers are used. Label insertion and molding techniques need special know-how. In-mold labeling allows the manufacture of the plastic item by injection molding (or other thermoforming procedure) and its labeling in the mold with a special label placed in the mold before the plastic is added. No supplemental postlabeling or postprinting of the finished item is necessary [31]. In-mold labels are not pressure-sensitive. The pressure-sensitive product manufacturer or user has to be informed about this technology because it is competitive with pressure-sensitive technology and the manufacture of in-mold adhesives is a hot melt formulation. For instance, for such labels paper of 100–120 g/m^2 is printed, overlacquered, and adhesive-coated (15–20 g/m^2) with a hot melt [115].

Other Labels. Decals used for clear overlays have to possess an adhesive that is water white before and after drying [110]. For advertising decals, weather resistance is required. These products are applied on buses, trucks, and trains and have to possess chemical resistance to washing agents also [116]. Baggage tags for airlines, promotional labels, tickets, and address labels are other label applications that have been developed [117].

3 TAPES AND THEIR APPLICATION

Tapes are weblike PSPs used to provide structural integrity, dimensional stability, shape retention, bonding, or another particular property (thermal, sound, chemical, or electrical isolation, etc.). In comparison with labels (where the role of the non-adhesive-coated carrier surface is more important for long-term end use, and the adhesive has to allow primarily only such instantaneous adhesion that ensures the contact between label and labeled item), for tapes the mechanical resistance of the carrier is the most important parameter and the adhesive acts as an assembly tool between the carrier and the taped product. Tapes function as an adhesive in film form, a prefabricated glue. As a glue in web form, on a solid-state carrier, they exhibit an exactly regulated thickness and allow fast processing. Taping requires only low energy use. No cleanup is required, and no pollution is produced.

The end use of tapes differs in various regions or countries. In the United States (1989), the most important tape qualities have been packaging, industrial, healthcare, and consumer tapes. In Japan (1987), the main application fields have been packaging, masking, electrical, application, stationery, and protective tapes [118]. In Europe (1987) the leading tape applications have been packaging, masking, anticorrosion, insulating, double-faced, hygienic, and reinforced products [119]. According to Roder [120], in the 1980s the main end uses for double-faced tapes were splicing, laminating, and mounting. The following fields have been considered the most important sectors: packaging/cardboard (70%), industrial packaging (20%), and household/office (10%).

3.1 Application Conditions

The tape market area includes packaging, electrical, industrial, surgical, masking, and consumer goods [121]. Most tapes are weblike fixing elements applied manually or with special winding machines. According to Schroeder [122], the winding up of tapes ensures the pressure necessary for the tape to adhere by lamination. It should be noted that some time ago, to achieve better wetting out and increase contact surface, tapes were applied with pressure and with solvents also. Product surface, geometry, and application climate influence the end use of tapes. It is evident that the application conditions for tackier adhesive-coated packaging tapes must differ from those for low tack adhesive-coated or adhesive-less masking tapes. Soft, conformable carrierless tapes need only low application forces. For medical tapes, deapplication is more important than application. Such examples illustrate the variable application conditions for tapes as a function of their end use.

Influence of the Product Surface on Tape Application

Application of PSPs generally consists of both delamination and lamination. The PSP is delaminated and builds up a new laminate with the adherend substrate. For labels the surface quality of the face stock material does not affect the application procedure. Labels are applied by delaminating a multiweb composite. The adhesive-coated surface (i.e., the PSA) and the release-coated surface (i.e., the liner) are the work surfaces in this process. For tapes, delamination consists of unwinding. The quality of the non-adhesive-coated surface of the carrier material is determinative for the delamination. This surface together with the adhesive-coated surface forms the working zone in delamination of the monoweb. After delaminating, both the label and the tape are laminated on a new substrate, whose surface characteristics affect the bonding.

Solid-state components with various surface qualities are used for tapes. Tapes for rubber profiles in automobiles have to adhere to difficult surfaces like EPDM, CR, and SBR, according to VW Specification TL 52018 [123]. Wire-wound tapes have to adhere to modified flame- and heat-resistant EPDM formulations [124]. EPDM roof systems use PSA tape for long-term performance [125]. Pipe isolation tapes should adhere to oiled surfaces also [126]. Pipe wrapping tapes have to adhere to metal surfaces, wood, rubber, or ceramics [85]. Repair tapes have to adhere to metal surfaces, wood, rubber, or ceramics [85]. A double-layered pressure-sensitive tape intended for industrial and commercial use has to bond to glass, concrete, steel, ceramic drywall, wood, tarpaulins, and covers [127]. Butt splicing tape has to adhere to PP, PE, PA, and coated paper [85]. Medical tapes must adhere to skin [128]. Diaper closure tapes are used as refastenable closure systems for disposable diapers, incontinence garments, and similar items [129]. Such systems allow reliable closure and refastening from an embossed, corona-treated polyethylene surface such as is typically used for a diaper cover sheet. Fluorescent adhesive tapes for highlighting have to be removable from written surfaces. These tapes are manufactured by coating plastic films with fluorescent inks and adhesives with low adhesion [130].

Mounting tapes must adhere to glass, concrete, steel, ceramic, drywall, wood, tarpaulins, and covers [127]. Silicone mounting tapes adhere to a variety of surfaces such as metals, glass, paper, fabric, plastic, silicone rubber, silicone varnished glass cloth, and silicone/glass laminates [131]. Foamlike transfer tapes are used for structural bonding of aluminum or copper-titanium-zinc roof panels [132]. When tapes are delaminated from anodized aluminum, cohesive breaks may produce adhesive deposits [133]. Cables are coated with coextruded 28% VAc [134]. Chlorinated polyethylene, acrylonitrile-butadiene rubber, liquid chorinated paraffin, and fire-retardant fillers (e.g., hydrated alumina, $CaCO_3$, zinc borate, and Sb_2O_3) have been compounded and formed into oil-resistant tapes (0.7 mm thick) for electrical wires and cables [135]. Insulation tapes have to adhere to this material. It is evident that various application conditions require different adhesion performance characteristics. Table 8.4 lists some typical peel values for different tapes; Table 8.5 illustrates the broad range of shear resistance values for tapes with various applications.

Table 8.4 Peel Resistance Values for Selected Tapes

Tape grade	Adhesive	Carrier	Peel resistance	Ref
Packaging	HMPSA, SEBS	—	800 g/25 mm	136
Packaging	HMPSA, AC, crosslinked	—	570 g/10 mm	137
Packaging	HMPSA.SIS	—	720 g/82 mm	138
Packaging	AC, crosslinked	—	925 g/25 mm	139
Packaging	RR	PP	640 g/15 mm	140
Packaging	RR	—	280 g/10 mm	141
Packaging	WB AC	—	1450 g/25 mm	142
Packaging	AC, crosslinked	—	1200 g/25 mm	143
Packaging	AC crosslinked	—	1000 g/25 mm	144
Removable	RR	—	40 g/10 mm	145
Removable	AC, EB crosslinked	—	410 g/25	146
Removable	—	—	200 g/25 mm	147
Removable	AC, crosslinked	PE	200 g/25 mm	148
Double-side coated	WB AC	PET	1300 g/20 mm	149
Temperature-resistant	Silicone	Al	3500 g/25 mm	150
Temperature-resistant	Silicone	PET	460 g/10 mm	151
Temperature-resistant	Fluoropolymer	—	45 g/20 mm	152
Insulating	HMPSA, SBC	—	370 g/15 mm	153
Masking	WB AC	Foam	300 g/25 mm	154
Masking	AC	PET	120 g/20 mm	155
Sealing	AC, UV-cured	—	1640 g/10 mm	156
Medical	RR, SB	PET	5 lb/in.	157

End Uses of PSPs 587

Table 8.5 Shear Properties of Selected Tapes

Code	Adhesive	Manufacturing method	Shear resistance	Ref.
Transfer tape	AC	Mechanical foaming of AC emulsion, impregnating, curing	4.5 kg/cm	158
Transfer tape	AC	AC oligomer, UV curing	>24 h	159
Insulating tape	Silicone	Coating	2.0 mm	150
Insulating tape	AC	Coating	48 h, 70°C	160
Packaging tape	RR SB	Coating	>4 h	161
Packaging tape	AC	Coating	5 h	137
Packaging tape	NR	Coating	2.6 mm, 2 h	140

Application Climate

The application temperature and atmospheric humidity during lamination and end use of the tapes may vary widely. Special conditions are given for medical, insulating, and sealing tapes. For instance, butt-splicing tapes have to resist temperatures up to 117°C [85]. Temperature-resistant silicone mounting tapes maintain their properties up to 288°C. They must exhibit good resistance to chemicals as well as to fungus, moisture, and weathering. Insulation tapes have to resist temperatures between 5 and 60°C [162]. Various tapes used in the automotive industry are often subjected to extremely high temperatures [163]. In a closed automobile in the sun, temperatures can reach 100–120°C. Car undercarriage protection and insulation tapes have to resist 20 h at 150°C [164]. Their resistance to water or solvents is tested by immersion in the liquid for periods of up to one week. Generally tapes should be applied between 20 and 30°C [133]. Products designed for short-term outdoor application should not be used for more than 3 days under such conditions. If the adhesive is to be used in a bathroom (e.g., for fastening soapdishes and utility racks to ceramic tile), the adhesive had to adhere well to tile, have high cohesive strength, withstand temperatures in the 110–120°F range, and endure 100% relative humidity. Because of the extreme working conditions of several tapes, their test methods use elevated temperatures also (Table 8.6).

3.2 Application Method

Tapes are applied manually or with a special apparatus. One-side coated tapes are unwound; double-side coated tapes are delaminated. For transfer tapes the adhesive is bonded to the sheet backing that is a release liner, so that the exposed adhesive surface of this tape is placed in contact with the desired surface, the release liner is stripped away, and the newly exposed adhesive surface is bonded to

Table 8.6 Aging Conditions for the Test of Adhesive Properties of PSPs

Product	Adhesive	Evaluation criteria	Storage temperature (°C)	Storage time	Ref.
Tape	AC, UV-curable	Retention of adhesive properties	70	3 wk	165
Tape	WB AC	Peel	40	60 days	149
Tape, cable	Polybutadiene, SBC	Peel, RB, creep	−5 3 40 70	30 days	153
Tape	AC, crosslinked	Retention of adhesive properties	80	—	143
Tape, masking	WB AC	Adhesion transfer	80	30 days	154
Sheet, removable	AC	Peel	60	100 h	147
Protective sheet	AC, crosslinked	Peel	60	3 days	148
Release liner	AC/SBR	Peel	70	28 days	166

a second substrate [167]. Self-adhesive tear tape designed as an easy-opening device for flexible packaging can be used with a motorized applicator [168]. Sealing tapes (without carrier) based on tackified butyl rubber have been applied with an extruder [169]. Butt-splicing tapes have high quick stick for high speed splicing applications [85]. According to [170], quick stick and high temperature shear are generally required for PSAs for tapes. According to Ref. 40, packaging tapes need more shear resistance than labels. Low shear values are sufficient for office tapes, doublefaced tapes, and foam mounting tapes. For closing polyethylene bags or nonwoven envelopes with a surface tape, removability/reclosability is required [171,172]. Delamination conditions may be various also. For instance, radiation-cured automotive tapes have to present an initial breakaway peel, and then the force needed to continue the breaking of the bond, the initial continuing peel, is measured (see Chapter 5 also). Initial continuing peel is lower (by about 30%) than breakaway peel [173].

Labels are discrete elements during lamination. Their transformation from continuous web form to a discrete label is carried out before labeling. Tapes are generally weblike products during their application. Their transformation to finite webs is carried out during final lamination or after lamination. For instance, cuttability with a hot knife and easy unwinding are required for bag closure tapes

[121]. Easy-tear tapes and certain masking tapes need low mechanical resistance to be torn by hand. On the other hand, transfer tapes are die-cut for some applications [119].

Rolldown properties of tapes are very important. Such characteristics depend on the release properties of the back of the carrier. As discussed in Chapter 6, there are various possible manufacturing ways to improve these properties. For instance, biaxially oriented multilayer PP films have been manufactured for adhesive-coated tapes. Generally, to prepare an adhesive tape that can be easily drawn from a roll without requiring an additional coating on the reverse side, at least two different layers having different compositions are coextruded and the thin back layer (having a thickness ratio of 1/3) contains an antiadhesive component [174]. Tapes with an olefinic carrier material are noisy during unwinding. Modification of the adhesive by addition of a mineral oil reduces the unwinding noise level, according to Galli [175]. Further improvement can be achieved by using a release coat and/or special treatment of the carrier back side. Polyvinyl carbamate, polyvinyl behenate, or other polymer can be used as release layer. It is supposed that the noise level is related to the mechanical properties and surface characteristics of the carrier material, to adhesive/abhesive properties, and to unrolling conditions. It has been demonstrated [176] that improving the surface tension (i.e., increasing it to more than 33 dyn/cm) reduces the noise level.

Application Equipment

Generally packaging tapes are applied with the aid of machines. Special (medical, application, etc.) tapes are used manually. Foamlike transfer tapes are applied under pressure [132]. A hydraulic press is used for bonding aluminum panels. Sealing tapes based on PVC are applied by hand or with a sealing machine [177]. Special application apparatus is suggested for folded tapes. Folding along the longitudinal axis is proposed to prevent the tapes from rewinding [178].

3.3 Main Tapes

The range of tape-like products is wide. They can be classified according to their construction or their use (see also Chapter 2). Carrierless tapes and paper, film, foam-based tapes, uncoated, and one-side or double-side coated products are manufactured. Friction tapes, splicing tapes, and many other types of tapes are produced [46]. The classic type of tape used for fixing and fastening is known as packaging tape.

Packaging Tapes

The classic tape construction, a mechanically resistant nondeformable but flexible carrier material coated with a high coating weight of an aggressive but inexpensive adhesive, is valid for packaging tapes without special requirements. The high processing speed for these tapes results in a low noise level in their use. Low noise, color-printable, PP-based packaging tapes have been developed [82]. Packaging tapes are applied either by hand or by automatic packaging machines for the fastening of goods. Generally packaging tapes are based on PVC or PP, and they may also have a reinforcing layer. Such products are some shade of brown [4]. Kraft paper, cloth, and oriented plastic carrier material are also used for these tapes. Packaging tapes accounted for 66% of European tape consumption in 1987 [119]. Although their volume is increasing, their proportion of the global PSP production is strongly decreasing.

Special packaging tape constructions are known also. A packaging tape described in one patent [179] is composed of a flexible member and a central tear-away section. The central tear-away portion is built up with two parallel lines of perforations extending the entire length of the flexible part. Adhesive tapes with the adhesive-coated area narrower than the substrate and with one edge with an adhesive-free strip have an easy-untie property and are useful for bundling electronic parts, building materials, vegetables, and other items [180].

For packaging tapes on a HMPSA basis, adhesion on cardboard is of special importance [181]. The desired adhesion value is 100 min. A flap test is carried out also; this is a combination of shear and peel adhesion and tack. According to Ref. 182, for such tapes one of the most important properties is open face aging.

Common HMPSA systems for tapes are based on SIS, a resin that is partially compatible with the midblock of the SBC and a plasticizer (oil). For such tapes there is generally no need for a very light color; a color number of 6 for the resin is adequate [182]. The hot melts for packaging tape are coated on prereleased BOPP (21–22 g/cm^2). According to Ref. 181, a peel value of 18 N/25 mm (on steel), loop tack values of 20–30 N/25 mm, and rolling ball values below 5 cm are preferred for packaging tapes. Typical SIS-based tape PSA performance characteristics are discussed in Ref. 183. For an 1100/100/10/2 resin/SIS/oil/antioxidant composition, rolling ball tack values of 11 cm, Polyken tack of 1400 g, loop tack of 16 N/25 mm, and peel value of 18 N/25 mm (on stainless steel) have been obtained. Packaging tapes need shear values higher than 680 min [40].

Tapes for Corrugated Board. Tapes for corrugated cardboard use PP as carrier material and have a silicone release coating [78]. Their application temperature is situated between −40 and +100°F [184]. Such products must have excellent tack and shear on cardboard. Although at normal temperatures rubber adhesives have better shear resistance than waterborne acrylics, acrylic adhesives display superior performance at 120°F [185]. There is a trend in the market to provide customers with printed packaging tapes. The text or graphics are imprinted on the nonadhesive side of the carrier material before the tape is made.

Box Closure Sealing Tapes. Tapes on cellulose hydrate or PVC as carrier material have been designed for box closures [186]. These are less heat- and aging-resistant and less tamperproof than wet tapes [187]. According to UN Recommendation R 001 [188], packages of dangerous goods should be closed with tapes called security tapes. These tapes have to ensure antitheft protection, drop destroying, and tear resistance. Reinforced wet glue tapes display 4% elongation, and nonreinforced plastic tapes, 40–70% [188]. In comparison with wet tapes, PSA tapes are more sensitive to dirt. Wet tapes do not need an unwinding force. These characteristics were the main advantages that allowed a fast exclusive application of wet tapes for security products some years ago. Sealing tapes based on PVC carriers are printable. Sandwich printing (flexo printing on the top side and gravure printing on the back between the film and the adhesive) is possible also. Such tapes are applied by hand or by using a sealing machine. These products are used for closing, manufacturer identification, antitheft protection, copyright protection, quality assurance, advertising, information, classification, and administration [177].

Diaper Closure Tapes. Special pressure-sensitive closure tapes are used for diapers [189]. Adhesives for diaper tapes have to display high shear [190–193]. Such systems have to allow reliable closure and refastenability. Tack values of 16–24 N/25 mm are acceptable. These tapes have to exhibit a maximum peel force at a peel rate between 10 and 400 cm/min and a log peel rate between 1.0 and 2.6 cm/min (see Chapter 5 also). Tapes exhibiting these values have been found to be strongly preferred by consumers.

Bag Lip Tapes. Double-coated PE tape has been designed for PE bag manufacturers [121]. The tape is applied to the lip of the bag. Bag closure tape has to be cut with the same hot knife (hot wire) that is used to form the bag. It should be easily unwound for automatic applications [82]. Tapes for closing PE bags, like those used for nonwoven envelopes must be reclosable. A clear PE carrier material is suggested [171]. A closure tape based on crosslinked acrylate terpolymer uses a reinforced web and a reinforcing filler. This adhesive is useful for bonding the edges of heat-recoverable sheets to one another to secure closure [194].

Tear Tapes. Self-adhesive tear tapes are designed to be used as an easy-opening device for flexible packaging. They can be used to remove the top of a pack or to form a lid, or in applications where a heavy duty opening system is required. These tapes provide tamper evidence, indicating when the packaging is intact. They are effective on overwrap, shrink sleeves, flow packs, and blister packs for foodstuffs and pharmaceutical products. Some are provided with holograms, and some combine easy opening, tamper evidence, and inherent security using a motorized applicator [168]. An adhesive tape used for fastening printer paper together is perforated along the centerline in the long direction to allow easy tear [195]. Easy-tear breakable tapes allow easy tearing or splitting of the tape during handling. Easy-tear breakable hank tapes are used in the same finished product as easy-tear paper tapes. The tape is applied by hand and is used for hanking and spot tape when fast easy removal is required. There is a difference between the tear behavior of paper-based and film-based tapes. Although a film-based tear tape can be used in the same finished product as paper tape, its physical properties, application method, and resultant assembly are different. Because of the higher tear resistance of film-based tear tapes, care must be taken to ensure a maximum of two laps [196]. Tapes for easy handling can comprise a continuous film–adhesive layer–perforated film laminate [197].

Freezer Tapes. Like freezer labels, freezer tapes are used at low temperatures (see Freezer Labels also). They are applied as packaging tapes. They can be coated with an adhesive based on natural rubber, styrene-butadiene rubber, regenerated rubber, rosin ester, or polyether [198]. They are based on primed soft PVC and have to resist temperatures between −29 and +113°C.

Pharmaceutical Sealing Tapes (Antitheft Tapes). Some years ago wet gummed products were applied almost exclusively as sealing tapes for pharmaceutical use. Now PSA tapes are applied (see Antitheft Labels) [188]. Such tapes are used as security closures for pharmaceuticals.

Repair Tapes. Repair tapes are flexible plastic tapes reinforced with nylon cord and have to adhere to metallic surfaces, wood, rubber, or ceramics. They are designed to offer puncture and tear resistance [85]. Such tapes are supplied as continuous web or die-cut finite elements. They have to be conformable and provide excellent bonding.

Insulating Tapes

Insulating tapes are used for a variety of applications, mainly for electrical or heat insulation or as sealants. They must have excellent mechanical, electrical, and thermal properties. Wire-wound tapes, electrically conductive or electrically insulating tapes, and thermally insulating tapes are the most important representatives of this product class.

Wire-Wound Tapes (Electrical Insulation Tapes). Electrical insulation tapes are used for taping generator motors and coils and transformers, where the tape serves as overwrap, layer insulation, or connection and lead-in tape [199]. The carrier is woven glass cloth impregnated with a high temperature resistant polyester resin or an impregnated woven polyester glass cloth. The latter variant is used where conformability is required. Oriented PP film (30–50 μm) is used for tapes, and oriented LDPE (80–110 μm) for insulation tapes [200]. Such nonflammable tapes can have an adhesive with 0–40% polychloroprene rubber [201]. An insulating tape with an acetate carrier is used for wire winding by a telephone manufacture [202]. It resists temperatures of up to 130°C [203]. It has to adhere to modified flame- and heat-resistant EPDM formulations [204]. High temperature resistant insulating tapes on a PET basis can be used at 180°C and an electrical tension of 5000 V [203]. Cable tapes have to resist 1 month at 70°C [205]. Fixing, transfer, carpet, and electrical insulating tapes are coated with an acrylic

End Uses of PSPs 593

PSA [206]. Certain electrical insulation tapes have to resist transformer oil; crosslinked acrylics are used for these products [207]. Electrical tapes must be approved by Underwriters Laboratories [208] and require elongation, shear, and peel resistance. Shear resistance should be measured after solvent exposure also. Electrical tapes have to possess high dielectric strength and good thermal dissipation. The edges of electrical insulating tapes used for taping must be straight and unbroken to provide distinct boundaries to the finished areas when the taping is complete [209]. Such tapes may be perforated.

Electrically Conductive Tapes. The principles of formulation of electrically conductive PSPs are discussed in Chapters 3–6. Tapes with adequate electrical conductivity are made by using a metal carrier (nickel foil) and an adhesive that has nickel particles as filler (3–20%) in an acrylic emulsion [210]. Electrically conductive adhesive tapes have been manufactured by coating an acrylic emulsion containing 3–20% nickel particles onto a rough flexible material (e.g., nickel foil) [210]. Such products can contain silicon carbide as filler also [211]. The adhesive may include a metallic network to ensure electrical conductivity [212].

Thermal Insulation Tapes. Double-side coated foam tapes are designed for industrial gasketing applications. They are based on closed-cell PE foams coated with a high tack medium shear adhesive [213]. Pipe insulation tape is used for gas pipelines [214]. Pipe-wrapping tapes are plastic tapes reinforced with nylon cord and have to adhere to metallic surfaces, wood, rubber, or ceramics. They are designed to offer puncture and tear resistance [85]. Pipe insulation tapes should also adhere to oiled surfaces [127]. An aluminum carrier and a tackified CSBR coated with 40 g/m^2 PSA are used for insulating tape in climatechnics [215]. Tapes applied as corrosion protection for steel pipes contain butyl rubber, crosslinked butyl rubber, regenerated rubber, tackifier (polybutene or resins), filler, and antioxidants [216]. Such adhesives are coated on a nonwoven carrier also. Airseal tape is made with butyl rubber on ethylene-propylene foam. Insulating tapes for gas and oil pipelines are manufactured by coextrusion of butyl rubber and a PE carrier. There are also foamable insulation tapes. For these products PUR foam is bonded in situ onto the (specially treated) carrier back [217]. Thermally conductive pressure-sensitive insulation tapes are manufactured as anticorrosion protection tapes also [218]. Pipe insulation tapes are wound by overlapping (50%) using special machines [203]. Bilayered heat-shrinkable insulating tapes for anticorrosion protection of petroleum and gas pipelines have been manufactured from photochemically cured low density polyethylene with an EVAc copolymer as an adhesive sublayer [219]. The liner dimensions of the two-layer insulating tape decrease by 10–50% depending on the degree of curing (application temperature 180°C). The application conditions and physicomechanical properties of the coating were determined for heat-shrinkable tapes with 5% shrinkage at a curing degree of 30% and shrinkage force of 0.07 MPa.

Special insulating tapes have to adhere to refrigerated surfaces. A self-sticking tape for thermal insulation used to secure a tight connection between heat exchange members (metal foils) and polyurethane foam based thermal insulation materials is prepared by applying an adhesive that is able to adhere to refrigerated surfaces on an olefin copolymer carrier material (containing release) that is bonded to the polyurethane [220]. Low temperature heat-curable silicone pressure-sensitive adhesives have been formulated from gumlike silicones containing alkenyl groups, a tackifying silicone resin, a curing agent, and a platinum catalyst [221]. Such adhesives for heat-resistant aluminum tapes are coated with a thickness of 50 μm and cure in 5 min at 80°C.

Foam Tapes. Classic foam tapes are foamlike PSPs with a foamed plastic carrier that is coated with a PSA. Carrierless foamlike tapes exist also. Such products are plastic foams impregnated with PSA or made from a foamed PSA. The foam carrier provides excellent conformability and stress distribution. In some applications the PSA is coated first on a flexible PUR, acrylic, or other foam and then laminated onto the carrier surface. The opposite surface of the foam may also be provided with a PSA layer [163]. Foam sealing tapes are made of a foam sealing tape layer and an interlayer strip rolled up together in a compressed form [222]. For such tapes there are standard thicknesses, roll lengths, and ranges of width given by the supplier [223]. Special foam mounting tapes are used for bonding soft printing plates for flexo printing [224]. Foam tapes can compensate for tolerances in the printing process and ensure that dot gain is minimized [225].

Automotive insert tapes are used as sealants. Such products are tapes based on a foam carrier and have to exhibit aging stability, weatherability, sealing, nonflammability. The tape roll is compressed before use (it is supplied in compressed 1/5 thickness). Double-side coated tapes with neoprene, PVC, PUR, or PE foam carrier have been developed [225]. When used as mounting tapes, these products have to satisfy requirements for stress relaxation, stress distribution, mechanical resistance, sealing and anticorrosion, and reinforcing and vibration/noise damping.

Mounting Tapes

Mounting tapes have been developed for a number of applications. They can be used for the temporary or permanent mounting of parts of an assembly (e.g., fixing films in a cartridge, mounting glass panes or windowpanes, etc.) [226]. Temperature-resistant silicone mounting tapes maintain their performance characteristics up to 288°C and adhere to a variety of surfaces, including metals, glass, paper, fabric, plastic, silicone rubber, silicone varnished glass cloth, and silicone glass laminates. They must exhibit good electrical properties and good resistance to chemicals as well as to fungus, moisture, and weathering [227].

Double-faced mounting tapes can be manufactured as double-side coated carrier based products or carrier-less, foamlike webs. Certain double-faced mounting tapes are built up with PET as the carrier material. They have a silicone adhesive on one side and a silicone release layer on the other. They are used in electronics [228].

Mounting tapes used in buildings must have the same life expectancy as the building elements (i.e., 10 years) [229]. Such mounting tapes can replace other, classic (mechanical) methods of fixing [230]. For instance, a glass-mounting tape is specially designed to bond two sheets of glass around the edges for a composite glass with a strengthening layer. The old method used black butyl tape to create a layer of air between two sheets [231]. Such tapes give a long-lasting waterproof seal. Another double-coated adhesive tape for mounting provides a cleaner watertight system and cuts labor costs. One side of the tape is applied to the window frame, and the glass pane is pressed against the other side of the tape. The pane is thus held securely when laths are nailed to the surrounding frame [232]. In a first step of mounting, one-side coated PE pressure-sensitive tape is applied [233]. Pressure-sensitive sealant tapes are soft, and their PSAs are tackified or plasticized formulations. They can be filled and crosslinked. Their quick stick values are low. Their peel resistance value (on aluminum) is about 120–140 oz/in.

Special sealant tapes for gasket applications are made by extrusion onto a release film. Such tapes must possess high tack and be resistant to gasoline.

Tissue tapes are applied to temporarily hold seat upholstery together. The double-faced tape holds leather in place prior to stitching and eliminates problems of leather creasing or slipping [232]. Pressure-sensitive tapes can be used to bond exterior trim to automobiles [234].

Certain mounting tapes have to be repositionable [235]. A special mounting tape for an instrument housing assembly is designed like a tear tape with a transfer adhesive [236]. The tape consists of a metallic foil strip of predetermined thickness and an acrylic transfer adhesive strip affixed to one side of the metallic foil and forming a tear band. Mounting tapes are used in the fashion and textile industries for attaching textile parts to each other. They are protected from telescoping during storage with a woven on nonwoven supplementary layer [237]. Fasteners are used in attaching automobile seat covers [162]. Certain mounting operations require high peel removable adhesives. In such cases postcrosslinking can be carried out. Postapplication crosslinking is proposed to achieve easier delamination. According to Charbonneau and Groff [238–240], the procedure allows delicate electronic components to be removed from the tape even though they were not removable before crosslinking.

Adhesives with excellent water resistance (resistance to humidity and condensed water), are required for technical tapes and pressure-sensitive assembly parts in the automotive industry [215]. Shear values of about 600–1000 min (25 $\times$ 25 mm, 1000 g, on steel) and peel values of 25–35 N/25 mm (20 min, on steel) are required. High shear, low tack applications are suggested for certain mounting tapes and double-coated foam tapes [47]. For these mounting tapes the target values of (180°, instantaneous) peel, loop tack, and 72°F shear are 34. 5/6.5 N/25 mm and 400 minutes. High coating weight is used. For mounting tapes a 88 μm thick adhesive layer has been proposed [47]. Peel was tested on steel and polycarbonate. A formulation of a PSA based on tackified SBS for double-faced carpet tape having 220 phr resin and 20 phr oil per 100 phr block copolymer (and a high coating weight of 40 g/m^2) exhibits relatively low adhesive characteristics (a peel strength of 16–21 N/25 mm), loop tack of 17–18 N/25 mm, rolling ball tack of 3–5 cm, and shear on steel of more than 100 h [182].

Foam-backed adhesive tapes are commonly used to cause an article to adhere to a substrate. The foam carrier is pigmented with carbon black to camouflage the presence of the tape [241]. Foam tapes used as structural bonding elements exhibit the following advantages [132]: (1) They do not damage the substrate; (2) no mechanical processing operations (drilling, screwing, riveting, etc.) are needed; (3) they give better and more uniform stress distribution; (4) they are lightweight; (5) they produce a watertight seal; and (6) they protect against corrosion. They might be used, for example, in the design of a thermal break. As defined in Ref. 132, a thermal break prevents differences in temperature within a structure, reducing the stresses that arise from different rates of expansion. Self-sticking, double-faced foam tapes should also be fire- and mildew-resistant [43]. According to Ref. 132, there are two main categories of double-faced tapes: double-side coated tapes with carrier and double-side coated tapes without carrier (foamlike tapes).

A foamlike pressure-sensitive tape is manufactured by using glass microbubbles as filler [241]. Dark glass microbubbles are embedded in a pigmented adhesive matrix. Transparent microbubbles are included in the composition of a foamlike tape [242]. Pressure-sensitive tapes filled with glass microbubbles have a foamlike appearance and character and are useful for purposes previously requiring a foam-backed PSA tape. The average diameter of the microbubbles should be between 5 and 200 μm. The thickness of the pressure-sensitive layer should exceed three times the average diameter of the microbubbles to enhance the flow of the bubbles in the matrix under applied pressure. This enables intimate contact to be built up with rough and uneven surfaces while retaining the foamlike character. Optimum performance is attained if the thickness of the PSA layer exceeds seven times the average diameter of the bubbles.

Plotter films are used as chablones that are cut from the film material. The negative lettering is applied on the substrate. Thus spray lacquering of the chablones is eliminated. The decorative and informational elements are placed and temporarily fixed with application tapes. Such tapes can be considered as temporary mounting tapes also.

Transfer Tapes. Transfer tapes are tapes without a carrier. They have a solid-state component only temporarily. The sheet backing is a release liner, and in use the exposed adhesive surface of this tape is placed in contact with the chosen substrate, the release liner is stripped away, and the newly exposed adhesive surface is bonded to the second surface. Their construction includes a release liner and the adhesive core. The adhesive core may be a continuous homogeneous adhesive layer or a semicontinuous heterogeneous adhesive layer. The heterogeneity of the adhesive layer in the latter case is due to embedded solid, liquid, or gasous particles (holes); i.e., the adhesive layer can be a foam [132].

Double-faced tapes were used some years ago only for low stress applications. Cohesive, high resistance transfer tapes were developed about a decade ago [243]. They can be used for structural bonding. Transfer tapes for structural and semistructural bonding have the following properties:

1. High temperature resistance (does not depend on the carrier)
2. Good environmental resistance
3. Conformability to the substrate shape
4. High transparency (for homogeneous tapes)
5. Low and constant adhesive thickness and variable adhesive thickness

According to the type of adhesive, carrierless tapes may be classified as (1) tapes having an acrylic film or (2) tapes having an adhesive foam core.

Traditional foam tapes have the disadvantage that the foam can split from the carrier (adhesive break) at too high a peel force. Foamlike transfer tapes have the advantages of (1) softness, thickness, and elasticity of the foam, which allow reliable bonding of uneven and textured surfaces; (2) the ability to dampen sound and vibration; and (3) resistance to temperatures of 90°C in the long term and 150°C in the short term.

The transfer of tape from the temporary liner can be regulated by means of the release, but it can also be controlled by using different adhesives or adhesives with different degrees of crosslinking. The radiation-curable compositions and curing procedure suggested in Refs. 238 and 239 can be used for transfer tapes also. The carrier can also be a foam coated on either one or both sides and having an overall thickness of 0.1–2.0 mm [242]. Both surfaces of the carrier may have low adhesion coatings, one of which is more effective than the other. When the tape is used as a transfer tape, when it is unwound the adhesive layer remains wholly adhered to the higher adhesion surface, from which it can be subsequently removed. To enhance immediate adhesion to rough and uneven surfaces a resilient foam backing can be used. Glass microbubbles can be incorporated to facilitate immediate adhesion to rough and uneven surfaces. For such tapes both a breakaway cleavage peel value and a continuing cleavage peel value have been measured. The continuing cleavage peel value is about 50–60% of the initial value. The cohesive (tensile) strength of the crosslinked adhesive (transferred core) may attain 6000 kPa (depending on the crosslinking conditions).

Transfer tapes with contoured adhesive have been proposed also. An uneven adhesive surface can be achieved by transferring adhesive during rewinding according to Ref. 244. This tape has a carrier with recesses on one side that are filled with adhesive. When the tape is unwound, the adhesive transfers from the recesses to the carrier. The use of EVAc copolymers for carrierless self-adhesive films (tapes) used to bond roofing insulation and laminate dissimilar materials is proposed in Ref. 245. Such formulations contain EVAc-polyolefin blends, rosin, waxes, and antioxidant. They are processed as hot melt and cast as 1–4 mm films (application temperature 180°C). The adhesive strength of such tapes decreases by 10–45% in 1 year.

Splicing Tapes. Splicing tapes have been used in the paper, converting, and printing industries. Papermaking and printing technology require splicing the end of one roll of paper to the beginning of another as well as splicing parts of a roll together after defective material has been cut out. Such splices have to be made quickly and easily with an adhesive that rapidly attains maximum strength [246].

The adhesive and end use requirements for splicing tapes are very severe. High tack and fast grabbing are required for high speed flying splicing. The tapes have to exhibit adhesion to various papers, liner board, foils, and films as well as high shear resistance and temperature resistance while traveling through the drying oven. The application time (i.e., bonding time) for splicing tapes is very short (less than 60 s according to Ref. 247). Reel changeover follows in less than 15 min. The joint length is less than 100 mm. The most difficult cases are characterized by high running speed, small reel diameter (for the next reel), maximum thickness of the web for the new, small diameter reel, and a short distance between the splicing point and the cutting line.

Butt splicing tapes, built up as single-coated PET, have to adhere to PP, PE, PA, and coated paper, resist temperatures up to 117°C, and possess high quick stick for high speed splicing applications [85]. For instance, the splice for flame-resistant tapes must be constructed in such a fashion that the roll can be unwound at a rate of 7.3 m/min [248]. Splicing tapes for paper manufacturing have to be recyclable [249] and resist temperatures up to 220–240°C. The paper used as carrier materials has to be dispersible, and the adhesive must be water-soluble but resistant to organic solvents. No migration is allowed. The adhesive has to bond at a running speed of 160 m/min. Such formulations must adhere to paper substrate running at 1200 m/min [250]. The tear resistance of the tape has to be as high as that of the paper to be bonded, and high instantaneous peel and shear resistance are required. For splicing tapes the corresponding tack, peel, and shear values are 2.2, 1.5, and 360 [47]. The shear value was measured at 250°F (50 μm adhesive thickness). The main types of splicing tapes are [249] one-side coated tape; double-side coated tape, and transfer tape.

In one-side coated splicing tapes the siliconized paper liner remains on the adhesive and has to be recyclable also. The carrier is water-dispersible or water-soluble [50]. Carrier-free splicing tapes have to be water-soluble also. The liner for transfer tapes has to allow detachment of the adhesive, i.e., it must display different degrees of release on its two sides. The requirement concerning the water solubility or dispersibility of the paper carrier used for splicing tapes is an old criterion [122,251]. Tack and heat stability together with water solubilty or dispersibility are required for splicing tapes in papermaking and printing [252]. Such tapes are used for bioelectrodes also because of their low electrical impedance and their conformability to skin. Water-soluble compositions used for splicing tapes contain polyvinylpyrrolidone with polyols or polyalkylglycol ethers as plasticizer, vinyl ether copolymers, neutralized acrylic acid-alkoxyalkyl acrylate copolymers, and water-soluble waxes [250], or acrylics [50]. Water solubility is given by a special composition containing vinyl carboxylic acid neutralized with an alkanolamine [252]. Other formulations contain acrylic acid polymerized in a water-soluble polyhydric alcohol and crosslinked with a multifunctional unsaturated monomer. Repulpable splicing tape especially adapted for splicing carbonless paper uses a water-dispersible PSA based on acrylate-acrylic acid copolymer, alkalies, and ethoxylated plasticizers. A polyamide-epichlorohydrin crosslinker may also be included [246] (see Chapter 6 also).

Splicing tapes have to be tested for the following properties: tack and adhesion on paper (on silicone raw paper also), dynamic and static shear strength at normal and high temperatures, resistance to grease and bleed-through, and water solubility at pH 3, 7, and 12 [50]. According to Ref. 250, water solubility between pH 3 and 9 should be given. As shown in Ref. 50, the adhesive properties of special acrylic-based, water-soluble adhesives for splicing tapes can vary as a function of atmospheric humidity (see Chapter 5 also).

Medical Tapes. Surgical pressure-sensitive adhesive sheet products include any product that has a flexible carrier material and a PSA [253]. Labels, tapes, adhesive bandages, adhesive plasters, adhesive surgical sheets, adhesive cornplaster, and adhesive absorbent dressings have been manufactured. Natural adhesives mixed with natural additives have been used [207]. Medical tapes have been developed for pharmaceutical companies, ostomy appliances, diagnostic apparatus, surgical grounding pads, transdermal drug delivery systems, and wound care products (see Medical Labels also).

Physiologically compatible adhesives are based on polyurethanes, silicones, acrylics, polyvinyl ether, and styrene copolymers. These should be breathable, cohesive, conformable, self-supporting compositions that cannot be degraded by body fluids [254]. They minimize skin irritation due to their ability to transmit air and moisture through the adhesive system.

Air- and moisture-permeable nonwovens (PET nonwoven, embossed nonwoven) or air-permeable tissue (breathable face stock) are used for medical tapes coated with a nonsensitizing acrylic porous adhesive for prolonged application of a medical device on the human skin. The air porosity rate has to fulfill a given value (50 s/100 cm^2 per square inch) [255]. Skin tolerance, no physiological effects, resistance to skin humidity, and sterilizing capability without color changes are required for medical tape adhesives [256]. Certain applications require water-soluble or crosslinked adhesives [128].

Surgical tapes are used for fastening cover materials during surgery. Cotton cloth and hydrophobic special textile materials coated with low energy polyfluorocarbon resins are used as carrier materials and coated with polyacrylate-based water-soluble PSAs. The adhesive has to be insensitive to moisture and insoluble in cold water and must display adequate skin adhesion and skin tolerance. The water solubility of the product is reached above 65°C at a pH above 9 [50,256]. Reusable, hydrophobized textile materials and extensible, deformable paper are required for medical tapes [227,257]. PVC can be used for self-adhesive medical tape or as an adhesive coating also [258]. Preferred carrier materials are those that permit transpiration and perspiration and/or the passage of tissue or wound exudate [257]. They should have a moisture vapor transmission of at least 500 g/m^2 over 24 h at 38°C with a humidity differential of at least about 1000 g/m^2. For medical tapes the conformability of PSA, its initial skin adhesion value, the skin adhesion value after 24–48 h, and adhesive transfer [260] are measured as the main performance characteristics. A desirable skin adhesive will generally exhibit an initial peel value of 50–100 g and final peel value (after 48 h) of 150–300 g.

Rubber-based and acrylate homo- and copolymers have been used for medical tapes [260,261] (see Chapter 6 also). For a composition containing polar comonomers and with built-in (carboxyl or hydroxyl group-containing) crosslinking comonomers, a coating weight of 34–68 g/m^2 has been suggested. Unfortunately, many of these compositions lose their adhesiveness in hot or humid environments and allow adhesive transfer during removal. The adhesives used for surgical tapes often exhibit a dynamic modulus too low for outstanding wear performance. A low storage and loss modulus result in a soft adhesive with adhesive transfer. A creep compliance of 1.2×10^{-5} cm^2/dyn is needed for a skin adhesive [262] (see also Chapter 3, Section 1). Irritation caused by removal of the tape was overcome by making the composition water-soluble [263]. So-called wet-stick adhesives (sterilized with ethylene oxide or gamma radiation) are used for surgical towels; they have to adhere to skin, wet textiles, and nonwovens [264].

Application Tape. Application tapes are special paper- or plastic-based mounting tapes. They are used together with plotter films as removable mounting aids. Application tapes are applied to ensure the transfer and temporary fixing of another permanent pressure-sensitive element (letters or written text) [265] (see also Chapter 6). The plastic-based tapes have PVC, PE, or PP carrier material that can be embossed to lower unwinding resistance. Plastic-based application tapes can also have paper release liners; these tapes are used for signs with small dimensions. Special application papers can be used as stencil paper also (see Plotter Film).

Protective Tapes

Protective tapes are used as temporary or permanent surface-protective or reinforcing elements. Most of them are really protective films with a tape-like geometry. They have various application domains.

Car Masking Tape. According to Ref. 132, the first masking tapes allowed the two-color painting of cars. These tapes are used during the spray painting of vehicles. Masking tapes may have special carrier constructions to give them conformability and temperature resistance. Some masking tapes have to be flexible, stretchable, and contractile, capable of maintaining the contour and the curvature of the position in which they are applied [266]. The adhesive is covered with a release liner. According to Lipson [267], a masking tape carrier has a stiffened longitudinal section extending from one edge, with an accordion-pleated structure to conform to small radii. The nonadhesive face of the tape is heat-reflective. Masking tape based on foam has been patented [268]. Masking tapes on paper basis (90 g/m^2) can be coated with 30 g/m^2 HMPSA. EVAc copolymers are used for HMPSA for weatherstripping tapes. A coating weight of 60 g/m^2 is applied for a bonding surface, with 50 g/m^2 on the opposite side to bond a wipe-clean surface.

Masking tapes used for spray varnishing in the automotive industry have been partially replaced by adherent, non-pressure-sensitive cover films that give less expensive, full surface, easily removed protection [269]. A pressure-sensitive adhesive tape for the protection of printed circuits, based on acrylics and phenolic terpene resin with a 100–200 g/m^2 siliconized crepe paper, is made by coating the adhesive mass first on a siliconized polyurethane release liner, then transferring it under pressure (10–30 N/cm^2) onto the final carrier material [270]. Special office masking tapes are applied to cover written text passages; the covered surface portion should give a shadow-free copy [271]. Generally masking tapes are rigorously tested for deposit-free removability at elevated temperatures also (see Table 8.7).

Mirror Tapes. Mirror tapes are large surface security films applied to the back of fragile substrates (see Protection Films).

Table 8.7 Conditions for Removability Test of Masking PSPs

PSP	Room temp.	Elevated temperature	Ref.
Masking tape	30 d	3 days, 80°C	154
Removable tapes or sheets	100 h	3 days, 60°C	147
Masking sheet	—	2 days, 60°C	155
Masking sheet, tape	—	3 days, 60°C	148

Storage conditions span the Room temp. and Elevated temperature columns.

Other Tapes

The number of other special tapes is large. Some of them have an ordinary construction, and others are more sophisticated. Certain products function mechanically to fix and fasten elements; others have dosage, insulation, or other special function.

Test Tapes. Special and standard tapes are used to test the adhesivity of varnish or printing ink [272]. Tapes can be used in testing laminates also. To evaluate a laminate section by light microscopy, a microtome is used to produce perfect cuts. A rapid method for fixing (reinforcing) the sample is to laminate it with a tape [273]. The surface treatment is tested according to ASTM 2141-68 by using a PSA tape [274–276].

4 PROTECTIVE FILMS

The use of protective films is a relatively new field. At the end of the 1980s, protective films were suggested for the posttreatment of metal coils as an alternative to oiling or wax coating [277]. The main criteria given by the German Association of Surface Coated Fine Coil (Fachverband oberflächenveredeltes Feinblech) for the choice of protective films are the nature, thickness, adhesion characteristics, forming, tear resistance, and light stability of the film, with the recommendation that "well-defined protective films should be applied outdoors for a given time only."

4.1 Application Conditions

Because they are used for only temporary bonding, application and deapplication are of equal importance for protective films. If the film is not detached completely, the different thermal dilatation of the metalic coil and plastic film may lead to stress cracking [277]. For film application, the chemical composition and processing quality of the surface play a special role.

Product Surface

As stated in Ref. 278, for protective films the adhesion of the PSA should be tailored to the structure of the surface to be protected. Protective films are used on very different substrates depending on the nature of the products to be protected. Generally supplier checklists concerning the application conditions of a protective film should contain the following data:

UV resistance
Desired adhesive strength
Nature of the surface to be protected
Thickness of the web to be protected
Future processing (forming) operations

It should be taken into account that, because of their higher elasticity, the processing of plastic surfaces for a given smoothness is more difficult [46]. The surface itself has a multilayered structure. There is a layer of dirt (3×10^{-6} mm), then an adsorption layer (about 3×10^{-7} mm) and a reaction layer (1×10^{-4} mm), followed by the material itself [279].

Various types of substrates have to be coated with protective films. Metallic or plastic surfaces are the most frequently encountered adherends. Various metals are used—unpainted and painted; powder and thermo painted; bright, glossy, and matte. Stainless steel, BA, cold-rolled, polished, or brushed, and aluminum, mill finished, anodized, or brushed, have been protected. Coil coating is the most important domain of metal coating. Here uncoated "pure" metal surfaces and precoated varnished surfaces are laminated with protective films. Coated steel materials include plastisol-coated, epoxy-modified polyester-coated, and hot dip galvanized steel [280]. Alkide, polyurethane, silicon-modified, and fluorine-containing macromolecular compounds have been suggested for coil coating also [281]. Acrylics, melamine, PVC, and lacquered, powder-coated, and thermolacquered surfaces may be used as substrates also. Coils are classified according to the metal, coating, and application. The main application fields are the automotive industry, construction, and household machines [282]. Coil-coating varnishes have to resist processing steps such as deep drawing, drilling, and cutting [283].

Quite a variety of surfaces have to be protected within the same application field. As shown in Ref. 284, a variety of lacquers are used for lacquering various plastic automotive components (depending on their elasticity). For instance, for low elasticity parts ($E > 3000$ N/mm^2), one-component acrylic-melamine, epoxide, and two-component PUR lacquers are applied. For parts having medium elasticity ($E = 1800$–3000 N/m^2), one- or two-component PUR-, acrylic-, or polyester-based varnishes are suggested. For elastic parts, two-component PUR lacquers have been proposed.

Customers refer to the typical stainless finishes as dull, bright, or annealed. Steel and copper surfaces may be chemically etched to produce a fibrous oxide growth on the surface [46]. The level of roughness is achieved by transferring a pattern from the working (calendering) rolls to the material, which then goes to the tempering mill to be (eventually) tinned or coated with chromium or other metal.

A dull finish is developed by final rolling followed by annealing and hot acid pickling. A low gloss finish results that is suitable for functional parts, where ultimate lubricant retention is required or where severe forming with possible intermediate annealing occurs. Extensive polishing and buffing is required to give the desired finish quality.

A bright finish is achieved via rolling with highly polished rolls followed by conventional annealing and electrolytic pickling. A relatively glossy gray surface results. This is a general-purpose finish, used where a minimum of polishing and buffing is employed to develop a bright luster. Typical uses include wheel covers and cookware.

For a bright annealed finish the final rolling is done by rolling with highly polished rolls and annealing in a protective atmosphere. It is used for applications where only a color coating is used to achieve the desired luster. A hardened finish is practicable only for certain martensitic, hardenable grades of steel. It is achieved by heat treating, annealing, and pickling. Additionally, there are many degrees of finish within each category [285]. For packaging steel, the measured roughness is expressed as the RA value (in micrometers). Roughness may vary from mirror finish (RA less than 0.17 μm) to matte finish (RA 0.70–1.00 μm) [286]. Generally a steel surface is passivated. Passivation is a chemical or electrochemical process that increases the metal surface's resistance to oxidation and facilitates lacquering. Tin, chromium, and chromium oxide may be used for top-coating. Usually coils are oiled to ensure better handling and chemical protection. Plasticizers such as dioctyl sebacate or acetyl tributyl citrate are used as the "oil." Their coating weight depends on the coil quality and use. Tin-coated coils are coated with 0.5–11.2 mg/m^2 oil. Uncoated packaging steel (black plate) has 25 mg/m^2 oil. For lacquering or printing, 4.4 g/m^2 oil is used [286]; for overseas shipments a larger quantity (10 mg/m^2) is used.

For a stainless steel mill finish, polished, BA, protective films with rubber resin adhesive are recommended. It would be difficult to achieve adequate contact between a hard acrylic adhesive and this surface. It would make the edging difficult, and the film could loosen. Furthermore, it is useful to have higher adhesive strength when forming special steel. As is known from the practice of dry laminating, rubber-based PSAs are used for Zellglas/PE lamination, where the mechanical properties of the components are quite different, or where a soft bonding layer is necessary [287]. The ability to bond through oil is essential in many processing applications. For metal protection for applications where chemical and water resistance are required, immersion bath compositions based on thermoplasts, chlorinated paraffin, and esters of phthalic acid have been suggested also [288].

The surface of aluminum with a mill finish is softer. Acrylic adhesive-coated protective films may be used for their protection. Various surface qualities—mill finish, polished, matte, and anodized—are manufactured. With anodized aluminum, the kind of treatment and its age influence the adhesion. The surface composition of aluminum changes during storage [289]. The concentration of alkali and alkaline earth metals increases. An anodized aluminum surface can have a very different oxide morphology [290]. Generally the surface of metals is rough and oxidized; the oxide layer contains water, and the water contains salts [291]. For electronic parts, brushed aluminum is used [292].

All lacquered surfaces (unpolished, polished, and bright polished) should be pretested before coating. Steel may be coated with epoxy-, polyester-, acrylic-, or melamine-based paint [293]. Flexible polyester is used for coil coating [294]. Polyester coil coating systems may contain nonvolatile ester alcohols [295]. Silicon acrylates are used for coil coating and for surface panels for buildings. Epoxide and polyester, zinc, or acrylic and melamine have used for coating cans and coils [296]. Cellulose, nylon, epoxy, vinyl, acrylic, and polyester powders may be used for metal coating [297]. For coil coating, high solids paints, powder paints, and systems capable of being hardened by radiation have been developed [298]. Some varnishes are based on plasticized PMMA.

Electrostatic powder spraying (EPS) is also used for coating coils and automotive parts. According to Ref. 299, in Europe the first powder coatings were developed in the early 1960s. The first powder coatings based on PE were introduced in the United States in 1971 [300]. The most used powder lacquers are based on epoxy, epoxy-polyester, polyester, and polyurethane derivatives. There are crosslinked systems cured at 140–220°C [301]. Epoxy-polyester systems are used particularly in the automotive industry, where acrylate powder lacquers are used also [302]. Powder-molded compounds (PMCs), where powder lacquering of the product occurs in the mold, is used for sanitary items and automotive parts [303].

It is well known that for certain acrylic coil coating formulations, 1–5% of special polyethylene dispersions is added to the recipe to impart better deformability and prevent blocking of the lacquered surface. Therefore, the application of a PE-based hot laminating film for such coils may cause PE/PE adhesion, i.e., a higher adhesion level. Such limitations exist for adhesive-coated protective films also. Rubber resin adhesive coated protective films cannot be used for copper or brass (bright or mill finish). Sulfur-containing copper reacts with the rubber.

Other surfaces such as glass, plastic profiles, acrylic sheets, PC plates, textiles, carpets, linoleum, parquetry, marble, wood, and melamine kitchen furniture also have to be protected. Most of these are manufactured in different qualities (grades). Furniture is lacquered with polyacrylates containing OH groups. Polyester acrylates in tripropylene glycol diacrylate are UV-cured on plastics [304]. Glass is covered with PVA/plasticizer [305]. Polycarbonate plates are produced with bright, glossy, and matte surfaces; melamines may be bright or matte; synthetic marble is produced with bright and matte surfaces. PVC plates of both soft and hard PVC are manufactured [306]. PVC plates may be coated to increase their abrasion resistance and gloss with a 0.003–0.004 in. layer of deck compound that displays different adhesion than common PVC [307].

Window frames are produced at a speed of 3 m/min from high impact hard PVC [308]. Other polymers like ABS, PP, PPE-HIPS blends, and PMMA have been tested also. PVC window profiles were first marketed 40 years ago [309]. Chlorinated PE or polyacryl ester is used as an impact modifier. SPVC, filler (TiO_2, $CaCO_3$), and chlorinated PVC are used as components in the formulation. After warm storage of the profiles, dimensional changes of 1.7% may occur. Screw rotation speed and the temperature in the extruder are also known to influence the shrinkage of the extruded product.

For furniture, UV-curable unsaturated polyester coatings and UV-curable polyester coatings are used [310]. Decorative laminates are manufactured from paper impregnated with melamine resins [306]. Modified polyethyl methacrylate may be used as a protective film for substrates such as PVC, ABS, polystyrene, wood, paper, and metal [311].

PMMA forming materials and semifinished products are produced from methyl methacrylate [310]. They are used for autotomotive, construction, light technical, advertising, household wares, and hygienic products. Semifinished products, cast and extruded PMMA plates, and extruded PC plates have been developed. So-called acrylic glass is manufactured with a minimum thickness of 2–3 mm [312]. Cast PMMA plates possess a thickness of 2–100 mm; extruded PMMA plates have a thickness of 1.5–6 mm. Extruded high impact PMMA plates have a gauge of 2–6 mm. Extruded PC plates have a gauge of 1–12 mm. According to Ref. 313, cast PMMA plates are manufactured with thicknesses of 2–25 and 30–100 mm; extruded plates with thicknesses of 1.5–6 mm; extruded HIPMMA, 2–6 mm; and extruded PC, 1–2 mm. Their abrasion resistance is improved by using special silicone-alkoxy condensation products [314]. Their surface quality may be quite different. As an example, polycarbonate plates can be patterned on both sides, with a pinspot pattern on one side, prism pattern on one side, both sides polished or patterned, and raindrop/haircell [315].

Polycarbonate with a thickness of 0.25 mm is used for overlays for keyboards [312]. Plastic-coated glass surfaces are used as building panels. The coating is an ionomer. The same system is used for bottle manufacturing (see Label End Use) [316].

As discussed earlier, there are both static (packaging) and dynamic (processing-related) uses for protective films. The requirements for these application fields are quite different.

Packaging Use of Protective Films

Certain protective films are used only as packaging materials. The protected item does not undergo work processes. Very often films are used for protection during transport, e.g., when stacking profiles; there is no paper between stacked items. Polycarbonate or PMMA sheets are provided with protective films on both sides to avoid damage to the mirror-polished surfaces during transport (and processing). Time of storage on the sheet or on the roll should not exceed 1–1.5 years. In packaging uses, the self-deformation of the coated plastic material has to be taken into account as a dynamic force. Plastics development reduces this risk. In the last decade the stability of plastic pipes at 80°C and 4 N/mm^2 force has increased in the main national norms from 170 to 1000 h [30]. Thermally induced extension of plates and profiles may produce tension and cause delamination of the protective film. PVC profiles display a coefficient of thermal expansion of 8×10^{-5} K^{-1}. Other protected films are processed together with the protected item and have to resist mechanical and thermal stresses during the technological cycle.

Processing Use

Processing (forming) of coated surfaces requires good adhesion between the protected surface and the protective film (see Table 8.8). The film must be able to undergo the processing. The main work processes are bending, punching, drilling, and deep drawing. Cutting, drilling, shearing, routing, and heat bending are the most important processing steps for protective films used for plastic plate

Table 8.8 Dependence of Protective Film Grade Used on the Characteristics of the Protected Product

	PSP characteristics			Product to be protected		
		Peel resistance (N/10 mm)				
	Coating weight	Dwell time				
Thickness	(g/m^2)	10 min	3 mo	Material	Roughness	Processing
50	0.7	0.3–1.0	0.15	SST, aluminum	High gloss and lacquered	Slight processing
70	5.0	0.8–1.8	0.25	SST, aluminum, plastics	Glossy and lacquered	Medium processing
75	3.0	>1.8	0.25	SST, aluminum, plastics	High gloss and lacquered	Medium processing
90	3.5	0.8–1.8	0.25	SST, aluminum, plastics	High gloss and lacquered	High degree of processing
100	2.0	0.8–1.8	0.25	SST, aluminum,	High gloss and lacquered	High degree of processing

protection. Polycarbonate plates may undergo sawing, shearing, punching (sometimes punching tools are heated to 140–180°C), drilling, turning, trapping, milling, grinding, and polishing [315]. During these operations, polyethylene protective films should not come into contact with oily or greasy substances. They should not be used at temperatures of more than 70–80°C. The film may be damaged by bending if the pressure of a tool is concentrated at a single point.

The thickness recommended for a protective film depends on the thickness of the protected item (e.g., a thin stainless steel sheet of 3 mm needs a protective film at least 90 μm thick. For drilling or punching aluminum or stainless steel sheets it is important to know that only a very few liquid processing aids can be used because polyethylene films are oil-permeable. The penetration of liquid between the sheet and film may produce residues. In this case 70–90 μm films should be used.

Protective films are also appropriate for deep drawing. It can happen that wrinkles will form on the edge (PVC would be better than PE). When the sheets are processed or embossed, wrinkles can be pressed into the surface. The processing conditions are quite different for different plastics and plastic plates. As an example, drilling speed during mounting is the highest (2000 m/min) for PMMA in comparison with PC (1000 m/min) or SAN (200–300 m/min) [317]. The high processing speed must be considered, with the low thermal conductivity of the material taken into account also. In comparison with steel (47–58 W/mK), polycarbonate possesses a coefficient of thermal conductivity of only 0.21–0.23 W/mK. Vibration welding is also used for plastic plates [318]. In some cases infrared radiant heating is used for warm-formed plastic plates. The halogen heating elements give their maximum radiation energy at 1.0 μm wavelength. However, at wavelengths of 0.5–1.5 μm both the processed plastic plate and the protective plastic film on it may absorb radiation energy [319]. According to the nature of the protected surface, self-adhesive films can be classified as

SAF for hot laminating uncoated and coated lacquered metal surfaces
SAF for hot laminating plastic surfaces
SAF for hot laminating paper
SAF for mold release casting
SAF for back protection of glass
SAF for special surfaces (fabric)

In choosing the SAF, the intrinsic adhesive properties of different raw material classes have to be taken into account also (Table 8.9).

It should be noted that coextruded SAFs have been used as hot laminating films for other applications for many years [326]. Peel strengths of 1–20 lb/in. have been achieved depending on the product surface and application conditions.

End Uses of PSPs 609

Table 8.9 Self-Adhesion of Elastomers/Plastomers Used for PSPs

Material	Adhesion Value	Comment	Ref.
SIS	70 g	—	320
Ethylene–α-olefin copolymer tackified with PIB, cold-stretchable adhesive film	65 g	ASTM 3354-74	321
Ethylene-propylene block copolymer grafted with maleic anhydride	8.5 kg/25 mm 3.8 kg/25 mm	On stainless steel On aluminum	322
Ethyl acrylate-ethylene-maleic anhydride copolymer, grafted with triethoxyvinylsilane	0.93 kg/cm 0.87 kg/cm	On glass On aluminum	323
PIB, APP, polybutadiene, butyl rubber	65 g	ASTM 3354-74	321
Plasticized PVC	0.07–0.25 N/cm	On siliconized release liner	324
NR, synthetic rubber, filler	?		325
PE, hot laminated	250 g/30 mm	On SST	—

For their application a laminater roll at a temperature of 300–500°F, pressure of up to 400 pli, and line speeds of 45–200 ft/min are suggested. Such products have a warranted shelf life of at least 6 months.

Application/Deapplication Climate

Polyethylene films become brittle at low temperatures, and it becomes very difficult to remove them. At temperatures below zero, films should be warmed before removal. The recommended temperature for their delamination is 8–22°C. According to Ref. 327, protective films should be applied between 15 and 50°C. Table 8.10 summarizes the application conditions for the most common protective films. As Table 8.10 indicates, very different application temperatures and pressures are used for protective films laminated on metallic or plastic surfaces. The application climate is strongly influenced by the nature of the substrate, substrate manufacturing procedures and laminating equipment, and the construction (grade) of the protective film.

Table 8.10 Application Conditions for Main Protective Film Grades

	Protective film		Application conditions	
Application	Adhesiveless	Adhesive-coated	Temp. (°C)	Pressure
Metal coils	Hot laminating	—	220–250	High
	—	AC or RR	RT	Low
Plastic plates	Warm laminating	—	50–70	Medium
	—	AC or RR	RT–70	Low
Finished plastic products	—	AC or RR	RT	Low
Carpet	—	AC or RR	RT	Low

Application Conditions According to Laminating Equipment. It is evident that laminating equipment determines the laminating temperature and pressure, which are the main laminating parameters [326]. The application temperature of the protective film depends on the manufacturing technology used for the protected item. Extruded plastic plates have an elevated temperature after extrusion. This is used to improve the adhesion of the protective film. Such plastic plates are laminated with adhesiveless protective films at high temperatures. As an example, extruded double-layered PMMA and PC plates are laminated with an SAF (40 μm, EVAc based) at 60–70°C. Cast items or plates are laminated with an SAF at room temperature, but they have to support the processing temperature (170°C for 30 min for PMMA, 160°C for 10 min for PC). In this case a greater film thickness (80 μm) is suggested. High temperature extruded plates may be laminated with adhesive-coated protective films also. In this case the adhesion of the film does not require the elevated laminating temperature. It is imposed by the manufacturing procedure used for the substrate to be protected. The laminating speed may vary with the equipment and the products to be protected. This influences the high temperature contact time of the film. For instance, the same coil coating production line may run at 20–120 m/min.

Application Conditions According to the Substrate Surface. Cast (chill roll) polyolefin films are used as self-adhesive protective films (without PSA). They have a thickness of 40–80 μm. They have to adhere to extruded and cast plastic plates. (After cooling the film has to be removed deposit-free and with low peel resistance from PC and AC plates.) For cast plates the application is carried out at room temperature; for extruded plates, at 60–70°C. In several cases such films have to survive thermal forming. In this process the protective film coated plastic plate is heated (for 5–15 min at 170–180°C).

The roughness of the surface (determined by manufacturing and treatment procedures) influences the type of protective film used also. For coil coating the gloss of the lacquer used is a decisive parameter. For instance, 30–40% gloss is recommended for hot laminating films. Generally protective films with higher adhesion (peel) are required for complex processing (multiple deep drawing) and for matte surfaces. Protective films used for deep drawing have to function as a lubricating agent also.

Application Method

Generally protective films are applied by lamination, which involves the use of laminating equipment. The properties of the protective film depend on the chemical and physical properties of the adherend, cleanness, elongation of the film during lamination (i.e., laminating method and equipment), and surface temperature of the adherend [327].

Application Equipment. Large-volume objects and profiles are applied with application lines. Such lines are supplied by specialized firms. Automatic lines with stacking and cutting devices are used. Generally the lines are built with an expander (banana) roller. The expander roller ensures a bubble- and wrinkle-free lamination. The brake should not be too strong, just strong enough to keep the films in a stretched position, in order to avoid wrinkles. In some cases where the film has to be removed and reapplied, the relamination is carried out by hand. This is the case in small firms, where short (100–200 m) rolls are used. In many cases (coil coating, plastic plate manufacturing, extrusion of plastic profiles) the lamination equipment is part of the production equipment for the web to be protected. Here the laminating speed and temperature are given by the main equipment. For instance, PVC window frames are produced at a speed of 3 m/min, so the lamination speed of the protective film has to be the same.

Table 8.11 Main Protective Films

Application	Product buildup
Coil coating	Adhesiveless hot laminating film
	Adhesive-coated film
Plastic plate protection	Adhesiveless warm laminating film
	Adhesive-coated film
Automotive storage and transport	Adhesive-coated film
	Adhesiveless warm laminating film
Deep drawing	Adhesive-coated film
	Adhesiveless film
Carpet protection	Adhesive-coated film
Product security	Adhesive-coated film
Furniture protection	Adhesive-coated film
	Adhesiveless warm laminating film
Building protection	Adhesive-coated film

4.2 Main Protective Films

Generally the classification of protective films according to their end use is related to the nature of the surface to be protected (Table 8.11). As discussed earlier, metals (coils) and plastics (profiles, plates, films, and other plastic-coated materials) are the most important application domains. Independently from their special use the following general requirements exist for protective films: adequate elongation at break, opacity, coefficient of friction (film-to-film), release of backing, stability of adhesion, low unwinding resistance, high tensile strength, and no telescoping. In the choice of film grades their adhesive properties must be taken into account relative to the surface in their application domain (Table 8.12).

Protective Films for Coil Coating

Coil coating is web coating, where calendered metal (steel and aluminum, copper, etc.) coils are coated with organic materials (lacquers, varnishes, or films). Liquid or dry coating materials can be used. Thin (10–20 μm) layers and thicker layers (60–100 μm for dry lacquers) are applied at a machine speed of 15–18 m/min. The protective film is laminated at the end of the varnishing process.

Because of outsourcing, the importance of coil coating has increased over recent years. This has led to the "finish first, fabricate later" principle, i.e., the prefinishing of the metal plate surface before the coil is processed. Steel plates 0.1–

Table 8.12 Adhesive Characteristics of Protective Films

Product	Adherend surface	Carrier Material	Thickness (μm)	Coating weight (g/m^2)	Peel resistance
Hot laminated film	Glossy (40%) varnished metal	PE	120	—	185.0 g/30 mm
Automotive protective film	Lacquered surface	PE/PP	50	5–10	5.0–8.0 N/25 mm
Thermally formed protective film	Polyamide	PE/PP	70	6–7	1.0 N/10 mm
Automotive technological protective film	PP, ABS, rough	PE/PP	100	14	1.5 N/10 mm
Carpet protective film	PP, PA, PET textured	PE	80	10–12	5.0 N/10 mm
Carpet underlay film	PA, PP, PET, EVAc	PE	25	—	5.0 N/10 mm
Plastic plate protective film	PC	EVAC	75–100	—	140–280 g/25 mm

End Uses of PSPs

Table 8.13 Manufacture and Application Characteristics of High Temperature Laminated Films

Polymer	BUR	MFI 190/2.16	Density (g/cm³)	Thickness (μm)	Surface tension (dyn/cm)	Tensile strength (N/mm²)	Peel	T (°C)
LDPE	1:1.8 1:2.4	0.6–0.8	0.923 0.02	50, 60	46–48	18/16	350 g/50 mm max 750 g/50 mm 250 ± 50 g/30 mm 250 g/30 mm	241 200
EVAc, EBA, PE+PB	—	—	—	—	—	—	80–120 g/10 mm	60–70

T = application temperature.

0.8 mm thick are used in the automotive, building, furniture, packaging, and other industries [328].

Adhesive-coated and adhesive-free protective films are used for coil coating. The best known adhesive-free protective films in this field are hot laminating films (Table 8.13). Low density PE (LDPE) may be sealed at 110°C. If the temperature is increased, a seal strength of 1–2 N/15 mm is achieved at 120°C [329]. According to Ref. 330, in thermal laminating, two thermoplastic substrates (with thermoplastic coatings) are combined with the application of heat and pressure. By the coating of coils the protective film is thermoplastic and it may or may not have an adhesive coating (or embedded PSA).

In many applications the coil (0.4–1.5 mm) to be protected by a film is first lacquered. Chromated, primed (5 μm) coils are lacquered (20 μm) continuously at a running speed of 50–80 m/min. The lamination temperature of the self-adhesive polyolefin protective film is between 200 and 250°C. After laminating, the film is cooled in a water bath. The debonding (90°) peel of such films is of about 180–250 g/30 cm (with tolerances of 50 g). The test conditions for such films are described in Ref. 331. Generally, relatively thick films (120, 150, or 170 μm) are used. Such films are corona-pretreated at 44–50 dyn/cm (see also Chapter 6).

The protection of aluminum, especially anodized aluminum, surfaces is a complex domain of coil coating. Standard tests of adhesion are generally made on stainless steel. Although the correlation between peel resistance on stainless steel and aluminum can be clear, the increase in adhesion with the dwell time on stainless steel and aluminum may be quite different (Table 8.14). Other problems

Table 8.14 Peel Buildup as a Function of the Substrate[a]

Peel resistance (N/10 mm) on SS after aging	Peel resistance (N/10 mm) on aluminum after aging
0.05 Instantaneous peel resistance	0.5 Instantaneous peel resistance
0.10	0.7
0.20	1.0
0.30	1.2
0.40	1.2
0.60	1.5
0.80	2.0
1.00 Peel resistance after 7 days	3.0 Peel resistance after 7 days

SS = stainless steel
[a] Peel adhesion was tested at intervals of 1 day.

End Uses of PSPs 615

Table 8.15 Dependence of Cleavage Peel on Coating Weight

Coating weight (g/m^2)	Cleavage peel (N/10 mm)
4	0.06
5	0.07
6	0.09
7	0.10
8	0.11
9	0.11
10	0.12
12	0.12

Table 8.16 Interdependence of Standard Peel and Cleavage Peel

Peel resistance (N/10 mm)	Cleavage peel (N/60 mm)
0.2	2.0
0.3	4.0
0.4	6.5
0.5	8.5
0.6	10.6
0.7	13.0

Table 8.17 Interdependence of Peel Adhesion on Stainless Steel and Aluminum

Peel on stainless steel (N/10 mm)	Peel on aluminum (N/10 mm)	Cleavage peel on SS (N/10 mm)
0.05	No adhesion	No adhesion
0.10	0.05	0.008
0.20	0.12	0.015
0.30	0.17	0.020
0.40	0.25	0.025
0.60	0.40	0.050
0.80	0.55	0.065
1.00	0.67	0.080

Instantaneous adhesion was measured for different coating weight values.

appear to be related to the use of protective films with "easy peel" and "cleavage peel" (see also Chapter 5). In this case the dependence of cleavage peel on the coating weight (Table 8.15), the interdependence between cleavage peel and standard peel (Table 8.16), and the interdependence between peel on SS and peel on aluminum must be clarified (Table 8.17).

Protective Films for Plastic Plates and Other Items

Various plastic plates and other items are laminated together with protective films during and after their manufacture to protect them during storage and processing. The main polymers used for plastic plates are PMMA, PC, polystyrene, styrene-acrylonitrile copolymers (SAN), PVC, and PP [332]. The protective films are generally applied via in-plant lamination. A hot roll laminater nip or a hot roll press is used, and heat and pressure cause the film to conform and adhere to the surface. As can be seen from Table 8.18, the grade of protective film used for a plastic plate depends on the plate material, manufacturing process, and geometry.

Generally, adhesive-coated and self-adhesive protective films are used for the protection of plastic films and plates. Relatively hard water-based acrylic adhesive coated protective films have been suggested as adhesive-coated protective films for plastic plates and films. Nontackified polyethylene (LDPE), polyethylene–EVAc blends, and polybutylene-tackified polyethylene (coextrudate) have been proposed as SAFs for plastic plates. Nonpolar common polyethylene films are corona-treated, and blends with polar monomers can be corona-treated also. Such films are generally laminated under pressure at elevated temperature. Generally the adhesion of the corona-treated polyethylene film to the plastic plate substrate increases with application temperature.

For instance, corona-treated LDPE (density of 0.923 g/cm^3) film (50–60 μm) is manufactured via a blow procedure, with a BUR of 1:1.8 to 1:2.4 depending on the dimensions of the film. It is pretreated for a surface tension of 46–48 dyn/cm. The adhesion of the film attains 350–750 g/50 mm.

The requirements concerning the dimensional stability and plasticity–elasticity balance of a protective film depend on the dimensional stability of the protected item. For instance, high molecular weight cast PMMA displays mainly thermoelastic behavior over a broad temperature range; therefore, the formed PMMA items do not display plastic deformation, i.e., dimensional changes [333]. The protective film used has to possess elastic properties also. On the other hand, low molecular weight extruded PMMA displays a narrow elastic domain during high temperature deformation. The transition from elastic to plastic deformation is not clear, and forming leads to dimensional changes. It should be taken into account that when the PMMA or PC items (plates) are first heated a manufacturing technology dependent shrinkage occurs also.

Table 8.18 Application and Processing Characteristics of Protective Films Used for Plastic Plates

Protective film		Protected plastic plate		Laminating temp. (°C)		Processing conditions (°C/min)	Adhesion (N/10 mm)
Grade	Thickness (μm)	Manuf. process	Material	Laminating cylinder	Plate		
Adhesive-coated	50	Extrusion, casting	PMMA	70	70	150/15	0.5–0.8
Adhesiveless	40	Extrusion	PMMA, PC	70	70	150/15 170/15	0.05–0.5
Adhesiveless	80	Casting	PMMA	70	30	170/30	0.01–0.04

To allow easier application, the plates are given protective films of different colors on their front and back, and plates on different chemical bases (e.g., PMMA or PC) or produced with different manufacturing methods are laminated with protective films of different colors. If these plates are used outdoors under severe weathering conditions, the polyethylene protective films have to be peeled off within 4 weeks in order to avoid the degradation (hardening) of the protective film and buildup of adhesion.

Another problem is the hygroscopy of plastic plates. They absorb humidity and reabsorb it after cooling below 100°C. The protective film seals the plate against water absorption also. Film-protected plates do not require elimination of water before use [333].

Cast PMMA has a broader processing temperature range than extruded PMMA, which has a more exactly controlled thickness tolerance and lower shrinkage than cast materials. Therefore extruded PMMA is easer to coat with a protective film. The product range for extruded grades is broader. Many surface qualities are manufactured into the product. Generally, cast PMMA is manufactured with gauges of 1.5–250 mm. Extruded PMMA can have a thickness of 1.5–18 mm. A quality required for food contact (XII.BGA Recommendation and FDA Regulation 177.1010) is available also.

PMMA is used for illuminated advertisement, light domes, displays, pictures, lamps, vehicles, noise-reducing panels, solar panels, furniture, and sanitary items, among other things. The plates are processed in various ways. Warm deep drawing is one of the main processing operations for PMMA plates. The maximal pressing force in heat pressing may attain 200 kN. Form stability after processing depends on the product. Cast PMMA achieves its form stability at 70°C, extruded PMMA by 60°C, and PC at 110°C. Polycarbonate needs a higher forming temperature (190–210°C); its thermal processing temperature range is narrow.

For extruded PC and PMMA plates, a 70/40 μm SAF with 18% EVAc is suggested. Such films are laminated at 60–70°C. For instance, a 40 μm coextrudate has an 8 μm EVAc adhesive layer (based on a copolymer with 7–9% VAc). Extruded PC or cast PMMA are processed as sanitary materials. The carrier film for such applications has to be deep drawable and has to support the processing conditions (160–170°C, 10–30 min) of the protected item. For such applications, self-adhesive or adhesive-coated films with a minimum thickness of 70–80 μm can be used.

Generally thin plates (less than 6–10 mm) are protected with SAF (70 μm). LDPE/EVAc films (80/20) with an ethylene-vinyl acetate copolymer having 9–12% VAc have been suggested. For thicker plates, adhesive-coated protective films of greater thickness (80–100 μm) have been proposed.

Polycarbonate is manufactured as film (5–750 μm) and as plates. Polycarbonate films are laminated with the protective film by using a laminating station (with rubber-coated cylinders) at 60°C and a lamination pressure of about 50 kPa/cm^2. Laminating pressure, cylinder hardness, cylinder cooling, temperature, running speed, and web tension influence the final adhesion level and film shrinkage. Extruded polycarbonate plates are laminated after processing and before cooling (at 60°C). Such plates are processed for 2–5 min at 180°C (e.g., for light domes, 2 min at 180°C). Protective films with a thickness of 25–100 μm (9–12% VAc) have been tested for PC plates. Passive protective films for PC plates have a thickness of 50 μm; processable protective films have a thickness of 70 μm. Such films are tested for adhesion buildup at normal and elevated temperatures. Both short-term and long-term adhesion performance have to be tested. Instantaneous adhesion at room temperature and adhesion after storage (2 days) at room temperature and adhesion buildup after storage at 80°C (4 h) are evaluated. As is known, PC plates are hygroscopic and are generally dried before use. Drying is carried out at 130°C for 0.5–48 h depending on the plate thickness (0.75–12 mm). Masking should be applied at both 50 and 100°C. Therefore, short-term adhesion buildup is tested after storage at 130°C also. Shrinkage at 130°C is very important. Long-term adhesion buildup is tested after storage at room temperature and at 40°C for 2–8 weeks. The end adhesion should be situated at about 4 N/20 cm. If adhesiveless films are used, the strong influence of the laminating temperature should be taken into account (Table 8.19). If adhesive-coated films are applied, attention should be paid to the dependence of the peel buildup on the coating weight (Table 8.20).

Laminating can ensure sufficient instantaneous adhesion. If the product stiffens when it is cooled and during storage at room temperature, surface contact decreases and delaminating may occur. On the other hand, for the same product, high temperature laminating (80°C) may lead to unacceptable adhesion buildup. For thick masking films used for PC or PMMA plates, shrinkage and lay flat are important quality criteria (Table 8.21).

Table 8.19 Dependence of Adhesion of Hot Laminated Film on Laminating Temperature[a]

Laminating temperature (°C)	Peel resistance (N/10 mm)
50	0.1
60	0.2
70	0.4
80	0.7
90	0.7
100	0.7

[a]The instantaneous peel resistance of a corona-treated (52 dyn/cm), 60 μm polyethylene film on extruded PMMA was measured.

Table 8.20 Dependence of Peel Buildup on Coating Weight[a]

Coating weight (g/m^2)	Peel resistance (N/10 mm) After 3 days	After 7 days
1.0	0.35	0.20
1.2	0.43	0.24
1.3	0.70	0.30
1.4	0.82	0.35
1.6	—	0.40

[a]A solvent-based acrylic adhesive was crosslinked with isocyanate curing agent (2.5%).

Table 8.21 End Use Characteristics of Masking Films for Plastic Plates[a]

Product code	Lay flat (mm) 90°C	120°C	Shrinkage (%)
1	1	1	1.1
2	1	1	1.0
3	1	3	1.0
4	1	2	1.1

[a]A 250 μm polyolefin film on PC was tested.

Protective Films for Automotive Storage and Transport

Protective films for storage and transport have been developed to replace conventional systems used to protect vehicle surfaces from scratches, iron dust, bird droppings, acid rain, etc. (see Table 2.13). The best known conventional protective systems are dispersions of low molecular weight, nonadhesive or slightly adhesive polymeric products like waxes or acrylic compounds (see SAFs Based on Other Compounds, Chapter 6, Section 3.1). Such products build up a protective layer of about 15 μm thickness on the protected surface. These compounds are applied in the liquid state and are eliminated as a liquid system also. Solvents are used for paraffin, and alkalies (WB solutions) for acrylics. Such systems giving full surface protection are more effective than carrier-based films. Their disadvantages include working time and pollution.

End Uses of PSPs 621

Protective coatings generally have to fulfill the following requirements:

Chemical resistance (resistance to organic solvents, grease, oils)
Versatility (suitable for all types of varnishes)
Weathering resistance (outdoor exposure at least 6 months, to pass the Florida test on cars, to resist harsh winter climates)
Minimum light transmission (lower than 0.04%)
Good environmental aspects (recycling, burning)
Low costs compared to waxes (material, labor)

As for their adhesive properties, an initial peel of 4 N/25 mm, easy removability (hand peel-off), and no visible changes to the protected surface are required. Table 8.22 lists the application requirements and test methods for automotive protective films. As can be seen from this table, these tests are related to the practical transport, storage, and end use conditions of vehicles and include mostly empirically stated conditions.

The surface to be protected must be clean; dust, stains, and waterdrops should be removed. The film should be stretched tight to minimize wrinkles and trapped

Table 8.22 Requirements for Automotive Protective Films

Application criteria		Test method	
Criterion	Parameter	Method	Conditions
Temperature resistance	High temperature	Storage stability	10–15 h/50°C
			10–15 h/80°C
	Low temperature		10–15 h/50°C
Chemical resistance	Battery acid	Contact	1 h, RT, pH 3
	Alkali		0.1 N NaOH, 80°C, 4 h
	Motor oil		1 h, 70°C
	Calcium chloride		Saturated solution
	Washer fluid		4 h, 80°C
	Gasoline		10 min
Resistance to air pollutants	Metal dust (iron)	Contact	Iron dust, 200 mesh, 80°C, 5 h
	Grain dust		80°C, 4 h
	Bird droppings		80°C, 4 h
	Smoke		80°C, 4 h
Sealing quality	—	—	Water leakage test
Humidity resistance	—	Storage	5 days
Aging resistance	External exposure	Storage	1, 2, and 3 months, Florida test

Table 8.23 Crosslinking Agent Concentration as a Function of Product Application

Product	Curing agent/adhesive ratio In adhesive	In primer
Automotive protective tape	1/76	1/250
Thermoformed protective film	1/1000	—
Common protective film	1/40	1/250

air. It is desirable that the boundary line between the wrapped and unwrapped areas be round and vertical. The film should be put on the substrate by slightly pressing it from the center area outwardly with a cloth by hand or with a sponge roller. Deapplication should be carried out by hand peel-off. Quality problems that may appear include air marks, a visible borderline, and adhesive deposits. Generally the chemical composition of the lacquers used strongly influences the applicability of the protective film. Soft lacquers, e.g., lacquers based on low T_g amino alkyd resins, are more sensitive than others. Highly crosslinked adhesives are suggested for automotive use (Table 8.23).

According to automotive suppliers, protective films for cars should meet special requirements concerning outward appearance, adhesive properties, stainability, and paint protection. Outward appearance is controlled by visual evaluation and concerns color, thickness, and light permeability. Adhesive properties are measured initially and after aging on the paint and by overlapping. Stainability is tested on automotive paint, but component stainability should also be tested, which means that the film should be tested on the various automotive parts (glass, sunroof, rubber seals, emblems, stainless steel moldings, headlight lens, etc.). Paint protection has to include resistance to acids, alkalies, rail dust, gasoline, and oil.

Automobile undercarriage protective coatings and films have been developed also. Such films have to display resistance to abrasion, light, weather, aging, washing, and gasoline. They have to be heat deformable (150°C). Certain films are based on a soft PVC carrier material [334].

Deep Drawable Protective Film

Deep drawability is a general requirement for protective films used as technological aids. Metals (coils) and plastics (films, plates, wovens, and nonwovens) are formed by deep drawing. Deep drawable films can be classified according to the surface protected into films for processing of metals and films for plastics (Fig. 8.1).

Deep drawing of metals is more complex than that of plastics because metals are less ductile and require much higher forces for cold drawing. According to the depth of the drawing, standard (depth of about 25 mm) and special deep drawing (more than 300 mm) are carried out. Advanced deep drawing may be carried out in one step or in multiple (one to four) steps. The forces acting on the film increase with the number of steps.

Common deep drawable film is used to protect metal plates (steel, copper, etc. with a thickness of 0.7–0.8 mm [335]) which are deep drawn in one or more steps. This procedure is carried out at room temperature. (Friction may cause a slight increase in temperature.) The protective film coated on the substrate has to display the same deformability as the adherend, i.e., a plastic deformation. The elastic forces in the film should be limited to prevent delamination. The film has

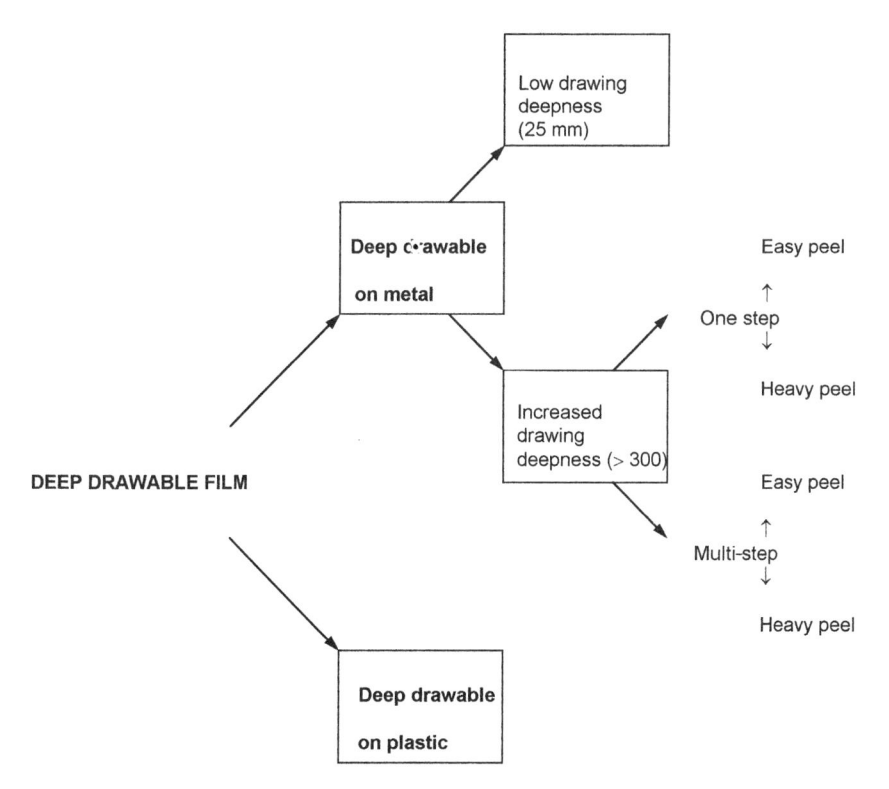

Figure 8.1 Classification of deep drawable protective films.

to resist the higher temperature produced by friction and must also play (partially) the role of lubricating agent. As discussed in Ref. 335, the friction (F) during processing is the sum of the solid-state friction between metal and metal (F_s), friction (F_L) in the thin lubricant layer at the boundary between the coil and the tool, and the hyrodynamic and hydrostatic friction (F_H) in the thick lubricant layer (Fig. 8.2):

$$F = F_S + F_L + F_H \qquad (8.1)$$

For a coil covered with a protective film, the polyethylene film participates in the friction between the solid-state components. In this process a high contact pressure (15–70 N/mm^2) is applied [335].

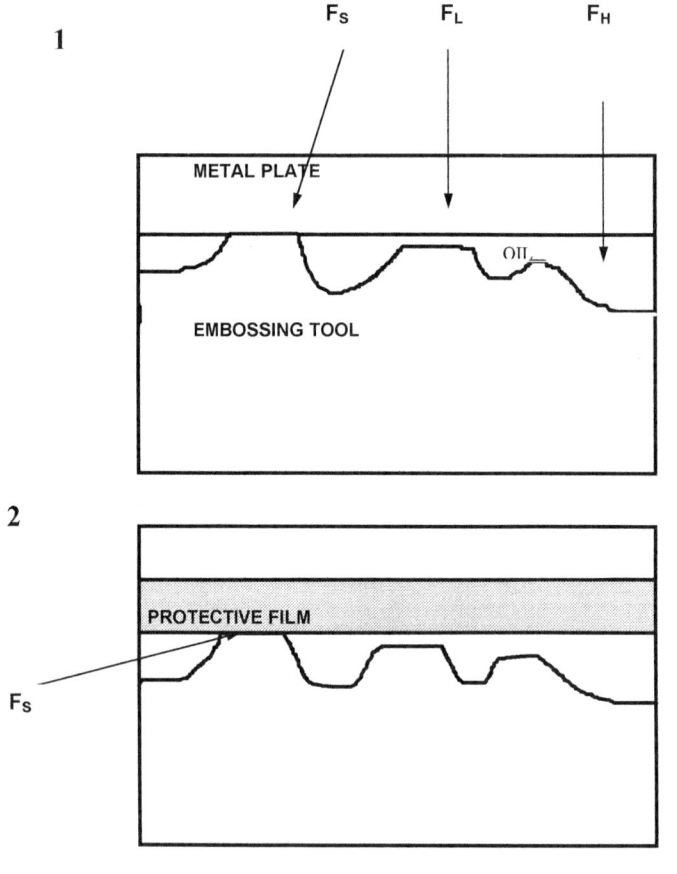

Figure 8.2 The role of the protective film as lubricating agent by deep drawing. (a) Processing without protective film; (b) processing with protective film.

The time-dependent interlayer phenomena during deep drawing are influenced by the compression stress caused by friction (τ_F), which depends on the temperature (T), normal stress (σ_N), relative speed (v_{rel}), and degree of forming ($\epsilon°$) [336]:

$$\tau_F = f(T, \sigma_N, v_{rel}, \epsilon°) \qquad (8.2)$$

The relative speed affects the temperature, and both depend on the drawing steps. The temperature influences the normal stress in the plastic component (protective film) also.

Plastics can be processed by deep drawing at room temperature or at elevated temperatures. Among the common polymers, PC, PVC, and ABS can be processed at room temperature via deep drawing [203]. As for PMMA, the deep drawing temperature of different PVC grades differs according to the polymerization procedure and film manufacturing process. Suspension-polymerized PVC has a lower and broader processing temperature range (125–165°C). High temperature deep drawing of PE can be carried out at a temperature as low as 30°C. The deep drawing of PC is carried out at 170–180°C. In such processes heat is also evolved by friction between the deep drawing tool and the material processed. Therefore water-based or organic lubricants are used. In the deep drawing of plastics the different thermal expansion coefficients of different materials should be taken into account. For instance, the coefficient of thermal expansion for polycarbonate is higher (65×10^{-6} K^{-1}) than that of steel (11×10^{-6} K^{-1}) [315].

The common deep drawable protective films are built up on a clear 50–70 μm thin film of PVC or on polyolefin. No slip agents are allowed in their composition. However, the protective film has to function as a lubricating agent also [333]. The carrier film for deep drawable protective films should have good coatability, good slip, excellent deformability, and long-term deformability. Good coatability refers to one-side wettability and adhesive anchorage. Good slip means that the bonded film should allow movement parallel to the drawing device. Deformability means that during embossing the film should follow the deformation of the metal plate. Metal plates are deep drawn up to 40–60 cm. Long-term deformability means that the film preserves its deformation after deep drawing and does not change form after the embossing device is removed. Such properties are given by soft PVC. Deformability is characteristic for LLDPE also, but its elastic forces do not permit its use alone. Excellent deep drawability is given by EVAc. Miura [337] states that the deep drawability of hydrocarbon-based films decreases as follows:

$$\text{EVAc} > \text{PE} > \text{PP} > \text{PSt} \qquad (8.3)$$

The processing temperature (°C) of the above polymers varies as follows:

$$PE\ (135) > PP\ (125) > PSt\ (105) > EVAc\ (90) \qquad (8.4)$$

Unfortunately, the high blocking tendency allows the use of EVAc only as a coextrudate. Special polymer compounds have been suggested also. For instance, AC copolymers have been suggested as additives for PP to improve its deep drawability [338]. Polyolefin-based deep drawable carrier film compositions contain an extensible component like EVAc, PP, or LLDPE, and other components that ensure adequate mechanical resistance, friction behavior, and machinability.

Deep drawable protective films are generally coated with a medium coating weight of 5.5–8.5 g/m^2 of a water-based acrylic PSA or 4.5–5.5 g/m^2 of a solvent-based rubber resin PSA. Because of the higher adhesion of the rubber resin formulation, such products are also release coated. For complex shapes and advanced deep drawing, rubber-based adhesives are preferred.

Protective Film for Carpets

Carpet-protecting films can be used for long fiber tufted carpets (high tack protective film), thermoformed embossed carpet (high tack deformable, deep drawable protective film), and processing protection. Carpet films applied for protection during storage differ from automotive carpet protection films, which undergo the forming processes of automotive carpets.

For the protection of textile webs, different films are used. Their characteristics have to be tailored to the manufacturing technology and buildup of the carpet, i.e., its adhesivity and postprocessing. Buildup includes fiber type, length, and construction. Tufted carpets generally have polyamide pile fibers and polyester-based fabric. Carpet processing depends on the end use. Automotive carpets are generally molded. Common carpet-protecting films for normal use are based on SBAC adhesives coated on an 80 μm PE carrier material. Automotive carpet films protect automotive carpets during the manufacture and transport of the vehicle. They have to possess deep drawability and good adhesion.

Mirror Tape

Mirror tape is a permanent security film used to avoid dangerous destruction (cracking) of large glass surfaces. This film is laminated as a reinforcing layer on the back of mirrors or on other impact-sensitive large surface items. The film has to have a high level of permanent adhesivity in order to avoid delamination of the glass by stress and dart drop resistance against glass splits.

Overlamination Protective Film

Self-wound overlamination protective films are used for overlapping and permanent protection of labels or other printed items. They have to display clarity, impact resistance, UV resistance, color stability, resistance to outdoor weathering, and printability. A matte finish, nonreflective character is needed also. Polypropylene and PET have been suggested as carrier materials for such PSPs. BOPP (25–50 μm, standard and battery approved, normal and heat-resistant lacquered) and PET (25 μm) are generally used.

For screen printers and litho printers, thermal laminating was used as the finishing process. For the overlamination of screenprinted labels a product buildup of 50/150, i.e., 12.5 μm PET/75 μm heat-activated adhesive, has been recommended. This combination gives a greater gloss depth, especially with embossed rough paper. UV-resistant thermolaminating films and inexpensive olefin copolymers have also been developed. At present, 10–15% of printed products are overlaminated. Heat laminating has been (partially) replaced by pressure-sensitive protective laminating [339]. PET and BOPP are used for pressure-sensitive overlaminating. Thermal laminaters vary in size and complexity. Tabletop models and machines with automatic feeders and automatically synchronized cutoff have been developed.

Separating Films

Separating films are used to protect web or sheetlike metallic, plastic, or varnished items during storage. They must have a very low peel. Therefore they are based on rubber resin formulations with a very low tackifier level (less than 10%) and a high degree of crosslinking. Because of the polar surface of the protected items (e.g., anodized aluminum or stainless steel) and the high laminating pressure, peel buildup occurs rapidly.

5 FORMS

As discussed earlier, forms are multilayered laminates that have exactly controlled adhesive forces in the laminate layers and are discontinuous, finite elements on or in a weblike product, which allow detachment of one or many parts of the forms, making simultaneous multipurpose information transfer possible. Piggyback mailing labels are a typical example of forms; they are sandwich labels. Such products can contain two or more peel-off layers over a single backing paper. The complete label, which is die cut and personalized in one pass, is affixed to the consignment, and two small numbered slips are peeled off and attached to control forms during the routing process [49]. Forms are applied by detaching, separating the parts of the form. This is carried out manually. Detachment and refixing of parts of forms (e.g., cards) is carried out by hand, but certain parts of a form (e.g., the main web) can be processed automatically also.

6 PLOTTER FILMS

Plotter films are weblike pressure-sensitive products that serve as raw materials for the manufacture of information-carrying items. Special equipment is used to cut letters, signs, and repromasks from plotter films [340]. The plotter film is used as the carrier material for self-adhesive signs, letters, graphics, typefaces, or decorative elements. It is a bulk or surface colored (coated) film with excellent dimensional stability. That means minimal deformation during processing, storage, and application. Plotter film has to display very good die cuttability, sharp contours, no plasticity or elasticity during cutting, and excellent lay flat and stiffness. It must also be weather-resistant. Such carrier materials display elongation values of about 300% MD and 300% CD and modulus values of 7–15 N/mm^2. High quality products exhibit elongation values of 24–200% (DIN 53455) [341]. The first U.S.-made pressure-sensitive PVC films were tested in Europe as plotter film in 1946 [342]. The first usable PVC films were manufactured in the 1950s. In 1984 computerized cutting plotters were launched.

Polyvinyl chloride (cast and calendered) and polyolefin films are manufactured as carriers for plotter film. Plasticized PVC and vinyl chloride copolymers have been developed for plotter films. Plasticized PVC films can be used outdoors for 3–5 years. Films based on vinyl chloride copolymers can withstand 5–7 years of external use [339]. The finished product must exhibit excellent chemical and environmental resistance. Resistance to water, acid, alkalies, surface-active agents, ethylene glycol, motor oil, methanol, gasoline, and certain other substances is required.

Such films are processed using special plotters to cut them. A procedure for the application of plotter films is described in Ref. 341. Lettercutting plotters have been developed [339]. Such cutting machines can use reels with a width of up to 1300 mm [342] and may run at speeds of up to 400–600 mm/s [342,343] and tolerances of 0.1 mm. The pressure of the cutter is adjustable for different film thicknesses. Roll plotter can process up to 1320 × 25,000 mm web with a minimum sign dimension of 3 mm [344]. Generally the working speed and acceleration are given as performance characteristics [345].

Plotter films are used together with a handling or fixing aid, i.e., the application tape (see Application Tape). Such plotter films have to be applied (laminated) at temperatures of 4–50°C [341]. Wet and dry application methods are used. In a wet technique warm water should be used for calendered film and cold water for cast film. Their end use temperature is in the range −56 to +107°C [341].

There are special application papers (or films) that can be used like a plotter film to produce logos or decorative elements. For such products the continuous, weblike part of the paper or film is used after the places to be lacquered or printed are cut off. Such stencil films or papers generally have a release liner because of their improved adhesivity. This is necessary to avoid ink penetration at the edges of the carrier [346].

7 OTHER PSPs

A number of pressure-sensitive imaging materials have been developed. Their number is growing with the increase in the number of new carrier materials, printing techniques, and packaging methods. In 1974, PSA stamps were introduced on a trial basis [347]. They cost five times as much as conventional stamps. Pressure-sensitive envelopes have been developed [348]. Message reply pads use low peel PSAs [349]. Cast vinyl marking films are used for producing labels, emblems, stripes, and other decorative markings for trucks, automobiles, and other equipment. They are durable, conformable, and dimensionally stable. Decal films have been developed for marking functional machinery, building parts, etc. [350]. Two-way decal films are used for push/pull glass doors, security decals, and association membership decals, etc. have been produced [351]. Collections are manufactured [352].

PSAs can be applied as laminating adhesives also. Decorative panels bearing PSA have been manufactured by pressing substrates (e.g., veneer) precoated with metallic alkoxides and/or chelates onto decorative sheets coated with a PSA containing reactive group (e.g., N-methylolacrylamide) [353]. Films based on cellulose have been applied by hot pressing on paper to protect it [354]. Electrically conductive PSAs are used for biomedical electrodes [355]. Laminating films for bookbinding are used for reinforcement, repair, binding, and lamination [356].

Decor films are used in household and technical domains [235,357–360]. They are manufactured as adhesive-coated or self-adhesive films. Self-adhesive decor films contain a high level of plasticizers; therefore it is difficult to print on them. For PVC decor films, water-based gravure printing is suggested [361]. Such products can have a printed release liner too.

Self-adhesive wall covering consists of a layer of fabric with a visible surface, a barrier of paper that has one surface fixed to another fabric layer, a pressure-sensitive adhesive coated on the barrier paper, and a release paper [362]. Low styrene content (15%) SEBS block copolymers have been proposed for adhesives for wall coverings [363]. Metallized film used for electromagnetic protection can be applied by using PSAs. Such products are suggested for low volume items with simple shapes and for optically low quality applications [364].

Self-sticking carpet tiles have a pressure-sensitive layer [365]. The system comprises tiles having adhesive and nonsticky material on the back. Self-adhesive wall carpets have been manufactured also. PSAs or PSPs may be used as formulation or constructional components in other products also. PSAs can be used for attachment of plastic cards (see also Business Forms) [366]. Diapers include standing gather, frontal tape, and foamed waist elastics [367]. A cleaning device based on PSPs has also been developed [368].

Pressure-sensitive aerosol adhesive is used to bond paper, cardboard, plastics, films, foil, felt, cloth, metal, glass, or wood [369]. The PSA market for do-it-yourself applications has a higher growth rate than the average for PSAs [370]. Sprayable HMPSAs are suitable for bonding laminating panels, thin plastics, fibrous paneling, and foam materials [371]. It is evident that new application fields and requirements, new raw materials, and advances in manufacturing technology will force an increase in the number of special pressure-sensitive products.

REFERENCES

1. I. Benedek and L. J. Heymans, *Pressure-Sensitive Adhesives Technology*, Marcel Dekker, New York, 1997, Chapter 5.
2. M. Fairley, *Labels Label.* 7/8:272 (1995).
3. *Adhes. Age* 7:34 (1983).
4. R. Jordan, *Adhäsion* 1/2:22 (1987).
5. *Finat Label. News* 3:12 (1996).
6. J. M. Casey, *Tappi J.* 6:42 (1998).
7. M. H. Mishne, *Screen Print.* 1:60 (1986).
8. P. Foreman and P. Mudge, EVA-based waterborne pressure sensitive adhesives, *Tech 12*, Technical Seminar Proceedings, Itasca, IL, May 3–5, 1989, p. 203.
9. K. Fust, *Coating* 2:65 (1988).
10. K. Heger, *Coating* 5:180 (1987).
11. *Druckwelt 14/15*:6 (1988).
12. *Kautsch. Gummi Kunstst.* 42(5):420 (1988).
13. *Adhäsion 11*:9 (1988).
14. *Siebdruck* 8:75 (1986).
15. Ball Bros Co., U.S. Patent 3352708/2; in *Coating* 7:210 (1969).
16. Owens Illinois Glass Company, Ohio, U.S. Patent 3147135; in *Adhäsion* 2:80 (1966).
17. *Adhäsion* 5:28 (1982).
18. Ball Brothers Co., Muncie, IN, DBP 1284585/1977.
19. *Coating* 9:269 (1972).
20. C. S. Maxwell, *Tappi J.* 8:1464 (1970).
21. L. Ridgeway and L. Pyle, *Tappi J.* 12:129 (1968).
22. *Papier Kunstst. Verarb.* 1:24 (1996).
23. *Etiketten-Labels* 5:22 (1995).
24. *Coating* 7:243 (1987).
25. *Coating* 5:151 (1969).
26. B. H. Gregory, Extrusion coating advances—Resins, processing, applications, markets, Polyethylene '93, The Global Challenge for Polyethylene in Film, Lamination, Extrusion, Coating Markets, Oct. 4, 1993, Maack Business Services, Zurich, Switzerland.
27. A. Ridgeway and L. K. Mengenhagen, Tappi Proceedings, *EAA/Polybutylene Blend for Packaging Applications*, Polymers, Laminations & Coatings Conference, 1992, p. 52.

28. *Coating* 12:450 (1986).
29. G. F. Bulian, Extrusion coating and adhesive laminating: Two techniques for the converter, Polyethylene 93, Maack Business Service, Conference, Zurich, 1993.
30. C. Gondro, *Kunststoffe* 10:1082 (1990).
31. U. Reichert, *Kunststoffe* 83(10):737 (1993).
32. *Verpack. Rundsch.* 12:1394 (1985).
33. C. Stöver, *Kunststoffe* 84(10):1426 (1994).
34. H. Maack, Coextruded and coinjected packaging developments and recycling aspects, Maack Business Service, Specialty Plastics Conference, 1989, Zurich, Switzerland.
35. *Adhes. Age* 11:10 (1986).
36. Nichimen Corporation, *Exceed* (Biaxially oriented EVOH film), Technical Data.
37. J. A. Fries, New developments in hot melt PSA and heat seals, Tappi Hot Melt Symposium, June 7–10, 1987, Monterey, CA.
38. *Coating* 6:214 (1996).
39. G. Bonneau and M. Baumassy, New tackifying dispersions for water based PSA for labels, 19th Münchener Klebstoff u. Veredelungsseminar, 1994, p. 82.
40. P. Dunckley, *Adhäsion* 11:19 (1989).
41. T. Kishi (Sekisui Chem. Co. Ltd.), Jpn. Patent 6369 879/29.03.1988; in *CAS Adhes.* 19:6 (1988).
42. Raflatac, *Raflatac Adhesives for Extreme Conditions*, Data Sheet, Raflatac OY, Tampere, Finnland, 1992.
43. *Kunstst. J.* 9:61 (1986).
44. *Labels Label.* 3/4:11 (1994).
45. A. Dobmann and J. Planje, *Papier Kunstst. Verarb.* 1:37 (1986).
46. *Adhes. Age* 12:38 (1987).
47. D. G. Pierson and J. J. Wilczynski, *Adhes. Age* 8:52 (1990).
48. *Adhes. Age* 3:10 (1974).
49. A. Prittie, *Finat News* 3:35 (1988).
50. Z. Czech, *Eur. Adhes. Sealants* 6:4 (1995).
51. W. Grebe, *Papier Kunstst. Verarb.* 2:49 (1985).
52. *Verpack. Rundsch.* 9:994 (1983).
53. T. G. Wood, *Adhes. Age* 7:19 (1987).
54. H. J. Teichmann, *Papier Kunstst. Verarb.* 11:10 (1994).
55. A. W. Norman, *Adhes. Age* 4:35 (1974).
56. *Labels Label. Int.* 5/6:24 (1997).
57. Lasertechnik in der Papierverarbeitung; in *Papier Kunstst. Verarb.* 11:57 (1994).
58. *Etiketten-Labels* 3:9 (1995).
59. *Labels Label. Int.* 5/6:20 (1997).
60. P. Thorne, *Finat News* 3:47 (1988).
61. A. Linder, Hauptmarkttendenzen im Einsatz von Selbstklebenden Etiketten, 19th Münchener Klebstoff und Veredelungsseminar, 1994, p. 118.
62. *Adhäsion* 5:36 (1988).
63. *Kunstst. J.* 9:82 (1986).
64. *Adhes. Age* 1:6 (1985).

65. F. T. R. Mecker, Low molecular weight isoprene based polymers—Modifiers for hot melts, Tappi Hot Melt Symposium 1984; in *Coating 11*:310 (1984).
66. *Coating 3*:65 (1974).
67. J. Lechat, European Tape and Label Conference, Exxon, Brussels, April 1989, p. 67.
68. *Adhäsion 11*:9 (1988).
69. *Verpack. Rundsch. 2*:154 (1986).
70. *Etiketten-Labels 5*:9 (1995).
71. *Papier Kunstst. Verarb. 12*:44 (1986).
72. H. Mueller and J. Tuerk (BASF A. G., Ludwigshafen, Germany), EP 0118726/ 19.09.1984.
73. *Papier Kunstst. Verarb. 11*:57 (1994).
74. *Druckwelt 14/15*:27 (1985).
75. *Tappi J. 6*:159 (1988).
76. *Etiketten-Labels 5*:100 (1995).
77. *Packlabel News*, Packlabel Europe 97, Ausgabe 2.
78. S. Benedek, *Adhäsion 3*:22 (1987).
79. *Etiketten-Labels 5*:24 (1995).
80. *Labels Label.*1:86 (1997).
81. A.C. Thomas and D.S. Culp (Avery International Co., Pasadena, CA), EP 0099087/ 23.12.1987.
82. *Neue Verpack. 3*:10 (1995).
83. *Labels Label. Int. 5/6*:18 (1997).
84. *Avery Rating Plates and Type Plates Last a Fairly Long Time*, Prospectus, Avery Etikettier-Logistik GmbH, Eching b. München, Germany.
85. *Adhes. Age 11*:10 (1986).
86. P. Broschk, Die Verklebte Information—Spezialaetiketten zur Fertigungssteuerung im Fahrzeugbau, *Adhäsion 6*:14 (1985).
87. J. Schaetti, Aufbügelbare chemisch reinigungsbeständige Etiketten, 9th Adhesive and Finishing Seminar, Munich, 1984.
88. D. Lacave, *Labels Label. 3/4*:54 (1994).
89. *Labels Label. 3*:82 (1994).
90. LOS, Lager Organisations System GmbH, Mannheim, Germany, Gebrauchsmuster, G 87 06 322.0/02.05.1987.
91. D. Allen, Jr. and E. Flam (Bard Inc., Murray Hill, NJ), U.S. Patent 4650817/ 17.03.1987.
92. Morstik 103 Adhesive, Morton Thiokol Inc., Adhesives and Coatings, MTD-MS 103-12/1988.
93. R. L. Sun and J. F. Kenney (Johnson and Johnson Products Inc.), U.S. Patent 4762688/09.08.1988; in *CAS Hot Melt Adhes. 26*:1 (1988).
94. S. E. Krampe and C. L. Moore (Minnesota Mining and Manuf. Co., St. Paul, MN), EP 0202831A2/26.11/1986.
95. U.S. Patent 3661874; in F. D. Blake (Minnesota Mining and Manuf., Co., St. Paul, MN), EP 0141504 A1/15.05.1985.
96. U.S. Patent 3865770; in F. C. Larimore and R. A. Sinclair (Minnesota Mining and Manuf. Co., St. Paul, MN), EP 0197662 A1/15.10.1986.

97. *Adhes. Age* 4:6 (1983).
98. DDR Patent 64111; in *Coating* 1:24 (1969).
99. J. R. Pennace, C. Ciuchta, D. Constantin, and T. Loftus (Flexcon Co. Inc., Spencer, MA), WO 8703477A/18.06.1987.
100. *Neue Verpack.* 1:156 (1991).
101. Z. Czech, *Eur. Adhes. Sealants* 6:4 (1995).
102. U.S. Patent 3321451; in S. E. Krampe and C. L. Moore (Minnesota Mining and Manuf. Co., St. Paul, MN), EP 0202831 A2/26.11.1986.
103. *Pack Rep.* 19:28 (86).
104. *Seifen Öle Fette Wachse* 112(17):613 (1986).
105. *Labels Label.* 3/4:44 (1994).
106. Tagsa, *Newsletter*, July 1995, p. 1, Tagsa Ltd., Newburry, U.K.
107. D. Boettger, *Labels Label.* 3/4:58 (1994).
108. S. W. Thompson, U.S. Patent 4674771/23.06.1987.
109. J. Nentwig, *Converter, Flessibili-Carta-Cartone* 1:66 (1991).
110. Oji Paper Co., Jpn. Patent 07325391/12.12.1995; in *Adhes. Age* 5:12 (1996).
111. *Labels Label.* 3/4:50 (1994).
112. *Labels Label.* 3/4:34 (1994).
113. *Adhes. Age* 8:8 (1983).
114. K. Unbehaun, *Deut. Drucker* 6:43 (1988).
115. *APR* 22:787 (1986).
116. R. Davis, *Fasson Facts Int.* 1:2 (1969).
117. *Etiketten-Labels* 5:91 (1995).
118. *Coating* 7:232 (1989).
119. *Adhäsion* 1/2:78 (1987).
120. H. Roder, *Adhäsion* 6:16 (1983).
121. W. E. Broxtermann, *Adhes. Age* 2:20 (1988).
122. K. F. Schroeder, *Adhäsion* 5:161 (1971).
123. W. Möhren, Doppelseitiges Haftklebeband zur Klebung von Gummiprofilen im Fahrzeugbau, *Adhäsion* 6:14 (1985).
124. H. Kato, H. Adachi, and H. Fujita, *Rubber Chem. Technol.* 56:287 (1983).
125. W. Schneider, D. Cotsakis, and R. Senderling, The novel use of pressure sensitive tape in EPDM single-ply roofing applications, PTSC XVII Technical Seminar, May 4, Woodfield Shaumburg, IL, 1986.
126. E. L. Scheinbart and J. E. Callan, *Adhes. Age* 3:17 (1973).
127. *Adhes. Age* 3:8 (1987).
128. E. H. Andrews, T. A. Khan, and K. A. Majid, *J. Mater. Sci.* 20:3621 (1985).
129. J. A. Miller and E. A. Jakusch (Minnesota Mining and Manuf. Co., St. Paul, MN), EP 0306232B1/07.04.1993.
130. G. Sala (International Carbon Solvent S.p.A.), EP 273997/13.07.1988; in *CAS Adhes.* 24:5 (1988).
131. *Adhes. Age* 9:8 (1986).
132. F. Altenfeld and D. Breker, Semi structural bonding with high performance pressure sensitive tapes, *3M Deutschland Neuss*, Feb. 2, 1993, p. 278.
133. TESA, *Technologie des Klebens*, Beiersdorf AG, Hamburg, Germany, p. 55, 1994.

134. M. Toshio (Furukawa Electric Co. Ltd.), Jpn. Patent 6263865/23.05.1988; in *CAS Adhes.* 24:6 (1988).
135. *Coating* 11:424 (1985).
136. K. Mizui (Mitsui Petrochemical Ind. Ltd.), PCT, WO 88 02767/21.04.1988; in *CAS Adhes.* 22:4 (1988).
137. R. M. Enanoza (Minnesota Mining and Manuf. Co.), EP 259968/11.08.1986; in *CAS Adhes.* 22:3 (1988).
138. *Coating* 1:8 (1986).
139. T. Sugiyama, N. Miyaji, I. Yoshihide, and T. Tange (Nippon Carbide Ind. Co. Inc.), Jpn. Patent 62199672/03.09.1987; in *CAS Adhes.* 14:3 (1988).
140. T. Hiroyoshi, H. Kuribayashi, and E. Usuda (Sumitomo Chem Co. Ltd.), EP 254002/27.01.1988; in *CAS Adhes.* 14:3 (1988).
141. C. Cervellati, G. Capaldi, and L. E. Jacob (Exxon Chemical Patents Inc.), EP 273585/06.07.1988.
142. R. G. Czerepinski and R. Gundermann (Dow. Chem. Co.), U.S. Patent 4713412/15.12.1987.
143. T. Tsubakimoto, K. Minami, A. Baba, and M. Yoshida (Nippon Shokubai Kagaku Kogyo Co., Ltd.), Jpn. Patent 6335676/16.02.1988; in *CAS Adhes.* 16:1 (1988).
144. I. Yorinobu, W. Yasuhhisa, and T. Hiroshi (Japan Synthetic Rubber Co. Ltd.), Jpn. Patent 6386777/18.04.1988; in *CAS Adhes.* 24:3 (1988).
145. C. H. Hill, N. C. Memmo, W. L. Phalen, Jr., and R. R. Suchanec (Hercules Inc.), EP 259697/16.03.1988; in *CAS Adhes.* 14:5 (1988).
146. D. Akihito, O. Tomohisa, and U. Toshishige (Hitachi Chem Co. Ltd), Jpn. Patent 6368683/28.03.1988; in *CAS Crosslinking React.* 14:7 (1988).
147. K. Akasha, S. Sanuki, and T. Matsuyama (Nippon Synth. Chem. Ind. Co. Ltd.), Jpn. Patent 6224336669/24.10.1987.
148. S. Shinji and Y. Yoshiuki (Nippon Synth. Chem. Co. Ind. Ltd.), Jpn. Patent 62243670/24.10.1987; in *CAS Adhes.* 13:4 (1988).
149. Y. Moroishi, T. Sugii, and K. Noda (Nitto Electric Ind. Co. Ltd.), Jpn. Patent 6381183/12.04.1988; in *CAS Adhes.* 22:6 (1988).
150. H. Yaguchi, H. Fukuda, Masayuki, and T. Ohashi (Bridgestone Corp.), Jpn. Patent 6327583/05.02.1988; in *CAS Adhes.* 19:6 (1988).
151. B. Copley and K. Melancon (Minnesota Mining and Manuf. Co.), U.S. Patent 878816/26.01.1986; in *CAS Siloxanes Silicones* 14:2 (1988).
152. A. Saburo (Central Glass Co. Ltd.), Jpn. Patent 63117085/21.05.1988.
153. L. E. Grunewald and D. J. Classen (Minnesota Mining and Manuf. Co., St. Paul, MN), EP 256662/24.02.1986.
154. S. Koriki, R. Mokino, K. Ito, and T. Unno (Nagoya Oil Chem. Co. Ltd.), EP 239348/30.09.1987; in *CAS Colloids* 4:5 (1988).
155. H. Kenjiro and S. Kotaro (Nitto Electric Ind. Co. Ltd.), Jpn. Patent 63118383/23.05.1988; in *CAS Adhes.* 24:5 (1988).
156. Minnesota Mining and Manuf. Co. Ltd., St. Paul, MN, Jpn. Patent 63178091/19.07.1988; in *CAS Adhes.* 25:7 (1988).
157. Morton Thiokol Inc., *Morstik 103 Adhesive*, Adhesives and Coatings, MTD-MS103-12/1988.

End Uses of PSPs

158. A. Nagasuka and M. Kobari (Sekisui Chem. Co., Ltd.), Jpn. Patent 6389585/20.04.1988; in *CAS Adhes.* 19:6 (1988).
159. J. P. Kealy and R. E. Zenk (Minnesota Mining and Manuf. Co., St. Paul, MN), Can. Patent 1224678/28.07.1987.
160. M. A. Johnson, *J. Plast. Film Sheeting* 1:50 (1988).
161. Ashland Chemicals, *Aroset* EPO 32216-APS-102, Data Sheet.
162. A. B. Wechsung, *Coating* 9:268 (1972).
163. EP 0120708, p. 2.
164. G. W. H. Lehmann and H. A. J. Curts (Beiersdorf AG, Hamburg, Germany), U.S. Patent 4038454/26.07.77.
165. Y. Sasaaki, D. L. Holguin, and R. Van Ham (Avery International Co., Pasadena, CA), EP 0252 717A2/13.01.1988.
166. *Coating* 1:12 (1984).
167. EP 0100146/1987.
168. *Openlines*, Vol. 8, Spring 1993, P. P. Payne Ltd., England.
169. R. G. Jahn, *Adhes. Age* 12:35 (1977).
170. *Adhes. Age* 3:6 (1985).
171. E. B. Richmann, S. Nemth, S. W. Tomlinson, and D. W. Wilson (Avery Int., San Marino, CA), U.S. Patent 624870/23.10.1975.
172. D. K. Fisher and B. J. Briddell (Adco Product Inc., Michigan Center, MI), EP 0426198A2/08.05.1991.
173. *Adhes. Age* 12:8 (1986).
174. EP 4673611.
175. G. Galli (Manuli Autoadesivi Spa., Cologno Monzese, Italy), EP 0191191A1/20.08.1986.
176. Ital. Patent 21842A/1982; in G. Galli (Manuli Autoadesivi Spa., Cologno Monzese, Italy), EP 0191191A1/20.08.1986.
177. Kalle Folien, *Siegel Band*, Hoechst., Mi 1984, 38T 5.84 LVI, Hoechst AG, Wiesbaden, Germany.
178. Better Packages Inc., U.S. Patent 3510037; in *Coating* 5:130 (1971).
179. R. E. Nelson, U.S. Patent 4647485/03.03.1987.
180. M. Satsuma (Nitto Electric Ind. Co.), Jpn. Patent 63 48381/01.03.1988; in *CAS Adhes.* 14:4 (1988).
181. C. Donker, R. Luth, and K. van Rijn, Hercules MBG 208 Hydrocarbon resin: A new resin for hot melt pressure sensitive (HMPSA) tapes, 19th Münchener Klebstoff und Veredelungsseminar, 1994, p. 64.
182. L. Jacob, New development of tackifiers for SBS copolymers, 19th Munich Adhesive and Converting Seminar, 1994, p. 107.
183. M. E. Ahner and M. L. Evans, Light colour aromatic containing resins, Tappi Hot Melt Symposium '85, June 16–19, 1985, Hilton Head, SC; in *Coating* 1:6 (1986).
184. K. Nakamura and Y. Miki (Nitto Electric Industrial Co. Ltd.), Jpn. Patent 6386786/18.04.1988; in *CAS Adhes.* 21:5 (1988).
185. F. M. Kuminski and T. D. Penn, Plastics Processing 42(2):25 1972.
186. B. Hanka, *Adhäsion* 10:342 (1971).

187. *Papier Kunstst. Verarb.* *1*:20 (1996).
188. *APR 18*:572 (1987).
189. C. Parodi, S. Giordano, A. Riva, and L. Vitalini, Styrene-butadiene block copolymers in hot melt adhesives for sanitary application, 19th Münchener Klebstoff und Veredelungsseminar, p. 119.
190. W. E. Lenney (Air Products and Chemical Inc.), Can. Patent 1225176/04.08.1987.
191. U.S. Patent 3257478; in W. E. Lenney (Air Products and Chemical Inc), Can. Patent 1225176/04.08.1987.
192. U.S. Patent 3697618; in W. E. Lenney (Air Products and Chemical Inc.), Can. Patent 1225176/04.08.1987.
193. U.S. Patent 3971766; in W. E. Lenney (Air Products and Chemical Inc.), Can. Patent 1225176/04.08.1987.
194. T. J. Bonk, T. I. Cheng, P. M. Olson, and D. E. Weiss, PCT, WO 87/00189/ 15.01.1987.
195. H. Nakahata, Jpn. Patent 07316510/05.12.1995; in *Adhes. Age* *5*:11 (1996).
196. Packard Electric, Engineering Specification ES-M-2359.
197. N. Yamazaki (Matsushita Electric Ind. Co., Ltd.), Jpn. Patent 63 86782/18.04.1988; in *CAS Adhes.* *12*:5 (1988).
198. H. K. Porter Co. Inc., Pittsburgh, PA, U.S. Patent 3149997; in *Adhäsion* *3*:79 (1966).
199. *Adhes. Age* *12*:352 (1984).
200. B. Kunze, S. Sommer, and G. Düsdorf, *Kunststoffe* *84*(10):1337 (1994).
201. Johns Manville Corp., U.S. Patent 3356635; in *Coating* *7*:210 (1969).
202. *Coating* *5*:151 (1969).
203. K. Pál, *Müanyag Fóliák*, Müszaki Könyvkiadó, Budapest, 1976, p. 45.
204. H. Kato, H. Adachi, and H. Fujita, *Rubber Chem. Technol.* *56*:287 (1983).
205. T. Kishi (Sekisui Chem. Co. Ltd), Jpn. Patent 6369879/29.03.1988; in *CAS Adhes.* *19*:6 (1988).
206. J. Andres, *APR 16*:444 (1986).
207. Minnesota Mining and Manuf. Co., St. Paul., MN, U.S. Patent 3307690; in *Coating* *4*:114 (1969).
208. *Adhes. Age* *8*:62 (1986).
209. Packard Electric, Engineering Specification ES-M-1881.
210. R. Shibata and H. Mixagawa (Hitachi Condenser Co. Ltd.), Jpn. Patent 6386785/ 18.04.1988; in *CAS Adhes.* *21*:5 (1988).
211. Allmänna Svenska Elekriska AB, DBP 1276771.
212. T. J. Kilduff and A. M. Biggar, U.S. Patent 3355545; in *Coating* *7*:210 (1969).
213. *Adhes. Age* *1*:6 (1985).
214. *Coating* *3*:210 (1985).
215. A. Dobmann and A. G. Viehofer, 19th Münchener Klebstoff und Veredelungsseminar, 1994, p. 168.
216. P. Penczek and B. Kujawa-Penczek, *Coating* *6*:232 (1991).
217. Minnesota Mining and Manuf. Co., St. Paul., MN, DBP 1486514.
218. Y. Torigoe (Dainichi Nippon Cable Ltd., Amagasaki, Japan), U.S. Patent 4645697/ 24.02.1987; in *Adhes. Age* *5*:26 (1987).

219. V. M. Ryabov, O. I. Chernikov and M. F. Nosova, *Plast. Massy* 7:58 (1988).
220. G. Camerini (Coverplast Italiana SpA), EP 248771/20.05.1986.
221. Y. Hamada and O. Takuman (Toray Silicone Co., Ltd.), EP 253601/20.01. 1988.
222. R. Tschudin-Mahrer (Irbit Research Consulting AG, Fribourg, Switzerland), U.S. Patent 4564550/14.01.1986.
223. K. Wabro, R. Milker and G. Krüger, *Haftklebstoffe und Haftklebebänder*, Astorplast GmbH, Afdorf, Germany, 1994, p. 129, 133 and 147.
224. B. Ellegard, *Etiketten Labels* 3:47 (1995).
225. *Labels Label.* 2:22 (1997).
226. Henkel & Cie., GmbH, Düsseldorf, Germany, DBP 1179660; in *Adhäsion* 2:84 (1966).
227. *Adhäsion* 6:14 (1985).
228. Johnson and Johnson, U.S. Patent 3403018; in *Coating* 1:24 (1969).
229. *Adhäsion* 11:37 (1994).
230. *Kaut. Gummi Kunstst.* 37(1):17 (1984).
231. *Eur. Adhes. J.* 6:23 (1995).
232. *Adhes. Age* 12:30 (1987).
233. W. Endlich, Klebende Dichtstoffe in der industriellen Fertigung, T.3, Seminar an der TA Wuppertal; in *Adhäsion* 3:26 (1982).
234. M. Gerace, *Adhes. Age* 8:15(1983).
235. J. W. Otter and G. R. Watts (Avery International Corp., Pasadena, CA), U.S. Patent 5346766/13.09.1994.
236. G. D. Bennett (Simmonds Precision, New York), U.S. Patent 20.01.1987; in *Adhes. Age* 5:28 (1988).
237. Kendall Co., Can. Patent 853145/1987.
238. R. R. Charbonneau and G. L. Groff (Minnesota Mining and Manuf. Co., St. Paul, MN), EP 0106559B1/25.04.1984.
239. U.S. Patent 2925174; in R. R. Charbonneau and G. L. Groff (Minnesota Mining and Manuf. Co., St. Paul, MN), EP 0106559 B1/25.04.1984.
240. U.S. Patent 4286047; in R. R. Charbonneau and G. L. Groff (Minnesota Mining and Manuf. Co.), EP 0106559 B1/25.04.1984.
241. G. F. Vesley, A. H. Paulson, and E. C. Barber, EP 0202938A2/26.11.1986.
242. U.S. Patent 4223067; in R. R. Charbonneau and G. L. Groff (Minnesota Mining and Manuf. Co.), EP 0106559B1/25.04.1984.
243. G. Bennett, P. Ludwig Geiss, J. Klingen, and T. Neeb, *Adhäsion* 7/8:19 (1996).
244. D. C. Stillwater, D. C. Kostenmaki, and M. H. Mazurek (Minnesota Mining and Manuf. Co.), U.S. Patent 5344681/06.09.1994.
245. J. Suchy, J. Hezina, and J. Matejka, Czech Patent 247802/15.121987; in *CAS Adhes.* 19:4 (1988).
246. F. D. Blake (Minnesota Mining and Manuf., Co.), EP 0141504 A1/15.05.1985.
247. *Coating* 9:33 (1988).
248. Packard Electric, Engineering Specification ES-M-2147.
249. Z. Czech, *Adhäsion* 11:26 (1994).

250. U.S. Patent 3096202; in P. Gleichenhagen and I. Wesselkamp (Beiersdorf AG, Hamburg, Germany), EP 0058 382B1/25.08.1982.
251. *Adhäsion* 5:161 (1971).
252. U.S. Patent 3865770; in F. C. Larimore and R. A. Sinclair (Minnesota Mining and Manuf. Co.), EP 0197662A1/15.10.1986.
253. T. H. Haddock (Johnson & Johnson Products Inc., New Brunswick, NJ), EP 0130080B1/02.01.1985.
254. D. Allen, Jr. and E. Flam (Bard Inc., Murray Hill, NJ), U.S. Patent 4650817/17.03.1987; in *Adhes. Age* 5:24 (1987).
255. *Adhes. Age* 4:6 (1983).
256. Z. Czech and D. Sander (Lohmann GmbH, Neuwied, Germany), DE 44 33 005/16.09.1994.
257. Johnson & Johnson, U.S. Patent 3483018; in *Coating* 1:23 (1971).
258. M. von Bittera, D. Schäpel, U. von Gizycki, and R. Rupp (Bayer AG, Leverkusen, Germany), EP 0147588B1/10.07.1985.
259. U.S. Patent 3321451; in S. E. Krampe and C. L. Moore (Minnesota Mining and Manuf. Co.), EP 0202831A2/26.11.1986.
260. R. L. Sun and J. F. Kennedy (Johnson and Johnson Products, Inc.), U.S. Patent 4762888/09.08.1988; in *CAS Hot Melt Adhes.* 26:1 (1988).
261. U.S. Patents 1760820, 3189581, 3371071, 3509111, and 3475363; in T. H. Haddock (Johnson & Johnson Products Inc., New Brunswick, NJ), EP 0130080B1/02.01.1985.
262. J. N. Kellen and C. W. Taylor (Minnesota Mining and Manuf. Co.), EP 0246A2/25.11.1987.
263. Martens et al., U.S. Patent 4181752; in J. N. Kellen and C. W. Taylor (Minnesota Mining and Manuf. Co.), EP 0246A2/25.11.1987.
264. *Adhäsion* 7/8:124 (1991).
265. *Verpack. Rundsch.* 9:994 (1983).
266. M. J. Huber (Quality Manufacturing Inc.), U.S. Patent 546692/07.11.1995; in *Adhes. Age* 5:12 (1996).
267. R. B. Lipson (Kwik Paint Products), U.S. Patent 5468533; in *Adhes. Age* 5:12 (1996).
268. H. Meichner (Rehau, Germany), Gebrauchsmuster, GM 8607368/18.03.1986.
269. I. Benedek, E. Frank, and G. Nicolaus (Poli-Film Verwaltungs GmbH, Wipperfürth, Germany), DE 4433626A1/21.09.1994.
270. T. Moldvai and N. Piatkowski (Inst. Cerc. Pielarie, Incaltaminte, Bucharest), Rom. Patent 93124/30.12.1987.
271. *Adhäsion* 11:481 (1967).
272. K. Taubert, *Adhäsion* 10:379 (1970).
273. *Papier Kunstst. Verarb.* 19:32 (1990).
274. G. Menges, W. Michaeli, R. Ludwig, and K. Scholl, *Kunststoffe* 80(11):1245 (1990).
275. E. Höflin and H. Breu, *Adhäsion* 10(6):252 (1966).
276. *APR* 29:798 (1988).
277. *Coating* 8:98 (1987).

278. BASF, Tl-2.2-21, November 1979, Teil 3, b.5, *Schutzfolien*, BASF AG, Ludwigshafen, Germany.
279. G. Fauner, *Adhäsion* 4:27 (1985).
280. *Eur. Adhes. J.* 6:22 (1995).
281. H. Bucholz, *Oberflächentechnik* 10:484 (1973).
282. *Coating* 3:66 (1984).
283. G. Allwyn, *Am. Paint J. Convention Daily Number* 15:33 (1969); in *Coating* 12:368 (1970).
284. *Kaut. Gummi Kunstst.* 40(10):981 (1987).
285. Poli Film-America, Cary, IL, *Product Application Guide*, 1994.
286. Hoogovens Packaging Steel, Ijmujden, The Netherlands, Technical Booklet, 1996.
287. W. E. Havercroft, *Paper, Film Foil Converter* 10:52 (1973).
288. J. W. Kühr, *Farbe Lack* 5:475 (1970).
289. *Coating* 2:78 (1984).
290. W. Brockman, O. D. Hennemann, H. Kollek, and C. Marx, *Adhäsion* 9:24 (1986).
291. W. S. D. Bascom and R. L. Patrick, *Adhes. Age* 19:25 (1974).
292. *Coating* 10:264 (1979).
293. V. Antonioli, *Rev. Colore* 58:79 (1973).
294. Dynamit Nobel A. G, Neth. Patent 72.07134; in *Coating* 5:122 (1974).
295. *Adhäsion* 1/2:28 (1983).
296. G. F. Bond and J. N. Ralston, *Ind. Finish. Surf. Coatings* 305:4 (1973).
297. Ch. W. Uezelmeier, *SPE-J.* 5:69 (1970).
298. G. Kuehl, *Metalloberflaeche* 42(3):139 (1988); in *CAS Crosslink. React.* 12:11 (1988).
299. K. H. Seifert, *Coating* 2:41 (1972).
300. W. Brushwell, *Farbe Lack* 1:34 (1974).
301. *Scope* 3:92 (1988).
302. *Coating* 3:66 (1984).
303. *Coating* 12:450 (1986).
304. *Farbe Lack* 92(4):296 (1986).
305. Ball Brothers, Muncie, IN, DBP 128585; in *Coating* 7:274 (1969).
306. *Eur. Plast. News* 1:28 (1996).
307. M. Saada and M. El-Moualled, *Chim. Ind.* 15:1917 (1970).
308. E. Röhrl and L. Schwiegk, *Plastverarbeiter* 42(10):39 (1991).
309. J. P. Weibel, W. Körmer, and S. Caporusso, *Kunststoffe* 83(7):541 (1993).
310. W. Kremer, *Farbe Lack* 94(3):205 (1988); in *CAS Crossl. React.* 12:11 (1988).
311. *KI 1020*:2 (1991).
312. P. Dippel, *Coating* 4:44 (1988).
313. *Plastverarbeiter* 9:9 (1994).
314. *Coating* 7:246 (1993).
315. Tielt, Belgium, ERTA Polycarbonate, Technical Booklet, Mar. 3, 1982.
316. *Coating* 5:170 (1996).
317. M. Heinze, *Kunststoffe* 83(8):630 (1993).
318. A. K. Schlarb and G. W. Ehrenstein, *Kunststoffe* 83(8):597 (1993).

319. W. Daum, *Kunststoffe* 84(10):1433 (1994).
320. D. J. P. Harrison, J. F. Johnson, and W. R. Yates, *Polym. Eng. Sci.* 22(14):865 (1982).
321. A. Haas (Soc. Chim. de Charbonnage-CdF), U.S. Patent 4624991/25.11.1986; in *Adhes. Age* 5:26 (1987).
322. H. Inada, T. Tomomori, and K. Kawakura (Tokuyama Soda Co., Ltd.), Jpn. Patent 62270642/25.11.1987; in *CAS Adhes.* 13:1 (1988).
323. O. Tadayuki, M. Kentaro, F. Kazunori, and K. Toshio (Sumitomo Chem. Co., Ltd.), EP 249442/16.12.1987.
324. Hoechst Folien, Datenblatt, Klebebandträger und Abdeckfolie, Metallisierungs und Dekorationsfolie, Ausgabe 07/92.
325. Bakelite Xylonite Ltd., Br. Patent 1081291; in *Coating* 4:114 (1969).
326. Dow, *DAF, Dow Adhesive Films*, Dow Chemical Co., Horgen, Switzerland, 1983.
327. Nitto, *Protective by Nitto*, Data Sheets, Nov. 4, 1986.
328. A. S. Jandel, *VDI* 37:29 (1996).
329. J. F. Kuik, *Papier Kunstst. Verarb.* 10:26 (1990).
330. R. M. Podhajny, *Convert. Packag.* 3:21 (1986).
331. ECCA, T 17/1985, *Adhérence des Films pelables*.
332. *Prodoc* 4:14 (1988).
333. Röhm AG, Technische Information, Produktbeschreibung, *Plexiglas GS*, XT, Ke. No. 212-1, November 1993, Darmstadt.
334. D. Symietz, *Kunststoffe* 11:11 (1994).
335. E. Doege and U. Hesberg, *Blech Rohre Profile* 39(1):25 (1992).
336. M. Herrmann, *Blech Rohre Profile* 40(2):164 (1993).
337. T. Miura, *Giesserei* 62(17):437 (1975).
338. *KI* 1169:5 (1994).
339. *Druckwelt* 5:51 (1988).
340. R. Uhlemayr (O. T. Drescher GmbH, Rutesheim, Germany), Gebrauchsmuster, PS 3508114/22.05.1986.
341. Grafityp, *Selbstklebefolien massgeschnitten*, Booklet, Grafityp, Houthalen, Belgium, October 1992.
342. *Grafityp Newslett.* 11:10 (1997).
343. Perigraf, Computer und Informationssysteme GmbH, Eibau, Germany, 1997.
344. Graphtec, RFC3100-60, *Tischschneideplotter*, Multiplot Grafiksysteme, Bad Emstal, Germany, 1997.
345. Cal Comp, Data Sheet, Sign Europe, Düsseldorf, 1997.
346. Regulus Schablonen Papiere, Application tape, Schutzfolie, Klaus Koenig KG, Germany, 1997, Booklet.
347. J. W. Prane, *Adhes. Age* 1:44 (1989).
348. *Papier Kunstst. Verarb.* 2:44 (1996).
349. Sealock, Andover, Hampshire, UK, Emulsion Adhesive E 959.
350. *Adhes. Age* 9:62 (1986).
351. *Labels Label.* 3/4:10 (1994).
352. *Papier Kunstst. Verarb.* 9:57 (1968).
353. Y. Ohata, K. Awano, M. Atsuji, and T. Hattori (Toa Gohsei Chemical Industry Co.,

Ltd., Aica Kogyo Co., Ltd.), Jpn. Patent 63,89345/20.04.1988; in *CAS Adhes. 21*:10 (1988).
354. Rasvino Predpriatie za organiziona Technika, Silistra, Bulgary, DBP 1777807; in *Coating 7*:210 (1969).
355. C. Zbigniew and H. D. Sander (Lohmann GmbH, Neuwied), DE 4219368/12.06.1992.
356. *Coating 1*:29 (1978).
357. *Adhäsion 6*:270 (1965).
358. G. Fuchs, *Adhäsion 3*:24 (1982).
359. *Coating 12*:342 (1984).
360. *APR 6*:159 (1986).
361. *APR 18*:686 (1986).
362. I. P. Rothernberg (Stik-Trim Industries Inc., New York), U.S. Patent 4650704/17.03.1987.
363. Hot Melts—The Future is Now, Tappi Hot Melt Symposium, June 2–4, Toronto; in *Coating 7*:175 (1980).
364. J. Weiss, *Metall 5*:367 (1994).
365. W. C. Zybco, W. Wald, and T. W. McCure (Burlington Industries Inc., Greensboro, NC), U.S. Patent 4680209/14.07.1987; in *Adhes. Age 2*:58 (1988).
366. J. Weidauer, *Eur. Adhes. Sealants 5*:26 (1995).
367. F. S. Thomas, U.S. Patent 3148/398; in *Adhäsion 2*:78 (1966).
368. *Adhes. Age 3*:10 (1987).
369. *Siebdruck 8*:7 (1988).
370. *Kunststoff J. 9*:15 (1986).
371. *Eur. Adhes. Sealants 6*:16 (1996).

Index

Abhesive (*see also* Release agent)
  built in, 236
  chemical composition of, 232
  coated, 233
    non-silicone, 233, 235
    silicone, 235
  coating properties of, 525
  formulation of,
    dependence of on chemical basis, 387
    dependence of on coating technology, 390
    dependence of on product classes, 392
    technology of, 394
  macromolecular basis of, 232
  manufacture of, 386
Acrylic adhesive, 13, 23
  of acrylic acid, 348, 362
    alkali soluble, 367
    water soluble, 368
  block copolymer for, 96, 197
    hard block of, 198
    chemical crosslinking of, 198

[Acrylic adhesive]
  for diaper tapes, 97, 380
  EB crosslinkable, 209, 364
  functionalization of for EB curing, 209
  for HMPSA, 206
  for labels, 39
  mechanical properties of, 122
  molecular weight of, 205
  photocrosslinked, 320
  postcured, 341
    T_g of, 341
  prepolymer, 205
    radiation cured, 205, 342
  for protective films, 60
  rubber for, 195
    formulation of, 197
    processibility of, 197
    solution polymerized, 206
    T_g of, 197
  skin compatible, 212 (*see also* Medical tapes)
  with solids, 34, 40, 341, 358
  solution polymerized, 357

643

[Acrylic adhesive]
  solvent based, 365
  use of, 365
  suspension polymerized, 209
  for tapes, 47
  T_g of, 203
Adherend surface, 85
Adhesion
  balance of, 260, 306, 348, 368 (see also Adhesive characteristics of PSPs)
  of masticated-crosslinked-tackified formulations, 353
  break, 350
  build up of, 257, 352
  on corona treated surface, 452
    build up of for tapes, 455
    dependence of on treatment energy level, 454, 455
  dependence of,
    on carboxylic functionality, 205
    on carrier strength, 433
    on time, 289
  with electric charges, 452
  influence of on removability, 289
  instantaneous, 275, 276, 277, 452
  mechanism of, 259
  parameters of, 305
  to skin, 368 (see also Medical tapes)
  subsequent, 446 (see also Release liner)
  surface-application surface ratio, 288,
  work of, 285
Adhesive
  acrylic, 100 (see also Acrylic adhesive)
    low cohesion, 173
    solvent based, 205
    UV light curable, 179, 341
  anchorage of, 24, 31, 289, 321
  application window of, 263
  breathable, 368 (see also Medical tapes)

[Adhesive]
  built-in, 33 (see also Adhesive)
  chemical composition of, 169
  choice of, 364
  coated, 33
  coating properties of, 524
    solvent based, 525
    water based, 525
  composite structure of, 263
  composition,
    influence of on adhesive properties, 305 (see also Adhesive characteristics of PSPs)
  for contact adhesion, 369
  contoured surface of, 22
  crack migration of, 320
  deformation work of, 105
  discontinous, 20, 22, 23 (see also Pattern coating)
  elastomeric components of, 169 (see also Elastomers)
  electrically conductive, 34, 223 (see also Electrical conductivity)
  enclosed, 24 (see also Tackification of carrier)
  failure nature, 290 (see also Removability)
  fluid permeable, 368 (see also Medical Tapes)
  geometry of, 313
    modification of, 313 (see also Pattern coating)
  for labels (see also Labels)
    performance parameters for, 373
    requirements for, 372
  formulation of, 343
    for synthesis, 338 (see also Polymacromerization)
  free flow region of, 267
  from the reel, 19 (see also Transfer tape)
  hardening of, 12
  for labels, 39
  choice of, 381

Index

[Adhesive]
    layer of, thick, 269 (*see also* Ultraviolet light induced polymerization)
    low temperature resistant, 369 (*see also* Freezer tape)
    macromolecular basis of, 169
    manufacturing possibilities of, 339
        in line, 339, 343
        off line, 339, 340, 343
    multilayer, 23, 34
    non degradable by body fluid, 368 (*see also* Medical tape)
    physiologically compatible, 368 (*see also* Medical tape)
    pigmented, 223 (*see also* Filler)
    for protective films, 60, 212, 363
        choice of, 381
        of EVAc, 212
    porous, 20, 21, 378 (*see also* Medical tape)
    powdered, 380
    properties, 94 (*see also* Adhesive characteristics of PSPs)
    re-adhering, 184
    reinforcing of, 13, 16
    removable, 19, 22, 23 (*see also* Removability)
    repositionable, 344, 429
    role of, 34
    self supporting, 368 (*see also* Transfer tape)
    softening of, 311, 364 (*see also* Tackification)
    solubility of,
        regulation of, 367 (*see also* Splicing tape)
    storable crosslinkable, 206
    strength of coatings,
        test of, 287
    synthesis of,
        for tapes, 47, 375, 376 (*see also* Ultraviolet light induced polymerization)
    thermally activatable, 223

[Adhesive]
    transfer index of, 31 (*see also* Removability)
    UV induced polymerization of, 341 (*see also* Ultraviolet light induced polymerization)
        in line, 341
    virtually non tacky, 17
    water activatable, 367,
    water soluble, of vinyl acetate-vinyl pyrrolidone, 210
    for wet surface, 369 (*see also* Bottle label)
Adhesive characteristics of PSPs, 255, 312
    balance of, 271 (*see also* Adhesion balance)
    characterization of, 256
    comparison of, 112, 114
    crosslinked, 318 (*see also* Crosslinking)
    definition of, 256
    general, 256
    influence of on other product characteristics, 325
    regulation of, 305
        adhesive, 305
        adhesive coating technology, 323
        adhesive geometry, 319 (*see also* Pattern coating)
        adhesive structure, 319
        carrier, 320, 321
        coating device, 324
        crosslinking, 315 (*see also* Crosslinking)
        direct/transfer coating, 27, 324
        manufacturing technology, 322
        product application technology, 324
    special, 263
    test of,
        aging conditions for, 588
        methods of, 263
Adhesive characteristics of protective films, 108, 298, 612

Index

Adhesive components, manufacture of, 337 (*see also* Adhesive manufacturing possibilities)
Agent
  abhesive, 387
  antiblock, 226
    for LLDPE, 155 (*see also* Linear low density polyethylene)
    for opacifying, 226 (*see also* Filler)
  antimicrobial, 33, 34
  antistatic, 35, 138, 139, 225, 443, 444 (*see also* Triboelectricity)
    gel forming with, 184
  slip, 226 (*see also* Slip)
    compatibility of, 227
    temperature resistance of, 227
Aging, 291 (*see also* Resistance)
Air-permeability
  of adhesive, 20 (*see also* Medical Tape)
  of carrier, 20 (*see also* Medical Tape)
Antitheft labels, 582, 592 (*see* also Tamper evident label)
Antioxidant, 15, 184, 223 (*see also* Aging)
  for EB cured formulations, 224
  for HMPSA, 224
  synergist for, 114
  yellowing of, 225
Antistatic
  characteristics, 138
    modification of, 443, 444
  protection, 139 (*see also* Triboelectricity)
Application
  pressure, 33 (*see also* Application conditions of protective films)
  windows for labels, 293
Application conditions of protective films, 60, 109, 271, 602, 603
  according to laminating equipment, 610
  according to substrate surface, 610
  for coil coating, 605

[Application conditions of protective films]
  dependence of on product surface, 603
  for laquered surface, 605
  for main grades, 610
  for metallic surface, 604
  for plastics, 606
Application tape, 484, 601, 628 (*see also* Plotter film)
  manufacture of, 485
Arrhenius equation, 78, 118
Autoadhesion, 85, 86 (*see also* Self adhesion)
  of PE, 454

Bag closure labels, 579
Bag lip tapes, 591
Barrier layer, 458 (*see also* Self adhesive film)
BGA approval, 372, 419
Binders for surface coating of paper, 396, 401
  influence of, on paper surface quality, 396
Bioelectrodes, 368 (*see also* Medical tape)
Bisequenced polymers, 188 (*see also* Thermoplastic elastomers)
Bleeding, 374 (*see also* Migration)
Block copolymers (*see also* Thermoplastic elastomers)
  ABA, 190
  ionic bonds in, 194
  multiarm, 190 (*see also* Styrene-isoprene-styrene block copolymers)
  radial (*see* Radial block copolymers)
  tackification of, 172 (*see also* Tackification)
  star, 190 (*see also* Star shaped block copolymers)
    crosslinking of, 192
    molecular weight of, 192
  synthesis of, 194

Index

Blocking, 226, 440 (*see also* Agent)
  of adhesive, 18
  of carrier, 82, 86, 155
  of HDPE, 155
  properties, modification of, 444
  resistance, 272
Blow procedure, 408, 410 (*see also* Manufacture of film carrier)
  for HDPE, 409, 410 (*see also* High density polyethylene)
  for LDPE, 409, 616 (*see also* Low density polyethylene)
  for LLDPE, 410 (*see also* Linear low density polyethylene)
  for PE film, 408
    influence of on molecular orientation, 408
    on shrinkage, 409 (*see also* Shrinkage)
  parameters of, 409
Blown film, 118, 451 (*see also* Manufacture of film carrier)
  equipment for, 155
Blow-up ratio, 117, 131, 155, 409, 410, 415, 616 (*see also* Manufacture of film carrier)
  influence of on cuttability, 554 (*see also* Cuttability)
Bond break character, changing of, 306 (*see also* Removability)
Bottle-labels, 578 (*see also* Label)
  wine, 578
Box closure sealing tapes, 590
Branching, 153
  of polyolefins, influence of on mechanical characteristics, 154, 156
Butadiene-styrene elastomer, 33 (*see also* Styrene block copolymers)
Butene polymers, 185
Butyl rubber, 13, 186 (*see also* Butene polymers)
  crosslinked, 186
  halogenated, 186
  latex of, 186

Calendering, dead, 172 (*see* Mastication)
Car masking tape, 601 (*see also* Tape)
Carboxylated styrene-butadiene rubber, 184, 187, 272
  gel content of, 187
  dispersion of, 39, 47, 121
  styrene content of, 187
  tackifier level for, 187
Carrier, 28
  adhesive free, 450 (*see also* Self adhesive film)
  chemical basis of, 150 (*see also* Manufacture of film carrier)
  of cloth, 31, 46, 54
  conformability of, 30, 285 (*see also* Conformability)
  deformability of, 28, 112
    dependence of on tickness, 110
    influence of on adhesive properties, 281, 322 (*see also* Adhesive characteristics of protective films)
  diffusion characteristics of, 321 (*see also* Medical tape)
  dimensional stability of, criterion of, 112 (*see also* Dimensional stability)
  discontinous, 20
  embossed, 20 (*see also* Medical tapes)
  foam-like, 269 (*see also* Foam-like tape)
  flexibility of, 135
  functionalization of, 109 (*see also* Corona treatment)
  geometry of, 22, 30
  impregnated, 16, 46, 319, 429, 431
  for labels, 28
    conformability of, 28
    destructibility of, 28
    dimensional stability of, 28
    hot melt coated, 359,
    mechanical properties of, 28
    surface quality of, 28

[Carrier]
  macromolecular basis of, 150 (*see also* Manufacture of film carrier)
  mechanical characteristics of, 127 (*see also* Mechanical properties of PSPs)
  metallic, 185, 431, 481
  non-adhesive, 75, 159
    chemical basis of, 159
    of PVC, 433
  nonwoven based, 20
  paper, 150
    top coat for, 150, 160 (*see also* Primer)
  of plastic, 150 (*see also* Film carrier)
    homogeneous, 160
  perforated, 23, 30
  pleated, 22, 31, 46, 472
    coating of, 479
  polar, 166
  polymeric, 160 (*see also* Film carrier)
  porous, 16, 364, 425, 472 (*see also* Medical label)
  for protective films, 60, 63, 107, 135 (*see also* Film carrier)
    mechanical characteristics of, 111
    T_g of, 107
  radiation cured, 321
  reinforced, 21
  role of, 28
    for dosage, 31, 32
    for information, 31
    for storage, 32
  self adhesive, 2, 32, 165 (*see also* Self adhesive films)
  slitting of, 30 (*see also* Converting properties)
  stain resistant, 31,
  for tapes, 30, 32, 46
    mechanical properties of, 28
  temporary, 137
  thickness of, 117
  Teflon, 31

Carrierless coatings, 56
  peelable, 56 (*see also* Protective coating)
Cellulose acetate carrier, 23, 163, 406, 429
  for labels, 39
    tamper evident, 164
  for tapes, 23, 46, 164, 353
Cellulose
  derivatives,
    cast film of, 406
    plasticization of, 151
  hydrate, 164, 406
    carrier of, 420, 429
Charge decay time, 139 (*see also* Surface resistivity)
Cling films, 4, 86 (*see also* Self adhesive film)
Chain mobility, 171 (*see also* Glass transition temperature)
Charge transfer complex, 183
Chemical treatment, 441 (*see also* Carrier)
Coalescence, 102
Coating
  components of, 336
    abhesive, 34
    adhesive, 34
    manufacture of, 336
  direct, 34, 270
  in line (*see* Manufacture of PSPs by adhesive coating)
  partial, 288 (*see also* Pattern coating)
  rheology, 102
    of printing inks, 103 (*see also* Printing ink)
  solventless, 21, 220 (*see also* Hot melt)
  spray, 21 (*see also* Hot melt)
  transfer, 27, 34, 270, 479
  viscosity, 379, 490, 491
  warm, 21 (*see also* Hot melt)
Coating device, 490
  choice of, 490,
  with doctor blade, 487

Index 649

[Coating device]
  with enclosed supply chamber, 491
  with gear in die, 464, 491
  gravure, 490
  for HMPSA, 468, 478, 491, 492
  influence of,
    on adhesive properties, 304
    on coating weight, 361
    on formulation of hot melts, 360
    on running speed, 361
  for in line coating of UV curable formulation, 341 (*see also* Ultraviolet light induced polymerization)
  for low coating weight, 494 (*see also* Manufacture of PSPs by adhesive coating)
  for primer, 465 (*see also* Primer)
  for protective films, 487 (*see also* Protective films)
    water based, 487
  reverse gravure, 490
    patterned, 493 (*see also* Pattern coating)
  roll, 468, 492
  rotary screen, 468, 478, 492
  screen printing, 478, 487, 492
  slot die, 468, 478, 492
  spray, 469, 492
    with controlled fiberization, 469
  viscosity for, 180, 361, 482 (*see also* Viscosity)
Coating image, 494
Coating machine (*see* Manufacture of PSPs by adhesive coating)
  for tapes, 51
Coating properties of PSPs, 524
Coating weight,
  critical, 267, 268, 275
  dependence of on carrier thickness, 267
  influence of,
    on adhesion energy, 289
    on adhesive properties, 264
    on peel resistance, 265

[Coating weight]
  tolerances of,
    for protective films, 487, 488
  values of, 267, 464
    for decorative films, 365
    for deep drawable films, 626 (*see also* Deep drawable protective film)
    for labels, 268, 464
    for laminating, 486
    for medical tape, 364
    for primer, 465,
    for protective films, 268, 271, 304, 464, 487 (*see* Manufacture of PSPs by adhesive coating)
    for readherable product, 291
    for tapes, 268, 468, 472, 479
    tolerances of, 271, 272, 464
    for water based PSA, 360, 478
Coextrusion, 411 (*see also* Manufacture of film carrier)
  for blown film, 412
  for flat film, 412
Coil coating, 1, 57, 451, 460 (*see also* Protective films for coil coating)
  with hot laminated PE, 452 (*see also* Hot laminating film)
  adhesion parameters of, 456
Cold flow, 12, 59, 63, 75, 80, 82, 83, 84, 86, 100, 107, 266
Cold seal adhesive, 38, 85, 211, 257, 394
Combination adhesive, 210
Comonomers,
  for hardening, 91
    influence of on cohesive strength, 178
  for softening, 91
Compatibility, 93, 171
  dependence of,
    on MWD, 172, 192
    on styrene content, 192 (*see also* Styrene block polymers)
    on temperature, 347

[Compatibility]
  of oils, 195 (*see also* Oil)
  of resins, 195, 213 (*see also* Resin)
    with carrier, 158
    dependence of on molecular
      weight, 213
    with multiphase systems, 195
  role of for tackifying, 172, 287, 346
Composite structure, 149
  reinforced, 190
Computer label, 37, 42 (*see also*
      Label)
Composition, spreadable, 341 (*see
      also* Manufacture of PSPs by
      adhesive coating)
Confectioning
  machine, for labels, 398
  properties, 548
Conformability, 95, 270, 295, 433
  dependence of on modulus, 133
  of ethylene-vinylacetate copolymers,
      167
  of medical tapes, 117, 425
  of tamper evident labels, 117 (*see
      also* Tamper evident labels)
Conformation time, 295 (*see also*
      Tape)
Conformational gas, 151 (*see also*
      Crosslinked network)
Contact angle, 439, 442
  for PET, 424
Contact
  build up, dependence of on
      pressure, 258
  hindrance, 306
  pressure, influence of on adhesive
      properties, 275
  site,
    network controlled, 319
    non-adhesive, 352
    spherical, 209, 319, 352
    surface reduction, 22, 34, 269, 319,
      351
      disadvantages of, 319
      methods of, 319
Controlled rheology, 154

Converting
  machines,
    for froms, 548
    for labels, 117, 398
  properties,
    of labels, 548
    of PSPs, 523, 568
Corona treatment, 32, 84, 85, 109,
      439, 452
  decay of, 439
  degree of,
    dependence of on treatment
      energy, 454
    parameters of, 455
  dependence of,
    on age of film, 454
    on crystallinity, 454
    on nature of film, 454
    on relative humidity of air, 454
    on slip, 439, 454 (*see also* Agent)
  deposition, 441
  disadvantages of, 440
  effects of, 439
  energy for, 439, 440, 445, 455
  mechanism of, 439
  ozone elimination of, 442
  of HDPE, 454
  of LDPE, 454, 616
  of paper, 20
  of PE, 343, 440, 452
  perforation with, 443
  of polar ethylene copolymers, 453
  of PP, 439
    shelf life of, 440
  of PSA, 439
  quality of, 452
    influence of on peel, 452
  saturation level of, 455
  spray, 439
  stationating time of, 455
Corrosion protective tapes, 35
Creep, 82 (*see also* Cold flow)
  compliance, 34, 82, 95
    values of, 95, 97
  dependence of,
    on crosslinking, 171

Index 651

[Creep]
  testing of, 100
  modulus, 118
Crosslinking, 95, 174
  of acrylates, 151, 177
    low cohesion, 173
  of adhesive, 268, 312 (see also Adhesive)
  anisotropy of, 270
  balance of, 179
  of block copolymers, 188
  chemical, 313, 351
    chemical basis for, 204
  density of, 179, 270
  dependence of on functionality, 179
  EB initiated, 173, 179, 351
    dependence of on the molecular weight, 177
  influence of on adhesive properties, 485 (see also Adhesive)
  monomer, 14, 177, 197, 341 (see also Monomer)
    T_g of, 209
  of natural rubber, 151
  penetration of, 362
  physical, 313 (see also Thermoplastic elastomers)
  of plastomers, 340
  possibilities of, comparison of, 352
  postapplication, 34, 206, 342
  postsynthesis, 173
  radiation induced, 339 (see also Radiation induced polymerization)
  of acrylates, 170
    dependence of on the molecular weight, 176
  of solvent based PSAs, 352
  superficial, 16, 270
  supplemental, 173
  thermal, of acrylics, 209
  UV light induced, 173, 180, 351 (see also Ultraviolet light induced polymerization)
Crosslinked network, 149
  functioning of, 149, 151

Crosslinking agent, 180, 291, 342, 382
  acrylic, 204, 205, 206
    for block copolymers, 204
    level of, 206, 342
  built in, 313, 352
  carboxylic, 205
  external, 313, 351, 352, 381
  influence of on peel, 314
  internal, 381
  latent, 34, 206, 269, 342
  level of, 292
    dependence of on application, 622
  monomer as, 175, 180 (see also Monomer)
    concentration of, 178
  multicomponent, 478
  for solvent based acrylic adhesives, 382
  for water based formulations, 209
Crystallinity, 87
  of blown film, 157
    influence of on tackifier level, 157 (see also Self adhesive film)
  of flat film, dependence of, on plasticizers, 158
  influence of,
    on density, 157
    on joint strength, 174
    on modulus, 157
    on optical properties, 157
    on self adhesivity
    on stiffness, 157
    on tensile modulus, 157
Curable prepolymer technology, 4 (see also Prepolymer)
Curl, 546 (see also Lay flat)
Cuttability, 262, 374 (see also Die-cutting)
  of adhesive, 104
  of carrier, 129
  film, 403, 425
  for labels, 30, 424
  of LLDPE, 403, 550

[Cuttability]
  paper, 401
  for protective films, 30
  dependence of,
    on BUR, 55 (*see also* Blow up ratio)
    on film thickness, 555
    on flexural resistance, 550
    on modulus, 552
    on product build up, 550
    on tack, 551
  evaluation of, 554
  of face stock materials, 554
  of labels, 531, 548
  of release liner materials, 554
Cutting, 549
  characteristics, 549
  laser, 554
  plotter, 552, 628 (*see also* Plotter film)
  thermal die-, 554 (*see also* Die-cutting)

Dahlquist criterion, 310
Dart drop impact strength, 131 (*see also* Mechanical properties of PSPs)
  dependence of, on MWD, 156
Debonding
  force of,
    distribution of, 290
    transfer of, 283
  energy of, 295
  time of, 281
Decalcomania, 8, 30, 292
Decals, 12 (*see also* Label)
Decoration
  film, 25, 35, 66, 131, 365, 428, 629
    automotive, 291
  tape, 35, 160
Deep drawability, 625 (*see also* Deep drawable protective film)
  of plastic films, 625
    dependence of on processibility, 626

Deep drawable protective film, 28, 130, 608, 622
  classification of, 623
  end use of, 623, 625
  as lubricant, 624
  functioning of, 625
  of PE, 432
Deep freeze
  applications, 383
  properties, 318
Degradation (*see also* Aging)
  by chain scission, 179
  of amorphous polypropylene, 201
  of SIS, 179 (*see also* Styrene block polymers)
Dehumidification (*see* Confectioning properties)
Depolymerization, 180 (*see also* Degradation)
Detackification, 344
Detackifier, formulation with, 346 (*see also* Removability)
Diaper tape, 97, 127, 128, 278, 427, 585 (*see also* Adhesive)
  for closure, 591
Diblock, 190 (*see also* Bisequenced polymers)
  for low temperaure use, 369,
  influence of,
    on peel, 189
    on tack, 190
    on viscosity, 190
Die-cutting, 524, 552 (*see also* Converting properties)
  clearance for, 553
  dependence of,
    carrier, 553
    release liner, 553
  machines for,
    flat bed, 553
    rotary, 553, 556
  of PE, 553
  of PET, 553
  tolerances for, 553
Die gap, 409 (*see also* Manufacture of film carrier)

Diene copolymers, 186 (*see also* Styrene block polymers)
Diluents, reactive, 180 (*see also* Ultraviolet light induced photopolymerization)
Dimensional stability, 129
  of paper, 396, 397
Dimethylsiloxane, 199, 237 (*see also* Silicones)
Dipole-dipole interaction, 193
Dispensing, 559 (*see* Labeling also)
Dissipation energy, 285
Down gauging, influence of on peel resistance, 126
Drying, 494
Differential Scanning Calorimetry,
  for carbamate based release agent, 390
  for olefinic raw materials, 417
Dwell time, 264, 272, 277, 295
  dynamic, 274
  static, 274
Dynamic mechanical analysis, 96, 263
  for evaluation of the adhesive properties, 308
Dynamic modulus, 95, 98

Edge curl, 545 (*see also* Printing properties)
Einstein equation, 121
Elasticity, 86
Elastic modulus, 79, 133
  dependence of on fillers, 135
  parameters of, 81
  values of, 134, 135
Elastomer, 79, 89 (*see also* Rubber)
  segregated, mechanical properties of, 196 (*see also* Thermoplastic elastomers)
  synthetic, 185
Electrical characteristics of PSPs, 137
Electrical conductivity
  of adhesive, 368
  of block copolymers, 194
  of PSPs, 139

Electrical double layer, 452 (*see also* Corona treatment)
Electrical insulation tapes, 592, 593 (*see also* Wire wound tape)
Electrical tapes, 45, 46, 428, 429, 430
Electrically conductive films, 140
  filler for, 161
  of carbon black, 444
Electrically conductive tapes, 222, 593
Electrolyte in polymer, 368 (*see also* Medical tape)
Electromagnetic-radio-frequency interference, 139
Electron beam induced polymerization, 24, 167, 179
  of acrylics, 343
  advantages of, 183
  of block copolymers, 179, 204
  dependence of,
    on molecular weight, 178
  formulation for, 342
Electrostatic
  control, 496
  discharge protection, 139
Elmendorf tear strength, 157, 423 (*see also* Mechanical properties of PSPs)
Elongation, 129 (*see also* Mechanical properties of PSPs)
  of carrier, 126 (*see also* Carrier)
  in printing, 547
Embossing, 555, 556 (*see also* Carrier)
Equipment
  for adhesive formulation, 371 (*see also* Manufacturing technology of adhesive)
  for HMPSA processing, 371 (*see also* Hot melt)
Ethylene-acrylic acid copolymers, 83 (*see also* Self adhesive film)
Ethylene copolymers, 154, 167
  acid, 154, 385 (*see also* Ethylene-acrylic acid copolymers)
  with acrylates, 154, 168

[Ethylene copolymers]
  butyl, 154
  methyl, 154
  for carrier, 32 (*see also* Self
    adhesive film)
  with maleic anhydride, 154
  with octene, 162 (*see also* Linear
    low density polyethylene)
  with polar vinyl monomers, 154,
    167 (*see also* Ethylene-vinyl
    acetate copolymers)
  terpolymers, 154
Ethylene-propylene
  copolymers, 13
  rubber, 159
    crosslinked, 159
    grafting of, 159
Ethylene-propylene-diene
  multipolymer, 83
Ethylene-vinyl acetate copolymers, 4,
    13, 15, 83, 130, 151, 167
  adhesives of, 40
    for tapes, 210
  adhesive carrier of, 168 (*see also*
    Self adhesive film)
  blocking of, 168, 272 (*see also*
    Blocking)
  branching of, 163
  coating of, 56, 59
  compatibility of, (*see also*
    Compatibility)
  with SBR, 211
  crystallization rate of, 185
  films of, 85
  HMPSAs of, 168 (*see also*
    Thermoplastic elastomers)
    formulation of, 169
  minimum film forming temperature
    of, 211
  non adhesive carrier of, 168
  plasticizer, 163
  processing temperature of, 154
  segmented, 163, 188 (*see also*
    Thermoplastic elastomers)
    adhesion of on soft PVC, 211
    adhesive characteristics of, 211

[Ethylene-vinyl acetate copolymers]
  for HMPSA, 211
  vinyl acetate content of, 168
  T_g of, 211
Extenders (*see* Fillers)
Extensibility,
  of the adhesive, 17
  of the carrier, 31, 135
Extrusion, 25 (*see also* Manufacture
  of film carrier)
  for adhesive coating, 369, 379

Face stock (*see also* Carrier)
  for labels, 38
  laminated, 25, 30 (*see also*
    Laminates)
Failure place, 288, 344 (*see also*
  Removability)
Fatty acid metal soaps, 15
FDA approval, 372, 419, 524, 573
Fiberlike carrier materials (*see also*
  Carrier)
  for labels, 425
  for tapes, 429
Fiberglass scrim carrier, 31, 430
Filler, 13, 14, 34, 122, 124, 135, 173,
    197, 206, 217, 342, 376
  carbon black as, 219 (*see also*
    Electrically conductive films)
  for carrierless tapes, 221
  characteristics of, influence of on
    mechanical properties, 120,
    124, 218
  concentration, influence of on
    mechanical properties, 120,
    121
  detackifying, 222 (*see also*
    Detackification)
  of electrolyte, 223 (*see also*
    Medical tape)
  expandable (*see also* Microspheres)
    influence on adhesive properties,
      485
    influence on peel build up, 486
  for EVAc hot melts, 211

Index 655

[Filler]
  fire retardant, 163, 218, 223 (*see also* Carrier)
  for electrical conductivity, 140, 218 (*see also* Electrical conductivity)
  influence of, 218
    on crosslinking, 223 (*see also* Crosslinking)
    on drying, 223 (*see also* Drying)
    on relaxation spectra, 321 (*see also* Relaxation)
  for medical tapes, 21, 219 (*see also* Medical tape)
  metallic, 219, 222 (*see also* Electrical conductivity)
  of microbubbles, 221 (*see also* Microspheres)
  monomer as, 175, 180
    concentration of, 178
  natural, 218
  of polymer particles, 222, 380
  for paper, 33 (*see also* Paper)
  for removability, 221 (*see also* Removability)
  theory, 100
  as viscosity regulating agents, 220
  water, 222
Film carrier, 2, 11, 402, 414 (*see also* Carrier)
  acrylic, 14
  cast, 15
  coextruded, tensile strenght of, 126 (*see also* Manufacture of film carrier)
  factors of development of, 403
  flat, advantages of, 412
  heterogeneous, 165
  homogeneous, 160
  general requirements for, 402
  for labels,
    formulation of, 421,
    main grades of, 413, 421
    manufacture of, 421 (*see also* Manufacture of film carrier)

[Film carrier]
  of polyolefins, 422
  for medical tape, 12 (*see also* Medical tape)
  non adhesive, 150
  oriented, 402 (*see also* Postextrusion processing of films)
  postextrusion modification of, 437 (*see also* Postextrusion processing of films)
  postprocessing procedures of, 407
  for protective films, 432
    formulation of, 432
    for outdoor use, 432
  reinforced, 124
  self adhesive, 14 (*see also* Self adhesive film)
  self sustaining, 14 (*see also* Transfer tape)
  special requirements for, 403, 404
    for labels, 404
  for tapes,
    formulation of, 426
    manufacture of, 426 (*see also* Manufacture of film carrier)
    requirements for, 405
Film release liner, 435 (*see also* Release liner)
  embossed, 436
  formulation of, 435 (*see also* Manufacture of film carrier)
  of HDPE, 435 (*see also* High density polyethylene)
  manufacture of, 435 (*see also* Manufacture of film carrier)
  of PET, 436
    advantages of, 437
  of polyolefins, 436
  of PP, 436
  of polystyrene, 436
  of PVC, 436
Flame treatment, 441 (*see also* Manufacture of film carrier)
Flap test, 309

Flat-die procedure, 410 (*see also* Manufacture of film carrier)
  postprocessing for, 411 (*see also* Postextrusion processing of films)
Flat film carrier, of polyolefin, 434 (*see also* Manufacture of film carrier)
Flexural crack resistance, 131
Flooring adhesive, 210
  T_g of, 211
Flory-Huggins theory, 346
Flow properties, improving of, 203
Fluorine treatment, 441 (*see also* Manufacture of film carrier)
Foam
  acrylic, 13
  elastic, 122
  -like tape, 16, 23, 379, 586, 596 (*see also* Tape)
  tape, 594, 596, 597
Foam generation, 361, 374
Folding, 555 (*see also* Converting properties)
Forms, 26, 27, 66, 474, 475, 477, 627
  printing machine for, 548
Formulation
  for adhesive properties, 344 (*see also* Adhesive characteristics of PSPs)
  dependence of,
    on carrier, 364 (*see also* Carrier)
    on coating technology, 358 (*see also* Manufacture of PSPs by adhesive coating)
    on crosslinking, 361 (*see also* Crosslinking)
    on end use, 364 (*see also* Pressure-sensitive products)
    on printing method, 364
    on product build up, 364
    on product class, 362
    on raw material, 354
  of HMPSA, 360 (*see also* Hot melt)
  of labels, 362 (*see also* Label)

[Formulation]
  of non-silicone release coating, 388 (*see also* Release liner)
    with acrylics, 389
    with fluorine compounds, 392
    with maleic acid copolymers, 392
    with polyamides, 390, 394
    with polyvinyl carbamate (*see also* Polyvinyl carbamate)
    with vinyl acetate polymers, 389, 394
    with vinyl ether copolymers, 390
  problems of, 371
  of protective films, 363 (*see also* Protective films)
  for removability, 350 (*see also* Manufacture of PSPs by adhesive coating)
  scope of, 343, 344
  for shear resistance, 350 (*see also* Shear resistance)
  for solids, 180 (*see also* Adhesive)
  of solvent based PSA, 360 (*see also* Solvent based systems)
  for tape, 350, 374 (*see also* Tape)
  of water based PSA, 360 (*see also* Water based formulation)
Freezer
  label, 97, 369, 577
  tape, 160, 186, 592
    with silicone, 198
Freezing height, 118, 155 (*see also* Manufacture of film carrier)

Gel effect, 206 (*see also* Solution acrylic, concentrated)
Gel content, 173
  of CSBR, 312
  influence of,
    on shear resistance, 187
    on tensile strength, 187
  of SBR, 318
    influence on processibility, 312
    influence on tack, 312, 318

Glass
  microbubbles (*see* Microspheres)
  powder, 352
Glass transition temperature, 78, 94
  dependence of,
    on comonomers, 91
    on tackifiers, 355
  parameters of, 263
  role of, 92, 171
  values of, 92, 96, 122, 177, 188, 190, 201, 381
Glue stick, 184, 291
Graft
  polymerization, 24, 122, 154, 167, 443 (*see also* Polymer-analogous reactions)
  for segregated structures, 194
  of styrene-butadiene, 187
  polymers,
    of LLDPE, 457
    of PE, 435, 460
    of vinyl esters, 367
Grit, in filler, 222 (*see also* Filler)
Guth-Gold equation, 129, 190
Guth-Smallwood equation, 88, 121, 135

Halpin-Tsai equation, 121
Health care tapes, 45 (*see also* Medical tapes)
Heat activated adhesive film, 59
Heat laminating, 85
High density polyethylene, 126, 415 (*see also* Polyethylene)
  advantages of, 414
  bimodal grades of, 155, 162
    for labels, 422
  blending of, 155
  blown film, 155, 409 (*see also* Film carrier)
  manufacture parameters of, 415
  carrier of, 405, 414
    for application tape, 485
    for labels, 405

[High density polyethylene]
  for tapes, 427, 430
  cross laminated, 422
  crosslinked, for labels, 422
  for flat film, 155
  oriented, for labels, 422
High solids, 478
Holding power, 193
  of crosslinked acrylics, 209
Holography, 535 (*see also* Printing)
Hot laminating film, 2, 32, 64, 85
  adhesion of,
    characteristics of, 452
    dependence of on corona treatment, 452 (*see also* Corona treatment)
    dependence of on laminating temperature, 619
  application conditions of, 32, 609, 613, 619 (*see also* Protective films for coil coating, 612, 614)
  bonding mechanismus of, 451
  manufacture characteristics of, 613 (*see also* Manufacture of film carrier)
  nonpolar, 451
    LDPE based, 451
  peel resistance of, 453, 608
    build up of, 453 (*see also* Peel)
    dependence of on adherend, 453
    dependence of on laminating conditions, 453
  suitability of, 456
Hot melt adhesive, 15, 93, 188, 358
  acrylic, 197, 270 (*see also* Adhesive)
    adhesive properties of, 197
    thermally reversible, 197
    viscosity of, 197
  coating technology of, 464, 468 (*see also* Manufacture of PSPs by adhesive coating)
    energy consumption of, 374

[Hot melt adhesive]
   in line with siliconizing, 469 (*see also* Manufacture of PSPs by adhesive coating)
   for labels, 469 (*see also* Label)
   crosslinking of, by radiation, 173, 339 (*see also* Styrene block copolymers)
   detackification of, 346
   of EVAc, 211 (*see also* Ethylene-vinyl acetate copolymers)
   formulation of, 211
   for labels, 39
   for low temperatures, 369
   formulation of, 309, 35 (*see also* Formulation)
   for tapes, 309 (*see also* Tape)
   without oil, 214 (*see also* Oil)
   oligomer based, viscosity of, 359
   for packaging tapes, 590 (*see also* Packaging tape)
   pattern coated, 417, 469 (*see also* Pattern coating)
   polypropylene based, 201 (*see also* Polypropylene)
   polyurethane acrylate, 198
   processing viscosity of, 214, 308 (*see also* Viscosity)
   reduction of, 214
   quality criteria for, 360
   repulpable, 366 (*see also* Solubilizers)
   SBS of, aging of, 192 (*see also* Styrene-butadiene)
   SIS of, 367 (*see also* Formulation)
   spray coated, 469 (*see also* Coating device)
   for tapes, 19, 31, 45, 47, 192
   thermoplastic components of, 367
   viscosity of, 468
   water dispersible, 367
   water soluble, 366, 367
Hot seal PSA, 155
Humidification (*see* Confectioning properties)

Humidity, influence of, on adhesive properties, 262, 293
Hydrogenated block copolymers, 188 (*see also* Styrene-ethylene-butylene-styrene block copolymers, 93, 122, 174, 190

Impact properties, 131 (*see also* Mechanical properties of PSPs)
   of EVAc, 131
   of LDPE, 131
   of LLDPE, 131
In-mold label, 43, 572, 575, 576, 583 (*see also* Pressure-sensitive products)
Instruction label, 580 (*see also* Label)
Insulation tape, 35, 52, 130, 272, 586, 592
   adhesive for, 371 (*see also* Formulation)
   anticorrosive, 461
   carrier for, 162, 426, 545 (*see also* Carrier)
   electric (*see also* Electrical insulation tapes)
   adhesive for, 140, 185
   resistivity of, 139, 375 (*see also* Electrical conductivity)
   formulation of, 170
   manufacture of, 480
   for pipe, 376, 585
   thermal, 32
Isocyanate, 197
   crosslinking agent, 199, 369, 378 (*see also* Crosslinking)

Krieger-Dougherty equation, 100

Label, 1, 2, 3, 36
   adhesive properties of, 294 (*see also* Adhesive for labels)
   aging of, 574 (*see also* Aging)

Index 659

[Label]
application of, 1, 17, 40, 570 (*see also* Pressure-sensitive products)
climate for, 574
conditions of, 570
domains of, 41, 577
machines for, 559 (*see also* Labeling)
methods for, 575 (*see also* Labeling)
automotive, 577
bar-coded, 576
blank, 540
blood bag, 574
build up of, 8, 38
copy, 42, 365, 403
classes of, 41, 576, 577
cutting of, 41 (*see also* Cutting)
end use of, 571 (*see also* Pressure-sensitive products)
dependence on product form, 573
dependence on product surface, 571
face stock of, 425 (*see also* Carrier)
general, 41
heat adhesion, 37 (*see also* Hot seal PSA)
lacquering of, 425, 541 (*see also* Printing)
coating device for, 540 (*see also* Coating device)
laser printable (*see* Printability)
mailing, 576
markets for, 571, 576
manufacture of, 449, 474 (*see also* Manufacture of PSPs by adhesive coating)
with hot melts, 474 (*see also* Hot melt)
multilayer, 38, 474 (*see also* Forms)
no label look, 397, 419, 425, 576
nonpaper, 39 (*see also* Carrier)
overprinting of (*see* Printing of labels)

[Label]
pharmaceutical, 39, 365 (*see also* Medical label)
postmodification of, 576 (*see also* Label printing)
principle of functioning, 40
removable, 185 (*see also* Removability)
repositionable, 41
repulpable, 44 (*see also* Recycling)
requirements for, 38, 40 (*see also* Pressure-sensitive products)
roll, 41
routing, 576
service temperature, 574 (*see also* Pressure-sensitive products)
sheet, 41
special, 41 (*see also* Pressure-sensitive products)
special characteristics of, 41
technological, 577
temperature stability of, 574 (*see also* Temperature resistant labels)
thermally printed, 365 (*see also* Label printing)
for thermometer, 13, 43, 577 (*see also* Temperature display labels)
water resistance of, 43 (*see also* Water resistant)
water solubility of, 43 (*see also* Warm water removable formulation)
wet adhesive, 37 (*see also* Bottle label)
Labeling, 559 (*see also* Converting properties)
characteristics of, 559
dependence of on stiffness, 421 (*see also* Stiffness)
high speed,
film for, 423
machines for, 575
non contact, 560
sleeve, 576

[Labeling]
  speed of, 559, 560, 575
  temperature of, 574 (*see also* Label)
  touch blow, 1, 27, 33, 560
Lacquering, 548 (*see also* Label)
Laminates, 24, 26, 27
  classification of, 35
  for labels, 425, 475 (*see also* Label)
  main characteristics of, 334
  roll, 575 (*see also* Label)
  sheet, 575 (*see also* Label)
Laminating, 334, 335, 496 (*see also* Manufacture of film carrier)
  cold, 575
  dry, 463
  for label manufacture, 475
  thermal, 627
Lap cling, 286, 287 (*see also* Adhesive characteristics of protective films)
  parameters of, 459
Latex foam, 21 (*see also* Tape)
Latex impregnated paper carrier (*see also* Carrier)
  for tapes, 20
Lay flat, 545 (*see also* Converting properties)
  dependence of on stiffness, 422 (*see also* Stiffness)
  of labels, 117 (*see also* Label)
  of paper, 396, 401
Legging adhesive, 210
  removable, 210
Linear low density polyethylene, 82, 161, 415 (*see also* Polyethylene)
  for blown film, 409 (*see also* Manufacture of film carrier)
  branching of, 162 (*see also* Branching)
  carrier of for tapes, 427 (*see also* Tape)
  as cling modifying component, 456 (*see also* Self adhesive film)

[Linear low density polyethylene]
  comparison of to LDPE, 415 (*see also* Low density polyethylene)
  cuttability of, 126 (*see also* Cuttability)
  impact resistance of, 157 (*see also* Impact properties)
  mechanical properties of, 124 (*see also* Film carrier)
  molecular weight distribution of, 90, 162
  optical properties of, 157, 161
  processing of, 161, 162 (*see also* Manufacture of film carrier)
  rheology of, 89
  self adhesivity of, 456
  shrinkage of, 161 (*see also* Shrinkage)
Loss modulus, 77, 90 (*see also* Dynamic mechanical analysis)
  influence of on adhesive properties, 310
  values of, for adhesive, 95
Low density polyethylene, 161, 451 (*see also* Polyethylene)
  for blown film, 409 (*see also* Manufacture of film carrier)
  carrier of, 413
  mechanical properties of, 124 (Mechanical properties of PSPs)
  rheology of, 90
Ludwick-Holloman equation, 131

Macrogel (*see* Styrene-butadiene rubber)
Macromer, 159, 202, 339, 340, 343, 465 (*see also* Adhesive)
Macromolecular compounds, non-Newtonian flow of, 257
Manufacture of abhesive components, 386 (*see also* Relese agent)
Manufacture of the carrier for PSPs, 394 (*see also* Manufacture of film carrier)

Manufacture of film carrier, 406, 407
  by calendering, 407
  by casting, 406
  by extrusion, 406, 408 (*see also* Blow procedure)
  for labels, 421 (*see also* Label)
  of polyolefins, 162 (*see also* Carrier)
  for protective films, 432 (*see also* Protective film)
Manufacture of PSPs by adhesive coating, 462
  advantages of, 463
  coating methods for, 466
  coating device for, 463 (*see also* Coating device)
  in line, 471 (*see also* Hot melt)
    advantages of, 472
  main equipment for, 490 (*see also* Equipment)
  with pattern, 493 (*see also* Pattern Coating)
  screen printing for, 463 (*see also* Printing)
    advantages of, 463
  technology of, 463
    as a function of chemical composition, 464 (*see also* Formulation)
    as a function of physical state of the adhesive, 465 (*see also* Hot melt)
    as a function of the product class, 471 (*see also* Pressure-sensitive products)
    as a function of product build up, 472 (*see also* Pressure-sensitive products)
Manufacturing technology of adhesive, 371
  for forms, 382 (*see also* Forms)
  for labels, 372 (*see also* Adhesive)
  for protective films, 381 (*see also* Adhesive)
  for tapes, 374 (*see also* Adhesive)

Masking tapes, 2, 19, 25, 28, 45, 46, 59, 589
  adhesive for, 20, 185, 364
  automotive, 377
  carrier for, 20, 30, 428, 431 (*see also* Carrier)
  manufacture of, 485
Mastication, 47, 104, 172, 180, 184, 295, 311, 323, 352
  for tapes, 352, 363, 368
Mechanical properties of PSPs, 116, 123, 124, 125
  dependence of,
    on chemical composition, 123, 153 (*see also* Formulation)
    on crystallinity, 154 (*see also* Crystallinity)
    on crosslinking, 120 (*see also* Crosslinking)
    on filling, 120 (*see also* Filler)
    on molecular order, 119, 123 (*see also* Ordered systems)
    on molecular orientation, 118, 124 (*see also* Postextrusion processing of films)
    on network modality, 120 (*see also* Network)
    on processing, 123 (*see also* Manufacture of film carrier)
    on rheology, 117 (*see also* Dynamic mechanical analysis)
    on sequence distribution, 120 (*see also* Sequence distribution)
  influence of,
    on adhesive performance, 122 (*see also* Adhesive characteristics of PSPs)
    on other performance characteristics, 136 (*see also* Converting properties)
  regulating of, 136
    by formulation, 136 (*see also* Formulation)
    by geometry, 136 (*see also* Manufacture of film carrier)

[Mechanical properties of PSPs]
   by manufacture, 136 (*see also* Manufacture of film carrier)
   structural background of, 118 (*see also* Ordered systems)
Mechanical resistance, of protective films, 62 (*see also* Mechanical properties of PSPs)
Mechanochemical destruction, 172, 352 (*see also* Mastication)
Medical plaster, 2, 365
Medical labels, 42, 580
Medical tape, 2, 4, 54, 127, 205, 295, 365, 581, 585, 599
   adhesive for, 21, 31, 34, 180, 185, 186, 342, 364, 368, 369, 377, 378
   carrier for, 31, 54, 95, 378, 425, 429, 431 (*see also* Plastic carrier)
   coating of, 21, 369
   for dosage, 54
   manufacture of, 370, 483
   raw materials for, 369, 370
   release liner for, 447
   requirements for, 368
   solvents for, 231
Melt index, 90, 133, 157 (*see also* Manufacture of film carrier)
Melt memory, 155 (*see also* Manufacture of film carrier)
Melting point depression, 156, 346 (*see also* Self adhesive film)
Metallocene resins, 162 (*see also* Polyolefin)
Microgel (*see* Styrene-butadiene rubber),
Microperforating (*see also* Perforating),
Microphase separation, 87 (*see also* Thermoplastic elastomers)
Microspheres, 132, 354, 380 (*see also* Filler)
   expanded, 23, 320, 354
   of glass, 13, 14, 47, 52, 100, 269, 379

[Microspheres]
   for paper, 132
   polymeric, 23, 34, 222, 319, 352
Midblock, saturated, 190 (*see also* Styrene-ethylene-butylene-styrene block copolymers)
   molecular weight of, 190
Migration, 110, 346
   resistance to, improving of, 93, 216
   characteristics of carrier,
      dependence of on crystallinity, 321
      dependence of on manufacturing method, 321
      dependence of on the resin, 360 (*see also* Resin)
      influence of on shrinkage, 547 (*see also* Shrinkage)
Mirror tape, 66, 432, 601, 626
Modulus, plateau of, 194
   dependence of on molecular weight, 203
Moisture permeability (*see also* Medical tape)
   for adhesive, 20
   for carrier, 20, 425 (*see also* Carrier)
Molecular association, 173, 193 (*see also* Thermoplastic elastomers)
Molecular flexibility, 171 (*see also* Ordered systems)
Molecular weight (*see also* Crosslinking)
   critical, 99, 203
   effective, 178, 292 (*see also* Crosslinking)
   of elastomers (*see also* Elastomers)
      for adhesives, 170
   of endblock, 188 (*see also* Thermoplastic elastomers)
   influence of,
      on debonding, 171 (*see also* Removability)
      on mechanical characteristics, 155 (*see also* Mechanical properties of PSPs)

[Molecular weight]
  on processibility, 155
  number average, 203
  postregulation of, 172, 203 (see also Polymacromerization)
  role of, 170 (see also Mechanical properties of PSPs)
  weight average, 203
Molecular weight distribution, 90
  influence of on mechanical characteristics, 154 (see also Mechanical properties of PSPs)
  of milled rubber, 184 (see also Mastication)
Monomer
  functional, 155, 270 (see also Crosslinking)
  functionality of, 179
  hard, 352, 465
    influence of on cohesion, 342
  low tack, 291
  for plastic carrier, 154 (see also Self adhesive film)
  reinforcing, 180, 299
Monoweb, 9 (see also Pressure-sensitive product)
  coated, 17, 18 (see also Manufacture of PSPs by adhesive coating)
  comparison of, 24
  labels, 16, 18 (see also Label)
  protective films, 58, 63 (see also Protective films)
  self supporting, 13 (see also Transfer tape)
  tapes, 17 (see also Tape)
  uncoated, 11 (see also Self adhesive film)
  comparison of, 23
Mooney-Rivlin equation, 129, 190
Mooney viscosity, 184
Mounting tape, 13, 28, 45, 46, 295, 297, 585, 594, 595
  adhesive for, 205, 222, 344
  build up of, 13
  carrier for, 427, 430, 431

[Mounting tape]
  double faced, 25, 429, 481
  manufacture of, 481
    by UV light induced polymerization, 482
  repositionable, 595
Multiphase structure, 154, 159, 173 (see also Ordered system)
  tackification of, 195 (see also Tackification)
Multiweb constructions, 24, 25, 27 (see also Laminates)

Nameplates, 405, 579 (see also Label)
  printing of, 580 (see also Printing)
Natural elastomer, 183 (see also Elastomer)
Natural rubber, 183, 369 (see also Natural elastomer),
  aging of, 301 (see also Aging)
  blends of, 129 (see also Formulation)
  crosslinking of, 172 (see also Crosslinking)
  degradation of, 94, 122, 172, 194 (see also Mastication)
  grafting of, 194 (see also Grafting)
  tackification of, 151 (see also Tackification)
Natural rubber latex, 22
  compatibility of, 184, 371
  influence of, on formulation, 184
Neoprene, carboxylated, 186
Network (see also Ordered System)
  covalently bonded, 175 (see also Crosslinking)
  physically bonded, 175, 192 (see also Molecular association)
  modulus of, dependence of on crosslinking, 178 (see also Crosslinking)
  thermally reversible, 194 (see also Thermoplastic elastomers)
Non flammability of halogenated diene polymers, 186 (see also Tape)

Office tapes, 35 (*see also* Tape)
Oils, 93, 263 (*see also* Thermoplastic elastomers)
   formulation with, 309, 310, 355
   influence of,
      on cohesion, 192
      on viscosity, 359
   plasticizer, 192, 214
      disadvantages of, 214
      naphtenic, 214, 355
      paraffinic, 350
Oligomer, 181, 182, 195 (*see also* Prepolymer)
   crosslinkable, 317, 358, 362 (*see also* Crosslinking)
   functionalized, 178
   influence of on peel build up, 318 (*see also* Formulation)
Oozing, 17 (*see also* Converting properties)
Operating tapes, 368 (*see also* Medical tape)
Optical characteristics of PSPs, 140 (*see also* Crystallinity)
Ordered systems, 87, 172, 173, 194 (*see also* Network)
   destruction of, 89 (*see also* Thermoplastic elastomers)
Orienting, influence of on the mechanical properties, 158 (*see also* Postextrusion processing of films)
Overcure, 179 (*see also* Crosslinking)
Overlamination protective film, 66, 626
Overprinting methods for labels, 536 (*see also* Label printing)

Packaging tape, 19, 45, 50, 588, 589
   application equipment for, 589
   adhesive for, 19, 96, 309, 377 (*see also* Adhesive)
      acrylic, 465 (*see also* Acrylic adhesive)
   carrier for, 377 (*see also* Carrier)

[Packaging tape]
   of polyolefin, 162, 590 (*see also* Film carrier)
   of PVC, 429, 590
   manufacture of, 478, 479 (*see also* Manufacture of PSPs by adhesive coating)
   printing of, 19 (*see also* Printing)
   tensile strength of, 125 (*see also* Mechanical properties of PSPs)
Paper carrier, 31, 436
   extensible, 31
   general requirements for, 396
   grades of, 398
   humidity content of, 262, 397, 544, 545, 547
   impregnated, 20, 31, 431
   for labels, 397 (*see also* Label)
      choice of, 399
      for direct thermal printing, 399
      requirements for, 398
      special, 398
   manufacture of, 395
   mechanical requirements for, 397
   for medical labels, 397 (*see also* Medical label)
   for mounting labels, 397
   overdrying of, 547
   pleated, 31
   PE coated, 446
   for protective films, 399 (*see also* Protective Film)
   for release liner, 397, 400, 436, 446 (*see also* Release liner)
   special requirements for, 397
   for tapes, 20, 31, 47, 431 (*see also* Tape)
   for splicing, 399
   thermally printable, 33 (*see also* Label printing)
Paper tear, influence of on removability, 291 (*see also* Removability)
Pattern coating, 22, 31, 184
   of hotmelts, 31

Peel resistance, 28, 256 (*see also* Adhesive characteristics of PSPs)
  build up of, 272, 273, 275
    dependence of on adherend surface quality, 274, 614 (*see also* Adherend surface)
    dependence of on coating weight, 620 (*see also* Coating weight)
    for protective films, 273, 614 (*see also* Adhesive characteristics of PSPs)
    for self adhesive films, 274, 608 (*see also* Self adhesive films)
    for separation films, 273 (*see also* Separating film)
  cleavage, 127, 135, 277, 278
    breakaway, 277
    continuing, 277, 279
    dependence of on coating weight, 615
    dependence of on standard peel, 615
  cling, 285, 286, 459 (*see also* Self adhesive film)
    parameters of, 459
  dependence of,
    on application conditions, 272, 275, 276 (*see also* Pressure-sensitive products)
    on break nature, 278 (*see also* Removability)
    on carrier deformation, 112, 113, 282, 281, 283, 284
    on carrier surface, 300 (*see also* Adhesion)
    on coating technology, 302 (*see also* Manufacture of PSPs by adhesive coating)
    on formulation, 278 (*see also* Formulation)
    on peel angle, 278 (*see also* Adhesive characteristics of PSPs)
    on stress rate, 277, 278
    on test methods, 262

[Peel resistance]
  gradient of, 272
  of heat seal copolymers, 275
  of protective films (*see also* Adhesive characteristics of protective films)
    dependence on aging, 301
    dependence on release agent, 489 (*see also* Release agent)
    parameters of, 301
    values of, 487
  regulation parameters of, 316, 352
  in slip-stick domain, 279, 280
  on steel, dependence of on peel on aluminum, 615
  tested as T-peel, 286
  values of, 97, 127, 197, 259, 261
    for protective films, 300
    for tapes, 586
  zip, 280
Perforating, 548, 555, 578 (*see also* Converting properties)
Pharmaceutical
  label, 580 (*see also* Medical label)
  sealing tapes, 592 (*see also* Tape)
Phase separation, 178, 192 (*see also* Multiphase structure)
  in radial block copolymers, 192, 311 (*see also* Radial block copolymer)
Photoinitiator of UV light induced polymerization, 182, 197, 206, 207, 208, 342, 356 (*see also* Ultraviolet light induced photopolymerization)
  benzophenone-containing system, 207
  built in, 182, 206, 208, 270, 317, 343, 356
Photoinitiated side reactions, 317 (*see also* Ultraviolet light induced photopolymerization)
Photopolymerization, 182, 206, 341 (*see also* Ultraviolet light induced photopolymerization)
  formulating freedom of, 317

[Photopolymerization]
  initiating mechanism of, 208
  of acrylics, 269 (*see also* Acrylic adhesive)
  on line, 206 (*see also* Polymacromerization)
  quantum yield of, 208
  for tapes, 207 (*see also* Tape)
Physical treatment, 450, 451 (*see also* Corona treatment)
Pigments, (*see* also Fillers)
  for surface coating of paper, 396, 401
Plasma treatment, 442 (*see also* Physical treatment)
  advantages of, 442
  effects of, 443
  inert, 443
  non-polymerization, 442
  oxidative, 442
Plastic anisotropy ratio, 129 (*see also* Mechanical properties of PSPs)
Plastic foam (*see also* Foam)
  carrier, for tapes, 431 (*see also* Foam tape)
  manufacture of, 421
Plasticing of PVC, 3, 13, 215 (*see also* Film carrier)
  external, 151
  internal, 151
Plasticizer, 93, 215, 216, 350 (*see also* Removability)
  alkali-dispersible, 367 (*see also* Water soluble formulation)
  choice of, 215
  for EVAc hotmelts, 211 (*see also* Hot melt)
  macromolecular, for PVC, 215
  migration of, 292
  for printing inks, 215
  for removability, 215
  tackifying effect of, 215
  for tapes, 216
    medical, 216
    splicing, 216
  water dispersibility of, dependence of on molecular weight, 216

[Plasticizer]
  water soluble, 216, 367 (*see also* Water soluble formulation)
Plastomers, 79, 89, 151
Plateau modulus, 94, 348 (*see also* Dynamic mechanical analysis)
  dependence of, on the molecular weight, 98
  values of, 94
  parameters of, 94
Plotter film, 66, 596, 627
Polarity of surface, 137 (*see also* Corona treatment)
  improving of, 137
Polyacrylate rubber, 376 (*see also* Acrylic adhesive)
Polyalkylenimine, 366 (*see also* Formulation)
Polyamide, 13
  carrier of, 164, 419
  thermoplastic elastomers of, 164
Polybutylene
  blending of, 458
    equipment for, 459
  compatibility of, 346
  derivatives of, for tackifier, 458
  level of,
    dependence of on film manufacture procedure, 459 (*see also* Manufacture of film carrier)
  for PE tackification, 458 (*see also* Self adhesive film)
  for medical tapes, 20 (*see also* Medical tape).
Polycarbonate carrier, 420 (*see also* Film carrier)
Polychloroprene rubber, 186 (*see also* Adhesive)
Polydimethylsiloxane, 194 (*see also* Release agent)
Polyester, 13, 31, 39, 126, 202
  carrier, 164, 402, 406 (*see also* Poly(ethyleneterephtalate) carrier)
  oligomers of, 202 (*see also* Adhesive)

Index 667

[Polyester]
    acrylated, 183 (*see also* Prepolymer)
    for HMPSA, 183 (*see also* Hot melt)
    radiation curable, 202
    T_g of, 203
    viscosity of, 202
Polyether, 184
Polyethylene
    bimodal, 131
    carrier of, 31, 151, 413 (*see also* Film carrier)
    manufacturing methods of, 161 (*see also* Manufacture of film carrier)
    for medical tapes, 20 (*see also* Medical tape)
    modulus of, 124 (*see also* Mechanical properties of PSPs)
    for protective films, 432 (*see also* Protective films)
    for tapes, 375 (*see also* Tape)
    compounding of with PP, 416
    foam of, 162, 431 (*see also* Foam tape)
    grades of, 161
    release liner, 20, 161 (*see also* Release liner)
    for tapes, 20
Poly(ethyleneterephtalate) carrier, 418, 429 (*see also* Printing)
    grades of, 418
    end use of, 419
    for labels, 424, 437 (*see also* Label)
    laquered, 424
    topcoated, 424
    writable, 424
    oriented, 419 (*see also* Postextrusion processing of films)
Poly(ethylene-vinyl acetate) carrier, 25, 428 (*see also* Ethylene-vinyl acetate copolymers)
    adhesive, 434 (*see also* Self adhesive film)

[Poly(ethylene-vinyl acetate) carrier]
    flame retardant properties of, 420
    non-adhesive, 420 (*see also* Film carrier)
Polyethylmethacrylate carrier, 31, 165
Polyethyloxazoline, 366 (*see also* Solubilizer)
Polyiimides, carrier of, 165, 426
Polyisobutylene
    base elastomer of, 170, 185 (*see also* Formulation)
    tackifier, 170, 350, 373 (*see also* Self adhesive film)
Polymer
    living, 193 (*see also* Thermoplastic elatomers)
    macrostructure of, 203 (*see also* Ordered system)
    microstructure of, 203
    photoinitiated degradation of, 224 (*see also* Aging)
    with water, for medical adhesives, 368 (*see also* Medical tape)
Polyisobutylene, 120
    tackifier, 197 (*see also* Self adhesive film)
    level of, 157
Polyisoprene (*see also* Elastomer)
    cis, 116
    low molecular weight, 93
Polymacromerization, 173, 179, 182, 339, 340 (*see also* Adhesive)
    water based, 340
Polymeric oxazoline derivatives, 368 (*see also* Polyethyloxazoline)
Polymer-analogous reactions, 24, 154
Polyolefin, 15
    carrier, 161
    foams of for tapes, 428 (*see also* Foam tape)
    for labels, 422 (*see also* Label)
    non-polar, 451 (*see also* Self adhesive film)
    coextruded, 451 (*see also* Coextrusion)
    polar, 451 (*see also* Ethylene copolymers)

[Polyolefin]
  for tapes, 426 (see also Film
    carrier)
Polypropylene (see also Polypropylene
    film carrier)
  amorphous, 200
    crosslinking of, 201
    synthesis of, 201
  atactic,
    adhesive of, 166, 200 (see also
      Semi pressure-sensitive
      adhesive)
    molecular weight of, 200, 201
    T_g of, 201
  calendered, 163 (see also
    Polypropylene film carrier)
  copolymers of, 163, 200, 201 (see
    also Ethylene-propylene)
Polypropylene film carrier, 39, 162,
  416 (see also Film carrier)
  advantages of, 416
  biaxially oriented, 31, 32, 162, 423
    (see also Postextrusion
    processing of films)
  calendered, 407
  comparison of,
    to PE, 416
    to PVC, 416, 423
  coextruded, 423, 424, 427
  embossed, 128
  end use of, 423 (see also
    Pressure-sensitive products)
  flat, 407
  for labels, 163 (see also Label)
    requirements for, 423
  lacquered, 424 (see also Converting
    properties)
  manufacturing of, 85, 166 (see also
    Manufacturing of film carrier)
  mechanical properties of, 423 (see
    also Mechanical properties of
    PSPs)
  metallized, 423
  oriented, 402, 416, 417, 423
    thermofixed, 417
  primer for, 163

[Polypropylene film carrier]
  printing of, 436
  surface tension of, 424
  tackification of, 24
  for tapes, 163, 427 (see also
    Packaging tape)
  warm laminated, 416 (see also
    Manufacture of film carrier)
Polypropylene film liner, 32 (see also
  Release liner)
  release for, 163
Polysegregation, 173 (see also
  Multiphase structure)
Polysiloxane (see also Release agent)
  crosslinked, 85, 199
  adhesive additive, 21
Polystyrene
  blends of, 424
  carrier of, 419 (see also Film
    carrier)
  for labels, 424 (see also Label)
  non-oriented, 419
  oriented, 419
    advantages of, 419
  UV curable, 183
Polyurethane, 120, 131, 199
  adhesives of, 199, 368
    medical, 342 (see also Medical
      tape)
  carrier film of, 420, 426
  foam carrier of, 23, 24, 26, 164
  foam of, 431 (see also Foam tape)
  gel, 369
  primer of, 25 (see also Primer)
  thermoplastic elastomers of, 200
    (see also Thermoplastic
    elastomers)
    hard segment of, 200
    soft segment of, 200
Polyvinylacetate, 14, 124
Polyvinylalcohol
  release layer of, 185
  for water removability, 210
Polyvinylbenzophenone, 317 (see also
  Photoinitiator of UV light
  induced polymerization)

Index

Polyvinylcarbamate, 234, 388, 389, 489, 589 (*see also* Release agent)
   handling of, 394
Polyvinylchloride, 59, 417
   antiblocking agent for, 418
   blends of, 131
   emulsion polymerized, 18, 417
   carrier, 25, 39, 160, 365, 402, 405 (*see also* Film carrier)
      advantages of, 418
      calendered, 407, 422
      cast, 408, 418
      embossed, 428
      hard, 417
      for labels, 422 (*see also* Label)
      oriented, 402, 418, 428
      primer for, 376, 429 (*see also* Primer)
      printability of, 160, 429, 544 (*see also* Printability)
      for protective films, 433 (*see also* Protective films)
      shrinkage of, 544 (*see also* Shrinkage)
      soft, 402, 422 (*see also* Plasticing of PVC)
      substitution of, 429
      for tapes, 26, 376, 428 (*see also* Packaging tape)
      use of, 160 (*see also* Film carrier)
   foam of, 160 (*see also* Foam tape)
   formulation of, 418
   opacizing agent for, 418 (*see also* Filler)
   plasticization of, 160, 428 (*see also* Plasticizing of PVC)
   plastisol of, 183 (*see also* Plasticing of PVC)
   suspension, 418
      end use of, 418 (*see also* Tape)
Polyvinylether, 212
   aging stability of, 212
   for curling resistant formulations, 212 (*see also* Curl)

[Polyvinylether]
   for masking tapes, 212, 377
   for medical tapes, 21, 212 (*see also* Medical tape)
   solubility of, 212 (*see also* Formulation)
   for splicing tapes, 212, 367 (*see also* Splicing tape)
Polyvinylpyrrolidone, adhesive formulations with, 366 (*see also* Solubilizer)
Postcoating synthesis, 339
Postcrosslinking, 173, 206, 352, 474 (*see also* Crosslinking)
Postextrusion processing of films, 118, 124, 158
Postlacquering (*see also* Converting properties)
Postpolymerization, 339 (*see also* Polymacromerization)
   raw materials for, 340
   thermal, 341
Postprinting (*see* Printing)
Pot life, 372, 374 (*see also* Crosslinking)
Prepolymers, 270, 358
   EB curable, 342 (*see also* Electron beam induced polymerization)
   functionalized, 183
   mixture of with monomers, 364
   molecular weight of, 342
   synthesis of, 206 (*see also* Formulation)
   UV curable, 342, 368 (*see also* Ultraviolet light induced photopolymerization)
   viscosity of, 342 (*see also* Adhesive)
Pressure of laminating for printing, 274 (*see also* Printing)
Pressure-sensitive adhesive, 1 (*see also* Adhesive)
   debonding of, 2
   definition of, 1, 2
   multilayered, 14 (*see also* Ultraviolet light induced photopolymerization)

[Pressure-sensitive adhesive]
  raw materials of, mechanical characteristics of, 128
Pressure-sensitive holograms, 534 (see also Printing of labels)
Pressure-sensitive imaging materials, 629
Pressure-sensitive products, 1, 7, 629
  adhesiveless, 1 (see also Self adhesive film)
    manufacture of, 448 (see also Manufature of film carrier)
    raw materials for, 448 (see also Plastomers)
  adhesive coated, 8 (see also Manufacture of PSPs by adhesive coating)
  application of,
    conditions of, 2, 337, 569
    fields of, 567
  build up of, 7, 449
  carrier for, 395, 445 (see also Film carrier)
    mechanical characteristics of, 432 (see also Mechanical properties of PSPs)
    paper based, 395 (see also Paper)
  carrierless, 270 (see also Transfer tape)
  chemical basis of, 149
  classes of, 35, 449
  components of, 28
    for coating, 33 (see also Manufacture of PSPs by adhesive coating)
    for printing, 35 (see also Converting properties)
    requirements for, 29
  deapplication conditions of, 2, 569
  end use of, 567
    properties for, 568
  macromer based, 340 (see also Polymacromerization)
  manufacture of 333, 448, 449
    auxiliary equipment for, 496 (see also Equipment)

[Pressure-sensitive products]
  by coating, 336, 462
  by extrusion, 450
  markets for, 567
  masking (see Masking tape), conditions for removability for, 602
  medical, 217 (see also Medical plaster, Medical label and Medical tape)
    end use requirements for, 581
  one component, 457 (see also Self adhesive film)
  performance characteristics of, 256
  physical basis of, 73
  rheology of, 73
  water resistant, 365
Pressure-sensitivity, 77, 82, 150
  macromolecular basis of, 150
  parameters of, 153
  by physical treatment, 1 (see also Physical treatment)
  of PSPs, comparison of, 114
Price-weight labels, 40, 373, 556, 578
  formulation of, 172
Primer, 35, 186, 231, 288, 384, 465
  chemical basis of, 232, 383
  for coated adhesive, 383
  coating of, 494
  coating weight of, 383 (see also Coating weight)
  coextrudable, 385
  for extruded layers, 385, 458
  formulation of, 383
  influence of on coating weight, 304
  for paper, 385
  for plastic films, 383, 385, 386
  as plasticizer barrier, 385
  for tapes, 20, 24, 25
    of PVC, 160, 231, 380, 383
    of EVAc, 210, 386
  concentration of, 178
  water based, 385, 386
Printability, 526 (see also Printing)
  computer, 576
  dependence of,

Index 671

[Printability]
    on carrier thickness, 529
    on product class, 529
    on surface wettability, 529
  of labels, 42, 526, 579
  laser, 576, 579
  of paper, 396
  of plastic films, 528, 529
    parameters of, 529
    non-polar, 529
  of PP,
    parameters of, 529
  requirements for, 527
  of tapes, 51 (*see also* Tape)
Printing, 526
  characteristics of carrier, 543 (*see also* Carrier)
  combined, 540 (*see also* Printing of label)
  digital, 42, 539
    offset, 539
    one shot color, 540
    postprinting, 42
  direct charge image, 537
  dot matrix, 539
  equipment for, 497
  electrography, 537
  electron beam, 538
  electrophotography, 537, 538
  electrostatic, 537
  flexography, 528, 529, 530, 531, 542
    narrow web, 397
  gravure, 426, 528, 530, 542
  impact, 536
  ink jet, 539
  laser, 537, 538
    line of, 538
  letterpress, 531
  litography, 528, 627
  magnetography, 539
  methods for PSPs, 530
    characteristics of, 527
    for PP, 529, 531
    for PET, 533, 534, 536, 538
    for polystyrene, 528

[Printing]
  for PVC, 528, 532, 536
  non-impact, 37, 396, 536
  offset, 397, 533
  of polypropylene tapes, 229, 542
  of protective films, 542
  quality of, 527
    parameters of, 527
  resolution for, 526
  screen, 528, 529, 532, 533, 546, 627
  of tapes, 542
  technological trends in, 529
  thermal, 33, 536
    direct, 536
    near edge, 537
    transfer, 536, 537
  toner based, 537
Printing inks, 35, 227
  for different printing methods, 528, 532
  reactive, 228
  special, 227
    photochromic, 228
  ultraviolet crosslinkable, 228, 542
  water based, 228, 530
Printing of labels, 28, 436, 531, 548 (*see also* Printing)
  cold foil transfer, 536
  flexo, 531, 532, 533, 548
    narrow web, 532
    postprinting, 532
    rotary, 532
  gravure, 531
    in mold, 531 (*see also* In-mold label)
  hot stamping, 534, 575
    advantages of, 534
    die-cutting for, 534 (*see also* Die-cutting)
    holographic, 535
    in-mold, 535
    transfer film for, 535
  letterpress, 533
    UV drying for, 533
  mask, 534

[Printing of labels]
   narrow web, 547
   offset litography, 533
      for laquering, 533
      for narrow web, 533
   screen, 532
      market segments for, 532
   of small series, 537
   tampon, 533
Processing
   aids, 227
   shear rate of, 545
Production speed, of tapes, 50
Propylene copolymers, 151
   heterophase, 159
Protective coatings, 36, 62, 434
   peelable, 197, 434
   requirements for, 621
Protective films, 1, 3, 55
   adhesive carrier for (*see also* Film carrier)
      of polyolefin, 433
   adhesive coated, 60, 64, 616
   adhesiveless, 64 (*see also* Self adhesive film)
   application of,
      climate of, 609
      domains of, 302, 303, 616
      equipment for, 611
      methods of, 611
   build up of, 8, 57
   classes of, 64
   coextruded, 461
   de-application of,
      climate of, 609
      conditions of, 61
   end use conditions of, 109
   formulation of, 184 (*see also* Formulation)
   grade of,
      dependence of on the protected product, 607
   history of, 59
   main grades of, 611, 612
   manufacture of by adhesive coating, 486

[Protective films]
   non adhesive carrier for
      of PVC, 433
   packaging use of, 607
   peel of, 127
   principle of functioning of, 60
   processing use of, 607, 616
   requirements for, 57
   self adhesive character of, 62
   special characteristics of, 64
   thickness of, 60
   uncoated, 59
   warm laminating of, 64 (*see also* Hot laminating films)
Protective films for automotive storage and transport, 620
   requirements for, 621
Protective films for carpet, 626
Protective films for coil coating, 612, 614 (*see also* Coil coating)
Protective films for plastic plates, 275, 434, 451, 460, 606, 607, 608
   application characteristics of, 617, 619
   end use characteristics of, 620
   for polycarbonate, 605, 606, 618
   for PMMA, 606, 607, 616, 618
   processing characteristics of, 617
   of PVC, 606
   self adhesive (*see also* Self adhesive film)
      EVAC based, 461
      polybutene based, 460 (*see also* Polybutylene)
Protective papers, 57, 460
Protective tapes, 601 (*see also* Masking tape)
Punching, 556 (*see also* Confectioning properties)

Quick stick, 262 (*see also* Tack)
   dependence of on storage modulus, 97, 311
   value of, 197

Index

Radiation dosage, 342, 343 (*see also* Radiation induced polymerization)
   dependence of,
      on resin level, 317
      on resin nature, 317, 356
Radiation induced polymerization, 316 (*see also* Electron beam induced polymerization)
Radial block copolymers, 97, 188, 190 (*see also* Thermoplastic elastomers)
   rubbery plateau of, 311 (*see also* Modulus)
Raw materials for PSA
   acrylic, 356 (*see also* Acrylic adhesive)
   elastic, 356 (*see also* Elastomers)
   viscoelastic, 356
   for coated adhesives for protective films, 382 (*see also* Formulation)
Re-adherability, 291 (*see also* Removability)
Recycling
   of carrier, 161
   of plastics, 193
Ready
   to coat polymerizable mixture, 339 (*see also* Polymacromerization)
   to use adhesive, 339 (*see also* Adhesive)
      hot melt, 359 (*see also* Hot melt)
Reinforced systems (*see* Ordered systems)
Relaxation, 95, 116, 279, 348
Release agent, 3 (*see also* Abhesive)
   coated layer of, 18, 185, 446
      discontinuous, 22
      fluorinated, 222
   embedded, 32, 158, 386, 392, 428, 446
   mechanism of functioning of, 234
   non silicone, 391
      water based, 392
      for paper, 392

[Release agent]
   for protective films, 63, 393
   radiation cured, 391
   for tapes, 393 (*see also* Tape)
   UV cured, 391 (*see also* Ultraviolet light induced photopolymerization)
   water based, 392
Release enchancing agent, 321 (*see also* Release agent)
Release layer, coextruded, 410 (*see also* Manufacture of film carrier)
Release liner, 8
   carrier for, 447
      manufacture of, 386 (*see also* Manufacture of film carrier)
   dimensional stability of, 117 (*see also* Shrinkage)
   double faced, 19, 25, 447 (*see also* Tape)
   films for, 436 (*see also* Film release liner)
   for labels, 40
      abhesive properties of, 35
   mechanical properties of, 30
   manufacture of, 445, 446
      coating device for, 446
      by extrusion, 448
   for protective films, 60 (*see also* Release agent)
   requirements for, 386
   role of, 117
   separate, 19, 25, 26
   siliconized, 19 (*see also* Silicone)
   for tapes, 19, 48, 429
Remoisturizing, 447 (*see also* Paper)
Removability, 93, 287, 291
   of acrylic PSAs, 346
   control of, 22, 291, 351
      by contact surface reduction, (*see also* contact surface reduction), 351
      by crosslinking, 351
      by discontinous adhesive, 22
      by silicone derivatives, 237

[Removability]
　of protective films, 63
　of self adhesive films (*see also* Self adhesive film)
　　dependence on aging, 291
　　dependence on stiffness, 285 (*see also* Stiffness)
Removable PSP, manufacture of, 351 (*see also* Self adhesive film)
Repair tapes, 592
Resin
　curing agent, 214
　detackifying, 222 (*see also* Detackification)
　end block compatible, 309 (*see also* Thermoplastic elastomer)
　end block reinforcing, 309
　　influence of on adhesive properties, 278 (*see also* Formulation)
　hydrocarbon based, 4, 93, 96, 310
　influence of on radiation curing, 214 (*see also* Radiation induced polymerization)
　liquid, for HMPSA formulation, 355
　midblock compatible, 309, 310 (*see also* Compatibility)
　partially compatible, 214 (*see also* Thermoplastic elastomer)
　phenolic, 184
　reactive, 184
　rosin, 15, 213
　　aging stability of, 214 (*see also* Aging)
　　comparison of acid/ester, 214
　saturated petroleum resins, for TPEs, 214
　solutions of, 213
　for tapes, 96
　T_g of, 92
　of UV curable systems, 317 (*see also* Ultraviolet light induced photopolymerization)
　　level of, 317
　water based dispersions of, 213

Resistance
　aging, 39
　to organic solvents, 367
　of carrier, 367
Rewindability, 55 (*see also* Converting properties)
Rheology
　of abhesive, 106 (*see also* Abhesive)
　of adhesive, 91 (*see also* Adhesive)
　　coated, 91
　　parameters of, 91
　　removable, 104
　of carrier material, 77 (*see also* Film carrier)
　of solid-state, 78
　during processing, 101
　energetic aspects of, 103
　of filled elastomers, 89
　of filled systems, 100 (*see also* Filler)
　of labels, 106
　of melts, 89
　of ordered systems, 87, 88 (*see also* Ordered systems)
　parameters of, 93
　polymer blends, 90
　of product, 106
　of protective films, 106 (*see also* Protective film)
　of reinforced systems, 100
　of tapes, 106
Rolldown properties, 589 (*see also* Unwinding of tapes)
Rolling adhesive moment tester, 298 (*see also* Adhesive characteristics of PSPs)
Rolling ball tack, 158, 288 (*see also* Tack)
　dependence of on adhesive anchorage, 288, 290
　values of,
　　for protective films, 276
　　for insulation tape, 481
Roll pressing for coating, 369, 378, 478, 524 (*see also* Medical tape)

Rosin resin, 15 (*see also* Resin)
  acid, 214, 279
    compatibility of, 280
    crystallization of, 214, 280
    influence on grit, 214
  ester, 184
Rubber
  carboxylated, latex of, 172
  crosslinked, 184
  regenerated, 369
Rubber elasticity, 86, 87
  theory of, 86
Rubber-resin adhesive, 183
  filled systems of, 175
  for labels, 39 (*see also* Adhesive)
  phase separation in, 159 (*see also* Multiphase structure)
  for protective films, 60 (*see also* Protective films)
  softness of, 301 (*see also* Modulus)
  for tapes, 20, 47, 161 (*see also* Adhesive)

Sealing tape, 12, 13, 54, 594, 595
  adhesives for, 47, 185, 197, 376
  of butyl rubber, 12 (*see also* Butyl rubber)
  carrier for, 429
  mechanical properties of, 129
Security labels, 27, 28, 30, 131, 398, 582 (*see also* Tamper evident label)
  destructibility of, 117
Segmentation, 175 (*see also* Segregated structure)
Segmented block copolymers, 122 (*see also* Styrene block copolymers)
Segmented polyurethane, 87 (*see also* Polyurethane)
Segregated structure, 98, 149 (*see also* Multiphase structure)
  of acrylic copolymers, 198 (*see also* Hot melt adhesive)
  build up of, 174

[Segregated structure]
  influence of, on mechanical properties, 157
  polyurethanes of, 174 (*see also* Polyurethane)
  crystallinity of, 174
Self-adhesion, 82
  of elastomers, for PSPs, 609
  of ethylene-vinyl acetate copolymers, 167
  of natural rubber, 151
  of plastomers for PSPs, 609
Self-adhesive pressure-sensitive product, 23 (*see also* Self adhesive film)
  application conditions of, 335
  bulky, 151
  raw materials for, 451 (*see also* Raw materials for PSA)
Self adhesive film, 158, 272 (*see also* Self adhesive protective film)
  classification of depending on protected surface, 608
  cold stretchable, 166 (*see also* Stretch film)
  non-polar, 451, 457 (*see also* Hot laminating film)
  polar, 434, 435, 451, 460
    of EVAc, 460 (*see also* Ethylene-vinyl acetate copolymers)
Self adhesive protective film, 60, 434
  of EVAc, 460 (*see also* Ethylene-vinyl acetate copolymers)
  manufacturing technology of, 462 (*see also* Manufacture of film carrier)
  polybutylene of, 458 (*see also* Polybutylene)
  polyisobutylene of, 458 (*see also* Polyisobutylene)
  polyolefin based, 433 (*see also* Hot laminating film)
  of PP, 462
  PVC of, 461

Self adhesivity, 24 (*see also* Pressure-sensitivity)
  of polybutadiene, 166
  time dependent, 174
Self-adhesive wall covering, 25, 399, 629
Self strengthening, 175 (*see also* Natural rubber)
Semi pressure-sensitive adhesives, 201, 269, 292
Separation energy, 103, 277 (*see also* Removability)
Separating film, 267, 627
Sequence distribution, 81
  influence of on tensile strength, 179
Shear adhesion to card board, 298 (*see also* Adhesive characteristics of PSPs)
Shear adhesion failure temperature, 158, 261
  dependence on sequence distribution, 189
Shear induced change, of the adhesive properties of hotmelts, 158
Shear modulus, 94
Shear resistance, 256
  dependence of on rheology, 101
  high temperature, 262
  improving of, 34
    by postcuring, 341
  of labels, 293
  of tapes, 295
  values of, 261
  for tapes, 587
Sheeter (*see* Confectioning properties)
Shrinkage, 129, 543
  of adherent, 616
  of adhesive, 544
  of carrier, 543, 544
    components of, 543
  dependence of on screen printing, 544
  humidity caused, 545 (*see also* Paper)
  test conditionf for, 544

Silage wrap films, 458 (*see also* Self adhesive film)
Silicone adhesive,
  for chemical resistant tapes, 378
  for double-faced tapes, 198
  elastomer, 198, 199
    T_g of, 198
  emulsified, 199
  heat resistant, 165, 198
  low remperature curable, 199
  polymers,
    electrical properties of, 140, 374
    for insulating tapes, 375 (*see also* Insulating tape)
  resin, 199
Silicone abhesive (*see also* Release agent)
  coating of, 447
  EB curing of, 447
  emulsion based, 446, 447
  low temperature curing, 445, 447
  solvent free, 445
  systems of, 445
  UV curing, 445, 447
Silicone release layer, 3 (*see also* Release agent)
  formulation of, 387, 445
  for silicone adhesive, 19
Silicone acrylates, 388, 447
Skin irritation, 20, 34, 368
Slayter paradox, 121
Slip, 18 (*see also* Agent)
  migration of, 287
Slitting, 30, 548, 549 (*see also* Converting properties)
  of tapes, 552, 558
Smoothness, 547 (*see also* Converting properties)
Solubilizer, 217 (*see also* Water soluble formulation)
  for HMPSA, 366
Solution acrylic, concentrated, 357 (*see also* Acrylic adhesive)
Solvent, 229, 496
  influence of,
    on adhesive properties, 230

Index

[Solvent]
    on pot life 231
    isoparaffinic, 356
Solvent based systems
    coating of, 468
    crosslinked, 468
    cuttability of, 552
Special tape (*see also* Tape)
    build up of, 19
    chemically resistant, 20
    for printed circuits, 20
    thermally resistant, 20
    water resistant, 20
Spherulite growth rate, 156 (*see also* Crystallinity)
Splicing tape, 50, 293, 295, 297, 598
    adhesive for, 205
        repulpable, 217, 365
        water soluble, 367, 599
    application conditions for, 367, 587
    carrier for, 367, 402
    requirements for, 599
    temperature resistance of, 367
Stiffness (*see also* Mechanical properties of PSPs)
    of block copolymers, 129
    of carrier, 132
        comparison of, 133
        factors of, 132
    dependence of,
        on crosslinking, 135 (*see also* Crosslinking)
        on filling, 135 (*see also* Filler)
        on modulus, 133 (*see also* Modulus)
        on molecular orientation, 133
    influence of, 132
        on conformability, 13 (*see also* Conformability)
        on cuttability, 133 (*see also* Cuttability)
        on lay flat, 546 (*see also* Lay flat)
        on machinability, 132
    of paper, 132, 401

[Stiffness]
    on printability, 547 (*see also* Printability)
    of tapes, 132
Star shaped block copolymers, 122, 178 (*see also* Thermoplastic elastomers)
    acrylic, 96 (*see also* Acrylic adhesive)
    epoxidized, 178
        UV curing of, 178 (*see also* Ultraviolet light induced photopolymerization)
    of polyamides, 194
    of styrene-diene, 129, 176
    of styrene-ethylene oxide, 194
Storage modulus, 75, 90, 263 (*see also* Dynamic mechanical Analysis)
    dependence of,
        on crosslinking, 95, 316
        stress frequency, 97, 311
    plateau of, 310
    values of, for the adhesive, 95, 96, 97, 293, 316
Storage time, of water based acrylics, 371 (*see also* Water based formulation)
Stress induced crystallization (*see also* Self strengthening)
    of natural rubber, 120
Stretch film, 459
Styrene-acrylate block copolymers, 193 (*see also* Thermoplastic elatomers)
Styrene block copolymers, 97, 174, 187
    compatibilizers, 193 (*see also* Compatibility)
    elastic recovery of, 174
    hard segment of, 195
    mechanical properties of,
        dependence of on composition, 129
    for labels, 40 (*see also* Label)
    modulus of, 195 (*see also* Modulus)

[Styrene block copolymers]
  morphology of, 98, 195
  radial, 129 (*see also* Radial block copolymers)
  mechanical properties of, 192
  radiation curable, 176 (*see also* Radiation induced polymerization)
  sequence distribution of, 187 (*see also* Sequence distribution)
  stiffness of, 159
  tackification of, 195 (*see also* Tackification)
  for tapes, 47 (*see also* Tape)
Styrene-butadiene
  random copolymers of, 186
  rubber, 85, 377
    aging stability of, 186
    gel content of, 377 (*see also* Gel content)
    latex of, 184, 360
    styrene content of, 186
    for tapes, 186, 377
Styrene-butadiene-styrene block copolymer, 87, 90, 122, 174
  aging of, 309
  carboxylated, 193
  hard block of, 87
  comparison of with SIS, 192, 309
  for hotmelts, 213 (*see also* Hot melt)
  screening recipe for, 224
  star, 191 (*see also* Star shaped block copolymers)
    molecular weight of, 191
    styrene content of, 191
  viscosity of, 309
Styrene-diene block copolymers, 158, 188 (*see also* Styrene block copolymers)
Styrene-ethylene-butylene-styrene block copolymers, 93, 122, 174, 190
  diblock, 179 (*see also* Diblock)
  styrene content of, 188

Styrene-isoprene-styrene block copolymers, 122, 174
  diblock, 120, 174 (*see also* Diblock)
  for labels, 189
  for hot melts (*see also* Hot melts)
    styrene content of, 213
  multiarm, 183
  radiation curing of, 193 (*see also* Radiation induced polymerization)
    dependence of on styrene content, 193
  rheology of, 91 (*see also* Rheology)
  screening recipe for, 224
  solid state morphology of, 191
  standard formulation for, 190
  star, 191 (*see also* Star shaped block copolymers)
    molecular weight of, 191
  for tapes, 22
  triblock, 188
Styrene-olefin block copolymers, 187 (*see also* Thermoplastic copolymers)
Sulfonation, 443 (*see also* Surface)
Surface
  chemical treatment of, 137
  contoured, 185
  geometry, modification of, 444
  modification of for carrier films, 438
  physical treatment of, 137 (*see also* Physical treatment)
  polarizability of, 140
  of product, influence of on label end use, 571
  resistivity of, 138
    dependence of on charge decay time, 139 (*see also* Charge decay time)
    influence of on converting, 139
  treatment, test of, 602
Surface polarity of carrier film, 438 (*see also* Carrier)
  modification of, 438

Index

[Surface polarity of carrier film]
  by corona treatment, 439 (*see also* Corona treatment)
  by physical treatment, 439 (*see also* Physical treatment)
  by tackifying, 439 (*see also* Tackification)
Surface tension, 452
  apparent, 454
  dependence of on corona treatment energy level, 455 (*see also* Corona treatment)
Surfactant, influence of on adhesive properties, 262 (*see also* Formulation)
Surgical application, 204 (*see* also Medical tapes)

Tack, 33, 94, 256, 275
  aged, 312
  application, 276, 277
  autoadhesive, 258
  dependence of on surface treatment, 277 (*see also* Surface)
  instantaneous, 291
  reduction of by UV light, 380 (*see also* Crosslinking)
  surface free of, 270, 362 (*see also* Crosslinking)
  as tear energy, 277
  test methods of, 261 (*see also* Adhesive characteristics of PSPs)
  values of, 261
Tackification, 213, 306, 344, 346, 349
  for anchorage, 32 (*see also* Adhesive)
  of carrier, 32, 346, 386 (*see also* Self adhesive film)
  of polypropylene, 32
  role of, 348
  steps of, 345 (*see also* Formulation)
  theory of, 310

Tackifier, 15, 32, 212, 357 (*see also* Resin)
  for acrylics, 203 (*see also* Formulation)
  level of, 203, 349
  for amorphous polypropylene, 201
  diffusion of, 287
  dispersion,
    formulation of, 224, 350
    storage stability of, 224
  embedded in carrier, 158 (*see also* Self adhesive Film)
  for EVAc hotmelts, 211
  for freezer tapes, 213 (*see also* Freezer tape)
  for hot melt adhesives, 213
    midblock compatible, 214
    low temperature, 369
    water releasable, 349
  influence of on the viscosity, 359
  level of, 307
    for acrylics, 348
    for adhesives for labels, 373
    for adhesives for protective films, 489
    for CSBR, 308
    for SBR, 348
  for postcrosslinked systems, 195, 357
  for SIS, 214, 310
  for SBS, 310
Tag, 574, 584
Tamper evident lables, 28, 42, 127, 580 (*see also* Security label)
Tan δ, 94, 263, 309 (*see also* Dynamic mechanical analysis)
  values of, 95
Tapes, 1, 2, 9, 10, 44, 122, 124
  adhesives for, 201, 361, 363
    choice of, 381
    crosslinked, 312
    removable, 22, 352
  adhesive properties of, 294, 296 (*see also* Adhesive characteristics of PSPs)
  application of, 584

[Tapes]
  climate of, 587
  conditions of, 584
  dependence of on product surface, 585
  equipment for, 589, 560
  fields of, 584
  method of, 587
  automotive, 589, 595 (*see also* Car masking tape)
  build up of, 8, 24, 45
  carrierless, 13, 15, 34, 46, 49, 52 (*see also* Transfer tape and Mounting tape)
    manufacture of, 481
  carrier for, 375 (*see also* Carrier)
  chemical basis of, 375
  classes of, 48, 50, 589
  coating technology for, 476
    hot melt based, 476
    solvent based, 476
    water based, 476
  cover, 429
  crepe, 35 (*see also* Paper)
  curl of, 546 (*see also* Converting properties)
  double faced, 19, 25, 26, 27, 31, 50, 162, 205, 230, 297, 436, 597
    for carpet, 310
    foam, 376, 379, 441
  end use of, 584
  extensible, 162
  fiber reinforced, 31, 35
  film based, 49
  folded, 589
  formulation of, 184
    dependence of on carrier material, 375
  functioning of, 48
  general, 48
  heat shrinkable, 130
  impregnated, 16
  low temperature, 20 (*see also* Freezer tape)
  manufacture of, 470, 472
    by coating, 476
    in line, 26

[Tapes]
  markets for, 45, 48
  medical, 2, 20 (*see also* Medical tape)
  noise reduction of, 388, 589
  non-flammable, 379, 592
  paper based, 49, 184
  prepreg, 16
  recyclable, 367
  removable,
    adhesives for, 23 (*see also* Formulation)
    filler for, 23 (*see also* Filler)
    for paper, 22
    requirements for, 45
  special, 52, 342, 602 (*see also* Special tape)
    adhesive characteristics of, 295
    mechanical characteristics of, 295
  textile based, 52
  temperature resistant, 198
  thick, 270, 342
  tissue, 595
  tropically resistant, 83, 378
  wet adhesive, 18
  wrapping, 585
  unwinding of, 46, 236, 585 (*see also* Unwinding)
Tapes for corrugated board, 590
Tear strength, 130 (*see also* Mechanical characteristics of PSPs)
  of carrier, 130
    bulk, 130
    surface, 130
  Elmendorf, 131 (*see also* Elmendorf tear strength)
  of labels, 124
  of polyethylene, for protective films, 432
  of polypropylene, 131
  of tapes, 131
Tear tape, 131, 380, 426, 588, 589, 591
Technological additives, 229 (*see also* Agent)
Telescoping, 558

Index

Temperature display labels, 577
Temperature resistant labels, 577
Temperature resistance of tapes, 49
Tensile strength, 123 (*see also* Mechanical characteristics of PSPs)
   of adhesive, 17, 20, 96
      dependence of on crosslinking, 270
   of carrier, 111, 124, 423
   of labels, 124 (*see also* Label)
   of self adhesive films, 285 (*see also* Self adhesive fim)
   of styrene-isoprene diblock copolymers, 120 (*see also* Diblock)
   of styrene isoprene radial block copolymers, 192 (*see also* Styrene block copolymers)
   of tapes, 124 (*see also* Tape)
Tensile modulus, 94 (*see also* Modulus)
   values of, 95
Test
   of release, high speed, 275
   tape, 602
Thermal insulation tape, 593
Thermodynamic adhesion work, 103, 105
Thermoplastic elastomers, 4, 32, 74, 79, 122
   blends of, 129
   with diblocks, 174 (*see also* Diblock)
   elasticity of, 151
   pressure-sensitive formulations of, adhesive characteristics of, 211
   with saturated midblock, 355 (*see also* Styrene-ethylene-butylene-styrene block copolymers)
   with triblocks, 174
Thermoplastic polyurethane, 13, 123, 202 (*see also* Polyurethane)
   crosslinking of, 202
   hard segment of, 202
Time-temperature superposition principle, 348 (*see also* Rheology)

Tissue tape, 21
Transdermal
   applications, 34 (*see also* Skin irritation)
   adhesive for, 34 (*see also* Medical tape)
   drug delivery system, 20, 198
Transfer
   films, 535
   printing materials, 19, 125
Transfer tape, 13, 16, 30, 50, 83, 100, 353, 597, 598
   application of, 16,
   foam-like, 275, 353 (*see also* Foam-like tape)
   manuafcture of, 482
      by foaming, 483
   patterned, 21 (*see also* Pattern coating)
   repositionable, 350
Triblock/diblock ratio, 189 (*see also* Thermoplastic elastomers)
Triboelectricity, 443
   reducing of, 444
Trisequenced polymers, 188 (*see also* Sequence distribution)
Two-phase morphology, 192 (*see also* Multiphase structure)

Ultraviolet light
   absorbers of, 224, 357,
   crosslinkable formulations, 277, 316
   surface treatment with, 442, 443
Ultraviolet light induced
   photopolymerization, 13, 270, 357, 379
   of acrylics, 13, 26, 34, 47, 83 (*see also* Acrylic adhesive)
   light intensity for, 341
   photoinitiator free, 341
   solvent based, 342
   for tapes, 128
Ultra low density polyethylene carrier, 415, 432 (*see also* Film carrier)
Unsaturation index, 195, 214 (*see also* Resin)

Unwinding (*see also* Converting properties)
  behavior, 18
  resistance of tapes, 557
  parameters of, 558

Very low density polyethylene carrier, 32, 415, 432 (*see also* Self adhesive film)
Vinyl acetate copolymers, 210
  with acrylate, 210
  with vinyl pyrrolidone, 210
Vinyl chloride copolymers, 151 (*see also* PVC)
Viscoelastic compounds, 150, 202, 306, 450
  acrylic, 202 (*see also* Raw materials for PSA)
Viscoelasticity, 75
  parameters of, 153
    macromolecular, 154
Viscoelastomer, 151 (*see also* Viscoelastic compounds)
Viscosity
  for air knife, 361
  for Meyer bar, 361
  for Rotogravure, 361
Viscosity of filled systems, 100
Viscosity regulators, 355
  polyisoprenes as, 93, 216
Viscous flow, 1 (*see also* Rheology)
Viscous components, 212 (*see also* Tackifier)

Warm coating, 323
Warm laminating, 2
  film, of VLDPE, 456 (*see also* Self adhesive film)
  mechanism of bonding of, 457
Warm water removable formulation, 365, 368
Water absorbent, 379 (*see also* Medical tape)

Water based formulation
  acrylic, 365 (*see also* Acrylic adhesive)
  for tapes, 377
  coating of, 468
  chloroprene, 374 (*see also* Neoprene)
  crosslinkable, 313 (*see also* Crosslinking)
  CSBR, for tapes, 377
  cuttability of, 552 (*see also* Cuttability)
  for labels, shelf life of, 372
Water soluble formulation, 365
  of polyvinylpyrrolidone, 367
Water resistant
  labels, 577 (*see also* Labels)
  tapes, 596
Wax, 15
  of polyethylene, 367
  of polypropylene, 367
Web
  cleaning device, 444
  coextruded, 322 (*see* also Coextrusion)
  control, equipment for, 497, 548
Wettability, 102, 137, 321, 443, 525 (*see also* Manufacture of PSPs by adhesive coating)
Williams-Landel-Ferry equation, 78, 101, 118
Williams Plasticity number, 197
Winding,
  equipment, 496
  properties, 557 (*see also* Unwinding)
Wire wound tapes, 28, 585, 592
Wrinkle build up, 547 (*see also* Converting properties)
Writability, 42, 51, 424

Ziegler-Natta catalyst, 193, 195
Zip peel (*see also* Peel), 281